Fortran for Scientists and Engineers

Fourth Edition

Fortran for Scientists and Engineers

Fourth Edition

Stephen J. Chapman
BAE Systems Australia

FORTRAN FOR SCIENTISTS AND ENGINEERS, FOURTH EDITION

Published by McGraw-Hill Education, 2 Penn Plaza, New York, NY 10121. Copyright © 2018 by McGraw-Hill Education. All rights reserved. Printed in the United States of America. Previous edition © 2008 and 2004. No part of this publication may be reproduced or distributed in any form or by any means, or stored in a database or retrieval system, without the prior written consent of McGraw-Hill Education, including, but not limited to, in any network or other electronic storage or transmission, or broadcast for distance learning.

Some ancillaries, including electronic and print components, may not be available to customers outside the United States.

This book is printed on acid-free paper.

1 2 3 4 5 6 7 8 9 LCR 21 20 19 18 17

ISBN 978–0–07–338589–1
MHID 0–07–338589–1

Chief Product Officer, SVP Products & Markets: *G. Scott Virkler*
Vice President, General Manager, Products & Markets: *Marty Lange*
Vice President, Content Design & Delivery: *Betsy Whalen*
Managing Director: *Thomas Timp*
Brand Manager: *Raghothaman Srinivasan/ Thomas M. Scaife, Ph.D*
Director, Product Development: *Rose Koos*
Product Developer: *Tina Bower*
Marketing Manager: *Shannon O'Donnell*

Director, Content Design & Delivery: *Linda Avenarius*
Program Manager: *Lora Neyens*
Content Project Managers: *Jane Mohr and Sandra Schnee*
Buyer: *Jennifer Pickel*
Design: *Studio Montage, St. Louis, MO*
Content Licensing Specialist: *DeAnna Dausener*
Cover Image: *hh5800/Getty Images*
Compositor: *Aptara®, Inc.*
Printer: *LSC Communications*

All credits appearing on page or at the end of the book are considered to be an extension of the copyright page.

Library of Congress Cataloging-in-Publication Data

Chapman, Stephen J., author.
Fortran for scientists and engineers / Stephen J. Chapman, BAE Systems
    Australia.
Fourth edition. | New York, NY : McGraw-Hill, a business unit of
    The McGraw-Hill Companies, Inc., [2017] | Includes index.
LCCN 2016052439 | ISBN 9780073385891 (alk. paper) | ISBN
    0073385891 (alk. paper)
LCSH: FORTRAN (Computer program language) | Science—Data
    processing. | Engineering—Data processing.
LCC QA76.73.F25 C425 2017 | DDC 005.13/3—dc23 LC record available at
    https://lccn.loc.gov/2016052439

The Internet addresses listed in the text were accurate at the time of publication. The inclusion of a website does not indicate an endorsement by the authors or McGraw-Hill Education, and McGraw-Hill Education does not guarantee the accuracy of the information presented at these sites.

mheducation.com/highered

This book is dedicated to my son Avi, who is the only one of our eight children actually making a living writing software!

ABOUT THE AUTHOR

STEPHEN J. CHAPMAN received a B.S. in Electrical Engineering from Louisiana State University (1975), an M.S.E. in Electrical Engineering from the University of Central Florida (1979), and pursued further graduate studies at Rice University.

From 1975 to 1980, he served as an officer in the U.S. Navy, assigned to teach Electrical Engineering at the U.S. Naval Nuclear Power School in Orlando, Florida. From 1980 to 1982, he was affiliated with the University of Houston, where he ran the power systems program in the College of Technology.

From 1982 to 1988 and from 1991 to 1995, he served as a Member of the Technical Staff of the Massachusetts Institute of Technology's Lincoln Laboratory, both at the main facility in Lexington, Massachusetts, and at the field site on Kwajalein Atoll in the Republic of the Marshall Islands. While there, he did research in radar signal processing systems. He ultimately became the leader of four large operational range instrumentation radars at the Kwajalein field site (TRADEX, ALTAIR, ALCOR, and MMW).

From 1988 to 1991, Mr. Chapman was a research engineer in Shell Development Company in Houston, Texas, where he did seismic signal processing research. He was also affiliated with the University of Houston, where he continued to teach on a part-time basis.

Mr. Chapman is currently Manager of Systems Modeling and Operational Analysis for BAE Systems Australia, in Melbourne, Australia. He is the leader of a team that has developed a model of how naval ships defend themselves against antiship missile attacks. This model contains more than 400,000 lines of MATLAB code written over more than a decade, so he has extensive practical experience applying MATLAB to real-world problems.

Mr. Chapman is a Senior Member of the Institute of Electrical and Electronic Engineers (and several of its component societies). He is also a member of the Association for Computing Machinery and the Institution of Engineers (Australia).

TABLE OF CONTENTS

	Preface	xix
1	**Introduction to Computers and The Fortran Language**	**1**
1.1	The Computer	2
	1.1.1. The CPU / 1.1.2. Memory / 1.1.3. Input and Output Devices	
1.2	Data Representation in a Computer	4
	1.2.1. The Binary Number System / 1.2.2. Octal and Hexadecimal Representations of Binary Numbers / 1.2.3. Types of Data Stored in Memory	
1.3	Computer Languages	12
1.4	The History of the Fortran Language	13
1.5	The Evolution of Fortran	16
1.6	Summary	19
	1.6.1. Exercises	
2	**Basic Elements of Fortran**	**22**
2.1	Introduction	22
2.2	The Fortran Character Set	23
2.3	The Structure of a Fortran Statement	23
2.4	The Structure of a Fortran Program	24
	2.4.1. The Declaration Section / 2.4.2. The Execution Section / 2.4.3. The Termination Section / 2.4.4. Program Style / 2.4.5. Compiling, Linking, and Executing the Fortran Program	
2.5	Constants and Variables	28
	2.5.1. Integer Constants and Variables / 2.5.2. Real Constants and Variables / 2.5.3. Character Constants and Variables / 2.5.4. Default and Explicit Variable Typing / 2.5.5. Keeping Constants Consistent in a Program	
2.6	Assignment Statements and Arithmetic Calculations	36
	2.6.1. Integer Arithmetic / 2.6.2. Real Arithmetic / 2.6.3. Hierarchy of Operations / 2.6.4. Mixed-Mode Arithmetic / 2.6.5. Mixed-Mode Arithmetic and Exponentiation	

	2.7	Intrinsic Functions	47
	2.8	List-Directed Input and Output Statements	49
	2.9	Initialization of Variables	55
	2.10	The `IMPLICIT NONE` Statement	57
	2.11	Program Examples	58
	2.12	Debugging Fortran Programs	66
	2.13	Summary	68
		2.13.1. Summary of Good Programming Practice / 2.13.2. Summary of Fortran Statements / 2.13.3. Exercises	
3	**Program Design and Branching Structures**		**81**
	3.1	Introduction to Top-Down Design Techniques	82
	3.2	Use of Pseudocode and Flowcharts	86
	3.3	Logical Constants, Variables, and Operators	89
		3.3.1. Logical Constants and Variables / 3.3.2. Assignment Statements and Logical Calculations / 3.3.3. Relational Operators / 3.3.4. Combinational Logic Operators / 3.3.5. Logical Values in Input and Output Statements / 3.3.6. The Significance of Logical Variables and Expressions	
	3.4	Control Constructs: Branches	94
		3.4.1. The Block IF Construct / 3.4.2. The ELSE and ELSE IF Clauses / 3.4.3. Examples Using Block IF Constructs / 3.4.4. Named Block IF Constructs / 3.4.5. Notes Concerning the Use of Block IF Constructs / 3.4.6. The Logical IF Statement / 3.4.7. The SELECT CASE Construct	
	3.5	More on Debugging Fortran Programs	118
	3.6	Summary	119
		3.6.1. Summary of Good Programming Practice / 3.6.2. Summary of Fortran Statements and Constructs / 3.6.3. Exercises	
4	**Loops and Character Manipulation**		**126**
	4.1	Control Constructs: Loops	126
		4.1.1 The While Loop / 4.1.2 The DO WHILE Loop / 4.1.3 The Iterative or Counting Loop / 4.1.4 The CYCLE and EXIT Statements / 4.1.5 Named Loops / 4.1.6 Nesting Loops and Block IF Constructs	
	4.2	Character Assignments and Character Manipulations	154
		4.2.1 Character Assignments / 4.2.2 Substring Specifications / 4.2.3 The Concatenation (//) Operator / 4.2.4 Relational Operators with Character Data / 4.2.5 Character Intrinsic Functions	
	4.3	Debugging Fortran Loops	168

	4.4	Summary	169
		4.4.1 Summary of Good Programming Practice / 4.4.2 Summary of Fortran Statements and Constructs / 4.4.3 Exercises	
5		Basic I/O Concepts	180
	5.1	Formats and Formatted `WRITE` Statements	180
	5.2	Output Devices	182
		5.2.1 Control Characters in Printer Output	
	5.3	Format Descriptors	184
		5.3.1 Integer Output—The I Descriptor / 5.3.2 Real Output—The F Descriptor / 5.3.3 Real Output—The E Descriptor / 5.3.4 True Scientific Notation—The ES Descriptor / 5.3.5 Logical Output—The L Descriptor / 5.3.6 Character Output—The A Descriptor / 5.3.7 Horizontal Positioning—The X and T Descriptor / 5.3.8 Repeating Groups of Format Descriptors / 5.3.9 Changing Output Lines—The Slash (/) Descriptor / 5.3.10 How Formats are Used During WRITEs	
	5.4	Formatted `READ` Statements	205
		5.4.1 Integer Input—The I Descriptor / 5.4.2 Real Input—The F Descriptor / 5.4.3 Logical Input—The L Descriptor / 5.4.4 Character Input—The A Descriptor / 5.4.5 Horizontal Positioning—The X and T Descriptors / 5.4.6 Vertical Positioning—The Slash (/) Descriptor / 5.4.7 How Formats are Used During READs	
	5.5	An Introduction to Files and File Processing	211
		5.5.1 The `OPEN` Statement / 5.5.2 The `CLOSE` Statement / 5.5.3 READs and WRITEs to Disk Files / 5.5.4 The `IOSTAT=` and `IOMSG=` Clauses in the READ Statement / 5.5.5 File Positioning	
	5.6	Summary	232
		5.6.1 Summary of Good Programming Practice / 5.6.2 Summary of Fortran Statements and Structures / 5.6.3 Exercises	
6		Introduction to Arrays	245
	6.1	Declaring Arrays	246
	6.2	Using Array Elements in Fortran Statements	247
		6.2.1 Array Elements are Just Ordinary Variables / 6.2.2 Initialization of Array Elements / 6.2.3 Changing the Subscript Range of an Array / 6.2.4 Out-of-Bounds Array Subscripts / 6.2.5 The Use of Named Constants with Array Declarations	
	6.3	Using Whole Arrays and Array Subsets in Fortran Statements	261
		6.3.1 Whole Array Operations / 6.3.2 Array Subsets	

	6.4	Input and Output	265
		6.4.1 Input and Output of Array Elements / 6.4.2 The Implied DO Loop / 6.4.3 Input and Output of Whole Arrays and Array Sections	
	6.5	Example Problems	271
	6.6	When Should You Use an Array?	287
	6.7	Summary	289
		6.7.1 Summary of Good Programming Practice / 6.7.2 Summary of Fortran Statements and Constructs / 6.7.3 Exercises	
7		**Introduction to Procedures**	**297**
	7.1	Subroutines	299
		7.1.1 Example Problem—Sorting / 7.1.2 The INTENT Attribute / 7.1.3 Variable Passing in Fortran: The Pass-By-Reference Scheme / 7.1.4 Passing Arrays to Subroutines / 7.1.5 Passing Character Variables to Subroutines / 7.1.6 Error Handling in Subroutines / 7.1.7 Examples	
	7.2	Sharing Data Using Modules	320
	7.3	Module Procedures	328
		7.3.1 Using Modules to Create Explicit Interfaces	
	7.4	Fortran Functions	331
		7.4.1 Unintended Side Effects in Functions / 7.4.2 Using Functions with Deliberate Side Effects	
	7.5	Passing Procedures as Arguments to Other Procedures	339
		7.5.1 Passing User-Defined Functions as Arguments / 7.5.2 Passing Subroutines as Arguments	
	7.6	Summary	344
		7.6.1 Summary of Good Programming Practice / 7.6.2 Summary of Fortran Statements and Structures / 7.6.3 Exercises	
8		**Additional Features of Arrays**	**360**
	8.1	2D or Rank 2 Arrays	360
		8.1.1 Declaring Rank 2 Arrays / 8.1.2 Rank 2 Array Storage / 8.1.3 Initializing Rank 2 Arrays / 8.1.4 Example Problem / 8.1.5 Whole Array Operations and Array Subsets	
	8.2	Multidimensional or Rank *n* Arrays	372
	8.3	Using Fortran Intrinsic Functions with Arrays	375
		8.3.1 Elemental Intrinsic Functions / 8.3.2 Inquiry Intrinsic Functions / 8.3.3 Transformational Intrinsic Functions	
	8.4	Masked Array Assignment: The WHERE Construct	378
		8.4.1 The WHERE Construct / 8.4.2 The WHERE Statement	
	8.5	The FORALL Construct	381
		8.5.1 The Form of the FORALL Construct / 8.5.2 The Significance of the FORALL Construct / 8.5.3 The FORALL Statement	

8.6	Allocatable Arrays	383
	8.6.1 Fortran Allocatable Arrays / 8.6.2 Using Fortran Allocatable Arrays in Assignment Statements	
8.7	Summary	393
	8.7.1 Summary of Good Programming Practice / 8.7.2 Summary of Fortran Statements and Constructs / 8.7.3 Exercises	

9 Additional Features of Procedures — 404

9.1	Passing Multidimensional Arrays to Subroutines and Functions	404
	9.1.1 Explicit Shape Dummy Arrays / 9.1.2 Assumed-Shape Dummy Arrays / 9.1.3 Assumed-Size Dummy Arrays	
9.2	The SAVE Attribute and Statement	417
9.3	Allocatable Arrays in Procedures	421
9.4	Automatic Arrays in Procedures	422
	9.4.1 Comparing Automatic Arrays and Allocatable Arrays / 9.4.2 Example Program	
9.5	Allocatable Arrays as Dummy Arguments in Procedures	430
	9.5.1 Allocatable Dummy Arguments / 9.5.2 Allocatable Functions	
9.6	Pure and Elemental Procedures	434
	9.6.1 Pure Procedures / 9.6.2 Elemental Procedures / 9.6.3 Impure Elemental Procedures	
9.7	Internal Procedures	436
9.8	Submodules	438
9.9	Summary	446
	9.9.1 Summary of Good Programming Practice / 9.9.2 Summary of Fortran Statements and Structures / 9.9.3 Exercises	

10 More about Character Variables — 457

10.1	Character Comparison Operations	458
	10.1.1 The Relational Operators with Character Data / 10.1.2 The Lexical Functions LLT, LLE, LGT, and LGE	
10.2	Intrinsic Character Functions	463
10.3	Passing Character Variables to Subroutines and Functions	465
10.4	Variable-Length Character Functions	471
10.5	Internal Files	473
10.6	Example Problems	474
10.7	Summary	479
	10.7.1 Summary of Good Programming Practice / 10.7.2 Summary of Fortran Statements and Structures / 10.7.3 Exercises	

11 Additional Intrinsic Data Types — 485

11.1 Alternate Kinds of the `REAL` Data Type — 485
11.1.1 Kinds of REAL Constants and Variables / 11.1.2 Determining the KIND of a Variable / 11.1.3 Selecting Precision in a Processor-Independent Manner / 11.1.4 Determining the KINDs of Data Types on a Particular Processor / 11.1.5 Mixed-Mode Arithmetic / 11.1.6 Higher Precision Intrinsic Functions / 11.1.7 When to Use High-Precision Real Values / 11.1.8 Solving Large Systems of Simultaneous Linear Equations

11.2 Alternate Lengths of the `INTEGER` Data Type — 509

11.3 Alternate Kinds of the `CHARACTER` Data Type — 511

11.4 The `COMPLEX` Data Type — 512
11.4.1 Complex Constants and Variables / 11.4.2 Initializing Complex Variables / 11.4.3 Mixed-Mode Arithmetic / 11.4.4 Using Complex Numbers with Relational Operators / 11.4.5 COMPLEX Intrinsic Functions

11.5 Summary — 522
11.5.1 Summary of Good Programming Practice / 11.5.2 Summary of Fortran Statements and Structures / 11.5.3 Exercises

12 Derived Data Types — 527

12.1 Introduction to Derived Data Types — 527

12.2 Working with Derived Data Types — 529

12.3 Input and Output of Derived Data Types — 529

12.4 Declaring Derived Data Types in Modules — 531

12.5 Returning Derived Types from Functions — 540

12.6 Dynamic Allocation of Derived Data Types — 544

12.7 Parameterized Derived Data Types — 545

12.8 Type Extension — 546

12.9 Type-Bound Procedures — 548

12.10 The `ASSOCIATE` Construct — 552

12.11 Summary — 553
12.11.1 Summary of Good Programming Practice / 12.11.2 Summary of Fortran Statements and Structures / 12.11.3 Exercises

13 Advanced Features of Procedures and Modules — 561

13.1 Scope and Scoping Units — 562

13.2 Blocks — 567

13.3 Recursive Procedures — 568

13.4 Keyword Arguments and Optional Arguments — 571

13.5	Procedure Interfaces and Interface Blocks	577
	13.5.1 Creating Interface Blocks / 13.5.2 Notes on the Use of Interface Blocks	
13.6	Generic Procedures	581
	13.6.1 User-Defined Generic Procedures / 13.6.2 Generic Interfaces for Procedures in Modules / 13.6.3 Generic Bound Procedures	
13.7	Extending Fortran with User-Defined Operators and Assignments	594
13.8	Bound Assignments and Operators	607
13.9	Restricting Access to the Contents of a Module	607
13.10	Advanced Options of the USE Statement	611
13.11	Intrinsic Modules	615
13.12	Access to Command Line Arguments and Environment Variables	615
	13.12.1 Access to Command Line Arguments / 13.12.2 Retrieving Environment Variables	
13.13	The VOLATILE Attribute and Statement	618
13.14	Summary	619
	13.14.1 Summary of Good Programming Practice / 13.14.2 Summary of Fortran Statements and Structures / 13.14.3 Exercises	

14 Advanced I/O Concepts — 633

14.1	Additional Format Descriptors	633
	14.1.1 Additional Forms of the E and ES Format Descriptors / 14.1.2 Engineering Notation—The EN Descriptor / 14.1.3 Double-Precision Data—The D Descriptor / 14.1.4 The Generalized (G) Format Descriptor / 14.1.5 The G0 Format Descriptor / 14.1.6 The Binary, Octal, and Hexadecimal (B, O, and Z) Descriptors / 14.1.7 The TAB Descriptors / 14.1.8 The Colon (:) Descriptor / 14.1.9 Scale Factors—The P Descriptor / 14.1.10 The SIGN Descriptors / 14.1.11 Blank Interpretation: The BN and BZ Descriptors / 14.1.12 Rounding Control: The RU, RD, RZ, RN, RC, and RP Descriptors / 14.1.13 Decimal Specifier: The DC and DP Descriptors	
14.2	Defaulting Values in List-Directed Input	642
14.3	Detailed Description of Fortran I/O Statements	644
	14.3.1 The OPEN Statement / 14.3.2 The CLOSE Statement / 14.3.3 The INQUIRE Statement / 14.3.4 The READ Statement / 14.3.5 Alternate Form of the READ Statement / 14.3.6 The WRITE Statement / 14.3.7 The PRINT Statement / 14.3.8 File Positioning Statements / 14.3.9 The ENDFILE Statement / 14.3.10 The WAIT Statement / 14.3.11 The FLUSH Statement	
14.4	Namelist I/O	668
14.5	Unformatted Files	671
14.6	Direct Access Files	673

14.7	Stream Access Mode	678
14.8	Nondefault I/O for Derived Types	678
14.9	Asynchronous I/O	687
	14.9.1. Performing Asynchronous I/O / 14.9.2. Problems with Asynchronous I/O	
14.10	Access to Processor-Specific I/O System Information	689
14.11	Summary	690
	14.11.1 Summary of Good Programming Practice / 14.11.2 Summary of Fortran Statements and Structures / 14.11.3 Exercises	

15 Pointers and Dynamic Data Structures — 698

15.1	Pointers and Targets	699
	15.1.1 Pointer Assignment Statements / 15.1.2 Pointer Association Status	
15.2	Using Pointers in Assignment Statements	705
15.3	Using Pointers with Arrays	707
15.4	Dynamic Memory Allocation with Pointers	709
15.5	Using Pointers as Components of Derived Data Types	712
15.6	Arrays of Pointers	725
15.7	Using Pointers in Procedures	727
	15.7.1 Using the INTENT Attribute with Pointers / 15.7.2 Pointer-valued Functions	
15.8	Procedure Pointers	733
15.9	Binary Tree Structures	736
	15.9.1 The Significance of Binary Tree Structures / 15.9.2 Building a Binary Tree Structure	
15.10	Summary	756
	15.10.1 Summary of Good Programming Practice / 15.10.2 Summary of Fortran Statements and Structures / 15.10.3 Exercises	

16 Object-Oriented Programming in Fortran — 763

16.1	An Introduction to Object-Oriented Programming	764
	16.1.1 Objects / 16.1.2 Messages / 16.1.3 Classes / 16.1.4 Class Hierarchy and Inheritance / 16.1.5 Object-Oriented Programming	
16.2	The Structure of a Fortran Class	769
16.3	The CLASS Keyword	770
16.4	Implementing Classes and Objects in Fortran	772
	16.4.1 Declaring Fields (Instance Variables) / 16.4.2 Creating Methods / 16.4.3 Creating (Instantiating) Objects from a Class	

16.5	First Example: A `timer` Class	775
	16.5.1 Implementing the `timer` Class / 16.5.2 Using the `timer` Class / 16.5.3 Comments on the `timer` Class	
16.6	Categories of Methods	780
16.7	Controlling Access to Class Members	789
16.8	Finalizers	790
16.9	Inheritance and Polymorphism	794
	16.9.1 Superclasses and Subclasses / 16.9.2 Defining and Using Subclasses / 16.9.3 The Relationship between Superclass Objects and Subclass Objects / 16.9.4 Polymorphism / 16.9.5 The `SELECT TYPE` Construct	
16.10	Preventing Methods from Being Overridden in Subclasses	809
16.11	Abstract Classes	809
16.12	Summary	831
	16.12.1 Summary of Good Programming Practice / 16.12.2 Summary of Fortran Statements and Structures / 16.12.3 Exercises	

17 Coarrays and Parallel Processing 837

17.1	Parallel Processing in Coarray Fortran	838
17.2	Creating a Simple Parallel Program	839
17.3	Coarrays	841
17.4	Synchronization between Images	843
17.5	Example: Sorting a Large Data Set	850
17.6	Allocatable Coarrays and Derived Data Types	856
17.7	Passing Coarrays to Procedures	857
17.8	Critical Sections	858
17.9	The Perils of parallel Programming	859
17.10	Summary	863
	17.10.1 Summary of Good Programming Practice / 17.10.2 Summary of Fortran Statements and Structures / 17.10.3 Exercises	

18 Redundant, Obsolescent, and Deleted Fortran Features 869

18.1	Pre-Fortran 90 Character Restrictions	870
18.2	Obsolescent Source Form	870
18.3	Redundant Data Type	871
18.4	Older, Obsolescent, and/or Undesirable Specification Statements	872
	18.4.1 Pre-Fortran 90 Specification Statements / 18.4.2 The `IMPLICIT` Statement / 18.4.3 The `DIMENSION` Statement / 18.4.4 The `DATA` Statement / 18.4.5 The `PARAMETER` Statement	

18.5 Sharing Memory Locations: COMMON and EQUIVALENCE 875
18.5.1 COMMON Blocks / 18.5.2 Initializing Data in COMMON Blocks: The BLOCK DATA Subprogram / 18.5.3 The Unlabeled COMMON Statement / 18.5.4 The EQUIVALENCE Statement

18.6 Undesirable Subprogram Features 882
18.6.1 Alternate Subroutine Returns / 18.6.2 Alternate Entry Points / 18.6.3 The Statement Function / 18.6.4 Passing Intrinsic Functions as Arguments

18.7 Miscellaneous Execution Control Features 889
18.7.1 The PAUSE Statement / 18.7.2 Arguments Associated with the STOP Statement / 18.7.3 The END Statement

18.8 Obsolete Branching and Looping Structures 892
18.8.1 The Arithmetic IF Statement / 18.8.2 The Unconditional GO TO Statement / 18.8.3 The Computed GO TO Statement / 18.8.4 The Assigned GO TO Statement / 18.8.5 Older Forms of DO Loops

18.9 Redundant Features of I/O Statements 896

18.10 Summary 897
18.10.1 Summary of Good Programming Practice / 18.10.2 Summary of Fortran Statements and Structures

Appendixes

A. The ASCII Character Set 903

B. Fortran/C Interoperability 904
B.1. Declaring Interoperable Data Types / B.2. Declaring Interoperable Procedures / B.3. Sample Programs—Fortran Calling C / B.4. Sample Programs—C Calling Fortran

C. Fortran Intrinsic Procedures 914
C.1. Classes of Intrinsic Procedures / C.2. Alphabetical List of Intrinsic Procedures / C.3. Mathematical and Type Conversion Intrinsic Procedures / C.4. Kind and Numeric Processor Intrinsic Functions / C.5. System Environment Procedures / C.6. Bit Intrinsic Procedures / C.7. Character Intrinsic Functions / C.8. Array and Pointer Intrinsic Functions / C.9. Miscellaneous Inquiry Functions / C.10. Miscellaneous Procedures / C.11. Coarray Functions

D. Order of Statements in a Fortran Program 961

E. Glossary 963

F. Answers to Quizzes 984

Index 1002
Summary of Selected Fortran Statements and Structures 1022

PREFACE

The first edition of this book was conceived as a result of my experience in writing and maintaining large Fortran programs in both the defense and geophysical fields. During my time in industry, it became obvious that the strategies and techniques required to write large, *maintainable* Fortran programs were quite different from what new engineers were learning in their Fortran programming classes at school. The incredible cost of maintaining and modifying large programs once they are placed into service absolutely demands that they be written to be easily understood and modified by people other than their original programmers. My goal for this book is to teach simultaneously both the fundamentals of the Fortran language and a programming style that results in good, maintainable programs. In addition, it is intended to serve as a reference for graduates working in industry.

It is quite difficult to teach undergraduates the importance of taking extra effort during the early stages of the program design process in order to make their programs more maintainable. Class programming assignments must by their very nature be simple enough for one person to complete in a short period of time, and they do not have to be maintained for years. Because the projects are simple, a student can often "wing it" and still produce working code. A student can take a course, perform all of the programming assignments, pass all of the tests, and still not learn the habits that are really needed when working on large projects in industry.

From the very beginning, this book teaches Fortran in a style suitable for use on large projects. It emphasizes the importance of going through a detailed design process before any code is written, using a top-down design technique to break the program up into logical portions that can be implemented separately. It stresses the use of procedures to implement those individual portions, and the importance of unit testing before the procedures are combined into a finished product. Finally, it emphasizes the importance of exhaustively testing the finished program with many different input data sets before it is released for use.

In addition, this book teaches Fortran as it is actually encountered by engineers and scientists working in industry and in laboratories. One fact of life is common in all programming environments: Large amounts of old legacy code that have to be maintained. The legacy code at a particular site may have been originally written in Fortran IV (or an even earlier version!), and it may use programming constructs that are no longer common today. For example, such code may use arithmetic IF statements, or computed or assigned GO TO statements. Chapter 18 is devoted to those older features of the language that are no longer commonly used, but that are encountered in legacy code.

The chapter emphasizes that these features should *never* be used in a new program, but also prepares the student to handle them when he or she encounters them.

CHANGES IN THIS EDITION

This edition builds directly on the success of *Fortran 95/2003 for Scientists and Engineers*, 3/e. It preserves the structure of the previous edition, while weaving the new Fortran 2008 material (and limited material from the proposed Fortran 2015 standard) throughout the text. It is amazing, but Fortran started life around 1954, and it is *still* evolving.

Most of the additions in Fortran 2008 are logical extensions of existing capabilities of Fortran 2003, and they are integrated into the text in the proper chapters. However, the use of parallel processing and Coarray Fortran is completely new, and Chapter 17 has been added to cover that material.

The vast majority of Fortran courses are limited to one-quarter or one semester, and the student is expected to pick up both the basics of the Fortran language and the concept of how to program. Such a course would cover Chapters 1 through 7 of this text, plus selected topics in Chapters 8 and 9 if there is time. This provides a good foundation for students to build on in their own time as they use the language in practical projects.

Advanced students and practicing scientists and engineers will need the material on COMPLEX numbers, derived data types, and pointers found in Chapters 11 through 15. Practicing scientists and engineers will almost certainly need the material on obsolete, redundant, and deleted Fortran features found in Chapter 18. These materials are rarely taught in the classroom, but they are included here to make the book a useful reference text when the language is actually used to solve real-world problems.

FEATURES OF THIS BOOK

Many features of this book are designed to emphasize the proper way to write reliable Fortran programs. These features should serve a student well as he or she is first learning Fortran, and should also be useful to the practitioner on the job. They include:

1. *Emphasis on Modern Fortran.*
   The book consistently teaches the best current practice in all of its examples. Many modern Fortran 2008 features duplicate and supersede older features of the Fortran language. In those cases, the proper usage of the modern language is presented. Examples of older usage are largely relegated to Chapter 18, where their old/undesirable nature is emphasized. Examples of modern Fortran features that supersede older features are the use of modules to share data instead of COMMON blocks, the use of DO . . . END DO loops instead of DO . . . CONTINUE loops, the use of internal procedures instead of statement functions, and the use of CASE constructs instead of computed GOTOs.

2. *Emphasis on Strong Typing.*

    The IMPLICIT NONE statement is used consistently throughout the book to force the explicit typing of every variable used in every program, and to catch common typographical errors at compilation time. In conjunction with the explicit declaration of every variable in a program, the book emphasizes the importance of creating a data dictionary that describes the purpose of each variable in a program unit.

3. *Emphasis on Top-Down Design Methodology.*

    The book introduces a top-down design methodology in Chapter 3, and then uses it consistently throughout the rest of the book. This methodology encourages a student to think about the proper design of a program *before* beginning to code. It emphasizes the importance of clearly defining the problem to be solved and the required inputs and outputs before any other work is begun. Once the problem is properly defined, it teaches the student to employ stepwise refinement to break the task down into successively smaller subtasks, and to implement the subtasks as separate subroutines or functions. Finally, it teaches the importance of testing at all stages of the process, both unit testing of the component routines and exhaustive testing of the final product. Several examples are given of programs that work properly for some data sets, and then fail for others.

    The formal design process taught by the book may be summarized as follows:

    - *Clearly state the problem that you are trying to solve.*
    - *Define the inputs required by the program and the outputs to be produced by the program.*
    - *Describe the algorithm that you intend to implement in the program. This step involves top-down design and stepwise decomposition, using pseudocode or flow charts.*
    - *Turn the algorithm into Fortran statements.*
    - *Test the Fortran program. This step includes unit testing of specific subprograms, and also exhaustive testing of the final program with many different data sets.*

4. *Emphasis on Procedures.*

    The book emphasizes the use of subroutines and functions to logically decompose tasks into smaller subtasks. It teaches the advantages of procedures for data hiding. It also emphasizes the importance of unit testing procedures before they are combined into the final program. In addition, the book teaches about the common mistakes made with procedures, and how to avoid them (argument type mismatches, array length mismatches, etc.). It emphasizes the advantages associated with explicit interfaces to procedures, which allow the Fortran compiler to catch most common programming errors at compilation time.

5. *Emphasis on Portability and Standard Fortran.*

    The book stresses the importance of writing portable Fortran code, so that a program can easily be moved from one type of computer to another one.

It teaches students to use only standard Fortran statements in their programs, so that they will be as portable as possible. In addition, it teaches the use of features such as the SELECTED_REAL_KIND function to avoid precision and kind differences when moving from computer to computer.

The book also teaches students to isolate machine-dependent code (such as code that calls machine-dependent system libraries) into a few specific procedures, so that only those procedures will have to be rewritten when a program is ported between computers.

6. *Good Programming Practice Boxes.*

These boxes highlight good programming practices when they are introduced for the convenience of the student. In addition, the good programming practices introduced in a chapter are summarized at the end of the chapter. An example Good Programming Practice Box is shown below:

Good Programming Practice
Always indent the body of an IF structure by two or more spaces to improve the readability of the code.

7. *Programming Pitfalls Boxes*

These boxes highlight common errors so that they can be avoided. An example Programming Pitfalls Box is shown below:

Programming Pitfalls
Beware of integer arithmetic. Integer division often gives unexpected results.

8. *Emphasis on Pointers and Dynamic Data Structures.*

Chapter 15 contains a detailed discussion of Fortran pointers, including possible problems resulting from the incorrect use of pointers such as memory leaks and pointers to deallocated memory. Examples of dynamic data structures in the chapter include linked lists and binary trees.

Chapter 16 contains a discussion of Fortran objects and object-oriented programming, including the use of dynamic pointers to achieve polymorphic behavior.

9. *Use of Sidebars.*

A number of sidebars are scattered throughout the book. These sidebars provide additional information of potential interest to the student. Some sidebars are historical in nature. For example, one sidebar in Chapter 1 describes the IBM Model 704, the first computer to ever run Fortran. Other sidebars

reinforce lessons from the main text. For example, Chapter 9 contains a sidebar reviewing and summarizing the many different types of arrays found in modern Fortran.
10. *Completeness.*
    Finally, the book endeavors to be a complete reference to the modern Fortran language, so that a practitioner can locate any required information quickly. Special attention has been paid to the index to make features easy to find. A special effort has also been made to cover such obscure and little understood features as passing procedure names by reference, and defaulting values in list-directed input statements.

PEDAGOGICAL FEATURES

The book includes several features designed to aid student comprehension. Each chapter begins with a list of the objectives that should be achieved in that chapter. A total of 27 quizzes appear scattered throughout the chapters, with answers to all questions included in Appendix F. These quizzes can serve as a useful self-test of comprehension. In addition, there are approximately 360 end-of-chapter exercises. Answers to selected exercises are available at the book's Web site, and of course answers to all exercises are included in the Instructor's Manual. Good programming practices are highlighted in all chapters with special Good Programming Practice boxes, and common errors are highlighted in Programming Pitfalls boxes. End-of-chapter materials include Summaries of Good Programming Practice and Summaries of Fortran Statements and Structures. Finally, a detailed description of every Fortran intrinsic procedure is included in Appendix C, and an extensive Glossary is included in Appendix E.

The book is accompanied by an Instructor's Manual, containing the solutions to all end-of-chapter exercises. Instructors can also download the solutions in the Instructor's Manual from the book's Web site. The source code for all examples in the book, plus other supplemental materials, can be downloaded by anyone from the book's Web site.

A NOTE ABOUT FORTRAN COMPILERS

Two Fortran compilers were used during the preparation of this book: the Intel Visual Fortran Compiler Version 16.0 and the GNU G95 Fortran compiler. Both compilers provide essentially complete implementations of Fortran 2008, with only a very few minor items not yet implemented. They are also both looking to the future, implementing features from the proposed Fortran 2015 standard.

I highly recommend both compilers to potential users. The great advantage of Intel Fortran is the very nice integrated debugging environment, and the great disadvantage is cost. The G95 compiler is free, but it is somewhat harder to debug.

A FINAL NOTE TO THE USER

No matter how hard I try to proofread a document like this book, it is inevitable that some typographical errors will slip through and appear in print. If you should spot any such errors, please drop me a note via the publisher, and I will do my best to get them eliminated from subsequent printings and editions. Thank you very much for your help in this matter.

I will maintain a complete list of errata and corrections at the book's World Wide Web site, which is www.mhhe.com/chapman4e. Please check that site for any updates and/or corrections.

ACKNOWLEDGMENTS

I would like to thank Raghu Srinivasan and the team at McGraw-Hill Education for making this revision possible. In addition, I would like to thank my wife Rosa and daughter Devorah for their support during the revision process. (In previous editions, I had thanked our other seven children as well, but they have all now flown the coop!)

Stephen J. Chapman
Melbourne, Victoria, Australia
August 7, 2016

1

Introduction to Computers and the Fortran Language

OBJECTIVES

- Know the basic components of a computer.
- Understand binary, octal, and hexadecimal numbers.
- Learn about the history of the Fortran language.

The computer was probably the most important invention of the twentieth century. It affects our lives profoundly in very many ways. When we go to the grocery store, the scanners that check out our groceries are run by computers. Our bank balances are maintained by computers, and the automatic teller machines and credit and debit cards that allow us to make banking transactions at any time of the day or night are run by more computers. Computers control our telephone and electric power systems, run our microwave ovens and other appliances, and control the engines in our cars. Almost any business in the developed world would collapse overnight if it were suddenly deprived of its computers. Considering their importance in our lives, it is almost impossible to believe that the first electronic computers were invented just about 75 years ago.

Just what is this device that has had such an impact on all of our lives? A **computer** is a special type of machine that stores information, and can perform mathematical calculations on that information at speeds much faster than human beings can think. A **program,** which is stored in the computer's memory, tells the computer what sequence of calculations is required, and which information to perform the calculations on. Most computers are very flexible. For example, the computer on which I write these words can also balance my checkbook, if I just execute a different program on it.

Computers can store huge amounts of information, and with proper programming, they can make that information instantly available when it is needed. For example, a bank's computer can hold the complete list of all the deposits and debits made by every one of its customers. On a larger scale, credit companies use their computers to hold the credit histories of every person in the United States—literally billions of pieces of information. When requested, they can search through those billions of pieces of information to recover the credit records of any single person, and present those records to the user in a matter of seconds.

It is important to realize that *computers do not think as humans understand thinking*. They merely follow the steps contained in their programs. When a computer appears to be doing something clever, it is because a clever person has written the program that it is executing. That is where we humans come into the act. It is our collective creativity that allows the computer to perform its seeming miracles. This book will help teach you how to write programs of your own, so that the computer will do what *you* want it to do.

1.1
THE COMPUTER

A block diagram of a typical computer is shown in Figure 1-1. The major components of the computer are the **central processing unit (CPU), main memory, secondary memory,** and **input** and **output devices.** These components are described in the paragraphs below.

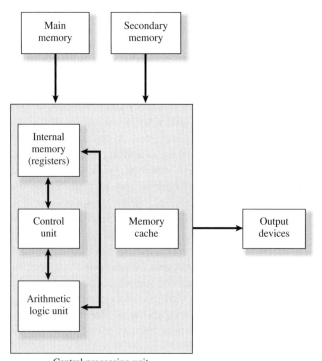

FIGURE 1-1
A block diagram of a typical computer.

1.1.1 The CPU

The central processing unit is the heart of any computer. It is divided into a *control unit*, an *arithmetic logic unit (ALU)*, and internal memory. The control unit within the CPU controls all of the other parts of the computer, while the ALU performs the actual mathematical calculations. The internal memory within a CPU consists of a series of *memory registers* used for the temporary storage of intermediate results during calculations, plus a *memory cache* to temporarily store data that will be needed in the near future.

The control unit of the CPU interprets the instructions of the computer program. It also fetches data values from main memory (or the memory cache) and stores them in the memory registers, and sends data values from memory registers to output devices or main memory. For example, if a program says to multiply two numbers together and save the result, the control unit will fetch the two numbers from main memory and store them in registers. Then, it will present the numbers in the registers to the ALU along with directions to multiply them and store the results in another register. Finally, after the ALU multiplies the numbers, the control unit will take the result from the destination register and store it back into the memory cache. (Other parts of the CPU copy the data from the memory cache to main memory in slower time.)

Modern CPUs have become dramatically faster by incorporating multiple ALUs running in parallel, allowing more operations to be performed in a given amount of time. They also incorporate larger memory caches on the CPU chip, allowing data to be fetched and saved very rapidly.

1.1.2 Memory

The memory of a computer is divided into three major types of memory: *cache* memory, *main* or *primary memory*, and *secondary memory*. Cache memory is memory stored on the CPU chip itself. This memory can be accessed very rapidly, allowing calculations to proceed at very high speed. The control unit looks ahead in the program to see what data will be needed, and pre-fetches it from main memory into the memory cache so that it can be used with minimal delay. The control unit also copies the results of calculations from the cache back to main memory when they are no longer needed.

Main memory usually consists of separate semiconductor chips connected to the CPU by conductors called a *memory bus*. It is very fast, and relatively inexpensive compared to the memory on the CPU itself. Data that is stored in main memory can be fetched for use in a few nanoseconds or less (sometimes *much* less) on a modern computer. Because it is so fast and cheap, main memory is used to temporarily store the program currently being executed by the computer, as well as the data that the program requires.

Main memory is not used for the permanent storage of programs or data. Most main memory is **volatile,** meaning that it is erased whenever the computer's power is turned off. Besides, main memory is relatively expensive, so we only buy enough to hold all of the programs actually being executed at any given time.

Secondary memory consists of devices that are slower and cheaper than main memory. They can store much more information for much less money than main memory can. In addition, most secondary memory devices are **nonvolatile,** meaning that they retain the programs and data stored in them whenever the computer's power is turned off. Typical secondary memory devices are **hard disks,** solid-state drives (SSD), USB memory sticks, and DVDs. Secondary storage devices are normally used to store programs and data that are not needed at the moment, but that may be needed some time in the future.

1.1.3 Input and Output Devices

Data is entered into a computer through an input device, and is output through an output device. The most common input devices on a modern computer are the keyboard and the mouse. We can type programs or data into a computer with a keyboard. Other types of input devices found on some computers include touchscreens, scanners, microphones, and cameras.

Output devices permit us to use the data stored in a computer. The most common output devices on today's computers are displays and printers. Other types of output devices include plotters and speakers.

1.2
DATA REPRESENTATION IN A COMPUTER

Computer memories are composed of billions of individual switches, each of which can be ON or OFF, but not at a state in between. Each switch represents one **binary digit** (also called a **bit**); the ON state is interpreted as a binary 1, and the OFF state is interpreted as a binary 0. Taken by itself, a single switch can only represent the numbers 0 and 1. Since we obviously need to work with numbers other than 0 and 1, a number of bits are grouped together to represent each number used in a computer. When several bits are grouped together, they can be used to represent numbers in the *binary* (base 2) *number system.*

The smallest common grouping of bits is called a **byte.** *A byte is a group of 8 bits that are used together to represent a binary number.* The byte is the fundamental unit used to measure the capacity of a computer's memory. For example, the personal computer on which I am writing these words has a main memory of 24 gigabytes (24,000,000,000 bytes) and a secondary memory (disk drive) with a storage of 2 terabytes (2,000,000,000,000 bytes).

The next larger grouping of bits in a computer is called a **word.** A word consists of 2, 4, or more consecutive bytes that are used to represent a single number in memory. The size of a word varies from computer to computer, so words are not a particularly good way to judge the size of computer memories. Modern CPUs tend to use words with lengths of either 32 or 64 bits.

1.2.1 The Binary Number System

In the familiar base 10 number system, the smallest (rightmost) digit of a number is the ones place (10^0). The next digit is in the tens place (10^1), and the next one is in the hundreds place (10^2), etc. Thus, the number 122_{10} is really $(1 \times 10^2) + (2 \times 10^1) + (2 \times 10^0)$. Each digit is worth a power of 10 more than the digit to the right of it in the base 10 system (see Figure 1-2a).

Introduction to Computers and the Fortran Language

FIGURE 1-2
(a) The base 10 number 122 is really $(1 \times 10^2) + (2 \times 10^1) + (2 \times 10^0)$. (b) Similarly, the base 2 number 101_2 is really $(1 \times 2^2) + (0 \times 2^1) + (1 \times 2^0)$.

Similarly, in the binary number system, the smallest (rightmost) digit is the ones place (2^0). The next digit is in the twos place (2^1), and the next one is in the fours place (2^2), etc. Each digit is worth a power of 2 more than the digit to the right of it in the base 2 system. For example, the binary number 101_2 is really $(1 \times 2^2) + (0 \times 2^1) + (1 \times 2^0) = 5$, and the binary number $111_2 = 7$ (see Figure 1-2b).

Note that three binary digits can be used to represent eight possible values: 0 ($= 000_2$) to 7 ($= 111_2$). In general, *if n bits are grouped together to form a binary number, then they can represent 2^n possible values.* Thus, a group of 8 bits (1 byte) can represent 256 possible values, a group of 16 bits (2 bytes) can be used to represent 65,536 possible values, and a group of 32 bits (4 bytes) can be used to represent 4,294,967,296 possible values.

In a typical implementation, half of all possible values are reserved for representing negative numbers, and half of the values are reserved for representing zero plus the positive numbers. Thus, a group of 8 bits (1 byte) is usually used to represent numbers between -128 and $+127$, including 0, and a group of 16 bits (2 bytes) is usually used to represent numbers between $-32,768$ and $+32,767$, including 0.[1]

TWO'S COMPLEMENT ARITHMETIC

The most common way to represent negative numbers in the binary number system is the two's complement representation. What is two's complement, and what is so special about it? Let's find out.

The Two's Complement Representation of Negative Numbers
In the two's complement representation, the leftmost bit of a number is the *sign bit*. If that bit is 0, then the number is positive; if it is 1, then the number is negative. To change a positive number into the corresponding negative number in the two's complement system, we perform two steps:

1. Complement the number (change all 1s to 0 and all 0s to 1).
2. Add 1 to the complemented number.

---

[1] The most common scheme for representing negative numbers in a computer's memory is the so-called *two's complement* representation, which is described in the sidebar.

Let's illustrate the process using simple 8-bit integers. As we already know, the 8-bit binary representation of the number 3 would be 00000011. The two's complement representation of the number −3 would be found as follows:

1. Complement the positive number: 11111100
2. Add 1 to the complemented number: 11111100 + 1 = 11111101

Exactly the same process is used to convert negative numbers back to positive numbers. To convert the number −3 (11111101) back to a positive 3, we would:

1. Complement the negative number: 00000010
2. Add 1 to the complemented number: 00000010 + 1 = 00000011

Two's Complement Arithmetic

Now we know how to represent numbers in two's complement representation, and to convert between positive and two's complement negative numbers. The special advantage of two's complement arithmetic is that *positive and negative numbers may be added together according to the rules of ordinary addition without regard to the sign, and the resulting answer will be correct, including the proper sign*. Because of this fact, a computer may add any two integers together without checking to see what the signs of the two integers are. This simplifies the design of computer circuits.

Let's do a few examples to illustrate this point.

1. Add 3 + 4 in two's complement arithmetic.

     3   00000011
  +4   00000100
     7   00000111

2. Add (−3) + (−4) in two's complement arithmetic.

     3   11111101
 +−4   11111100
  −7   111111001

In a case like this, we ignore the extra ninth bit resulting from the sum, and the answer is 11111001. The two's complement of 11111001 is 00000111 or 7, so the result of the addition was −7!

3. Add 3 + (−4) in two's complement arithmetic.

  −3   00000011
 +−4   11111100
  −1   11111111

The answer is 11111111. The two's complement of 11111111 is 00000001 or 1, so the result of the addition was −1.

With two's complement numbers, binary addition comes up with the correct answer regardless of whether the numbers being added are both positive, both negative, or mixed.

1.2.2 Octal and Hexadecimal Representations of Binary Numbers

Computers work in the binary number system, but people think in the decimal number system. Fortunately, we can program the computer to accept inputs and give its outputs in the decimal system, converting them internally to binary form for processing. Most of the time, the fact that computers work with binary numbers is irrelevant to the programmer.

However, there are some cases in which a scientist or engineer has to work directly with the binary representations coded into the computer. For example, individual bits or groups of bits within a word might contain status information about the operation of some machine. If so, the programmer will have to consider the individual bits of the word, and work in the binary number system.

A scientist or engineer who has to work in the binary number system immediately faces the problem that binary numbers are unwieldy. For example, a number like 1100_{10} in the decimal system is 010001001100_2 in the binary system. It is easy to get lost working with such a number! To avoid this problem, we customarily break binary numbers down into groups of 3 or 4 bits, and represent those bits by a single base 8 (octal) or base 16 (hexadecimal) number.

To understand this idea, note that a group of 3 bits can represent any number between 0 (= 000_2) and 7 (= 111_2). These are the numbers found in an **octal** or base 8 arithmetic system. An octal number system has seven digits: 0 through 7. We can break a binary number up into groups of 3 bits, and substitute the appropriate octal digit for each group. Let's use the number 010001001100_2 as an example. Breaking the number into groups of three digits yields $010|001|001|100_2$. If each group of 3 bits is replaced by the appropriate octal number, the value can be written as 2114_8. The octal number represents exactly the same pattern of bits as the binary number, but it is more compact.

Similarly, a group of 4 bits can represent any number between 0 (= 0000_2) and 15 (= 1111_2). These are the numbers found in a **hexadecimal** or base 16 arithmetic system. A hexadecimal number system has 16 digits: 0 through 9 and A through F. Since the hexadecimal system needs 16 digits, we use digits 0 through 9 for the first 10 of them, and then letters A through F for the remaining 6. Thus, $9_{16} = 9_{10}$, $A_{16} = 10_{10}$, $B_{16} = 11_{10}$, and so forth. We can break a binary number up into groups of 4 bits, and substitute the appropriate hexadecimal digit for each group. Let's use the number 010001001100_2 again as an example. Breaking the number into groups of four digits yields $0100|0100|1100_2$. If each group of 4 bits is replaced by the appropriate hexadecimal number, the value can be written as $44C_{16}$. The hexadecimal number represents exactly the same pattern of bits as the binary number, but more compactly.

Some computer vendors prefer to use octal numbers to represent bit patterns, while other computer vendors prefer to use hexadecimal numbers to represent bit patterns. Both representations are equivalent, in that they represent the pattern of bits in a compact form. A Fortran language program can input or output numbers in any of the four formats (decimal, binary, octal, or hexadecimal). Table 1-1 lists the decimal, binary, octal, and hexadecimal forms of the numbers 0 to 15.

TABLE 1-1
Table of decimal, binary, octal, and hexadecimal numbers

Decimal	Binary	Octal	Hexadecimal
0	0000	0	0
1	0001	1	1
2	0010	2	2
3	0011	3	3
4	0100	4	4
5	0101	5	5
6	0110	6	6
7	0111	7	7
8	1000	10	8
9	1001	11	9
10	1010	12	A
11	1011	13	B
12	1100	14	C
13	1101	15	D
14	1110	16	E
15	1111	17	F

1.2.3 Types of Data Stored in Memory

Three common types of data are stored in a computer's memory: **character data, integer data,** and **real data** (numbers with a decimal point). Each type of data has different characteristics, and takes up a different amount of memory in the computer.

Character Data

The **character data** type consists of characters and symbols. A typical system for representing character data in a non-Oriental language must include the following symbols:

1. The 26 uppercase letters A through Z
2. The 26 lowercase letters a through z
3. The 10 digits 0 through 9
4. Miscellaneous common symbols, such as ",(), {}, [], !, ~, @, #, $, %, ^, &, and *.
5. Any special letters or symbols required by the language, such as à, ç, ë, and £.

Since the total number of characters and symbols required to write Western languages is less than 256, *it is customary to use 1 byte of memory to store each character*. Therefore, 10,000 characters would occupy 10,000 bytes of the computer's memory.

The particular bit values corresponding to each letter or symbol may vary from computer to computer, depending upon the coding system used for the characters. The most important coding system is ASCII, which stands for the American Standard Code for Information Interchange (ANSI X3.4 1986, or ISO/IEC 646:1991). The ASCII coding system defines the values to associate with the first 128 of the 256 possible values that can be stored in a 1-byte character. The 8-bit codes corresponding to each letter and number in the ASCII coding system are given in Appendix A.

The second 128 characters that can be stored in a 1-byte character are *not* defined by the ASCII character set, and they used to be defined differently depending on the language used in a particular country or region. These definitions are a part of the ISO 8859 standard series, and they are sometimes referred to as "code pages." For example, the ISO 8859-1 (Latin 1) character set is the version used in Western European countries. There are similar code pages available for Eastern European languages, Arabic, Greek, Hebrew, and so forth. Unfortunately, the use of different code pages made the output of programs and the contents of files appear different in different countries. As a result, these code pages are falling out of favor, and being replaced by the Unicode system described below.

Some Oriental languages such as Chinese and Japanese contain more than 256 characters (in fact, about 4000 characters are needed to represent each of these languages). To accommodate these languages and all of the other languages in the world, a coding system called Unicode[2] has been developed. In the Unicode coding system, each character is stored in 2 bytes of memory, so the Unicode system supports 65,536 possible different characters. The first 128 Unicode characters are identical to the ASCII character set, and other blocks of characters are devoted to various languages such as Chinese, Japanese, Hebrew, Arabic, and Hindi. When the Unicode coding system is used, character data can be represented in any language.

Integer Data

The **integer data** type consists of the positive integers, the negative integers, and zero. The amount of memory devoted to storing an integer will vary from computer to computer, but will usually be 1, 2, 4, or 8 bytes. Four-byte integers are the most common type in modern computers.

Since a finite number of bits are used to store each value, only integers that fall within a certain range can be represented on a computer. Usually, the smallest number that can be stored in an *n*-bit integer is

$$\text{Smallest integer value} = -2^{n-1} \tag{1-1}$$

and the largest number that can be stored in an *n*-bit integer is

$$\text{Largest integer value} = 2^{n-1} - 1 \tag{1-2}$$

For a 4-byte integer, the smallest and largest possible values are −2,147,483,648 and 2,147,483,647, respectively. Attempts to use an integer larger than the largest possible

---

[2] Also referred to by the corresponding standard number, ISO/IEC 10646:2014.

value or smaller than the smallest (most negative) possible value result in an error called an *overflow condition*.[3]

Real Data

The integer data type has two fundamental limitations:

1. It is not possible to represent numbers with fractional parts (0.25, 1.5, 3.14159, etc.) as integer data.
2. It is not possible to represent very large positive integers or very small negative integers, because there are not enough bits available to represent the value. The largest and smallest possible integers that can be stored in a given memory location will be given by Equations (1-1) and (1-2).

To get around these limitations, computers include a **real** or **floating-point** data type.

The real data type stores numbers in a type of scientific notation. We all know that very large or very small numbers can be most conveniently written in scientific notation. For example, the speed of light in a vacuum is about 299,800,000 m/s. This number is easier to work with in scientific notation: 2.998×10^8 m/s. The two parts of a number expressed in scientific notation are called the **mantissa** and the **exponent.** The mantissa of the number above is 2.998, and the exponent (in the base 10 system) is 8.

The real numbers in a computer are similar to the scientific notation above, except that a computer works in the base 2 system instead of the base 10 system. Real numbers usually occupy 32 bits (4 bytes) of computer memory, divided into two components: a 24-bit mantissa and an 8-bit exponent (Figure 1-3).[4] The mantissa contains a number between −1.0 and 1.0, and the exponent contains the power of 2 required to scale the number to its actual value.

Real numbers are characterized by two quantities: **precision** and **range.** *Precision* is the number of significant digits that can be preserved in a number, and *range* is the difference between the largest and smallest numbers that can be represented. The precision of a real number depends on the number of bits in its mantissa, while the range of the number depends on the number of bits in its exponent. A 24-bit mantissa can represent approximately $\pm 2^{23}$ numbers, or about seven significant decimal digits, so the precision of real numbers is about seven significant digits. An 8-bit exponent can represent multipliers between 2^{-128} and 2^{127}, so the range of real numbers is from about 10^{-38} to 10^{38}. Note that the real data type can represent numbers much larger or much smaller than integers can, but only with seven significant digits of precision.

---

[3] When an overflow condition occurs, some processors will abort the program causing the overflow condition. Other processors will "wrap around" from the most positive integer to the most negative integer without giving the user a warning that anything has happened. This behavior varies for different types of computers.

[4] This discussion is based on the IEEE Standard 754 for floating-point numbers, which is representative of most modern computers.

Introduction to Computers and the Fortran Language

FIGURE 1-3
This floating-point number includes a 24-bit mantissa and an 8-bit exponent.

When a value with more than seven digits of precision is stored in a real variable, *only the most significant 7 bits of the number will be preserved*. The remaining information will be lost forever. For example, if the value 12,345,678.9 is stored in a real variable on a PC, it will be rounded off to 12,345,680.0. This difference between the original value and the number stored in the computer is known as **round-off error**.

You will use the real data type in many places throughout this book and in your programs after you finish this course. It is quite useful, but you must always remember the limitations associated with round-off error, or your programs might give you an unpleasant surprise. For example, if your program must be able to distinguish between the numbers 1,000,000.0 and 1,000,000.1, then you cannot use the standard real data type.[5] It simply does not have enough precision to tell the difference between these two numbers!

Programming Pitfalls
Always remember the precision and range of the data types that you are working with. Failure to do so can result in subtle programming errors that are very hard to find.

Quiz 1-1

This quiz provides a quick check to see if you have understood the concepts introduced in Section 1.2. If you have trouble with the quiz, reread the section, ask your instructor, or discuss the material with a fellow student. The answers to this quiz are found in the back of the book.

1. Express the following decimal numbers as their binary equivalents:
   (a) 27_{10}
   (b) 11_{10}
   (c) 35_{10}
   (d) 127_{10}

(continued)

---

[5] We will learn how to use high-precision floating-point numbers in Chapter 11.

(concluded)

2. Express the following binary numbers as their decimal equivalents:
   (a) 1110_2
   (b) 01010101_2
   (c) 1001_2

3. Express the following binary numbers as octal and hexadecimal numbers:
   (a) 1110010110101101_2
   (b) 1110111101_2
   (c) 1001011100111111_2

4. Is the fourth bit of the number 131_{10} a 1 or a 0?

5. Assume that the following numbers are the contents of a character variable. Find the character corresponding to each number according to the ASCII encoding scheme (The character codes in the ASCII encoding scheme are defined in Appendix A):
   (a) 77_{10}
   (b) 01111011_2
   (c) 249_{10}

6. Find the maximum and minimum values that can be stored in a 2-byte integer variable.

7. Can a 4-byte variable of the real data type be used to store larger numbers than a 4-byte variable of the integer data type? Why or why not? If it can, what is given up by the real variable to make this possible?

1.3
COMPUTER LANGUAGES

When a computer executes a program, it executes a string of very simple operations such as load, store, add, subtract, multiply, and so on. Each such operation has a unique binary pattern called an *operation code (op code)* to specify it. The program that a computer executes is just a string of op codes (and the data associated with the op codes[6]) in the order necessary to achieve a purpose. Op codes are collectively called **machine language,** since they are the actual language that a computer recognizes and executes.

Unfortunately, we humans find machine language very hard to work with. We prefer to work with English-like statements and algebraic equations that are expressed in forms familiar to us, instead of arbitrary patterns of zeros and ones. We like to program computers with **high-level languages.** We write out our instructions in a high-level language, and then use special programs called **compilers** and **linkers** to convert the instructions into the machine language that the computer understands.

---

[6] The data associated with op codes is called *operands.*

There are many different high-level languages, with different characteristics. Some of them are designed to work well for business problems, while others are designed for general scientific use. Still others are especially suited for applications like operating systems programming. It is important to pick a proper language to match the problem that you are trying to solve.

Some common high-level computer languages today include Ada, C, C++, Fortran, and Java. Historically, Fortran has been the pre-eminent language for general scientific computations. It has been around in one form or another for more than 60 years, and has been used to implement everything from computer models of nuclear power plants to aircraft design programs to seismic signal processing systems, including some projects requiring literally millions of lines of code. The language is especially useful for numerical analysis and technical calculations. In addition, Fortran is the dominant language in the world of supercomputers and massively parallel computers.

1.4
THE HISTORY OF THE FORTRAN LANGUAGE

Fortran is the grandfather of all scientific computer languages. The name Fortran is derived from FORmula TRANslation, indicating that the language was intended from the start for translating scientific equations into computer code. The first version of the FORTRAN[7] language was developed during the years 1954–1957 by IBM for use with its Type 704 computer (see Figure 1-4). Before that time, essentially all computer programs were generated by hand in machine language, which was a slow, tedious, and error-prone process. FORTRAN was a truly revolutionary product. For the first time, a programmer could write a desired algorithm as a series of standard algebraic equations, and the FORTRAN compiler would convert the statements into the machine language that the computer could recognize and execute.

THE IBM TYPE 704 COMPUTER

The IBM Type 704 computer was the first computer ever to use the FORTRAN language. It was released in 1954, and was widely used from then until about 1960, when it was replaced by the Model 709. As you can see from Figure 1-4, the computer occupied a whole room.

What could a computer like that do in 1954? Not much, by today's standards. Any PC sitting on a desktop can run rings around it. The 704 could perform about 4000 integer multiplications and divisions per second, and an average of about 8000 floating-point operations per second. It could read data from magnetic drums

---

[7] Versions of the language before Fortran 90 were known as FORTRAN (written with all capital letters), while Fortran 90 and later versions are known as Fortran (with only the first letter capitalized).

FIGURE 1-4
The IBM Type 704 computer. *(© Bettmann/Getty Images)*

(the equivalent of a disk drive) into memory at a rate of about 50,000 bytes per second. The amount of data storage available on a magnetic drum was also very small, so most programs that were not currently in use were stored as decks of punched cards.

By comparison, a typical modern personal computer (circa 2006) performed more than 20,000,000,000 integer multiplications and divisions per second, and hundreds of millions of floating-point operations per second. Some of today's workstations are small enough to sit on a desktop, and yet can perform more than 5,000,000,000 floating-point operations per second! Reads from disk into memory occur at rates greater than 25,000,000 bytes per second, and a typical PC disk drive can store more than 200,000,000,000 bytes of data.

The limited resources available in the 704 and other machines of that generation placed a great premium on efficient programming. The structured programming techniques that we use today were simply not possible, because there was not enough speed or memory to support them. The earliest versions of FORTRAN were designed with those limitations in mind, which is why we find many archaic features preserved as living fossils in modern versions of Fortran.

FORTRAN was a wonderful idea! People began using it as soon as it was available, because it made programming so much easier than machine language did. The language was officially released in April 1957, and by the fall of 1958, more than half of all IBM 704 computer programs were being written in Fortran.

The original FORTRAN language was very small compared to our modern versions of Fortran. It contained only a limited number of statement types, and supported only the integer and real data types. There were also no subroutines in the first FORTRAN. It was a first effort at writing a high-level computer language, and naturally many deficiencies were found as people started using the language regularly. IBM addressed those problems, releasing FORTRAN II in the spring of 1958.

Further developments continued through 1962, when FORTRAN IV was released. FORTRAN IV was a great improvement, and it became the standard version of Fortran for the next 15 years. In 1966, FORTRAN IV was adopted as an ANSI standard, and it came to be known as FORTRAN 66.

The Fortran language received another major update in 1977. FORTRAN 77 included many new features designed to make structured programs easier to write and maintain, and it quickly became "the" Fortran. FORTRAN 77 introduced such structures as the block IF, and was the first version of Fortran in which character variables were truly easy to manipulate.

The next major update of Fortran was Fortran 90.[8] Fortran 90 included all of FORTRAN 77 as a subset, and extended the language in many important new directions. Among the major improvements introduced to the language in Fortran 90 were a new free source format, array sections, whole-array operations, parameterized data types, derived data types, and explicit interfaces. Fortran 90 was a dramatic improvement over earlier versions of the language.

Fortran 90 was followed in 1996 by a minor update called Fortran 95. Fortran 95 added a number of new features to the language such as the FORALL construct, pure functions, and some new intrinsic procedures. In addition, it clarified numerous ambiguities in the Fortran 90 standard.

Fortran 2003 was the next update.[9] This is a more major change from Fortran 95, including new features such as enhanced derived types, object-oriented programming support, Unicode character set support, data manipulation enhancements, procedure pointers, and interoperability with the C language. It was followed by a more minor update called Fortran 2008.

The subject of this book is the Fortran 2008 language. The designers of Fortran 2008 were careful to make the language backward compatible with FORTRAN 77 and earlier versions. Because of this backward compatibility, most of the millions of programs written in FORTRAN 77 also work with Fortran 2008. Unfortunately, being backward compatible with earlier versions of Fortran required that Fortran 2008 retain some archaic features that should never be used in any modern program. *In this book,*

---

[8] American National Standard Programming Language Fortran, ANSI X3.198-1992; and International Standards Organization ISO/IEC 1539: 1991, Information Technology—Programming Languages—Fortran.

[9] International Standards Organization ISO/IEC 1539: 2004, Information Technology—Programming Languages—Fortran.

we will learn to program in Fortran *using only its modern features.* The older features that are retained for backward compatibility are relegated to Chapter 18 of this book. They are described there in case you run into any of them in older programs, but they should never be used in any new program.

1.5
THE EVOLUTION OF FORTRAN

The Fortran language is a dynamic language that is constantly evolving to keep up with advances in programming practice and computing technology. A major new version appears about once per decade.

The responsibility for developing new versions of the Fortran language lies with the International Organization for Standardization's (ISO) Fortran Working Group, WG5. That organization has delegated authority to the J3 Committee of the InterNational Committee for Information Technology Standards (INCITS) to actually prepare new versions of the language. The preparation of each new version is an extended process involving first asking for suggestions for inclusion in the language, deciding which suggestions are feasible to implement, writing and circulating drafts to all interested parties throughout the world, and correcting the drafts and trying again until general agreement is reached. Eventually, a worldwide vote is held and the standard is adopted.

The designers of new versions of the Fortran language must strike a delicate balance between backward compatibility with the existing base of Fortran programs and the introduction of desirable new features. Although modern structured programming features and approaches have been introduced into the language, many undesirable features from earlier versions of Fortran have been retained for backward compatibility.

The designers have developed a mechanism for identifying undesirable and obsolete features of the Fortran language that should no longer be used, and for eventually eliminating them from the language. Those parts of the language that have been superseded by new and better methods are declared to be **obsolescent features.** *Features that have been declared obsolescent should never be used in any new programs.* As the use of these features declines in the existing Fortran code base, they will then be considered for deletion from the language. No feature will ever be deleted from a version of the language unless it was on the obsolescent list in at least one previous version, and unless the usage of the feature has dropped off to negligible levels. In this fashion, the language can evolve without threatening the existing Fortran code base.

The redundant, obsolescent, and deleted features of Fortran 2008 are described in Chapter 18 in case a programmer runs into them in existing programs, but they should never be used in any new programs.

We can get a feeling for just how much the Fortran language has evolved over the years by examining Figures 1-5 through 1-7. These three figures show programs for calculating the solutions to the quadratic equation $ax^2 + bx + c = 0$ in the styles of the original FORTRAN I, of FORTRAN 77, and of Fortran 2008. It is obvious that the language has become more readable and structured over the years. Amazingly, though, Fortran 2008 compilers will still compile the FORTRAN I program with just a few minor changes![10]

FIGURE 1-5
A FORTRAN I program to solve for the roots of the quadratic equation $ax^2 + bx + c = 0$.

```
C       SOLVE QUADRATIC EQUATION IN FORTRAN I
        READ 100,A,B,C
100     FORMAT(3F12.4)
        DISCR = B**2-4*A*C
        IF (DISCR) 10,20,30
10      X1=(-B)/(2.*A)
        X2=SQRTF(ABSF(DISCR))/(2.*A)
        PRINT 110,X1,X2
110     FORMAT(5H X = ,F12.3,4H +i ,F12.3)
        PRINT 120,X1,X2
120     FORMAT(5H X = ,F12.3,4H -i ,F12.3)
        GOTO 40
20      X1=(-B)/(2.*A)
        PRINT 130,X1
130     FORMAT(11H X1 = X2 = ,F12.3)
        GOTO 40
30      X1=((-B)+SQRTF(ABSF(DISCR)))/(2.*A)
        X2=((-B)-SQRTF(ABSF(DISCR)))/(2.*A)
        PRINT 140,X1
140     FORMAT(6H X1 = ,F12.3)
        PRINT 150,X2
150     FORMAT(6H X2 = ,F12.3)
40      CONTINUE
        STOP 25252
```

FIGURE 1-6
A FORTRAN 77 program to solve for the roots of the quadratic equation $ax^2 + bx + c = 0$.

```
        PROGRAM QUAD4
C
C       This program reads the coefficients of a quadratic equation of
C       the form
C             A * X**2 + B * X + C = 0,
C       and solves for the roots of the equation (FORTRAN 77 style).
C
C       Get the coefficients of the quadratic equation.
C
        WRITE (*,*) 'Enter the coefficients A, B and C: '
        READ (*,*) A, B, C
C
C       Echo the coefficients to make sure they are entered correctly.
C
```

(continued)

---

[10] Change SQRTF to SQRT, ABSF to ABS, and add an END statement.

(*concluded*)

```
      WRITE (*,100) 'The coefficients are : ', A, B, C
100   FORMAT (1X,A,3F10.4)
C
C     Check the discriminant and calculate its roots.
C
      DISCR = B**2 - 4.*A*C
      IF ( DISCR .LT. 0) THEN
         WRITE (*,*) ' This equation has complex roots:'
         WRITE (*,*) ' X = ', -B/(2.*A), ' +i ', SQRT(ABS(DISCR))/(2.*A)
         WRITE (*,*) ' X = ', -B/(2.*A), ' -i ', SQRT(ABS(DISCR))/(2.*A)
      ELSE IF ( (B**2 - 4.*A*C) .EQ. 0) THEN
         WRITE (*,*) ' This equation has a single repeated real root:'
         WRITE (*,*) ' X = ', -B/(2.*A)
      ELSE
         WRITE (*,*) ' This equation has two distinct real roots:'
         WRITE (*,*) ' X = ', (-B + SQRT(ABS(DISCR)))/(2.*A)
         WRITE (*,*) ' X = ', (-B - SQRT(ABS(DISCR)))/(2.*A)
      END IF
C
      END
```

FIGURE 1-7
A Fortran 2008 program to solve for the roots of the quadratic equation $ax^2 + bx + c = 0$.

```
PROGRAM roots

!  Purpose:
!    This program solves for the roots of a quadratic equation of the form
!    A * X**2 + B * X + C = 0.  It calculates the answers regardless of the
!    type of roots that the equation possesses (Fortran 95/2003 style).
!
IMPLICIT NONE

!  Declare the variables used in this program

REAL :: a                 ! Coefficient of X**2 term of equation
REAL :: b                 ! Coefficient of X term of equation
REAL :: c                 ! Constant term of equation
REAL :: discriminant      ! Discriminant of the equation
REAL :: imag_part         ! Imaginary part of equation (for complex roots)
REAL :: real_part         ! Real part of equation (for complex roots)
REAL :: x1                ! First solution of equation (for real roots)
REAL :: x2                ! Second solution of equation (for real roots)

!  Prompt the user for the coefficients of the equation
WRITE (*,*) 'This program solves for the roots of a quadratic '
WRITE (*,*) 'equation of the form A * X**2 + B * X + C = 0. '
WRITE (*,*) 'Enter the coefficients A, B, and C:'
READ  (*,*) a, b, c

!  Echo back coefficients
WRITE (*,*) 'The coefficients A, B, and C are: ', a, b, c
```

(*continued*)

(*concluded*)

```
! Calculate discriminant
discriminant = b**2 - 4. * a * c

! Solve for the roots, depending upon the value of the discriminant

IF ( discriminant > 0. ) THEN ! there are two real roots, so...

    X1 = ( -b + sqrt(discriminant) ) / ( 2. * a )
    X2 = ( -b - sqrt(discriminant) ) / ( 2. * a )
    WRITE (*,*) 'This equation has two real roots:'
    WRITE (*,*) 'X1 = ', x1
    WRITE (*,*) 'X2 = ', x2

ELSE IF ( discriminant == 0. ) THEN ! there is one repeated root, so...

    x1 = ( -b ) / ( 2. * a )
    WRITE (*,*) 'This equation has two identical real roots:'
    WRITE (*,*) 'X1 = X2 = ', x1

ELSE ! there are complex roots, so ...

    real_part = ( -b ) / ( 2. * a )
    imag_part = sqrt ( abs ( discriminant ) ) / ( 2. * a )
    WRITE (*,*) 'This equation has complex roots:'
    WRITE (*,*) 'X1 = ', real_part, ' +i ', imag_part
    WRITE (*,*) 'X2 = ', real_part, ' -i ', imag_part

END IF

END PROGRAM roots
```

1.6
SUMMARY

A computer is a special type of machine that stores information, and can perform mathematical calculations on that information at speeds much faster than human beings can think. A program, which is stored in the computer's memory, tells the computer what sequence of calculations is required, and which information to perform the calculations on.

The major components of a computer are the central processing unit (CPU), cache memory, main memory, secondary memory, and input and output devices. The CPU performs all of the control and calculation functions of the computer. Cache memory is very fact memory integrated directly on the CPU chip. Main memory is somewhat slower memory that is used to store the program being executed and its associated data. Main memory is volatile, meaning that its contents are lost whenever power is turned off. Secondary memory is slower and cheaper than main memory. It is nonvolatile. Hard disks are common secondary memory devices. Input and output devices are used to read data into the computer and to output data from the computer. The most common input device is a keyboard, and the most common output devices are displays or printers.

Computer memories are composed of millions of individual switches, each of which can be ON or OFF, but not at a state in between. These individual switches are binary devices called bits. Eight bits are grouped together to form a *byte* of memory, and 2 or more bytes (depending on the computer) are grouped together to form a *word* of memory.

Computer memories can be used to store *character*, *integer*, or *real* data. Each character in most character data sets occupies 1 byte of memory. The 256 possible values in the byte allow for 256 possible character codes. (Characters in the Unicode character set occupy 2 bytes, allowing for 65,536 possible character codes.) Integer values occupy 1, 2, 4, or 8 bytes of memory, and store integer quantities. Real values store numbers in a kind of scientific notation. They usually occupy 4 bytes of memory. The bits are divided into a separate mantissa and exponent. The *precision* of the number depends upon the number of bits in the mantissa, and the *range* of the number depends upon the number of bits in the exponent.

The earliest computers were programmed in *machine language*. This process was slow, cumbersome, and error-prone. High-level languages began to appear in about 1954, and they quickly replaced machine language coding for most uses. FORTRAN was one of the first high-level languages ever created.

The FORTRAN I computer language and compiler were originally developed in 1954–1957. The language has since gone through many revisions, and a standard mechanism has been created to evolve the language. This book teaches good programming practices using the modern Fortran language.

1.6.1. Exercises

1-1. Express the following decimal numbers as their binary equivalents:

(a) 10_{10}

(b) 32_{10}

(c) 77_{10}

(d) 63_{10}

1-2. Express the following binary numbers as their decimal equivalents:

(a) 01001000_2

(b) 10001001_2

(c) 11111111_2

(d) 0101_2

1-3. Express the following numbers in both octal and hexadecimal forms:

(a) 1010111011110001_2

(b) 330_{10}

(c) 111_{10}

(d) 11111101101_2

1-4. Express the following numbers in binary and decimal forms:

(a) 377_8

(b) $1A8_{16}$

(c) 111_8

(d) $1FF_{16}$

1-5. Some computers (such as IBM mainframes) used to implement real data using a 23-bit mantissa and a 9-bit exponent. What precision and range can we expect from real data on these machines?

1-6. Some Cray supercomputers used to support 46-bit and 64-bit integer data types. What are the maximum and minimum values that we could express in a 46-bit integer? in a 64-bit integer?

1-7. Find the 16-bit two's complement representation of the following decimal numbers:

(a) 55_{10}

(b) -5_{10}

(c) 1024_{10}

(d) -1024_{10}

1-8. Add the two's complement numbers 0010010010010010_2 and 1111110011111100_2 using binary arithmetic. Convert the two numbers to decimal form, and add them as decimals. Do the two answers agree?

1-9. The largest possible 8-bit two's complement number is 01111111_2, and the smallest possible 8-bit two's complement number is 10000000_2. Convert these numbers to decimal form. How do they compare to the results of Equations (1-1) and (1-2)?

1-10. The Fortran language includes a second type of floating-point data known as double precision. A double-precision number usually occupies 8 bytes (64 bits), instead of the 4 bytes occupied by a real number. In the most common implementation, 53 bits are used for the mantissa and 11 bits are used for the exponent. How many significant digits does a double-precision value have? What is the range of double-precision numbers?

2

Basic Elements of Fortran

OBJECTIVES

- Know which characters are legal in a Fortran statement.
- Know the basic structure of a Fortran statement and a Fortran program.
- Know the difference between executable and nonexecutable statements.
- Know the difference between constants and variables.
- Understand the differences among the INTEGER, REAL, and CHARACTER data types.
- Learn the difference between default and explicit typing, and understand why explicit typing should always be used.
- Know the structure of a Fortran assignment statement.
- Learn the differences between integer arithmetic and real arithmetic, and when each one should be used.
- Know the Fortran hierarchy of operations.
- Learn how Fortran handles mixed-mode arithmetic expressions.
- Learn what intrinsic functions are, and how to use them.
- Know how to use list-directed input and output statements.
- Know why it is important to always use the IMPLICIT NONE statement.

2.1
INTRODUCTION

As engineers and scientists, we design and execute computer programs to accomplish a goal. The goal typically involves technical calculations that would be too difficult or take too long to be performed by hand. Fortran is one of the computer languages commonly used for these technical calculations.

This chapter introduces the basic elements of the Fortran language. By the end of the chapter, we will be able to write simple but functional Fortran programs.

2.2
THE FORTRAN CHARACTER SET

Every language, whether it is a natural language such as English or a computer language such as Fortran, Java, or C++, has its own special alphabet. Only the characters in this alphabet may be used with the language.

The special alphabet used with the Fortran language is known as the **Fortran character set.** The Fortran character set consists of 97 characters, as shown in Table 2-1.

TABLE 2-1
The Fortran character set

Number of symbols	Type	Values
26	Uppercase letters	A-Z
26	Lowercase letters	a-z
10	Digits	0-9
1	Underscore character	_
5	Arithmetic symbols	+ - * / **
28	Miscellaneous symbols	( ) . = , ' $: ! " % & ; < > ? ~ \ [ ] ` ^ { } | # @ and blank

Note that the uppercase letters of the alphabet are equivalent to the lowercase ones in the Fortran character set. (For example, the uppercase letter A is equivalent to the lowercase letter a.) In other words, Fortran is *case insensitive*. This behavior is in contrast with such case sensitive languages as C++ and Java, in which A and a are two totally different things.

2.3
THE STRUCTURE OF A FORTRAN STATEMENT

A Fortran program consists of a series of *statements* designed to accomplish the goal of the programmer. There are two basic types of statements: **executable statements** and **nonexecutable statements.** Executable statements describe the actions taken by the program when it is executed (additions, subtractions, multiplications, divisions, etc.), while nonexecutable statements provide information necessary for the proper operation of the program. We will see many examples of each type of statement as we learn more about the Fortran language.

Fortran statements may be entered anywhere on a line, and each line may be up to 132 characters long. If a statement is too long to fit onto a single line, then it may be continued on the next line by ending the current line (and optionally starting the next line) with an ampersand (&) character. For example, the following three Fortran statements are identical:

```
output = input1 + input2 ! Sum the inputs
output = input1 &
       + input2           ! Sum the inputs
```

```
999 output = input1 &        ! Sum the inputs
           & + input2
```

Each of the statements specifies that the computer should add the two quantities stored in `input1` and `input2` and save the result in `output`. A Fortran statement can be continued over up to 256 lines, if required.

The last statement shown above starts with a number, known as a **statement label.** A statement label can be any number between 1 and 99,999. It is the "name" of a Fortran statement, and may be used to refer to the statement in other parts of the program. Note that a statement label has no significance other than as a "name" for the statement. It is *not* a line number, and it tells nothing about the order in which statements are executed. *Statement labels are rare in modern Fortran, and most statements will not have one.* If a statement label is used, it must be unique within a given program unit. For example, if 100 is used as a statement label on one line, it cannot be used again as a statement label on any other line in the same program unit.

Any characters following an exclamation point are **comments,** and are ignored by the Fortran compiler. All text from the exclamation point to the end of the line will be ignored, so comments may appear on the same line as an executable statement. Comments are very important, because they help us document the proper operation of a program. In the third example above, the comment is ignored, so the ampersand is treated by the compiler as the last character on the line.

2.4
THE STRUCTURE OF A FORTRAN PROGRAM

Each Fortran program consists of a mixture of executable and nonexecutable statements, which must occur in a specific order. An example Fortran program is shown in Figure 2-1. This program reads in two numbers, multiplies them together, and prints out the result. Let's examine the significant features of this program.

FIGURE 2-1
A simple Fortran program.

```
PROGRAM my_first_program

!  Purpose:
!    To illustrate some of the basic features of a Fortran program.
!

! Declare the variables used in this program.
INTEGER :: i, j, k           ! All variables are integers

! Get two values to store in variables i and j
WRITE (*,*) 'Enter the numbers to multiply: '
READ (*,*) i, j
```

(*continued*)

(*concluded*)

```
! Multiply the numbers together
k = i * j

!  Write out the result.
WRITE (*,*) 'Result = ', k

!  Finish up.
STOP
END PROGRAM my_first_program
```

This Fortran program, like all Fortran program units,[1] is divided into three sections:

1. *The declaration section.* This section consists of a group of nonexecutable statements at the beginning of the program that define the name of the program and the number and types of variables referenced in the program.
2. *The execution section.* This section consists of one or more statements describing the actions to be performed by the program.
3. *The termination section.* This section consists of a statement or statements stopping the execution of the program and telling the compiler that the program is complete.

Note that comments may be inserted freely anywhere within, before, or after the program.

2.4.1 The Declaration Section

The declaration section consists of the nonexecutable statements at the beginning of the program that define the name of the program and the number and types of variables referenced in the program.

The first statement in this section is the PROGRAM statement. It is a nonexecutable statement that specifies the name of the program to the Fortran compiler. Fortran program names may be up to 63 characters long and contain any combination of alphabetic characters, digits, and the underscore (_) character. However, the first character in a program name must always be alphabetic. If present, the PROGRAM statement must be the first line of the program. In this example, the program has been named my_first_program.

The next several lines in the program are comments that describe the purpose of the program. Next comes the INTEGER type declaration statement. This nonexecutable statement will be described later in this chapter. Here, it declares that three integer variables called i, j, and k will be used in this program.

---

[1] A *program unit* is a separately-compiled piece of Fortran code. We will meet several other types of program units beginning in Chapter 7.

2.4.2 The Execution Section

The execution section consists of one or more executable statements describing the actions to be performed by the program.

The first executable statement in this program is the `WRITE` statement, which writes out a message prompting the user to enter the two numbers to be multiplied together. The next executable statement is a `READ` statement, which reads in the two integers supplied by the user. The third executable statement instructs the computer to multiply the two numbers i and j together, and to store the result in variable k. The final `WRITE` statement prints out the result for the user to see. Comments may be embedded anywhere throughout the execution section.

All of these statements will be explained in detail later in this chapter.

2.4.3 The Termination Section

The termination section consists of the `STOP` and `END PROGRAM` statements. The `STOP` statement is a statement that tells the computer to stop running the program. The `END PROGRAM` statement is a statement that tells the compiler that there are no more statements to be compiled in the program.

The `STOP` statement takes one of the following forms:

```
STOP
STOP 3
STOP 'Error stop'
```

If the `STOP` statement is used by itself, execution will stop. If the `STOP` statement is used with a number, that number will be printed out when the program stops, and will normally be returned to the operating system as an error code. If the `STOP` statement is used with a character string, that string will be printed out when the program stops.

When the `STOP` statement immediately precedes the `END PROGRAM` statement as in this example, it is optional. The compiler will automatically generate a `STOP` command when the `END PROGRAM` statement is reached. The `STOP` statement is therefore rarely used.[2]

There is an alternate version of the `STOP` statement called `ERROR STOP`. This version stops the program, but it also notifies the operating system that the program failed to execute properly. An example might be:

```
ERROR STOP 'Cannot access database'
```

This version of the `STOP` statement was added in Fortran 2008, and it might be useful if you need to inform an operating system script that a program failed abnormally.

---

[2] There is a philosophical disagreement among Fortran programmers about the use of the `STOP` statement. Some programming instructors believe that it should always be used, even though it is redundant when located before an `END PROGRAM` statement. They argue that the `STOP` statement makes the end of execution explicit. The author of this book is of the school that believes that a good program should only have *one* starting point and *one* ending point, with no additional stopping points anywhere along the way. In that case, a `STOP` is totally redundant and will never be used. Depending on the philosophy of your instructor, you may or may not be encouraged to use this statement.

2.4.4 Program Style

This example program follows a commonly used Fortran convention of capitalizing keywords such as `PROGRAM`, `READ`, and `WRITE`, while using lowercase for the program variables. Names are written with underscores between the words, as in `my_first_program` above. It also uses capital letters for named constants such as `PI` (π). This is *not* a Fortran requirement; the program would have worked just as well if all capital letters or all lowercase letters were used. Since uppercase and lowercase letters are equivalent in Fortran, the program functions identically in either case.

Throughout this book, we will follow this convention of capitalizing Fortran keywords and constants, and using lowercase for variables, procedure names, etc.

Some programmers use other styles to write Fortran programs. For example, Java programmers who also work with Fortran might adopt a Java-like convention in which keywords and names are in lowercase, with capital letters at the beginning of each word (sometimes called "camel case"). Such a programmer might give this program the name `myFirstProgram`. This is an equally valid way to write a Fortran program.

It is not necessary for you to follow any specific convention to write a Fortran program, but *you should always be consistent* in your programming style. Establish a standard practice, or adopt the standard practice of the organization in which you work, and then follow it consistently in all of your programs.

Good Programming Practice
Adopt a programming style, and then follow it consistently in all of your programs.

2.4.5 Compiling, Linking, and Executing the Fortran Program

Before the sample program can be run, it must be compiled into object code with a Fortran compiler, and then linked with a computer's system libraries to produce an executable program (Figure 2-2). These two steps are usually done together in response to a single programmer command. *The details of compiling and linking are different for every compiler and operating system.* You should ask your instructor or consult the appropriate manuals to determine the proper procedure for your system.

FIGURE 2-2
Creating an executable Fortran program involves two steps, compiling and linking.

Fortran programs can be compiled, linked, and executed in one of two possible modes: **batch** and **interactive.** In batch mode, a program is executed without an input from or interaction with a user. This is the way most Fortran programs worked in the early days. A program would be submitted as a deck of punched cards or in a file, and it would be compiled, linked, and executed without any user interaction. All input data for the program had to be placed on cards or put in files before the job was started, and all output went to output files or to a line printer.

By contrast, a program that is run in interactive mode is compiled, linked, and executed while a user is waiting at an input device such as the computer keyboard or a terminal. Since the program executes with the human present, it can ask for input data from the user as it is executing, and it can display intermediate and final results as soon as they are computed.

Today, most Fortran programs are executed in interactive mode. However, some very large Fortran programs that execute for days at a time are still run in batch mode.

2.5
CONSTANTS AND VARIABLES

A **constant** is a data object that is defined before a program is executed, and that does not change value during the execution of the program. When a Fortran compiler encounters a constant, it places the value of the constant in a known location in memory, and then references that memory location whenever the constant is used in the program. A **variable** is a data object that can change value during the execution of a program. (The value of a Fortran variable may or may not be initialized before a program is executed.) When a Fortran compiler encounters a variable, it reserves a known location in memory for the variable, and then references that memory location whenever the variable is used in the program.

Each Fortran variable in a program unit must have a unique name. The variable name is a label for a specific location in memory that is easy for humans to remember and use. Fortran variable names may be up to 63 characters long, and may contain any combination of alphabetic characters, digits, and the underscore (_) character. However, the first character in a name must always be alphabetic. The following examples are valid variable names:

```
time
distance
z123456789
I_want_to_go_home
```

The following examples are invalid variable names:

```
this_is_a_very_very_very_very_very_very_very_very_long_variable_name
```
(Name is too long.)

```
3_days
```
(First character is a number.)

```
A$
```
($ is an illegal character.)

When writing a program, it is important to pick meaningful names for the variables. Meaningful names make a program *much* easier to read and to maintain. Names such as day, month, and year are quite clear even to a person seeing a program for the first time. Since spaces cannot be used in Fortran variable names, underscore characters can be substituted to create meaningful names. For example, *exchange rate* might become exchange_rate.

Good Programming Practice
Use meaningful variable names whenever possible.

It is also important to include a **data dictionary** in the header of any program that you write. A data dictionary lists the definition of each variable used in a program. The definition should include both a description of the contents of the item and the units in which it is measured. A data dictionary may seem unnecessary while the program is being written, but it is invaluable when you or another person have to go back and modify the program at a later time.

Good Programming Practice
Create a data dictionary for each program to make program maintenance easier.

There are five intrinsic or "built-in" types of Fortran constants and variables. Three of them are numeric (types INTEGER, REAL, and COMPLEX), one is logical (type LOGICAL), and one consists of strings of characters (type CHARACTER). The simplest forms of the INTEGER, REAL, and CHARACTER data types will be discussed now. The LOGICAL data type is included in Chapter 3. More advanced forms of various data types will be discussed in Chapter 11.

In addition to the intrinsic data types, Fortran permits a programmer to define **derived data types,** which are special data types intended to solve particular problems. Derived data types will also be discussed in Chapter 12.

2.5.1 Integer Constants and Variables

The integer data type consists of integer constants and variables. This data type can only store integer values—it cannot represent numbers with fractional parts.

An **integer constant** is any number that does not contain a decimal point. If a constant is positive, it may be written either with or without a + sign. No commas may be embedded within an integer constant. The following examples are valid integer constants:

```
       0
    -999
123456789
     +17
```

The following examples are *not* valid integer constants:

    1,000,000   (Embedded commas are illegal.)
    -100.   (If it has a decimal point, it is not an integer constant!)

An **integer variable** is a variable containing a value of the integer data type.

Constants and variables of the integer data type are usually stored in a single word on a computer. Since the length of a word varies from 32 bits to 64 bits on different computers, the largest integer that can be stored in a computer also varies. The largest and smallest integers that can be stored in a particular computer can be determined from the word size by applying Equations (1-1) and (1-2).

Almost all Fortran compilers support integers with more than one length. For example, most PC compilers support 16-bit, 32-bit, and 64-bit integers. These different lengths of integers are known as different **kinds** of integers. Fortran has an explicit mechanism for choosing which kind of integer is used for a given value. This mechanism is explained in Chapter 11.

2.5.2 Real Constants and Variables

The real data type consists of numbers stored in real or floating-point format. Unlike integers, the real data type can represent numbers with fractional components.

A **real constant** is a constant written with a decimal point. It may be written with or without an exponent. If the constant is positive, it may be written either with or without a + sign. No commas may be embedded within a real constant.

Real constants may be written with or without an exponent. If used, the exponent consists of the letter E followed by a positive or negative integer, which corresponds to the power of 10 used when the number is written in scientific notation. If the exponent is positive, the + sign may be omitted. The mantissa of the number (the part of the number that precedes the exponent) should contain a decimal point. The following examples are valid real constants:

    10.
    -999.9
    +1.0E-3   ($= 1.0 \times 10^{-3}$, or 0.001)
    123.45E20   ($= 123.45 \times 10^{20}$, or 1.2345×10^{22})
    0.12E+1   ($= 0.12 \times 10^{1}$, or 1.2)

The following examples are *not* valid real constants:

    1,000,000.   (Embedded commas are illegal.)
    111E3   (A decimal point is required in the mantissa.)
    -12.0E1.5   (Decimal points are not allowed in exponents.)

A **real variable** is a variable containing a value of the real data type.

A real value is stored in two parts: the **mantissa** and the **exponent**. The number of bits allocated to the mantissa determines the *precision* of the constant (that is, the number of significant digits to which the constant is known), while the number of bits allocated to the exponent determines the *range* of the constant (that is, the largest and

Basic Elements of Fortran

TABLE 2-2
Precision and range of real numbers

Computer standard	Total number of bits	Number of bits in mantissa	Precision in decimal digits	Number of bits in exponent	Exponent range
IEEE 754	32	24	7	8	10^{-38} to 10^{38}
	64	53	15	11	10^{-308} to 10^{308}
	128	112	34	16	10^{-4932} to 10^{4932}

the smallest values that can be represented). For a given word size, the more precise a real number is, the smaller its range is, and vice versa, as described in the previous chapter.

Over the last 25 years, almost all computers have switched to using floating-point numbers that conform to IEEE Standard 754. Table 2-2 shows the precision and the range of typical real constants and variables on IEEE Standard 754 compliant computers.

All Fortran compilers support real numbers with more than one length. For example, PC compilers support both 32-bit real numbers and 64-bit real numbers. These different lengths of real numbers are known as different **kinds**. By selecting the proper kind, it is possible to increase the precision and range of a real constant or variable. Fortran has an explicit mechanism for choosing which kind of real is used for a given value. This mechanism is explained in detail in Chapter 11.

2.5.3 Character Constants and Variables

The character data type consists of strings of alphanumeric characters. A **character constant** is a string of characters enclosed in single (') or double (") quotes. The minimum number of characters in a string is 0, while the maximum number of characters in a string varies from compiler to compiler.

The characters between the two single or double quotes are said to be in a **character context.** Any characters representable on a computer are legal in a character *context, not just the 97 characters forming the Fortran character set.*

The following are valid character constants:

```
'This is a test!'
'b'                     (a single blank)[3]
'{^}'                   (These characters are legal in a character
                        context even though they are not a part of
                        the Fortran character set.)
"3.141593"              (This is a character string, not a number.)
```

---

[3] In places where the difference matters, the symbol ƀ is used to indicate a blank character, so that the student can tell the difference between a string containing *no* characters ('') and one containing a single blank character ('ƀ').

The following are not valid character constants:

```
This is a test!          (No single or double quotes)
'This is a test!"        (Mismatched quotes)
"Try this one.'          (Unbalanced single quotes)
```

If a character string must include an apostrophe, then that apostrophe may be represented by two consecutive single quotes. For example, the string "Man's best friend" would be written in a character constant as

```
'Man''s best friend'
```

Alternatively, the character string containing a single quote can be surrounded by double quotes. For example, the string "Man's best friend" could be written as

```
"Man's best friend"
```

Similarly, a character string containing double quotes can be surrounded by single quotes. The character string "Who cares?" could be written in a character constant as

```
'"Who cares?"'
```

Character constants are most often used to print descriptive information using the WRITE statement. For example, the string `'Result = '` in Figure 2-1 is a valid character constant:

```
WRITE (*,*) 'Result = ', k
```

A **character variable** is a variable containing a value of the character data type.

2.5.4 Default and Explicit Variable Typing

When we look at a constant, it is easy to see whether it is of type integer, real, or character. If a number does not have a decimal point, it is of type integer; if it has a decimal point, it is of type real. If the constant is enclosed in single or double quotes, it is of type character. With variables, the situation is not so clear. How do we (or the compiler) know if the variable junk contains an integer, real, or character value?

There are two possible ways in which the type of a variable can be defined: **default typing** and **explicit typing**. If the type of a variable is not explicitly specified in the program, then default typing is used. By default:

Any variable names beginning with the letters i, j, k, l, m, or n are assumed to be of type INTEGER. Any variable names starting with another letter are assumed to be of type REAL.

Therefore, a variable called `incr` is assumed to be of type integer by default, while a variable called `big` is assumed to be of type real by default. This default typing convention goes all the way back to the original Fortran I in 1954. Note that no variable names are of type character by default, because this data type didn't exist in Fortran I!

The type of a variable may also be explicitly defined in the declaration section at the beginning of a program. The following Fortran statements can be used to specify the type of variables:[4]

```
INTEGER :: var1 [, var2, var3, ...]
REAL ::    var1 [, var2, var3, ...]
```

where the values inside the *[ ]* are optional. In this case, the values inside the brackets show that more than two variables may be declared on a single line if they are separated by commas.

These nonexecutable statements are called **type declaration statements.** They should be placed after the `PROGRAM` statement and before the first executable statement in the program, as shown in the example below.

```
PROGRAM example
INTEGER :: day, month, year
REAL :: second
...
(Executable statements follow here...)
```

There are no default names associated with the character data type, so all character variables must be explicitly typed using the `CHARACTER` type declaration statement. This statement is a bit more complicated than the previous ones, since character variables may have different lengths. Its form is:

```
CHARACTER(len=<len>) :: var1 [, var2, var3, ...]
```

where `<len>` is the number of characters in the variables. The `(len=<len>)` portion of the statement is optional. If only a number appears in the parentheses, then the character variables declared by the statement are of that length. If the parentheses are entirely absent, then the character variables declared by the statement have length 1. For example, the type declaration statements

```
CHARACTER(len=10) :: first, last
CHARACTER :: initial
CHARACTER(15) :: id
```

define two 10-character variables called *first* and *last*, a 1-character variable called *initial*, and a 15-character variable called *id*.

---

[4] The double colon `::` is optional in the above statements for backward compatibility with earlier versions of Fortran. Thus, the following two statements are equivalent

```
INTEGER count
INTEGER :: count
```

The form with the double colon is preferred, because the double colons are not optional in more advanced forms of the type specification statement that we will see later.

2.5.5 Keeping Constants Consistent in a Program

It is important to always keep your physical constants consistent throughout a program. For example, do not use the value 3.14 for π at one point in a program, and 3.141593 at another point in the program. Also, you should always write your constants with at least as much precision as your computer will accept. If the real data type on your computer has seven significant digits of precision, then π should be written as 3.141593, *not* as 3.14!

The best way to achieve consistency and precision throughout a program is to *assign a name to a constant, and then to use that name to refer to the constant throughout the program.* If we assign the name PI to the constant 3.141593, then we can refer to PI by name throughout the program, and be certain that we are getting the same value everywhere. Furthermore, assigning meaningful names to constants improves the overall readability of our programs, because a programmer can tell at a glance just what the constant represents.

Named constants are created using the PARAMETER attribute of a type declaration statement. The form of a type declaration statement with a PARAMETER attribute is

```
type, PARAMETER :: name = value [, name2 = value2, ...]
```

where type is the type of the constant (integer, real, logical, or character), and name is the name assigned to constant *value*. More than one parameter may be declared on a single line if they are separated by commas. For example, the following statement assigns the name pi to the constant 3.141593.

```
REAL, PARAMETER :: PI = 3.141593
```

If the named constant is of type character, then it is not necessary to declare the length of the character string. Since the named constant is being defined on the same line as its type declaration, the Fortran compiler can directly count the number of characters in the string. For example, the following statements declare a named constant error_message to be the 14-character string 'Unknown error!'.

```
CHARACTER, PARAMETER :: ERROR_MESSAGE = 'Unknown error!'
```

In languages such as C, C++, and Java, named constants are usually written in all capital letters. Many Fortran programmers are also familiar with these languages, and they have adopted the convention of writing named constants in capital letters in Fortran as well. We will follow that practice in this book.

Good Programming Practice
Keep your physical constants consistent and precise throughout a program. To improve the consistency and understandability of your code, assign a name to any important constants, and refer to them by name in the program.

Quiz 2-1

This quiz provides a quick check to see if you have understood the concepts introduced in Section 2.5. If you have trouble with the quiz, reread the section, ask your instructor, or discuss the material with a fellow student. The answers to this quiz are found in the back of the book.

Questions 1 to 12 contain a list of valid and invalid constants. State whether or not each constant is valid. If the constant is valid, specify its type. If it is invalid, say why it is invalid.

1. 10.0
2. -100,000
3. 123E-5
4. 'That's ok!'
5. -32768
6. 3.14159
7. "Who are you?"
8. '3.14159'
9. 'Distance =
10. "That's ok!"
11. 17.877E+6
12. 13.0^2

Questions 13 to 16 contain two real constants each. Tell whether or not the two constants represent the same value within the computer:

13. 4650.; 4.65E+3
14. -12.71; -1.27E1
15. 0.0001; 1.0E4
16. 3.14159E0; 314.159E-3

Questions 17 and 18 contain a list of valid and invalid Fortran program names. State whether or not each program name is valid. If it is invalid, say why it is invalid.

17. PROGRAM new_program
18. PROGRAM 3rd

Questions 19 to 23 contain a list of valid and invalid Fortran variable names. State whether or not each variable name is valid. If the variable name is valid, specify its type (assume default typing). If it is invalid, say why it is invalid.

(continued)

> (*concluded*)
>
> 19. `length`
> 20. `distance`
> 21. `1problem`
> 22. `when_does_school_end`
> 23. `_ok`
>
> Are the following PARAMETER declarations correct or incorrect? If a statement is incorrect, state why it is invalid.
>
> 24. `REAL, PARAMETER BEGIN = -30`
> 25. `CHARACTER, PARAMETER :: NAME = 'Rosa'`

2.6

ASSIGNMENT STATEMENTS AND ARITHMETIC CALCULATIONS

Calculations are specified in Fortran with an **assignment statement,** whose general form is

`variable_name = expression`

The assignment statement calculates the value of the expression to the right of the equal sign, and *assigns* that value to the variable named on the left of the equal sign. Note that the equal sign does not mean equality in the usual sense of the word. Instead, it means: *store the value of* `expression` *into location* `variable_name`. For this reason, the equal sign is called the **assignment operator.** A statement like

`i = i + 1`

is complete nonsense in ordinary algebra, but makes perfect sense in Fortran. In Fortran, it means: Take the current value stored in variable `i`, add one to it, and store the result back into variable `i`.

The expression to the right of the assignment operator can be any valid combination of constants, variables, parentheses, and arithmetic or logical operators. The standard arithmetic operators included in Fortran are:

- `+` Addition
- `-` Subtraction
- `*` Multiplication
- `/` Division
- `**` Exponentiation

Note that the symbols for multiplication (`*`), division (`/`), and exponentiation (`**`) are not the ones used in ordinary mathematical expressions. These special symbols were

Basic Elements of Fortran

chosen because they were available in 1950s-era computer character sets, and because they were different from the characters being used in variable names.

The five arithmetic operators described above are **binary operators,** which means that they should occur between and apply to two variables or constants, as shown:

```
a + b
a - b
a ** b
a * b
a / b
```

In addition, the + and − symbols can occur as **unary operators,** which means that they apply to one variable or constant, as shown:

```
+23
-a
```

The following rules apply when using Fortran arithmetic operators:

1. No two operators may occur side by side. Thus, the expression a * -b is illegal. In Fortran, it must be written as a * (-b). Similarly, a ** -2 is illegal, and should be written as a ** (-2).
2. Implied multiplication is illegal in Fortran. An expression like x(y + z) means that we should add y and z, and then multiply the result by x. The implied multiplication must be written explicitly in Fortran as x * (y + z).
3. Parentheses may be used to group terms whenever desired. When parentheses are used, the expressions inside the parentheses are evaluated before the expressions outside the parentheses. For example, the expression 2 ** ((8+2)/5) is evaluated as shown below

```
2 ** ((8+2)/5) = 2 ** (10/5)
               = 2 ** 2
               = 4
```

2.6.1 Integer Arithmetic

Integer arithmetic is arithmetic involving only integer data. Integer arithmetic always produces an integer result. This is especially important to remember when an expression involves division, since there can be no fractional part in the answer. If the division of two integers is not itself an integer, *the computer automatically truncates the fractional part of the answer*. This behavior can lead to surprising and unexpected answers. For example, integer arithmetic produces the following strange results:

$$\frac{3}{4} = 0 \quad \frac{4}{4} = 1 \quad \frac{5}{4} = 1 \quad \frac{6}{4} = 1$$

$$\frac{7}{4} = 1 \quad \frac{8}{4} = 2 \quad \frac{9}{4} = 2$$

Because of this behavior, integers should *never* be used to calculate real-world quantities that vary continuously, such as distance, speed, and time. They should only be used for things that are intrinsically integer in nature, such as counters and indices.

Programming Pitfalls
Beware of integer arithmetic. Integer division often gives unexpected results.

2.6.2 Real Arithmetic

Real arithmetic (or **floating-point arithmetic**) is arithmetic involving real constants and variables. Real arithmetic always produces a real result that is essentially what we would expect. For example, real arithmetic produces the following results:

$$\frac{3.}{4.} = 0.75 \qquad \frac{4.}{4.} = 1. \qquad \frac{5.}{4.} = 1.25 \qquad \frac{6.}{4.} = 1.50$$

$$\frac{7.}{4.} = 1.75 \qquad \frac{8.}{4.} = 2. \qquad \frac{9.}{4.} = 2.25 \qquad \frac{1.}{3.} = 0.3333333$$

However, real numbers do have peculiarities of their own. Because of the finite word length of a computer, some real numbers cannot be represented exactly. For example, the number 1/3 is equal to 0.33333333333..., but since the numbers stored in the computer have limited precision, the representation of 1/3 in the computer might be 0.3333333. As a result of this limitation in precision, some quantities that are theoretically equal will not be equal when evaluated by the computer. For example, on some computers

$$3. * (1. / 3.) \neq 1. \, ,$$

but

$$2. * (1. / 2.) = 1.$$

Tests for equality must be performed very cautiously when working with real numbers.

Programming Pitfalls
Beware of real arithmetic: Due to limited precision, two theoretically identical expressions often give slightly different results.

2.6.3 Hierarchy of Operations

Often, many arithmetic operations are combined into a single expression. For example, consider the equation for the distance traveled by an object starting from rest and subjected to a constant acceleration:

```
distance = 0.5 * accel * time ** 2
```

There are two multiplications and an exponentiation in this expression. In such an expression, it is important to know the order in which the operations are evaluated. If exponentiation is evaluated before multiplication, this expression is equivalent to

```
distance = 0.5 * accel * (time ** 2)
```

But if multiplication is evaluated before exponentiation, this expression is equivalent to

```
distance = (0.5 * accel * time) ** 2
```

These two equations have different results, and we must be able to unambiguously distinguish between them.

To make the evaluation of expressions unambiguous, Fortran has established a series of rules governing the hierarchy or order in which operations are evaluated within an expression. The Fortran rules generally follow the normal rules of algebra. The order in which the arithmetic operations are evaluated is:

1. The contents of all parentheses are evaluated first, starting from the innermost parentheses and working outward.
2. All exponentials are evaluated, working from right to left.
3. All multiplications and divisions are evaluated, working from left to right.
4. All additions and subtractions are evaluated, working from left to right.

Following these rules, we see that the first of our two possible interpretations is correct—time is squared before the multiplications are performed.

Some people use simple phrases to help them remember the order of operations. For example, try "Please excuse my dear Aunt Sally". The first letters of these words give the order of evaluation: parentheses, exponents, multiplication, division, addition, subtraction.

EXAMPLE 2-1

Variables a, b, c, d, e, f, and g have been initialized to the following values:

```
a = 3. b = 2. c = 5. d = 4. e = 10. f = 2. g = 3.
```

Evaluate the following Fortran assignment statements:

(a) output = a*b+c*d+e/f**g
(b) output = a*(b+c)*d+(e/f)**g
(c) output = a*(b+c)*(d+e)/f**g

Solution

(a) Expression to evaluate: output = a*b+c*d+e/f**g
Fill in numbers: output = 3.*2.+5.*4.+10./2.**3.
First, evaluate 2.**3.:   output = 3.*2.+5.*4.+10./8.
Now, evaluate multiplications
and divisions from left to right: output = 6. +5.*4.+10./8.
                                output = 6. +20. +10./8.
                                output = 6. +20. + 1.25
Now evaluate additions: output = 27.25

(b) Expression to evaluate: output = a* (b+c)*d+(e/f)**g
Fill in numbers: output = 3.*(2.+5.)*4.+(10./2.)**3.
First, evaluate parentheses: output = 3.*7.*4.+5.**3.
Now, evaluate exponents: output = 3.*7.*4.+125.
Evaluate multiplications and
divisions from left to right: output = 21.*4.+125.
                                output = 84. + 125.
Evaluate additions: output = 209.

(c) Expression to evaluate: output = a*(b+c)*(d+e)/f**g
Fill in numbers: output = 3.*(2.+5.)*(4.+10.)/2.**3.
First, evaluate parentheses: output = 3.*7.*14./2.**3.
Now, evaluate exponents: output = 3.*7.*14./8.
Evaluate multiplications and
divisions from left to right: output = 21.*14./8.
                                output = 294./8.
                                output = 36.75

As we saw above, the order in which operations are performed has a major effect on the final result of an algebraic expression.

EXAMPLE 2-2

Variables a, b, and c have been initialized to the following values:

$$a = 3. \quad b = 2. \quad c = 3.$$

Evaluate the following Fortran assignment statements:

(a) output = a**(b**c)
(b) output = (a**b)**c
(c) output = a**b**c

Solution

(a) Expression to evaluate: output = a**(b**c)
Fill in numbers: output = 3.**(2.**3.)

Basic Elements of Fortran

    Evaluate expression in parentheses: `output = 3.**8.`
    Evaluate remaining expression: `output = 6561.`

  (b) Expression to evaluate: `output = (a**b)**c`
    Fill in numbers: `output = (3.**2.)**3.`
    Evaluate expression in parentheses: `output = 9.**3.`
    Evaluate remaining expression: `output = 729.`

  (c) Expression to evaluate: `output = a**b**c`
    Fill in numbers: `output = 3.**2.**3.`
    First, evaluate rightmost exponent: `output = 3.**8.`
    Now, evaluate remaining exponent: `output = 6561.`

    The results of (a) and (c) are identical, but the expression in (a) is easier to understand and less ambiguous than the expression in (c).

    It is important that every expression in a program be made as clear as possible. Any program of value must not only be written but also be maintained and modified when necessary. You should always ask yourself: "Will I easily understand this expression if I come back to it in six months? Can another programmer look at my code and easily understand what I am doing?" If there is any doubt in your mind, use extra parentheses in the expression to make it as clear as possible.

Good Programming Practice

Use parentheses as necessary to make your equations clear and easy to understand.

    If parentheses are used within an expression, then the parentheses must be balanced. That is, there must be an equal number of open parentheses and close parentheses within the expression. It is an error to have more of one type than the other. Errors of this sort are usually typographical, and the Fortran compiler catches them. For example, the expression

$$(2. + 4.) / 2.)$$

produces an error during compilation because of the mismatched parentheses.

2.6.4 Mixed-Mode Arithmetic

When an arithmetic operation is performed using two real numbers, its immediate result is of type real. Similarly, when an arithmetic operation is performed using two integers, the result is of type integer. In general, arithmetic operations are only defined between numbers of the same type. For example, the addition of two real numbers is a valid operation, and the addition of two integers is a valid operation, but the addition of a real number and an integer is *not* a valid operation. This is true because real numbers and integers are stored in completely different forms in the computer.

What happens if an operation is between a real number and an integer? Expressions containing both real numbers and integers are called **mixed-mode expressions,** and arithmetic involving both real numbers and integers is called *mixed-mode arithmetic*. In the case of an operation between a real number and an integer, the integer is converted by the computer into a real number, and real arithmetic is used on the numbers. The result is of type real. For example, consider the following equations:

Integer expression: $\dfrac{3}{2}$ is evaluated to be 1 (integer result)

Real expression: $\dfrac{3.}{2.}$ is evaluated to be 1.5 (real result)

Mixed-mode expression: $\dfrac{3.}{2}$ is evaluated to be 1.5 (real result)

The rules governing mixed-mode arithmetic can be confusing to beginning programmers, and even experienced programmers may trip up on them from time to time. This is especially true when the mixed-mode expression involves division. Consider the following expressions:

	Expression	Result
1.	1 + 1/4	1
2.	1. + 1/4	1.
3.	1 + 1./4	1.25

Expression 1 contains only integers, so it is evaluated by integer arithmetic. In integer arithmetic, 1/4 = 0 and 1 + 0 = 1, so the final result is 1 (an integer). Expression 2 is a mixed-mode expression containing both real numbers and integers. However, the first operation to be performed is a division, since division comes before addition in the hierarchy of operations. The division is between integers, so the result is 1/4 = 0. Next comes an addition between a real 1. and an integer 0, so the compiler converts the integer 0 into a real number, and then performs the addition. The resulting number is 1. (a real number). Expression 3 is also a mixed-mode expression containing both real numbers and integers. The first operation to be performed is a division between a real number and an integer, so the compiler converts the integer 4 into a real number, and then performs the division. The result is a real 0.25. The next operation to be performed is an addition between an integer 1 and a real 0.25, so the compiler converts the integer 1 into a real number, and then performs the addition. The resulting number is 1.25 (a real number).

Basic Elements of Fortran

To summarize,

1. An operation between an integer and a real number is called a mixed-mode operation, and an expression containing one or more such operations is called a mixed-mode expression.
2. When a mixed-mode operation is encountered, Fortran converts the integer into a real number, and then performs the operation to get a real result.
3. The automatic mode conversion does not occur until a real number and an integer both appear in the *same* operation. Therefore, it is possible for a portion of an expression to be evaluated in integer arithmetic, followed by another portion evaluated in real arithmetic.

Automatic type conversion also occurs when the variable to which the expression is assigned is of a different type than the result of the expression. For example, consider the following assignment statement:

$$\text{nres} = 1.25 + 9 / 4$$

where nres is an integer. The expression to the right of the equal sign evaluates to 3.25, which is a real number. Since nres is an integer, the 3.25 is automatically converted into the integer number 3 before being stored in nres.

Programming Pitfalls

Mixed-mode expressions are dangerous because they are hard to understand and may produce misleading results. Avoid them whenever possible.

Fortran includes five type conversion functions that allow us to explicitly control the conversion between integer and real values. These functions are described in Table 2-3.

The REAL, INT, NINT, CEILING, and FLOOR functions may be used to avoid undesirable mixed-mode expressions by explicitly converting data types from one form to another. The REAL function converts an integer into a real number, and the INT, NINT, CEILING, and FLOOR functions convert real numbers into integers. The INT function truncates the real number, while the NINT function rounds it to the nearest integer value. The CEILING function returns the nearest integer greater than or equal

TABLE 2-3
Type conversion functions

Function name and arguments	Argument type	Result type	Comments
INT(X)	REAL	INTEGER	Integer part of x (x is truncated)
NINT(X)	REAL	INTEGER	Nearest integer to x (x is rounded)
CEILING(X)	REAL	INTEGER	Nearest integer above or equal to the value of x
FLOOR(X)	REAL	INTEGER	Nearest integer below or equal to the value of x
REAL(I)	INTEGER	REAL	Converts integer value to real

to the real number and the FLOOR function returns the nearest integer less than or equal to the real number.

To understand the distinction between these two operations, let's consider the real numbers 2.9995 and −2.9995. The results of each function with these inputs are shown below:

Function	Result	Description
INT(2.9995)	2	Truncates 2.9995 to 2
NINT(2.9995)	3	Rounds 2.9995 to 3
CEILING(2.9995)	3	Selects nearest integer above 2.9995
FLOOR(2.9995)	2	Selects nearest integer below 2.9995
INT(-2.9995)	−2	Truncates −2.9995 to −2
NINT(-2.9995)	−3	Rounds −2.9995 to −3
CEILING(-2.9995)	−2	Selects nearest integer above −2.9995
FLOOR(-2.9995)	−3	Selects nearest integer below −2.9995

The NINT function is especially useful when converting back from real to integer form, since the small round-off errors occurring in real calculations will not affect the resulting integer value.

2.6.5 Mixed-Mode Arithmetic and Exponentiation

As a general rule, mixed-mode arithmetic operations are undesirable because they are hard to understand and can sometimes lead to unexpected results. However, there is one exception to this rule: exponentiation. For exponentiation, mixed-mode operation is actually *desirable*.

To understand why this is so, consider the assignment statement

$$\text{result} = y ** n$$

where result and y are real, and n is an integer. The expression y ** n is shorthand for "use y as a factor n times", and that is exactly what the computer does when it encounters this expression. Since y is a real number and the computer is multiplying y by itself, the computer is really doing real arithmetic and not mixed-mode arithmetic!

Now consider the assignment statement

$$\text{result} = y ** x$$

where result, y, and x are real. The expression y ** x is shorthand for "use y as a factor x times", but this time x is not an integer. Instead, x might be a number like 2.5. It is not physically possible to multiply a number by itself 2.5 times, so we have to rely on indirect methods to calculate y ** x in this case. The most common approach is to use the standard algebraic formula that says that

$$y^x = e^{x \ln y} \tag{2-1}$$

Using this equation, we can evaluate y ** x by taking the natural logarithm of y, multiplying by x, and then calculating *e* to the resulting power. While this technique certainly works, it takes longer to perform and is less accurate than an ordinary series of multiplications. Therefore, if given a choice, we should try to raise real numbers to integer powers instead of real powers.

Good Programming Practice
Use integer exponents instead of real exponents whenever possible.

Also, note that *it is not possible to raise a negative number to a negative real power*. Raising a negative number to an integer power is a perfectly legal operation. For example, (-2.0)**2 = 4. However, raising a negative number to a real power will not work, since the natural logarithm of a negative number is undefined. Therefore, the expression (-2.0)**2.0 will produce a runtime error.

Programming Pitfalls
Never raise a negative number to a real power.

Quiz 2-2

This quiz provides a quick check to see if you have understood the concepts introduced in Section 2.6. If you have trouble with the quiz, reread the section, ask your instructor, or discuss the material with a fellow student. The answers to this quiz are found in the back of the book.

1. In what order are the arithmetic and logical operations evaluated if they appear within an arithmetic expression? How do parentheses modify this order?
2. Are the following expressions legal or illegal? If they are legal, what is their result? If they are illegal, what is wrong with them?
   (*a*) 37 / 3
   (*b*) 37 + 17 / 3
   (*c*) 28 / 3 / 4
   (*d*) (28 / 3) / 4
   (*e*) 28 / (3 / 4)

(*continued*)

(*concluded*)

   (*f*) -3. ** 4. / 2.
   (*g*) 3. ** (-4. / 2.)
   (*h*) 4. ** -3

3. Evaluate the following expressions:
   (*a*) 2 + 5 * 2 - 5
   (*b*) (2 + 5) * (2 - 5)
   (*c*) 2 + (5 * 2) - 5
   (*d*) (2 + 5) * 2 - 5

4. Are the following expressions legal or illegal? If they are legal, what is their result? If they are illegal, what is wrong with them?
   (*a*) 2. ** 2. ** 3.
   (*b*) 2. ** (-2.)
   (*c*) (-2) ** 2
   (*d*) (-2.) ** (-2.2)
   (*e*) (-2.) ** NINT(-2.2)
   (*f*) (-2.) ** FLOOR(-2.2)

5. Are the following statements legal or illegal? If they are legal, what is their result? If they are illegal, what is wrong with them?

   ```
   INTEGER :: i, j
   INTEGER, PARAMETER :: K = 4
   i = K ** 2
   j = i / K
   K = i + j
   ```

6. What value is stored in `result` after the following statements are executed?

   ```
   REAL :: a, b, c, result
   a = 10.
   b = 1.5
   c = 5.
   result = FLOOR(a / b) + b * c ** 2
   ```

7. What values are stored in a, b, and n after the following statements are executed?

   ```
   REAL :: a, b
   INTEGER :: n, i, j
   i = 10.
   j = 3
   n = i / j
   a = i / j
   b = REAL(i) / j
   ```

2.7
INTRINSIC FUNCTIONS

In mathematics, a **function** is an expression that accepts one or more input values and calculates a single result from them. Scientific and technical calculations usually require functions that are more complex than the simple addition, subtraction, multiplication, division, and exponentiation operations that we have discussed so far. Some of these functions are very common, and are used in many different technical disciplines. Others are rarer and specific to a single problem or a small number of problems. Examples of very common functions are the trigonometric functions, logarithms, and square roots. Examples of rarer functions include the hyperbolic functions, Bessel functions, and so forth.

The Fortran language has mechanisms to support both the very common functions and the less common functions. Many of the most common ones are built directly into the Fortran language. They are called **intrinsic functions.** Less common functions are not included in the Fortran language, but the user can supply any function needed to solve a particular problem as either an **external function** or an **internal function.** External functions will be described in Chapter 7, and internal functions will be described in Chapter 9.

A Fortran function takes one or more input values, and calculates a *single* output value from them. The input values to the function are known as **arguments;** they appear in parentheses immediately after the function name. The output of a function is a single number, logical value, or character string, which can be used together with other functions, constants, and variables in Fortran expressions. When a function appears in a Fortran statement, the arguments of the function are passed to a separate routine that computes the result of the function, and then the result is used in place of the function in the original calculation (see Figure 2-3). Intrinsic functions are supplied with the Fortran compiler. For external and internal functions, the routine must be supplied by the user.

A list of some common intrinsic functions is given in Table 2-4. A more complete list of Fortran intrinsic functions is given in Appendix B, along with a brief description of each one.

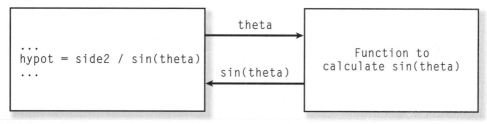

FIGURE 2-3
When a function is included in a Fortran statement, the argument(s) of the function are passed to a separate routine that computes the result of the function, and then the result is used in place of the function in the original calculation.

TABLE 2-4
Some common intrinsic functions

Function name and arguments	Function value	Argument type	Result type	Comments
SQRT(X)	$\sqrt{x}$	REAL	REAL	Square root of x for $x \geq 0$
ABS(X)		REAL/INTEGER	*	Absolute value of x
ACHAR(I)		INTEGER	CHAR(1)	Returns the character at position I in the ASCII collating sequence
SIN(X)	$\sin(x)$	REAL	REAL	Sine of x (x must be in *radians*)
SIND(X)	$\sin(x)$	REAL	REAL	Sine of x (x must be in *degrees*)
COS(X)	$\cos(x)$	REAL	REAL	Cosine of x (x must be in *radians*)
COSD(X)	$\cos(x)$	REAL	REAL	Cosine of x (x must be in *degrees*)
TAN(X)	$\tan(x)$	REAL	REAL	Tangent of x (x must be in *radians*)
TAND(X)	$\tan(x)$	REAL	REAL	Tangent of x (x must be in *degrees*)
EXP(X)	e^x	REAL	REAL	e raised to the xth Power
LOG(X)	$\log_e(x)$	REAL	REAL	Natural logarithm of x for $x > 0$
LOG10(X)	$\log_{10}(x)$	REAL	REAL	Base-10 logarithm of x for $x > 0$
IACHAR(C)		CHAR(1)	INTEGER	Returns the position of the character C in the ASCII collating sequence
MOD(A,B)		REAL/INTEGER	*	Remainder or Modulo Function
MAX(A,B)		REAL/INTEGER	*	Picks the larger of a and b
MIN(A,B)		REAL/INTEGER	*	Picks the smaller of a and b
ASIN(X)	$\sin^{-1}(x)$	REAL	REAL	Inverse sine of x for $-1 \leq x \leq 1$ (results in *radians*)
ASIND(X)	$\sin^{-1}(x)$	REAL	REAL	Inverse sine of x for $-1 \leq x \leq 1$ (results in *degrees*)
ACOS(X)	$\cos^{-1}(x)$	REAL	REAL	Inverse cosine of x for $-1 \leq x \leq 1$ (results in *radians*)
ACOSD(X)	$\cos^{-1}(x)$	REAL	REAL	Inverse cosine of x for $-1 \leq x \leq 1$ (results in *degrees*)
ATAN(X)	$\tan^{-1}(x)$	REAL	REAL	Inverse tangent of x (results in *radians* in the range $-\frac{\pi}{2} \leq x \leq \frac{\pi}{2}$)
ATAND(X)	$\tan^{-1}(x)$	REAL	REAL	Inverse tangent of x (results in *radians* in the range $-90 \leq x \leq 90$)
ATAN2(Y/X)	$\tan^{-1}(y/x)$	REAL	REAL	Four quadrant inverse tangent of x (results in *radians* in the range $-\pi \leq x \leq \pi$)
ATAN2D(Y,X)	$\tan^{-1}(y/x)$	REAL	REAL	Four quadrant inverse tangent of x (results in *radians* in the range $-180 \leq x \leq 180$)

Note:
* = Result is of the same type as the input argument(s).

Fortran functions are used by naming them in an expression. For example, the intrinsic function SIN can be used to calculate the sine of a number as follows:

$$y = \text{SIN(theta)}$$

where theta is the argument of the function SIN. After this statement is executed, the variable y contains the sine of the value stored in variable theta. Note from Table 2-4 that the trigonometric functions without a "D" in their name expect their arguments to be in radians. If the variable theta is in degrees, then we must convert degrees to radians ($180° = \pi$ radians) before computing the sine. This conversion can be done in the same statement as the sine calculation:

$$y = \text{SIN (theta*(3.141593/180.))}$$

Alternately, we could create a named constant containing the conversion factor, and refer to that constant when the function is executed:

```
INTEGER, PARAMETER :: DEG_2_RAD = 3.141593 / 180.
...
y = SIN (theta * DEG_TO_RAD)
```

The argument of a function can be a constant, a variable, an expression, or even the result of another function. All of the following statements are legal:

```
y = SIN(3.141593)    (argument is a constant)
y = SIN(x)           (argument is a variable)
y = SIN(PI*x)        (argument is an expression)
y = SIN(SQRT(x))     (argument is the result of another function)
```

Functions may be used in expressions anywhere that a constant or variable may be used. However, functions may never appear on the left side of the assignment operator (equal sign), since they are not memory locations, and nothing can be stored in them.

The type of argument required by a function and the type of value returned by it are specified in Table 2-4 for the intrinsic functions listed there. Some of these intrinsic functions are **generic functions,** which means that they can use more than one type of input data. The absolute value function ABS is a generic function. If X is a real number, then the type of ABS(X) is real. If X is an integer, then the type of ABS(X) is integer. Some functions are called **specific functions,** because they can use only one specific type of input data, and produce only one specific type of output value. For example, the function IABS requires an integer argument and returns an integer result. A list of all intrinsic functions (both generic and specific) is provided in Appendix B.

2.8
LIST-DIRECTED INPUT AND OUTPUT STATEMENTS

An **input statement** reads one or more values from an input device and stores them into variables specified by the programmer. The input device could be a keyboard in an interactive environment, or an input disk file in a batch environment. An **output statement**

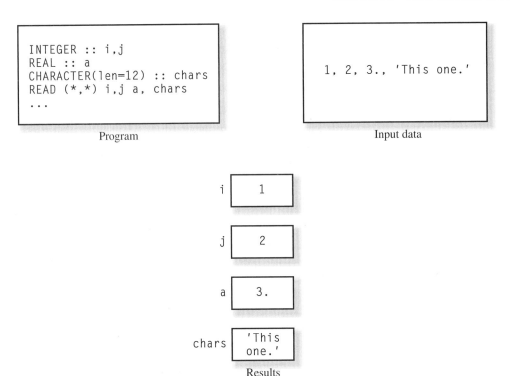

FIGURE 2-4
For list-directed input, the type and order of the input data values must match the type and order of the supplied input data.

writes one or more values to an output device. The output device could be a CRT screen in an interactive environment, or an output listing file in a batch environment.

We have already seen input and output statements in `my_first_program`, which is shown in Figure 2-1. The input statement in the figure is of the form

```
READ (*,*) input_list
```

where *input_list* is the list of variables into which the values being read are placed. If there is more than one variable in the list, they should be separated by commas. The parentheses (*,*) in the statement contains control information for the read. The first field in the parentheses specifies the *input/output unit* (or i/o unit) from which the data is to be read (the concept of an input/output unit will be explained in Chapter 5). An asterisk in this field means that the data is to be read from the standard input device for the computer—usually the keyboard when running in interactive mode. The second field in the parentheses specifies the format in which the data is to be read (formats will also be explained in Chapter 5). An asterisk in this field means that list-directed input (sometimes called free-format input) is to be used.

The term **list-directed input** means that *the types of the variables in the variable list determine the required format of the input data* (Figure 2-4). For example, consider the following statements:

Basic Elements of Fortran

```
PROGRAM input_example
INTEGER :: i, j
REAL :: a
CHARACTER(len=12) :: chars
READ (*,*) i, j, a, chars
END PROGRAM input_example
```

The input data supplied to the program must consist of two integers, a real number, and a character string. Furthermore, they must be in that order. The values may be all on one line separated by commas or blanks, or they may be on separate lines. The list-directed READ statement will continue to read input data until values have been found for all of the variables in the list. If the input data supplied to the program at execution time is

1, 2, 3.,'This one.'

then the variable i will be filled with a 1, j will be filled with a 2, a will be filled with a 3.0, and chars with be filled with 'This one. '. Since the input character string is only 9 characters long, while the variable chars has room for 12 characters, the string is *left justified* in the character variable, and three blanks are automatically added at the end of it to fill out the remaining space. Also note that for list-directed reads, input character strings must be enclosed in single or double quotes if they contain spaces.

When using list-directed input, the values to be read must match the variables in the input list both in order and type. If the input data had been

1, 2, 'This one.', 3.

then a runtime error would have occurred when the program tried to read the data.

Each READ statement in a program begins reading from a new line of input data. If any data was left over on the previous input line, that data is discarded. For example, consider the following program:

```
PROGRAM input_example_2
INTEGER :: i, j, k, l
READ (*,*) i, j
READ (*,*) k, l
END PROGRAM input_example_2
```

If the input data to this program is:

1, 2, 3, 4
5, 6, 7, 8

then after the READ statements, i will contain a 1, j will contain a 2, k will contain a 5, and l will contain a 6 (Figure 2-5).

It is a good idea to always *echo* any value that you read into a program from a keyboard. Echoing a value means displaying the value with a WRITE statement after it has been read. If you do not do so, a typing error in the input data might cause a wrong answer, and the user of the program would never know that anything was wrong. You may echo the data either immediately after it is read or somewhere further down in the program output, but *every input variable should be echoed somewhere in the program's output.*

```
INTEGER :: i,j,k,l
READ (*,*) i,j
READ (*,*) k,l
...
```

Program

```
1, 2, 3, 4
5, 6, 7, 8
```

Input data

i	1
j	2
k	5
l	6

Results

FIGURE 2-5
Each list-directed READ statement begins reading from a new line of input data, and any unused data left on the previous line is discarded. Here, the values 3 and 4 on the first line of input data are never used.

Good Programming Practice
Echo any variables that a user enters into a program from a keyboard, so that the user can be certain that they were typed and processed correctly.

The *list-directed output statement* is of the form

WRITE (*,*) output_list

where *output_list* is the list of data items (variables, constants, or expressions) that are to be written. If there is more than one item in the list, then the items should be separated by commas. The parentheses (*,*) in the statement contains control information for the write, where the two asterisks have the same meaning as for a list-directed read statement.[5]

---

[5] There is another form of list-directed output statement:

PRINT *, output_list

This statement is equivalent to the list-directed WRITE statement discussed above, and is used by some programmers. The PRINT statement is never used in this book, but it is discussed in Chapter 14 Section 14.3.7.

Basic Elements of Fortran

The term **list-directed output** means that *the types of the values in the output list of the write statement determine the format of the output data*. For example, consider the following statements:

```
PROGRAM output_example
INTEGER :: ix
REAL :: theta
ix = 1
test = .TRUE.
theta = 3.141593
WRITE (*,*) ' IX =            ', ix
WRITE (*,*) ' THETA =         ', theta
WRITE (*,*) ' COS(THETA) =    ', COS(theta)
WRITE (*,*) REAL(ix), NINT(theta)
END PROGRAM output_example
```

The output resulting from these statements is:

```
 IX =                  1
 THETA =               3.141593
 COS(THETA) =         -1.000000
   1.000000            3
```

This example illustrates several points about the list-directed write statement:

1. The output list may contain constants (' IX = ' is a constant), variables, functions, and expressions. In each case, the value of the constant, variable, function, or expression is output to the standard output device.
2. The format of the output data matches the type of the value being output. For example, even though theta is of type real, NINT(theta) is of type integer. Therefore, the sixth write statement produces an output of 3 (the nearest integer to 3.141593).
3. The output of list-directed write statements is not very pretty. The values printed out do not line up in neat columns, and there is no way to control the number of significant digits displayed for real numbers. We will learn how to produce neatly formatted output in Chapter 5.

Quiz 2-3

This quiz provides a quick check to see if you have understood the concepts introduced in Sections 2.7 and 2.8. If you have trouble with the quiz, reread the sections, ask your instructor, or discuss the material with a fellow student. The answers to this quiz are found in the back of the book.

Convert the following algebraic equations into Fortran assignment statements:

1. The equivalent resistance R_{eq} of four resistors R_1, R_2, R_3, and R_4 connected in series:

$$R_{eq} = R_1 + R_2 + R_3 + R_4$$

(continued)

(*continued*)

2. The equivalent resistance R_{eq} of four resistors R_1, R_2, R_3, and R_4 connected in parallel:

$$R_{eq} = \frac{1}{\dfrac{1}{R_1} + \dfrac{1}{R_2} + \dfrac{1}{R_3} + \dfrac{1}{R_4}}$$

3. The period T of an oscillating pendulum:

$$T = 2\pi\sqrt{\frac{L}{g}}$$

where L is the length of the pendulum, and g is the acceleration due to gravity.

4. The equation for damped sinusoidal oscillation:

$$v(t) = V_M e^{-\alpha t} \cos \omega t$$

where V_M is the maximum value of the oscillation, α is the exponential damping factor, and ω is the angular velocity of the oscillation.

Convert the following Fortran assignment statements into algebraic equations:

5. The motion of an object in a constant gravitational field:

```
distance = 0.5 * accel * t**2 + vel_0 * t + pos_0
```

6. The oscillating frequency of a damped RLC circuit:

```
freq = 1. / (2. * PI * SQRT(l * c))
```

where `PI` is the constant π (3.141592...).

7. Energy storage in an inductor:

```
energy = 1.0 / 2.0 * inductance * current**2
```

8. What values will be printed out when the following statements are executed?

```
PROGRAM quiz_1
INTEGER :: i
REAL :: a
a = 0.05
i = NINT(2. * 3.141493 / a)
a = a * (5 / 3)
WRITE (*,*) i, a
END PROGRAM quiz_1
```

9. If the input data is as shown, what will be printed out by the following program?

```
PROGRAM quiz_2
INTEGER :: i, j, k
```

(*continued*)

Basic Elements of Fortran

> (*concluded*)
> ```
> REAL :: a, b, c
> READ (*,*) i, j, a
> READ (*,*) b, k
> c = SIN ((3.141593 / 180) * a)
> WRITE (*,*) i, j, k, a, b, c
> END PROGRAM quiz_2
> ```
> The input data is :
> ```
> 1, 3
> 2., 45., 17.
> 30., 180, 6.
> ```

2.9
INITIALIZATION OF VARIABLES

Consider the following program:

```
PROGRAM init
INTEGER :: i
WRITE (*,*) i
END PROGRAM init
```

What is the value stored in the variable i? What will be printed out by the WRITE statement? The answer is: We don't know!

The variable i is an example of an **uninitialized variable.** It has been defined by the INTEGER :: i statement, but no value has been placed into it yet. The value of an uninitialized variable is not defined by the Fortran standard. Some compilers automatically set uninitialized variables to zero, and some set them to different arbitrary patterns. Some compilers for older version of Fortran leave whatever values previously existed at the memory location of the variables. Some compilers even produce a runtime error if a variable is used without first being initialized.

Uninitialized variables can present a serious problem. Since they are handled differently on different machines, a program that works fine on one computer may fail when transported to another one. On some machines, the same program could work sometimes and fail sometimes, depending on the data left behind by the previous program occupying the same memory. Such a situation is totally unacceptable, and we must avoid it by always initializing all of the variables in our programs.

Good Programming Practice
Always initialize all variables in a program before using them.

There are three techniques available to initialize variables in a Fortran program: assignment statements, `READ` statements, and initialization in type declaration statements.[6] An assignment statement assigns the value of the expression to the right of the equal sign to the variable on the left of the equal sign. In the following code, the variable i is initialized to 1, and we know that a 1 will be printed out by the `WRITE` statement.

```
PROGRAM init_1
INTEGER :: i
i = 1
WRITE (*,*) i
END PROGRAM init_1
```

A `READ` statement may be used to initialize variables with values input by the user. Unlike initialization with assignment statements, the user can change the value stored in the variable each time that the program is run. For example, the following code will initialize variable i with whatever value the user desires, and that value will be printed out by the `WRITE` statement.

```
PROGRAM init_2
INTEGER :: i
READ (*,*) i
WRITE (*,*) i
END PROGRAM init_2
```

The third technique available to initialize variables in a Fortran program is to specify their initial values in the type declaration statement that defines them. This declaration specifies that *a value should be pre-loaded into a variable during the compilation and linking process*. Note the fundamental difference between initialization in a type declaration statement and initialization in an assignment statement: A type declaration statement initializes the variable before the program begins to run, whereas an assignment statement initializes the variable during execution.

The form of a type declaration statement used to initialize variables is

```
type :: var1 = value, [var2 = value, ... ]
```

Any number of variables may be declared and initialized in a single type declaration statement provided that they are separated by commas. An example of type declaration statements used to initialize a series of variables is

```
REAL :: time = 0.0, distance = 5128.
INTEGER :: loop = 10
```

Before program execution, time is initialized to 0.0, distance is initialized to 5128., and loop is initialized to 10.

In the following code, the variable i is initialized by the type declaration statement, so we know that when execution starts, the variable i will contain the value 1. Therefore, the `WRITE` statement will print out a 1.

---

[6]A fourth, older technique uses the `DATA` statement. This statement is kept for backward compatibility with earlier versions of Fortran, but it has been superseded by initialization in type declaration statements. `DATA` statements should not be used in new programs. The `DATA` statement is described in Chapter 18.

```
PROGRAM init_3
INTEGER :: i = 1
WRITE (*,*) i
END PROGRAM init_3
```

2.10
THE IMPLICIT NONE STATEMENT

There is another very important nonexecutable statement: the IMPLICIT NONE statement. When it is used, the IMPLICIT NONE statement disables the default typing provisions of Fortran. When the IMPLICIT NONE statement is included in a program, *any variable that does not appear in an explicit type declaration statement is considered an error*. The IMPLICIT NONE statement should appear after the PROGRAM statement and before any type declaration statements.

When the IMPLICIT NONE statement is included in a program, the programmer must explicitly declare the type of every variable in the program. On first thought, this might seem to be a disadvantage, since the programmer must do more work when he or she first writes a program. This initial impression couldn't be more wrong. In fact, there are several advantages to using this statement.

The majority of programming errors are simple typographical errors. The IMPLICIT NONE statement catches these errors at compilation time, before they can produce subtle errors during execution. For example, consider the following simple program:

```
PROGRAM test_1
REAL :: time = 10.0
WRITE (*,*) 'Time = ', tmie
END PROGRAM test_1
```

In this program, the variable time is misspelled tmie at one point. When this program is compiled with a Fortran compiler and executed, the output is "Time = 0.000000E+00", which is the wrong answer! In contrast, consider the same program with the IMPLICIT NONE statement present:

```
PROGRAM test_1
IMPLICIT NONE
REAL :: time = 10.0
WRITE (*,*) 'Time = ', tmie
END PROGRAM test_1
```

When compiled with the same compiler, this program produces the following compile-time error:[7]

```
1 PROGRAM test_1
2 IMPLICIT NONE
3 REAL :: time = 10.0
4 WRITE (*,*) 'Time = ', tmie
                         .....................1
(1) Error: This name does not have a type, and must have an explicit type. [TMIE]

5 END PROGRAM
```

---

[7] The exact error message will vary in different Fortran compilers.

Instead of having a wrong answer in an otherwise-working program, we have an explicit error message flagging the problem at compilation time. This is an enormous advantage when working with longer programs containing many variables.

Another advantage of the IMPLICIT NONE statement is that it makes the code more maintainable. Any program using the statement must have a complete list of all variables included in the declaration section of the program. If the program must be modified, a programmer can check the list to avoid using variable names that are already defined in the program. This checking helps to eliminate a very common error, in which the modifications to the program inadvertently change the values of some variables used elsewhere in the program.

In general, the use of the IMPLICIT NONE statement becomes more and more advantageous as the size of a programming project increases. The use of IMPLICIT NONE is so important to the designing of good programs that we will use it consistently everywhere throughout this book.

Good Programming Practice
Always explicitly define every variable in your programs, and use the IMPLICIT NONE statement to help you spot and correct typographical errors before they become program execution errors.

2.11
PROGRAM EXAMPLES

In this chapter, we have presented the fundamental concepts required to write simple but functional Fortran programs. We will now present a few example problems in which these concepts are used.

EXAMPLE 2-3 *Temperature Conversion:*

Design a Fortran program that reads an input temperature in degrees Fahrenheit, converts it to an absolute temperature in kelvins, and writes out the result.

SOLUTION
The relationship between temperature in degrees Fahrenheit (°F) and temperature in kelvins (K) can be found in any physics textbook. It is

$$T \text{ (in kelvin)} = \left[\frac{5}{9}T \text{ (in °F)} - 32.0\right] + 273.15 \qquad (2\text{-}2)$$

The physics books also give us sample values on both temperature scales, which we can use to check the operation of our program. Two such values are:

The boiling point of water	212° F	373.15 K
The sublimation point of dry ice	-110° F	194.26 K

Our program must perform the following steps:

1. Prompt the user to enter an input temperature in °F.
2. Read the input temperature.
3. Calculate the temperature in kelvins from Equation (2-2).
4. Write out the result, and stop.

The resulting program is shown in Figure 2-6.

FIGURE 2-6
Program to convert degrees Fahrenheit into kelvins.

```
PROGRAM temp_conversion
!  Purpose:
!    To convert an input temperature from degrees Fahrenheit to
!    an output temperature in kelvins.
!
!  Record of revisions:
!      Date       Programmer         Description of change
!      ====       ==========         =====================
!   11/03/15 -- S. J. Chapman        Original code
!
IMPLICIT NONE        ! Force explicit declaration of variables

! Data dictionary: declare variable types, definitions, & units
REAL :: temp_f       ! Temperature in degrees Fahrenheit
REAL :: temp_k       ! Temperature in kelvins

! Prompt the user for the input temperature.
WRITE (*,*) 'Enter the temperature in degrees Fahrenheit: '
READ  (*,*) temp_f

! Convert to kelvins.
temp_k = (5. / 9.) * (temp_f - 32.) + 273.15

! Write out the result.
WRITE (*,*) temp_f, ' degrees Fahrenheit = ', temp_k, ' kelvins'

! Finish up.
END PROGRAM temp_conversion
```

To test the completed program, we will run it with the known input values given above. Note that user inputs appear in bold face below.[8]

```
C:\book\fortran\chap2>temp_conversion
Enter the temperature in degrees Fahrenheit:
212
     212.000000 degrees Fahrenheit =    373.150000 kelvins
```

---

[8] Fortran programs such as this are normally executed from a **command line**. In Windows, a Command Window can be opened by clicking the Start button, selecting the Run option, and typing "cmd" as the program to start. When the Command Window is running, the prompt shows the name of the current working directory (C:\book\fortran\chap2 in this example), and a program is executed by typing its name on the command line. Note that the prompt would look different on other operating systems such as Linux or Unix.

```
C:\book\fortran\chap2>temp_conversion
Enter the temperature in degrees Fahrenheit:
-110
    -110.000000 degrees Fahrenheit =    194.261100 kelvins
```

The results of the program match the values from the physics book.

In the above program, we echoed the input values and printed the output values together with their units. The results of this program only make sense if the units (degrees Fahrenheit and kelvins) are included together with their values. As a general rule, the units associated with any input value should always be printed along with the prompt that requests the value, and the units associated with any output value should always be printed along with that value.

Good Programming Practice
Always include the appropriate units with any values that you read or write in a program.

The above program exhibits many of the good programming practices that we have described in this chapter. It uses the `IMPLICIT NONE` statement to force the explicit typing of all variables in the program. It includes a data dictionary as a part of the declaration section, with each variable being given a type, definition, and units. It also uses descriptive variable names. The variable `temp_f` is initialized by a `READ` statement before it is used. All input values are echoed, and appropriate units are attached to all printed values.

EXAMPLE 2-4 *Electrical Engineering: Calculating Real, Reactive, and Apparent Power:*

Figure 2-7 shows a sinusoidal AC voltage source with voltage V supplying a load of impedance $Z \angle \theta\ \Omega$. From simple circuit theory, the rms current I, the real power P, reactive power Q, apparent power S, and power factor PF supplied to the load are given by the equations

$$V = IR \qquad (2\text{-}3)$$

$$P = VI \cos \theta \qquad (2\text{-}4)$$

$$Q = VI \cos \theta \qquad (2\text{-}5)$$

$$S = VI \qquad (2\text{-}6)$$

$$PF = \cos \theta \qquad (2\text{-}7)$$

Basic Elements of Fortran

FIGURE 2-7
A sinusoidal AC voltage source with voltage V supplying a load of impedance $Z \angle \theta \; \Omega$.

where V is the rms voltage of the power source in units of volts (V). The units of current are amperes (A), of real power are watts (W), of reactive power are volt-amperes-reactive (VAR), and of apparent power are volt-amperes (VA). The power factor has no units associated with it.

Given the rms voltage of the power source and the magnitude and angle of the impedance Z, write a program that calculates the rms current I, the real power P, reactive power Q, apparent power S, and power factor PF of the load.

SOLUTION

In this program, we need to read in the rms voltage V of the voltage source and the magnitude Z and angle θ of the impedance. The input voltage source will be measured in volts, the magnitude of the impedance Z in ohms, and the angle of the impedance θ in degrees. Once the data is read, we must convert the angle θ into radians for use with the Fortran trigonometric functions. Next, the desired values must be calculated, and the results must be printed out.

The program must perform the following steps:

1. Prompt the user to enter the source voltage in volts.
2. Read the source voltage.
3. Prompt the user to enter the magnitude and angle of the impedance in ohms and degrees.
4. Read the magnitude and angle of the impedance.
5. Calculate the current I from Equation (2-3).
6. Calculate the real power P from Equation (2-4).
7. Calculate the reactive power Q from Equation (2-5).
8. Calculate the apparent power S from Equation (2-6).
9. Calculate the power factor PF from Equation (2-7).
10. Write out the results, and stop.

The final Fortran program is shown in Figure 2-8.

FIGURE 2-8
Program to calculate the real power, reactive power, apparent power, and power factor supplied to a load.

```fortran
PROGRAM power
!
!   Purpose:
!     To calculate the current, real, reactive, and apparent power,
!     and the power factor supplied to a load.
!
!   Record of revisions:
!       Date          Programmer          Description of change
!       ====          ==========          =====================
!     11/03/15       S. J. Chapman        Original code
!
IMPLICIT NONE

! Data dictionary: declare constants
REAL,PARAMETER :: DEG_2_RAD = 0.01745329  ! Deg to radians factor

! Data dictionary: declare variable types, definitions, & units
REAL :: amps         ! Current in the load (A)
REAL :: p            ! Real power of load (W)
REAL :: pf           ! Power factor of load (no units)
REAL :: q            ! Reactive power of the load (VAR)
REAL :: s            ! Apparent power of the load (VA)
REAL :: theta        ! Impedance angle of the load (deg)
REAL :: volts        ! Rms voltage of the power source (V)
REAL :: z            ! Magnitude of the load impedance (ohms)

! Prompt the user for the rms voltage.
WRITE (*,*) 'Enter the rms voltage of the source: '
READ  (*,*) volts

! Prompt the user for the magnitude and angle of the impedance.
WRITE (*,*) 'Enter the magnitude and angle of the impedance '
WRITE (*,*) 'in ohms and degrees: '
READ  (*,*) z, theta

! Perform calculations
amps = volts / z                                ! Rms current
p = volts * amps * cos (theta * DEG_2_RAD)      ! Real power
q = volts * amps * sin (theta * DEG_2_RAD)      ! Reactive power
s = volts * amps                                ! Apparent power
pf = cos ( theta * DEG_2_RAD)                   ! Power factor

! Write out the results.
WRITE (*,*) 'Voltage        = ', volts, ' volts'
WRITE (*,*) 'Impedance      = ', z, ' ohms at ', theta,' degrees'
WRITE (*,*) 'Current        = ', amps, ' amps'
WRITE (*,*) 'Real Power     = ', p, ' watts'
WRITE (*,*) 'Reactive Power = ', q, ' VAR'
WRITE (*,*) 'Apparent Power = ', s, ' VA'
WRITE (*,*) 'Power Factor   = ', pf

! Finish up.
END PROGRAM power
```

Basic Elements of Fortran

This program also exhibits many of the good programming practices that we have described. It uses the IMPLICIT NONE statement to force the explicit typing of all variables in the program. It includes a variable dictionary defining the uses of all of the variables in the program. It also uses descriptive variable names (although the variable names are short, P, Q, S, and PF are the standard accepted abbreviations for the corresponding quantities). All variables are initialized before they are used. The program defines a named constant for the degrees-to-radians conversion factor, and then uses that name everywhere throughout the program when the conversion factor is required. All input values are echoed, and appropriate units are attached to all printed values.

To verify the operation of program power, we will do a sample calculation by hand and compare the results with the output of the program. If the rms voltage V is 120 V, the magnitude of the impedance Z is 5 Ω, and the angle θ is 30°, then the values are

$$I = \frac{V}{Z} = \frac{120 \text{ V}}{5 \Omega} = 24 \text{ A} \tag{2-3}$$

$$P = VI \cos \theta = (120 \text{ V})(24 \text{ A}) \cos 30° = 2494 \text{ W} \tag{2-4}$$

$$Q = VI \sin \theta = (120 \text{ V})(24 \text{ A}) \sin 30° = 1440 \text{ VAR} \tag{2-5}$$

$$S = VI = (120 \text{ V})(24 \text{ A}) = 2880 \text{ VA} \tag{2-6}$$

$$PF = \cos \theta = \cos 30° = 0.86603 \tag{2-7}$$

When we run program power with the specified input data, the results are identical with our hand calculations:

```
C:\book\fortran\chap2>power
Enter the rms voltage of the source:
120
Enter the magnitude and angle of the impedance
in ohms and degrees:
5., 30.
Voltage         =      120.000000 volts
Impedance       =        5.000000 ohms at    30.000000 degrees
Current         =       24.000000 amps
Real Power      =     2494.153000 watts
Reactive Power  =     1440.000000 VAR
Apparent Power  =     2880.000000 VA
Power Factor    =    8.660254E-01
```

EXAMPLE 2-5 *Carbon 14 Dating:*

A radioactive isotope of an element is a form of the element that is not stable. Instead, it spontaneously decays into another element over a period of time. Radioactive decay is an exponential process. If Q_0 is the initial quantity of a radioactive substance at time

FIGURE 2-9
The radioactive decay of carbon 14 as a function of time. Notice that 50 percent of the original carbon 14 is left after about 5730 years have elapsed.

$t = 0$, then the amount of that substance that will be present at any time t in the future is given by

$$Q(t) = Q_0 e^{-\lambda t} \qquad (2\text{-}8)$$

where λ is the radioactive decay constant (see Figure 2-9).

Because radioactive decay occurs at a known rate, it can be used as a clock to measure the time since the decay started. If we know the initial amount of the radioactive material Q_0 present in a sample, and the amount of the material Q left at the current time, we can solve for t in Equation (2-8) to determine how long the decay has been going on. The resulting equation is

$$t_{\text{decay}} = -\frac{1}{\lambda} \log_e \frac{Q}{Q_0} \qquad (2\text{-}9)$$

Equation (2-8) has practical applications in many areas of science. For example, archaeologists use a radioactive clock based on carbon 14 to determine the time that has passed since a once-living thing died. Carbon 14 is continually taken into the body while a plant or animal is living, so the amount of it present in the body at the time of death is assumed to be known. The decay constant λ of carbon 14 is well known to be 0.00012097/year, so if the amount of carbon 14 remaining now can be accurately measured, then Equation (2-9) can be used to determine how long ago the living thing died.

Write a program that reads the percentage of carbon 14 remaining in a sample, calculates the age of the sample from it, and prints out the result with proper units.

SOLUTION
Our program must perform the following steps:

1. Prompt the user to enter the percentage of carbon 14 remaining in the sample.

2. Read in the percentage.
3. Convert the percentage into the fraction $\frac{Q}{Q_o}$.
4. Calculate the age of the sample in years using Equation (2-8).
5. Write out the result, and stop.

The resulting code is shown in Figure 2-10.

FIGURE 2-10
Program to calculate the age of a sample from the percentage of carbon 14 remaining in it.

```
PROGRAM c14_date
!
!  Purpose:
!    To calculate the age of an organic sample from the percentage
!    of the original carbon 14 remaining in the sample.
!
!  Record of revisions:
!      Date        Programmer          Description of change
!      ====        ==========          =====================
!    11/03/15     S. J. Chapman         Original code
!
IMPLICIT NONE

! Data dictionary: declare constants
REAL,PARAMETER :: LAMDA = 0.00012097  ! The radioactive decay
                                      ! constant of carbon 14,
                                      ! in units of 1/years.

! Data dictionary: declare variable types, definitions, & units
REAL :: age        ! The age of the sample (years)
REAL :: percent    ! The percentage of carbon 14 remaining at the time
                   ! of the measurement (%)
REAL :: ratio      ! The ratio of the carbon 14 remaining at the time
                   ! of the measurement to the original amount of
                   ! carbon 14 (no units)

! Prompt the user for the percentage of C-14 remaining.
WRITE (*,*) 'Enter the percentage of carbon 14 remaining:'
READ  (*,*) percent

! Echo the user's input value.
WRITE (*,*) 'The remaining carbon 14 = ', percent, ' %.'

! Perform calculations
ratio = percent / 100.                ! Convert to fractional ratio
age = (-1.0 / LAMDA) * log(ratio)     ! Get age in years

! Tell the user about the age of the sample.
WRITE (*,*) 'The age of the sample is ', age, ' years.'

! Finish up.
END PROGRAM c14_date
```

To test the completed program, we will calculate the time it takes for half of the carbon 14 to disappear. This time is known as the *half-life* of carbon 14.

```
C:\book\fortran\chap2>c14_date
Enter the percentage of carbon 14 remaining:
50.
The remaining carbon 14 =        50.000000 %.
The age of the sample is         5729.910000 years.
```

The *CRC Handbook of Chemistry and Physics* states that the half-life of carbon 14 is 5730 years, so output of the program agrees with the reference book.

2.12
DEBUGGING FORTRAN PROGRAMS

There is an old saying that the only sure things in life are death and taxes. We can add one more certainty to that list: if you write a program of any significant size, it won't work the first time you try it! Errors in programs are known as **bugs,** and the process of locating and eliminating them is known as **debugging.** Given that we have written a program and it is not working, how do we debug it?

Three types of errors are found in Fortran programs. The first type of error is a **syntax error.** Syntax errors are errors in the Fortran statement itself, such as spelling errors or punctuation errors. These errors are detected by the compiler during compilation. The second type of error is the **runtime error.** A runtime error occurs when an illegal mathematical operation is attempted during program execution (for example, attempting to divide by zero). These errors cause the program to abort during execution. The third type of error is a **logical error.** Logical errors occur when the program compiles and runs successfully but produces the wrong answer.

The most common mistakes made during programming are *typographical errors*. Some typographical errors create invalid Fortran statements. These errors produce syntax errors that are caught by the compiler. Other typographical errors occur in variable names. For example, the letters in some variable names might have been transposed. If you have used the `IMPLICIT NONE` statement, then the compiler will also catch most of these errors. However, if one legal variable name is substituted for another legal variable name, the compiler cannot detect the error. This sort of substitution might occur if you have two similar variable names. For example, if variables `vel1` and `vel2` are both used for velocities in the program, then one of them might be inadvertently used instead of the other one at some point. This sort of typographical error will produce a logical error. You must check for that sort of error by manually inspecting the code, since the compiler cannot catch it.

Sometimes it is possible to successfully compile and link the program, but there are runtime errors or logical errors when the program is executed. In this case, there is something wrong either with the input data or with the logical structure of the program. The first step in locating this sort of bug should be to *check the input data to the program*. Your program should have been designed to echo its input data. If not, go back and add WRITE statements to verify that the input values are what you expect them to be.

If the variable names seem to be correct and the input data is correct, then you are probably dealing with a logical error. You should check each of your assignment statements.

1. If an assignment statement is very long, break it into several smaller assignment statements. Smaller statements are easier to verify.
2. Check the placement of parentheses in your assignment statements. It is a very common error to have the operations in an assignment statement evaluated in the wrong order. If you have any doubts as to the order in which the variables are being evaluated, add extra sets of parentheses to make your intentions clear.
3. Make sure that you have initialized all of your variables properly.
4. Be sure that any functions you use are in the correct units. For example, the input to trigonometric functions must be in units of radians, not degrees.
5. Check for possible errors due to integer or mixed-mode arithmetic.

If you are still getting the wrong answer, add WRITE statements at various points in your program to see the results of intermediate calculations. If you can locate the point where the calculations go bad, then you know just where to look for the problem, which is 95% of the battle.

If you still cannot find the problem after all of the above steps, explain what you are doing to another student or to your instructor, and let them look at the code. It is very common for a person to see just what he or she expects to see when they look at their own code. Another person can often quickly spot an error that you have overlooked time after time.

Good Programming Practice
To reduce your debugging effort, make sure that during your program design you:

1. Use the IMPLICIT NONE statement.
2. Echo all input values.
3. Initialize all variables.
4. Use parentheses to make the functions of assignment statements clear.

All modern compilers have special debugging tools called *symbolic debuggers*. A symbolic debugger is a tool that allows you to walk through the execution of your program one statement at a time, and to examine the values of any variables at each step along the way. Symbolic debuggers allow you to see all of the intermediate results without having to insert a lot of WRITE statements into your code. They are powerful and flexible, but unfortunately they are different for every type of compiler. If you will be using a symbolic debugger in your class, your instructor will introduce you to the debugger appropriate for your compiler and computer.

2.13
SUMMARY

In this chapter, we have presented many of the fundamental concepts required to write functional Fortran programs. We described the basic structure of Fortran programs, and introduced four data types: integer, real, logical, and character. We introduced the assignment statement, arithmetic calculations, intrinsic functions, and list-directed input/output statements. Throughout the chapter, we have emphasized those features of the language that are important for writing understandable and maintainable Fortran code.

The Fortran statements introduced in this chapter must appear in a specific order in a Fortran program. The proper order is summarized in Table 2-5.

The order in which Fortran expressions are evaluated follows a fixed hierarchy, with operations at a higher level evaluated before operations at lower levels. The hierarchy of operations is summarized in Table 2-6.

The Fortran language includes a number of built-in functions to help us solve problems. These functions are called intrinsic functions, since they are intrinsic to the Fortran language itself. Some common intrinsic functions are summarized in Tables 2-3 and 2-4, and a complete listing of intrinsic functions is contained in Appendix B.

There are two varieties of intrinsic functions: specific functions and generic functions. Specific functions require that their input data be of a specific type; if data of the

TABLE 2-5
The order of Fortran statements in a program

1. `PROGRAM` Statement
2. `IMPLICIT NONE` Statement
3. **Type Declaration Statements:**
   `REAL` Statement(s)   ( )
   `INTEGER` Statement(s)   (Any number in any order)
   `CHARACTER` Statement(s)   ( )
4. **Executable Statements:**
   Assignment Statement(s)   ( )
   `READ` Statement(s)   (Any number in the order)
   `WRITE` Statement(s)   (required to accomplish the)
   `STOP` Statement(s)   (desired task.)
5. `END PROGRAM` Statement

TABLE 2-6
Fortran hierarchy of operations

1. Operations within parentheses are evaluated first, starting with the innermost parentheses and working outward.
2. All exponential operations are evaluated next, working from *right* to *left*.
3. All multiplications and divisions are evaluated, working from left to right.
4. All additions and subtractions are evaluated, working from left to right.

Basic Elements of Fortran

wrong type is supplied to a specific function, the result will be meaningless. In contrast, generic functions can accept input data of more than one type and produce correct results.

2.13.1 Summary of Good Programming Practice

Every Fortran program should be designed so that another person who is familiar with Fortran can easily understand it. This is very important, since a good program may be used for a long period of time. Over that time, conditions will change, and the program will need to be modified to reflect the changes. The program modifications may be done by someone other than the original programmer. The programmer making the modifications must understand the original program well before attempting to change it.

It is much harder to design clear, understandable, and maintainable programs than it is to simply write programs. To do so, a programmer must develop the discipline to properly document his or her work. In addition, the programmer must be careful to avoid known pitfalls along the path to good programs. The following guidelines will help you to develop good programs:

1. Use meaningful variable names whenever possible. Use names that can be understood at a glance, like `day`, `month`, and `year`.
2. Always use the `IMPLICIT NONE` statement to catch typographical errors in your program at compilation time.
3. Create a data dictionary in each program that you write. The data dictionary should explicitly declare and define each variable in the program. Be sure to include the physical units associated with each variable, if applicable.
4. Use a consistent number of significant digits in constants. For example, do not use 3.14 for π in one part of your program, and 3.141593 in another part of the program. To ensure consistency, a constant may be named, and the constant may be referenced by name wherever it is needed.
5. Be sure to specify all constants with as much precision as your computer will support. For example, specify π as 3.141593, *not* 3.14.
6. Do not use integer arithmetic to calculate continuously varying real-world quantities such as distance and time. Use integer arithmetic only for things that are intrinsically integer, such as counters.
7. Avoid mixed-mode arithmetic except for exponentiation. If it is necessary to mix integer and real variables in a single expression, use the intrinsic functions `REAL`, `INT`, `NINT`, `CEILING`, and `FLOOR` to make the type conversions explicit.
8. Use extra parentheses whenever necessary to improve the readability of your expressions.
9. Always echo any variables that you enter into a program from a keyboard to make sure that they were typed and processed correctly.
10. Initialize all variables in a program before using them. The variables may be initialized with assignment statements, with `READ` statements, or directly in type declaration statements.
11. Always print the physical units associated with any value being written out. The units are important for the proper interpretation of a program's results.

2.13.2 Summary of Fortran Statements

The following summary describes the Fortran statements introduced in this chapter.

Assignment Statement:

$$\text{variable} = \text{expression}$$

Examples:

```
pi = 3.141593
distance = 0.5 * acceleration * time ** 2
side = hypot * cos(theta)
```

Description:
The left side of the assignment statement must be a variable name. The right side of the assignment statement can be any constant, variable, function, or expression. The value of the quantity on the right-hand side of the equal sign is stored into the variable named on the left-hand side of the equal sign.

CHARACTER Statement:

```
CHARACTER(len=<len>) :: variable_name1[, variable_name2, ...]
CHARACTER(<len>) :: variable_name1[, variable_name2, ...]
CHARACTER :: variable_name1[, variable_name2, ...]
```

Examples:

```
CHARACTER(len=10) :: first, last, middle
CHARACTER(10) :: first = 'My Name'
CHARACTER :: middle_initial
```

Description:
The CHARACTER statement is a type declaration statement that declares variables of the character data type. The length in characters of each variable is specified by the (len=<len>), or by <len>. If the length is absent, then the length of the variables defaults to 1.
　　The value of a CHARACTER variable may be initialized with a string when it is declared, as shown in the second example above.

END PROGRAM Statement:

```
END PROGRAM [name]
```

Description:
The END PROGRAM statement must be the last statement in a Fortran program segment. It tells the compiler that there are no further statements to process. Program execution is stopped when the END PROGRAM statement is reached. The name of the program may optionally be included in the END PROGRAM statement.

Basic Elements of Fortran

ERROR STOP Statement:

```
ERROR STOP
ERROR STOP n
ERROR STOP 'message'
```

Description:
The ERROR STOP statement stops the execution of a Fortran program, and notifies the operating system that an execution error occurred.

IMPLICIT NONE Statement:

```
IMPLICIT NONE
```

Description:
The IMPLICIT NONE statement turns off default typing in Fortran. When it is used in a program, every variable in the program must be explicitly declared in a type declaration statement.

INTEGER Statement:

```
INTEGER :: variable_name1[, variable_name2, ...]
```

Examples:

```
INTEGER :: i, j, count
INTEGER :: day = 4
```

Description:
The INTEGER statement is a type declaration statement that declares variables of the integer data type. This statement overrides the default typing specified in Fortran. The value of an INTEGER variable may be initialized when it is declared, as shown in the second example above.

PROGRAM Statement:

```
PROGRAM program_name
```

Example:

```
PROGRAM my_program
```

Description:
The PROGRAM statement specifies the name of a Fortran program. It must be the first statement in the program. The name must be unique, and cannot be used as a variable name within the program. A program name may consist of 1 to 31 alphabetic, numeric, and underscore characters, but the first character in the program name must be alphabetic.

READ Statement (List-Directed READ):

```
READ (*,*) variable_name1[, variable_name2, ...]
```

Examples:

```
READ (*,*) stress
READ (*,*) distance, time
```

Description:
The list-directed READ statement reads one or more values from the standard input device and loads them into the variables in the list. The values are stored in the order in which the variables are listed. Data values must be separated by blanks or by commas. As many lines as necessary will be read. Each READ statement begins searching for values with a new line.

REAL Statement:

```
REAL :: variable_name1[, variable_name2, ...]
REAL :: variable_name = value
```

Examples:

```
REAL :: distance, time
REAL :: distance = 100
```

Description:
The REAL statement is a type declaration statement that declares variables of the real data type. This statement overrides the default typing specified in Fortran. The value of a REAL variable may be initialized when it is declared, as shown in the second example above.

STOP Statement:

```
STOP
STOP n
STOP 'message'
```

Description:
The STOP statement stops the execution of a Fortran program. There may be more than one STOP statement within a program. A STOP statement that immediately precedes an END PROGRAM statement may be omitted, since execution is also stopped when the END PROGRAM statement is reached.

Basic Elements of Fortran

> **WRITE Statement (List-Directed WRITE):**
>
>   WRITE (*,*) expression1 [,expression2, etc.]
>
> Examples:
>
>   WRITE (*,*) stress
>   WRITE (*,*) distance, time
>   WRITE (*,*) 'SIN(theta) = ', SIN(theta)
>
> Description:
> The list-directed WRITE statement writes the values of one or more expressions to the standard output device. The values are written in the order in which the expressions are listed.

2.13.3 Exercises

2-1. State whether or not each of the following Fortran constants is valid. If valid, state what type of constant it is. If not, state why it is invalid.

(a) 3.14159

(b) '.TRUE.'

(c) -123,456.789

(d) +1E-12

(e) 'Who's coming for dinner?'

(f) "Pass / Fail'

(g) "Enter name:"

2-2. For each of the following pairs of numbers, state whether they represent the same value or different values within the computer.

(a) 123.E+0; 123

(b) 1234.E-3; 1.234E3

(c) 1.41421; 1.41421E0

(d) 0.000005E+6; 5.

2-3. State whether each of the following program names is valid or not. If not, state why the name is invalid.

(a) junk

(b) 3rd

(c) Who_are_you?

(d) time_to_intercept

2-4. Which of the following expressions are legal in Fortran? If an expression is legal, evaluate it.

(a) 2.**3 / 3**2

(b) 2 * 6 + 6 ** 2 / 2

(c) 2 * (-10.)**-3.

(d) 2 / (-10.) ** 3.

(e) 23 / (4 / 8)

2-5. Which of the following expressions are legal in Fortran? If an expression is legal, evaluate it.

(a) ((58/4)*(4/58))

(b) ((58/4)*(4/58.))

(c) ((58./4)*(4/58.))

(d) ((58./4*(4/58.))

2-6. Evaluate each of the following expressions.

(a) 13 / 5 * 6

(b) (13 / 5) * 6

(c) 13 / (5 * 6)

(d) 13. / 5 * 6

(e) 13 / 5 * 6.

(f) INT(13. / 5) * 6

(g) NINT(13. / 5) * 6

(h) CEILING(13. / 5) * 6

(i) FLOOR(13. / 5) * 6

2-7. Evaluate each of the following expressions.

(a) 3 ** 3 ** 2

(b) (3 ** 3) ** 2

(c) 3 ** (3 ** 2)

2-8. What values will be output from the following program?

```
PROGRAM sample_1
INTEGER :: i1, i2, i3, i4
REAL :: a1 = 2.4, a2
i1 = a1
i2 = INT( -a1 * i1 )
i3 = NINT( -a1 * i1 )
i4 = FLOOR( -a1 * i1 )
a2 = a1**i1
WRITE (*,*) i1, i2, i3, i4, a1, a2
END PROGRAM sample_1
```

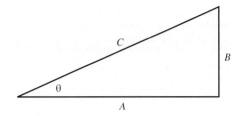

FIGURE 2-11
The right triangle of Exercise 2-9.

2-9. Figure 2-11 shows a right triangle with a hypotenuse of length C and angle θ. From elementary trigonometry, the length of sides A and B are given by

$$A = C \cos \theta$$
$$B = C \sin \theta$$

The following program is intended to calculate the lengths of sides A and B given the hypotenuse C and angle θ. Will this program run? Will it produce the correct result? Why or why not?

```
PROGRAM triangle
REAL :: a, b, c, theta
WRITE (*,*) 'Enter the length of the hypotenuse C:'
READ (*,*) c
WRITE (*,*) 'Enter the angle THETA in degrees:'
READ (*,*) theta
a = c * COS ( theta )
b = c * SIN ( theta )
WRITE (*,*) 'The length of the adjacent side is ', a
WRITE (*,*) 'The length of the opposite side is ', b
END PROGRAM triangle
```

2-10. What output will be produced by the following program?

```
PROGRAM example
REAL :: a, b, c
INTEGER :: k, l, m
READ (*,*) a, b, c, k
READ (*,*) l, m
WRITE (*,*) a, b, c, k, l, m
END PROGRAM example
```

The input data to the program is:

```
-3.141592
100, 200., 300, 400
-100, -200, -300
-400
```

2-11. Write a Fortran program that calculates an hourly employee's weekly pay. The program should ask the user for the person's pay rate and the number of hours worked during the week. It should then calculate the total pay from the formula

Total Pay = Hourly Pay Rate × Hours Worked

Finally, it should display the total weekly pay. Check your program by computing the weekly pay for a person earning $7.90 per hour and working for 42 hours.

2-12. The potential energy of an object due to its height above the surface of the Earth is given by the equation

$$PE = mgh \tag{2-10}$$

where m is the mass of the object, g is the acceleration due to gravity, and h is the height above the surface of the Earth. The kinetic energy of a moving object is given by the equation

$$KE = \frac{1}{2}mv^2 \tag{2-11}$$

where m is the mass of the object and v is the velocity of the object. Write a Fortran statement for the total energy (potential plus kinetic) possessed by an object in the Earth's gravitational field.

2-13. If a stationary ball is released at a height h above the surface of the Earth, the velocity of the ball v when it hits the Earth is given by the equation

$$v = \sqrt{2gh} \tag{2-12}$$

where g is the acceleration due to gravity, and h is the height above the surface of the Earth (assuming no air friction). Write a Fortran equation for the velocity of the ball when it hits the Earth.

2-14. Write a Fortran program that calculates the velocity of the ball v when it hits the Earth from a given height h, using Equation (2-12) *equation reference goes here*. Use the program to calculate the velocity for a height of (*a*) 1 meter; (*b*) 10 meters; and (*c*) 100 meters.

2-15. In Einstein's Theory of Relativity, the rest mass of matter is related to an equivalent energy by the equation

$$E = mc^2 \tag{2-13}$$

where E is the energy in joules, m is mass in kilograms, and c is the speed of light in meters per second ($c = 2.9979 \times 10^8$ m/s). Suppose that a 400 MW (= 400 million joules per second) nuclear power generating station supplies full power to the electrical grid for a year. Write a program that calculates the amount of mass consumed in the course of the year. Use good programming practices in your program. (*Note:* Assume that the generating station is 100% efficient in producing electrical energy.)

2-16. Generalize the program of the previous exercise to calculate the mass consumed by a generating station with a user-specified output power for a user-specified period of months.

2-17. Period of a Pendulum The period of an oscillating pendulum T (in seconds) is given by the equation

$$T = 2\pi\sqrt{\frac{L}{g}} \tag{2-14}$$

where L is the length of the pendulum in meters, and g is the acceleration due to gravity in meters per second squared. Write a Fortran program to calculate the period of a

Basic Elements of Fortran

pendulum of length L. The length of the pendulum will be specified by the user when the program is run. Use good programming practices in your program. (The acceleration due to gravity at the Earth's surface is 9.81 m/s².)

2-18. Write a program to calculate the hypotenuse of a right triangle, given the lengths of its two sides. Use good programming practices in your program.

2-19. Logarithms to an Arbitrary Base Write a program to calculate the logarithm of a number x to an arbitrary base b ($\log_b x$). Use the following equation for the calculation

$$\log_b x = \frac{\log_{10} x}{\log_{10} b} \tag{2-15}$$

Test the program by calculating the logarithm to the base e of 100. (Note that you can check your answer using the LOG(X) function, which calculates $\log_e x$.)

2-20. Write a program using the IMPLICIT NONE statement, and do not declare one of the variables in the program. What sort of error message is generated by your compiler?

2-21. The distance between two points (x_1, y_1) and (x_2, y_2) on a Cartesian coordinate plane (see Figure (2-12)) is given by the equation

$$d = \sqrt{(x_1 - x_2)^2 + (y_1 - y_2)^2} \tag{2-16}$$

Write a Fortran program to calculate the distance between any two points (x_1, y_1) and (x_2, y_2) specified by the user. Use good programming practices in your program. Use the program to calculate the distance between the points $(-1,1)$ and $(6,2)$.

2-22. Decibels Engineers often measure the ratio of two power measurements in *decibels*, or dB. The equation for the ratio of two power measurements in decibels is

$$dB = 10 \log_{10} \frac{P_2}{P_1} \tag{2-17}$$

where P_2 is the power level being measured, and P_1 is some reference power level. Assume that the reference power level P_1 is 1 mW, and write a program that accepts an input power P_2 and converts it into dB with respect to the 1 mW reference level.

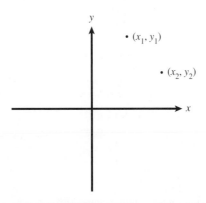

FIGURE 2-12
A Cartesian plane containing two points (x_1, y_1) and (x_2, y_2).

2-23. Hyperbolic cosine The hyperbolic cosine function is defined by the equation

$$\cosh x = \frac{e^x + e^{-x}}{2} \tag{2-18}$$

Write a Fortran program to calculate the hyperbolic cosine of a user-supplied value x. Use the program to calculate the hyperbolic cosine of 3.0. Compare the answer that your program produces to the answer produced by the Fortran intrinsic function COSH(x).

2-24. Compound Interest Suppose that you deposit a sum of money P in an interest-bearing account at a local bank (P stands for *present value*). If the bank pays you interest on the money at a rate of i percent per year and compounds the interest m times a year, the amount of money that you will have in the bank after n years is given by the equation

$$F = P\left(1 + \frac{APR}{100m}\right)^{mn} \tag{2-19}$$

where F is the future value of the account and APR is the annual percentage rate on the account. The quantity $\frac{APR}{100m}$ is the fraction of interest earned in one compounding period (the extra factor of 100 in the denominator converts the rate from percentages to fractional amounts). Write a Fortran program that will read an initial amount of money P, an annual interest rate APR, the number of times m that the interest is compounded in a year, and the number of years n that the money is left in the account. The program should calculate the future value F of this account.

Use this program to calculate the future value of the bank account if $1000.00 is deposited in an account with an APR of 5% for a period of 1 year, and the interest is compounded (*a*) annually, (*b*) semiannually, or (*c*) monthly. How much difference does the rate of compounding make on the amount in the account?

2-25. Radio Receiver A simplified version of the front end of an AM radio receiver is shown in Figure 2-13. This receiver consists of an *RLC* tuned circuit containing a resistor, capacitor, and an inductor connected in series. The *RLC* circuit is connected to an external antenna and ground as shown in the picture.

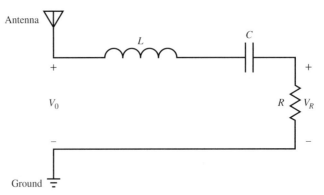

FIGURE 2-13
A simplified representation of an AM radio set.

The tuned circuit allows the radio to select a specific station out of all the stations transmitting on the AM band. At the resonant frequency of the circuit, essentially all of the signal V_0 appearing at the antenna appears across the resistor, which represents the rest of the radio. In other words, the radio receives its strongest signal at the resonant frequency. The resonant frequency of the LC circuit is given by the equation

$$f_0 = \frac{1}{2\pi\sqrt{LC}} \tag{2-20}$$

where L is inductance in henrys (H) and C is capacitance in farads (F). Write a program that calculates the resonant frequency of this radio set given specific values of L and C. Test your program by calculating the frequency of the radio when $L = 0.1$ mH and $C = 0.25$ nF.

2-26. **Aircraft Turning Radius** An object moving in a circular path at a constant tangential velocity v is shown in Figure 2-14. The radial acceleration required for the object to move in the circular path is given by Equation (2-21)

$$a = \frac{v^2}{r} \tag{2-21}$$

where a is the centripetal acceleration of the object in m/s², v is the tangential velocity of the object in m/s, and r is the turning radius in meters. Suppose that the object is an aircraft, and write a program to answer the following questions about it:

(a) Suppose that the aircraft is moving at Mach 0.80, or 80% of the speed of sound. If the centripetal acceleration is 2.5g, what is the turning radius of the aircraft? (Note: For this problem, you may assume that Mach 1 is equal to 340 m/s, and that $1g = 9.81$ m/s².)

(b) Suppose that the speed of the aircraft increases to Mach 1.5. What is the turning radius of the aircraft now?

(c) Suppose that the maximum acceleration that the pilot can stand is 7g. What is the minimum possible turning radius of the aircraft at Mach 1.5?

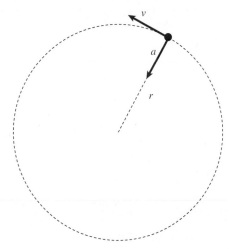

FIGURE 2-14
An object moving in uniform circular motion due to the centripetal acceleration a.

2-27. Escape Velocity The escape velocity from the surface of a planet or moon (ignoring the effects of atmosphere) is given by Equation (2-22)

$$v_{esc} = \frac{\sqrt{2GM}}{R} \qquad (2\text{-}22)$$

where v_{esc} is the escape velocity in meters per second, G is the gravitational constant $(6.673 \times 10^{-11} \text{ Nm}^{-2}\text{kg}^{-2})$, M is the mass of the planet in kilograms, and R is the radius of the planet in meters. Write a program that will calculate the escape velocity as a function of mass and radius, and use the program to calculate the escape velocity for the bodies given below.

Body	Mass (kg)	Radius (m)
Earth	6.0×10^{24}	6.4×10^{6}
Moon	7.4×10^{22}	1.7×10^{6}
Ceres	8.7×10^{20}	4.7×10^{5}
Jupiter	1.9×10^{27}	7.1×10^{7}

3

Program Design and Branching Structures

OBJECTIVES

- Learn the concepts of top-down design and decomposition.
- Learn about pseudocode and flowcharts, and why they should be used.
- Know how to create and use `LOGICAL` constants and variables.
- Learn about relational and combinational logical operators, and how they fit into the hierarchy of operations.
- Know how to use the `IF` construct.
- Know how to use the `SELECT CASE` construct.

In the previous chapter, we developed several complete working Fortran programs. However, all of the programs were very simple, consisting of a series of Fortran statements that were executed one after another in a fixed order. Such programs are called *sequential* programs. They read input data, process it to produce a desired answer, print out the answer, and quit. There is no way to repeat sections of the program more than once, and there is no way to selectively execute only certain portions of the program depending on values of the input data.

In the next two chapters, we will introduce a number of Fortran statements that allow us to control the order in which statements are executed in a program. There are two broad categories of control statements: **branches,** which select specific sections of the code to execute, and **loops,** which cause specific sections of the code to be repeated. Branches will be introduced in this chapter, and loops will be covered in Chapter 4.

With the introduction of branches and loops, our programs are going to become more complex, and it will get easier to make mistakes. To help avoid programming errors, we will introduce a formal program design procedure based upon the technique known as top-down design. We will also introduce two common algorithm development tools, flowcharts and pseudocode.

After introducing the program design process, we will introduce the logical data type and the operations that produce them. Logical expressions are used to control many branching statements, so we will learn about them before studying branches.

Finally, we will study the various types of Fortran branching statements.

3.1
INTRODUCTION TO TOP-DOWN DESIGN TECHNIQUES

Suppose that you are an engineer working in industry, and that you need to write a Fortran program to solve some problem. How do you begin?

When given a new problem, there is a natural tendency to sit down at a terminal and start programming without "wasting" a lot of time thinking about it first. It is often possible to get away with this "on the fly" approach to programming for very small problems, such as many of the examples in this book. In the real world, however, problems are larger, and a programmer attempting this approach will become hopelessly bogged down. For larger problems, it pays to completely think out the problem and the approach you are going to take to it before writing a single line of code.

We will introduce a formal program design process in this section, and then apply that process to every major application developed in the remainder of the book. For some of the simple examples that we will be doing, the design process will seem like overkill. However, as the problems that we solve get larger and larger, the process becomes more and more essential to successful programming.

When I was an undergraduate, one of my professors was fond of saying, "Programming is easy. It's knowing what to program that's hard." His point was forcefully driven home to me after I left university and began working in industry on larger-scale software projects. I found that the most difficult part of my job was to *understand the problem* I was trying to solve. Once I really understood the problem, it became easy to break the problem apart into smaller, more easily manageable pieces with well-defined functions, and then to tackle those pieces one at a time.

Top-down design is the process of starting with a large task and breaking it down into smaller, more easily understandable pieces (subtasks) that perform a portion of the desired task. Each subtask may in turn be subdivided into smaller subtasks if necessary. Once the program is divided into small pieces, each piece can be coded and tested independently. We do not attempt to combine the subtasks into a complete task until each of the subtasks has been verified to work properly by itself.

The concept of top-down design is the basis of our formal program design process. We will now introduce the details of the process, which is illustrated in Figure 3-1. The steps involved are:

1. *Clearly state the problem that you are trying to solve.*

    Programs are usually written to fill some perceived need, but that need may not be articulated clearly by the person requesting the program. For example, a user may ask for a program to solve a system of simultaneous linear equations. This request is not clear enough to allow a programmer to design a program to meet the need; he or she must first know much more about the problem to be solved. Is the system of equations to be solved real or complex? What is the maximum number of equations and unknowns that the program must handle? Are there any symmetries in the equations that might be exploited to make the task easier? The program designer will have to talk with the user requesting the program, and the two of them will have to come up with a clear statement of exactly what they are trying to accomplish. A clear statement of the

Program Design and Branching Structures

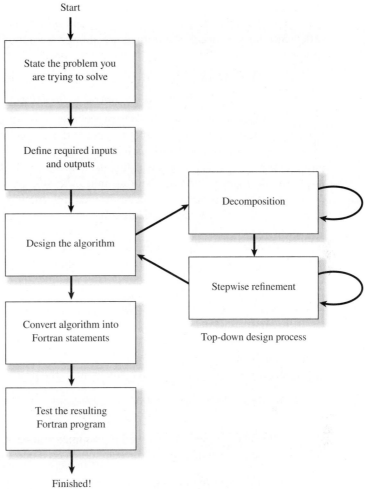

FIGURE 3-1
The program design process used in this book.

problem will prevent misunderstandings, and it will also help the program designer to properly organize his or her thoughts. In the example we were describing, a proper statement of the problem might have been:

Design and write a program to solve a system of simultaneous linear equations having real coefficients and with up to 20 equations in 20 unknowns.

2. *Define the inputs required by the program and the outputs to be produced by the program.*

The inputs to the program and the outputs produced by the program must be specified so that the new program will properly fit into the overall processing scheme.

In the above example, the coefficients of the equations to be solved are probably in some pre-existing order, and our new program needs to be able to read them in that order. Similarly, it needs to produce the answers required by the programs that may follow it in the overall processing scheme, and to write out those answers in the format needed by the programs following it.

3. *Design the algorithm that you intend to implement in the program.*

An **algorithm** is a step-by-step procedure for finding the solution to a problem. It is at this stage in the process that top-down design techniques come into play. The designer looks for logical divisions within the problem, and divides it up into subtasks along those lines. This process is called *decomposition*. If the subtasks are themselves large, the designer can break them up into even smaller sub-subtasks. This process continues until the problem has been divided into many small pieces, each of which does a simple, clearly understandable job.

After the problem has been decomposed into small pieces, each piece is further refined through a process called *stepwise refinement*. In stepwise refinement, a designer starts with a general description of what the piece of code should do, and then defines the functions of the piece in greater and greater detail until they are specific enough to be turned into Fortran statements. Stepwise refinement is usually done with **pseudocode,** which will be described in the next section.

It is often helpful to solve a simple example of the problem by hand during the algorithm development process. If the designer understands the steps that he or she went through in solving the problem by hand, then he or she will be better able to apply decomposition and stepwise refinement to the problem.

4. *Turn the algorithm into* Fortran *statements.*

If the decomposition and refinement process were carried out properly, this step will be very simple. All that the programmer will have to do is to replace pseudocode with the corresponding Fortran statements on a one-for-one basis.

5. *Test the resulting* Fortran *program.*

This step is the real killer. The components of the program must first be tested individually, if possible, and then the program as a whole must be tested. When testing a program, we must verify that it works correctly for *all legal input data sets*. It is very common for a program to be written, tested with some standard data set, and released for use, only to find that it produces the wrong answers (or crashes) with a different input data set. If the algorithm implemented in a program includes different branches, we must test all of the possible branches to confirm that the program operates correctly under every possible circumstance.

Large programs typically go through a series of tests before they are released for general use (see Figure 3-2). The first stage of testing is sometimes called **unit testing.** During unit testing, the individual subtasks of the program are tested separately to confirm that they work correctly. The programmer usually writes small programs called "stubs" or "test drivers" to execute the code under test, and to see if the code is returning the proper results. This verifies the operation of the subtasks at a basic level before they are combined into larger groups.

Program Design and Branching Structures

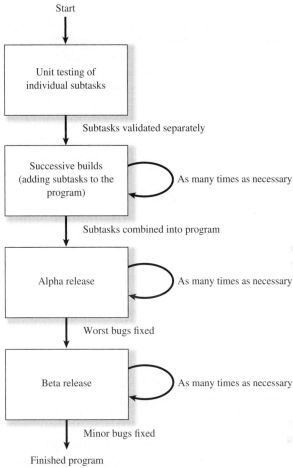

FIGURE 3-2
A typical testing process for a large program.

After the unit testing is completed, the program goes through a series of *builds* during which the individual subtasks are combined to produce the final program. The first build of the program typically includes only a few of the subtasks. It is used to check the interactions among those subtasks and the functions performed by the combinations of the subtasks. In successive builds, more and more subtasks are added, until the entire program is complete. Testing is performed on each build, and any errors (bugs) that are detected are corrected before moving on to the next build.

Testing continues even after the program is complete. The first complete version of the program is usually called the **alpha release.** It is exercised by the programmers and others very close to them in as many different ways as possible, and the bugs discovered during the testing are corrected. When the most serious bugs have been removed from the program, a new version called the **beta release** is prepared. The beta release is normally given to "friendly" outside users who have a need for the program in their normal day-to-day jobs. These users put the program through its paces under many different conditions and with many different input data sets, and they report any bugs that they find to the programmers. When those bugs have been corrected, the program is ready to be released for general use.

Because the programs in this book are fairly small, we will not go through the sort of extensive testing described above. However, we will follow the basic principles in testing all of our programs.

The program design process may be summarized as follows:

1. Clearly state the problem that you are trying to solve.
2. Define the inputs required by the program and the outputs to be produced by the program.
3. Design the algorithm that you intend to implement in the program.
4. Turn the algorithm into Fortran statements.
5. Test the Fortran program.

Good Programming Practice

Follow the steps of the program design process to produce reliable, understandable Fortran programs.

In a large programming project, the time actually spent in programming is surprisingly small. In his book *The Mythical Man-Month*,[1] Frederick P. Brooks, Jr., suggests that in a typical large software project, 1/3 of the time is spent planning what to do (steps 1 through 3), 1/6 of the time is spent actually writing the program (step 4), and fully 1/2 of the time is spent in testing and debugging the program! Clearly, anything that we can do to reduce the testing and debugging time will be very helpful. We can best reduce the testing and debugging time by doing a very careful job in the planning phase, and by using good programming practices. Good programming practices will reduce the number of bugs in the program, and will make the ones that do creep in easier to find.

3.2
USE OF PSEUDOCODE AND FLOWCHARTS

As a part of the design process, it is necessary to describe the algorithm that you intend to implement. The description of the algorithm should be in a standard form that is easy for both you and other people to understand, and the description should aid you in turning your concept into Fortran code. The standard forms that we use to describe algorithms are called **constructs,** and an algorithm described using these constructs is called a structured algorithm. When the algorithm is implemented in a Fortran program, the resulting program is called a **structured program.**

---

[1] *The Mythical Man-Month, Anniversary Edition*, by Frederick P. Brooks, Jr., Addison-Wesley, 1995.

Program Design and Branching Structures

The constructs used to build algorithms can be described in two different ways: pseudocode and flowcharts. **Pseudocode** is a hybrid mixture of Fortran and English. It is structured like Fortran, with a separate line for each distinct idea or segment of code, but the descriptions on each line are in English. Each line of the pseudocode should describe its idea in plain, easily understandable English. Pseudocode is very useful for developing algorithms, since it is flexible and easy to modify. It is especially useful since pseudocode can be written and modified on the same computer terminal used to write the Fortran program—no special graphical capabilities are required.

For example, the pseudocode for the algorithm in Example 2-3 is:

```
Prompt user to enter temperature in degrees Fahrenheit
Read temperature in degrees Fahrenheit (temp_f)
temp_k in kelvins ← (5./9.) * (temp_f - 32) + 273.15
Write temperature in kelvins
```

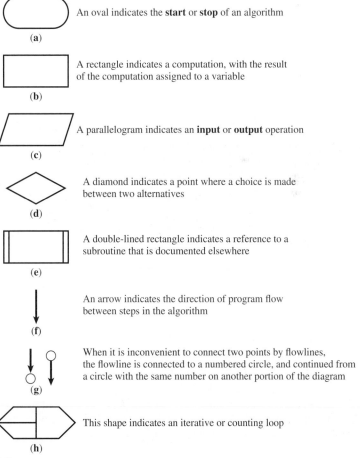

(a) An oval indicates the **start** or **stop** of an algorithm

(b) A rectangle indicates a computation, with the result of the computation assigned to a variable

(c) A parallelogram indicates an **input** or **output** operation

(d) A diamond indicates a point where a choice is made between two alternatives

(e) A double-lined rectangle indicates a reference to a subroutine that is documented elsewhere

(f) An arrow indicates the direction of program flow between steps in the algorithm

(g) When it is inconvenient to connect two points by flowlines, the flowline is connected to a numbered circle, and continued from a circle with the same number on another portion of the diagram

(h) This shape indicates an iterative or counting loop

FIGURE 3-3
Common symbols used in flowcharts.

Notice that a left arrow (←) is used instead of an equal sign (=) to indicate that a value is stored in a variable, since this avoids any confusion between assignment and equality. Pseudocode is intended to aid you in organizing your thoughts before converting them into Fortran code.

Flowcharts are a way to describe algorithms graphically. In a flowchart, different graphical symbols represent the different operations in the algorithm, and our standard constructs are made up of collections of one or more of these symbols. Flowcharts are very useful for describing the algorithm implemented in a program after it is completed. However, since they are graphical, flowcharts tend to be cumbersome to modify, and they are not very useful during the preliminary stages of algorithm definition when rapid changes are occurring. The most common graphical symbols used in flowcharts are shown in Figure 3-3, and the flowchart for the algorithm in Example 2-3 is shown in Figure 3-4.

Throughout the examples in this book, we will illustrate the use of both pseudocode and flowcharts. You are welcome to use whichever one of these tools gives you the best results in your own programming projects.

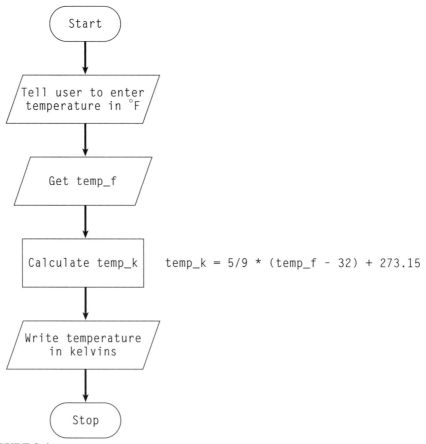

FIGURE 3-4
Flowchart for the algorithm in Example 2-3.

3.3
LOGICAL CONSTANTS, VARIABLES, AND OPERATORS

As we mentioned in the introduction to this chapter, most Fortran branching structures are controlled by logical values. Before studying the branching structures, we will introduce the data types that control them.

3.3.1 Logical Constants and Variables

The logical data type contains one of only two possible values: TRUE or FALSE. A **logical constant** can have one of the following values: .TRUE. or .FALSE. (note that the periods are required on either side of the values to distinguish them from variable names). Thus, the following are valid logical constants:

```
.TRUE.
.FALSE.
```

The following are not valid logical constants:

```
TRUE      (No periods—this is a variable name)
.FALSE    (Unbalanced periods)
```

Logical constants are rarely used, but logical expressions and variables are commonly used to control program execution, as we will see later in the chapter.

A **logical variable** is a variable containing a value of the logical data type. A logical variable is declared using the LOGICAL statement:

```
LOGICAL :: var1 [, var2, var3, ...]
```

This type declaration statement should be placed after the PROGRAM statement and before the first executable statement in the program, as shown in the example below:

```
PROGRAM example
LOGICAL :: test1, test2
...
(Executable statements follow)
```

3.3.2 Assignment Statements and Logical Calculations

Like arithmetic calculations, logical calculations are performed with an assignment statement, whose form is

```
logical_variable_name = logical_expression
```

The expression to the right of the equal sign can be any combination of valid logical constants, logical variables, and logical operators. A **logical operator** is an operator on numeric, character, or logical data that yields a logical result. There are two basic type of logical operators: **relational operators** and **combinational operators.**

TABLE 3-1
Relational logic operators

Operation		Meaning
New style	Older style	
==	.EQ.	Equal to
/=	.NE.	Not equal to
>	.GT.	Greater than
>=	.GE.	Greater than or equal to
<	.LT.	Less than
<=	.LE.	Less than or equal to

3.3.3 Relational Operators

Relational logic operators are operators with two numerical or character operands that yield a logical result. The result depends on the *relationship* between the two values being compared, so these operators are called relational. The general form of a relational operator is

$$a_1 \text{ op } a_2$$

where a_1 and a_2 are arithmetic expressions, variables, constants, or character strings, and op is one of the relational logic operators listed in Table 3-1.

There are two forms of each relational operator. The first one is composed of symbols, and the second one is composed of characters surrounded by periods. In the second form, the periods are a part of the operator and must always be present. The first form of the operators was introduced in Fortran 90, while the second form is a holdover from earlier versions of Fortran. You may use either form of the operators in your program, but the first form is preferred in new programs.

If the relationship between a_1 and a_2 expressed by the operator is true, then the operation returns a value of .TRUE.; otherwise, the operation returns a value of .FALSE..

Some relational operations and their results are given below:

Operation	Result
3 < 4	.TRUE.
3 <= 4	.TRUE.
3 == 4	.FALSE.
3 > 4	.FALSE.
4 <= 4	.TRUE.
'A' < 'B'	.TRUE.

The last logical expression is .TRUE. because characters are evaluated in alphabetical order.

The equivalence relational operator is written with two equal signs, while the assignment operator is written with a single equal sign. These are very different operators that beginning programmers often confuse. The == symbol is a *comparison* operation that returns a logical result, while the = symbol *assigns* the value of the expression to the right of the equal sign to the variable on the left of the equal sign. It is a very common mistake for beginning programmers to use a single equal sign when trying to do a comparison.

Programming Pitfalls
Be careful not to confuse the equivalence relational operator (==) with the assignment operator (=).

In the hierarchy of operations, relational operators are evaluated after all arithmetic operators have been evaluated. Therefore, the following two expressions are equivalent (both are .TRUE.):

$$7 + 3 < 2 + 11$$
$$(7 + 3) < (2 + 11)$$

If the comparison is between real and integer values, then the integer value is converted to a real value before the comparison is performed. Comparisons between numerical data and character data are illegal and will cause a compile-time error:

```
4 == 4.    .TRUE.    (Integer is converted to real and comparison is made)
4 <= 'A'             (Illegal—produces a compile-time error)
```

3.3.4 Combinational Logic Operators

Combinational logic operators are operators with one or two logical operands that yield a logical result. There are four binary operators, .AND., .OR., .EQV., and .NEQV., and one unary operator, .NOT.. The general form of a binary combinational logic operation is

$$l_1 \text{ .op. } l_2$$

where l_1 and l_2 are logical expressions, variables, or constants, and .op. is one of the combinational operators listed in Table 3-2.

The periods are a part of the operator and must always be present. If the relationship between l_1 and l_2 expressed by the operator is true, then the operation returns a value of .TRUE.; otherwise, the operation returns a value of .FALSE..

The results of the operators are summarized in the **truth tables** in Table 3-3(A) and (B), which show the result of each operation for all possible combinations of l_1 and l_2.

TABLE 3-2
Combinational logic operators

Operator	Function	Definition
l_1 .AND. l_2	Logical AND	Result is TRUE if both l_1 and l_2 are TRUE
l_1 .OR. l_2	Logical OR	Result is TRUE if either or both of l_1 and l_2 are TRUE
l_1 .EQV. l_2	Logical equivalence	Result is TRUE if l_1 is the same as l_2 (either both TRUE or both FALSE)
l_1 .NEQV. l_2	Logical nonequivalence	Result is TRUE if one of l_1 and l_2 is TRUE and the other one is FALSE
.NOT. l_1	Logical NOT	Result is TRUE if l_1 is FALSE, and FALSE if l_1 is TRUE

TABLE 3-3A
Truth tables for binary combinational logic operators

l_1	l_2	l_1 .AND. l_2	l_1 .OR. l_2	l_1 .EQV. l_2	l_1 .NEQV. l_2
.FALSE.	.FALSE.	.FALSE.	.FALSE.	.TRUE.	.FALSE.
.FALSE.	.TRUE.	.FALSE.	.TRUE.	.FALSE.	.TRUE.
.TRUE.	.FALSE.	.FALSE.	.TRUE.	.FALSE.	.TRUE.
.TRUE.	.TRUE.	.TRUE.	.TRUE.	.TRUE.	.FALSE.

TABLE 3-3B
Truth table for .NOT. operator

l_1	.NOT. l_1
.FALSE.	.TRUE.
.TRUE.	.FALSE.

In the hierarchy of operations, combinational logic operators are evaluated *after all arithmetic operations and all relational operators have been evaluated*. The order in which the operators in an expression are evaluated is:

1. All arithmetic operators are evaluated first in the order previously described.
2. All relational operators (==, /=, >, >=, <, <=) are evaluated, working from left to right.
3. All .NOT. operators are evaluated.
4. All .AND. operators are evaluated, working from left to right.
5. All .OR. operators are evaluated, working from left to right.
6. All .EQV. and .NEQV. operators are evaluated, working from left to right.

As with arithmetic operations, parentheses can be used to change the default order of evaluation. Examples of some combinational logic operators and their results are given below.

Program Design and Branching Structures

EXAMPLE 3-1 Assume that the following variables are initialized with the values shown, and calculate the result of the specified expressions:

$$log1 = .TRUE.$$
$$log2 = .TRUE.$$
$$log3 = .FALSE.$$

Logical Expression	Result
(a) .NOT. log1	.FALSE.
(b) log1 .OR. log3	.TRUE.
(c) log1 .AND. log3	.FALSE.
(d) log2 .NEQV. log3	.TRUE.
(e) log1 .AND. log2 .OR. log3	.TRUE.
(f) log1 .OR. log2 .AND. log3	.TRUE.
(g) .NOT. (log1 .EQV. log2)	.FALSE.

The .NOT. operator is evaluated before other combinational logic operators. Therefore, the parentheses in part (g) of the above example were required. If they had been absent, the expression in part (g) would have been evaluated in the order (.NOT. L1) .EQV. L2.

Combinational logic operations involving numerical or character data are illegal and will cause a compile-time error:

$$4 \text{ .AND. } 3 \qquad \text{Error}$$

3.3.5 Logical Values in Input and Output Statements

If a logical variable appears in a list-directed READ statement, then the corresponding input value must either be the constants .TRUE. or .FALSE., or else a character or a group of characters beginning with a T or an F. If the input value is .TRUE., or the first character of the input value is T, then the logical variable will be set to .TRUE.. If the input value is .FALSE., or the first character of the input value is F, then the logical variable will be set to .FALSE.. Any input value beginning with another character will produce a runtime error.

If a logical variable or expression appears in a list-directed WRITE statement, then the corresponding output value will be the single character T if the value of the variable is .TRUE., and F if the value of the variable is .FALSE..

3.3.6 The Significance of Logical Variables and Expressions

Logical variables and expressions are rarely the final product of a Fortran program. Nevertheless, they are absolutely essential to the proper operation of most programs. Most of the major branching and looping structures of Fortran are controlled by logical values, so must be able to read and write logical expressions to understand and use Fortran control statements.

Quiz 3-1

This quiz provides a quick check to see if you have understood the concepts introduced in Section 3.3. If you have trouble with the quiz, reread the sections, ask your instructor, or discuss the material with a fellow student. The answers to this quiz are found in the back of the book.

1. Suppose that the real variables a, b, and c contain the values $-10.$, 0.1, and 2.1, respectively, and that the logical variable l1, l2, and l3 contain the values .TRUE., .FALSE., and .FALSE., respectively. Is each of the following expressions legal or illegal? If an expression is legal, what will its result be?

   (a) a > b .OR. b > c
   (b) (.NOT. a) .OR. l1
   (c) l1 .AND. .NOT. l2
   (d) a < b .EQV. b < c
   (e) l1 .OR. l2 .AND. l3
   (f) l1 .OR. (l2 .AND. l3)
   (g) (l1 .OR. l2) .AND. l3
   (h) a .OR. b .AND. l1

2. If the input data is as shown, what will be printed out by the following program?

   ```
   PROGRAM quiz_31
   INTEGER :: i, j, k
   LOGICAL :: l
   READ (*,*) i, j
   READ (*,*) k
   l = i + j == k
   WRITE (*,*) l
   END PROGRAM quiz_31
   ```

   The input data is :

   1, 3, 5
   2, 4, 6

3.4
CONTROL CONSTRUCTS: BRANCHES

Branches are Fortran statements that permit us to select and execute specific sections of code (called *blocks*) while skipping other sections of code. They are variations of the IF statement, plus the SELECT CASE.

3.4.1 The Block IF Construct

The commonest form of the IF statement is the block IF construct. This construct specifies that a block of code will be executed if and only if a certain logical expression is true. The block IF construct has the form

```
IF (logical_expr) THEN
   Statement 1
   Statement 2                    }   Block 1
   ...
END IF
```

If the logical expression is true, the program executes the statements in the block between the IF and END IF statements. If the logical expression is false, then the program skips all of the statements in the block between the IF and END IF statements, and executes the next statement after the END IF. The flowchart for a block IF construct is shown in Figure 3-5.

The IF (...) THEN is a single Fortran statement that must be written together on the same line, and the statements to be executed must occupy separate lines below the IF (...) THEN statement. An END IF statement must follow them on a separate line. There should not be a statement number on the line containing the END IF statement. For readability, the block of code between the IF and END IF statements is usually indented by two or three spaces, but this is not actually required.

Good Programming Practice
Always indent the body of a block IF construct by two or more spaces to improve the readability of the code.

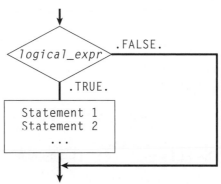

FIGURE 3-5
Flowchart for a simple block IF construct.

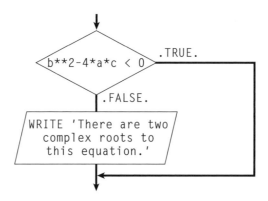

FIGURE 3-6
Flowchart showing structure to determine if a quadratic equation has two complex roots.

As an example of a block IF construct, consider the solution of a quadratic equation of the form

$$ax^2 + bx + c = 0 \tag{3-1}$$

The solution to this equation is

$$x = \frac{-b \pm \sqrt{b^2 - 4ac}}{2a} \tag{3-2}$$

The term $b^2 - 4ac$ is known as the *discriminant* of the equation. If $b^2 - 4ac > 0$, then there are two distinct real roots to the quadratic equation. If $b^2 - 4ac = 0$, then there is a single repeated root to the equation, and if $b^2 - 4ac < 0$, then there are two complex roots to the quadratic equation.

Suppose that we wanted to examine the discriminant of the quadratic equation and tell a user if the equation has complex roots. In pseudocode, the block IF construct to do this would take the form

```
IF (b**2 - 4.*a*c) < 0. THEN
    Write message that equation has two complex roots.
END of IF
```

In Fortran, the block IF construct is

```
IF ( (b**2 - 4.*a*c) < 0.) THEN
    WRITE (*,*) 'There are two complex roots to this equation.'
END IF
```

The flowchart for this construct is shown in Figure 3-6.

3.4.2 The ELSE and ELSE IF Clauses

In the simple block IF construct, a block of code is executed if the controlling logical expression is true. If the controlling logical expression is false, all of the statements in the construct are skipped.

Sometimes we may want to execute one set of statements if some condition is true, and different sets of statements if other conditions are true. In fact, there might be many different options to consider. An ELSE clause and one or more ELSE IF clauses may be added to the block IF construct for this purpose. The block IF construct with an ELSE clause and an ELSE IF clause has the form

```
IF (logical_expr_1) THEN
    Statement 1
    Statement 2                    } Block 1
    ...
ELSE IF (logical_expr_2) THEN
    Statement 1
    Statement 2                    } Block 2
    ...
ELSE
    Statement 1
    Statement 2                    } Block 3
    ...
END IF
```

If *logical_expr_1* is true, then the program executes the statements in Block 1, and skips to the first executable statement following the END IF. Otherwise, the program checks for the status of *logical_expr_2*. If *logical_expr_2* is true, then the program executes the statements in Block 2, and skips to the first executable statement following the END IF. If both logical expressions are false, then the program executes the statements in Block 3.

The ELSE and ELSE IF statements must occupy lines by themselves. There should not be a statement number on a line containing an ELSE or ELSE IF statement.

There can be any number of ELSE IF clauses in a block IF construct. The logical expression in each clause will be tested only if the logical expressions in every clause above it are false. Once one of the expressions proves to be true and the corresponding code block is executed, the program skips to the first executable statement following the END IF.

The flowchart for a block IF construct with an ELSE IF and an ELSE clause is shown in Figure 3-7.

To illustrate the use of the ELSE and ELSE IF clauses, let's reconsider the quadratic equation once more. Suppose that we wanted to examine the discriminant of a quadratic equation and to tell a user whether the equation has two complex roots, two identical real roots, or two distinct real roots. In pseudocode, this construct would take the form

```
IF (b**2 - 4.*a*c) < 0.0 THEN
    Write message that equation has two complex roots.
ELSE IF (b**2 - 4.*a*c) > 0.0 THEN
    Write message that equation has two distinct real roots.
ELSE
    Write message that equation has two identical real roots.
END IF
```

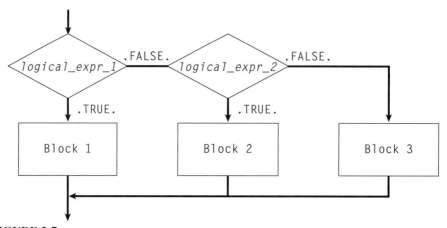

FIGURE 3-7
Flowchart for a block IF construct with an ELSE IF (...) THEN clause and an ELSE clause.

The Fortran statements to do this are

```
IF ( (b**2 - 4.*a*c) < 0.0 ) THEN
   WRITE (*,*) 'This equation has two complex roots.'
ELSE IF ( (b**2 - 4.*a*c) > 0.0 ) THEN
   WRITE (*,*) 'This equation has two distinct real roots.'
ELSE
   WRITE (*,*) 'This equation has two identical real roots.'
END IF
```

The flowchart for this construct is shown in Figure 3-8.

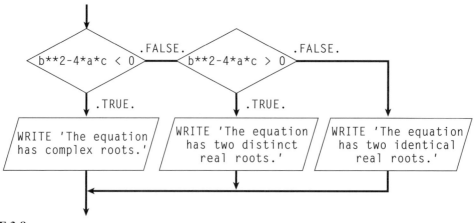

FIGURE 3-8
Flowchart showing structure to determine whether a quadratic equation has two complex roots, two identical real roots, or two distinct real roots.

3.4.3 Examples Using Block IF Constructs

We will now look at two examples that illustrate the use of block IF constructs.

EXAMPLE 3-2

The Quadratic Equation:

Design and write a program to solve for the roots of a quadratic equation, regardless of type.

SOLUTION
We will follow the design steps outlined earlier in the chapter.

1. **State the problem.**
   The problem statement for this example is very simple. We want to write a program that will solve for the roots of a quadratic equation, whether they are distinct real roots, repeated real roots, or complex roots.

2. **Define the inputs and outputs.**
   The inputs required by this program are the coefficients a, b, and c of the quadratic equation

$$ax^2 + bx + c = 0 \tag{3-1}$$

   The output from the program will be the roots of the quadratic equation, whether they are distinct real roots, repeated real roots, or complex roots.

3. **Design the algorithm.**
   This task can be broken down into three major sections, whose functions are input, processing, and output:

   ```
   Read the input data
   Calculate the roots
   Write out the roots
   ```

   We will now break each of the above major sections into smaller, more detailed pieces. There are three possible ways to calculate the roots, depending on the value of the discriminant, so it is logical to implement this algorithm with a three-branched IF statement. The resulting pseudocode is:

   ```
   Prompt the user for the coefficients a, b, and c.
      Read a, b, and c
      Echo the input coefficients
      discriminant ← b**2 - 4. * a * c

      IF discriminant > 0 THEN
         x1 ← ( -b + sqrt(discriminant) ) / ( 2. * a )
         x2 ← ( -b - sqrt(discriminant) ) / ( 2. * a )
         Write message that equation has two distinct real roots.
         Write out the two roots.
   ```

(continued)

(concluded)

```
    ELSE IF discriminant < 0 THEN
        real_part ← -b / ( 2. * a )
        imag_part ← sqrt ( abs ( discriminant ) ) / ( 2. * a )
        Write message that equation has two complex roots.
        Write out the two roots.
    ELSE
        x1 ← -b / ( 2. * a )
        Write message that equation has two identical real roots.
        Write out the repeated root.
    END IF
```

The flowchart for this program is shown in Figure 3-9.

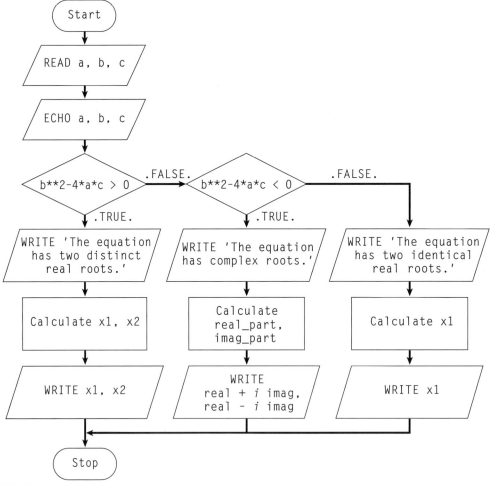

FIGURE 3-9
Flowchart of program roots.

4. **Turn the algorithm into Fortran statements.**
   The final Fortran code is shown in Figure 3-10.

FIGURE 3-10
Program to solve for the roots of a quadratic equation.

```
PROGRAM roots
! Purpose:
!   This program solves for the roots of a quadratic equation of the
!   form a*x**2 + b*x + c = 0. It calculates the answers regardless
!   of the type of roots that the equation possesses.
!
! Record of revisions:
!   Date        Programmer          Description of change
!   ====        ==========          =====================
!   11/06/15    S. J. Chapman       Original code
!
IMPLICIT NONE

! Data dictionary: declare variable types, definitions, & units
REAL :: a                ! Coefficient of x**2 term of equation
REAL :: b                ! Coefficient of x term of equation
REAL :: c                ! Constant term of equation
REAL :: discriminant     ! Discriminant of the equation
REAL :: imag_part        ! Imaginary part of equation (for complex roots)
REAL :: real_part        ! Real part of equation (for complex roots)
REAL :: x1               ! First solution of equation (for real roots)
REAL :: x2               ! Second solution of equation (for real roots)

! Prompt the user for the coefficients of the equation
WRITE (*,*) 'This program solves for the roots of a quadratic '
WRITE (*,*) 'equation of the form A * X**2 + B * X + C = 0. '
WRITE (*,*) 'Enter the coefficients A, B, and C: '
READ (*,*) a, b, c

! Echo back coefficients
WRITE (*,*) 'The coefficients A, B, and C are: ', a, b, c

! Calculate discriminant
discriminant = b**2 - 4. * a * c

! Solve for the roots, depending upon the value of the discriminant
IF ( discriminant > 0. ) THEN ! there are two real roots, so...

   x1 = ( -b + sqrt(discriminant) ) / ( 2. * a )
   x2 = ( -b - sqrt(discriminant) ) / ( 2. * a )
   WRITE (*,*) 'This equation has two real roots:'
   WRITE (*,*) 'X1 = ', x1
   WRITE (*,*) 'X2 = ', x2

ELSE ( discriminant < 0. ) THEN ! there are complex roots, so ...

   real_part = ( -b ) / ( 2. * a )
```

(continued)

(concluded)

```
        imag_part = sqrt ( abs ( discriminant ) ) / ( 2. * a )
        WRITE (*,*) 'This equation has complex roots:'
        WRITE (*,*) 'X1 = ', real_part, ' +i ', imag_part
        WRITE (*,*) 'X2 = ', real_part, ' -i ', imag_part

    ELSE IF ( discriminant == 0. ) THEN ! there is one repeated root, so...

        x1 = ( -b ) / ( 2. * a )
        WRITE (*,*) 'This equation has two identical real roots:'
        WRITE (*,*) 'X1 = X2 = ', x1

    END IF

END PROGRAM roots
```

5. **Test the program.**

Next, we must test the program using real input data. Since there are three possible paths through the program, we must test all three paths before we can be certain that the program is working properly. From Equation (3-2), it is possible to verify the solutions to the equations given below:

$$x^2 + 5x + 6 = 0 \qquad x = -2 \text{ and } x = -3$$

$$x^2 + 4x + 4 = 0 \qquad x = -2$$

$$x^2 + 2x + 5 = 0 \qquad x = -1 \pm i2$$

If this program is compiled, and then run three times with the above coefficients, the results are as shown below (user inputs are shown in bold face):

```
C:\book\fortran\chap3>roots
This program solves for the roots of a quadratic
equation of the form A * X**2 + B * X + C = 0.
Enter the coefficients A, B, and C:
1., 5., 6.
The coefficients A, B, and C are:    1.000000    5.000000
    6.000000
This equation has two real roots:
X1 =       -2.000000
X2 =       -3.000000

C:\book\fortran\chap3>roots
This program solves for the roots of a quadratic
equation of the form A * X**2 + B * X + C = 0.
Enter the coefficients A, B, and C:
1., 4., 4.
The coefficients A, B, and C are:    1.000000    4.000000
    4.000000
This equation has two identical real roots:
X1 = X2 =      -2.000000
```

```
C:\book\fortran\chap3>roots
This program solves for the roots of a quadratic
equation of the form A * X**2 + B * X + C = 0.
Enter the coefficients A, B, and C:
1., 2., 5.
The coefficients A, B, and C are:   1.000000      2.000000
   5.000000
This equation has complex roots:
X1 =    -1.000000 +i     2.000000
X2 =    -1.000000 -i     2.000000
```

The program gives the correct answers for our test data in all three possible cases.

EXAMPLE 3-3

Evaluating a Function of Two Variables:

Write a Fortran program to evaluate a function $f(x,y)$ for any two user-specified values x and y. The function $f(x,y)$ is defined as follows:

$$f(x, y) = \begin{cases} x + y & x \geq 0 \text{ and } y \geq 0 \\ x + y^2 & x \geq 0 \text{ and } y < 0 \\ x^2 + y & x < 0 \text{ and } y \geq 0 \\ x^2 + y^2 & x < 0 \text{ and } y < 0 \end{cases}$$

SOLUTION

The function $f(x,y)$ is evaluated differently depending on the signs of the two independent variables x and y. To determine the proper equation to apply, it will be necessary to check for the signs of the x and y values supplied by the user.

1. **State the problem.**

    This problem statement is very simple: Evaluate the function $f(x,y)$ for any user-supplied values of x and y.

2. **Define the inputs and outputs.**

    The inputs required by this program are the values of the independent variables x and y. The output from the program will be the value of the function $f(x,y)$.

3. **Design the algorithm.**

    This task can be broken down into three major sections, whose functions are input, processing, and output:

    ```
    Read the input values x and y
    Calculate f(x,y)
    Write out f(x,y)
    ```

We will now break each of the above major sections into smaller, more detailed pieces. There are four possible ways to calculate the function $f(x,y)$, depending upon the values of x and y, so it is logical to implement this algorithm with a four-branched IF statement. The resulting pseudocode is:

```
Prompt the user for the values x and y.
Read x and y
Echo the input coefficients
IF x ≥ 0 and y ≥ 0 THEN
    fun ← x + y
ELSE IF x ≥ 0 and y < 0 THEN
    fun ← x + y**2
ELSE IF x < 0 and y ≥ 0 THEN
    fun ← x**2 + y
ELSE
    fun ← x**2 + y**2
END IF
Write out f(x,y)
```

The flowchart for this program is shown in Figure 3-11.

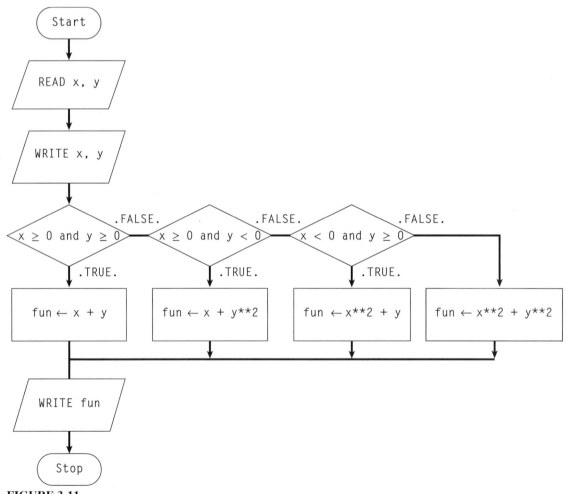

FIGURE 3-11
Flowchart of program funxy.

4. **Turn the algorithm into Fortran statements.**

The final Fortran code is shown in Figure 3-12.

FIGURE 3-12
Program `funxy` from Example 3-3.

```
! Purpose:
!   This program solves the function f(x,y) for a user-specified x and y,
!   where f(x,y) is defined as:
!
!            _
!           |
!           |  X + Y         X >= 0 and Y >= 0
!           |  X + Y**2      X >= 0 and Y < 0
!   F(X,Y) =|  X**2 + Y      X < 0 and Y >= 0
!           |  X**2 + Y**2   X < 0 and Y < 0
!           |
!            _
!
! Record of revisions:
!     Date         Programmer          Description of change
!    ========     ==============      ======================
!   11/06/15      S. J. Chapman        Original code
!
IMPLICIT NONE

! Data dictionary: declare variable types, definitions, & units
REAL :: x          ! First independent variable
REAL :: y          ! Second independent variable
REAL :: fun        ! Resulting function

! Prompt the user for the values x and y
WRITE (*,*) 'Enter the coefficients x and y: '
READ (*,*) x, y

! Write the coefficients of x and y.
WRITE (*,*) 'The coefficients x and y are: ', x, y

! Calculate the function f(x,y) based upon the signs of x and y.
IF ( ( x >= 0. ) .AND. ( y >= 0. ) ) THEN
   fun = x + y
ELSE IF ( ( x >= 0. ) .AND. ( y < 0. ) ) THEN
   fun = x + y**2
ELSE IF ( ( x < 0. ) .AND. ( y >= 0. ) ) THEN
   fun = x**2 + y
ELSE
   fun = x**2 + y**2
END IF

! Write the value of the function.
WRITE (*,*) 'The value of the function is: ', fun

END PROGRAM funxy
```

5. **Test the program.**

Next, we must test the program using real input data. Since there are four possible paths through the program, we must test all four paths before we can be certain that the program is working properly. To test all four possible paths, we will execute the program with the four sets of input values $(x,y) = (2,3), (2,-3), (-2,3)$, and $(-2, -3)$. Calculating by hand, we see that

$$f(2,3) = 2 + 3 = 5$$
$$f(2,-3) = 2 + (-3)^2 = 11$$
$$f(-2,3) = (-2)^2 + 3 = 7$$
$$f(-2,-3) = (-2)^2 + (-3)^2 = 13$$

If this program is compiled, and then run four times with the above values, the results are:

```
C:\book\fortran\chap3>funxy
Enter the coefficients X and Y:
2. 3.
The coefficients X and Y are:       2.000000        3.000000
The value of the function is:       5.000000

C:\book\fortran\chap3>funxy
Enter the coefficients X and Y:
2. -3.
The coefficients X and Y are:       2.000000       -3.000000
The value of the function is:      11.000000

C:\book\fortran\chap3>funxy
Enter the coefficients X and Y:
-2. 3.
The coefficients X and Y are:      -2.000000        3.000000
The value of the function is:       7.000000

C:\book\fortran\chap3>funxy
Enter the coefficients X and Y:
-2. -3.
The coefficients X and Y are:      -2.000000       -3.000000
The value of the function is:      13.000000
```

The program gives the correct answers for our test values in all four possible cases.

3.4.4 Named Block IF Constructs

It is possible to assign a name to a block IF construct. The general form of the construct with a name attached is

```
[name:] IF (logical_expr_1) THEN
   Statement 1
   Statement 2                        Block 1
   ...
```

```
        ELSE IF (logical_expr_2) THEN [name]
            Statement 1                    ⎫
            Statement 2                    ⎬    Block 2
            ...                            ⎭
        ELSE [name]
            Statement 1                    ⎫
            Statement 2                    ⎬    Block 3
            ...                            ⎭
        END IF [name]
```

where `name` may be up to 63 alphanumeric characters long, beginning with a letter. The name given to the IF construct must be unique within each program unit, and must not be the same as any constant or variable name within the program unit. If a name is assigned to an IF, then the same name must appear on the associated END IF. Names are optional on the ELSE and ELSE IF statements of the construct, but if they are used, they must be the same as the name on the IF.

Why would we want to name an IF construct? For simple examples like the ones we have seen so far, there is no particular reason to do so. The principal reason for using names is to help us (and the compiler) keep IF constructs straight in our own minds when they get very complicated. For example, suppose that we have a complex IF construct that is hundreds of lines long, spanning many pages of listings. If we name all of the parts of such a construct, then we can tell at a glance which construct a particular ELSE or ELSE IF statement belongs to. They make a programmer's intentions explicitly clear. In addition, names on constructs can help the compiler flag the specific location of an error when one occurs.

Good Programming Practice

Assign a name to any large and complicated IF constructs in your program to help you keep the parts of the construct associated together in your own mind.

3.4.5 Notes Concerning the Use of Block IF Constructs

The block IF construct is very flexible. It must have one IF (...) THEN statement and one END IF statement. In between, it can have any number of ELSE IF clauses, and may also have one ELSE clause. With this combination of features, it is possible to implement any desired branching construct.

In addition, block IF constructs may be **nested**. Two block IF constructs are said to be nested if one of them lies entirely within a single code block of the other one. The following two IF constructs are properly nested.

```
            outer: IF (x > 0.) THEN
                ...
                inner: IF (y < 0.) THEN
                    ...
                END IF inner
                ...
            END IF outer
```

It is a good idea to name IF constructs when they are being nested, since the name explicitly indicates which IF a particular END IF is associated with. If the constructs are not named, the Fortran compiler always associates a given END IF with the most recent IF statement. This works well for a properly written program, but can cause the compiler to produce confusing error messages in cases where the programmer makes a coding error. For example, suppose we have a large program containing a construct like the one shown below:

```
PROGRAM mixup
   ...
   IF (test1) THEN
      ...
      IF (test2) THEN
         ...
         IF (test3) THEN
            ...
         END IF
         ...
      END IF
      ...
   END IF
   ...
END PROGRAM mixup
```

This program contains three nested IF constructs that may span hundreds of lines of code. Now suppose that the first END IF statement is accidentally deleted during an editing session. When that happens, the compiler will automatically associate the second END IF with the innermost IF (test3) construct, and the third END IF with the middle IF (test2). When the compiler reaches the END PROGRAM statement, it will notice that the first IF (test1) construct was never ended, and it will generate an error message saying that there is a missing END IF. Unfortunately, it can't tell *where* the problem occurred, so we will have to go back and manually search the entire program to locate the problem.

In contrast, consider what happens if we assign names to each IF construct. The resulting program would be:

```
PROGRAM mixup_1
   ...
   outer: IF (test1) THEN
      ...
      ...
      middle: IF (test2) THEN
         ...
         ...
         inner: IF (test3) THEN
            ...
            ...
         END IF inner
         ...
      END IF middle
      ...
   END IF outer
   ...
END PROGRAM mixup_1
```

Suppose that the first END IF statement is again accidentally deleted during an editing session. When that happens, the compiler will notice that there is no END IF associated with the inner IF, and it will generate an error message as soon as it encounters the END IF middle statement. Furthermore, the error message will explicitly state that the problem is associated with the inner IF construct, so we know just where to go to fix it.

It is sometimes possible to implement an algorithm using either ELSE IF clauses or nested IF statements. In that case, a programmer may choose whichever style he or she prefers.

EXAMPLE 3-4 *Assigning Letter Grades:*

Suppose that we are writing a program that reads in a numerical grade and assigns a letter grade to it according to the following table:

95 < GRADE	A
86 < GRADE ≤ 95	B
76 < GRADE ≤ 86	C
66 < GRADE ≤ 76	D
0 < GRADE ≤ 66	F

Write an IF construct that will assign the grades as described above using (*a*) multiple ELSE IF clauses and (*b*) nested IF constructs.

SOLUTION

(*a*) One possible structure using ELSE IF clauses is

```
IF ( grade > 95.0 ) THEN
    WRITE (*,*) 'The grade is A.'
ELSE IF ( grade > 86.0 ) THEN
    WRITE (*,*) 'The grade is B.'
ELSE IF ( grade > 76.0 ) THEN
    WRITE (*,*) 'The grade is C.'
ELSE IF ( grade > 66.0 ) THEN
    WRITE (*,*) 'The grade is D.'
ELSE
    WRITE (*,*) 'The grade is F.'
END IF
```

(*b*) One possible structure using nested IF constructs is

```
if1: IF ( grade > 95.0 ) THEN
        WRITE (*,*) 'The grade is A.'
     ELSE
        if2: IF ( grade > 86.0 ) THEN
                WRITE (*,*) 'The grade is B.'
             ELSE
                if3: IF ( grade > 76.0 ) THEN
                        WRITE (*,*) 'The grade is C.'
                     ELSE
```

```
            if4: IF ( grade > 66.0 ) THEN
                    WRITE (*,*) 'The grade is D.'
                 ELSE
                    WRITE (*,*) 'The grade is F.'
                 END IF if4
              END IF if3
           END IF if2
        END IF if1
```

It should be clear from the above example that if there are a lot of mutually exclusive options, a single `IF` construct with `ELSE IF` clauses will be simpler than a nested `IF` construct.

Good Programming Practice
For branches in which there are many mutually exclusive options, use a single `IF` construct with `ELSE IF` clauses in preference to nested `IF` constructs.

3.4.6 The Logical IF Statement

There is an alternative form of the block `IF` construct described above. It is just a single statement of the form

              `IF (logical_expr) Statement`

where `Statement` is an executable Fortran statement. If the logical expression is true, the program executes the statement on the same line with it. Otherwise, the program skips to the next executable statement in the program. This form of the logical `IF` is equivalent to a block `IF` construct with only one statement in the `IF` block.

3.4.7 The SELECT CASE Construct

The `SELECT CASE` construct is another form of branching construct. It permits a programmer to select a particular code block to execute based on the value of a single integer, character, or logical expression. The general form of a `CASE` construct is:

```
[name:] SELECT CASE (case_expr)
CASE (case_selector_1) [name]
   Statement 1
   Statement 2                          }         Block 1
   ...
```

Program Design and Branching Structures

```
              CASE (case_selector_2) [name]
                  Statement 1
                  Statement 2          } Block 2
                  ...
              ...
              CASE DEFAULT [name]
                  Statement 1
                  Statement 2          } Block n
                  ...
              END SELECT [name]
```

If the value of *case_expr* is in the range of values included in *case_selector_1*, then the first code block will be executed. Similarly, if the value of *case_expr* is in the range of values included in *case_selector_2*, then the second code block will be executed. The same idea applies for any other cases in the construct. The default code block is optional. If it is present, the default code block will be executed whenever the value of *case_expr* is outside the range of all of the case selectors. If it is not present and the value of *case_expr* is outside the range of all of the case selectors, then none of the code blocks will be executed. The pseudocode for the case construct looks just like its Fortran implementation; a flowchart for this construct is shown in Figure 3-13.

A name may be assigned to a CASE construct, if desired. The name must be unique within each program unit. If a name is assigned to a SELECT CASE statement, then the same name must appear on the associated END SELECT. Names are optional on the CASE statements of the construct, but if they are used, they must be the same as the name on the SELECT CASE statement.

The *case_expr* may be any integer, character, or logical expression. Each case selector must be an integer, character, or logical value *or a range of values*. All case selectors must be *mutually exclusive*; no single value can appear in more than one case selector.

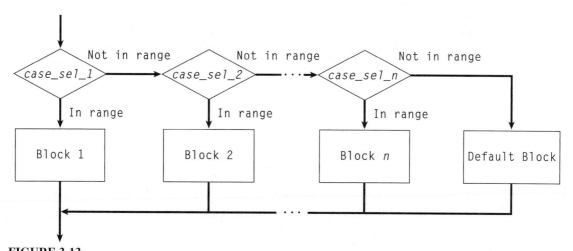

FIGURE 3-13
Flowchart for a CASE construct.

Let's look at a simple example of a CASE construct. This example prints out a message based on the value of an integer variable.

```
INTEGER :: temp_c          ! Temperature in degrees C
...
temp: SELECT CASE (temp_c)
CASE (:-1)
   WRITE (*,*) "It's below freezing today!"
CASE (0)
   WRITE (*,*) "It's exactly at the freezing point."
CASE (1:20)
   WRITE (*,*) "It's cool today."
CASE (21:33)
   WRITE (*,*) "It's warm today."
CASE (34:)
   WRITE (*,*) "It's hot today."
END SELECT temp
```

The value of temp_c controls which case is selected. If the temperature is less than zero, then the first case will be selected, and the message printed out will be "It's below freezing today!". If the temperature is exactly zero, then the second case will be selected, and so forth. Note that the cases do not overlap—a given temperature can appear in only one of the cases.

The *case_selector* can take one of four forms:

case_value	Execute block if *case_value* == *case_expr*
low_value:	Execute block if *low_value* <= *case_expr*
:high_value	Execute block if *case_expr* <= *high_value*
low_value:high_value	Execute block if *low_value* <= *case_expr* <= *high_value*

or it can be a list of any combination of these forms separated by commas.

The following statements determine whether an integer between 1 and 10 is even or odd, and print out an appropriate message. It illustrates the use of a list of values as case selectors, and also the use of the CASE DEFAULT block.

```
INTEGER :: value
...
SELECT CASE (value)
CASE (1,3,5,7,9)
   WRITE (*,*) 'The value is odd.'
CASE (2,4,6,8,10)
   WRITE (*,*) 'The value is even.'
CASE (11:)
   WRITE (*,*) 'The value is too high.'
CASE DEFAULT
   WRITE (*,*) 'The value is negative or zero.'
END SELECT
```

Program Design and Branching Structures

The CASE DEFAULT block is extremely important for good programming design. If an input value in a SELECT CASE statement does not match any of the cases, none of the cases will be executed. In a well-designed program, this is usually the result of an error in the logical design or an illegal input. You should always include a default case, and have that case create a warning message for the user.

Good Programming Practice
Always include a DEFAULT CASE clause in your case constructs to trap any logical errors or illegal inputs that might occur in a program.

EXAMPLE 3-5

Selecting the Day of the Week with a SELECT CASE Construct:

Write a program that reads an integer from the keyboard, and displays the day of the week corresponding to that integer. Be sure to handle the case of an illegal input value.

SOLUTION
In this example, we will prompt the user to enter an integer between 1 and 7, and then use a SELECT CASE construct to select the day of the week corresponding to that number, using the convention that Sunday is the first day of the week. The SELECT CASE construct will also include a default case to handle illegal days of the week.

The resulting program is shown in Figure 3-14.

FIGURE 3-14
Program day_of_week from Example 3-5.

```
PROGRAM day_of_week
!
! Purpose:
!   This program displays the day of week corresponding to
!   a input integer value.
!
! Record of revisions:
!    Date          Programmer       Description of change
!    ====          ==========       =====================
!  11/06/15     S. J. Chapman       Original code
!
IMPLICIT NONE

! Data dictionary: declare variable types, definitions, & units
CHARACTER(len=11) :: c_day   ! Character string containing day
INTEGER :: i_day             ! Integer day of week
```

(continued)

(*concluded*)

```
! Prompt the user for the numeric day of the week
WRITE (*,*) 'Enter the day of the week (1-7): '
READ (*,*) i_day

! Get the corresponding day of the week.
SELECT CASE (i_day)
CASE (1)
   c_day = 'Sunday'
CASE (2)
   c_day = 'Monday'
CASE (3)
   c_day = 'Tuesday'
CASE (4)
   c_day = 'Wednesday'
CASE (5)
   c_day = 'Thursday'
CASE (6)
   c_day = 'Friday'
CASE (7)
   c_day = 'Saturday'
CASE DEFAULT
   c_day = 'Invalid day'
END SELECT

! Write the resulting day
WRITE (*,*) 'Day = ', c_day

END PROGRAM day_of_week
```

If this program is compiled, and then executed three times with various values, the results are:

```
C:\book\fortran\chap3>day_of_week
Enter the day of the week (1-7):
1
Day = Sunday

C:\book\fortran\chap3>day_of_week
Enter the day of the week (1-7):
5
Day = Thursday

C:\book\fortran\chap3>day_of_week
Enter the day of the week (1-7):
-2
Day = Invalid day
```

Note that this program gave correct values for valid days of the week, and also displayed an error message for an invalid day.

EXAMPLE 3-6

Using Characters in a SELECT CASE Construct:

Write a program that reads a character string from the keyboard containing a day of the week, and displays "Weekday" if the day falls between Monday and Friday, and "weekend" if the day is Saturday or Sunday. Be sure to handle the case of an illegal input value.

SOLUTION

In this example, we will prompt the user to enter a day of the week, and then use a SELECT CASE construct to select whether the day is a weekday or it falls on the weekend. The SELECT CASE construct will also include a default case to handle illegal days of the week.

The resulting program is shown in Figure 3-15.

FIGURE 3-15

Program weekday_weekend from Example 3-6.

```
PROGRAM weekday_weekend
!
! Purpose:
!   This program accepts a character string containing a
!   day of the week, and responds with a message specifying
!   whether the day is a weekday or falls on the weekend.
!
! Record of revisions:
!    Date          Programmer        Description of change
!    ====          ==========        =====================
!  11/06/15      S. J. Chapman       Original code
!
IMPLICIT NONE

! Declare the variables used in this program.
CHARACTER(len=11) :: c_day   ! Character string containing day
CHARACTER(len=11) :: c_type  ! Character string with day type

! Prompt the user for the day of the week
WRITE (*,*) 'Enter the name of the day: '
READ (*,*) c_day

! Get the corresponding day of the week.
SELECT CASE (c_day)
CASE ('Monday','Tuesday','Wednesday','Thursday','Friday')
   c_type = 'Weekday'
CASE ('Saturday','Sunday')
   c_type = 'Weekend'
CASE DEFAULT
   c_type = 'Invalid day'
END SELECT

! Write the resulting day type
WRITE (*,*) 'Day Type = ', c_type
END PROGRAM weekday_weekend
```

If this program is compiled, and then executed three times with various values, the results are:

```
C:\book\fortran\chap3>weekday_weekend
Enter the name of the day:
Tuesday
Day Type = Weekday

C:\book\fortran\chap3>weekday_weekend
Enter the name of the day:
Sunday
Day Type = Weekend

C:\book\fortran\chap3>weekday_weekend
Enter the name of the day:
Holiday
Day Type = Invalid day
```

Note that this program gave correct values for valid days of the week, and also displayed an error message for an invalid day. This program illustrates the use of a list of possible case values in each CASE clause.

Quiz 3-2

This quiz provides a quick check to see if you have understood the concepts introduced in Section 3.5. If you have trouble with the quiz, reread the section, ask your instructor, or discuss the material with a fellow student. The answers to this quiz are found in the back of the book.

Write Fortran statements that perform the functions described below:

1. If x is greater than or equal to zero, then assign the square root of x to variable sqrt_x and print out the result. Otherwise, print out an error message about the argument of the square root function, and set sqrt_x to zero.

2. A variable fun is calculated as numerator / denominator. If the absolute value of denominator is less than 1.0E-10, write "Divide by 0 error." Otherwise, calculate and print out fun.

3. The cost per mile for a rented vehicle is $0.30 for the first 100 miles, $0.20 for the next 200 miles, and $0.15 for all miles in excess of 300 miles. Write Fortran statements that determine the total cost and the average cost per mile for a given number of miles (stored in variable distance).

Examine the following Fortran statements. Are they correct or incorrect? If they are correct, what is output by them? If they are incorrect, what is wrong with them?

(continued)

(*concluded*)

4.  ```
    IF ( volts > 125. ) THEN
        WRITE (*,*) 'WARNING: High voltage on line. '
    IF ( volts < 105. ) THEN
        WRITE (*,*) 'WARNING: Low voltage on line. '
    ELSE
        WRITE (*,*) 'Line voltage is within tolerances. '
    END IF
    ```

5.  ```
    PROGRAM test
    LOGICAL :: warn
    REAL :: distance
    REAL, PARAMETER :: LIMIT = 100.
    warn = .TRUE.
    distance = 55. + 10.
    IF ( distance > LIMIT .OR. warn ) THEN
        WRITE (*,*) 'Warning: Distance exceeds limit.'
    ELSE
        WRITE (*,*) 'Distance = ', distance
    END IF
    ```

6.  ```
    REAL, PARAMETER :: PI = 3.141593
    REAL :: a = 10.
    SELECT CASE ( a * sqrt(PI) )
    CASE (0:)
        WRITE (*,*) 'a > 0'
    CASE (:0)
        WRITE (*,*) 'a < 0'
    CASE DEFAULT
        WRITE (*,*) 'a = 0'
    END SELECT
    ```

7.  ```
    CHARACTER(len=6) :: color = 'yellow'
    SELECT CASE ( color )
    CASE ('red')
        WRITE (*,*) 'Stop now!'
    CASE ('yellow')
        WRITE (*,*) 'Prepare to stop.'
    CASE ('green')
        WRITE (*,*) 'Proceed through intersection.'
    CASE DEFAULT
        WRITE (*,*) 'Illegal color encountered.'
    END SELECT
    ```

8.  ```
    IF ( temperature > 37. ) THEN
        WRITE (*,*) 'Human body temperature exceeded. '
    ELSE IF ( temperature > 100. )
        WRITE (*,*) 'Boiling point of water exceeded. '
    END IF
    ```
    ```

3.5
MORE ON DEBUGGING FORTRAN PROGRAMS

It is much easier to make a mistake when writing a program containing branches and loops than it is when writing simple sequential programs. Even after going through the full design process, a program of any size is almost guaranteed not to be completely correct the first time it is used. Suppose that we have built the program and tested it, only to find that the output values are in error. How do we go about finding the bugs and fixing them?

The best approach to locating the error is to use a symbolic debugger, if one is supplied with your compiler. You must ask your instructor or else check with your system's manuals to determine how to use the symbolic debugger supplied with your particular compiler, because they all differ from one another.

An alternate approach to locating the error is to insert `WRITE` statements into the code to print out important variables at key points in the program. When the program is run, the `WRITE` statements will print out the values of the key variables. These values can be compared to the ones you expect, and the places where the actual and expected values differ will serve as a clue to help you locate the problem. For example, to verify the operation of a block `IF` construct:

```
WRITE (*,*) 'At if1: var1 = ', var1
if1: IF ( sqrt(var1) > 1. ) THEN
    WRITE (*,*) 'At if1: sqrt(var1) > 1.'
    ...
ELSE IF ( sqrt(var1) < 1. ) THEN
    WRITE (*,*) 'At if1: sqrt(var1) < 1.'
    ...
ELSE
    WRITE (*,*) 'At if1: sqrt(var1) == 1.'
    ...
END IF if1
```

When the program is executed, its output listing will contain detailed information about the variables controlling the block `IF` construct and just which branch was executed.

Once you have located the portion of the code in which the error occurs, you can take a look at the specific statements in that area to locate the problem. Two common errors are described below. Be sure to check for them in your code.

1. *If the problem is in an* `IF` *construct, check to see if you used the proper relational operator in your logical expressions.* Did you use > when you really intended >=, etc.? Logical errors of this sort can be very hard to spot, since the compiler will not give an error message for them. Be especially careful of logical expressions that are very complex, since they will be hard to understand, and very easy to mess up. You should use extra parentheses to make them easier to understand. If the logical expressions are really large, consider breaking them down into simpler expressions that are easier to follow.

2. *Another common problem with* IF *statements occurs when real variables are tested for equality.* Because of small round-off errors during floating-point arithmetic operations, two numbers that theoretically should be equal will differ by a tiny amount, and the test for equality will fail. When working with real variables, it is often a good idea to replace a test for equality with a test for *near equality*. For example, instead of testing to see if x is equal to 10., you should test to see if |x - 10.| < 0.0001. Any value of x between 9.9999 and 10.0001 will satisfy the latter test, so round-off error will not cause problems. In Fortran statements,

```
IF ( x == 10. ) THEN
```

would be replaced by

```
IF ( abs(x - 10.) <= 0.0001 ) THEN
```

Good Programming Practice
Be cautious about testing for equality with real variables in an IF construct, since round-off errors may cause two variables that should be equal to fail a test for equality. Instead, test to see if the variables are nearly equal within the round-off error to be expected on the computer you are working with.

3.6
SUMMARY

In this chapter, we presented the top-down approach to program design, including pseudocode and flowcharts.

Next, we discussed the logical data type and more details of the character data type, which can be used to control Fortran branching structures. This material included relational operators, which compare two numbers or character expressions to produce a logical result, and combinational logic operators, which produce a logical result from one or two logical input values.

The Fortran hierarchy of operations, expanded to include the relational and combinational logic operators, is summarized in Table 3-4.

Finally, we have presented the basic types of Fortran branches and loops. The principal types of branch is the block IF–ELSE IF–ELSE–END IF construct. This construct is very flexible. It can have as many ELSE IF clauses as needed to construct any desired test. Furthermore, block IF constructs can be nested to produce more complex tests. A second type of branch is the CASE construct. It may be used to select among mutually exclusive alternatives specified by an integer, character, or logical control expression.

TABLE 3-4
Fortran hierarchy of operations

1. Operations within parentheses are evaluated first, starting with the innermost parentheses and working outward.
2. All exponential operations are evaluated next, working from *right* to *left*.
3. All multiplications and divisions are evaluated, working from left to right.
4. All additions and subtractions are evaluated, working from left to right.
5. All relational operators (`==`, `/=`, `>`, `>=`, `<`, `<=`) are evaluated, working from left to right.
6. All `.NOT.` operators are evaluated.
7. All `.AND.` operators are evaluated, working from left to right.
8. All `.OR.` operators are evaluated, working from left to right.
9. All `.EQV.` and `.NEQV.` operators are evaluated, working from left to right.

3.6.1 Summary of Good Programming Practice

The following guidelines should be adhered to when programming with branch or loop constructs. By following them consistently, your code will contain fewer bugs, will be easier to debug, and will be more understandable to others who may need to work with it in the future.

1. Always indent code blocks in block `IF` and `CASE` constructs to make them more readable.
2. Be cautious about testing for equality with real variables in an `IF` construct, since round-off errors may cause two variables that should be equal to fail a test for equality. Instead, test to see if the variables are nearly equal within the round-off error to be expected on the computer you are working with.
3. Always include a `DEFAULT CASE` clause in your case constructs to trap any logical errors or illegal inputs that might occur in a program.

3.6.2 Summary of Fortran Statements and Constructs

The following summary describes the Fortran statements and constructs introduced in this chapter.

Block IF Construct

```
[name:] IF ( logical_expr_1 ) THEN
   Block 1
ELSE IF ( logical_expr_2 ) THEN [name]
   Block 2
ELSE [name]
   Block 3
END IF [name]
```

(continued)

Program Design and Branching Structures

(concluded)

Description:
The block IF construct permits the execution of a code block based on the results of one or more logical expressions. If `logical_expr_1` is true, the first code block will be executed. If `logical_expr_1` is false and `logical_expr_2` is true, the second code block will be executed. If both logical expressions are false, the third code block will be executed. After any block is executed, control jumps the first statement after the construct.

There must be one and only one IF ( ) THEN statement in a block IF construct. There may be any number of ELSE IF clauses (zero or more), and there may be at most one ELSE clause in the construct. The name is optional, but if it is used on the IF statement, then it must be used on the END IF statement. The name is optional on the ELSE IF and ELSE statements even if it is used on the IF and END IF statements.

CASE Construct

```
[name:] SELECT CASE (case_expr)
CASE (case_selector_1) [name]
    Block 1
CASE (case_selector_2) [name]
    Block 2
CASE DEFAULT [name]
    Block n
END SELECT [name]
```

Description:
The CASE construct executes a specific block of statements based on the value of the *case_expr*, which can be an integer, character, or logical value. Each case selector specifies one or more possible values for the case expression. If the *case_expr* is a value included in a given case selector, then the corresponding block of statements is executed, and control will jump to the first executable statement after the end of the construct. If no case selector is executed, then the CASE DEFAULT block will be executed if present, and control will jump to the first executable statement after the end of the construct. If CASE DEFAULT is not present, the construct does nothing.

There must be one SELECT CASE statement and one END SELECT statement in a CASE construct. There will be one or more CASE statements. At most one CASE DEFAULT statement may be included. Note that all case selectors must be *mutually exclusive*. The name is optional, but if it is used on the SELECT CASE statement, then it must also be used on the END SELECT statement. The name is optional on the CASE statements even if it is used on the SELECT CASE and END SELECT statements.

LOGICAL Statement:

```
LOGICAL :: variable_name1[, variable_name2, etc.]
```

Examples:
```
LOGICAL :: initialize, debug
LOGICAL :: debug = .false.
```

Description:
The LOGICAL statement is a type declaration statement that declares variables of the logical data type. The value of a LOGICAL variable may be initialized when it is declared, as shown in the second example above.

Logical IF Statement:

```
IF ( logical_expr ) statement
```

Description:
The Logical IF statement is a special case of the block IF construct. If `logical_expr` is true, then the statement on the line with the IF is executed. Execution continues at the next line after the IF statement.
  This statement may be used instead of the block IF construct if only one statement needs to be executed as a result of the logical condition.

3.6.3. Exercises

3-1. Which of the following expressions are legal in Fortran? If an expression is legal, evaluate it.

(a) 5.5 >= 5

(b) 20 > 20

(c) .NOT. 6 > 5

(d) .TRUE. > .FALSE.

(e) 35 / 17. > 35 / 17

(f) 7 <= 8 .EQV. 3 / 2 == 1

(g) 17.5 .AND. (3.3 > 2.)

3-2 The tangent function is defined as tan θ = sin θ/cos θ. This expression can be evaluated to solve for the tangent as long as the magnitude of cos θ is not too near to 0. (If cos θ is 0, evaluating the equation for tan θ will produce a divide-by-zero error.) Assume that θ is given in degrees, and write Fortran statements to evaluate tan θ as long as the magnitude of cos θ is greater than or equal to 10^{-20}. If the magnitude of cos θ is less than 10^{-20}, write out an error message instead.

3-3 Write the Fortran statements required to calculate $y(t)$ from the equation

$$y(t) = \begin{cases} -3t^2 + 5 & t \geq 0 \\ 3t^2 + 5 & t < 0 \end{cases}$$

for a user-supplied value of t.

3-4 The following Fortran statements are intended to alert a user to dangerously high oral thermometer readings (values are in degrees Fahrenheit). Are they correct or incorrect? If they are incorrect, explain why and correct them.

```
IF ( temp < 97.5 ) THEN
    WRITE (*,*) 'Temperature below normal'
ELSE IF ( temp > 97.5 ) THEN
    WRITE (*,*) 'Temperature normal'
```

```
        ELSE IF ( temp > 99.5 ) THEN
            WRITE (*,*) 'Temperature slightly high'
        ELSE IF ( temp > 103.0 ) THEN
            WRITE (*,*) 'Temperature dangerously high'
        END IF
```

3-5 The cost of sending a package by an express delivery service is $15.00 for the first two pounds, and $5.00 for each pound or fraction thereof over two pounds. If the package weighs more than 70 pounds, a $15.00 excess weight surcharge is added to the cost. No package over 100 pounds will be accepted. Write a program that accepts the weight of a package in pounds and computes the cost of mailing the package. Be sure to handle the case of overweight packages.

3-6 The inverse sine function ASIN(x) is only defined for the range $-1.0 \le x \le 1.0$. If x is outside this range, the value NaN (not a number) occurs when the function is evaluated. The following Fortran statements calculate the inverse sine of a number if it is in the proper range, and print an error message if it is not. Assume that x and inverse_sine are real. Is this code correct or incorrect? If it is incorrect, explain why and correct it.

```
        test: IF ( ABS(x) <= 1. ) THEN
            inverse_sine = ASIN(x)
        ELSE test
            WRITE (*,*) x, ' is out of range!'
        END IF test
```

3-7 In Example 3-3, we wrote a program to evaluate the function $f(x,y)$ for any two user-specified values x and y, where the function $f(x,y)$ was defined as follows.

$$f(x, y) = \begin{cases} x + y & x \ge 0 \text{ and } y \ge 0 \\ x + y^2 & x \ge 0 \text{ and } y < 0 \\ x^2 + y & x < 0 \text{ and } y \ge 0 \\ x^2 + y^2 & x < 0 \text{ and } y < 0 \end{cases}$$

The problem was solved by using a single block IF construct with four code blocks to calculate $f(x,y)$ for all possible combinations of x and y. Rewrite program funxy to use nested IF constructs, where the outer construct evaluates the value of x and the inner constructs evaluate the value of y. Be sure to assign names to each of your constructs.

3-8 Write a program to evaluate the function

$$y(x) = \ln \frac{1}{1-x}$$

for any user-specified value of x, where x is a number <1.0 (note that ln is the natural logarithm, the logarithm to the base e). Use an if structure to verify that the value passed to the program is legal. If the value of x is legal, calculate y(x). If not, write a suitable error message and quit.

3-9 Suppose that a student has the option of enrolling for a single elective during a term. The student must select a course from a limited list of options: "English", "History", "Astronomy", or "Literature". Construct a fragment of Fortran code that will prompt the student for his or her choice, read in the choice, and use the answer as the case expression for a `CASE` construct. Be sure to include a default case to handle invalid inputs.

3-10 The author of this book now lives in Australia. In 2009, individual citizens and residents of Australia paid the following income taxes:

Taxable Income (in A$)	Tax on This Income
$0–$6000	None
$6001–$34,000	15¢ for each $1 over $6000
$34,001–$80,000	$4200 plus 30¢ for each $1 over $34,000
$80,001–$180,000	$18,000 plus 40¢ for each $1 over $80,000
Over $180,000	$58,000 plus 45¢ for each $1 over $180,000

In addition, a flat 1.5 percent Medicare levy is charged on all income. Write a program to calculate how much income tax a person will owe based on this information. The program should accept a total income figure from the user, and calculate the income tax, Medicare Levy, and total tax payable by the individual.

3-11 In 2002, individual citizens and residents of Australia paid the following income taxes:

Taxable Income (in A$)	Tax on This Income
$0–$6,000	None
$6,001–$20,000	17¢ for each $1 over $6,000
$20,001–$50,000	$2,380 plus 30¢ for each $1 over $20,000
$50,001–$60,000	$11,380 plus 42¢ for each $1 over $50,000
Over $60,000	$15,580 plus 47¢ for each $1 over $60,000

In addition, a flat 1.5 percent Medicare levy was charged on all income. Write a program to calculate how much *less* income tax a person paid on a given amount of income in 2009 than he or she would have paid in 2002.

3-12 It is often hard to compare the value of two items if they are priced in different currencies. Write a program that will allow a user to enter the cost of a purchase in US dollars, Australian dollars, Euros, or UK pounds, and then convert the cost into any of the other currencies, as specified by the user. Use the following conversion factors in your program:

$$A\$ 1.00 = US\ \$ 0.71$$
$$€1.00 = US\ \$ 1.12$$
$$UK£ 1.00 = US\ \$1.42$$

3-13 Decibels In Exercise 2-22, we wrote a program to calculate a power level in decibels with respect to a 1 mW reference level. The equation implemented was

$$dB = 10\log_{10}\frac{P_2}{P_1} \qquad (2\text{-}16)$$

where P_2 is the power level being measured, and P_1 is reference power level (1 milliwatt). This equation uses the logarithm to the base 10, which is undefined for negative or zero values. Modify the program to trap negative or zero input values, and inform the user of the invalid input values.

3-14 Refraction When a ray of light passes from a region with an index of refraction n_1 into a region with a different index of refraction n_2, the light ray is bent (see Figure 3-16). The angle at which the light is bent is given by *Snell's law*

$$n_1 \sin \theta_1 = n_2 \sin \theta_2 \qquad (3\text{-}3)$$

where θ_1 is the angle of incidence of the light in the first region, and θ_2 is the angle of incidence of the light in the second region. Using Snell's law, it is possible to predict the angle of incidence of a light ray in Region 2 if the angle of incidence θ_1 in Region 1 and the indices of refraction n_1 and n_2 are known. The equation to perform this calculation is

$$\theta_2 = \sin^{-1}\left(\frac{n_1}{n_2} \sin \theta_1\right) \qquad (3\text{-}4)$$

Write a Fortran program to calculate the angle of incidence (in degrees) of a light ray in Region 2 given the angle of incidence θ_1 in Region 1 and the indices of refraction n_1 and n_2. (*Note*: If $n_1 > n_2$, then for some angles θ_1, Equation (3-4) will have no real solution because the absolute value of the quantity $\left(\frac{n_2}{n_1} \sin \theta_1\right)$ will be greater than 1.0. When this occurs, all light is reflected back into Region 1, and no light passes into Region 2 at all. Your program must be able to recognize and properly handle this condition.) Test your program by running it for the following two cases: (*a*) $n_1 = 1.0$, $n_2 = 1.7$, and $\theta_1 = 45°$; (*b*) $n_1 = 1.7$, $n_2 = 1.0$, and $\theta_1 = 45°$.

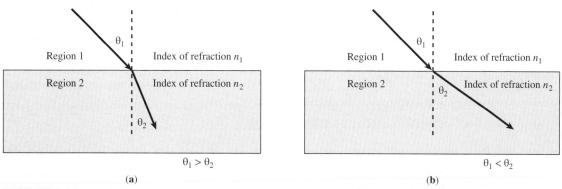

FIGURE 3-16
A ray of light bends as it passes from one medium into another one. (a) If the ray of light passes from a region with a low index of refraction into a region with a higher index of refraction, the ray of light bends more toward the vertical. (b) If the ray of light passes from a region with a high index of refraction into a region with a lower index of refraction, the ray of light bends away from the vertical.

4

Loops and Character Manipulation

OBJECTIVES

- Know how to create and use while loops.
- Know how to create and use counting loops.
- Know when you should use while loops, and when you should use counting loops.
- Know the purpose of the `CONTINUE` and `EXIT` statements, and how to use them.
- Understand loop names, and why they are used.
- Learn about character assignments and character operators.
- Learn about substrings and string manipulations.

In the previous chapter, we introduced branching structures, which allowed a program to select and execute one of several possible sets of statements, depending on the value of some control expression. In this chapter, we will introduce **loops,** which cause specific sections of the code to be repeated.

We will also learn more about how to manipulate character variables in this chapter. Many of the manipulations will involve loops, and we will use the character manipulations as practice in using loops.

4.1
CONTROL CONSTRUCTS: LOOPS

Loops are Fortran constructs that permit us to execute a sequence of statements more than once. There are two basic forms of loop constructs: **while loops** and **iterative loops** (or **counting loops**). The major difference between these two types of loops is in how the repetition is controlled. The code in a while loop is repeated an indefinite number of times until some user-specified condition is satisfied. By contrast, the code in an iterative loop is repeated a specified number of times, and the number of repetitions is known before the loop starts.

4.1.1 The While Loop

A **while loop** is a block of statements that are repeated indefinitely as long as some condition is satisfied. The general form of a while loop in Fortran is

```
DO
    ...
    IF (logical_expr) EXIT       } Code Block
    ...
END DO
```

The block of statements between the DO and END DO are repeated indefinitely until the *logical_expr* becomes true and the EXIT statement is executed. After the EXIT statement is executed, control transfers to the first statement after the END DO.

A while loop may contain one or more EXIT statements to terminate its execution. Each EXIT statement is usually a part of an IF statement or block of construct. If the *logical_expr* in the IF is false when the statement is executed, the loop continues to execute. If the *logical_expr* in the IF is true when the statement is executed, control transfers immediately to the first statement after the END DO. If the logical expression is true the first time we reach the while loop, the statements in the loop below the IF will never be executed at all!

The pseudocode corresponding to a while loop is

```
WHILE
    ...
    IF logical_expr EXIT
    ...
End of WHILE
```

and the flowchart for this construct is shown in Figure 4-1.

In a good structured program, every while loop should have a *single entry point* and a *single exit point*. The entry point for a while loop is the DO statement, and the exit point is the EXIT statement. Having only a single exit point from a loop helps us to confirm that the loop operates properly under all circumstances. Therefore, each while loop should have only one EXIT statement.

Good Programming Practice
Each while loop should contain only one EXIT statement.

We will now show an example statistical analysis program that is implemented using a while loop.

EXAMPLE 4-1

Statistical Analysis:

It is very common in science and engineering to work with large sets of numbers, each of which is a measurement of some particular property that we are interested in.

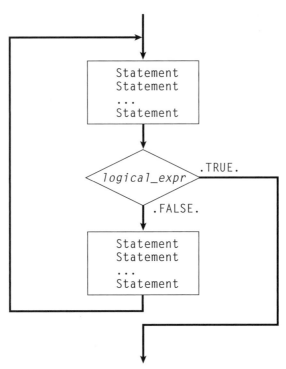

FIGURE 4-1
Flowchart for a while loop.

A simple example would be the grades on the first test in this course. Each grade would be a measurement of how much a particular student has learned in the course to date.

Much of the time, we are not interested in looking closely at every single measurement that we make. Instead, we want to summarize the results of a set of measurements with a few numbers that tell us a lot about the overall data set. Two such numbers are the *average* (or *arithmetic mean*) and the *standard deviation* of the set of measurements. The average or arithmetic mean of a set of numbers is defined as

$$\bar{x} = \frac{1}{N}\sum_{i=1}^{N} x_i \tag{4-1}$$

where x_i is sample i out of N samples. The standard deviation of a set of numbers is defined as

$$s = \sqrt{\frac{N\sum_{i=1}^{N} x_i^2 - \left(\sum_{i=1}^{N} x_i\right)^2}{N(N-1)}} \tag{4-2}$$

Standard deviation is a measure of the amount of scatter on the measurements; the greater the standard deviation, the more scattered the points in the data set are.

Loops and Character Manipulation

Implement an algorithm that reads in a set of measurements and calculates the mean and the standard deviation of the input data set.

SOLUTION

This program must be able to read in an arbitrary number of measurements, and then calculate the mean and standard deviation of those measurements. We will use a while loop to accumulate the input measurements before performing the calculations.

When all of the measurements have been read, we must have some way of telling the program that there is no more data to enter. For now, we will assume that all the input measurements are either positive or zero, and we will use a negative input value as a *flag* to indicate that there is no more data to read. If a negative value is entered, then the program will stop reading input values and will calculate the mean and standard deviation of the data set.

1. **State the problem.**

    Since we assume that the input numbers must be positive or zero, a proper statement of this problem would be: *calculate the average and the standard deviation of a set of measurements, assuming that all of the measurements are either positive or zero, and assuming that we do not know in advance how many measurements are included in the data set. A negative input value will mark the end of the set of measurements.*

2. **Define the inputs and outputs.**

    The inputs required by this program are an unknown number of positive or zero real (floating-point) numbers. The outputs from this program are a printout of the mean and the standard deviation of the input data set. In addition, we will print out the number of data points input to the program, since this is a useful check that the input data was read correctly.

3. **Design the algorithm.**

    This program can be broken down into three major steps:

    ```
    Accumulate the input data
    Calculate the mean and standard deviation
    Write out the mean, standard deviation, and number of points
    ```

    The first major step of the program is to accumulate the input data. To do this, we will have to prompt the user to enter the desired numbers. When the numbers are entered, we will have to keep track of the number of values entered, plus the sum and the sum of the squares of those values. The pseudocode for these steps is:

    ```
    Initialize n, sum_x, and sum_x2 to 0
    WHILE
        Prompt user for next number
        Read in next x
        IF x < 0. EXIT
        n ← n + 1
        sum_x ← sum_x + x
        sum_x2 ← sum_x2 + x**2
    End of WHILE
    ```

Note that we have to read in the first value before the IF ( ) EXIT test so that the while loop can have a value to test the first time it executes.

Next, we must calculate the mean and standard deviation. The pseudocode for this step is just the Fortran versions of Equations (4.1) and (4.2).

```
x_bar ← sum_x / REAL(n)
std_dev ← SQRT((REAL(n)*sum_x2 - sum_x**2) / (REAL(n)*REAL(n-1)))
```

Finally, we must write out the results.

```
Write out the mean value x_bar
Write out the standard deviation std_dev
Write out the number of input data points n
```

The flowchart for this program is shown in Figure 4-2.

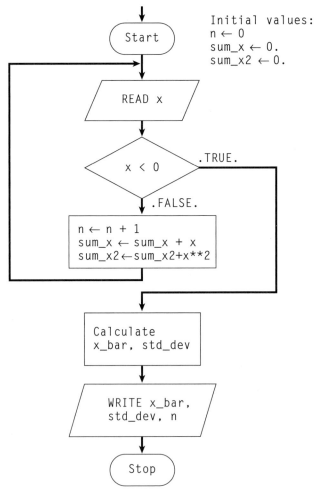

FIGURE 4-2
Flowchart for the statistical analysis program of Example 4-1.

4. **Turn the algorithm into Fortran statements.**
   The final Fortran program is shown in Figure 4-3.

FIGURE 4-3
Program to calculate the mean and standard deviation of a set of nonnegative real numbers.

```
PROGRAM stats_1
!
!   Purpose:
!     To calculate mean and the standard deviation of an input
!     data set containing an arbitrary number of input values.
!
!   Record of revisions:
!       Date        Programmer          Description of change
!       ====        ==========          =====================
!     11/10/15     S. J. Chapman        Original code
!
IMPLICIT NONE

! Data dictionary: declare variable types, definitions, & units
INTEGER :: n = 0      ! The number of input samples.
REAL :: std_dev = 0.  ! The standard deviation of the input samples.
REAL :: sum_x = 0.    ! The sum of the input values.
REAL :: sum_x2 = 0.   ! The sum of the squares of the input values.
REAL :: x = 0.        ! An input data value.
REAL :: x_bar         ! The average of the input samples.

! While Loop to read input values.
DO
   ! Read in next value
   WRITE (*,*) 'Enter number: '
   READ  (*,*) x
   WRITE (*,*) 'The number is ', x

   ! Test for loop exit
   IF ( x < 0 ) EXIT

   ! Otherwise, accumulate sums.
   n      = n + 1
   sum_x  = sum_x + x
   sum_x2 = sum_x2 + x**2
END DO

! Calculate the mean and standard deviation
x_bar = sum_x / real(n)
std_dev = sqrt( (real(n) * sum_x2 - sum_x**2) / (real(n) * real(n-1)) )

! Tell user.
WRITE (*,*) 'The mean of this data set is:', x_bar
WRITE (*,*) 'The standard deviation is:   ', std_dev
WRITE (*,*) 'The number of data points is:', n

END PROGRAM stats_1
```

5. **Test the program.**

To test this program, we will calculate the answers by hand for a simple data set, and then compare the answers to the results of the program. If we used three input values: 3, 4, and 5, then the mean and standard deviation would be

$$\bar{x} = \frac{1}{N}\sum_{i=1}^{N} x_i = \frac{1}{3}(12) = 4$$

$$s = \sqrt{\frac{N\sum_{i=1}^{N} x_i^2 - \left(\sum_{i=1}^{N} x_i\right)^2}{N(N-1)}} = 1$$

When the above values are fed into the program, the results are

```
C:\book\fortran\chap4>stats_1
Enter number:
3.
The number is          3.000000
Enter number:
4.
The number is          4.000000
Enter number:
5.
The number is          5.000000
Enter number:
-1.
The number is         -1.000000
The mean of this data set is:        4.000000
The standard deviation is:           1.000000
The number of data points is:               3
```

The program gives the correct answers for our test data set.

In the example above, we failed to follow the design process completely. This failure has left the program with a fatal flaw! Did you spot it?

We have failed because *we did not completely test the program for all possible types of inputs.* Look at the example once again. If we enter either no numbers or only one number, then we will be dividing by zero in the above equations! The division-by-zero error will cause the program to abort. We need to modify the program to detect this problem, inform the user of it, and stop gracefully.

A modified version of the program called `stats_2` is shown in Figure 4-4, with the changes shown in bold face. Here, we check to see if there are enough input values before performing the calculations. If not, the program will print out an intelligent error message and quit. Test the modified program for yourself.

FIGURE 4-4
A modified statistical analysis program that avoids the divide-by-zero problems inherent in program stats_1.

```
PROGRAM stats_2
!
!  Purpose:
!    To calculate mean and the standard deviation of an input
!    data set containing an arbitrary number of input values.
!
!  Record of revisions:
!      Date        Programmer          Description of change
!      ====        ==========          =====================
!    11/10/15    S. J. Chapman         Original code
! 1. 11/12/15    S. J. Chapman         Correct divide-by-0 error if
!                                      0 or 1 input values given.
!
IMPLICIT NONE

! Data dictionary: declare variable types, definitions, & units
INTEGER :: n = 0      ! The number of input samples.
REAL :: std_dev = 0.  ! The standard deviation of the input samples.
REAL :: sum_x = 0.    ! The sum of the input values.
REAL :: sum_x2 = 0.   ! The sum of the squares of the input values.
REAL :: x = 0.        ! An input data value.
REAL :: x_bar         ! The average of the input samples.

! While Loop to read input values.
DO
   ! Read in next value
   WRITE (*,*) 'Enter number: '
   READ (*,*) x
   WRITE (*,*) 'The number is ', x

   ! Test for loop exit
   IF ( x < 0 ) EXIT

   ! Otherwise, accumulate sums.
   n       = n + 1
   sum_x   = sum_x + x
   sum_x2  = sum_x2 + x**2
END DO

! Check to see if we have enough input data.
IF ( n < 2 ) THEN ! Insufficient information

   WRITE (*,*) 'At least 2 values must be entered!'

ELSE ! There is enough information, so
     ! calculate the mean and standard deviation

   x_bar = sum_x / real(n)
   std_dev = sqrt( (real(n) * sum_x2 - sum_x**2) / (real(n)*real(n-1)))
```

(*continued*)

(*concluded*)

```
      ! Tell user.
      WRITE (*,*) 'The mean of this data set is:', x_bar
      WRITE (*,*) 'The standard deviation is:   ', std_dev
      WRITE (*,*) 'The number of data points is:', n
   END IF

END PROGRAM stats_2
```

4.1.2 The DO WHILE Loop

There is an alternate form of the while loop in Fortran, called the DO WHILE loop. The DO WHILE construct has the form

```
DO WHILE (logical_expr)
   ...                    ⎫  Statement 1
   ...                    ⎬  Statement 2
   ...                    ⎪  ...
   ...                    ⎭  Statement n
END DO
```

If the logical expression is true, statements 1 through *n* will be executed, and then control will return to the DO WHILE statement. If the logical expression is still true, the statements will be executed again. This process will be repeated until the logical expression becomes false. When control returns to the DO WHILE statement and the logical expression is false, the program will execute the first statement after the END DO.

This construct is a special case of the more general while loop, in which the exit test must always occur at the top of the loop. There is no reason to ever use it, since the general while loop does the same job with more flexibility.

Good Programming Practice

Do not use DO WHILE loops in new programs. Use the more general while loop instead.

4.1.3 The Iterative or Counting Loop

In the Fortran language, a loop that executes a block of statements a specified number of times is called an **iterative DO loop** or a **counting loop.** The counting loop construct has the form

```
DO index = istart, iend, incr
   Statement 1              ⎫
   ...                      ⎬  Body
   Statement n              ⎭
END DO
```

where index is an integer variable used as the loop counter (also known as the **loop index**). The integer quantities istart, iend, and incr are the *parameters* of the counting loop; they control the values of the variable `index` during execution. The parameter `incr` is optional; if it is missing, it is assumed to be 1.

The statements between the `DO` statement and the `END DO` statement are known as the *body* of the loop. They are executed repeatedly during each pass of the DO loop.

The counting loop construct functions as follows:

1. Each of the three DO loop parameters `istart`, `iend`, and `incr` may be a constant, a variable, or an expression. If they are variables or expressions, then their values are calculated before the start of the loop, and the resulting values are used to control the loop.
2. At the beginning of the execution of the DO loop, the program assigns the value `istart` to control variable `index`. If `index*incr` ≤ `iend*incr`, the program executes the statements within the body of the loop.
3. After the statements in the body of the loop have been executed, the control variable is recalculated as

$$index = index + incr$$

If `index*incr` is still ≤ `iend*incr`, the program executes the statements within the body again.
4. Step 2 is repeated over and over as long as `index*incr` ≤ `iend*incr`. When this condition is no longer true, execution skips to the first statement following the end of the DO loop.

The number of iterations to be performed by the DO loop may be calculated using the following equation

$$iter = \frac{iend - istart + incr}{incr} \tag{4-3}$$

Let's look at a number of specific examples to make the operation of the counting loop clearer. First, consider the following example:

```
DO i = 1, 10
    Statement 1
    ...
    Statement n
END DO
```

In this case, statements 1 through *n* will be executed 10 times. The index variable `i` will be 1 on the first time, 2 on the second time, and so on. The index variable will be 10 on the last pass through the statements. When control is returned to the DO statement after the tenth pass, the index variable `i` will be increased to 11. Since 11 × 1 > 10 × 1, control will transfer to the first statement after the END DO statement.

Second, consider the following example:

```
DO i = 1, 10, 2
    Statement 1
    ...
    Statement n
END DO
```

In this case, statements 1 through *n* will be executed five times. The index variable `i` will be 1 on the first time, 3 on the second time, and so on. The index variable will be 9 on the fifth and last pass through the statements. When control is returned to the DO statement after the fifth pass, the index variable i will be increased to 11. Since $11 \times 2 > 10 \times 2$, control will transfer to the first statement after the END DO statement.

Third, consider the following example:

```
DO i = 1, 10, -1
   Statement 1
   ...
   Statement n
END DO
```

Here, *statements 1 through n will never be executed*, since `index*incr > iend*incr` on the very first time that the DO statement is reached. Instead, control will transfer to the first statement after the END DO statement.

Finally, consider the example:

```
DO i = 3, -3, -2
   Statement 1
   ...
   Statement n
END DO
```

In this case, statements 1 through *n* will be executed four times. The index variable i will be 3 on the first time, 1 on the second time, -1 on the third time, and -3 on the fourth time. When control is returned to the DO statement after the fourth pass, the index variable i will be decreased to -5. Since $-5 \times -2 > -3 \times -2$, control will transfer to the first statement after the END DO statement.

The pseudocode corresponding to a counting loop is

```
DO for index = istart to iend by incr
   Statement 1
   ...
   Statement n
End of DO
```

and the flowchart for this construct is shown in Figure 4-5.

EXAMPLE 4-2

The Factorial Function:

To illustrate the operation of a counting loop, we will use a DO loop to calculate the factorial function. The factorial function is defined as

$$n! = \begin{cases} 1 & n = 0 \\ n \times (n-1) \times (n-2) \times \cdots \times 2 \times 1 & n > 0 \end{cases} \quad (4\text{-}4)$$

The Fortran code to calculate N factorial for positive value of N would be

```
n_factorial = 1
DO i = 1, n
   n_factorial = n_factorial * i
END DO
```

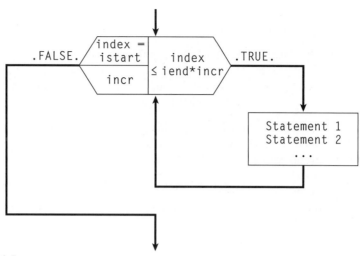

FIGURE 4-5
Flowchart for a DO loop construct.

Suppose that we wish to calculate the value of 5!. If n is 5, the DO loop parameters will be istart = 1, iend = 5, and incr = 1. This loop will be executed five times, with the variable i taking on values of 1, 2, 3, 4, and 5 in the successive loops. The resulting value of n_factorial will be $1 \times 2 \times 3 \times 4 \times 5 = 120$.

EXAMPLE 4-3 *Calculating the Day of Year:*

The *day of year* is the number of days (including the current day) that have elapsed since the beginning of a given year. It is a number in the range 1 to 365 for ordinary years, and 1 to 366 for leap years. Write a Fortran program that accepts a day, month, and year, and calculates the day of year corresponding to that date.

SOLUTION
To determine the day of year, this program will need to sum up the number of days in each month preceding the current month, plus the number of elapsed days in the current month. A DO loop will be used to perform this sum. Since the number of days in each month varies, it is necessary to determine the correct number of days to add for each month. A SELECT CASE construct will be used to determine the proper number of days to add for each month.

During a leap year, an extra day must be added to the day of year for any month after February. This extra day accounts for the presence of February 29 in the leap year. Therefore, to perform the day of year calculation correctly, we must determine which years are leap years. In the Gregorian calendar, leap years are determined by the following rules:

1. Years evenly divisible by 400 are leap years.
2. Years evenly divisible by 100 but *not* by 400 are not leap years.
3. All years divisible by 4 but *not* by 100 are leap years.
4. All other years are not leap years.

We will use the MOD (for modulo) function to determine whether or not a year is evenly divisible by a given number. If the result of the MOD function is zero, then the year is evenly divisible.

A program to calculate the day of year is shown in Figure 4-6. Note that the program sums up the number of days in each month before the current month, and that it uses a SELECT CASE construct to determine the number of days in each month.

FIGURE 4-6
A program to calculate the equivalent day of year from a given day, month, and year.

```
PROGRAM doy
!  Purpose:
!    This program calculates the day of year corresponding to a
!    specified date.  It illustrates the use of counting loops
!    and the SELECT CASE construct.
!
!  Record of revisions:
!      Date           Programmer            Description of change
!      ====           ==========            =====================
!    11/13/15        S. J. Chapman          Original code
!
IMPLICIT NONE

! Data dictionary: declare variable types, definitions, & units
INTEGER :: day            ! Day (dd)
INTEGER :: day_of_year    ! Day of year
INTEGER :: i              ! Index variable
INTEGER :: leap_day       ! Extra day for leap year
INTEGER :: month          ! Month (mm)
INTEGER :: year           ! Year (yyyy)

! Get day, month, and year to convert
WRITE (*,*) 'This program calculates the day of year given the '
WRITE (*,*) 'current date.  Enter current month (1-12), day(1-31),'
WRITE (*,*) 'and year in that order: '
READ (*,*) month, day, year

! Check for leap year, and add extra day if necessary
IF ( MOD(year,400) == 0 ) THEN
   leap_day = 1              ! Years divisible by 400 are leap years
ELSE IF ( MOD(year,100) == 0 ) THEN
   leap_day = 0              ! Other centuries are not leap years
```

(*continued*)

Loops and Character Manipulation

(concluded)

```
ELSE IF ( MOD(year,4) == 0 ) THEN
   leap_day = 1            ! Otherwise every 4th year is a leap year
ELSE
   leap_day = 0            ! Other years are not leap years
END IF

! Calculate day of year
day_of_year = day
DO i = 1, month-1

   ! Add days in months from January to last month
   SELECT CASE (i)
   CASE (1,3,5,7,8,10,12)
      day_of_year = day_of_year + 31
   CASE (4,6,9,11)
      day_of_year = day_of_year + 30
   CASE (2)
      day_of_year = day_of_year + 28 + leap_day
   END SELECT

END DO

! Tell user
WRITE (*,*) 'Day          = ', day
WRITE (*,*) 'Month        = ', month
WRITE (*,*) 'Year         = ', year
WRITE (*,*) 'day of year  = ', day_of_year

END PROGRAM doy
```

We will use the following known results to test the program:

1. Year 1999 is not a leap year. January 1 must be day of year 1, and December 31 must be day of year 365.
2. Year 2000 is a leap year. January 1 must be day of year 1, and December 31 must be day of year 366.
3. Year 2001 is not a leap year. March 1 must be day of year 60, since January has 31 days, February has 28 days, and this is the first day of March.

If this program is compiled, and then run five times with the above dates, the results are

```
C:\book\fortran\chap4>doy

This program calculates the day of year given the
current date. Enter current month (1-12), day(1-31),
and year in that order: 1 1 1999
Day          =           1
Month        =           1
Year         =        1999
day of year  =           1
```

(continued)

(*concluded*)

```
C:\book\fortran\chap4>doy

This program calculates the day of year given the
current date. Enter current month (1-12), day(1-31),
and year in that order: 12 31 1999

Day           =           31
Month         =           12
Year          =         1999
day of year   =          365

C:\book\fortran\chap4>doy

This program calculates the day of year given the
current date. Enter current month (1-12), day(1-31),
and year in that order: 1 1 2000

Day           =            1
Month         =            1
Year          =         2000
day of year   =            1

C:\book\fortran\chap4>doy

This program calculates the day of year given the
current date. Enter current month (1-12), day(1-31),
and year in that order: 12 31 2000

Day           =           31
Month         =           12
Year          =         2000
day of year   =          366

C:\book\fortran\chap4>doy

This program calculates the day of year given the
current date. Enter current month (1-12), day(1-31),
and year in that order: 3 1 2001

Day           =            1
Month         =            3
Year          =         2001
day of year   =           60
```

The program gives the correct answers for our test dates in all five test cases.

EXAMPLE 4-4 *Statistical Analysis:*

Implement an algorithm that reads in a set of measurements and calculates the mean and the standard deviation of the input data set, when any value in the data set can be positive, negative, or zero.

Loops and Character Manipulation

SOLUTION

This program must be able to read in an arbitrary number of measurements, and then calculate the mean and standard deviation of those measurements. Each measurement can be positive, negative, or zero.

Since we cannot use a data value as a flag this time, we will ask the user for the number of input values, and then use a DO loop to read in those values. A flowchart for this program is shown in Figure 4-7. Note that the while loop has been replaced by a

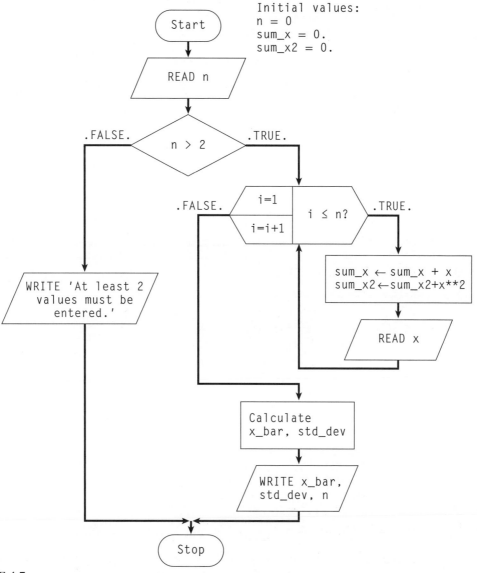

FIGURE 4-7
Flowchart for modified statistical analysis program using a DO loop.

counting loop. The modified program that permits the use of any input value is shown in Figure 4-8. Verify its operation for yourself by finding the mean and standard deviation of the following five input values: 3., −1., 0., 1., and −2.

FIGURE 4-8
Modified statistical analysis program that works with both positive and input values.

```
PROGRAM stats_3
!
! Purpose:
!   To calculate mean and the standard deviation of an input
!   data set, where each input value can be positive, negative,
!   or zero.
!
! Record of revisions:
!     Date       Programmer       Description of change
!     ====       ==========       =====================
!   11/13/15    S. J. Chapman     Original code
!
IMPLICIT NONE

! Data dictionary: declare variable types, definitions, & units
INTEGER :: i              ! Loop index
INTEGER :: n = 0          ! The number of input samples.
REAL :: std_dev           ! The standard deviation of the input samples.
REAL :: sum_x = 0.        ! The sum of the input values.
REAL :: sum_x2 = 0.       ! The sum of the squares of the input values.
REAL :: x = 0.            ! An input data value.
REAL :: x_bar             ! The average of the input samples.

! Get the number of points to input.
WRITE (*,*) 'Enter number of points: '
READ  (*,*) n

! Check to see if we have enough input data.
IF ( n < 2 ) THEN ! Insufficient data

   WRITE (*,*) 'At least 2 values must be entered.'

ELSE ! we will have enough data, so let's get it.

   ! Loop to read input values.
   DO i = 1, n

      ! Read values
      WRITE (*,*) 'Enter number: '
      READ  (*,*) x
      WRITE (*,*) 'The number is ', x

      ! Accumulate sums.
      sum_x  = sum_x + x
      sum_x2 = sum_x2 + x**2

   END DO
```

(continued)

Loops and Character Manipulation

(*concluded*)

```
    ! Now calculate statistics.
    x_bar = sum_x / real(n)
    std_dev = sqrt((real(n)*sum_x2 - sum_x**2) / (real(n)*real(n-1)))

    ! Tell user.
    WRITE (*,*) 'The mean of this data set is:', x_bar
    WRITE (*,*) 'The standard deviation is:  ', std_dev
    WRITE (*,*) 'The number of data points is:', n
END IF

END PROGRAM stats_3
```

Details of Operation

Now that we have seen examples of a counting DO loop in operation, we will examine some of the important details required to use DO loops properly.

1. It is not necessary to indent the body of the DO loop as we have shown above. The Fortran compiler will recognize the loop even if every statement in it starts in column 1. However, the code is much more readable if the body of the DO loop is indented, so you should always indent the bodies of your DO loops.

Good Programming Practice
Always indent the body of a DO loop by two or more spaces to improve the readability of the code.

2. The index variable of a DO loop *must not be modified anywhere within the* DO *loop*. Since the index variable is used to control the repetitions in the DO loop, changing it could produce unexpected results. In the worst case, modifying the index variable could produce an *infinite loop* that never completes. Consider the following example:

```
PROGRAM bad_1
INTEGER :: i
DO i = 1, 4
    i = 2
END DO
END PROGRAM bad_1
```

If i is reset to 2 every time through the loop, the loop will never end, because the index variable can never be greater than 4! This loop will run forever unless the program containing it is killed. Almost all Fortran compilers will recognize this problem, and will generate a compile-time error if a program attempts to modify an index variable within a loop.

> **Programming Pitfalls**
> Never modify the value of a DO loop index variable while inside the loop.

3. If the number of iterations calculated from Equation 4-3 is less than or equal to zero, the statements within the DO loop are never executed at all. For example, the statements in the following DO loop will never be executed

   ```
   DO i = 3, 2
   ...
   END DO
   ```

   since

   $$\text{iter} = \frac{\text{iend} - \text{istart} + \text{incr}}{\text{incr}} = \frac{2 - 3 + 1}{1} = 0$$

4. It is possible to design counting DO loops that count down as well as up. The following DO loop executes three times with i being 3, 2, and 1 in the successive loops.

   ```
   DO i = 3, 1, -1
   ...
   END DO
   ```

5. The index variable and control parameters of a DO loop should always be of type integer.

   The use of real variables as DO loop indices and DO loop control parameters used to be a legal but undesirable feature of Fortran. It was declared obsolescent in Fortran 90, and has been completely deleted from Fortran 95.

6. It is possible to branch out of a DO loop at any time while the loop is executing. If program execution does branch out of a DO loop before it would otherwise finish, the loop index variable retains the value that it had when the branch occurs. Consider the following example:

   ```
   INTEGER :: i
   DO i = 1, 5
   ...
   IF (i >= 3) EXIT
   ...
   END DO
   WRITE (*,*) i
   ```

   Execution will branch out of the DO loop and go to the WRITE statement on the third pass through the loop. When execution gets to the WRITE statement, variable i will contain a value of 3.

7. If a DO loop completes normally, *the value of the index variable is undefined when the loop is completed*. In the example shown below, the value written out by the WRITE statement is not defined in the Fortran standard.

```
INTEGER :: i
DO i = 1, 5
   ...
END DO
WRITE (*,*) i
```

On many computers, after the loop has completed, the index variable i will contain the first value of the index variable to fail the index*incr ≤ iend*incr test. In the above code, the result would usually contain a 6 after the loop is finished. However, don't count on it! Since the value is officially undefined in the Fortran standard, some compilers may produce a different result. If your code depends on the value of the index variable after the loop is completed, you may get different results as the program is moved between computers.

Good Programming Practice
Never depend on an index variable to retain a specific value after a DO loop completes normally.

4.1.4 The CYCLE and EXIT Statements

There are two additional statements that can be used to control the operation of while loops and counting DO loops: CYCLE and EXIT.

If the CYCLE statement is executed in the body of a DO loop, the execution of the *current iteration* of the loop will stop, and control will be returned to the top of the loop. The loop index will be incremented, and execution will resume again if the index has not reached its limit. An example of the CYCLE statement in a counting DO loop is shown below.

```
PROGRAM test_cycle
INTEGER :: i
DO i = 1, 5
   IF ( i == 3 ) CYCLE
   WRITE (*,*) i
END DO
WRITE (*,*) 'End of loop!'
END PROGRAM test_cycle
```

The flowchart for this loop is shown in Figure 4-9a. When this program is executed, the output is:

```
C:\book\fortran\chap4>test_cycle
 1
 2
 4
 5
 End of loop!
```

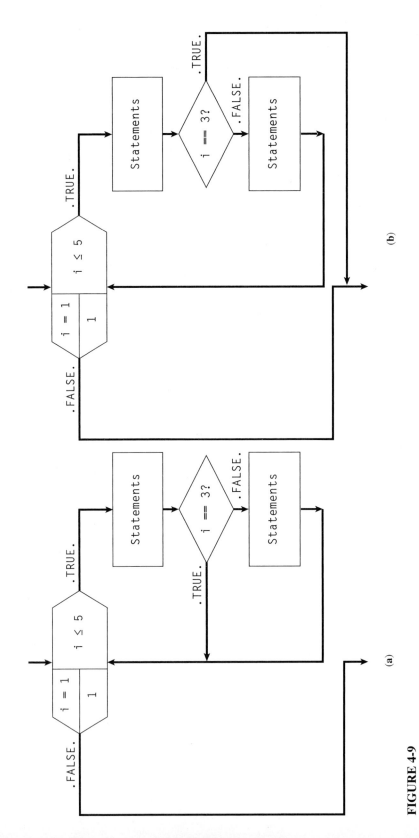

FIGURE 4-9
(a) Flowchart of a DO loop containing a CYCLE statement. (b) Flowchart of a DO loop containing an EXIT statement.

Note that the `CYCLE` statement was executed on the iteration when i was 3, and control returned to the top of the loop without executing the `WRITE` statement. After control returned to the top of the loop, the loop index was incremented and the loop continued to execute.

If the `EXIT` statement is executed in the body of a loop, the execution of the loop will stop and control will be transferred to the first executable statement after the loop. An example of the `EXIT` statement in a DO loop is shown below.

```
PROGRAM test_exit
INTEGER :: i
DO i = 1, 5
   IF ( i == 3 ) EXIT
   WRITE (*,*) i
END DO
WRITE (*,*) 'End of loop!'
END PROGRAM test_exit
```

The flowchart for this loop is shown in Figure 4-9*b*. When this program is executed, the output is:

```
C:\book\fortran\chap4>test_exit
   1
   2
End of loop!
```

Note that the `EXIT` statement was executed on the iteration when i was 3, and control transferred to the first executable statement after the loop without executing the `WRITE` statement.

Both the `CYCLE` and `EXIT` statements work with both while loops and counting DO loops.

4.1.5 Named Loops

It is possible to assign a name to a loop. The general form of a while loop with a name attached is

```
[name:] DO
   Statement
   Statement
   Statement
   IF ( logical_expr ) CYCLE [name]
   ...
   IF ( logical_expr ) EXIT [name]
END DO [name]
```

and the general form of a counting loop with a name attached is

```
[name:] DO index = istart, iend, incr
   Statement
   Statement
   IF ( logical_expr ) CYCLE [name]
   ...
END DO [name]
```

where *name* may be up to 63 alphanumeric characters long, beginning with a letter. The name given to the loop must be unique within each program unit. If a name is assigned to a loop, then the same name must appear on the associated END DO. Names are optional on any CYCLE and EXIT statements associated with the loop, but if they are used, they must be the same as the name on the DO statement.

Why would we want to name a loop? For simple examples like the ones we have seen so far, there is no particular reason to do so. The principal reason for using names is to help us (and the compiler) keep loops straight in our own minds when they get very complicated. For example, suppose that we have a complex loop that is hundreds of lines long, spanning many pages of listings. There may be many smaller loops inside body of that loop. If we name all of the parts of the loop, then we can tell at a glance which construct a particular END DO, CYCLE, or EXIT statement belongs to. They make our intentions explicitly clear. In addition, names on constructs can help the compiler flag the specific location of an error when one occurs.

Good Programming Practice
Assign a name to any large and complicated loops in your program to help you keep the parts of the construct associated together in your own mind.

4.1.6 Nesting Loops and Block IF Constructs

Nesting loops

It is possible for one loop to be completely inside another loop. If one loop is completely inside another one, the two loops are called **nested** loops. The following example shows two nested DO loops used to calculate and write out the product of two integers.

```
PROGRAM nested_loops
INTEGER :: i, j, product
DO i = 1, 3
   DO j = 1, 3
      product = i * j
      WRITE (*,*) i, ' * ', j, ' = ', product
   END DO
END DO
END PROGRAM nested_loops
```

In this example, the outer DO loop will assign a value of 1 to index variable i, and then the inner DO loop will be executed. The inner DO loop will be executed three times with index variable j having values 1, 2, and 3. When the entire inner DO loop has been completed, the outer DO loop will assign a value of 2 to index variable i, and the inner

DO loop will be executed again. This process repeats until the outer DO loop has executed three times, and the resulting output is

```
1 *   1 =   1
1 *   2 =   2
1 *   3 =   3
2 *   1 =   2
2 *   2 =   4
2 *   3 =   6
3 *   1 =   3
3 *   2 =   6
3 *   3 =   9
```

Note that the inner DO loop executes completely before the index variable of the outer DO loop is incremented.

When a Fortran compiler encounters an END DO *statement, it associates that statement with the innermost currently open loop*. Therefore, the first END DO statement above closes the "DO j = 1, 3" loop, and the second END DO statement above closes the "DO i = 1, 3" loop. This fact can produce hard-to-find errors if an END DO statement is accidentally deleted somewhere within a nested loop construct. If each of the nested loops are named, then the error will be much easier to find.

To illustrate this problem, let's "accidentally" delete the inner END DO statement in the previous example, and compile the program with the Intel Visual Fortran compiler.

```
PROGRAM bad_nested_loops_1
INTEGER :: i, j, product
DO i = 1, 3
   DO j = 1, 3
      product = i * j
      WRITE (*,*) i, ' * ', j, ' = ', product
END DO
END PROGRAM bad_nested_loops_1
```

The output of the compiler is:

```
C:\book\fortran\chap4>ifort bad_nested_loops_1.f90
Intel(R) Visual Fortran Intel(R) 64 Compiler for applications running on
Intel(R) 64, Version 16.0.2.180 Build 20160204
Copyright (C) 1985-2016 Intel Corporation. All rights reserved.

bad_nested_loops_1.f90(3): error #6321: An unterminated block exists.
DO i = 1, 3
^
compilation aborted for bad_nested_loops_1.f90 (code 1)
```

The compiler reports that there is a problem with the loop construct, but it could not detect the problem until the END PROGRAM statement is reached, and it cannot tell where the problem occurred. If the program is very large, we would be faced with a difficult task when we tried to locate the problem.

Now let's name each loop and "accidentally" delete the inner END DO statement.

```
PROGRAM bad_nested_loops_2
INTEGER :: i, j, product
outer: DO i = 1, 3
   inner: DO j = 1, 3
      product = i * j
      WRITE (*,*) i, ' * ', j, ' = ', product
END DO outer
END PROGRAM bad_nested_loops_2
```

When we compile the program with the Intel Visual Fortran compiler, the output is:

```
C:\book\fortran\chap4>df bad_nested_loops_2.f90
Intel(R) Visual Fortran Intel(R) 64 Compiler for applications running on
Intel(R) 64, Version 16.0.2.180 Build 20160204
Copyright (C) 1985-2016 Intel Corporation. All rights reserved.

bad_nested_loops_2.f90(7): error #6606: The block construct names must
match, and they do not.   [OUTER]
END DO outer
-------^
bad_nested_loops_2.f90(3): error #8147: DO construct with a construct name
must be terminated by an ENDDO statement with the same name.
outer: DO i = 1, 3
^

compilation aborted for bad_nested_loops_2.f90 (code 1)
```

The compiler reports that there is a problem with the loop construct, and it reports which loops were involved in the problem. This can be a major aid in debugging the program.

Good Programming Practice

Assign names to all nested loops so that they will be easier to understand and debug.

If DO *loops are nested, they must have independent index variables.* Remember that it is not possible to change an index variable within the body of a DO loop. Therefore, it is not possible to use the same index variable for two nested DO loops, since the inner loop would be attempting to change the index variable of the outer loop within the body of the outer loop.

Also, *if two loops are to be nested, one of them must lie completely within the other one.* The following DO loops are incorrectly nested, and a compile-time error will be generated for this code.

```
outer: DO i = 1, 3
   ...
   inner: DO j = 1, 3
      ...
END DO outer
   ...
END DO inner
```

Loops and Character Manipulation

The CYCLE and EXIT statements in nested loops

If a CYCLE or EXIT statement appears inside an *unnamed* set of nested loops, then the CYCLE or EXIT statement refers to the *innermost* of the loops containing it. For example, consider the following program

```
PROGRAM test_cycle_1
INTEGER :: i, j, product
DO i = 1, 3
   DO j = 1, 3
      IF ( j == 2) CYCLE
      product = i * j
      WRITE (*,*) i, ' * ', j, ' = ', product
   END DO
END DO
END PROGRAM test_cycle_1
```

If the inner loop counter j is equal to 2, then the CYCLE statement will be executed. This will cause the remainder of the code block of the *innermost* DO loop to be skipped, and execution of the innermost loop will start over with j increased by 1. The resulting output values are

```
1 *   1 =   1
1 *   3 =   3
2 *   1 =   2
2 *   3 =   6
3 *   1 =   3
3 *   3 =   9
```

Each time the inner loop variable had the value 2, execution of the inner loop was skipped.

It is also possible to make the CYCLE or EXIT statement refer to the *outer* loop of a nested construct of named loops by specifying a loop name in the statement. In the following example, when the inner loop counter j is equal to 2, the CYCLE outer statement will be executed. This will cause the remainder of the code block of the *outer* DO loop to be skipped, and execution of the outer loop will start over with i increased by 1.

```
PROGRAM test_cycle_2
INTEGER :: i, j, product
outer: DO i = 1, 3
   inner: DO j = 1, 3
      IF ( j == 2) CYCLE outer
      product = i * j
      WRITE (*,*) i, ' * ', j, ' = ', product
   END DO inner
END DO outer
END PROGRAM test_cycle_2
```

The resulting output values are

```
1 *   1 =   1
2 *   1 =   2
3 *   1 =   3
```

You should always use loop names with CYCLE or EXIT statements in nested loops to make sure that the proper loop is affected by the statements.

Good Programming Practice
Use loop names with `CYCLE` or `EXIT` statements in nested loops to make sure that the proper loop is affected by the statements.

Nesting loops within IF constructs and vice versa

It is possible to nest loops within block `IF` constructs or block `IF` constructs within loops. If a loop is nested within a block `IF` construct, *the loop must lie entirely within a single code block* of the `IF` construct. For example, the following statements are illegal since the loop stretches between the `IF` and the `ELSE` code blocks of the `IF` construct.

```
outer: IF ( a < b ) THEN
   ...
   inner: DO i = 1, 3
      ...
ELSE
   ...
   END DO inner
   ...
END IF outer
```

In contrast, the following statements are legal, since the loop lies entirely within a single code block of the `IF` construct.

```
outer: IF ( a < b ) THEN
   ...
   inner: DO i = 1, 3
      ...
   END DO inner
   ...
ELSE
   ...
END IF outer
```

Exiting from loops inside nested structures

In Fortran 2008 and later, the `EXIT` statement can exit to any label on *any* structure that contains the `DO` loop. For example, in the code below, when `i` is equal to 3, execution will transfer to the first statement after the end of the `IF` structure.[1]

```
if1: IF ( i > 0 ) THEN
   loop_1: DO i = 1, 5
      IF ( i == 3 ) EXIT if1
      WRITE (*,*) i
   END DO loop_1

ELSE if1
   ...

END IF if1
```

---

[1] At the time of writing, this feature has not been fully implemented on all common Fortran compilers.

Quiz 4-1

This quiz provides a quick check to see if you have understood the concepts introduced in Section 4.1. If you have trouble with the quiz, reread the section, ask your instructor, or discuss the material with a fellow student. The answers to this quiz are found in the back of the book.

Examine the following DO loops and determine how many times each loop will be executed. Assume that all of the index variables shown are of type integer.

1. `DO index = 5, 10`
2. `DO j = 7, 10, -1`
3. `DO index = 1, 10, 10`
4. `DO loop_counter = -2, 10, 2`
5. `DO time = -5, -10, -1`
6. `DO i = -10, -7, -3`

Examine the following loops and determine the value in ires at the end of each of the loops. Assume that ires, incr, and all index variables are integers.

7. ```
   ires = 0
   DO index = 1, 10
      ires = ires + 1
   END DO
   ```

8. ```
   ires = 0
   DO index = 1, 10
      ires = ires + index
   END DO
   ```

9. ```
   ires = 0
   DO index = 1, 10
      IF ( ires == 10 ) CYCLE
      ires = ires + index
   END DO
   ```

10. ```
    ires = 0
    DO index1 = 1, 10
       DO index2 = 1, 10
          ires = ires + 1
       END DO
    END DO
    ```

11. ```
    ires = 0
    DO index1 = 1, 10
       DO index2 = index1, 10
          IF ( index2 > 6 ) EXIT
          ires = ires + 1
       END DO
    END DO
    ```

(*continued*)
    ```

(*concluded*)

Examine the following Fortran statements and tell whether or not they are valid. If they are invalid, indicate the reason why they are invalid.

12. ```
    loop1: DO i = 1, 10
       loop2: DO j = 1, 10
          loop3: DO i = i, j
             ...
          END DO loop3
       END DO loop2
    END DO loop1
    ```

13. ```
    loop1: DO i = 1, 10
       loop2: DO j = i, 10
          loop3: DO k = i, j
             ...
          END DO loop3
       END DO loop2
    END DO loop1
    ```

14. ```
    loopx: DO i = 1, 10
       ...
       loopy: DO j = 1, 10
          ...
       END DO loopx
    END DO loopy
    ```

4.2

CHARACTER ASSIGNMENTS AND CHARACTER MANIPULATIONS

Character data can be manipulated using **character expressions.** A character expression can be any combination of valid character constants, character variables, character operators, and character functions. A **character operator** is an operator on character data that yields a character result. There are two basic types of character operators: **substring specifications** and **concatenation.** Character functions are functions that yield a character result.

4.2.1 Character Assignments

A character expression may be assigned to a character variable with an assignment statement. If the character expression is *shorter* than the length of the character variable to which it is assigned, then the rest of the variable is padded out with blanks. For example, the statements

```
CHARACTER(len=3) :: file_ext
file_ext = 'f'
```
```

Loops and Character Manipulation

store the value `'f`∅`'` into variable file_ext (∅ denotes a blank character). If the character expression is *longer* than the length of the character variable to which it is assigned, then the excess portion of the character variable is discarded. For example, the statements

```
CHARACTER(len=3) :: file_ext_2
file_extent_2 = 'FILE01'
```

will store the value `'FIL'` into variable file_ext_2, and the characters `'E01'` are discarded.

4.2.2 Substring Specifications

A **substring specification** selects a portion of a character variable, and treats that portion as if it were an independent character variable. For example, if the variable str1 is a six-character variable containing the string `'123456'`, then the substring str1(2:4) would be a three-character variable containing the string `'234'`. The substring str1(2:4) really refers to the same memory locations as characters 2 through 4 of str1, so if the contents of str1(2:4) are changed, the characters in the middle of variable str1 will also be changed.

A character substring is denoted by placing integer values representing the starting and ending character numbers separated by a colon in parentheses following the variable name. If the ending character number is less than the starting number, a zero-length character string will be produced.

The following example illustrates the use of substrings.

EXAMPLE 4-5

What will the contents of variables a, b, and c be at the end of the following program?

```
PROGRAM test_char1
CHARACTER(len=8) :: a, b, c
a = 'ABCDEFGHIJ'
b = '12345678'
c = a(5:7)
b(7:8) = a(2:6)
END PROGRAM test_char1
```

SOLUTION
The character manipulations in this program are:

1. Line 3 assigns the string `'ABCDEFGHIJ'` to a, but only the first eight characters are saved since a is only eight characters long. Therefore, a will contain `'ABCDEFGH'`.
2. Line 4 statement assigns the string `'12345678'` to b.
3. Line 5 assigns the character substring a(5:7) to c. Since c is eight characters long, five blanks will be padded onto variable c, and c will contain `'EFG`∅∅∅∅∅`'`.
4. Line 6 assigns substring a(2:6) to substring b(7:8). Since b(7:8) is only two characters long, only the first two characters of a(2:6) will be used. Therefore, variable b will contain `'123456BC'`.

4.2.3 The Concatenation (//) Operator

It is possible to combine two or more strings or substrings into a single large string. This operation is known as **concatenation.** The concatenation operator in Fortran is a double slash with no space between the slashes (//). For example, after the following lines are executed,

```
PROGRAM test_char2
CHARACTER(len=10) :: a
CHARACTER(len=8) :: b, c
a = 'ABCDEFGHIJ'
b = '12345678'
c = a(1:3) // b(4:5) // a(6:8)
END PROGRAM test_char2
```

variable c will contain the string 'ABC45FGH'.

4.2.4 Relational Operators with Character Data

Character strings can be compared in logical expressions using the relational operators ==, /=, <, <=, >, and >=. The result of the comparison is a logical value that is either true or false. For instance, the expression '123' == '123' is true, while the expression '123' == '1234' is false. In standard Fortran, character strings may be compared with character strings, and numbers may be compared with numbers, but *character strings may not be compared to numbers.*

How are two characters compared to determine if one is greater than the other? The comparison is based on the **collating sequence** of the characters on the computer where the program is being executed. The collating sequence of the characters is the order in which they occur within a specific character set. For example, the character 'A' is character number 65 in the ASCII character set, while the character 'B' is character number 66 in the set (see Appendix A). Therefore, the logical expression 'A' < 'B' is true in the ASCII character set. On the other hand, the character 'a' is character number 97 in the ASCII set, so 'a' < 'A' is false in the ASCII character set. Note that during character comparisons, a lowercase letter is different than the corresponding uppercase letter.

How are two strings compared to determine if one is greater than the other? The comparison begins with the first character in each string. If they are the same, then the second two characters are compared. This process continues until the first difference is found between the strings. For example, 'AAAAAB' > 'AAAAAA'.

What happens if the strings are of different lengths? The comparison begins with the first letter in each string, and progresses through each letter until a difference is found. If the two strings are the same all the way to the end of one of them, then the other string is considered the larger of the two. Therefore,

'AB' > 'AAAA'   and   'AAAAA' > 'AAAA'

Loops and Character Manipulation

TABLE 4-1
Some common character intrinsic functions

Function name and argument(s)	Argument types	Result type	Comments
ACHAR(ival)	INT	CHAR	Returns the character corresponding to ival in the ASCII collating sequence.
IACHAR(char)	CHAR	INT	Returns the integer corresponding to char in the ASCII collating sequence.
LEN(str1)	CHAR	INT	Returns length of str1 in characters.
LEN_TRIM(str1)	CHAR	INT	Returns length of str1, excluding any trailing blanks.
TRIM(str1)	CHAR	CHAR	Returns str1 with trailing blanks removed.

4.2.5 Character Intrinsic Functions

A few common character intrinsic functions are listed in Table 4-1. Function IACHAR(c) accepts a single character c, and returns the integer corresponding to its position in the ASCII character set. For example, the function IACHAR('A') returns the integer 65, because 'A' is the 65th character in the ASCII character set.

Function ACHAR(i) accepts an integer value i, and returns the character at that position in the ASCII character set. For example, the function ACHAR(65) returns the character 'A', because 'A' is the 65th character in the ASCII character set.

Function LEN(str) and LEN_TRIM(str) return the length of the specified character string. Function LEN(str) returns the length including any trailing blanks, while function LEN_TRIM(str) returns the string with any trailing blanks stripped off.

Function TRIM(str) accepts a character string, and returns the string with any trailing blanks stripped off.

Quiz 4-2

This quiz provides a quick check to see if you have understood the concepts introduced in Section 4.2. If you have trouble with the quiz, reread the sections, ask your instructor, or discuss the material with a fellow student. The answers to this quiz are found in the back of the book.

1. Assume that a computer uses the ASCII character set. Is each of the following expressions legal or illegal? If an expression is legal, what will its result be? (Note that ᑭ denotes a blank character.)

    (a) 'AAA' >= 'aaa'
    (b) '1A' < 'A1'
    (c) 'Helloᑭᑭᑭ ' // 'there'
    (d) TRIM('Helloᑭᑭᑭ ') // 'there'

(continued)

(*concluded*)

2. Suppose that character variables `str1`, `str2`, and `str3` contain the values `'abc'`, `'abcd'`, `'ABC'`, respectively, and that a computer uses the ASCII character set. Is each of the following expressions legal or illegal? If an expression is legal, what will its result be?

   (*a*) `str2(2:4)`
   (*b*) `str3 // str2(4:4)`
   (*c*) `str1 > str2`
   (*d*) `str1 > str3`
   (*e*) `str2 > 0`
   (*f*) `IACHAR('C') == 67`
   (*g*) `'Z' >= ACHAR(100)`

3. What is written out by each of the `WRITE` statements below?

```
PROGRAM test_char
CHARACTER(len=10) :: str1 = 'Hello'
CHARACTER(len=10) :: str2 = 'World'
CHARACTER(len=20) :: str3
str3 = str1 // str2
WRITE (*,*) LEN(str3)
WRITE (*,*) LEN_TRIM(str3)
str3 = TRIM(str1) // TRIM(str2)
WRITE (*,*) LEN(str3)
WRITE (*,*) LEN_TRIM(str3)
END PROGRAM test_char
```

EXAMPLE 4-6

Shifting Strings to Uppercase:

As we learned in this chapter, uppercase and lowercase letters are different inside strings. This difference between upper- and lowercase letters can cause a problem when we are attempting to match or compare two character strings, since `'STRING'` is not the same as `'string'` or `'String'`. If we wanted to compare two strings to see if they contained the same words, we would not get the correct answer if the capitalization of the words differed.

When making comparisons, it is often desirable to shift all characters to uppercase, so that identical strings will always match. Write a program that accepts two strings from a user, and compares them to determine if they are equal, ignoring case. To do the comparison, convert a copy of each string to uppercase, and compare the copies. Tell the user whether or not the two strings are the same.

SOLUTION
We will assume that the computer executing the program uses the ASCII character set, or a superset of it such as ISO 8859 or ISO 10646 (Unicode).

Loops and Character Manipulation

Appendix A shows the ASCII collating sequence. If we look at Appendix A, we can see that *there is a fixed offset of 32 characters* between an uppercase letter and the corresponding lowercase letter in each collating sequence. All letters are in order, and there are no nonalphabetic characters mixed into the middle of the alphabet.

1. **State the problem.**

   Write a program that reads two character strings, converts all of the lowercase letters in a copy of each character string to uppercase, and compares the strings for equality. The conversion process should not affect numeric and special characters. The program should write out a message indicating whether the two strings are equal or not, ignoring case.

2. **Define the inputs and outputs.**

   The inputs to the program are two strings str1 and str2. The output from the program is a message stating whether or not the two strings are equal, ignoring case.

3. **Describe the algorithm.**

   Looking at the ASCII table in Appendix A, we note that the uppercase letters begin at sequence number 65, while the lowercase letters begin at sequence number 97. There are exactly 32 numbers between each uppercase letter and its lowercase equivalent. Furthermore, there are no other symbols mixed into the middle of the alphabet.

   These facts give us our basic algorithm for shifting strings to uppercase. We will determine if a character is lowercase by deciding if it is between 'a' and 'z' in the ASCII character set. If it is, then we will subtract 32 from its sequence number to convert it to uppercase using the ACHAR and IACHAR functions. The initial pseudocode for this algorithm is

```
Prompt for str1 and str2
READ str1, str2

Make a copy of str1 and str2 in str1a and str2a
DO for each character in str1
   Determine if character is lowercase. If so,
      Convert to integer form
      Subtract 32 from the integer
      Convert back to character form
   End of IF
END of DO
DO for each character in str2
   Determine if character is lowercase. If so,
      Convert to integer form
      Subtract 32 from the integer
      Convert back to character form
   End of IF
END of DO

Compare shifted strings
Write out results
```

The final pseudocode for this program is

```
Prompt for str1 and str2
READ str1, str2

str1a ← str1
str2a ← str2

DO for i = 1 to LEN(str1a)
   IF str1a(i:i) >= 'a') .AND. str1a(i:i) <= 'z' THEN
      str1a(i:i) ← ACHAR ( IACHAR (str1a(i:i) - 32 ) )
   END of IF
END of DO

DO for i = 1 to LEN(str2a)
   IF str2a(i:i) >= 'a') .AND. str2a(i:i) <= 'z' THEN
      str2a(i:i) ← ACHAR ( IACHAR (str2a(i:i) - 32 ) )
   END of IF
END of DO

IF str1a == str2a
   WRITE that the strings are equal
ELSE
   WRITE that the strings are not equal
END IF
```

where length is the length of the input character string.

4. **Turn the algorithm into Fortran statements.**
   The resulting Fortran program is shown in Figure 4-10.

FIGURE 4-10
Program compare.

```
PROGRAM compare
!
!  Purpose:
!    To compare two strings to see if they are equivalent,
!    ignoring case.
!
!  Record of revisions:
!      Date         Programmer         Description of change
!      ====         ==========         =====================
!    11/14/15     S. J. Chapman        Original code
!
IMPLICIT NONE

! Data dictionary: declare variable types, definitions, & units
INTEGER :: i                   ! Loop index
CHARACTER(len=20) :: str1      ! First string to compare
CHARACTER(len=20) :: str1a     ! Copy of first string to compare
CHARACTER(len=20) :: str2      ! Second string to compare
CHARACTER(len=20) :: str2a     ! Copy of second string to compare
```

(*continued*)

(*concluded*)

```
! Prompt for the strings
WRITE (*,*) 'Enter first string to compare:'
READ (*,*) str1
WRITE (*,*) 'Enter second string to compare:'
READ (*,*) str2

! Make copies so that the original strings are not modified
str1a = str1
str2a = str2

! Now shift lowercase letters to uppercase.
DO i = 1, LEN(str1a)
   IF ( str1a(i:i) >= 'a' .AND. str1a(i:i) <= 'z' ) THEN
      str1a(i:i) = ACHAR ( IACHAR ( str1a(i:i) ) - 32 )
   END IF
END DO
DO i = 1, LEN(str2a)
   IF ( str2a(i:i) >= 'a' .AND. str2a(i:i) <= 'z' ) THEN
      str2a(i:i) = ACHAR ( IACHAR ( str2a(i:i) ) - 32 )
   END IF
END DO

! Compare strings and write result
IF ( str1a == str2a ) THEN
   WRITE (*,*) "'", str1, "' = '", str2, "' ignoring case."
ELSE
   WRITE (*,*) "'", str1, "' /= '", str2, "' ignoring case."
END IF

END PROGRAM compare
```

5. **Test the resulting Fortran program.**

We will test this program by passing it two pairs of strings to compare. One pair is identical except for case, and the other pair is not. The results from the program for two sets of input strings are:

```
C:\book\fortran\chap4>compare
Enter first string to compare:
'This is a test.'
Enter second string to compare:
'THIS IS A TEST.'
'This is a test.   ' = 'THIS IS A TEST.   ' ignoring case.

C:\book\fortran\chap4>compare
Enter first string to compare:
'This is a test.'
Enter second string to compare:
'This is another test.'
'This is a test.   ' /= 'This is another test' ignoring case.
```

The program appears to be working correctly.

EXAMPLE 4-7

Physics—The Flight of a Ball:

If we assume negligible air friction and ignore the curvature of the Earth, a ball that is thrown into the air from any point on the Earth's surface will follow a parabolic flight path (see Figure 4-11a). The height of the ball at any time t after it is thrown is given by Equation (4-5)

$$y(t) = y_0 + v_{y0}t + \frac{1}{2}gt^2 \qquad (4\text{-}5)$$

where y_0 is the initial height of the object above the ground, v_{y0} is the initial vertical velocity of the object, and g is the acceleration due to the Earth's gravity. The horizontal distance (range) traveled by the ball as a function of time after it is thrown is given by Equation (4-6)

$$x(t) = x_0 + v_{x0}t \qquad (4\text{-}6)$$

where x_0 is the initial horizontal position of the ball on the ground, and v_{x0} is the initial horizontal velocity of the ball.

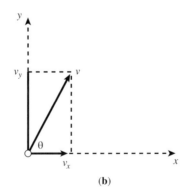

FIGURE 4-11
(a) When a ball is thrown upward, it follows a parabolic trajectory. (b) The horizontal and vertical components of a velocity vector v at an angle θ with respect to the horizontal.

Loops and Character Manipulation

If the ball is thrown with some initial velocity v_0 at an angle of θ degrees with respect to the Earth's surface, then the initial horizontal and vertical components of velocity will be

$$v_{x0} = v_0 \cos \theta \qquad (4\text{-}7)$$

$$v_{y0} = v_0 \sin \theta \qquad (4\text{-}8)$$

Assume that the ball is initially thrown from position $(x_0, y_0) = (0,0)$ with an initial velocity v of 20 m/s at an initial angle of θ degrees. Design, write, and test a program that will determine the horizontal distance traveled by the ball from the time it was thrown until it touches the ground again. The program should do this for all angles θ from 0° to 90° in 1° steps. Determine the angle θ that maximizes the range of the ball.

SOLUTION

In order to solve this problem, we must determine an equation for the range of the thrown ball. We can do this by first finding the time that the ball remains in the air, and then finding the horizontal distance that the ball can travel during that time.

The time that the ball will remain in the air after it is thrown may be calculated from Equation (4.5). The ball will touch the ground at the time t for which $y(t) = 0$. Remembering that the ball will start from ground level ($y(0) = 0$), and solving for t, we get:

$$y(t) = y_o + v_{yo}t + \frac{1}{2}gt^2 \qquad (4\text{-}5)$$

$$0 = 0 + v_{yo}t + \frac{1}{2}gt^2$$

$$0 = \left(v_{yo} + \frac{1}{2}gt\right)t$$

so the ball will be at ground level at time $t_1 = 0$ (when we threw it), and at time

$$t_2 = -\frac{2v_{y0}}{g} \qquad (4\text{-}9)$$

The horizontal distance that the ball will travel in time t_2 is found using Equation (4-6):

$$\text{Range} = x(t_2) = x_o + v_{xo}t_2 \qquad (4\text{-}6)$$

$$\text{Range} = 0 + v_{x0}\left(-\frac{2v_{y0}}{g}\right)$$

$$\text{Range} = -\frac{2v_{x0}v_{y0}}{g}$$

We can substitute Equations (4.7) and (4.8) for v_{xo} and v_{yo} to get an equation expressed in terms of the initial velocity v and initial angle θ:

$$\text{Range} = -\frac{2(v_0 \cos \theta)(v_0 \sin \theta)}{g}$$

$$\text{Range} = -\frac{2v_0^2}{g} \cos \theta \sin \theta \tag{4-10}$$

From the problem statement, we know that the initial velocity v_0 is 20 m/s, and that the ball will be thrown at all angles from 0° to 90° in 1° steps. Finally, any elementary physics textbook will tell us that the acceleration due to the Earth's gravity is -9.81 m/s².

Now let's apply our design technique to this problem.

1. **State the problem.**

    A proper statement of this problem would be: *Calculate the range that a ball would travel when it is thrown with an initial velocity of v_0 at an initial angle θ. Calculate this range for a v_o of 20 m/s and all angles between 0° and 90°, in 1° increments. Determine the angle θ that will result in the maximum range for the ball. Assume that there is no air friction.*

2. **Define the inputs and outputs.**

    As the problem is defined above, no inputs are required. We know from the problem statement what v_0 and θ will be, so there is no need to read them in. The outputs from this program will be a table showing the range of the ball for each angle θ, and the angle θ for which the range is maximum.

3. **Design the algorithm.**

    This program can be broken down into the following major steps:

    ```
    DO for theta = 0 to 90 degrees
       Calculate the range of the ball for each angle theta
       Determine if this theta yields the maximum range so far
       Write out the range as a function of theta
    END of DO
    WRITE out the theta yielding maximum range
    ```

    An iterative DO loop is appropriate for this algorithm, since we are calculating the range of the ball for a specified number of angles. We will calculate the range for each value of θ, and compare each range with the maximum range found so far to determine which angle yields the maximum range. Note that the trigonometric functions work in radians, so the angles in degrees must be converted to radians before the range is calculated. The detailed pseudocode for this algorithm is

    ```
    Initialize max_range and max_degrees to 0
    Initialize v0 to 20 meters/second
    DO for theta = 0 to 90 degrees
       radian ← theta * degrees_2_rad      (Convert degrees to radians)
       angle ← (-2. * v0**2 / gravity ) * sin(radian) * cos(radian)
       Write out theta and range
    ```

Loops and Character Manipulation

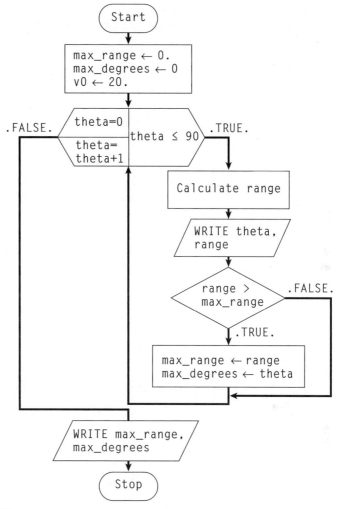

FIGURE 4-12
Flowchart for a program to determine the angle θ at which a ball thrown with an initial velocity v_0 of 20 m/s will travel the farthest.

```
    IF range > max_range then
       max_range   ← range
       max_degrees ← theta
    END of IF
END of DO
Write out max_degrees, max_range
```

The flowchart for this program is shown in Figure 4-12.

4. **Turn the algorithm into Fortran statements.**
   The final Fortran program is shown in Figure 4-13.

FIGURE 4-13
Program `ball` to determine the angle that maximizes the range of a thrown ball.

```
PROGRAM ball
!
!  Purpose:
!    To calculate distance traveled by a ball thrown at a specified
!    angle THETA and at a specified velocity V0 from a point on the
!    surface of the earth, ignoring the effects of air friction and
!    the earth's curvature.
!
!  Record of revisions:
!      Date       Programmer        Description of change
!      ====       ==========        =====================
!    11/14/15    S. J. Chapman       Original code
!
IMPLICIT NONE

! Data dictionary: declare constants
REAL, PARAMETER :: DEGREES_2_RAD = 0.01745329 ! Deg ==> rad conv.
REAL, PARAMETER :: GRAVITY = -9.81            ! Accel. due to gravity (m/s)

! Data dictionary: declare variable types, definitions, & units
INTEGER :: max_degrees   ! angle at which the max rng occurs (degrees)
REAL :: max_range        ! Maximum range for the ball at vel v0 (meters)
REAL :: range            ! Range of the ball at a particular angle (meters)
REAL :: radian           ! Angle at which the ball was thrown (in radians)
INTEGER :: theta         ! Angle at which the ball was thrown (in degrees)
REAL :: v0               ! Velocity of the ball (in m/s)

! Initialize variables.
max_range = 0.
max_degrees = 0
v0 = 20.

! Loop over all specified angles.

loop: DO theta = 0, 90

   ! Get angle in radians
   radian = real(theta) * DEGREES_2_RAD

   ! Calculate range in meters.
   range = (-2. * v0**2 / GRAVITY) * SIN(radian) * COS(radian)

   ! Write out the range for this angle.
   WRITE (*,*) 'Theta = ', theta, ' degrees; Range = ', range, &
               ' meters'

   ! Compare the range to the previous maximum range.  If this
   ! range is larger, save it and the angle at which it occurred.
   IF ( range > max_range ) THEN
      max_range = range
      max_degrees = theta
```

(*continued*)

Loops and Character Manipulation

(concluded)

```
        END IF

     END DO loop

     ! Skip a line, and then write out the maximum range and the angle
     ! at which it occurred.
     WRITE (*,*) ' '
     WRITE (*,*) 'Max range = ', max_range, ' at ', max_degrees, ' degrees'
     END PROGRAM ball
```

The degrees-to-radians conversion factor is always a constant, so in the program it is given a name using the PARAMETER attribute, and all references to the constant within the program use that name. The acceleration due to gravity at sea level can be found in any physics text. It is about 9.81 m/sec², directed downward.

5. Test the program.

To test this program, we will calculate the answers by hand for a few of the angles, and compare the results with the output of the program.

θ	$v_{xo} = v_o \cos\theta$	$v_{yo} = v_o \sin\theta$	$t_2 = -\dfrac{2v_{yo}}{g}$	$x = v_{xo}t_2$
0°	20 m/s	0 m/s	0 s	0 m
5°	19.92 m/s	1.74 m/s	0.355 s	7.08 m
40°	15.32 m/s	12.86 m/s	2.621 s	40.15 m
45°	14.14 m/s	14.14 m/s	2.883 s	40.77 m

When program ball is executed, a 90-line table of angles and ranges is produced. To save space, only a portion of the table is reproduced below.

```
C:\book\fortran\chap4>ball
     Theta =      0 degrees; Range =       0.000000E+00 meters
     Theta =      1 degrees; Range =           1.423017 meters
     Theta =      2 degrees; Range =           2.844300 meters
     Theta =      3 degrees; Range =           4.262118 meters
     Theta =      4 degrees; Range =           5.674743 meters
     Theta =      5 degrees; Range =           7.080455 meters
          ...
     Theta =     40 degrees; Range =          40.155260 meters
     Theta =     41 degrees; Range =          40.377900 meters
     Theta =     42 degrees; Range =          40.551350 meters
     Theta =     43 degrees; Range =          40.675390 meters
     Theta =     44 degrees; Range =          40.749880 meters
     Theta =     45 degrees; Range =          40.774720 meters
     Theta =     46 degrees; Range =          40.749880 meters
     Theta =     47 degrees; Range =          40.675390 meters
     Theta =     48 degrees; Range =          40.551350 meters
```

```
Theta =      49 degrees; Range =      40.377900 meters
Theta =      50 degrees; Range =      40.155260 meters
   ...
Theta =      85 degrees; Range =       7.080470 meters
Theta =      86 degrees; Range =       5.674757 meters
Theta =      87 degrees; Range =       4.262130 meters
Theta =      88 degrees; Range =       2.844310 meters
Theta =      89 degrees; Range =       1.423035 meters
Theta =      90 degrees; Range =       1.587826E-05 meters

Max range =      40.774720 at     45 degrees
```

The program output matches our hand calculation for the angles calculated above to the four-digit accuracy of the hand calculation. Note that the maximum range occurred at an angle of 45°.

4.3
DEBUGGING FORTRAN LOOPS

The best approach to locating an error in a program containing loops is to use a symbolic debugger, if one is supplied with your compiler. You must ask your instructor or else check with your system's manuals to determine how to use the symbolic debugger supplied with your particular compiler.

An alternate approach to locating the error is to insert WRITE statements into the code to print out important variables at key points in the program. When the program is run, the WRITE statements will print out the values of the key variables. These values can be compared to the ones you expect, and the places where the actual and expected values differ will serve as a clue to help you locate the problem. For example, to verify the operation of a counting loop, the following WRITE statements could be added to the program.

```
      WRITE (*,*) 'At loop1: ist, ien, inc = ', ist, ien, inc
loop1: DO i = ist, ien, inc
         WRITE (*,*) 'In loop1: i = ', i
         ...
      END DO loop1
      WRITE (*,*) 'loop1 completed'
```

When the program is executed, its output listing will contain detailed information about the variables controlling the DO loop and just how many times the loop was executed.

Once you have located the portion of the code in which the error occurs, you can take a look at the specific statements in that area to locate the problem. A list of some common errors is given below. Be sure to check for them in your code.

1. *Most errors in counting* DO *loops involve mistakes with the loop parameters.* If you add WRITE statements to the DO loop as shown above, the problem should be fairly clear. Did the DO loop start with the correct value? Did it end

with the correct value? Did it increment at the proper step? If not, check the parameters of the `DO` loop closely. You will probably spot an error in the control parameters.
2. Errors in while loops are usually related to errors in the logical expression used to control their function. These errors may be detected by examining the `IF (logical_expr) EXIT` statement of the while loop with `WRITE` statements.

4.4
SUMMARY

In this chapter, we have presented the basic types of Fortran loops, plus some additional details about manipulating character data.

There are two basic types of loops in Fortran, the while loop and the iterative or counting `DO` loop. The while loop is used to repeat a section of code in cases where we do not know in advance how many times the loop must be repeated. The counting `DO` loop is used to repeat a section of code in cases where we know in advance how many times the loop should be repeated.

It is possible to exit from a loop at any time using the `EXIT` statement. It is also possible to jump back to the top of a loop using the `CYCLE` statement. If loops are nested, an `EXIT` or `CYCLE` statement refers by default to the innermost loop.

4.4.1 Summary of Good Programming Practice

The following guidelines should be adhered to when programming with branch or loop constructs. By following them consistently, your code will contain fewer bugs, will be easier to debug, and will be more understandable to others who may need to work with it in the future.

1. Always indent code blocks in `DO` loops to make them more readable.
2. Use a while loop to repeat sections of code when you don't know in advance how often the loop will be executed.
3. Make sure that there is only one exit from a while loop.
4. Use a counting `DO` loop to repeat sections of code when you know in advance how often the loop will be executed.
5. Never attempt to modify the values of `DO` loop index while inside the loop.
6. Assign names to large and complicated loops or `IF` constructs, especially if they are nested.
7. Use loop names with `CYCLE` and `EXIT` statements in nested loops to make certain that the proper loop is affected by the action of the `CYCLE` or `EXIT` statement.

4.4.2 Summary of Fortran Statements and Constructs

The following summary describes the Fortran statements and constructs introduced in this chapter.

CYCLE Statement:

                CYCLE [name]

Example:

                CYCLE inner

Description:
The CYCLE statement may appear within any DO loop. When the statement is executed, all of the statements below it within the loop are skipped, and control returns to the top of the loop. In while loops, execution resumes from the top of the loop. In counting loops, the loop index is incremented, and if the index is still less than its limit, execution resumes from the top of the loop.

An unnamed CYCLE statement always causes the *innermost* loop containing the statement to cycle. A named CYCLE statement causes the named loop to cycle, even if it is not the innermost loop.

DO Loop (Iterative or Counting Loop) Construct:

                [name:] DO index = istart, iend, incr
                    ...
                END DO [name]

Example:

                loop: DO index = 1, last_value, 3
                    ...
                END DO loop

Description:
The iterative DO loop is used to repeat a block of code a known number of times. During the first iteration of the DO loop, the variable *index* is set to the value *istart*. index is incremented by incr in each successive loop until its *index*incr* > *iend*incr*, at which time the loop terminates. The loop name is optional, but if it is used on the DO statement, then it must be used on the END DO statement. The loop variable index is incremented and tested *before* each loop, so the DO loop code will never be executed at all if *istart*incr* > *iend*incr*.

EXIT Statement:

                EXIT [name]

Example:

                EXIT loop1

Description:
The EXIT statement may appear within any DO loop. When an EXIT statement is encountered, the program stops executing the loop and jumps to the first executable statement after the END DO.

An unnamed EXIT statement always causes the *innermost* loop containing the statement to exit. A named EXIT statement causes the named loop to exit, even if it is not the innermost loop.

Loops and Character Manipulation

WHILE Loop Construct:

```
[name:] DO
    ...
    IF ( logical_expr ) EXIT [name]
    ...
END DO [name]
```

Example:

```
loop1: DO
    ...
    IF ( istatus /= 0 ) EXIT loop1
    ...
END DO loop1
```

Description:
The while loop is used to repeat a block of code until a specified `logical_expr` becomes true. It differs from a counting DO loop in that we do not know in advance how many times the loop will be repeated. When the IF statement of the loop is executed with the `logical_expr` true, execution skips to the next statement following the end of the loop.

The name of the loop is optional, but if a name is included on the DO statement, then the same name must appear on the END DO statement. The name on the EXIT statement is optional; it may be left out even if the DO and END DO are named.

4.4.3 Exercises

4-1. Which of the following expressions are legal in Fortran? If an expression is legal, evaluate it. Assume the ASCII collating sequence.

(a) `'123' > 'abc'`

(b) `'9478' == 9478`

(c) `ACHAR(65) // ACHAR(95) // ACHAR(72)`

(d) `ACHAR(IACHAR('j') + 5)`

4-2. Write the Fortran statements required to calculate and print out the squares of all the even integers between 0 and 50.

4-3. Write a Fortran program to evaluate the equation $y(x) = x^2 - 3x + 2$ for all values of x between -1 and 3, in steps of 0.1.

4-4. Write the Fortran statements required to calculate $y(t)$ from the equation:

$$y(t) = \begin{cases} -3t^2 + 5 & t \geq 0 \\ 3t^2 + 5 & t < 0 \end{cases}$$

4-5. Write a Fortran program to calculate the factorial function, as defined in Example 4-2. Be sure to handle the special cases of 0! and of illegal input values.

4-6. What is the difference in behavior between a `CYCLE` statement and an `EXIT` statement?

4-7. Modify program `stats_2` to use the `DO WHILE` construct instead of the while construct currently in the program.

4-8. Examine the following `DO` statements and determine how many times each loop will be executed. (Assume that all loop index variables are integers.)

(a) `DO irange = -32768, 32767`

(b) `DO j = 100, 1, -10`

(c) `DO kount = 2, 3, 4`

(d) `DO index = -4, -7`

(e) `DO i = -10, 10, 10`

(f) `DO i = 10, -2, 0`

(g) `DO`

4-9. Examine the following iterative `DO` loops and determine the value of `ires` at the end of each of the loops, and also the number of times each loop executes. Assume that all variables are integers.

(a)
```
ires = 0
DO index = -10, 10
   ires = ires + 1
END DO
```

(b)
```
ires = 0
loop1: DO index1 = 1, 20, 5
   IF ( index1 <= 10 ) CYCLE
   loop2: DO index2 = index1, 20, 5
      ires = ires + index2
   END DO loop2
END DO loop1
```

(c)
```
ires = 0
loop1: DO index1 = 10, 4, -2
   loop2: DO index2 = 2, index1, 2
      IF ( index2 > 6 ) EXIT loop2
      ires = ires + index2
   END DO loop2
END DO loop1
```

(d)
```
ires = 0
loop1: DO index1 = 10, 4, -2
   loop2: DO index2 = 2, index1, 2
      IF ( index2 > 6 ) EXIT loop1
      ires = ires + index2
   END DO loop2
END DO loop1
```

4-10. Examine the following while loops and determine the value of `ires` at the end of each of the loops, and the number of times each loop executes. Assume that all variables are integers.

(a)
```
ires = 0
loop1: DO
    ires = ires + 1
    IF ( (ires / 10 ) * 10 == ires ) EXIT
END DO loop1
```

(b)
```
ires = 2
loop2: DO
    ires = ires**2
    IF ( ires > 200 ) EXIT
END DO loop2
```

(c)
```
ires = 2
DO WHILE ( ires > 200 )
    ires = ires**2
END DO
```

4-11. Modify program `ball` from Example 4-7 to read in the acceleration due to gravity at a particular location, and to calculate the maximum range of the ball for that acceleration. After modifying the program, run it with accelerations of -9.8 m/sec^2, -9.7 m/sec^2, and -9.6 m/sec^2. What effect does the reduction in gravitational attraction have on the range of the ball? What effect does the reduction in gravitational attraction have on the best angle θ at which to throw the ball?

4-12. Modify program `ball` from Example 4-7 to read in the initial velocity with which the ball is thrown. After modifying the program, run it with initial velocities of 10 m/sec, 20 m/sec, and 30 m/sec. What effect does changing the initial velocity v_o have on the range of the ball? What effect does it have on the best angle θ at which to throw the ball?

4-13. Program `doy` in Example 4-3 calculates the day of year associated with any given month, day, and year. As written, this program does not check to see if the data entered by the user is valid. It will accept nonsense values for months and days, and do calculations with them to produce meaningless results. Modify the program so that it checks the input values for validity before using them. If the inputs are invalid, the program should tell the user what is wrong, and quit. The year should be number greater than zero, the month should be a number between 1 and 12, and the day should be a number between 1 and a maximum that depends on the month. Use a `SELECT CASE` construct to implement the bounds checking performed on the day.

4-14. Write a Fortran program to evaluate the function

$$y(x) = \ln \frac{1}{1-x} \tag{4-11}$$

for any user-specified value of x, where ln is the natural logarithm (logarithm to the base e). Write the program with a while loop, so that the program repeats the calculation for each legal value of x entered into the program. When an illegal value of x is entered, terminate the program. (Note that values of $x \leq 1$ are illegal, because the natural log of a negative real number is not defined.)

4-15. Write a Fortran program to convert all uppercase characters in a user-supplied character string to lowercase, without changing the uppercase and nonalphabetic characters in the string. Assume that your computer uses the ASCII collating sequence.

4-16. Calculating Orbits When a satellite orbits the Earth, the satellite's orbit will form an ellipse with the Earth located at one of the focal points of the ellipse. The satellite's orbit can be expressed in polar coordinates as

$$r = \frac{p}{1 - \varepsilon \cos \theta} \tag{4-12}$$

where r and θ are the distance and angle of the satellite from the center of the Earth, p is a parameter specifying the size of the orbit, and ε is a parameter representing the eccentricity of the orbit. A circular orbit has an eccentricity ε of zero. An elliptical orbit has an eccentricity of $0 \leq \varepsilon \leq 1$. If $\varepsilon > 1$, the satellite follows a hyperbolic path and escapes from the Earth's gravitational field.

Consider a satellite with a size parameter $p = 1200$ km. Write a program to calculate the distance of the satellite from the center of the Earth as a function of θ if the satellite has an eccentricity of (a) $\varepsilon = 0$; (b) $\varepsilon = 0.25$; (c) $\varepsilon = 0.5$. Write a single program in which r and ε are both input values.

How close does each orbit come to the Earth? How far away does each orbit get from the Earth?

4-17. Write a program `caps` that reads in a character string, searches for all of the words within the string, and capitalizes the first letter of each word, while shifting the remainder of the word to lowercase. Assume that all nonalphabetic and nonnumeric characters can mark the boundaries of a word within the character variable (e.g., periods, commas, etc.). Nonalphabetic characters should be left unchanged.

4-18. Current through a Diode The current flowing through the semiconductor diode shown in Figure 4-14 is given by the equation

$$i_D = I_o \left(e^{\frac{qv_D}{kT}} - 1 \right) \tag{4-13}$$

where

v_D = the voltage across the diode, in volts
i_D = the current flow through the diode, in amperes
I_o = the leakage current of the diode, in amperes
q = the charge on an electron, 1.602×10^{-19} C
k = Boltzmann's constant, 1.38×10^{-23} J/K
T = temperature, in kelvins (K)

The leakage current I_o of the diode is 2.0 µA. Write a computer program to calculate the current flowing through this diode for all voltages from -1.0 V to $+0.6$ V, in 0.1 V steps. Repeat this process for the following temperatures: 75 °F, 100 °F, and 125 °F. Use the program of Example 2-3 to convert the temperatures from °F to kelvins.

FIGURE 4-14
A semiconductor diode.

4-19. Binary to Decimal Conversion Write a program that prompts a user for a binary number, which will be entered as a string of 0s and 1s in a character variable. For example, the user might enter 01000101 as a character string. The program should then convert the input binary number into a decimal number, and display the corresponding decimal number to the user.

This program should be able to handle numbers from 0000000000 to 1111111111, converting them into the equivalent decimal values 0 to 1023. It should also test for and handle an invalid value among the input characters (a letter, symbol, or a number greater than one). Test your program with the following binary numbers.

(a) 0010010010

(b) 1111111111

(c) 10000000001

(d) 01111111110

4-20. **Decimal to Binary Conversion** Write a program that prompts a user for a decimal integer in the range 0 to 1023, and converts the number into the equivalent binary number. The binary number should consist of 0s and 1s in a character string. The program should display the corresponding binary number to the user. Test your program with the following decimal numbers.

(a) 256

(b) 63

(c) 140

(d) 768

4-21. **Octal to Decimal Conversion** Write a program that prompts a user for an octal number, which will be entered as a string of 0s to 7s in a character variable. For example, the user might enter 377 as a character string. The program should then convert the input octal number into a decimal number, and display the corresponding decimal number to the user. Design the program to handle up to five octal digits. (*Hint:* This might be a great place for a SELECT CASE structure.) Test your program with the following binary numbers.

(a) 377

(b) 11111

(c) 70000

(d) 77777

4-22. **Fibonacci Numbers** The *n*th Fibonacci number is defined by the following recursive equations:

$$f(1) = 1$$
$$f(2) = 2 \qquad\qquad (4\text{-}14)$$
$$f(n) = f(n-1) + f(n-2)$$

Therefore, $f(3) = f(2) + f(1) = 2 + 1 = 3$, and so forth for higher numbers. Write a program to calculate and write out the *n*th Fibonacci number for $n > 2$, where n is input by the user. Use a while loop to perform the calculation.

4-23. **Tension on a Cable** A 200 kilogram object is to be hung from the end of a rigid 3-m horizontal pole of negligible weight, as shown in Figure 4-15. The pole is attached to a wall by a pivot and is supported by a 3-m cable that is attached to the wall at a higher point. The tension on this cable is given by the equation

$$T = \frac{W \cdot lc \cdot lp}{d\sqrt{lp^2 - d^2}} \qquad\qquad (4\text{-}15)$$

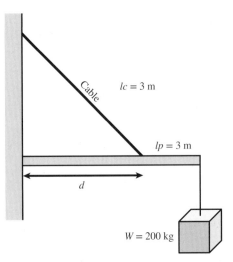

FIGURE 4-15
A 200 pound weight suspended from a rigid bar supported by a cable.

where T is the tension on the cable, W is the weight of the object, lc is the length of the cable, lp is the length of the pole, and d is the distance along the pole at which the cable is attached. Write a program to determine the distance d at which to attach the cable to the pole in order to minimize the tension on the cable. To do this, the program should calculate the tension on the cable at 0.1 m intervals from $d = 0.5$ m to $d = 2.8$ m, and should locate the position d that produces the minimum tension.

4-24. If the maximum tension on the cable in the previous exercise is 350, over what range of distances d is it safe to attach the cable to the pole?

4-25. Bacterial Growth Suppose that a biologist performs an experiment in which he or she measures the rate at which a specific type of bacterium reproduces asexually in different culture media. The experiment shows that in Medium A the bacteria reproduce once every 90 minutes, and in Medium B the bacteria reproduce once every 120 minutes. Assume that a single bacterium is placed on each culture medium at the beginning of the experiment. Write a Fortran program that calculates and writes out the number of bacteria present in each culture at intervals of 6 hours from the beginning of the experiment until 24 hours have elapsed. How do the numbers of bacteria compare on the two media after 24 hours?

4-26. Decibels Engineers often measure the ratio of two power measurements in *decibels*, or dB. The equation for the ratio of two power measurements in decibels is

$$\text{dB} = \log_{10}\frac{P_2}{P_1} \qquad (4\text{-}16)$$

where P_2 is the power level being measured, and P_1 is some reference power level. Assume that the reference power level P_1 is 1 W, and write a program that calculates the decibel level corresponding to power levels between 1 and 20W, in 0.5 W steps.

4-27. Infinite Series Trigonometric functions are usually calculated on computers by using a *truncated infinite series*. An *infinite series* is an infinite set of terms that together add up to the value of a particular function or expression. For example, one infinite series used to evaluate the sine of a number is

$$\sin x = x - \frac{x^3}{3!} + \frac{x^5}{5!} - \frac{x^7}{7!} + \frac{x^9}{9!} + \cdots \tag{4-17}$$

or

$$\sin x = \sum_{n=1}^{\infty} (-1)^{n-1} \frac{x^{2n-1}}{(2n-1)!} \tag{4-18}$$

where x is in units of radians.

Since a computer does not have enough time to add an infinite number of terms for every sine that is calculated, the infinite series is *truncated* after a finite number of terms. The number of terms that should be kept in the series is just enough to calculate the function to the precision of the floating-point numbers on the computer on which the function is being evaluated. The truncated infinite series for $\sin x$ is

$$\sin x = \sum_{n=1}^{N} (-1)^{n-1} \frac{x^{2n-1}}{(2n-1)!} \tag{4-19}$$

where N is the number of terms to retain in the series.

Write a Fortran program that reads in a value for x in degrees, and then calculates the sine of x using the sine intrinsic function. Next, calculate the sine of x using Equation (4.19), with $N = 1, 2, 3, \ldots, 10$. Compare the true value of $\sin x$ with the values calculated using the truncated infinite series. How many terms are required to calculate $\sin x$ to the full accuracy of your computer?

4-28. Geometric Mean The *geometric mean* of a set of numbers x_1 through x_n is defined as the nth root of the product of the numbers:

$$\text{geometric mean} = \sqrt[n]{x_1 x_2 x_3 \ldots x_n} \tag{4-20}$$

Write a Fortran program that will accept an arbitrary number of positive input values and calculate both the arithmetic mean (i.e., the average) and the geometric mean of the numbers. Use a while loop to get the input values, and terminate the inputs a user enters a negative number. Test your program by calculating the average and geometric mean of the four numbers 10, 5, 4, and 5.

4-29. RMS Average The *root-mean-square* (rms) *average* is another way of calculating a mean for a set of numbers. The rms average of a series of numbers is the square root of the arithmetic mean of the squares of the numbers:

$$\text{rms average} = \sqrt{\frac{1}{N} \sum_{i=1}^{N} x_i^2} \tag{4-21}$$

Write a Fortran program that will accept an arbitrary number of positive input values and calculate the rms average of the numbers. Prompt the user for the number of values to be entered, and use a DO loop to read in the numbers. Test your program by calculating the rms average of the four numbers 10, 5, 4, and 5.

4-30. Harmonic Mean The *harmonic mean* is yet another way of calculating a mean for a set of numbers. The harmonic mean of a set of numbers is given by the equation:

$$\text{harmonic mean} = \frac{N}{\frac{1}{x_1} + \frac{1}{x_2} + \frac{1}{x_3} + \cdots + \frac{1}{x_N}} \quad (4\text{-}22)$$

Write a Fortran program that will read in an arbitrary number of positive input values and calculate the harmonic mean of the numbers. Use any method that you desire to read in the input values. Test your program by calculating the harmonic mean of the four numbers 10, 5, 4, and 5.

4-31. Write a single Fortran program that calculates the arithmetic mean (average), rms average, geometric mean, and harmonic mean for a set of positive numbers. Use any method that you desire to read in the input values. Compare these values for each of the following sets of numbers:

(a) 4, 4, 4, 4, 4, 4, 4

(b) 5, 2, 3, 6, 3, 2, 6

(c) 4, 1, 4, 7, 4, 1, 7

(d) 1, 2, 3, 4, 5, 6, 7

4-32. Mean Time Between Failure Calculations The reliability of a piece of electronic equipment is usually measured in terms of Mean Time Between Failures (MTBF), where MTBF is the average time that the piece of equipment can operate before a failure occurs in it. For large systems containing many pieces of electronic equipment, it is customary to determine the MTBFs of each component, and to calculate the overall MTBF of the system from the failure rates of the individual components. If the system is structured like the one shown in Figure 4-16, every component must work in order for the whole system to work, and the overall system MTBF can be calculated as

$$\text{MBTF}_{sys} = \frac{1}{\frac{1}{\text{MTBF}_1} + \frac{1}{\text{MTBF}_2} + \cdots + \frac{1}{\text{MTBF}_N}} \quad (4\text{-}23)$$

Write a program that reads in the number of series components in a system and the MTBFs for each component, and then calculates the overall MTBF for the system. To

FIGURE 4-16
An electronic system containing three subsystems with known MTBFs.

test your program, determine the MTBF for a radar system consisting of an antenna subsystem with an MTBF of 2000 hours, a transmitter with an MTBF of 800 hours, a receiver with an MTBF of 3000 hours, and a computer with an MTBF of 5000 hours.

4-33. Ideal Gas Law An ideal gas is one in which all collisions between molecules are perfectly elastic. It is possible to think of the molecules in an ideal gas as perfectly hard billiard balls that collide and bounce off of each other without losing kinetic energy.

Such a gas can be characterized by three quantities: absolute pressure (P), volume (V), and absolute temperature (T). The relationship among these quantities in an ideal gas is known as the Ideal Gas Law:

$$PV = nRT \tag{4-24}$$

where P is the pressure of the gas in kilopascals (kPa), V is the volume of the gas in liters (L), n is the number of molecules of the gas in units of moles (mol), R is the universal gas constant (8.314 L · kPa/mol · K), and T is the absolute temperature in kelvins (K). (*Note:* 1 mol = 6.02×10^{23} molecules.)

Assume that a sample of an ideal gas contains 1 mole of molecules at a temperature of 273 K, and answer the following questions.

(*a*) Write a program to calculate and print out the volume of this gas as its pressure varies from 1 to 1001 kPa in steps of 100 kPa.

(*b*) Suppose that the temperature of the gas is increased to 300 K. How does the volume of this gas vary with pressure over the same range now?

4-34. Assume that the volume of 1 mole of an ideal gas has a fixed volume of 10 L, and calculate and print out the pressure of the gas as a function of temperature as the temperature is changed from 250 to 400 kelvins.

4-35. The Lever The lever (Figure 4-17) is the simplest possible machine. It is used to lift loads that would otherwise be too heavy to lift. If there is no friction, the relationship between the force applied to the lever and the weight that can be lifted is given by the equation

$$F_{APP} \times d_1 = Weight \times d_2 \tag{4-25}$$

where F_{APP} is the applied force in newtons, d_1 is the distance from the fulcrum to the point where the force is applied, d_2 is the distance from the fulcrum to the location of the load, and *Weight* is the weight (= downward force) of the load.

Assume that the applied force consists of weights that can be stacked onto one end of the lever. Write a program that will calculate weight required to lift a load of 600 kg if the distance d_2 from the fulcrum to the location of the load is fixed at 1 m, and the distance d_1 from the fulcrum to the point where the weights are applied varies from 0.5 m to 3.0 m in 0.1 m steps. Assuming that we only have 400 kg of weights available, what is the shortest distance d_1 that could be used in this lever?

FIGURE 4-17
A lever.

5

Basic I/O Concepts

OBJECTIVES

- Know how to use formatted `WRITE` statements to create neatly formatted output from a program.
- Learn how to use the I, F, E, ES, L, A, X, T, and / format descriptors.
- Know how to use formatted `READ` statements to read data into a program.
- Know how to open, read, write, navigate through, and close files.

In the previous chapters, we have read values into and written them out of our programs using list-directed `READ` and `WRITE` statements. List-directed I/O statements are said to be in **free format.** Free format is specified by the second asterisk in the `READ (*,*)` and `WRITE (*,*)` statements. As we saw, the results of writing out data in free format are not always pretty. There are often a large number of extra spaces in the output. In this chapter, we will learn how to write out data using **formats** that specify the exact way in which the numbers should be printed out.

Formats may be used either when writing or when reading data. Since they are most useful during output, we will examine formatted `WRITE` statements first, and postpone formatted `READ` statements until a later section in the chapter.

The second major topic introduced in this chapter is disk file processing. We will learn the basics of how to read from and write to disk files. Advanced disk file processing will be postponed to Chapter 14.

5.1
FORMATS AND FORMATTED WRITE STATEMENTS

A **format** may be used to specify the exact manner in which variables are to be printed out by a program. In general, a format can specify both the horizontal and the vertical position of the variables on the paper, and also the number of significant digits to be printed out. A typical formatted `WRITE` statement for an integer `i` and a real variable `result` is shown below:

```
WRITE (*,100) i, result
100 FORMAT (' The result for iteration ', I3,' is ', F7.3)
```

The FORMAT statement contains the formatting information used by the WRITE statement. The number 100 that appears within the parentheses in the WRITE statement is the statement label of the FORMAT statement describing how the values contained in i and result are to be printed out. I3 and F7.3 are the **format descriptors** associated with variables i and result, respectively. In this case, the FORMAT statement specifies that the program should first write out the phrase 'The result for iteration ', followed by the value of variable i. The format descriptor I3 specifies that a space three characters wide should be used to print out the value of variable i. The value of i will be followed by the phrase ' is ' and then the value of the variable result. The format descriptor F7.3 specifies that a space seven characters wide should be used to print out the value of variable result, and that it should be printed with three digits to the right of the decimal point. The resulting output line is shown below, compared to the same line printed with free format.

```
The result for iteration 21 is   3.142                  (formatted)
The result for iteration      21 is    3.141593   (free format)
```

Note that we are able to eliminate both extra blank spaces and undesired decimal places by using format statements. Note also that the value in variable result was rounded before it was printed out in F7.3 format. (Only the value printed out has been rounded; the contents of variable result are unchanged.) Formatted I/O will permit us to create neat output listings from our programs.

In addition to FORMAT statements, formats may be specified in character constants or variables. If a character constant or variable is used to contain the format, then the constant or the name of the variable appears within the parentheses in the WRITE statement. For example, the following three WRITE statements are equivalent:

```
WRITE (*,100) i, x                  ! Format in FORMAT statement
100 FORMAT (1X,I6,F10.2)

CHARACTER(len=20) :: string         ! Format in character variable
string = '(1X,I6,F10.2)'
WRITE (*,string) i, x
WRITE (*,'(1X,I6,F10.2)') i, x      ! Format in character constant
```

We will mix formats in FORMAT statements, character constants, and character variables in examples throughout this chapter.

In the above example, each format descriptor was separated from its neighbors by commas. With a few exceptions, *multiple format descriptors in a single format must be separated by commas.*[1]

---

[1] There is another form of formatted output statement:

PRINT fmt, *output_list*

This statement is equivalent to the formatted WRITE statement discussed above, where *fmt* is either the number of a format statement or a character constant or variable. The PRINT statement is never used in this book, but it is discussed in Section 14.3.7.

5.2
OUTPUT DEVICES

To understand the structure of a FORMAT statement, we must know something about the **output devices** on which our data will be displayed. The output from a Fortran program is displayed on an output device. There are many types of output devices that are used with computers. Some output devices produce permanent paper copies of the data, while others just display it temporarily for us to see. Common output devices include laser printers, line printers, and monitors.

The traditional way to get a paper copy of the output of a Fortran program was on a **line printer.** A line printer was a type of printer that originally got its name from the fact that it printed output data a line at a time. Since it was the first common computer output device, Fortran output specifications were designed with it in mind. Other more modern output devices are generally built to be backward compatible with the line printer, so that the same output statement can be used for any of the devices.

A line printer printed on computer paper that was divided into pages on a continuous roll. There were perforations between the pages so that it was easy to separate them. The most common size of line printer paper in the United States was 11 inches high by $14\frac{7}{8}$ inches wide. Each page was divided into a number of lines, and each line is divided into 132 columns, with one character per column. Since most line printers printed either 6 lines per vertical inch or 8 lines per vertical inch, the printers could print either 60 or 72 lines per page (note that this assumes a 0.5-inch margin at the top and the bottom of each page; if the margin is made larger, fewer lines could be printed).

Most modern printers are laser printers, which print on separate sheets of paper instead of on a connected roll of paper. The paper size is usually "Letter or Legal" in the North America, and A4 or A3 in the rest of the world. Laser printers can be set to print either 80 or 132 columns depending on text size, so they can be compatible with line printers and respond the same way to output from Fortran programs.

The format specifies where a line is to be printed on a line printer or laser printer page (vertical position), and also where each variable is to be printed within the line (horizontal position).

5.2.1 Control Characters in Printer Output

The computer builds up a complete image of each line in memory before sending it to an output device. The computer memory containing the image of the line is called the **output buffer** (see Figure 5-1). In the days of line printers, the first character in a line had a special function and was known as a **control character.** The control character specified the vertical spacing for the line. The remaining 132 characters in the buffer contain the data to be printed on that line. All versions of Fortran up to and including Fortran 95 included special behavior for control characters.

Basic I/O Concepts

FIGURE 5-1
The output buffer is usually 133 characters long. The first character is the control character, and the next 132 characters are an image of what is to be printed on the line.

The control character was not printed on the page by the line printer. Instead, it provided vertical positioning control information to the printer. Table 5-1 shows the vertical spacing resulting from different control characters.

A '1' character caused the printer to skip the remainder of the current page and print the current line at the top of the next page. A blank character caused the printer to print the current line right below the previous one, while a '0' character caused the printer to skip a line before the current line is printed. A '+' character specified no spacing; in this case, the new line overwrote the previous line. If any other character was used as the control character, the result should be the same as for a blank.

For list-directed output [WRITE (*,*)], a blank control character was automatically inserted at the beginning of each output buffer. Therefore, list-directed output was always printed in single-spaced lines.

The following FORMAT statements illustrate the use of the control character. They will print a heading at the top of a new page, skip one line, and then print column headings for Table 5-1 below it.

```
WRITE (*,100)
100 FORMAT ('1','This heading is at the top of a new page.')
WRITE (*,110)
110 FORMAT ('0','  Control Character     Action ')
WRITE (*,120)
120 FORMAT (' ','  =================     ====== ')
```

The results of executing these Fortran statements are shown below.

Control characters were a special mechanism designed to work with line printers. Line printers are effectively extinct, and have been for many years, so the use of the column 1 as a control character has been deleted from the Fortran 2003 standard. According to the new standard, column 1 of the output buffer is an ordinary character that has no special purpose. It is printed out like any other character.

TABLE 5-1
Fortran control characters

Control character	Action
1	Skip to new page
Blank	Single spacing
0	Double spacing
+	No spacing (print over previous line)

```
This heading is at the top of a new page.


Control Character Action
================== ======
```

FIGURE 5-2
Results printing Table 5-1 column headings using the old control character mechanism.

Programming Pitfalls
Be aware of control characters in older Fortran programs and in compilers that support and modify older programs.

Fortran compilers still support this mechanism for backward compatibility, but it is normally turned off by default. In modern Fortran programs, the first character in a line no longer has a special meaning.

5.3
FORMAT DESCRIPTORS

There are many different format descriptors. They fall into four basic categories:

1. Format descriptors that describe the *vertical position* of a line of text.
2. Format descriptors that describe the *horizontal position* of data in a line.
3. Format descriptors that describe the output format of a particular value.
4. Format descriptors that control the repetition of portions of a format.

We will deal with some common examples of format descriptors in this chapter. Other less common format descriptors will be postponed to Chapter 14. Table 5-2 contains a list of symbols used with format descriptors, together with their meanings.

5.3.1 Integer Output—The I Descriptor

The descriptor used to describe the display format of integer data is the I descriptor. It has the general form

$$rIw \quad \text{or} \quad rIw.m$$

Basic I/O Concepts

■ TABLE 5-2
Symbols used with format descriptors

Symbol	Meaning
c	Column number
d	Number of digits to right of decimal place for real input or output
m	Minimum number of digits to be displayed
n	Number of spaces to skip
r	**Repeat count**—the number of times to use a descriptor or group of descriptors
w	**Field width**—the number of characters to use for the input or output

where r, w, and m have the meanings given in Table 5-2. Integer values are *right justified* in their fields. This means that integers are printed out so that the last digit of the integer occupies the rightmost column of the field. If an integer is too large to fit into the field in which it is to be printed, then the field is filled with asterisks. For example, the following statements:

```
INTEGER :: index = -12, junk = 4, number = -12345
WRITE (*,200) index, index+12, junk, number
WRITE (*,210) index, index+12, junk, number
WRITE (*,220) index, index+12, junk, number
200 FORMAT (' ', 2I5,   I6, I10 )
210 FORMAT (' ', 2I5.0, I6, I10.8 )
220 FORMAT (' ', 2I5.3, I6, I5 )
```

will produce the output

```
  -12    0    4    -12345
  -12         4 -00012345
 -012  000    4*****
----|----|----|----|----|----|
    5   10   15   20   25   30
```

The special case of the zero length descriptor I0 causes the integer to be written out with a *variable field width* sufficient to hold the information contained in the integer. For example, the following statements:

```
INTEGER :: index = -12, junk = 4, number = -12345
WRITE (*,100) index, junk, number
100 FORMAT (I0,1X,I0,1X,I0)
```

will produce the output

```
-12 4 -12345
----|----|----|----|----|----|
    5   10   15   20   25   30
```

This form of the format descriptor is especially useful for ensuring that the data will always be displayed, but it is not suitable for creating tables of data, because the columns of data will not be aligned properly.

5.3.2 Real Output—The F Descriptor

One format descriptor used to describe the display format of real data is the F descriptor. It has the form

$$rFw.d$$

where r, w, and d have the meanings given in Table 5-2. Real values are printed *right justified* within their fields. If necessary, the number will be rounded off before it is displayed. For example, suppose that the variable `pi` contains the value 3.141593. If this variable is displayed using the F7.3 format descriptor, the displayed value will be ƅƅ3.142. On the other hand, if the displayed number includes more significant digits than the internal representation of the number, extra zeros will be appended to the right of the decimal point. If the variable `pi` is displayed with an F10.8 format descriptor, the resulting value will be 3.14159300. If a real number is too large to fit into the field in which it is to be printed, then the field is filled with asterisks.

For example, the following statements:

```
REAL :: a = -12.3, b = .123, c = 123.456
WRITE (*,200) a, b, c
WRITE (*,210) a, b, c
200 FORMAT (2F6.3, F8.3 )
210 FORMAT (3F10.2 )
```

will produce the output

```
****** 0.123 123.456
    -12.30      0.12      123.46
----|----|----|----|----|----|
    5   10   15   20   25   30
```

5.3.3 Real Output—The E Descriptor

Real data can also be printed in **exponential notation** using the E descriptor. Scientific notation is a popular way for scientists and engineers to display very large or very small numbers. It consists of expressing a number as a normalized value between 1 and 10 multiplied by 10 raised to a power.

To understand the convenience of scientific notation, let's consider the following two examples from chemistry and physics. *Avogadro's number* is the number of atoms in a mole of a substance. It can be written out as 602,000,000,000,000,000,000,000 or it can be expressed in scientific notation as 6.02×10^{23}. On the other hand, the charge on an electron is 0.0000000000000000001602 coulombs. This number can be expressed in scientific notation as 1.602×10^{-19}. Scientific notation is clearly a much more convenient way to write these numbers!

The E format descriptor has the form

$$rEw.d$$

where *r*, *w*, and *d* have the meanings given in Table 5-2. Unlike normal scientific notation, the real values displayed in exponential notation with the E descriptor are normalized to a range between 0.1 and 1.0. That is, they are displayed as a number between 0.1 and 1.0 multiplied by a power of 10. For example, the standard scientific notation for the number 4096.0 would be 4.096×10^3, while the computer output with the E descriptor would be 0.4096×10^4. Since it is not easy to represent exponents on a line printer, the computer output would appear on the printer as 0.4096E+04.

If a real number cannot fit into the field in which it is to be printed, then the field is filled with asterisks. You should be especially careful with field sizes when working with the E format descriptor, since many items must be considered when sizing the output field. For example, suppose that we want to print out a variable in the E format with four significant digits of accuracy. Then a field width of **11** characters is required, as shown below: 1 for the sign of the mantissa, 2 for the zero and decimal point, 4 for the actual mantissa, 1 for the E, 1 for the sign of the exponent, and 2 for the exponent itself.

$$\pm 0.\text{ddddE}\pm\text{ee}$$

In general, the width of an E format descriptor field must satisfy the expression

$$w \geq d + 7 \tag{5-1}$$

or the field may be filled with asterisks.[2] The seven extra characters required are used as follows: 1 for the sign of the mantissa, 2 for the zero and decimal point, 1 for the E, 1 for the sign of the exponent, and 2 for the exponent itself.

For example, the following statements:

```
REAL :: a = 1.2346E6, b = 0.001, c = -77.7E10 , d = -77.7E10
WRITE (*,200) a, b, c, d
200 FORMAT (2E14.4, E13.6, E11.6 )
```

will produce the output[3]

```
     0.1235E+07     0.1000E-02-0.777000E+12**********
----|----|----|----|----|----|----|----|----|----|----|
    5   10   15   20   25   30   35   40   45   50   55
```

---

[2] If the number to be displayed in the field is positive, then the field width *w* need only be six characters larger than *d*. If the number is negative, an extra character is needed for the minus sign. Hence, in general, *w* must be $\geq d + 7$. Also, note that some compilers suppress the leading zero, so that one less column is required.

[3] The presence of the leading zero in an E format descriptor is optional, and whether or not it is there will differ among compiler vendors. Some compilers display leading zeros, while others do not. The following two lines show the output that could be produced by two different compilers for this example, and both would be considered correct.

```
         0.1235E+07     0.1000E-02-0.777000E+12**********
          .1235E+07      .1000E-02 -.777000E+12**********
    ----|----|----|----|----|----|----|----|----|----|----|
        5   10   15   20   25   30   35   40   45   50   55
```

Notice that the fourth field is all asterisks, since the format descriptor does not satisfy Equation (5-1).

5.3.4 True Scientific Notation—The ES Descriptor

As mentioned above, the output of the E format descriptor doesn't exactly match conventional scientific notation. Conventional scientific notation expresses a number as a value between 1.0 and 10.0 times a power of 10, while the E format expresses the number as a value between 0.1 and 1.0 times a power of 10.

We can make the computer output match conventional scientific notation by using a slightly modified version of the E descriptor called the ES descriptor. The ES descriptor is exactly the same as the E descriptor, except that the number to be output will be displayed with a mantissa in the range between 1 and 10. The ES format descriptor has the form

$$rESw.d$$

where r, w, and d have the meanings given in Table 5-2. The formula for the minimum width of an ES format descriptor is the same as the formula for the width of an E format descriptor, but the ES descriptor can display one more significant digit in a given width because the leading zero is replaced by a significant digit. The ES field must satisfy the expression

$$w \geq d + 7 \qquad (5\text{-}1)$$

or the field may be filled with asterisks.[4]

For example, the following statements:

```
REAL :: a = 1.2346E6, b = 0.001, c = -77.7E10
WRITE (*,200) a, b, c
200 FORMAT (2ES14.4, ES12.6)
```

will produce the output

```
      1.2346E+06    1.0000E-03************
----|----|----|----|----|----|----|----|
    5   10   15   20   25   30   35   40
```

The third field is all asterisks, since the format descriptor does not satisfy Equation (5-1).

Good Programming Practice
When displaying very large or very small numbers, use the ES format descriptor to cause them to be displayed in conventional scientific notation. This display will help a reader to quickly understand the output numbers.

---

[4] If the number to be displayed in the field is positive, then the field width w need only be six characters larger than d. If the number is negative, an extra character is needed for the minus sign. Hence, in general, $w \geq d + 7$.

Basic I/O Concepts

5.3.5 Logical Output—The L Descriptor

The descriptor used to display logical data has the form

$$rLw$$

where r and w have the meanings given in Table 5-2. The value of a logical variable can only be .TRUE. or .FALSE.. The output of logical variable is either a T or an F, right justified in the output field.

For example, the following statements:

```
LOGICAL :: output = .TRUE., debug = .FALSE.
WRITE (*,"(2L5 )") output, debug
```

will produce the output

```
     T    F
----|----|----|
    5    10   15
```

5.3.6 Character Output—The A Descriptor

Character data is displayed using the A format descriptor.

$$rA \text{ or } rAw$$

where r and w have the meanings given in Table 5-2. The rA descriptor displays character data in a field whose width is the same as the number of characters being displayed, while the rAw descriptor displays character data in a field of fixed width w. If the width w of the field is longer than the length of the character variable, the variable is printed out *right justified* in the field. If the width of the field is shorter than the length of the character variable, only the first w characters of the variable will be printed out in the field.

For example, the following statements:

```
CHARACTER(len=17) :: string = 'This is a string.'
WRITE (*,10) string
WRITE (*,11) string
WRITE (*,12) string
10 FORMAT (' ', A)
11 FORMAT (' ', A20)
12 FORMAT (' ', A6)
```

will produce the output

```
          This is a string.
             This is a string.
          This i
          ----|----|----|----|----|
              5   10   15   20   25
```

5.3.7 Horizontal Positioning—The X and T Descriptor

Two format descriptors are available to control the spacing of data in the output buffer, and therefore on the final output line. They are the X descriptor, which inserts spaces into the buffer, and the T descriptor, which "tabs" over to a specific column in the buffer. The X descriptor has the form

$$nX$$

where n is the number of blanks to insert. It is used to *add one or more blanks* between two values on the output line. The T descriptor has the form

$$Tc$$

where c is the column number to go to. It is used to *jump directly to a specific column* in the output buffer. The T descriptor works much like a "tab" character on a typewriter, except that it is possible to jump to any position in the output line, even if we are already past that position in the FORMAT statement.

For example, the following statements:

```
CHARACTER(len=10) :: first_name = 'James     '
CHARACTER :: initial = 'R'
CHARACTER(len=16) :: last_name = 'Johnson         '
CHARACTER(len=9) :: class = 'COSC 2301'
INTEGER :: grade = 92
WRITE (*,100) first_name, initial, last_name, grade, class
100 FORMAT (A10, 1X, A1, 1X, A10, 4X, I3, T51, A9)
```

will produce the output

```
 James     R Johnson        92                COSC 2301
----|----|----|----|----|----|----|----|----|----|----|----|
    5   10   15   20   25   30   35   40   45   50   55   60
```

The first 1X descriptor produces a blank control character, so this output line is printed on the next line of the printer. The first name begins in column 1, the middle initial begins in column 12, the last name begins in column 14, the grade begins in column 28, and course name begins in column 50. (The course name begins in column 51 of the buffer, but it is printed in column 50, since the first character in the output buffer is the control character.) This same output structure could have been created with the following statements:

```
WRITE (*,110) first_name, initial, last_name, class, grade
110 FORMAT (A10, T13, A1, T15, A10, T51, A9, T29, I3)
```

In this example, we are actually jumping backward in the output line when we print out the grade.

Since you may freely move anywhere in the output buffer with the T descriptor, it is possible to accidentally overwrite portions of your output data before the line is printed. For example, if we change the tab descriptor for class from T51 to T17,

Basic I/O Concepts

```
      WRITE (*,120) first_name, initial, last_name, class, grade
  120 FORMAT (A10, T13, A1, T15, A10, T17, A9, T29, I3)
```

the program will produce the following output:

```
JAMES      R JOCOSC 2301    92
----|----|----|----|----|----|----|----|----|----|----|----|
    5   10   15   20   25   30   35   40   45   50   55   60
```

Programming Pitfalls

When using the T descriptor, be careful to make certain that your fields do not overlap.

5.3.8 Repeating Groups of Format Descriptors

We have seen that many individual format descriptors can be repeated by preceding them with a repeat count. For example, the format descriptor 2I10 is the same as the pair of descriptors I10, I10.

It is also possible to repeat *whole groups* of format descriptors by enclosing the whole group within parentheses and placing a repetition count in front of the parentheses. For example, the following two FORMAT statements are equivalent:

```
  320 FORMAT (I6, I6, F10.2, F10.2, I6, F10.2, F10.2 )
  320 FORMAT (I6, 2(I6, 2F10.2) )
```

Groups of format descriptors may be nested if desired. For example, the following two FORMAT statements are equivalent:

```
  330 FORMAT (I6, F10.2, A, F10.2, A, I6, F10.2, A, F10.2, A )
  330 FORMAT (2(I6, 2(F10.2,A)) )
```

However, don't go overboard with nesting. The more complicated you make your FORMAT statements, the harder it will be for you or someone else to understand and debug them.

If an asterisk is used instead of a number for the repetition count, then the contents of the parentheses will be repeated indefinitely as long as there is additional data to write out. A FORMAT statement such as

```
  340 FORMAT (I6, *(I6, 2F10.2) )
```

will reuse the (I6, 2F10.2) descriptors indefinitely as long as there is more data to print out.

5.3.9 Changing Output Lines—The Slash (/) Descriptor

The slash (/) descriptor causes the current output buffer to be sent to the printer, and a new output buffer to be started. With slash descriptors, a single WRITE statement can display output values on more than one line. Several slashes can be used together to skip several lines. The slash is one of the special descriptors that does not have to be separated from other descriptors by commas. However, you may use commas if you wish.

For example, suppose that we need to print out the results of an experiment in which we have measured the amplitude and phase of a signal at a certain time and depth. Assume that the integer variable `index` is 10 and the real variables `time`, `depth`, `amplitude`, and `phase` are 300., 330., 850.65, and 30., respectively. Then the statements

```
WRITE (*,100) index, time, depth, amplitude, phase
100 FORMAT (T20,'Results for Test Number ',I3,///, &
    'Time      = ',F7.0/, &
    'Depth     = ',F7.1,' meters',/, &
    'Amplitude = ',F8.2/ &,
    'Phase     = ',F7.1)
```

generate six separate output buffers. The first buffer puts a title on the output. The next two output buffers are empty, so two blank lines are printed. The final four output buffers contain the output for one variable each, so the four values for `time`, `depth`, `amplitude`, and `phase` are printed on successive lines. The resulting output is shown in Figure 5-3.

Notice the 1X descriptors after each slash. These descriptors place a blank in the character of each output buffer, so that each subsequent line starts in column 2.

5.3.10 How Formats are Used during WRITEs

Most Fortran compilers verify the syntax of FORMAT statements and character constants containing formats at compilation time, but do not otherwise process them. Character variables containing formats are not even checked at compilation time for valid syntax, since the format may be modified dynamically during program execution. In all cases, formats are saved unchanged as character strings within the compiled program. When

```
            Results for Test Number 10


    Time      =   300.
    Depth     =   330.0 meters
    Amplitude =   850.65
    Phase     =   30.2
```

FIGURE 5-3
Results printing amplitude and phase.

Basic I/O Concepts

the program is executed, the characters in a format are used as a template to guide the operation of the formatted WRITE.

At execution time, the list of output variables associated with the WRITE statement is processed together with the format of the statement. The program begins at the left end of the variable list and the left end of the format, and scans from left to right, associating the first variable in the output list with the first format descriptor in the format, and so forth. The variables in the output list must be of the same type and in the same order as the format descriptors in the format, or a runtime error will occur. For example, the program in Figure 5-4 will compile and link correctly, since all the statements in it are legal Fortran statements, and the program doesn't check for correspondence between the format descriptors and the data types until it runs. However, it will abort at runtime, when the check shows a logical format descriptor corresponding to a character variable.

FIGURE 5-4
A Fortran program showing a runtime error resulting from a data/format descriptor mismatch. Note that the Fortran compiler did not check for format correspondence, so it missed the error.

```
C:\book\fortran\chap5>type bad_format.f90
PROGRAM bad_format
IMPLICIT NONE
INTEGER :: i = 10
CHARACTER(len=6) :: j = 'ABCDEF'
WRITE (*,100) i, j
100 FORMAT ( I10, L10 )
END PROGRAM

C:\book\fortran\chap5>ifort bad_format.f90
Intel(R) Visual Fortran Intel(R) 64 Compiler for applications running on
Intel(R) 64, Version 16.0.2.180 Build 20160204
Copyright (C) 1985-2016 Intel Corporation.  All rights reserved.

Microsoft (R) Incremental Linker Version 12.00.40629.0
Copyright (C) Microsoft Corporation.  All rights reserved.

-out:bad_format.exe
-subsystem:console
bad_format.obj

C:\book\fortran\chap5>bad_format
forrtl: severe (61): format/variable-type mismatch, unit -1, file CONOUT$
Image              PC                Routine        Line        Source
bad_format.exe     00007FF7512BE7AB  Unknown        Unknown     Unknown
bad_format.exe     00007FF7512B619D  Unknown        Unknown     Unknown
bad_format.exe     00007FF7512B109C  Unknown        Unknown     Unknown
bad_format.exe     00007FF75130124E  Unknown        Unknown     Unknown
bad_format.exe     00007FF751301524  Unknown        Unknown     Unknown
KERNEL32.DLL       00007FFA56B38102  Unknown        Unknown     Unknown
ntdll.dll          00007FFA594DC5B4  Unknown        Unknown     Unknown
```

Programming Pitfalls
Make sure that there is a one-to-one correspondence between the types of the data in a WRITE statement and the types of the format descriptors in the associated FORMAT statement, or your program will fail at execution time.

As the program moves from left to right through the variable list of a WRITE statement, it also scans from left to right through the associated format. However, the order in which the contents of a format are used may be modified by the inclusion of repetition counters and parentheses. Formats are scanned according to the following rules:

1. *Formats are scanned in order from left to right*. The first variable format descriptor in the format is associated with the first value in the output list of the WRITE statement, and so forth. The type of each format descriptor must match the type of the data being output. In the example shown below, descriptor I5 is associated with variable i, I10 with variable j, I15 with variable k, and F10.2 with variable a.

         WRITE (*,10) i, j, k, a
      10 FORMAT (I5, I10, I15, F10.2)

2. *If a format descriptor has a repetition count associated with it, the descriptor will be used the number of times specified in the repetition count before the next descriptor will be used*. In the example shown below, descriptor I5 is associated with variable i, and again with variable j. After it has been used twice, I10 is associated with variable k, and F10.2 is associated with variable a.

         WRITE (*,20) i, j, k, a
      20 FORMAT (2I5, I10, F10.2)

3. *If a group of format descriptors included within parentheses has a repetition count associated with it, the entire group will be used the number of times specified in the repetition count before the next descriptor will be used*. Each descriptor within the group will be used in order from left to right during each repetition. In the example shown below, descriptor F10.2 is associated with variable a. Next, the group in parentheses is used twice, so I5 is associated with i, E14.6 is associated with b, I5 is associated with j, and E14.6 is associated with c. Finally, F10.2 is associated with d.

         WRITE (*,30) a, i, b, j, c, d
      30 FORMAT (F10.2, 2(I5, E14.6), F10.2)

4. If the WRITE statement runs out of variables before the end of the format, *the use of the format stops at the first format descriptor without a corresponding variable, or at the end of the format, whichever comes first*. For example, the statements

Basic I/O Concepts

```
        INTEGER :: m = 1
        WRITE (*,40) m
     40 FORMAT ('M = ', I3, 'N = ', I4, 'O = ', F7.2)
```

will produce the output

```
            M =    1 N =
        ----|----|----|----|----|----|
            5   10   15   20   25   30
```

since the use of the format stops at I4, which is the first unmatched format descriptor. The statements

```
        REAL :: voltage = 13800.
        WRITE (*,50) voltage / 1000.
     50 FORMAT ('Voltage = ', F8.1, ' kV')
```

will produce the output

```
        Voltage =     13.8 kV
        ----|----|----|----|----|----|
            5   10   15   20   25   30
```

since there are no unmatched descriptors, and the use of the format stops at the end of the statement.

5. If the scan reaches the end of the format before the WRITE statement runs out of values, the program sends the current output buffer to the printer, and starts over *at the rightmost open parenthesis in the format that is not preceded by a repetition count*. For example, the statements

```
        INTEGER :: j = 1, k = 2, l = 3, m = 4, n = 5
        WRITE (*,60) j, k, l, m, n
     60 FORMAT ('value = ', I3)
```

will produce the output

```
            value =    1
            value =    2
            value =    3
            value =    4
            value =    5
        ----|----|----|----|----|----|
            5   10   15   20   25   30
```

When the program reaches the end of the FORMAT statement after it prints j with the I3 descriptor, it sends that output buffer to the printer and goes back to the rightmost open parenthesis not preceded by a repetition count. In this case, the rightmost open parenthesis without a repetition count is the opening parenthesis of the statement, so the entire statement is used again to print k, l, m, and n. By contrast, the statements

```
        INTEGER :: j = 1, k = 2, l = 3, m = 4, n = 5
        WRITE (*,60) j, k, l, m, n
     60 FORMAT ('Value = ',/, ('New Line',2(3X,I5)))
```

will produce the output

```
Value =
New Line     1     2
New Line     3     4
New Line     5
----|----|----|----|----|----|
    5   10   15   20   25   30
```

In this case, the entire FORMAT statement is used to print values j and k. Since the rightmost open parenthesis not preceded by a repetition count is the one just before 'New Line', that part of the statement is used again to print l, m, and n. Note that the open parenthesis associated with (3X,I5) was ignored because it had a repetition count associated with it.

EXAMPLE 5-1

Generating a Table of Information:

A good way to illustrate the use of formatted WRITE statements is to generate and print out a table of data. The example program shown in Figure 5-5 generates the square roots, squares, and cubes of all integers between 1 and 10, and presents the data in a table with appropriate headings.

FIGURE 5-5
A Fortran program to generate a table of square roots, squares, and cubes.

```
PROGRAM table
!
!  Purpose:
!    To illustrate the use of formatted WRITE statements.  This
!    program generates a table containing the square roots, squares,
!    and cubes of all integers between 1 and 10.  The table includes
!    a title and column headings.
!
!  Record of revisions:
!      Date       Programmer          Description of change
!      ====       ==========          =====================
!    11/18/15    S. J. Chapman        Original code
!
IMPLICIT NONE

INTEGER :: cube           ! The cube of i
INTEGER :: i              ! Index variable
INTEGER :: square         ! The square of i
REAL    :: square_root    ! The square root of i

! Print the title of the table on a new page.
WRITE (*,100)
100 FORMAT (T3, 'Table of Square Roots, Squares, and Cubes'/)

! Print the column headings after skipping one line.
WRITE (*,110)
```

(continued)

Basic I/O Concepts

(*concluded*)

```
110 FORMAT (T4,'Number',T13,'Square Root',T29,'Square',T39,'Cube')
    WRITE (*,120)
120 FORMAT (T4,'======',T13,'============',T29,'======',T39,'===='/)

! Generate the required values, and print them out.
DO i = 1, 10
   square_root = SQRT ( REAL(i) )
   square = i**2
   cube = i**3
   WRITE (*,130) i, square_root, square, cube
130 FORMAT (1X, T4, I4, T13, F10.6, T27, I6, T37, I6)
END DO

END PROGRAM table
```

This program uses the tab format descriptor to set up neat columns of data for the table. When this program is compiled and executed using the Intel Fortran compiler, the results are

```
C:\book\fortran\chap5>table
    Table of Square Roots, Squares, and Cubes

   Number    Square Root      Square    Cube
   ======    ===========      ======    ====

      1       1.000000           1         1
      2       1.414214           4         8
      3       1.732051           9        27
      4       2.000000          16        64
      5       2.236068          25       125
      6       2.449490          36       216
      7       2.645751          49       343
      8       2.828427          64       512
      9       3.000000          81       729
     10       3.162278         100      1000
```

EXAMPLE 5-2 *Charge on a Capacitor:*

A *capacitor* is an electrical device that stores electric charge. It essentially consists of two flat plates with an insulating material (the *dielectric*) between them (see Figure 5-6). The capacitance of a capacitor is defined as

$$C = \frac{Q}{V} \tag{5-2}$$

where Q is the amount of charge stored in a capacitor in units of coulombs and V is the voltage between the two plates of the capacitor in volts. The units of capacitance are farads (F), with 1 farad = 1 coulomb per volt. When a charge is present on the plates

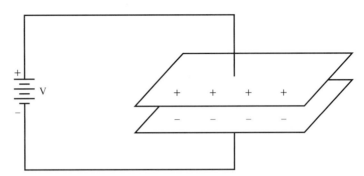

FIGURE 5-6
A capacitor consists of two metal plates separated by an insulating material.

of the capacitor, there is an electric field between the two plates. The energy stored in this electric field is given by the equation

$$E = \frac{1}{2}CV^2 \qquad (5\text{-}3)$$

where E is the energy in joules. Write a program that will perform one of the following calculations:

1. For a known capacitance and voltage, calculate the charge on the plates, the number of electrons on the plates, and the energy stored in the electric field.
2. For a known charge and voltage, calculate the capacitance of the capacitor, the number of electrons on the plates, and the energy stored in the electric field.

SOLUTION
This program must be able to ask the user which calculation he or she wishes to perform, read in the appropriate values for that calculation, and write out the results in a reasonable format. Note that this problem will require us to work with very small and very large numbers, so we will have to pay special attention to the FORMAT statements in the program. For example, capacitors are typically rated in microfarads (μF or 10^{-6} F) or picofarads (pF or 10^{-12} F), and there are 6.241461×10^{18} electrons per coulomb of charge.

1. **State the problem.**
   The problem may be succinctly stated as follows:
   (a) For a known capacitance and voltage, calculate the charge on a capacitor, the number of electrons stored, and the energy stored in its electric field.
   (b) For a known charge and voltage, calculate the capacitance of the capacitor, the number of electrons stored, and the energy stored in its electric field.

2. **Define the inputs and outputs.**
   There are two possible sets of input values to this program:
   (a) Capacitance in farads and voltage in volts.
   (b) Charge in coulombs and voltage in volts.

The outputs from the program in either mode will be the capacitance of the capacitor, the voltage across the capacitor, the charge on the plates of the capacitor, and the

Basic I/O Concepts

number of electrons on the plates of the capacitor. The output must be printed out in a reasonable and understandable format.

3. **Describe the algorithm.**

This program can be broken down into four major steps:

```
Decide which calculation is required
Get the input data for that calculation
Calculate the unknown quantities
Write out the capacitance, voltage, charge and number of electrons
```

The first major step of the program is to decide which calculation is required. There are two types of calculations: Type 1 requires capacitance and voltage, while Type 2 requires charge and voltage. We must prompt the user for the type of input data, read his or her answer, and then read in the appropriate data. The pseudocode for these steps is:

```
Prompt user for the type of calculation "type"
WHILE
    Read type
    IF type == 1 or type == 2 EXIT
    Tell user of invalid value
End of WHILE

IF type == 1 THEN
    Prompt the user for the capacitance c in farads
    Read capacitance c
    Prompt the user for the voltage v in volts
    Read voltage v
ELSE IF type == 2 THEN
    Prompt the user for the charge "charge" in coulombs
    Read "charge"
    Prompt the user for the voltage v in volts
    Read voltage v
END IF
```

Next, we must calculate unknown values. For Type 1 calculations, the unknown values are charge, the number of electrons, and the energy in the electric field, while for Type 2 calculations, the unknown values are capacitance, the number of electrons, and the energy in the electric field. The pseudocode for this step is shown below:

```
IF type == 1 THEN
    charge ← c * v
ELSE
    c ← charge / v
END IF
electrons ← charge * electrons_per_coulomb
energy ← 0.5 * c * v**2
```

where `electrons_per_coulomb` is the number of electrons per coulomb of charge (6.241461×10^{18}). Finally, we must write out the results in a useful format.

```
                WRITE v, c, charge, electrons, energy
```

The flowchart for this program is shown in Figure 5-7.

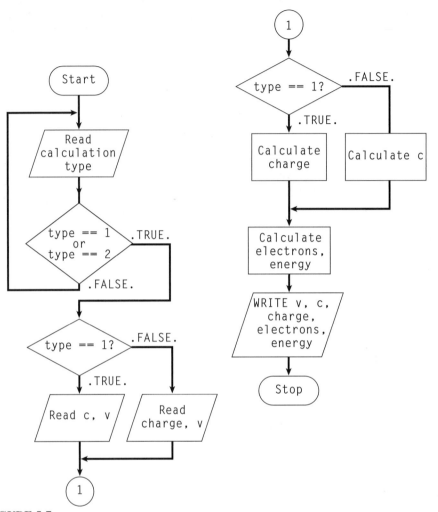

FIGURE 5-7
Flowchart for the program to calculate information about a capacitor.

4. **Turn the algorithm into Fortran statements.**
The final Fortran program is shown in Figure 5-8.

FIGURE 5-8
Program to perform capacitor calculations.

```
PROGRAM capacitor
!
! Purpose:
!   To calculate the behavior of a capacitor as follows:
!   1. If capacitance and voltage are known, calculate
!      charge, number of electrons, and energy stored.
```

(*continued*)

(*continued*)

```
!       2.  If charge and voltage are known, calculate capa-
!           citance, number of electrons, and energy stored.
!
! Record of revisions:
!      Date        Programmer          Description of change
!      ====        ==========          =====================
!    11/18/15     S. J. Chapman        Original code
!
IMPLICIT NONE

! Data dictionary: declare constants
REAL, PARAMETER :: ELECTRONS_PER_COULOMB = 6.241461E18

! Data dictionary: declare variable types, definitions, & units
REAL :: c            ! Capacitance of the capacitor (farads).
REAL :: charge       ! Charge on the capacitor (coulombs).
REAL :: electrons    ! Number of electrons on the plates of the capacitor
REAL :: energy       ! Energy stored in the electric field (joules)
INTEGER :: type      ! Type of input data available for the calculation:
                     !   1:   C and V
                     !   2:   CHARGE and V
REAL :: v            ! Voltage on the capacitor (volts).

! Prompt user for the type of input data available.
WRITE (*, 100)
100 FORMAT (' This program calculates information about a ' &
            'capacitor.',/, ' Please specify the type of information',&
            ' available from the following list:',/,&
            '    1 -- capacitance and voltage ',/,&
            '    2 -- charge and voltage ',//,&
            ' Select options 1 or 2: ')

! Get response and validate it.
DO
   READ (*,*) type
   IF ( (type == 1) .OR. (type == 2) ) EXIT
   WRITE (*,110) type
   110 FORMAT (' Invalid response: ', I6, '.  Please enter 1 or 2:')
END DO

! Get additional data based upon the type of calculation.
input: IF ( type == 1 ) THEN

   ! Get capacitance.
   WRITE (*,'Enter capacitance in farads: ')
   READ (*,*) c

   ! Get voltage.
   WRITE (*,'Enter voltage in volts: ')
   READ (*,*) v

ELSE
```

(*continued*)

(*concluded*)

```
            ! Get charge.
            WRITE (*,'Enter charge in coulombs: ')
            READ (*,*) charge

            ! Get voltage.
            WRITE (*,'Enter voltage in volts: ')
            READ (*,*) v
      END IF input

      ! Calculate the unknown quantities.
      calculate: IF ( type == 1 ) THEN
            charge = c * v                              ! Charge
      ELSE
            c = charge / v                              ! Capacitance
      END IF calculate
      electrons = charge * ELECTRONS_PER_COULOMB        ! Electrons
      energy = 0.5 * c * v**2                           ! Energy

      ! Write out answers.
      WRITE (*,120) v, c, charge, electrons, energy
  120 FORMAT ('For this capacitor: ',/, &
              ' Voltage              = ', F10.2, ' V',/, &
              ' Capacitance          = ', ES10.3, ' F',/, &
              ' Total charge         = ', ES10.3, ' C',/, &
              ' Number of electrons  = ', ES10.3,/, &
              ' Total energy         = ', F10.4, ' joules' )

      END PROGRAM capacitor
```

5. Test the program.

To test this program, we will calculate the answers by hand for a simple data set, and then compare the answers to the results of the program. If we use a voltage of 100 V and a capacitance of 100 μF, the resulting charge on the plates of the capacitor is 0.01 C, there are 6.241×10^{16} electrons on the capacitor, and the energy stored is 0.5 joules.

Running these values through the program using both options 1 and 2 yields the following results:

```
C:\book\fortran\chap5>capacitor

This program calculates information about a capacitor.
Please specify the type of information available from the following list:
   1 -- capacitance and voltage
   2 -- charge and voltage

Select options 1 or 2:
1
Enter capacitance in farads:
100.e-6
Enter voltage in volts:
100.
```

(*continued*)

(*concluded*)

```
For this capacitor:
  Voltage             =    100.00 V
  Capacitance         = 1.000E-04 F
  Total charge        = 1.000E-02 C
  Number of electrons = 6.241E+16
  Total energy        =     .5000 joules
C:\book\fortran\chap5>capacitor
This program calculates information about a capacitor.
Please specify the type of information available from the following list:
  1 -- capacitance and voltage
  2 -- charge and voltage
Select options 1 or 2:
2
Enter charge in coulombs:
0.01
Enter voltage in volts:
100.
For this capacitor:
  Voltage             =    100.00 V
  Capacitance         = 1.000E-04 F
  Total charge        = 1.000E-02 C
  Number of electrons = 6.241E+16
  Total energy        =     .5000 joules
```

The program gives the correct answers for our test data set.

In Example 5-2, sometimes formats appeared in FORMAT statements, and sometimes they appeared as character constants within WRITE statements. Since these two forms of formats are equivalent, either one could be used to provide a format for any WRITE statement. If that is so, when should we use a FORMAT statement, and when should we use a character constant? This author usually lets common sense be a guide: If a format is small and fits conveniently, I place it in a character constant within the WRITE statement. If the format is large and complicated, I place it in separate FORMAT statement.

Quiz 5-1

This quiz provides a quick check to see if you have understood the concepts introduced in Sections 5.1 through 5.3. If you have trouble with the quiz, reread the sections, ask your instructor, or discuss the material with a fellow student. The answers to this quiz are found in the back of the book. Unless otherwise stated, assume that variables beginning with the letters I - N are integers, and all other variables are reals.

(*continued*)

(*continued*)

Write Fortran statements that perform the operations described below.

1. Print the title 'This is a test!' starting in column 25.
2. Skip a line, then display the values of i, j, and data_1 in fields 10 characters wide. Allow two decimal points for the real variable.
3. Beginning in column 12, write out the string 'The result is' followed by the value of result expressed to five significant digits in correct scientific notation.

Assume that real variables a, b, and c are initialized with -0.0001, 6.02×10^{23}, and 3.141593, respectively, and that integer variables i, j, and k are initialized with 32767, 24, and -1010101, respectively. What will be printed out by each of the following sets of statements?

4. WRITE (*,10) a, b, c
   10 FORMAT (3F10.4)

5. WRITE (*,20) a, b, c
   20 FORMAT (F10.3, 2X, E10.3, 2X, F10.5)

6. WRITE (*,40) a, b, c
   40 FORMAT (ES10.4, ES11.4, F10.4)

7. WRITE (*,'(I5)') i, j, k

8. CHARACTER(len=30) :: fmt
   fmt = "(I0, 2X, I8.8, 2X, I8)"
   WRITE (*,fmt) i, j, k

Assume that string_1 is a 10-character variable initialized with the string 'ABCDEFGHIJ', and that string_2 is a 5-character variable initialized with the string '12345'. What will be printed out by each of the following sets of statements?

9. WRITE (*,"(2A10)") string_1, string_2

10. WRITE (*,80) string_1, string_2
    80 FORMAT (T21,A10,T24,A5)

11. WRITE (*,100) string_1, string_2
    100 FORMAT (A5,2X,A5)

Examine the following Fortran statements. Are they correct or incorrect? If they are incorrect, why are they incorrect? Assume default typing for variable names where they are not otherwise defined.

12. WRITE (*,'(2I6,F10.4)') istart, istop, step

(*continued*)

Basic I/O Concepts

(concluded)

13.  ```
     LOGICAL :: test
     CHARACTER(len=6) :: name
     INTEGER :: ierror
     WRITE (*,200) name, test, ierror
     200 FORMAT ('Test name: ',A,/,' Completion status : ',&
     I6, ' Test results: ', L6 )
     ```

What output will be generated by the following program? Describe the output from the program, including both the horizontal and vertical position of each output item.

14.  ```
     INTEGER :: index1 = 1, index2 = 2
     REAL :: x1 = 1.2, y1 = 2.4, x2 = 2.4, y2 = 4.8
     WRITE (*,120) index1, x1, y1, index2, x2, y2
     120 FORMAT (T11,'Output Data',/, &
                 T11,'===========',//,&
                 ('POINT(',I2,') = ',2F14.6))
     ```

5.4
FORMATTED READ STATEMENTS

An *input device* is a piece of equipment that can enter data into a computer. The most common input device on a modern computer is a keyboard. As data is entered into the input device, it is stored in an **input buffer** in the computer's memory. Once an entire line has been typed into the input buffer, the user hits the ENTER key on his or her keyboard, and the input buffer is made available for processing by the computer.

A READ statement reads one or more data values from the input buffer associated with an input device. The particular input device to read from is specified by the i/o unit number in the READ statement, as we will explain later in the chapter. It is possible to use a **formatted READ statement** to specify the exact manner in which the contents of an input buffer are to be interpreted.

In general, a format specifies which columns of the input buffer are to be associated with a particular variable and how those columns are to be interpreted. A typical formatted READ statement is shown below:

```
READ (*,100) increment
100 FORMAT (6X,I6)
```

This statement specifies that the first six columns of the input buffer are to be skipped, and then the contents of columns 7 through 12 are to be interpreted as an integer, with the resulting value stored in variable increment. As with WRITEs, formats may be stored in FORMAT statements, character constants, or character variables.

Formats associated with READs use many of the same format descriptors as formats associated with WRITEs. However, the interpretation of those descriptors is somewhat different. The meanings of the format descriptors commonly found with READs are described below.

5.4.1 Integer Input—The I Descriptor

The descriptor used to read integer data is the I descriptor. It has the general form

$$rIw$$

where r and w have the meanings given in Table 5-2. An integer value may be placed anywhere within its field, and it will be read and interpreted correctly.

5.4.2 Real Input—The F Descriptor

The format descriptor used to describe the input format of real data is the F descriptor. It has the form

$$rFw.d$$

where r, w, and d have the meanings given in Table 5-2. The interpretation of real data in a formatted READ statement is rather complicated. The input value in an F input field may be a real number with a decimal point, a real number in exponential notation, or a number without a decimal point. If a real number with a decimal point or a real number in exponential notation is present in the field, then the number is always interpreted correctly. For example, consider the following statement:

```
READ (*,'(3F10.4)') a, b, c
```

Assume that the input data for this statement is

```
  1.5         0.15E+01   15.0E-01
----|----|----|----|----|----|
    5   10   15   20   25   30
```

After the statement is executed, all three variables will contain the number 1.5.

If a number *without* a decimal point appears in the field, then a decimal point is assumed to be in the position specified by the d term of the format descriptor. For example, if the format descriptor is F10.4, then the four rightmost digits of the number are assumed to be the fractional part of the input value, and the remaining digits are assumed to be the integer part of the input value. Consider the following Fortran statements

```
READ (*,'(3F10.4)') a, b, c
```

Assume that the input data for these statements is

```
             15    150       15000
----|----|----|----|----|----|
    5   10   15   20   25   30
```

Then after these statements are executed, a will contain 0.0015, b will contain 0.0150, and c will contain 1.5000. The use of values without decimal points in a real input field is very confusing. It is a relic from an earlier version of Fortran that should never be used in your programs.

> **Good Programming Practice**
> Always include a decimal point in any real values used with a formatted READ statement.

The E and ES format descriptors are completely identical to the F descriptor for inputting data. They may be used in the place of the F descriptor, if desired.

5.4.3 Logical Input—The L Descriptor

The descriptor used to read logical data has the form

$$rLw$$

where r and w have the meanings given in Table 5-2. The value of a logical variable can only be .TRUE. or .FALSE.. The input value must be either the values .TRUE. or .FALSE., or else a block of characters beginning with a T or an F as the first nonblank character in the input field. If any other character is the first nonblank character in the field, a runtime error will occur. The logical input format descriptor is rarely used.

5.4.4 Character Input—The A Descriptor

Character data is read using the A format descriptor.

$$rA \quad \text{or} \quad rAw$$

where r and w have the meanings given in Table 5-2. The rA descriptor reads character data in a field whose width is the same as the length of the character variable being read, while the rAw descriptor reads character data in a field of fixed width w. If the width w of the field is larger than the length of the character variable, the data from the *rightmost* portion of the field is loaded into the character variable. If the width of the field is smaller than the length of the character variable, the characters in the field will be stored in the leftmost characters of the variable, and the remainder of the variable will be padded with blanks.

For example, consider the following statements

```
CHARACTER(len=10) :: string_1, string_2
CHARACTER(len=5)  :: string_3
CHARACTER(len=15) :: string_4, string_5
READ (*,'(A)')   string_1
READ (*,'(A10)') string_2
READ (*,'(A10)') string_3
READ (*,'(A10)') string_4
READ (*,'(A)')   string_5
```

Assume that the input data for these statements is

```
ABCDEFGHIJKLMNO
ABCDEFGHIJKLMNO
ABCDEFGHIJKLMNO
ABCDEFGHIJKLMNO
ABCDEFGHIJKLMNO
----|----|----|
    5   10   15
```

After the statements are executed, variable `string_1` will contain `'ABCDEFGHIJ'`, since `string_1` is 10 characters long, and the A descriptor will read as many characters as the length of variable. Variable `string_2` will contain `'ABCDEFGHIJ'`, since `string_2` is 10 characters long, and the A10 descriptor will read 10 characters. Variable `string_3` is only 5 characters long, and the A10 descriptor is 10 characters long, so `string_3` will contain the 5 rightmost of the 10 characters in the field: `'FGHIJ'`. Variable `string_4` will contain `'ABCDEFGHIJbbbbb'`, since `string_4` is 15 characters long, and the A10 descriptor will only read 10 characters. Finally `string_5` will contain `'ABCDEFGHIJKLMNO'`, since `string_5` is 15 characters long, and the A descriptor will read as many characters as the length of the variable.

5.4.5 Horizontal Positioning—The X and T Descriptors

The X and T format descriptors may be used when reading formatted input data. The chief use of the X descriptor is to skip over fields in the input data that we do not wish to read. The T descriptor may be used for the same purpose, but it may also be used to read the same data twice in two different formats. For example, the following code reads the values in characters 1 through 6 of the input buffer twice—once as an integer and once as a character string.

```
CHARACTER(len=6) :: string
INTEGER :: input
READ (*,'(I6,T1,A6)') input, string
```

5.4.6 Vertical Positioning—The Slash (/) Descriptor

The slash (/) format descriptor causes a formatted READ statement to discard the current input buffer, get another one from the input device, and start processing from the beginning of the new input buffer. For example, the following formatted READ statement reads the values of variables a and b from the first input line, skips down two lines, and reads the values of variables c and d from the third input line.

```
REAL :: a, b, c, d
READ (*,300) a, b, c, d
300 FORMAT (2F10.2,//,2F10.2)
```

Basic I/O Concepts

If the input data for these statements is

```
        1.0       2.0       3.0
        4.0       5.0       6.0
        7.0       8.0       9.0
----|----|----|----|----|----|
    5   10   15   20   25   30
```

then the contents of variables a, b, c, and d will be 1.0, 2.0, 7.0, and 8.0, respectively.

5.4.7 How Formats are Used during READs

Most Fortran compilers verify the syntax of FORMAT statements and character constants containing formats at compilation time, but do not otherwise process them. Character variables containing formats are not even checked at compilation time for valid syntax, since the format may be modified dynamically during program execution. In all cases, formats are saved unchanged as character strings within the compiled program. When the program is executed, the characters in a format are used as a template to guide the operation of the formatted READ.

At execution time, the list of input variables associated with the READ statement is processed together with the format of the statement. The rules for scanning a format are essentially the same for READs as they are for WRITEs. The order of scanning, repetition counts, and the use of parentheses are identical.

When the number of variables to be read and the number of descriptors in the format differ, formatted READs behave as follows:

1. If the READ statement runs out of variables before the end of the format, the use of the format stops after the last variable has been read. The next READ statement will start with a new input buffer, and all of the other data in the original input buffer will be lost. For example, consider the following statements

   ```
   READ (*,30) i, j
   READ (*,30) k, l, m
   30 FORMAT (5I5)
   ```

   and the following input data

   ```
        1    2    3    4    5
        6    7    8    9   10
   ----|----|----|----|----|
       5   10   15   20   25
   ```

   After the first statement is executed, the values of i and j will be 1 and 2, respectively. The first READ ends at that point, so that input buffer is thrown away without ever using the remainder of the buffer. The next READ uses the second input buffer, so the values of k, l, and m will be 6, 7, and 8.

2. If the scan reaches the end of the format before the READ statement runs out of variables, the program discards the current input buffer. It gets a new input buffer

and resumes in the format *at the rightmost open parenthesis that is not preceded by a repetition count*. For example, consider the statements

```
READ (*,40) i, j, k, l, m
40 FORMAT (I5,(T6,2I5))
```

and the input data

```
       1    2    3    4    5
       6    7    8    9   10
----|----|----|----|----|
    5   10   15   20   25
```

When the READ statement is executed, variables i, j, and k will be read from the first input buffer. They will contain 1, 2, and 3, respectively. The FORMAT statement ends at that point, so the first input buffer is discarded and the next one is used. The FORMAT statement starts over at the rightmost open parentheses not preceded by a repetition count, so variables l and m will contain 7 and 8, respectively.

Quiz 5-2

This quiz provides a quick check to see if you have understood the concepts introduced in Section 5.4. If you have trouble with the quiz, reread the section, ask your instructor, or discuss the material with a fellow student. The answers to this quiz are found in the back of the book. Unless otherwise stated, assume that variables beginning with the letters I-N are integers, and all other variables are reals.

Write Fortran statements that perform the functions described below.

1. Read the values of a real variable amplitude from columns 10 to 20, an integer variable count from columns 30 to 35, and a character variable identity from columns 60 to 72 of the current input buffer.
2. Read a 25-character variable called title from columns 10 to 34 of the first input line, and then read 5 integer variables i1 through i5 from columns 5 to 12 on each of the next 5 lines.
3. Read columns 11 to 20 from the current input line into a character variable string, skip two lines, and read columns 11 to 20 into an integer variable number. Do this with a single formatted READ statement.

What will be stored in each of the following variables?

4. READ (*,'(3F10.4)') a, b, c

   With the input data:

   ```
      1.65E-10    17.       -11.7
   ----|----|----|----|----|----|----|
       5   10   15   20   25   30   35
   ```

(continued)

Basic I/O Concepts

(concluded)

5. ```
   READ (*,20) a, b, c
   20 FORMAT (E10.2,F10.2,/,20X,F10.2)
   ```
   With the input data:
   ```
   -3.1415932.7182818210.1E10
           -11.         -5.        37.5532
   ----|----|----|----|----|----|----|
       5   10   15   20   25   30   35
   ```

6. ```
   READ (*,'(3I5)') i, j, k
   ```
   With the input data:
   ```
   -35    67053687
   ----|----|----|----|----|----|----|
       5   10   15   20   25   30   35
   ```

7. ```
   CHARACTER(len=5)  :: string_1
   CHARACTER(len=10) :: string_2, string_4
   CHARACTER(len=15) :: string_3
   READ (*,'(4A10)') string_1, string_2, string_3, string_4
   ```
   With the input data:
   ```
   ABCDEFGHIJLKMNOPQRSTUVWXYZ0123 _TEST_ 1
   ----|----|----|----|----|----|----|----|
       5   10   15   20   25   30   35   40
   ```

Examine the following Fortran statements. Are they correct or incorrect? If they are incorrect, why are they incorrect? If they are correct, what do they do?

8. ```
   READ (*,100) nvals, time1, time2
   100 FORMAT (10X,I10,F10.2,F10.4)
   ```

9. ```
   READ (*,220) junk, scratch
   220 FORMAT (T60,I15,/,E15.3)
   ```

10. ```
    READ (*,220) icount, range, azimuth, elevation
    220 FORMAT (I6, 4X, F20.2)
    ```

5.5

AN INTRODUCTION TO FILES AND FILE PROCESSING

The programs that we have written up to now have involved relatively small amounts of input and output data. We have typed in the input data from the keyboard each time that a program has been run, and the output data has gone directly to a terminal or printer. This is acceptable for small data sets, but it rapidly becomes prohibitive when working with large volumes of data. Imagine having to type in 100,000 input values each time a program is run! Such a process would be both time consuming and prone to typing errors. We need a convenient way to read in and write out large data sets, and to be able to use them repeatedly without retyping.

Fortunately, computers have a standard structure for holding data that we will be able to use in our programs. This structure is called a **file.** A file consists of many lines of data that are related to each other, and that can be accessed as a unit. Each line of information in a file is called a **record.** Fortran can read information from a file or write information to a file one record at a time.

The files on a computer can be stored on various types of devices, which are collectively know as *secondary memory*. (The computer's RAM is its primary memory.) Secondary memory is slower than the computer's main memory, but it still allows relatively quick access to the data. Common secondary storage devices include hard disk drives, USB memory sticks, and CDs or DVDs.

In the early days of computers, magnetic tapes were the most common type of secondary storage device. Computer magnetic tapes store data in a manner similar to the audio cassette tapes that were used to play music. Like them, computer magnetic tapes must be read (or "played") in order from the beginning of the tape to the end of it. When we read data in consecutive order one record after another in this manner, we are using **sequential access.** Other devices such as hard disks have the ability to jump from one record to another anywhere within a file. When we jump freely from one record to another following no specific order, we are using **direct access.** For historical reasons, sequential access is the default access technique in Fortran, even if we are working with devices capable of direct access.

To use files within a Fortran program, we will need some way to select the desired file and to read from or write to it. Fortunately, Fortran has a wonderfully flexible method to read from and write to files, whether they are on disk, magnetic tape, or some other device attached to the computer. This mechanism is known as the **input/output unit** (i/o unit, sometimes called a "logical unit", or simply a "unit"). The i/o unit corresponds to the first asterisk in the READ (*,*) and WRITE (*,*) statements. If that asterisk is replaced by an i/o unit number, then the *corresponding read or write will be to the device assigned to that unit* instead of to the standard input or output device. The statements to read or write any file or device attached to the computer are exactly the same except for the i/o unit number in the first position, so we already know most of what we need to know to use file i/o. An i/o unit number must be of type INTEGER.

Several Fortran statements may be used to control disk file input and output. The ones discussed in this chapter are summarized in Table 5-3.

TABLE 5-3
Fortran file control statements

I/O statement	Function
OPEN	Associate a specific disk file with a specific i/o unit number.
CLOSE	End the association of a specific disk file with a specific i/o unit number.
READ	Read data from a specified i/o unit number.
WRITE	Write data to a specified i/o unit number.
REWIND	Move to the beginning of a file.
BACKSPACE	Move back one record in a file.

Basic I/O Concepts

I/O unit numbers are assigned to disk files or devices using the OPEN statement, and detached from them using the CLOSE statement. Once a file is attached to an i/o unit using the OPEN statement, we can read and write in exactly the same manner that we have already learned. When we are through with the file, the CLOSE statement closes the file and releases the i/o unit to be assigned to some other file. The REWIND and BACKSPACE statements may be used to change the current reading or writing position in a file while it is open.

Certain unit numbers are pre-defined to be connected to certain input or output devices, so that we don't need an OPEN statement to use these devices. These pre-defined units vary from processor to processor.[5] Typically, i/o unit 5 is pre-defined to be the *standard input device* for your program (i.e., the keyboard if you are running at a terminal, or the input batch file if you are running in batch mode). Similarly, i/o unit 6 is usually pre-defined to be the *standard output device* for your program (the screen if you are running at a terminal, or the line printer if you are running in batch mode). These assignments date back to the early days of Fortran on IBM computers, so they have been copied by most other vendors in their Fortran compilers. Another common association is i/o unit 0 for the *standard error device* for your program. This assignment goes back to the C language and Unix-based computers.

However, you cannot count on any of these associations always being true for every processor. If you need to read from and write to the standard devices, always use the asterisk instead of the standard unit number for that device. The asterisk is guaranteed to work correctly on any computer system.

Good Programming Practice
Always use asterisks instead of i/o unit numbers when referring to the standard input or standard output devices. The standard i/o unit numbers vary from processor to processor, but the asterisk works correctly on all processors.

If we want to access any files or devices other than the pre-defined standard devices, we must first use an OPEN statement to associate the file or device with a specific i/o unit number. Once the association has been established, we can use ordinary Fortran READs and WRITEs with that unit to work with the data in the file.[6]

5.5.1 The OPEN Statement

The OPEN statement associates a file with a given i/o unit number. Its form is

OPEN (*open_list*)

---

[5] A processor is defined as the combination of a specific computer with a specific compiler.
[6] Some Fortran compilers attach default files to logical units that have not been opened. For example, in Intel Fortran, a write to an unopened i/o unit 26 will automatically go into a file called fort.26. You should never use this feature, since it is non-standard and varies from processor to processor. Your programs will be much more portable if you always use an OPEN statement before writing to a file.

where `open_list` contains a series of clauses specifying the i/o unit number, the file name, and information about how to access the file. The clauses in the list are separated by commas. The full list of possible clauses in the OPEN statement will be postponed until Chapter 14. For now, we will introduce only the six most important items from the list. They are

1. A `UNIT=` clause indicating the i/o unit number to associate with this file. This clause has the form,

    UNIT= *int_expr*

    where *int_expr* can be a nonnegative integer value.

2. A `FILE=` clause specifying the name of the file to be opened. This clause has the form,

    FILE= *char_expr*

    where *char_expr* is a character value containing name of the file to be opened.

3. A `STATUS=` clause specifying the status of the file to be opened. This clause has the form,

    STATUS= *char_expr*

    where *char_expr* is one of the following: `'OLD'`, `'NEW'`, `'REPLACE'`, `'SCRATCH'`, or `'UNKNOWN'`.

4. An `ACTION=` clause specifying whether a file is to be opened for reading only, for writing only, or for both reading and writing. This clause has the form,

    ACTION= *char_expr*

    where *char_expr* is one of the following: `'READ'`, `'WRITE'`, or `'READWRITE'`. If no action is specified, the file is opened for both reading and writing.

5. An `IOSTAT=` clause specifying the name of an integer variable in which the status of the open operation can be returned. This clause has the form,

    IOSTAT= *int_var*

    where *int_var* is an integer variable. If the OPEN statement is successful, a 0 will be returned in the integer variable. If it is not successful, a positive number corresponding to a system error message will be returned in the variable. The system error messages vary from processor to processor, but a zero always means success.

6. An `IOMSG=` clause specifying the name of a character variable that will contain a message if an error occurs. This clause has the form,

    IOMSG= *chart_var*

    where *char_var* is a character variable. If the OPEN statement is successful, the contents of the character variable will be unchanged. If it is not successful, a descriptive error message will be returned in this string.

Basic I/O Concepts

The above clauses may appear in any order in the OPEN statement. Some examples of correct OPEN statements are shown below.

Case 1: Opening a File for Input

The statement below opens a file named EXAMPLE.DAT and attaches it to i/o unit 8.

```
INTEGER :: ierror
OPEN (UNIT=8, FILE='EXAMPLE.DAT', STATUS='OLD', ACTION='READ', &
      IOSTAT=ierror, IOMSG=err_string)
```

The STATUS='OLD' clause specifies that the file already exists; if it does not exist, then the OPEN statement will return an error code in variable ierror, and an error message in character string err_string. This is the proper form of the OPEN statement for an *input file*. If we are opening a file to read input data from, then the file had better be present with data in it! If it is not there, something is obviously wrong. By checking the returned value in ierror, we can tell that there is a problem and take appropriate action.

The ACTION='READ' clause specifies that the file should be read-only. If an attempt is made to write to the file, an error will occur. This behavior is appropriate for an input file.

Case 2: Opening a File for Output

The statements below open a file named OUTDAT and attach it to i/o unit 25.

```
INTEGER :: unit, ierror
CHARACTER(len=6) :: filename
unit = 25
filename = 'OUTDAT'
OPEN (UNIT=unit, FILE=filename, STATUS='NEW', ACTION='WRITE', &
      IOSTAT=ierror, IOMSG=err_string)
```

or

```
OPEN (UNIT=unit, FILE=filename, STATUS='REPLACE', ACTION='WRITE', &
      IOSTAT=ierror, IOMSG=err_string)
```

The STATUS='NEW' clause specifies that the file is a new file; if it already exists, then the OPEN statement will return an error code in variable ierror. This is the proper form of the OPEN statement for an *output file* if we want to make sure that we don't overwrite the data in a file that already exists.

The STATUS='REPLACE' clause specifies that a new file should be opened for output whether a file by the same name exists or not. If the file already exists, the program will delete it, create a new file, and open it for output. The old contents of the file will be lost. If it does not exist, the program will create a new file by that name and open it. This is the proper form of the OPEN statement for an output file if we want to open the file whether or not a previous file exists with the same name.

The ACTION='WRITE' clause specifies that the file should be write-only. If an attempt is made to read from the file, an error will occur. This behavior is appropriate for an output file.

Case 3: Opening a Scratch File

The statement below opens a *scratch file* and attaches it to i/o unit 12.

```
OPEN (UNIT=12, STATUS='SCRATCH', IOSTAT=ierror)
```

A scratch file is a temporary file that is created by the program, and that will be deleted automatically when the file is closed or when the program terminates. This type of file may be used for saving intermediate results while a program is running, but it may not be used to save anything that we want to keep after the program finishes. Notice that no file name is specified in the OPEN statement. In fact, it is an error to specify a file name with a scratch file. Since no ACTION= clause is included, the file has been opened for both reading and writing.

Good Programming Practice
Always be careful to specify the proper status in OPEN statements, depending on whether you are reading from or writing to a file. This practice will help prevent errors such as accidentally overwriting data files that you want to keep.

5.5.2 The CLOSE Statement

The CLOSE statement closes a file and releases the i/o unit number associated with it. Its form is

```
CLOSE (close_list)
```

where *close_list* must contain a clause specifying the i/o number, and may specify other options that will be discussed with the advanced i/o material in Chapter 14. If no CLOSE statement is included in the program for a given file, that file will be closed automatically when the program terminates.

After a nonscratch file is closed, it may be reopened at any time using a new OPEN statement. When it is reopened, it may be associated with the same i/o unit or with a different i/o unit. After the file is closed, the i/o unit that was associated with it is free to be reassigned to any other file in a new OPEN statement.

5.5.3 READs and WRITEs to Disk Files

Once a file has been connected to an i/o unit via the OPEN statement, it is possible to read from or write to the file using the same READ and WRITE statements that we have been using. For example, the statements

```
OPEN (UNIT=8, FILE='INPUT.DAT',STATUS='OLD',IOSTAT=ierror)
READ (8,*) x, y, z
```

will read the values of variables x, y, and z in free format from the file INPUT.DAT, and the statements

```
OPEN (UNIT=9, FILE='OUTPUT.DAT',STATUS='REPLACE',IOSTAT=ierror)
WRITE (9,100) x, y, z
100 FORMAT (' X = ', F10.2, ' Y = ', F10.2, ' Z = ', F10.2 )
```

will write the values of variables x, y, and z to the file OUTPUT.DAT in the specified format.

5.5.4 The IOSTAT= and IOMSG= Clauses in the READ Statement

The IOSTAT= and IOMSG= clauses are important additional features that may be added to the READ statement when working with disk files. The form of the IOSTAT= clause is

$$\text{IOSTAT}= \textit{int\_var}$$

where *int_var* is an integer variable. If the READ statement is successful, a 0 will be returned in the integer variable. If it is not successful due to a file or format error, a positive number corresponding to a system error message will be returned in the variable. If it is not successful because the end of the input data file has been reached, a negative number will be returned in the variable.[7]

If an IOMSG= clause is included in a READ statement and the returned i/o status is nonzero, then the character string returned by the IOMSG= clause will explain in words what went wrong. The program should be designed to display this message to the user.

If no IOSTAT= clause is present in a READ statement, *any attempt to read a line beyond the end of a file will abort the program.* This behavior is unacceptable in a well-designed program. We often want to read all of the data from a file until the end is reached, and then perform some sort of processing on that data. This is where the IOSTAT= clause comes in: If an IOSTAT= clause is present, the program will not abort on an attempt to read a line beyond the end of a file. Instead, the READ will complete with the IOSTAT variable set to a negative number. We can then test the value of the variable, and process the data accordingly.

Good Programming Practice

Always include the IOSTAT= clause when reading from a disk file. This clause provides a graceful way to detect end-of-data conditions on the input files.

---

[7] There is an alternate method of detecting file read errors and end-of-file conditions using ERR= and END= clauses. These clauses of the READ statement will be described in Chapter 14. The IOSTAT= clause and IOMSG= clause lend themselves better to structured programming than the other clauses do, so they are being postponed to the later chapter.

EXAMPLE 5-3

Reading Data from a File:

It is very common to read a large data set into a program from a file, and then to process the data in some fashion. Often, the program will have no way of knowing in advance just how much data is present in the file. In that case, the program needs to read the data in a while loop until the end of the data set is reached, and then must detect that there is no more data to read. Once it has read in all of the data, the program can process it in whatever manner is required.

Let's illustrate this process by writing a program that can read in an unknown number of real values from a disk file, and detect the end of the data in the disk file.

SOLUTION

This program must open the input disk file, and then read the values from it using the IOSTAT= clause to detect problems. If the IOSTAT variable contains a negative number after a READ, then the end of the file has been reached. If the IOSTAT variable contains 0 after a READ, then everything was ok. If the IOSTAT variable contains a positive number after a READ, then a READ error occurred. In this example, the program should stop if a READ error occurs.

1. **State the problem.**

   The problem may be succinctly stated as follows:

   Write a program that can read an unknown number of real values from a user-specified input data file, detecting the end of the data file as it occurs.

2. **Define the inputs and outputs.**

   The inputs to this program consist of:

   (*a*) The name of the file to be opened.
   (*b*) The data contained in that file.

   The outputs from the program will be the input values in the data file. At the end of the file, an informative message will be written out telling how many valid input values were found.

3. **Describe the algorithm.**

   This pseudocode for this program is

   ```
   Initialize nvals to 0
   Prompt user for file name
   Get the name of the input file
   OPEN the input file
   Check for errors on OPEN

   If no OPEN error THEN
      ! Read input data
      WHILE
         READ value
         IF status /= 0 EXIT
         nvals ← nvals + 1
   ```

Basic I/O Concepts

```
            WRITE valid data to screen
         END of WHILE

         ! Check to see if the WHILE terminated due to end of file
         ! or READ error
         IF status > 0
            WRITE 'READ error occurred on line', nvals
         ELSE
            WRITE number of valid input values nvals
         END of IF ( status > 0 )
      END of IF (no OPEN error)
   END PROGRAM
```

A flowchart for the program is shown in Figure 5-9.

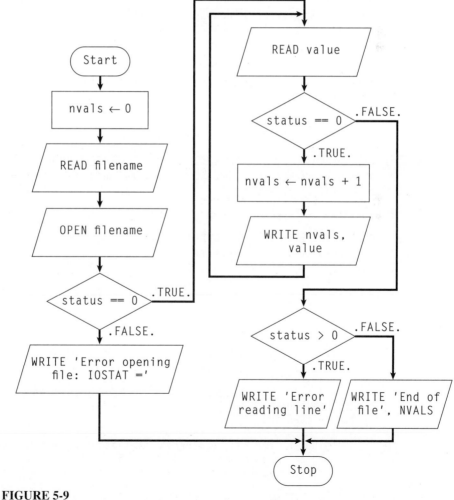

FIGURE 5-9
Flowchart for a program to read an unknown number of values from an input data file.

4. **Turn the algorithm into Fortran statements.**
   The final Fortran program is shown in Figure 5-10.

FIGURE 5-10
Program to read an unknown number of values from a user-specified input disk file.

```
PROGRAM read_file
!
!  Purpose:
!    To illustrate how to read an unknown number of values from
!    an input data file, detecting both any formatting errors and
!    the end of file.
!
!  Record of revisions:
!      Date        Programmer         Description of change
!      ====        ==========         =====================
!    11/18/15     S. J. Chapman         Original code
!
IMPLICIT NONE

! Data dictionary: declare variable types, definitions, & units
CHARACTER(len=20) :: filename    ! Name of file to open
CHARACTER(len=80) :: msg         ! Error message
INTEGER :: nvals = 0             ! Number of values read in
INTEGER :: status                ! I/O status
REAL :: value                    ! The real value read in

! Get the file name, and echo it back to the user.
WRITE (*,*) 'Please enter input file name: '
READ (*,*) filename
WRITE (*,1000) filename
1000 FORMAT ('The input file name is: ', A)

! Open the file, and check for errors on open.
OPEN (UNIT=3, FILE=filename, STATUS='OLD', ACTION='READ', &
      IOSTAT=status, IOMSG=msg )
openif: IF ( status == 0 ) THEN

   ! OPEN was ok.  Read values.
   readloop: DO
      READ (3,*,IOSTAT=status) value    ! Get next value
      IF ( status /= 0 ) EXIT           ! EXIT if not valid.
      nvals = nvals + 1                 ! Valid: increase count
      WRITE (*,1010) nvals, value       ! Echo to screen
      1010 FORMAT ('Line ', I6, ': Value = ',F10.4 )
   END DO readloop

   ! The WHILE loop has terminated.  Was it because of a READ
   ! error or because of the end of the input file?
   readif: IF ( status > 0 ) THEN   ! a READ error occurred.  Tell user.
```

(continued)

Basic I/O Concepts

(*concluded*)

```
      WRITE (*,1020) nvals + 1
      1020 FORMAT ('An error occurred reading line ', I6)
   ELSE ! the end of the data was reached.  Tell user.

      WRITE (*,1030) nvals
      1030 FORMAT ('End of file reached.  There were ', I6, &
                   ' values in the file.')
      END IF readif

ELSE openif
   WRITE (*,1040) status
   1040 FORMAT ('Error opening file: IOSTAT = ', I6 )
   WRITE (*,1050) TRIM(msg)
   1050 FORMAT (A)
END IF openif

! Close file
CLOSE ( UNIT=3 )

END PROGRAM read_file
```

Note that the input file is opened with STATUS='OLD', since we are reading from the file, and the input data must already exist before the program is executed.

5. **Test the program.**

To test this program, we will create two input files, one with valid data and one with an input data error. We will run the program with both input files, and verify that it works correctly both for valid data and for data containing input errors. Also, we will run the program with an invalid file name to show that it can properly handle missing input files.

The valid input file is called READ1.DAT. It contains the following lines:

```
         -17.0
         30.001
         1.0
         12000.
         -0.012
```

The invalid input file is called READ2.DAT. It contains the following lines:

```
         -17.0
         30.001
         ABCDEF
         12000.
         -0.012
```

Running these files through the program yields the following results:

```
C:\book\fortran\chap5>read_file
Please enter input file name:
read1.dat
The input file name is: read1.dat
Line     1: Value =   -17.0000
Line     2: Value =    30.0010
```

```
Line     3: Value =     1.0000
Line     4: Value = 12000.0000
Line     5: Value =    -.0120
```

```
End of file reached.  There were     5 values in the file.

C:\book\fortran\chap5>read_file
Please enter input file name:
read2.dat
The input file name is: read2.dat
Line     1: Value =   -17.0000
Line     2: Value =    30.0010

An error occurred reading line    3
```

Finally, let's test the program with an invalid input file name.

```
C:\book\fortran\chap5>read_file
Please enter input file name:
xxx
The input file name is: xxx
Error opening file: IOSTAT =   29
file not found, unit 3, file C:\Data\book\fortran\chap5\xxx
```

The number of the IOSTAT error reported by this program will vary from processor to processor, but it will always be positive. You must consult a listing of the runtime error codes for your particular compiler to find the exact meaning of the error code that your computer reports. For the Fortran compiler used here, IOSTAT = 29 means "File not found." Note that the error message returned from the IOMSG clause is clear to the user, without having to look up the meaning of status 29!

This program correctly read all of the values in the input file, and detected the end of the data set when it occurred.

5.5.5 File Positioning

As we stated previously, ordinary Fortran files are sequential—they are read in order from the first record in the file to the last record in the file. However, we sometimes need to read a piece of data more than once, or to process a whole file more than once during a program. How can we skip around within a sequential file?

Fortran provides two statements to help us move around within a sequential file. They are the BACKSPACE statement, which moves back one record each time it is called, and the REWIND statement, which restarts the file at its beginning. The forms of these statements are

$$\text{BACKSPACE (UNIT=unit)}$$

and

$$\text{REWIND (UNIT=unit)}$$

Basic I/O Concepts

where `unit` is the i/o unit number associated with the file that we want to work with.[8]

Both statements can also include `IOSTAT=` and `IOMSG=` clauses to detect errors during the backspace or rewind operation without causing the program to abort.

EXAMPLE 5-4

Using File Positioning Commands:

We will now illustrate the use of scratch files and file positioning commands in a simple example problem. Write a program that accepts a series of nonnegative real values and stores them in a scratch file. After the data is input, the program should ask the user what data record he or she is interested in, and then recover and display that value from the disk file.

SOLUTION

Since the program is expected to read only positive or zero values, we can use a negative value as a flag to terminate the input to the program. A Fortran program that does this is shown in Figure 5-11. This program opens a scratch file, and then reads input values from the user. If a value is nonnegative, it is written to the scratch file. When a negative value is encountered, the program asks the user for the record to display. It checks to see if a valid record number was entered. If the record number is valid, it rewinds the file and reads forward to that record number. Finally, it displays the contents of that record to the user.

FIGURE 5-11
Sample program illustrating the use of file positioning commands.

```
PROGRAM scratch_file
!
!  Purpose:
!    To illustrate the use of a scratch file and positioning
!    commands as follows:
!    1.  Read in an arbitrary number of positive or zero
!        values, saving them in a scratch file.  Stop
!        reading when a negative value is encountered.
!    2.  Ask the user for a record number to display.
!    3.  Rewind the file, get that value, and display it.
!
!  Record of revisions:
!      Date       Programmer          Description of change
!      ====       ==========          =====================
!   11/19/15    S. J. Chapman         Original code
!
IMPLICIT NONE

! Data dictionary: declare constants
INTEGER, PARAMETER :: LU = 8   ! i/o unit for scratch file
```

(continued)

---

[8] Alternate forms of these statements are described in Chapter 14.

(*continued*)

```
! Data dictionary: declare variable types, definitions, & units
REAL :: data            ! Data value stored in a disk file
INTEGER :: icount = 0   ! The number of input data records
INTEGER :: irec         ! Record number to recover and display
INTEGER :: j            ! Loop index

! Open the scratch file
OPEN (UNIT=LU, STATUS='SCRATCH')

! Prompt user and get input data.
WRITE (*, 100)
100 FORMAT ('Enter positive or zero input values. ',/, &
            'A negative value terminates input.' )

! Get the input values, and write them to the scratch file
DO
   WRITE (*, 110) icount + 1     ! Prompt for next value
   110 FORMAT ('Enter sample ',I4,':' )
   READ (*,*) data               ! Read value
   IF ( data < 0. ) EXIT         ! Exit on negative numbers
   icount = icount + 1           ! Valid value: bump count
   WRITE (LU,120) data           ! Write data to scratch file
   120 FORMAT (ES16.6)
END DO

! Now we have all of the records.  Ask which record to see.
! icount records are in the file.
WRITE (*,130) icount
130 FORMAT ('Which record do you want to see (1 to ',I4, ')? ')
READ (*,*) irec

! Do we have a legal record number?  If so, get the record.
! If not, tell the user and stop.
IF ( (irec >= 1) .AND. (irec <= icount) ) THEN

   ! This is a legal record.  Rewind the scratch file.
   REWIND (UNIT=LU)

   ! Read forward to the desired record.
   DO j = 1, irec
      READ (LU,*) data
   END DO

   ! Tell user.
   WRITE (*,140) irec, data
   140 FORMAT ('The value of record ', I4, ' is ', ES14.5 )

ELSE

   ! We have an illegal record number.  Tell user.
   WRITE (*,150) irec
   150 FORMAT ('Illegal record number entered: ', I8)
END IF
```

(*continued*)

(*concluded*)

```
! Close file
CLOSE(LU)
END PROGRAM scratch_file
```

Let us test the program with valid data:

```
C:\book\fortran\chap5>scratch_file
Enter positive or zero input values.
A negative input value terminates input.

Enter sample  1:
234.
Enter sample  2:
12.34
Enter sample  3:
0.
Enter sample  4:
16.
Enter sample  5:
11.235
Enter sample  6:
2.
Enter sample  7:
-1
Which record do you want to see (1 to  6)?
5
The value of record   5 is   1.12350E+01
```

Next, we should test the program with an invalid record number to see that the error condition is handled properly.

```
C:\book\fortran\chap5>scratch_file
Enter positive or zero input values.
A negative input value terminates input.
Enter sample    1:
234.
Enter sample    2:
12.34
Enter sample    3:
0.
Enter sample    4:
16.
Enter sample    5:
11.235
Enter sample    6:
2.
Enter sample    7:
-1
Which record do you want to see (1 to 6):
7
Illegal record number entered: 7
```

The program appears to be functioning correctly.

EXAMPLE 5-5

Fitting a Line to a Set of Noisy Measurements:

The velocity of a falling object in the presence of a constant gravitational field is given by the equation

$$v(t) = at + v_0 \quad (5\text{-}4)$$

where $v(t)$ is the velocity at any time t, a is the acceleration due to gravity, and v_0 is the velocity at time 0. This equation is derived from elementary physics—it is known to every freshman physics student. If we plot velocity versus time for the falling object, our (v, t) measurement points should fall along a straight line. However, the same freshman physics student also knows that if we go out into the laboratory and attempt to *measure* the velocity versus time of an object, our measurements will *not* fall along a straight line. They may come close, but they will never line up perfectly. Why not? Because we can never make perfect measurements. There is always some *noise* included in the measurements, which distorts them.

There are many cases in science and engineering where there are noisy sets of data such as this, and we wish to estimate the straight line that "best fits" the data. This problem is called the *linear regression* problem. Given a noisy set of measurements (x, y) that appear to fall along a straight line, how can we find the equation of the line

$$y = mx + b \quad (5\text{-}5)$$

that "best fits" the measurements? If we can determine the regression coefficients m and b, then we can use this equation to predict the value of y at any given x by evaluating Equation 5-5 for that value of x.

A standard method for finding the regression coefficients m and b is the *method of least squares*. This method is named "least squares" because it produces the line $y = mx + b$ for which the sum of the squares of the differences between the observed y values and the predicted y values is as small as possible. The slope of the least squares line is given by

$$m = \frac{(\Sigma xy) - (\Sigma x)\bar{y}}{(\Sigma x^2) - (\Sigma x)\bar{x}} \quad (5\text{-}6)$$

and the intercept of the least squares line is given by

$$b = \bar{y} - m\bar{x} \quad (5\text{-}7)$$

where

Σx is the sum of the x values
Σx^2 is the sum of the squares of the x values
Σxy is the sum of the products of the corresponding x and y values
$\bar{x}$ is the mean (average) of the x values
$\bar{y}$ is the mean (average) of the y values

Basic I/O Concepts

Write a program that will calculate the least squares slope m and y-axis intercept b for a given set of noisy measured data points (x, y) which are to be found in an input data file.

SOLUTION

1. **State the problem.**

   Calculate the slope m and intercept b of a least squares line that best fits an input data set consisting of an arbitrary number of (x, y) pairs. The input (x, y) data resides in a user-specified input file.

2. **Define the inputs and outputs.**

   The inputs required by this program are pairs of points (x, y), where x and y are real quantities. Each pair of points will be located on a separate line in the input disk file. The number of points in the disk file is not known in advance.

   The outputs from this program are the slope and intercept of the least squares fitted line, plus the number of points going into the fit.

3. **Describe the algorithm.**

   This program can be broken down into four major steps:

   ```
   Get the name of the input file and open it
   Accumulate the input statistics
   Calculate the slope and intercept
   Write out the slope and intercept
   ```

   The first major step of the program is to get the name of the input file and to open the file. To do this, we will have to prompt the user to enter the name of the input file. After the file is opened, we must check to see that the open was successful. Next, we must read the file and keep track of the number of values entered, plus the sums Σx, Σy, Σx^2, and Σxy. The pseudocode for these steps is:

   ```
   Initialize n, sum_x, sum_x2, sum_y, and sum_xy to 0
   Prompt user for input file name
   Open file "filename"
   Check for error on OPEN

   WHILE
       READ x, y from file "filename"
       IF (end of file) EXIT
       n ← n + 1
       sum_x ← sum_x + x
       sum_y ← sum_y + y
       sum_x2 ← sum_x2 + x**2
       sum_xy ← sum_xy + x*y
   End of WHILE
   ```

   Next, we must calculate the slope and intercept of the least squares line. The pseudocode for this step is just the Fortran versions of Equations (5-6) and (5-7).

   ```
   x_bar ← sum_x / real(n)
   y_bar ← sum_y / real(n)
   slope ← (sum_xy - sum_x * y_bar) / ( sum_x2 - sum_x * x_bar)
   y_int ← y_bar - slope * x_bar
   ```

Finally, we must write out the results.

> Write out slope "slope" and intercept "y_int".

4. **Turn the algorithm into Fortran statements.**
The final Fortran program is shown in Figure 5-12.

FIGURE 5-12
The least squares fit program of Example 5-5.

```
PROGRAM least_squares_fit
!
!  Purpose:
!    To perform a least-squares fit of an input data set
!    to a straight line, and print out the resulting slope
!    and intercept values.  The input data for this fit
!    comes from a user-specified input data file.
!
!  Record of revisions:
!      Date        Programmer         Description of change
!      ====        ==========         =====================
!    11/19/15     S. J. Chapman       Original code
!
IMPLICIT NONE

! Data dictionary: declare constants
INTEGER, PARAMETER :: LU = 18 ! I/O unit for disk I/O

! Data dictionary: declare variable types, definitions, & units
! Note that cumulative variables are all initialized to zero.
CHARACTER(len=24) :: filename ! Input file name (<= 24 chars)
INTEGER :: ierror             ! Status flag from I/O statements
CHARACTER(len=80) :: msg      ! Error message
INTEGER :: n = 0              ! Number of input data pairs (x,y)
REAL :: slope                 ! Slope of the line
REAL :: sum_x = 0.            ! Sum of all input X values
REAL :: sum_x2 = 0.           ! Sum of all input X values squared
REAL :: sum_xy = 0.           ! Sum of all input X*Y values
REAL :: sum_y = 0.            ! Sum of all input Y values
REAL :: x                     ! An input X value
REAL :: x_bar                 ! Average X value
REAL :: y                     ! An input Y value
REAL :: y_bar                 ! Average Y value
REAL :: y_int                 ! Y-axis intercept of the line

! Prompt user and get the name of the input file.
WRITE (*,1000)
1000 FORMAT ('This program performs a least-squares fit of an ',/, &
             'input data set to a straight line. Enter the name',/ &
             'of the file containing the input (x,y) pairs: ' )
```

(continued)

Basic I/O Concepts

(*concluded*)

```
READ (*,'(A)') filename
! Open the input file
OPEN (UNIT=LU, FILE=filename, STATUS='OLD', IOSTAT=ierror, IOMSG=msg)

! Check to see of the OPEN failed.
errorcheck: IF ( ierror > 0 ) THEN

   WRITE (*,1010) filename
   1010 FORMAT ('ERROR: File ',A,' does not exist!')
   WRITE (*,'(A)') TRIM(msg)

ELSE

   ! File opened successfully. Read the (x,y) pairs from
   ! the input file.
   DO
      READ (LU,*,IOSTAT=ierror) x, y    ! Get pair
      IF ( ierror /= 0 ) EXIT
      n      = n + 1                    !
      sum_x  = sum_x + x                ! Calculate
      sum_y  = sum_y + y                !   statistics
      sum_x2 = sum_x2 + x**2            !
      sum_xy = sum_xy + x * y           !
   END DO

   ! Now calculate the slope and intercept.
   x_bar = sum_x / real(n)
   y_bar = sum_y / real(n)
   slope = (sum_xy - sum_x * y_bar) / ( sum_x2 - sum_x * x_bar)
   y_int = y_bar - slope * x_bar

   ! Tell user.
   WRITE (*, 1020 ) slope, y_int, N
   1020 FORMAT ('Regression coefficients for the least-squares line:',&
           /,' slope (m)     = ', F12.3,&
           /,' Intercept (b) = ', F12.3,&
           /,' No of points  = ', I12)

   ! Close input file, and quit.
   CLOSE (UNIT=LU)

END IF errorcheck

END PROGRAM least_squares_fit
```

5. **Test the program.**

To test this program, we will try a simple data set. For example, if every point in the input data set actually falls along a line, then the resulting slope and intercept should be exactly the slope and intercept of that line. Thus, the data set

1.1, 1.1
2.2, 2.2

```
3.3, 3.3
4.4, 4.4
5.5, 5.5
6.6, 6.6
7.7, 7.7
```

should produce a slope of 1.0 and an intercept of 0.0. If we place these values in a file called INPUT, and run the program, the results are:

```
C:\book\fortran\chap5>least_squares_fit

This program performs a least-squares fit of an
input data set to a straight line. Enter the name
of the file containing the input (x,y) pairs:
INPUT
Regression coefficients for the least-squares line:
  slope (m)      =    1.000
  Intercept (b)  =     .000
  No of points   =       7
```

Now let's add some noise to the measurements. The data set becomes

```
1.1, 1.01
2.2, 2.30
3.3, 3.05
4.4, 4.28
5.5, 5.75
6.6, 6.48
7.7, 7.84
```

If these values are placed in a file called INPUT1, and the program is run on that file, the results are:

```
C:\book\fortran\chap5>least_squares_fit

This program performs a least-squares fit of an
input data set to a straight line. Enter the name
of the file containing the input (x,y) pairs:
INPUT1
Regression coefficients for the least-squares line:
  slope (m)      =    1.024
  Intercept (b)  =    -.120
  No of points   =       7
```

If we calculate the answer by hand, it is easy to show that the program gives the correct answers for our two test data sets. The noisy input data set and the resulting least squares fitted line are shown in Figure 5-13.

The program in this example has a problem—it cannot distinguish between the end of an input file and a read error (such as character data instead of real data) in the input file. How would you modify the program to distinguish between these two possible cases?

Also, note that this program stored two simple formats in character constants instead of defining a separate format statement for each of them. This is good practice when the formats are simple.

Basic I/O Concepts

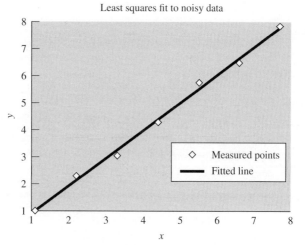

FIGURE 5-13
A noisy input data set and the resulting least squares fitted line.

Quiz 5-3

This quiz provides a quick check to see if you have understood the concepts introduced in Section 5.5. If you have trouble with the quiz, reread the section, ask your instructor, or discuss the material with a fellow student. The answers to this quiz are found in the back of the book.

Write Fortran statements that perform the functions described below. Unless otherwise stated, assume that variables beginning with the letters I - N are integers, and all other variables are reals.

1. Open an existing file named IN052691 on i/o unit 25 for read-only input, and check the status to see if the OPEN was successful.
2. Open a new output file, making sure that you do not overwrite any existing file by the same name. The name of the output file is stored in character variable out_name.
3. Close the file attached to unit 24.
4. Read variables first and last from i/o unit 8 in free format, checking for end of data during the READ.
5. Backspace eight lines in the file attached to i/o unit 13.

Examine the following Fortran statements. Are they correct or incorrect? If they are incorrect, why are they incorrect? Unless otherwise stated, assume that variables beginning with the letters I - N are integers, and all other variables are reals.

6. `OPEN (UNIT=35, FILE='DATA1', STATUS='REPLACE',IOSTAT=ierror)`
   `READ (35,*) n, data1, data2`

(continued)

(*concluded*)

```
7.  CHARACTER(len=80) :: str
    OPEN (UNIT=11, FILE='DATA1', STATUS='SCRATCH',IOSTAT=ierror, &
    IOMSG=str)
8.  OPEN (UNIT=15,STATUS='SCRATCH',ACTION='READ', IOSTAT=ierror)
9.  OPEN (UNIT=x, FILE='JUNK', STATUS='NEW',IOSTAT=ierror)
10. OPEN (UNIT=9, FILE='TEMP.DAT', STATUS='OLD', ACTION='READ', &
    IOSTAT=ierror)
    READ (9,*) x, y
```

5.6
SUMMARY

In this chapter, we presented a basic introduction to formatted WRITE and READ statements, and to the use of disk files for input and output of data.

In a formatted WRITE statement, the second asterisk of the unformatted WRITE statement (WRITE (*,*)) is replaced by a FORMAT statement number or a character constant or variable containing the format. The format describes how the output data is to be displayed. It consists of format descriptors that describe the vertical and horizontal position of the data on a page, as well as display format for integer, real, logical, and character data types.

The format descriptors discussed in this chapter are summarized in Table 5-4.

TABLE 5-4
Fortran format descriptors discussed in Chapter 5

FORMAT descriptors		Usage
Aw	A	Character data
E$w.d$		Real data in exponential notation
ES$w.d$		Real data in scientific notation
F$w.d$		Real data in decimal notation
Iw	I$w.m$	Integer data
I0		Integer data with variable field width
Lw		Logical data
Tc		TAB: move to column c of current line
nX		Horizontal spacing: skip n spaces
/		Vertical spacing: move down one line

where:
c column number
d number of digits to right of decimal place
m minimum number of digits to be displayed
n number of spaces to skip
w field width in characters

Basic I/O Concepts

Formatted `READ` statements use a format to describe how the input data is to be interpreted. All of the above format descriptors are also legal in formatted `READ` statements.

A disk file is opened using the `OPEN` statement, read and written using `READ` and `WRITE` statements, and closed using the `CLOSE` statement. The `OPEN` statement associates a file with an i/o unit number, and that i/o unit number is used by the `READ` statements and `WRITE` statements in the program to access the file. When the file is closed, the association is broken.

It is possible to move around within a sequential disk file using the `BACKSPACE` and `REWIND` statements. The `BACKSPACE` statement moves the current position in the file backward by one record whenever it is executed, and the `REWIND` statement moves the current position back to the first record in the file.

5.6.1 Summary of Good Programming Practice

The following guidelines should be adhered to when programming with formatted output statements or with disk i/o. By following them consistently, your code will contain fewer bugs, will be easier to debug, and will be more understandable to others who may need to work with it in the future.

1. Always be careful to match the type of data in a `WRITE` statement to the type of descriptors in the corresponding format. Integers should be associated with `I` format descriptors; reals with `E`, `ES`, or `F` format descriptors; logicals with `L` descriptors; and characters with `A` descriptors. A mismatch between data types and format descriptors will result in an error at execution time.
2. Use the `ES` format descriptor instead of the `E` descriptor when displaying data in exponential format to make the output data appear to be in conventional scientific notation.
3. Use an asterisk instead of an i/o unit number when reading from the standard input device or writing to the standard output device. This makes your code more portable, since the asterisk is the same on all systems, while the actual unit numbers assigned to standard input and standard output devices may vary from system to system.
4. Always open input files with `STATUS='OLD'`. By definition, an input file must already exist if we are to read data from it. If the file does not exist, this is an error, and the `STATUS='OLD'` will catch that error. Input files should also be opened with `ACTION='READ'` to prevent accidental overwriting of the input data.
5. Open output files with `STATUS='NEW'` or `STATUS='REPLACE'`, depending on whether or not you want to preserve the existing contents of the output file. If the file is opened with `STATUS='NEW'`, it should be impossible to overwrite an existing file, so the program cannot accidentally destroy data. If you don't care about the existing data in the output file, open the file with `STATUS='REPLACE'`, and the file will be overwritten if it exists. Open scratch files with `STATUS='SCRATCH'`, so that they will be automatically deleted upon closing.
6. Always include the `IOSTAT=` clause when reading from disk files to detect an end-of-file or error condition.

5.6.2 Summary of Fortran Statements and Structures

The following summary describes the Fortran statements and structures introduced in this chapter.

BACKSPACE Statement:

>   BACKSPACE (UNIT=*unit*)

Example:

>   BACKSPACE (UNIT=8)

Description:
The BACKSPACE statement moves the current position of a file back by one record.

CLOSE Statement:

>   CLOSE (*close_list*)

Example:

>   CLOSE (UNIT=8)

Description:
The CLOSE statement closes the file associated with a i/o unit number.

FORMAT Statement:

>   *label* FORMAT (*format descriptor*, ...)

Example:

>   100 FORMAT (' This is a test: ', I6)

Description:
The FORMAT statement describes the position and format of the data being read or written.

Formatted READ Statement:

>   READ (*unit,format*) *input_list*

Examples:

>   READ (1,100) time, speed
>   100 FORMAT (F10.4, F18.4)
>   READ (1,'(I6)') index

(continued)

Basic I/O Concepts

(*concluded*)

Description:
The formatted READ statement reads data from an input buffer according to the format descriptors specified in the format. The format is a character string that may be specified in a FORMAT statement, a character constant, or a character variable.

Formatted WRITE Statement:

        WRITE (*unit,format*) *output_list*

Examples:

```
WRITE (*,100) i, j, slope
100 FORMAT (2I10, F10.2)
WRITE (*,'(2I10, F10.2)') i, j, slope
```

Description:
The formatted WRITE statement outputs the data in the output list according to the format descriptors specified in the format. The format is a character string that may be specified in a FORMAT statement, a character constant, or a character variable.

OPEN Statement:

        OPEN (*open_list*)

Example:

```
OPEN (UNIT=8, FILE='IN', STATUS='OLD' ACTION='READ', &
      IOSTAT=ierror,IOMSG=msg)
```

Description:
The OPEN statement associates a file with an i/o unit number, so that it can be accessed by READ or WRITE statements.

REWIND Statement:

        REWIND (UNIT=*lu*)

Example:

        REWIND (UNIT=8)

Description:
The REWIND statement moves the current position of a file back to the beginning.

5.6.3 Exercises

5-1. What is the purpose of a format? In what three ways can formats be specified?

5-2. What is printed out by the following Fortran statements?

(a)
```
INTEGER :: i
CHARACTER(len=20) :: fmt
fmt = "('i = ', I6.5)"
i = -123
WRITE (*,fmt) i
WRITE (*,'(I0)') i
```

(b)
```
REAL :: a, b, sum, difference
a = 1.0020E6
b = 1.0001E6
sum = a + b
difference = a - b
WRITE (*,101) a, b, sum, difference
101 FORMAT ('A = ',ES14.6,' B = ',E14.6, &
    ' Sum = ',E14.6,' Diff = ', F14.6)
```

(c)
```
INTEGER :: i1, i2
i1 = 10
i2 = 4**2
WRITE (*,300) i1 > i2
300 FORMAT ('Result = ', L6)
```

5-3. What is printed out by the following Fortran statements?

```
REAL :: a = 1.602E-19, b = 57.2957795, c = -1.
WRITE (*,'(ES14.7,2(1X,E13.7))') a, b, c
```

5-4. For the Fortran statements and input data given below, state what the values of each variable will be when the READ statement has been completed.

Statements:
```
CHARACTER(5)  :: a
CHARACTER(10) :: b
CHARACTER(15) :: c
READ (*,'(3A10)') a, b, c
```

Input Data:
```
This is a test of reading characters.
----|----|----|----|----|----|----|----|----|
    5   10   15   20   25   30   35   40   45
```

5-5. For the Fortran statements and input data given below, state what the values of each variable will be when the READ statements have been completed.

(a) Statements:
```
INTEGER :: item1, item2, item3, item4, item5
INTEGER :: item6, item7, item8, item9, item10
READ (*,*) item1, item2, item3, item4, item5, item6
READ (*,*) item7, item8, item9, item10
```

Input Data:

```
      -300    -250    -210    -160    -135
      -105    - 70    - 55    - 28    - 11
        17      55     102     165     225
----|----|----|----|----|----|----|----|----|----|
    5   10   15   20   25   30   35   40   45   50
```

(b) Statements:

```
INTEGER :: item1, item2, item3, item4, item5
INTEGER :: item6, item7, item8, item9, item10
READ (*,8) item1, item2, item3, item4, item5, item6
READ (*,8) item7, item8, item9, item10
8 FORMAT (4I10)
```

Input Data: Same as for (a) above.

5-6. Table of Logarithms Write a Fortran program to generate a table of the base 10 logarithms between 1 and 10 in steps of 0.1. The table should include a title describing the table and row and column headings. This table should be organized as shown below:

	X.0	X.1	X.2	X.3	X.4	X.5	X.6	X.7	X.8	X.9
1.0	0.000	0.041	0.079	0.114	...					
2.0	0.301	0.322	0.342	0.362	...					
3.0	...									
4.0	...									
5.0	...									
6.0	...									
7.0	...									
8.0	...									
9.0	...									
10.0	...									

5-7. Example 5-3 illustrates the technique of reading an arbitrary amount of real data from an input data file. Modify that program to read in the data from an input data file and to calculate the mean and standard deviation of the samples in the file.

5-8. A real number `length` is to be displayed in F$w.d$ format with four digits to the right of the decimal point ($d = 4$). If the number is known to lie within the range $-10000.0 \leq$ `length` ≤ 10000.0, what is the minimum field width w that will always be able to display the value of `length`?

5-9. In what columns will the following characters be printed? Why?

```
WRITE (*,'(T30,A)') 'Rubbish!'
```

5-10. Write Fortran statements to perform the functions described below. Assume that variables beginning with I-N are integers, and all other variables are reals.

(a) Skip to a new line and print the title 'INPUT DATA' starting in column 40.

(b) Skip a line, and then display the data point number ipoint in columns 6 to 10, and the data point value data_1 in columns 15 to 26. Display the data value in scientific notation with seven significant digits.

5-11. What is the minimum field width necessary to display any real data value in E or ES format with six significant bits of accuracy?

5-12. Write a Fortran program that reads in a time in seconds since the start of the day (this value will be somewhere between 0. and 86400.), and writes out the time in the form HH:MM:SS using the 24-hour-clock convention. Use the I$w.m$ format descriptor to ensure that leading zeros are preserved in the MM and SS fields. Also, be sure to check the input number of seconds for validity, and write an appropriate error message if an invalid number is entered.

5-13. Gravitational Acceleration The acceleration due to the Earth's gravity at any height h above the surface of the Earth is given by the equation

$$g = -G \frac{M}{(R+h)^2} \quad (5\text{-}8)$$

where G is the gravitational constant (6.672×10^{-11} N m^2/kg^2), M is the mass of the Earth (5.98×10^{24} kg), R is the mean radius of the Earth (6371 km), and h is the height above the Earth's surface. If M is measured in kg and R and h in meters, then the resulting acceleration will be in units of meters per second squared. Write a program to calculate the acceleration due to the Earth's gravity in 500 km increments at heights from 0 km to 40,000 km above the surface of the Earth. Print out the results in a table of height versus acceleration with appropriate labels, including the units of the output values.

5-14. What is the proper STATUS to use when opening a file for reading input data? What is the proper STATUS to use when opening a file for writing output data? What is the proper STATUS to use when opening a temporary storage file?

5-15. What is the proper ACTION to use when opening a file for reading input data? What is the proper ACTION to use when opening a file for writing output data? What is the proper ACTION to use when opening a temporary storage file?

5-16. Is a CLOSE statement always required in a Fortran program that uses disk files? Why or why not?

5-17. Write Fortran statements to perform the functions described below. Assume that file INPUT.DAT contains a series of real values organized with one value per record.

(a) Open an existing file named INPUT.DAT on i/o unit 98 for input, and a new file named NEWOUT.DAT on i/o unit 99 for output.

(b) Read data values from file INPUT.DAT until the end of file is reached. Write all positive data values to the output file.

(c) Close the input and output data files.

FIGURE 5-14
A rectangle.

5-18. Write a program that reads an arbitrary number of real values from a user-specified input data file, rounds the values to the nearest integer, and writes the integers out to a user-specified output file. Open the input and output files with the appropriate status, and be sure to handle end of file and error conditions properly.

5-19. **Area of a Rectangle** The area of the rectangle in Figure 5-14 is given by Equation (5-9) and the perimeter of the rectangle is given by Equation (5-10).

$$area = W \times H \qquad (5\text{-}9)$$

$$perimeter = 2W + 2H \qquad (5\text{-}10)$$

Assume that the total perimeter of a rectangle is limited to 10, and write a program that calculates and plots the area of the rectangle as its width is varied from the smallest possible value to the largest possible value. Use format statements to create a neat output table. At what width is the area of the rectangle maximized?

5-20. Write a program that opens a scratch file and writes the integers 1 through 10 in the first 10 records. Next, move back six records in the file, and read the value stored in that record. Save that value in variable x. Next, move back three records in the file, and read the value stored in that record. Save that value in variable y. Multiply the two values x and y together. What is their product?

5-21. Examine the following Fortran statements. Are they correct or incorrect? If they are incorrect, why are they incorrect? (Unless otherwise indicated, assume that variables beginning with I-N are integers, and all other variables are reals.)

(a) `OPEN (UNIT=1, FILE='INFO.DAT', STATUS='NEW', IOSTAT=ierror)`
`READ (1,*) i, j, k`

(b) `OPEN (UNIT=17, FILE='TEMP.DAT', STATUS='SCRATCH', IOSTAT=ierror)`

(c) `OPEN (UNIT=99, FILE='INFO.DAT', STATUS='NEW', &`
`     ACTION='READWRITE', IOSTAT=ierror)`
`WRITE (99,*) i, j, k`

(d) `INTEGER :: unit = 8`
`OPEN (UNIT=unit, FILE='INFO.DAT', STATUS='OLD', IOSTAT=ierror)`
`READ (8,*) unit`
`CLOSE (UNIT=unit)`

(e) `OPEN (UNIT=9, FILE='OUTPUT.DAT', STATUS='NEW', ACTION='WRITE', &`
`     IOSTAT=ierror)`

```
        WRITE (9,*) mydat1, mydat2
        WRITE (9,*) mydat3, mydat4
        CLOSE (UNIT=9)
```

5-22. Table of Sines and Cosines Write a program to generate a table containing the sine and cosine of θ for θ between 0° and 90°, in 1° increments. The program should properly label each of the columns in the table.

5-23. Table of Speed versus Height The velocity of an initially stationary ball can be calculated as a function of the distance it has fallen from the equation

$$v = \sqrt{2g\Delta h} \quad (5\text{-}11)$$

where g is the acceleration due to gravity and Δh is the distance that the ball has fallen. If g is in units of m/s² and Δh is in units of meters, then the velocity will be in units of m/s. Write a program to create a table of the velocity of the ball as a function of how far it has fallen for distances from 0 to 200 meters in steps of 10 m. The program should properly label each of the columns in the table.

5-24. Potential versus Kinetic Energy The potential energy of a ball due to its height above ground is given by the equation

$$PE = mgh \quad (5\text{-}12)$$

where m is the mass of the ball in kilograms, g is the acceleration due to gravity in m/s², and h is the height of the ball about the surface of the Earth in meters. The kinetic energy of a ball due to its speed is given by the equation

$$KE = \frac{1}{2}mv^2 \quad (5\text{-}13)$$

where m is the mass of the ball in kilograms, and v is the velocity of the ball in m/s. Assume that a ball is initially stationary at a height of 100 m. When this ball is released, it will start to fall. Calculate the potential energy and the kinetic energy of the ball at 10 m increments as it falls from the initial height of 100 m to the ground, and create a table containing height, PE, KE, and the total energy (PE + KE) of the ball at each step. The program should properly label each of the columns in the table. What happens to the total energy as the ball falls? (*Note:* You can use Equation (5-11) to calculate the velocity at a given height, and then use that velocity to calculate the KE.)

5-25. Interest Calculations Suppose that you have a sum of money P in an interest-bearing account at a local bank (P stands for *present value*). If the bank pays you interest on the money at a rate of i percent per year and compounds the interest monthly, the amount of money that you will have in the bank after n months is given by the equation

$$F = P\left(1 + \frac{i}{1200}\right)^n \quad (5\text{-}14)$$

where F is the future value of the account and $i/12$ is the monthly percentage interest rate (the extra factor of 100 in the denominator converts the interest rate from percentages to fractional amounts). Write a Fortran program that will read an initial amount of money P and an annual interest rate i, and will calculate and write out a table showing the future value of the account every month for the next 4 years. The table should be written to an output file called 'interest'. Be sure to properly label the columns of your table.

5-26. Write a program to read a set of integers from an input data file, and locate the largest and smallest values within the data file. Print out the largest and smallest values, together with the lines on which they were found. Assume that you do not know the number of values in the file before the file is read.

5-27. Means In Exercise 4-31, we wrote a Fortran program that calculated the arithmetic mean (average), rms average, geometric mean, and harmonic mean for a set of numbers. Modify that program to read an arbitrary number of values from an input data file, and calculate the means of those numbers. To test the program, place the following values into an input data file and run the program on that file: 1.0, 2.0, 5.0, 4.0, 3.0, 2.1, 4.7, 3.0.

5-28. Converting Radians to Degrees/Minutes/Seconds Angles are often measured in degrees (°), minutes ('), and seconds ("), with 360 degrees in a circle, 60 minutes in a degree, and 60 seconds in a minute. Write a program that reads angles in radians from an input disk file, and converts them into degrees, minutes, and seconds. Test your program by placing the following four angles expressed in radians into an input file, and reading that file into the program: 0.0, 1.0, 3.141593, 6.0.

5-29. There is a logical error in program `least_squares_fit` from Example 5-5. The error can cause the program to abort with a divide-by-zero error. It slipped through the example because we did not test the program exhaustively for all possible inputs. Find the error, and rewrite the program to eliminate it.

5-30. Ideal Gas Law Modify the Ideal Gas Law programs in Exercise 4-33 to print their output in neat columns, with appropriate column headings.

5-31. Antenna Gain Pattern The gain G of a certain microwave dish antenna can be expressed as a function of angle by the equation

$$G(\theta) = |\text{sinc } 6\theta| \quad \text{for } -\frac{\pi}{2} \leq \theta \leq \frac{\pi}{2} \tag{5-15}$$

where θ is measured in radians from the boresite of the dish, and the sinc function is defined as follows:

$$\text{sinc } x = \begin{cases} \dfrac{\sin x}{x} & x \neq 0 \\ 1 & x = 0 \end{cases} \tag{5-16}$$

Calculate a table of gain versus the angle off boresite *in degrees* for this antenna for the range $0° \leq \theta \leq 90°$ in 1° steps. Label this table with the title "Antenna Gain vs Angle (deg)", and include column headings on the output.

5-32. Bacterial Growth Modify the bacterial growth problem of Exercise 4-25 to produce a neat table containing the number of bacteria as a function of time.

5-33. Output Power from a Motor The output power produced by a rotating motor is given by the equation

$$P = \tau_{\text{IND}}\, \omega_m \tag{5-17}$$

where τ_{IND} is the induced torque on the shaft in newton-meters, ω_m is the rotational speed of the shaft in radians per second, and P is in watts. Assume that the rotational speed of a particular motor shaft is given by the equation

$$\omega_m = 377(1 - e^{-0.25t}) \text{ rad/s} \tag{5-18}$$

and the induced torque on the shaft is given by

$$\tau_{IND} = 10e^{-0.25t} \text{ N} \cdot \text{m} \tag{5-19}$$

Calculate the torque, speed, and power supplied by this shaft versus time for $0 \leq t \leq 10$ s at intervals of 0.25 s, and display the results in a table. Be sure to label your table and provide column headings.

5-34. Calculating Orbits When a satellite orbits the Earth, the satellite's orbit will form an ellipse with the Earth located at one of the focal points of the ellipse. The satellite's orbit can be expressed in polar coordinates as

$$r = \frac{p}{1 - \varepsilon \cos \theta} \tag{5-20}$$

where r and θ are the distance and angle of the satellite from the center of the Earth, p is a parameter specifying the size of the orbit, and ε is a parameter representing the eccentricity of the orbit. A circular orbit has an eccentricity ε of 0. An elliptical orbit has an eccentricity of $0 \leq \varepsilon \leq 1$. If $\varepsilon > 1$, the satellite follows a hyperbolic path and escapes from the Earth's gravitational field.

Consider a satellite with a size parameter $p = 10{,}000$ km. Calculate and create a table of the height of this satellite versus θ if (a) $\varepsilon = 0$; (b) $\varepsilon = 0.25$; (c) $\varepsilon = 0.5$. How close does each orbit come to the center of the Earth? How far away does each orbit get from the center of the Earth?

5-35. Apogee and Perigee The term r in Equation (5-20) refers to the range from a satellite to the *center of the Earth*. If the radius of the Earth $R = 6.371 \times 10^6$ m, then we can calculate the satellite height above the Earth from the equation

$$h = r - R \tag{5-21}$$

where h is the height in meters, and r is the range to the center of the Earth calculated from Equation (5-20).

The *apogee* of an orbit is the maximum height of the orbit above the surface of the Earth, and the *perigee* of an orbit is the minimum height of the orbit above the surface of the Earth. We can use Equations (5-20) and (5-21) to calculate the apogee and perigee of an orbit.

Consider a satellite with a size parameter $p = 10{,}000$ km. Calculate and create a table of the apogee and perigee of this satellite versus eccentricity for $0 \leq \varepsilon \leq 0.5$, in steps of 0.05.

5-36. Dynamically Modifying Format Descriptors Write a program to read a set of four real values in free format from each line of an input data file, and print them out on the standard output device. Each value should be printed in F14.6 format if it is exactly zero or if it lies in the range $0.01 \leq |\text{value}| < 1000.0$, and in ES14.6 format otherwise. (*Hint:* Define the output format in a character variable, and modify it to match each line of data as it is printed.) Test your program on the following data set:

```
         0.00012         -250.        6.02E23       -0.012
             0.0       12345.6        1.6E-19     -1000.
----|----|----|----|----|----|----|----|----|----|
    5   10   15   20   25   30   35   40   45   50
```

5-37. Correlation Coefficient The method of least squares is used to fit a straight line to a noisy input data set consisting of pairs of values (x, y). As we saw in Example 5-5, the best fit to equation

$$y = mx + b \tag{5-5}$$

is given by

$$m = \frac{(\Sigma xy) - (\Sigma x)\bar{y}}{(\Sigma x^2) - (\Sigma x)\bar{x}} \tag{5-6}$$

and

$$b = \bar{y} - m\bar{x} \tag{5-7}$$

where

Σx is the sum of the x values
Σx^2 is the sum of the squares of the x values
Σxy is the sum of the products of the corresponding x and y values
$\bar{x}$ is the mean (average) of the x values
$\bar{y}$ is the mean (average) of the y values

Figure 5-15 shows two data sets and the least squares fits associated with each one. As you can see, the low-noise data fits the least squares line much better than the noisy data does. It would be useful to have some quantitative way to describe how well the data fits the least squares line given by Equations (5-5) through (5-7).

(a)

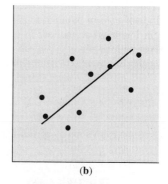
(b)

FIGURE 5-15
Two different least squares fits: (a) with good, low-noise data; (b) with very noisy data.

There is a standard statistical measure of the "goodness of fit" of a data set to a least squares line. It is called a *correlation coefficient*. The correlation coefficient is equal to 1.0 when there is a perfect positive linear relationship between data x and y, and it is equal to -1.0 when there is a perfect negative linear relationship between data x and y. The correlation coefficient is 0.0 when there is no linear relationship between x and y at all. The correlation coefficient is given by the equation

$$r = \frac{n(\Sigma xy) - (\Sigma x)(\Sigma y)}{\sqrt{[(n\Sigma x^2) - (\Sigma x)^2][(n\Sigma y^2) - (\Sigma y)^2]}} \qquad (5\text{-}22)$$

where r is the correlation coefficient and n is the number of data points included in the fit.

Write a program to read an arbitrary number of (x, y) data pairs from an input data file, and to calculate and print out both the least squares fit to the data and the correlation coefficient for the fit. If the correlation coefficient is small ($|r| < 0.3$), write out a warning message to the user.

5-38. Aircraft Turning Radius An object moving in a circular path at a constant tangential velocity v is shown in Figure 5-16. The radial acceleration required for the object to move in the circular path was given by the Equation (5-15)

$$a = \frac{v^2}{r} \qquad (5\text{-}23)$$

where a is the centripetal acceleration of the object in m/s², v is the tangential velocity of the object in m/s, and r is the turning radius in meters. Suppose that the object is an aircraft, and write a program to answer the following questions about it:

(a) Print a table of the aircraft turning radius as a function of aircraft speed for speeds between Mach 0.5 and Mach 2.0 in Mach 0.1 steps, assuming that the acceleration remains 2 g. Be sure to include proper labels on your table.

(b) Print a table of the aircraft turning radius as a function of centripetal acceleration for accelerations between 2 g and 8 g in 0.5 g steps, assuming a constant speed of Mach 0.85. Be sure to include proper labels on your table.

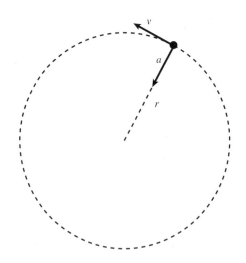

FIGURE 5-16
An object moving in uniform circular motion due to the centripetal acceleration a.

6

Introduction to Arrays

OBJECTIVES

- Know how to define, initialize, and use arrays.
- Know how to use whole array operations to operate on entire arrays of data in a single statement.
- Know how to use array sections.
- Learn how to read and write arrays and array sections.

An **array** is a group of variables or constants, all of the same type, which are referred to by a single name. The values in the group occupy consecutive locations in the computer's memory (see Figure 6-1). An individual value within the array is called an **array element;** it is identified by the name of the array together with a **subscript** pointing to the particular location within the array. For example, the first variable shown in Figure 6-1 is referred to as a(1), and the fifth variable shown in the figure is referred to as a(5). The subscript of an array is of type INTEGER. Either constants or variables may be used for array subscripts.

As we shall see, arrays can be extremely powerful tools. They permit us to apply the same algorithm over and over again to many different data items with a simple DO loop. For example, suppose that we need to take the square root of 100 different real numbers. If the numbers are stored as elements of an array a consisting of 100 real values, then the code

```
DO i = 1, 100
   a(i) = SQRT(a(i))
END DO
```

will take the square root of each real number, and store it back into the memory location that it came from. If we wanted to take the square root of 100 real numbers without using arrays, we would have to write out

```
a1   = SQRT(a1)
a2   = SQRT(a2)
    . . .
a100 = SQRT(a100)
```

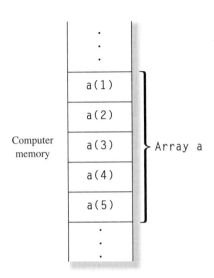

FIGURE 6-1
The elements of an array occupy successive locations in a computer's memory.

as 100 separate statements! Arrays are obviously a *much* cleaner and shorter way to handle repeated similar operations.

Arrays are very powerful tools for manipulating data in Fortran. As we shall see, it is possible to manipulate and perform calculations with individual elements of arrays one by one, with whole arrays at once, or with various subsets of arrays. We will first learn how to declare arrays in Fortran programs. Then, we will learn how to use individual array elements in Fortran statements, and afterward we will learn to use whole arrays or array subsets in Fortran statements.

6.1
DECLARING ARRAYS

Before an array can be used, its type and the number of elements it contains must be declared to the compiler in a type declaration statement, so that the compiler will know what sort of data is to be stored in the array, and how much memory is required to hold it. For example, a real array `voltage` containing 16 elements could be declared as follows:[1]

```
REAL, DIMENSION(16) :: voltage
```

The `DIMENSION` **attribute** in the type declaration statement declares the size of the array being defined. The elements in array `voltage` would be addressed as

---

[1] An alternate way to declare an array is to attach the dimension information directly to the array name:

```
REAL :: voltage(16)
```

This declaration style is provided for backward compatibility with earlier version of Fortran. It is fully equivalent to the array declaration shown above.

Introduction to Arrays

`voltage(1)`, `voltage(2)`, etc., up to `voltage(16)`. Similarly, an array of fifty 20-character-long variables could be declared as follows:

```
CHARACTER(len = 20), DIMENSION(50) :: last_name
```

Each of the elements in array `last_name` would be a 20-character-long variable, and the elements would be addressed as `last_name(1)`, `last_name(2)`, etc.

Arrays may be declared with more than one subscript, so they may be organized into two or more dimensions. These arrays are convenient for representing data that is normally organized into multiple dimensions, such as map information. The number of subscripts declared for a given array is called the **rank** of the array. Both array `voltage` and array `last_name` are rank 1 arrays, since they have only one subscript. We will see more complex arrays later in Chapter 8.

The number of elements in a given dimension of an array is called the **extent** of the array in that dimension. The extent of the first (and only) subscript of array `voltage` is 20, and the extent of the first (and only) subscript of array `last_name` is 50. The **shape** of an array is defined as the combination of its rank and the extent of the array in each dimension. Thus, two arrays have the same shape if they have the same rank and the same extent in each dimension. Finally, the **size** of an array is the total number of elements declared in that array. For simple rank 1 arrays, the size of the array is the same as the extent of its single subscript. Therefore, the size of array `voltage` is 20, and the size of array `last_name` is 50.

Array constants may also be defined. An array constant is an array consisting entirely of constants. It is defined by placing the constant values between special delimiters, called **array constructors.** The starting delimiter of a Fortran array constructor is (/ or [, and the ending delimiter of an array constructor is /) or]. For example, each of the two expressions shown below defines an array constant containing five integer elements:

```
(/ 1, 2, 3, 4, 5 /)
[ 1, 2, 3, 4, 5 ]
```

The form of the array constructor using (/ /) is older than the array constructor using [ ], so more existing programs use it. You should recognize both forms of array constructor. We will use them both throughout the rest of this book, with a preference for the newer form.

■ 6.2
USING ARRAY ELEMENTS IN FORTRAN STATEMENTS

This section contains some of the practical details involved in using arrays in Fortran programs.

6.2.1 Array Elements are Just Ordinary Variables

Each element of an array is a variable just like any other variable, and *an array element may be used in any place where an ordinary variable of the same type may be used.*

Array elements may be included in arithmetic and logical expressions, and the results of an expression may be assigned to an array element. For example, assume that arrays index and temp are declared as:

```
INTEGER, DIMENSION(10) :: index
REAL, DIMENSION(3) :: temp
```

Then the following Fortran statements are perfectly valid:

```
index(1) = 5
temp(3) = REAL(index(1)) / 4.
WRITE (*,*) ' index(1) = ', index(1)
```

Under certain circumstances, entire arrays or subsets of arrays can be used in expressions and assignment statements. These circumstances will be explained in Section 6.3.

6.2.2 Initialization of Array Elements

Just as with ordinary variables, the values in an array must be initialized before use. If an array is not initialized, the contents of the array elements are undefined. In the following Fortran statements, array j is an example of an **uninitialized array.**

```
INTEGER, DIMENSION(10) :: j
WRITE (*,*) 'j(1) = ', j(1)
```

The array j has been declared by the type declaration statement, but no values have been placed into it yet. Since the contents of an uninitialized array are unknown and can vary from computer to computer, *the elements of the array should never be used until they are initialized to known values.*

Good Programming Practice

Always initialize the elements in an array before they are used.

The elements in an array may be initialized by one of three techniques:

1. Arrays may be initialized using assignment statements.
2. Arrays may be initialized in type declaration statements at compilation time.
3. Arrays may be initialized using READ statements.

Initializing arrays with assignment statements

Initial values may be assigned to the array using assignment statements, either element-by-element in a DO loop or all at once with an array constructor. For example, the following DO loop will initialize the elements of array array1 to 0.0, 2.0, 3.0, etc., one element at a time:

```
REAL, DIMENSION(10) :: array1
DO i = 1, 10
```

Introduction to Arrays

```
            array1(i) = REAL(i)
        END DO
```

The following assignment statement accomplishes the same function all at once using an array constructor:

```
            REAL, DIMENSION(10) :: array1
            array1 = [1.,2.,3.,4.,5.,6.,7.,8.,9.,10.]
```

It is also possible to initialize all of the elements of an array to a single value with a simple assignment statement. For example, the following statement initializes all of the elements of array1 to zero:

```
            REAL, DIMENSION(10) :: array1
            array1 = 0.
```

The simple program shown in Figure 6-2 calculates the squares of the numbers in array number, and then prints out the numbers and their squares. Note that the values in array number are initialized element-by-element with a DO loop.

FIGURE 6-2
A program to calculate the squares of the integers from 1 to 10, using assignment statements to initialize the values in array number.

```
PROGRAM squares

IMPLICIT NONE

INTEGER :: i
INTEGER, DIMENSION(10) :: number, square

! Initialize number and calculate square.
DO i = 1, 10
    number(i) = i              ! Initialize number
    square(i) = number(i)**2   ! Calculate square
END DO

! Write out each number and its square.
DO i = 1, 10
    WRITE (*,100) number(i), square(i)
    100 FORMAT ('Number = ',I6,' Square = ',I6)
END DO

END PROGRAM squares
```

Initializing arrays in type declaration statements

Initial values may be loaded into an array at compilation time by declaring their values in a type declaration statement. To initialize an array in a type declaration statement, we use an array constructor to declare its initial values in that statement. For example, the following statement declares a five-element integer array array2, and initializes the elements of array2 to 1, 2, 3, 4, and 5:

```
            INTEGER, DIMENSION(5) :: array2 = [ 1, 2, 3, 4, 5 ]
```

The five-element array constant [1, 2, 3, 4, 5] was used to initialize the five-element array array2. In general, *the number of elements in the constant must match the number of elements in the array being initialized.* Either too few or too many elements will result in a compiler error.

This method works well to initialize small arrays, but what do we do if the array has 100 (or even 1000) elements? Writing out the initial values for a 100-element array would be very tedious and repetitive. To initialize larger arrays, we can use an **implied DO loop.** An implied DO loop has the general form

```
(arg1, arg2, ... , index = istart, iend, incr)
```

where `arg1`, `arg2`, etc., are values evaluated each time the loop is executed, and `index`, `istart`, `iend`, and `incr` function in exactly the same way as they do for ordinary counting DO loops. For example, the `array2` declaration above could be written using an implied DO loop as:

```
INTEGER, DIMENSION(5) :: array2 = [ (i, i=1,5) ]
```

and a 1000-element array could be initialized to have the values 1, 2, ..., 1000 using an implied DO loop as follows:

```
INTEGER, DIMENSION(1000) :: array3 = [ (i, i=1,1000) ]
```

Implied DO loops can be nested or mixed with constants to produce complex patterns. For example, the following statements initialize the elements of `array4` to zero if they are not divisible by 5, and to the element number if they are divisible by 5.

```
INTEGER, DIMENSION(25) :: array4 = [ ((0,i=1,4),5*j, j=1,5) ]
```

The inner DO loop `(0,i=1,4)` executes completely for each step of the outer DO loop, so for each value of the outer loop index `j`, we will have four zeros (from the inner loop) followed by the number `5*j`. The resulting pattern of values produced by these nested loops is:

```
0, 0, 0, 0, 5, 0, 0, 0, 0, 10, 0, 0, 0, 0, 15, ...
```

Finally, all the elements of an array can be initialized to a single constant value by simply including the constant in the type declaration statement. In the following example, all of the elements of `array5` are initialized to 1.0:

```
REAL, DIMENSION(100) :: array5 = 1.0
```

The program in Figure 6-3 illustrates the use of type declaration statements to initialize the values in an array. It calculates the square roots of the numbers in array `value`, and then prints out the numbers and their square roots.

FIGURE 6-3
A program to calculate the square roots of the integers from 1 to 10, using a type declaration statement to initialize the values in array `value`.

```
PROGRAM square_roots

IMPLICIT NONE

INTEGER :: i
REAL, DIMENSION(10) :: value = [ (i, i=1,10) ]
```

(continued)

(concluded)

```
REAL, DIMENSION(10) :: square_root

! Calculate the square roots of the numbers.
DO i = 1, 10
   square_root(i) = SQRT(value(i))
END DO

! Write out each number and its square root.
DO i = 1, 10
   WRITE (*,100) value(i), square_root(i)
   100 FORMAT ('Value = ',F5.1,' Square Root = ',F10.4)
END DO

END PROGRAM square_roots
```

Initializing arrays with READ statements

Array may also be initialized with READ statements. The use of arrays in I/O statements will be described in detail in Section 6.4.

6.2.3 Changing the Subscript Range of an Array

The elements of an *N*-element array are normally addressed using the subscripts 1, 2, ..., *N*. Thus, the elements of array arr declared with the statement

```
REAL, DIMENSION(5) :: arr
```

would be addressed as arr(1), arr(2), arr(3), arr(4), and arr(5). In some problems, however, it is more convenient to address the array elements with other subscripts. For example, the possible grades on an exam might range from 0 to 100. If we wished to accumulate statistics on the number of people scoring any given grade, it would be convenient to have a 101-element array whose subscripts ranged from 0 to 100 instead of 1 to 101. If the subscripts ranged from 0 to 100, each student's exam grade could be used directly as an index into the array.

For such problems, Fortran provides a way to specify the range of numbers that will be used to address the elements of an array. To specify the subscript range, we include the starting and ending subscript numbers in the declaration statement, with the two numbers separated by a colon.

```
REAL, DIMENSION(lower_bound:upper_bound) :: array
```

For example, the following three arrays all consist of five elements:

```
REAL, DIMENSION(5) :: a1
REAL, DIMENSION(-2:2) :: b1
REAL, DIMENSION(5:9) :: c1
```

Array a1 is addressed with subscripts 1 through 5, array b1 is addressed with subscripts −2 through 2, and array c1 is addressed with subscripts 5 through 9. *All three arrays have the same shape*, since they have the same number of dimensions and the same extent in each dimension.

In general, the number of elements in a given dimension of an array can be found from the equation

$$\text{extent} = \text{upper\_bound} - \text{lower\_bound} + 1 \tag{6-1}$$

The simple program squares_2 shown in Figure 6-4 calculates the squares of the numbers in array number, and then prints out the numbers and their squares. The arrays in this example contain 11 elements, addressed by the subscripts −5, −4, . . ., 0, . . ., 4, 5.

FIGURE 6-4
A program to calculate the squares of the integers from −5 to 5, using array elements addressed by subscripts −5 through 5.

```
PROGRAM squares_2

IMPLICIT NONE

INTEGER :: i
INTEGER, DIMENSION(-5:5) :: number, square

! Initialize number and calculate square.
DO i = -5, 5
   number(i) = i                   ! Initialize number
   square(i) = number(i)**2        ! Calculate square
END DO

! Write out each number and its square.
DO i = -5, 5
   WRITE (*,100) number(i), square(i)
   100 FORMAT ('Number = ',I6,' Square = ',I6)
END DO

END PROGRAM squares_2
```

When program squares_2 is executed, the results are

```
C:\book\fortran\chap6>squares_2
Number =     -5 Square =     25
Number =     -4 Square =     16
Number =     -3 Square =      9
Number =     -2 Square =      4
Number =     -1 Square =      1
Number =      0 Square =      0
Number =      1 Square =      1
Number =      2 Square =      4
Number =      3 Square =      9
Number =      4 Square =     16
Number =      5 Square =     25
```

6.2.4 Out-of-Bounds Array Subscripts

Each element of an array is addressed using an integer subscript. The range of integers that can be used to address array elements depends on the declared extent of the array.

Introduction to Arrays

For a real array declared as

```
REAL, DIMENSION(5) :: a
```

the integer subscripts 1 through 5 address elements in the array. *Any other integers (less than 1 or greater than 5) could not be used as subscripts, since they do not correspond to allocated memory locations.* Such integer subscripts are said to be **out of bounds** for the array. But what would happen if we make a mistake and try to access the out-of-bounds element a(6) in a program?

The answer to this question is very complicated, since it varies from compiler to compiler, and also on the compilation options selected for the compiler. In some cases, a running Fortran program will check every subscript used to reference an array to see if it is in bounds. If an out-of-bounds subscript is detected, the program will issue an informative error message and stop. Unfortunately, such **bounds checking** requires a lot of computer time, and the program will run more slowly. To make programs run faster, most Fortran compilers make bounds checking optional. If it is turned on, programs run slower, but they are protected from out-of-bounds references. If it is turned off, programs will run much faster, but out-of-bounds references will not be checked. If your Fortran compiler has a bounds checking option, you should always turn it on during debugging to help detect programming errors. Once the program has been debugged, bounds checking can be turned off if necessary to increase the execution speed of the final program.

Good Programming Practice
Always turn on the bounds checking option on your Fortran compiler during program development and debugging to help you catch programming errors producing out-of-bounds references. The bounds checking option may be turned off if necessary for greater speed in the final program.

What happens in a program if an out-of-bounds reference occurs and the bounds checking option is not turned on? Sometimes, the program will abort. Much of the time, though, the computer will simply go to the location in memory *at which the referenced array element would have been if it had been allocated*, and use that memory location (see Figure 6-5). For example, the array a declared above has five elements in it. If a(6) were used in a program, the computer would access the first word beyond the end of array a. Since that memory location will be allocated for a totally different purpose, the program can fail in subtle and bizarre ways that can be almost impossible to track down. Be careful with your array subscripts, and always use the bounds checker when you are debugging!

The program shown in Figure 6-6 illustrates the behavior of a Fortran program containing incorrect array references with and without bounds checking turned on. This simple program declares a five-element real array a and a five-element real array b. The array a is initialized with the values 1., 2., 3., 4., and 5., and array b is initialized with the values 10., 20., 30., 40., and 50. Many Fortran compilers will allocate the memory for array b immediately after and the memory for array a, as shown in Figure 6-5.[2]

---

[2] But they are not required to do so. The Fortran standard does not constrict how the compilers choose to allocate data in memory.

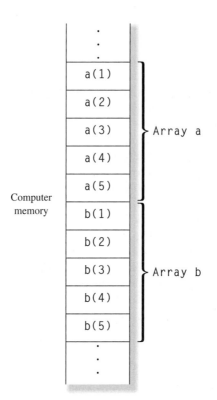

FIGURE 6-5
A computer memory showing a five-element array a immediately followed by a five-element array b. If bounds checking is turned off, some processors may not recognize the end of array a, and may treat the memory location after the end of a as a(6).

The program in Figure 6-6 uses a DO loop to write out the values in the elements 1 through 6 of array a, despite the fact that array a only has five elements. Therefore, it will attempt to access the out-of-bounds array element a(6).

FIGURE 6-6
A simple program to illustrate the effect of out-of-bounds array references with and without bounds checking turned on.

```
PROGRAM bounds
!
! Purpose:
!   To illustrate the effect of accessing an out-of-bounds
!   array element.
!
! Record of revisions:
!     Date        Programmer          Description of change
!     ====        ==========          =====================
!   11/15/15    S. J. Chapman         Original code
!
IMPLICIT NONE
```

(continued)

Introduction to Arrays

(*concluded*)

```
! Declare and initialize the variables used in this program.
INTEGER :: i                   ! Loop index
REAL, DIMENSION(5) :: a = (/ 1., 2., 3., 4., 5./)
REAL, DIMENSION(5) :: b = (/10.,20.,30.,40.,50./)

! Write out the values of array a
DO i = 1, 6
   WRITE (*,100) i, a(i)
   100 FORMAT ( 'a(', I1, ') = ', F6.2 )
END DO

END PROGRAM bounds
```

If this program is compiled with the Intel Visual Fortran compiler on a PC-compatible computer with bounds checking turned *on* (the -check option), the result is

```
C:\book\fortran\chap6>ifort -check bounds.f90
Intel(R) Visual Fortran Intel(R) 64 Compiler for applications running on
Intel(R) 64, Version 16.0.2.180 Build 20160204
Copyright (C) 1985-2016 Intel Corporation. All rights reserved.

Microsoft (R) Incremental Linker Version 12.00.40629.0
Copyright (C) Microsoft Corporation. All rights reserved.

-out:bounds.exe
-subsystem:console
bounds.obj

C:\book\fortran\chap6>bounds

a(1) =   1.00
a(2) =   2.00
a(3) =   3.00
a(4) =   4.00
a(5) =   5.00
forrtl: severe (408): fort: (10): Subscript #1 of the array A has value 6
which is greater than the upper bound of 5

Image              PC                 Routine         Line         Source
bounds.exe         00007FF62EEAB66E   Unknown         Unknown      Unknown
bounds.exe         00007FF62EEA117A   Unknown         Unknown      Unknown
bounds.exe         00007FF62EEF116E   Unknown         Unknown      Unknown
bounds.exe         00007FF62EEF1A28   Unknown         Unknown      Unknown
KERNEL32.DLL       00007FFA56B38102   Unknown         Unknown      Unknown
ntdll.dll          00007FFA594DC5B4   Unknown         Unknown      Unknown
```

The program checked each array reference, and aborted when an out-of-bounds expression was encountered. Note that the error message tells us what is wrong, and even the line number at which it occurred. If the program is compiled with bounds checking turned *off*, the result is

```
C:\book\fortran\chap6>bounds

a(1) =   1.00
```

```
a(2) =    2.00
a(3) =    3.00
a(4) =    4.00
a(5) =    5.00
a(6) =   10.00
```

When the program tried to write out a(6), it wrote out the contents of the first memory location after the end of the array. This location just happened to be the first element of array b.

6.2.5 The Use of Named Constants with Array Declarations

In many Fortran programs, arrays are used to store large amounts of information. The amount of information that a program can process depends on the size of the arrays it contains. If the arrays are relatively small, the program will be small and will not require much memory to run, but it will only be able to handle a small amount of data. On the other hand, if the arrays are large, the program will be able to handle a lot of information, but it will require a lot of memory to run. The array sizes in such a program are frequently changed to make it run better for different problems or on different processors.

It is good practice to always declare the array sizes using named constants. Named constants make it easy to resize the arrays in a Fortran program. In the following code, the sizes of all arrays can be changed by simply changing the single named constant MAX_SIZE.

```
INTEGER, PARAMETER :: MAX_SIZE = 1000
REAL :: array1(MAX_SIZE)
REAL :: array2(MAX_SIZE)
REAL :: array3(2*MAX_SIZE)
```

This may seem like a small point, but it is *very* important to the proper maintenance of large Fortran programs. If all related array sizes in a program are declared using named constants, and if those same named constants are used in any size tests in the program, then it will be much simpler to modify the program later. Imagine what it would be like if you had to locate and change every reference to array sizes within a 50,000 line program! The process could take weeks to complete and debug. By contrast, the size of a well-designed program could be modified in five minutes by changing only one statement in the code.

Good Programming Practice

Always declare the sizes of arrays in a Fortran program using parameters to make them easy to change.

EXAMPLE 6-1

Finding the Largest and Smallest Values in a Data Set:

To illustrate the use of arrays, we will write a simple program that reads in data values, and finds the largest and smallest numbers in the data set. The program will then write out the values, with the word 'LARGEST' printed by the largest value and the word 'SMALLEST' printed by the smallest value in the data set.

Introduction to Arrays

SOLUTION

This program must ask the user for the number of values to read, and then read the input values into an array. Once the values are all read, it must go through the data to find the largest and smallest values in the data set. Finally, it must print out the values, with the appropriate annotations beside the largest and smallest values in the data set.

1. **State the problem.**

    We have not yet specified the type of data to be processed. If we are processing integer data, then the problem may be stated as follows:

    Develop a program to read a user-specified number of integer values from the standard input device, locate the largest and smallest values in the data set, and write out all of the values with the words 'LARGEST' and 'SMALLEST' printed by the largest and smallest values in the data set.

2. **Define the inputs and outputs.**

    There are two types of inputs to this program:

    (a) An integer containing the number of integer values to read. This value will come from the standard input device.
    (b) The integer values in the data set. These values will also come from the standard input device.

    The outputs from this program are the values in the data set, with the word 'LARGEST' printed by the largest value, and the word 'SMALLEST' printed by the smallest value.

3. **Describe the algorithm.**

    The program can be broken down into four major steps

    ```
    Get the number of values to read
    Read the input values into an array
    Find the largest and smallest values in the array
    Write out the data with the words 'LARGEST' and 'SMALLEST' at
       the appropriate places
    ```

    The first two major steps of the program are to get the number of values to read in and to read the values into an input array. We must prompt the user for the number of values to read. If that number is less than or equal to the size of the input array, then we should read in the data values. Otherwise, we should warn the user and quit. The detailed pseudocode for these steps is:

    ```
    Prompt user for the number of input values nvals
    Read in nvals
    IF nvals <= max_size then
       DO for j = 1 to nvals
          Read in input values
       End of DO
       ...
       ... (Further processing here)
       ...
    ELSE
       Tell user that there are too many values for array size
    End of IF
    END PROGRAM
    ```

Next we must locate the largest and smallest values in the data set. We will use variables `ilarge` and `ismall` as pointers to the array elements having the largest and smallest values. The pseudocode to find the largest and smallest values is:

```
! Find largest value
temp ← input(1)
ilarge ← 1
DO for j = 2 to nvals
   IF input(j) > temp then
      temp ← input(j)
      ilarge ← j
   End of IF
End of DO

! Find smallest value
temp ← input(1)
ismall ← 1
DO for j = 2 to nvals
   IF input(j) < temp then
      temp ← input(j)
      ismall ← j
   End of IF
End of DO
```

The final step is writing out the values with the largest and smallest numbers labeled:

```
DO for j = 1 to nvals
   IF ismall == j then
      Write input(j) and 'SMALLEST'
   ELSE IF ilarge == j then
      Write input(j) and 'LARGEST'
   ELSE
      Write input(j)
   END of IF
End of DO
```

4. **Turn the algorithm into Fortran statements.**

   The resulting Fortran program is shown in Figure 6-7.

FIGURE 6-7
A program to read in a data set from the standard input device, find the largest and smallest values, and print the values with the largest and smallest values labeled.

```
PROGRAM extremes
!
!  Purpose:
!    To find the largest and smallest values in a data set,
!    and to print out the data set with the largest and smallest
!    values labeled.
!
!  Record of revisions:
!      Date        Programmer       Description of change
!      ====        ==========       =====================
!    11/16/15    S. J. Chapman       Original code
!
```

(continued)

(*continued*)

```
IMPLICIT NONE

! Data dictionary: declare constants
INTEGER, PARAMETER :: MAX_SIZE = 10   ! Max size of data set

! Data dictionary: declare variable types, definitions, & units
INTEGER, DIMENSION(MAX_SIZE) :: input  ! Input values
INTEGER :: ilarge                      ! Pointer to largest value
INTEGER :: ismall                      ! Pointer to smallest value
INTEGER :: j                           ! DO loop index
INTEGER :: nvals                       ! Number of vals in data set
INTEGER :: temp                        ! Temporary variable

! Get number of values in data set
WRITE (*,*) 'Enter number of values in data set:'
READ (*,*) nvals

! Is the number <= MAX_SIZE?
size: IF ( nvals <= MAX_SIZE ) THEN

   ! Get input values.
   in: DO J = 1, nvals
      WRITE (*,100) 'Enter value ', j
      100 FORMAT (A,I3,': ')
      READ (*,*) input(j)
   END DO in

   ! Find the largest value.
   temp = input(1)
   ilarge = 1
   large: DO j = 2, nvals
      IF ( input(j) > temp ) THEN
         temp = input(j)
         ilarge = j
      END IF
   END DO large

   ! Find the smallest value.
   temp = input(1)
   ismall = 1
   small: DO j = 2, nvals
      IF ( input(j) < temp ) THEN
         temp = input(j)
         ismall = j
      END IF
   END DO small

   ! Write out list.
   WRITE (*,110)
   110 FORMAT ('The values are:')
   out: DO j = 1, nvals
      IF ( j == ilarge ) THEN
         WRITE (*,'(I6,2X,A)') input(j), 'LARGEST'
```

(*continued*)

(*concluded*)

```
      ELSE IF ( J == ismall ) THEN
         WRITE (*,'(I6,2X,A)') input(j), 'SMALLEST'
      ELSE
         WRITE (*,'(I6)') input(j)
      END IF
   END DO out

ELSE size

   ! nvals > max_size. Tell user and quit.
   WRITE (*,120) nvals, MAX_SIZE
   120 FORMAT ('Too many input values: ', I6, ' > ', I6)

END IF size

END PROGRAM extremes
```

5. **Test the program.**

To test this program, we will use two data sets, one with 6 values and one with 12 values. Running this program with six values yields the following result:

```
C:\book\fortran\chap6>extremes
Enter number of values in data set:
6
Enter value  1:
-6
Enter value  2:
5
Enter value  3:
-11
Enter value  4:
16
Enter value  5:
9
Enter value  6:
0

The values are:
    -6
     5
   -11 SMALLEST
    16 LARGEST
     9
     0
```

The program correctly labeled the largest and smallest values in the data set. Running this program with 12 values yields the following result:

```
C:\book\fortran\chap6>extremes
Enter number of values in data set:
12
Too many input values:       12 >      10
```

Introduction to Arrays

The program recognized that there were too many input values, and quit. Thus, the program gives the correct answers for both of our test data sets.

This program used the named constant MAX_SIZE to declare the size of the array, and also in all comparisons related to the array. As a result, we could change this program to process up to 1000 values by simply changing the value of MAX_SIZE from 10 to 1000.

6.3
USING WHOLE ARRAYS AND ARRAY SUBSETS IN FORTRAN STATEMENTS

Both whole arrays and array subsets may be used in Fortran statements. When they are, the operations are performed on all of the specified array elements simultaneously. This section teaches us how to use whole arrays and array subsets in Fortran statements.

6.3.1 Whole Array Operations

Under certain circumstances, **whole arrays** may be used in arithmetic calculations as though they were ordinary variables. If two arrays are the same **shape,** then they can be used in ordinary arithmetic operations, and the operation will be applied on an element-by-element basis (Figure 6-8). Consider the example program in Figure 6-9. Here, arrays a, b, c, and d are all four elements long. Each element in array c is calculated as the sum of the corresponding elements in arrays a and b, using a DO loop. Array d is calculated as the sum of arrays a and b in a single assignment statement.

a(1)	1.		b(1)	5.		d(1)	6.
a(2)	2.		b(2)	6.		d(2)	8.
a(3)	3.		b(3)	7.		d(3)	10.
a(4)	4.		b(4)	8.		d(4)	12.
a		+	b		=	d	

FIGURE 6-8
When an operation is applied to two arrays of the same shape, the operation is performed on the arrays on an element-by-element basis.

FIGURE 6-9
A program illustrating both element-by-element addition and whole array addition.

```
PROGRAM add_arrays

IMPLICIT NONE

INTEGER :: i
REAL, DIMENSION(4) :: a = [ 1., 2., 3., 4.]
REAL, DIMENSION(4) :: b = [ 5., 6., 7., 8.]
REAL, DIMENSION(4) :: c, d

! Element by element addition
DO i = 1, 4
   c(i) = a(i) + b(i)
END DO

! Whole array addition
d = a + b

! Write out results
WRITE (*,100) 'c', c
WRITE (*,100) 'd', d
100 FORMAT (A,' = ',5(F6.1,1X))

END PROGRAM add_arrays
```

When this program is executed, the results are exactly the same for both calculations:

```
C:\book\fortran\chap6>add_arrays
c =   6.0   8.0  10.0  12.0
d =   6.0   8.0  10.0  12.0
```

Two arrays can be used as operands in an intrinsic operation (addition, etc.) if and only if they have the *same shape*. This means that they must have the *same number of dimensions* (the same **rank**), and *the same number of elements in each dimension* (the same **extent**). Two arrays of the same shape are said to be **conformable**. Note that although the two arrays must be the same shape, they do *not* have to have the same subscript range in each dimension. The following arrays can be added freely even though the subscript ranges used to address their elements are different.

```
REAL, DIMENSION(1:4)   :: a = [ 1., 2., 3., 4.]
REAL, DIMENSION(5:8)   :: b = [ 5., 6., 7., 8.]
REAL, DIMENSION(101:104) :: c
c = a + b
```

If two arrays are not conformable, then any attempt to perform arithmetic operations with them will produce a compile-time error.

Scalar values are also conformable with arrays. In that case, the scalar value is applied equally to every element of the array. For example, after the following piece of code is executed, array c will contain the values [10., 20., 30., 40.].

```
REAL, DIMENSION(4) :: a = [ 1., 2., 3., 4.], c
REAL :: b = 10
c = a * b
```

Introduction to Arrays

Many Fortran intrinsic functions that are used with scalar values will also accept arrays as input arguments, and return arrays as results. The returned arrays will contain the result of applying the function to the input array on an element-by-element basis. These functions are called **elemental intrinsic functions,** since they operate on arrays on an element-by-element basis. Most common functions are elemental, including `ABS`, `SIN`, `COS`, `EXP`, `LOG`, etc. A complete list of elemental functions is contained in Appendix B. For example, consider an array a defined as

```
REAL, DIMENSION(4) :: a = [ -1., 2., -3., 4.]
```

Then, the function `ABS(a)` would return [1., 2., 3., 4.].

6.3.2 Array Subsets

We have already seen that it is possible to use either array elements or entire arrays in calculations. In addition, it is possible to use subsets of arrays in calculations. A subset of an array is called an **array section.** It is specified by replacing an array subscript with a **subscript triplet** or **vector subscript.**

A subscript triplet has the general form

subscript_1 : subscript_2 : stride

where *subscript_1* is the first subscript to be included in the array subset, *subscript_2* is the last subscript to be included in the array subset, and *stride* is the subscript increment through the data set. It works much like an implied `DO` loop. A subscript triplet specifies the ordered set of all array subscripts starting with *subscript_1* and ending with *subscript_2*, advancing at a rate of *stride* between values. For example, let's define an array array as

```
INTEGER, DIMENSION(10) :: array = [1,2,3,4,5,6,7,8,9,10]
```

Then the array subset `array(1:10:2)` would be an array containing only elements `array(1)`, `array(3)`, `array(5)`, `array(7)`, and `array(9)`.

Any or all of the components of a subscript triplet may be defaulted. If *subscript_1* is missing from the triplet, it defaults to the subscript of the first element in the array. If *subscript_2* is missing from the triplet, it defaults to the subscript of the last element in the array. If *stride* is missing from the triplet, it defaults to one. All of the following possibilities are examples of legal triplets:

subscript_1 : subscript_2 : stride
subscript_1 : subscript_2
subscript_1 :
subscript_1 : : stride
: subscript_2
: subscript_2 : stride
: : stride
:

EXAMPLE 6-2

Specifying Array Sections with Subscript Triplets:

Assume the following type declarations statements:

```
INTEGER :: i = 3, j = 7
REAL, DIMENSION(10) :: a = [1.,-2.,3.,-4.,5.,-6.,7.,-8.,9.,-10.]
```

Determine the number of elements in and the contents of the array sections specified by each of the following subscript triplets:

(a) a(:)
(b) a(i:j)
(c) a(i:j:i)
(d) a(i:j:j)
(e) a(i:)
(f) a(:j)
(g) a(::i)

SOLUTION

(a) a(:) is identical to the original array: [1., −2., 3., −4., 5., −6., 7., −8., 9., −10.]
(b) a(i:j) is the array subset starting at element 3 and ending at element 7, with a default stride of 1: [3., −4., 5., −6., 7.]
(c) a(i:j:i) is the array subset starting at element 3 and ending at element 7, with a stride of 3: [3., −6.]
(d) a(i:j:j) is the array subset starting at element 3 and ending at element 7, with a stride of 7: [3.]
(e) a(i:) is the array subset starting at element 3 and by default ending at element 10 (the end of the array), with a default stride of 1: [3., −4., 5., −6., 7., −8., 9., −10.]
(f) a(:j) is the array subset starting by default at element 1 and ending at element 7, with a default stride of 1: [1., −2., 3., −4., 5., −6., 7.]
(g) a(::i) is the array subset starting by default at element 1 and ending by default at element 10, with a stride of 3: [1., −4., 7., −10.]

Subscript triplets select ordered subsets of array elements for use in calculations. In contrast, **vector subscripts** allow arbitrary combinations of array elements to be selected for use in an operation. *A vector subscript is a one-dimensional integer array specifying the array elements to be used in a calculation.* The array elements may be specified in any order, and more than once. The resulting array will contain one element for each subscript specified in the vector. For example, consider the following type declaration statements:

```
INTEGER, DIMENSION(5) :: vec = [1, 6, 4, 1, 9 ]
REAL, DIMENSION(10) :: a = [1., -2., 3., -4., 5., -6., 7., -8., 9., -10.]
```

With these definitions, a(vec) would be the array [1., −6., −4., 1., 9.].

If a vector subscript includes any array element more than once, then the resulting array section is called a **many-one array section.** Such an array section *cannot be used on the left side of an assignment statement*, because it would specify that two or more different values should be assigned to the same array element at the same time! For example, consider the following Fortran statements:

```
INTEGER, DIMENSION(5) :: vec = [1, 2, 1 ]
REAL, DIMENSION(10) :: a = [10.,20.,30.]
REAL, DIMENSION(2) :: b
b(vec) = a
```

The assignment statement attempts to assign both the value 10. and the value 30. to array element b(1), which is impossible.

6.4
INPUT AND OUTPUT

It is possible to perform I/O operations on either individual array elements or entire arrays. Both types of I/O operations are described in this section.

6.4.1 Input and Output of Array Elements

We previously stated that an *array element* is a variable just like any other variable, and that an array element may be used in any place where an ordinary variable of the same type may be used. Therefore, READ and WRITE statements containing array elements are just like READ and WRITE statements for any other variables. To write out specific elements from an array, just name them in the argument list of the WRITE statement. For example, the following code writes out the first five elements of the real array a.

```
WRITE (*,100) a(1), a(2), a(3), a(4), a(5)
100 FORMAT ('a = ', 5F10.2)
```

6.4.2 The Implied DO Loop

The implied DO loop is also permitted in I/O statements. It allows an argument list to be written many times as a function of an index variable. Every argument in the argument list is written once for each value of the index variable in the implied DO loop. With an implied DO loop, the previous statement becomes:

```
WRITE (*,100) ( a(i), i = 1, 5 )
100 FORMAT ('a = ', 5F10.2)
```

The argument list in this case contains only one item: a(i). This list is repeated once for each value of the index variable i. Since i takes on the values from 1 to 5, the array elements a(1), a(2), a(3), a(4), and a(5) will be written.

The general form of a WRITE or READ statement with an implied DO loop is:

```
WRITE (unit,format) (arg1, arg2, ... , index = istart, iend, incr)
READ (unit,format) (arg1, arg2, ... , index = istart, iend, incr)
```

where *arg1, arg2,* etc., are the values to be written or read. The variable *index* is the DO loop index, and *istart, iend,* and *incr* are respectively the starting value, ending value, and increment of the loop index variable. The index and all of the loop control parameters should be of type INTEGER.

For a WRITE statement containing an implied DO loop, each argument in the argument list is written once each time the loop is executed. Therefore, a statement like

```
WRITE (*,1000) (i, 2*i, 3*i, i = 1, 3)
1000 FORMAT (9I6)
```

will write out nine values on a single line:

```
    1     2     3     2     4     6     3     6     9
```

Now let's look at a slightly more complicated example using arrays with an implied DO loop. Figure 6-10 shows a program that calculates the square root and cube root of a set of numbers, and prints out a table of square and cube roots. The program computes square roots and cube roots for all numbers between 1 and MAX_SIZE, where MAX_SIZE is a parameter. What will the output of this program look like?

FIGURE 6-10
A program that computes the square and cube roots of a set of number, and writes them out using an implied DO loop.

```
PROGRAM square_and_cube_roots
!
!   Purpose:
!     To calculate a table of numbers, square roots, and cube roots
!     using an implied DO loop to output the table.
!
!   Record of revisions:
!       Date          Programmer         Description of change
!       ====          ==========         =====================
!     11/16/15       S. J. Chapman        Original code
!
IMPLICIT NONE

! Data dictionary: declare constants
INTEGER, PARAMETER :: MAX_SIZE = 10      ! Max values in array

! Data dictionary: declare variable types, definitions, & units
INTEGER :: j                             ! Loop index
REAL, DIMENSION(MAX_SIZE) :: value        ! Array of numbers
REAL, DIMENSION(MAX_SIZE) :: square_root  ! Array of square roots
REAL, DIMENSION(MAX_SIZE) :: cube_root    ! Array of cube roots
```

(continued)

Introduction to Arrays

(*concluded*)

```
! Calculate the square roots & cube roots of the numbers.
DO j = 1, MAX_SIZE
   value(j) = real(j)
   square_root(j) = sqrt(value(j))
   cube_root(j) = value(j)**(1.0/3.0)
END DO

! Write out each number, its square root, and its cube root.
WRITE (*,100)
100 FORMAT (20X,'Table of Square and Cube Roots',/, &
            4X,' Number   Square Root Cube Root', &
            3X,' Number   Square Root Cube Root',/, &
            4X,' ======   =========== =========', &
            3X,' ======   =========== =========')
WRITE (*,110) (value(j), square_root(j), cube_root(j), j = 1, MAX_SIZE)
110 FORMAT (2(4X,F6.0,9X,F6.4,6X,F6.4))

END PROGRAM square_and_cube_roots
```

The implied DO loop in this example will be executed 10 times, with j taking on every value between 1 and 10 (the loop increment is defaulted to 1 here). During each iteration of the loop, the entire argument list will be written out. Therefore, this WRITE statement will write out 30 values, six per line. The resulting output is

```
                     Table of Square and Cube Roots
    Number    Square Root     Cube Root     Number    Square Root    Cube Root
    ======    ===========     =========     ======    ===========    =========
      1.         1.0000         1.0000        2.        1.4142         1.2599
      3.         1.7321         1.4422        4.        2.0000         1.5874
      5.         2.2361         1.7100        6.        2.4495         1.8171
      7.         2.6458         1.9129        8.        2.8284         2.0000
      9.         3.0000         2.0801       10.        3.1623         2.1544
```

Nested implied DO loops

Like ordinary DO loops, implied DO loops may also be *nested*. If they are nested, the inner loop will execute completely for each step in the outer loop. As a simple example, consider the following statements

```
WRITE (*,100) ((i, j, j = 1, 3), i = 1, 2)
100 FORMAT (I5,1X,I5)
```

There are two implicit DO loops in this WRITE statement. The index variable of the inner loop is j, and the index variable of the outer loop is i. When the WRITE statement is executed, variable j will take on values 1, 2, and 3 while i is 1, and then 1, 2, and 3 while i is 2. The output from this statement will be

```
    1    1
    1    2
    1    3
    2    1
    2    2
    2    3
```

Nested implied DO loops are important when working with arrays having two or more dimensions, as we will see later in Chapter 8.

The difference between I/O with standard DO loops and I/O with implied DO loops

Array input and output can be performed either with a standard DO loop containing I/O statements or with an implied DO loop. However, *there are subtle differences between the two types of loops.* To better understand those differences, let's compare the same output statement written with both types of loops. We will assume that integer array arr is initialized as follows

```
INTEGER, DIMENSION(5) :: arr = [ 1, 2, 3, 4, 5 ]
```

and compare output using a regular DO loop with output using an implied DO loop. An output statement using an ordinary DO loop is shown below

```
DO i = 1, 5
   WRITE (*,1000) arr(i), 2.*arr(i). 3*arr(i)
   1000 FORMAT (6I6)
END DO
```

In this loop, the WRITE statement is executed *five times*. In fact, this loop is equivalent to the following statements

```
WRITE (*,1000) arr(1), 2.*arr(1). 3*arr(1)
WRITE (*,1000) arr(2), 2.*arr(2). 3*arr(2)
WRITE (*,1000) arr(3), 2.*arr(3). 3*arr(3)
WRITE (*,1000) arr(4), 2.*arr(4). 3*arr(4)
WRITE (*,1000) arr(5), 2.*arr(5). 3*arr(5)
1000 FORMAT (6I6)
```

An output statement using an implied DO loop is shown below

```
WRITE (*,1000) (arr(i), 2.*arr(i). 3*arr(i), i = 1, 5)
1000 FORMAT (6I6)
```

Here, there is only *one* WRITE statement, but the WRITE statement has 15 arguments. In fact, the WRITE statement with the implied DO loop is equivalent to

```
WRITE (*,1000) arr(1), 2.*arr(1). 3*arr(1), &
               arr(2), 2.*arr(2). 3*arr(2), &
               arr(3), 2.*arr(3). 3*arr(3), &
               arr(4), 2.*arr(4). 3*arr(4), &
               arr(5), 2.*arr(5). 3*arr(5)
1000 FORMAT (6I6)
```

The main difference between having many WRITE statements with few arguments and one WRITE statement with many arguments is in the behavior of its associated format. Remember that each WRITE statement starts at the beginning of the format. Therefore, each of the five WRITE statements in the standard DO loop will start over at the beginning of the FORMAT statement, and only the first three of the six I6 descriptors will be used. The output of the standard DO loop will be

Introduction to Arrays

```
1    2    3
2    4    6
3    6    9
4    8   12
5   10   15
```

On the other hand, the implied DO loop produces a single WRITE statement with 15 arguments, so the associated format will be used completely 2½ times. The output of the implied DO loop will be

```
1    2    3    2    4    6
3    6    9    4    8   12
5   10   15
```

The same concept applies to a comparison of READ statements using standard DO loops with READ statements using implied DO loops. (See Exercise 6-9 at the end of the chapter.)

6.4.3 Input and Output of Whole Arrays and Array Sections

Entire arrays or array sections may also be read or written with READ and WRITE statements. If an array name is mentioned without subscripts in a Fortran I/O statement, then the compiler assumes that every element in the array is to be read in or written out. If an array section is mentioned in a Fortran I/O statement, then the compiler assumes that the entire section is to be read in or written out. Figure 6-11 shows a simple example of using an array and two array sections in I/O statements.

FIGURE 6-11
An example program illustrating array I/O.

```
PROGRAM array_io
!
! Purpose:
!   To illustrate array I/O.
!
! Record of revisions:
!      Date       Programmer          Description of change
!      ====       ==========          =====================
!   11/17/15     S. J. Chapman        Original code
!
IMPLICIT NONE

! Data dictionary: declare variable types & definitions
REAL, DIMENSION(5) :: a = [1.,2.,3.,20.,10.] ! 5-element test array
INTEGER, DIMENSION(4) :: vec = [4,3,4,5]     ! vector subscript

! Output entire array.
WRITE (*,100) a
100 FORMAT ( 6F8.3 )
```

(continued)

(*concluded*)

```
! Output array section selected by a triplet.
WRITE (*,100) a(2::2)

! Output array section selected by a vector subscript.
WRITE (*,100) a(vec)

END PROGRAM array_io
```

The output from this program is:

```
   1.000     2.000     3.000    20.000    10.000
   2.000    20.000
  20.000     3.000    20.000    10.000
```

Quiz 6-1

This quiz provides a quick check to see if you have understood the concepts introduced in Sections 6.1 through 6.4. If you have trouble with the quiz, reread the sections, ask your instructor, or discuss the material with a fellow student. The answers to this quiz are found in the back of the book.

For questions 1 to 3, determine the length of the array specified by each of the following declaration statements and the valid subscript range for each array.

1. `INTEGER :: itemp(15)`
2. `LOGICAL :: test(0:255)`
3. ```
   INTEGER, PARAMETER :: I1 = -20
   INTEGER, PARAMETER :: I2 = -1
   REAL, DIMENSION(I1:I1*I2) :: a
   ```

Determine which of the following Fortran statements are valid. For each valid statement, specify what will happen in the program. Assume default typing for any variable not explicitly typed.

4. ```
   REAL:: phase(0:11) = [ 0., 1., 2., 3., 3., 3., &
                          3., 3., 3., 2., 1., 0. ]
   ```
5. `REAL, DIMENSION(10) :: phase = 0.`
6. ```
   INTEGER :: data1(256)
   data1 = 0
   data1(10:256:10) = 1000
   WRITE (*,'(10I8)') data1
   ```
7. ```
   REAL, DIMENSION(21:31) :: array1 = 10.
   REAL, DIMENSION(10) :: array2 = 3.
   WRITE (*,'(1X,10I8)') array1 + array2
   ```
8. ```
   INTEGER :: i, j
   INTEGER, DIMENSION(10) :: sub1
   INTEGER, DIMENSION(0:9) :: sub2
   ```

(*continued*)
   ```

Introduction to Arrays

```
    (concluded)
        INTEGER, DIMENSION(100) :: in = [ ((0,i=1,9), j*10,j=1,10) ]
        sub1 = in(10:100:10)
        sub2 = sub1 / 10
        WRITE (*,100) sub1 * sub2
    100 FORMAT (10I8)
 9. REAL, DIMENSION(-3:0) :: error
    error(-3) = 0.00012
    error(-2) = 0.0152
    error(-1) = 0.0
    WRITE (*,500) error
    500 FORMAT (T6,error = ,/,(3X,I6))
10. INTEGER, PARAMETER :: MAX = 10
    INTEGER :: i
    INTEGER, DIMENSION(MAX) :: ivec1 = (/(i,i=1,10)/)
    INTEGER, DIMENSION(MAX) :: ivec2 = (/(i,i=10,1,-1)/)
    REAL, DIMENSION(MAX) :: data1
    data1 = real(ivec1)**2
    WRITE (*,500) data1(ivec2)
    500 FORMAT ('Output = ',/,5(3X,F7.1))
11. INTEGER, PARAMETER :: NPOINT = 10
    REAL, DIMENSION(NPOINT) :: mydata
    DO i = 1, NPOINT
       READ (*,*) mydata
    END DO
```

6.5
EXAMPLE PROBLEMS

Now we will examine two example problems that illustrate the use of arrays.

EXAMPLE 6-3   *Sorting Data:*

In many scientific and engineering applications, it is necessary to take a random input data set and to sort it so that the numbers in the data set are either all in *ascending order* (lowest-to-highest) or all in *descending order* (highest-to-lowest). For example, suppose that you were a zoologist studying a large population of animals, and that you wanted to identify the largest 5% of the animals in the population. The most straightforward way to approach this problem would be to sort the sizes of all of the animals in the population into ascending order, and take the top 5% of the values.

Sorting data into ascending or descending order seems to be an easy job. After all, we do it all the time. It is simple matter for us to sort the data (10, 3, 6, 4, 9) into the order (3, 4, 6, 9, 10). How do we do it? We first scan the input data list (10, 3, 6, 4, 9) to find the smallest value in the list (3), and then scan the remaining input data (10, 6, 4, 9) to find the next smallest value (4), etc., until the complete list is sorted.

In fact, sorting can be a very difficult job. As the number of values to be sorted increases, the time required to perform the simple sort described above increases rapidly, since we must scan the input data set once for each value sorted. For very large data sets, this technique just takes too long to be practical. Even worse, how would we sort the data if there were too many numbers to fit into the main memory of the computer? The development of efficient sorting techniques for large data sets is an active area of research, and is the subject of whole courses all by itself.

In this example, we will confine ourselves to the simplest possible algorithm to illustrate the concept of sorting. This simplest algorithm is called the selection sort. It is just a computer implementation of the mental math described above. The basic algorithm for the selection sort is:

1. Scan the list of numbers to be sorted to locate the smallest value in the list. Place that value at the front of the list by swapping it with the value currently at the front of the list. If the value at the front of the list is already the smallest value, then do nothing.
2. Scan the list of numbers from position 2 to the end to locate the next smallest value in the list. Place that value in position 2 of the list by swapping it with the value currently at that position. If the value in position 2 is already the next smallest value, then do nothing.
3. Scan the list of numbers from position 3 to the end to locate the third smallest value in the list. Place that value in position 3 of the list by swapping it with the value currently at that position. If the value in position 3 is already the third smallest value, then do nothing.
4. Repeat this process until the next-to-last position in the list is reached. After the next-to-last position in the list has been processed, the sort is complete.

Note that if we are sorting N values, this sorting algorithm requires $N - 1$ scans through the data to accomplish the sort.

This process is illustrated in Figure 6-12. Since there are five values in the data set to be sorted, we will make four scans through the data. During the first pass through the entire data set, the minimum value is 3, so the 3 is swapped with the 10 that was in position 1. Pass 2 searches for the minimum value in positions 2 through 5. That minimum is 4, so the 4 is swapped with the 10 in position 2. Pass 3 searches for the minimum value in positions 3 through 5. That minimum is 6, which is already in position 3, so no swapping is required. Finally, pass 4 searches for the minimum value in positions 4 through 5. That minimum is 9, so the 9 is swapped with the 10 in position 4, and the sort is completed.

Programming Pitfalls:
The selection sort algorithm is the easiest sorting algorithm to understand, but it is computationally inefficient. *It should never be applied to sort really large data sets* (say, sets with more than 1000 elements). Over the years, computer scientists have developed much more efficient sorting algorithms. We will encounter one such algorithm (the *heapsort algorithm*) in Exercise 7-35.

Introduction to Arrays

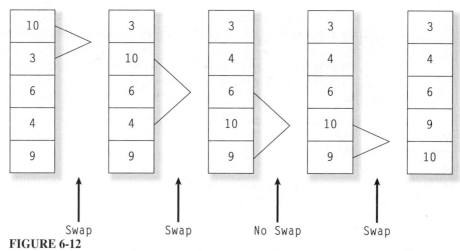

FIGURE 6-12
An example problem demonstrating the selection sort algorithm.

We will now develop a program to read in a data set from a file, sort it into ascending order, and display the sorted data set.

SOLUTION
This program must be able to ask the user for the name of the file to be sorted, open that file, read the input data, sort the data, and write out the sorted data. The design process for this problem is given below.

1. **State the problem.**
    We have not yet specified the type of data to be sorted. If the data is real, then the problem may be stated as follows:

    Develop a program to read an arbitrary number of real input data values from a user-supplied file, sort the data into ascending order, and write the sorted data to the standard output device.

2. **Define the inputs and outputs.**
    There are two types of inputs to this program:
    (*a*) A character string containing the file name of the input data file. This string will come from the standard input device.
    (*b*) The real data values in the file.
    The outputs from this program are the sorted real data values written to the standard output device.

3. **Describe the algorithm.**
    This program can be broken down into five major steps:

    ```
    Get the input file name
    Open the input file
    Read the input data into an array
    Sort the data in ascending order
    Write the sorted data
    ```

The first three major steps of the program are to get the name of the input file, to open the file, and to read in the data. We must prompt the user for the input file name, read in the name, and open the file. If the file open is successful, we must read in the data, keeping track of the number of values that have been read. Since we don't know how many data values to expect, a while loop is appropriate for the READ. A flowchart for these steps is shown in Figure 6-13, and the detailed pseudocode is shown below:

```
Prompt user for the input file name "filename"
Read the file name "filename"
OPEN file "filename"
IF OPEN is successful THEN
    WHILE
        Read value into temp
        IF read not successful EXIT
        nvals ← nvals + 1
```

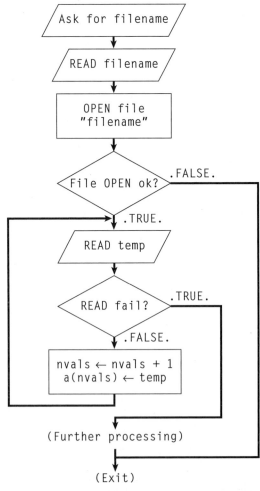

FIGURE 6-13
Flowchart for reading values to sort from an input file.

Introduction to Arrays

```
                a(nvals) ← temp
         End of WHILE
            ...
            ...                  (Insert sorting step here)
            ...                  (Insert writing step here)
         End of IF
```

Next we have to sort the data. We will need to make `nvals-1` passes through the data, finding the smallest remaining value each time. We will use a pointer to locate the smallest value in each pass. Once the smallest value is found, it will be swapped to the top of the list if it is not already there. A flowchart for these steps is shown in Figure 6-14, and the detailed pseudocode is shown below:

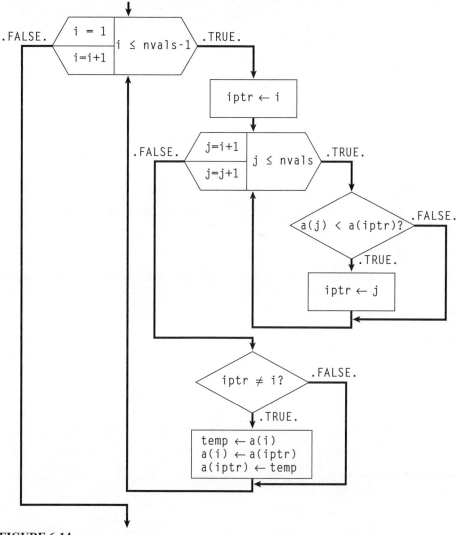

FIGURE 6-14
Flowchart for sorting values with a selection sort.

```
DO for i = 1 to nvals-1

   ! Find the minimum value in a(i) through a(nvals)
   iptr ← i
   DO for j == i+1 to nvals
      IF a(j) < a(iptr) THEN
         iptr ← j
      END of IF
   END of DO

   ! iptr now points to the min value, so swap a(iptr) with
   ! a(i) if iptr /= i.
   IF i /= iptr THEN
      temp ← a(i)
      a(i) ← a(iptr)
      a(iptr) ← temp
   END of IF
END of DO
```

The final step is writing out the sorted values. No refinement of the pseudocode is required for that step. The final pseudocode is the combination of the reading, sorting, and writing steps.

4. **Turn the algorithm into Fortran statements.**

The resulting Fortran program is shown in Figure 6-15.

FIGURE 6-15
A program to read values from an input data file, and to sort them into ascending order.

```
PROGRAM sort1
!
!  Purpose:
!     To read in a real input data set, sort it into ascending order
!     using the selection sort algorithm, and to write the sorted
!     data to the standard output device.
!
!  Record of revisions:
!      Date         Programmer          Description of change
!      ====         ==========          =====================
!    11/17/15     S. J. Chapman         Original code
!
IMPLICIT NONE

! Data dictionary: declare constants
INTEGER, PARAMETER :: MAX_SIZE = 10 ! Maximum input data set size

! Data dictionary: declare variable types & definitions
REAL, DIMENSION(MAX_SIZE) :: a     ! Data array to sort
CHARACTER(len=20) :: filename      ! Input data file name
INTEGER :: i                       ! Loop index
INTEGER :: iptr                    ! Pointer to smallest value
INTEGER :: j                       ! Loop index
CHARACTER(len=80) :: msg           ! Error message
```

(*continued*)

(continued)

```
INTEGER :: nvals = 0          ! Number of data values to sort
INTEGER :: status             ! I/O status: 0 for success
REAL :: temp                  ! Temporary variable for swapping

! Get the name of the file containing the input data.
WRITE (*,1000)
1000 FORMAT ('Enter the file name with the data to be sorted: ')
READ (*,'(A20)') filename

! Open input data file. Status is OLD because the input data must
! already exist.
OPEN ( UNIT=9, FILE=filename, STATUS='OLD', ACTION='READ', &
       IOSTAT=status, IOMSG=msg )

! Was the OPEN successful?
fileopen: IF ( status == 0 ) THEN    ! Open successful

   ! The file was opened successfully, so read the data to sort
   ! from it, sort the data, and write out the results.
   ! First read in data.
   DO
      READ (9, *, IOSTAT=status) temp    ! Get value
      IF ( status /= 0 ) EXIT            ! Exit on end of data
      nvals = nvals + 1                  ! Bump count
      a(nvals) = temp                    ! Save value in array
   END DO

   ! Now, sort the data.
   outer: DO i = 1, nvals-1

      ! Find the minimum value in a(i) through a(nvals)
      iptr = i
      inner: DO j = i+1, nvals
         minval: IF ( a(j) < a(iptr) ) THEN
            iptr = j
         END IF minval
      END DO inner

      ! iptr now points to the minimum value, so swap a(iptr) with
      ! a(i) if i /= iptr.
      swap: IF ( i /= iptr ) THEN
         temp    = a(i)
         a(i)    = a(iptr)
         A(iptr) = temp
      END IF swap

   END DO outer

   ! Now write out the sorted data.
   WRITE (*,'(A)') 'The sorted output data values are: '
   WRITE (*,'(3X,F10.4)') ( a(i), i = 1, nvals )

ELSE fileopen
```

(continued)

(concluded)

```
    ! Else file open failed. Tell user.
    WRITE (*,1050) TRIM(msg)
    1050 FORMAT ('File open failed--error = ', A)
END IF fileopen

END PROGRAM sort1
```

5. **Test the program.**

To test this program, we will create an input data file and run the program with it. The data set will contain a mixture of positive and negative numbers as well as at least one duplicated value to see if the program works properly under those conditions. The following data set will be placed in file INPUT2:

```
    13.3
    12.
    -3.0
    0.
    4.0
    6.6
    4.
    -6.
```

Running this file values through the program yields the following result:

```
C:\book\fortran\chap6>sort1
Enter the file name containing the data to be sorted:
input2
The sorted output data values are:
    -6.0000
    -3.0000
      .0000
     4.0000
     4.0000
     6.6000
    12.0000
    13.3000
```

The program gives the correct answers for our test data set. Note that it works for both positive and negative numbers as well as for repeated numbers.

To be certain that our program works properly, we must test it for every possible type of input data. This program worked properly for the test input data set, but will it work for *all* input data sets? Study the code now and see if you can spot any flaws before continuing to the next paragraph.

The program has a major flaw that must be corrected. If there are more than 10 values in the input data file, this program will attempt to store input data in memory locations a(11), a(12), etc., that have not been allocated in the program (this is an

Introduction to Arrays

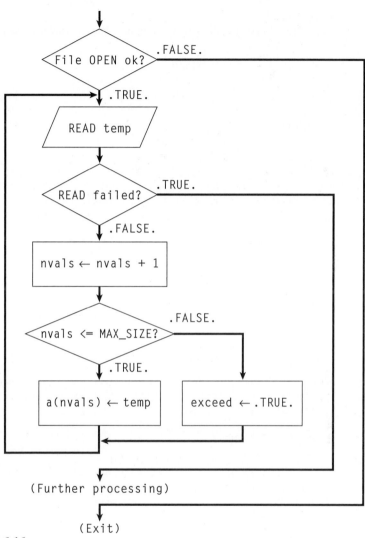

FIGURE 6-16
Corrected flowchart for reading the values to sort from an input file without causing an array overflow.

out-of-bounds or **array overflow** condition). If bounds checking is turned on, the program will abort when we try to write to a(11). If bounds checking is not turned on, the results are unpredictable and vary from computer to computer. This program must be rewritten to prevent it from attempting to write into locations beyond the end of the allocated array. This can be done by checking to see if the number of values exceeds max_size before storing each number into array a. The corrected flowchart for reading in the data is shown in Figure 6-16, and the corrected program is shown in Figure 6-17.

FIGURE 6-17
A corrected version of the sort program that detects array overflows.

```
PROGRAM sort2
!
! Purpose:
!   To read in a real input data set, sort it into ascending order
!   using the selection sort algorithm, and to write the sorted
!   data to the standard output device.
!
! Record of revisions:
!      Date          Programmer        Description of change
!      ====          ==========        =====================
!    11/15/05      S. J. Chapman       Original code
! 1. 11/16/05      S. J. Chapman       Modified to protect against
!                                      array overflow.
!
IMPLICIT NONE

! Data dictionary: declare constants
INTEGER, PARAMETER :: MAX_SIZE = 10 ! Maximum input data set size

! Data dictionary: declare variable types & definitions
REAL, DIMENSION(MAX_SIZE) :: a       ! Data array to sort
LOGICAL :: exceed = .FALSE.          ! Logical indicating that array
                                     ! limits are exceeded.
CHARACTER(len=20) :: filename        ! Input data file name
INTEGER :: i                         ! Loop index
INTEGER :: iptr                      ! Pointer to smallest value
INTEGER :: j                         ! Loop index
CHARACTER(len=80) :: msg             ! Error message
INTEGER :: nvals = 0                 ! Number of data values to sort
INTEGER :: status                    ! I/O status: 0 for success
REAL :: temp                         ! Temporary variable for swapping

! Get the name of the file containing the input data.
WRITE (*,1000)
1000 FORMAT ('Enter the file name with the data to be sorted: ')
READ (*,'(A20)') filename

! Open input data file. Status is OLD because the input data must
! already exist.
OPEN ( UNIT=9, FILE=filename, STATUS='OLD', ACTION='READ', &
       IOSTAT=status, IOMSG=msg )

! Was the OPEN successful?
fileopen: IF ( status == 0 ) THEN    ! Open successful

   ! The file was opened successfully, so read the data to sort
   ! from it, sort the data, and write out the results.
   ! First read in data.
   DO
      READ (9, *, IOSTAT=status) temp    ! Get value
      IF ( status /= 0 ) EXIT            ! Exit on end of data
      nvals = nvals + 1                  ! Bump count
```

(continued)

Introduction to Arrays

(*concluded*)

```
      size: IF ( nvals <= MAX_SIZE ) THEN   ! Too many values?
         a(nvals) = temp                    ! No: Save value in array
      ELSE
         exceed = .TRUE.                    ! Yes: Array overflow
      END IF size
END DO

! Was the array size exceeded? If so, tell user and quit.
toobig: IF ( exceed ) THEN
   WRITE (*,1010) nvals, MAX_SIZE
   1010 FORMAT (' Maximum array size exceeded: ', I6, ' > ', I6 )
ELSE toobig

   ! Limit not exceeded: sort the data.
   outer: DO i = 1, nvals-1

      ! Find the minimum value in a(i) through a(nvals)
      iptr = i
      inner: DO j = i+1, nvals
         minval: IF ( a(j) < a(iptr) ) THEN
            iptr = j
         END IF minval
      END DO inner

      ! iptr now points to the minimum value, so swap a(iptr) with
      ! a(i) if i /= iptr.
      swap: IF ( i /= iptr ) THEN
         temp    = a(i)
         a(i)    = a(iptr)
         a(iptr) = temp
      END IF swap

   END DO outer

   ! Now write out the sorted data.
   WRITE (*,'(A)') ' The sorted output data values are: '
   WRITE (*,'(3X,F10.4)') ( a(i), i = 1, nvals )

   END IF toobig

ELSE fileopen

   ! Else file open failed. Tell user.
   WRITE (*,1050) TRIM(msg)
   1050 FORMAT ('File open failed--error = ', A)

END IF fileopen

END PROGRAM sort2
```

  In the test for array overflow conditions, we have used a logical variable exceed. If the next value to be read into the array would result on an array overflow, then exceed is set to true, and the value is not stored. When all values have been read from the input file, the program checks to see if the array size would have been exceeded. If so, it writes out an error message and quits. If not, it reads in and sorts the numbers.

This program also illustrates the proper use of named constants to allow the size of a program to be changed easily. The size of array a is set by parameter `MAX_SIZE`, and the test for array overflow within the code also uses parameter `MAX_SIZE`. The maximum sorting capacity of this program could be changed from 10 to 1000 by simply modifying the definition of the named constant `MAX_SIZE` at the top of the program.

EXAMPLE 6-4

The Median:

In Chapter 4, we examined two common statistical measures of data: averages (or means) and standard deviations. Another common statistical measure of data is the median. The median of a data set is the value such that half of the numbers in the data set are larger than the value and half of the numbers in the data set are smaller than the value. If there are an even number of values in the data set, then there cannot be a value exactly in the middle. In that case, the median is usually defined as the average of the two elements in the middle. The median value of a data set is often close to the average value of the data set, but not always. For example, consider the following data set:

```
1
2
3
4
100
```

The average or mean of this data set is 22, while the median of this data set is 3!

An easy way to compute the median of a data set is to sort it into ascending order, and then to select the value in the middle of the data set as the median. If there are an even number of values in the data set, then average the two middle values to get the median.

Write a program to calculate the mean, median, and standard deviation of an input data set that is read from a user-specified file.

SOLUTION

This program must be able to read in an arbitrary number of measurements from a file, and then calculate the mean and standard deviation of those measurements.

1. **State the problem.**

    Calculate the average, median, and standard deviation of a set of measurements that are read from a user-specified input file, and write those values out on the standard output device.

2. **Define the inputs and outputs.**

    There are two types of inputs to this program:

    (*a*) A character string containing the file name of the input data file. This string will come from the standard input device.

    (*b*) The real data values in the file.

Introduction to Arrays

The outputs from this program are the average, median, and standard deviation of the input data set. They are written to the standard output device.

3. **Describe the algorithm.**

   This program can be broken down into six major steps:

   ```
   Get the input file name
   Open the input file
   Read the input data into an array
   Sort the data in ascending order
   Calculate the average, mean, and standard deviation
   Write average, median, and standard deviation
   ```

   The detailed pseudocode for the first four steps is similar to that of the previous example:

   ```
   Initialize variables.
   Prompt user for the input file name "filename"
   Read the file name "filename"
   OPEN file "filename"
   IF OPEN is successful THEN
       WHILE
           Read value into temp
           IF read not successful EXIT
           nvals ← nvals + 1
           IF nvals <= max_size then
               a(nvals) ← temp
           ELSE
               exceed ← .TRUE.
       End of IF
   End of WHILE

   ! Notify user if array size exceeded.
   IF array size exceeded then
       Write out message to user
   ELSE
       ! Sort the data
       DO for i = 1 to nvals-1

           ! Find the minimum value in a(i) through a(nvals)
           iptr ← i
           DO for j = i+1 to nvals
               IF a(j) < a(iptr) THEN
                   iptr ← j
               END of IF
           END of DO (for j = i+1 to nvals)

           ! iptr now points to the min value, so swap A(iptr)
           ! with a(i) if iptr /= i.
           IF i /= iptr THEN
               temp ← a(i)
               a(i) ← a(iptr)
               a(iptr) ← temp
           END of IF
       END of DO (for i = 1 to nvals-1)
   ```

```
        (Add code here)

    End of IF (array size exceeded...)

End of IF (open successful...)
```

The fifth step is to calculate the required average, median, and standard deviation. To do this, we must first accumulate some statistics on the data (Σx and Σx^2), and then apply the definitions of average, median, and standard deviation given previously. The pseudocode for this step is:

```
DO for i = 1 to nvals
   sum_x ← sum_x + a(i)
   sum_x2 ← sum_x2 + a(i)**2
End of DO
IF nvals >= 2 THEN
   x_bar ← sum_x / real(nvals)
   std_dev ← sqrt((real(nvals)*sum_x2-
             sum_x**2)/(real(nvals)*real(nvals-1)))
   IF nvals is an even number THEN
      median ← (a(nvals/2) + a(nvals/2+1)) / 2.
   ELSE
      median ← a(nvals/2+1)
   END of IF
END of IF
```

We will decide if `nvals` is an even number by using the modulo function `mod(nvals,2)`. If `nvals` is even, this function will return a 0; if `nvals` is odd, it will return a 1. Finally, we must write out the results.

```
Write out average, median, standard deviation, and no. of points
```

4. **Turn the algorithm into Fortran statements.**

   The resulting Fortran program is shown in Figure 6-18.

FIGURE 6-18
A program to read in values from an input data file, and to calculate their mean, median, and standard deviation.

```
PROGRAM stats_4
!
!   Purpose:
!     To calculate mean, median, and standard deviation of an input
!     data set read from a file.
!
!   Record of revisions:
!       Date        Programmer       Description of change
!       ====        ==========       =====================
!     11/18/15    S. J. Chapman      Original code
!
IMPLICIT NONE

! Data dictionary: declare constants
```

(continued)

Introduction to Arrays

(*continued*)

```fortran
INTEGER, PARAMETER :: MAX_SIZE = 100 ! Max data size

! Data dictionary: declare variable types & definitions
REAL, DIMENSION(MAX_SIZE) :: a    ! Data array to sort
LOGICAL :: exceed = .FALSE.       ! Logical indicating that array
                                  ! limits are exceeded.
CHARACTER(len=20) :: filename     ! Input data file name
INTEGER :: i                      ! Loop index
INTEGER :: iptr                   ! Pointer to smallest value
INTEGER :: j                      ! Loop index
REAL :: median                    ! The median of the input samples
CHARACTER(len=80) :: msg          ! Error message
INTEGER :: nvals = 0              ! Number of data values to sort
INTEGER :: status                 ! I/O status: 0 for success
REAL :: std_dev                   ! Standard deviation of input samples
REAL :: sum_x = 0.                ! Sum of input values
REAL :: sum_x2 = 0.               ! Sum of input values squared
REAL :: temp                      ! Temporary variable for swapping
REAL :: x_bar                     ! Average of input values

! Get the name of the file containing the input data.
WRITE (*,1000)
1000 FORMAT ('Enter the file name with the data to be processed: ')
READ (*,'(A20)') filename

! Open input data file. Status is OLD because the input data must
! already exist.
OPEN ( UNIT=9, FILE=filename, STATUS='OLD', ACTION='READ', &
       IOSTAT=status, IOMSG=msg )

! Was the OPEN successful?
fileopen: IF ( status == 0 ) THEN         ! Open successful

   ! The file was opened successfully, so read the data to sort
   ! from it, sort the data, and write out the results.
   ! First read in data.
   DO
      READ (9, *, IOSTAT=status) temp   ! Get value
      IF ( status /= 0 ) EXIT           ! Exit on end of data
      nvals = nvals + 1                 ! Bump count
      size: IF ( nvals <= MAX_SIZE ) THEN ! Too many values?
         a(nvals) = temp                ! No: Save value in array
      ELSE
         exceed = .TRUE.                ! Yes: Array overflow
      END IF size
   END DO

   ! Was the array size exceeded? If so, tell user and quit.
   toobig: IF ( exceed ) THEN
      WRITE (*,1010) nvals, MAX_SIZE
      1010 FORMAT ('Maximum array size exceeded: ', I0, ' > ', I0 )
   ELSE

      ! Limit not exceeded: sort the data.
```

(*continued*)

(*continued*)

```
        outer: DO i = 1, nvals-1

           ! Find the minimum value in a(i) through a(nvals)
           iptr = i
           inner: DO j = i+1, nvals
              minval: IF ( a(j) < a(iptr) ) THEN
                 iptr = j
              END IF minval
           END DO inner

           ! iptr now points to the minimum value, so swap A(iptr)
           ! with a(i) if i /= iptr.
           swap: IF ( i /= iptr ) THEN
              temp     = a(i)
              a(i)     = a(iptr)
              a(iptr)  = temp
           END IF swap

        END DO outer

        ! The data is now sorted. Accumulate sums to calculate
        ! statistics.
        sums: DO i = 1, nvals
           sum_x  = sum_x + a(i)
           sum_x2 = sum_x2 + a(i)**2
        END DO sums

        ! Check to see if we have enough input data.
        enough: IF ( nvals < 2 ) THEN

           ! Insufficient data.
           WRITE (*,*) ' At least 2 values must be entered.'

        ELSE

           ! Calculate the mean, median, and standard deviation
           x_bar   = sum_x / real(nvals)
           std_dev = sqrt( (real(nvals) * sum_x2 - sum_x**2) &
                   / (real(nvals) * real(nvals-1)) )
           even: IF ( mod(nvals,2) == 0 ) THEN
              median = ( a(nvals/2) + a(nvals/2+1) ) / 2.
           ELSE
              median = a(nvals/2+1)
           END IF even

           ! Tell user.
           WRITE (*,*) 'The mean of this data set is:    ', x_bar
           WRITE (*,*) 'The median of this data set is:', median
           WRITE (*,*) 'The standard deviation is:       ', std_dev
           WRITE (*,*) 'The number of data points is:  ', nvals

        END IF enough

     END IF toobig

  ELSE fileopen
```

(*continued*)

(*concluded*)

```
        ! Else file open failed. Tell user.
        WRITE (*,1050) TRIM(msg)
        1050 FORMAT ('File open failed--error = ', A)
    END IF fileopen
END PROGRAM stats_4
```

5. **Test the program.**

To test this program, we will calculate the answers by hand for a simple data set, and then compare the answers to the results of the program. If we use five input values: 5, 3, 4, 1, and 9, then the mean and standard deviation would be

$$\bar{x} = \frac{1}{N}\sum_{i=1}^{N} x_i = \frac{1}{5}(22) = 4.4 \qquad (4\text{-}1)$$

$$s = \sqrt{\frac{N\sum_{i=1}^{N} x_i^2 - \left(\sum_{i=1}^{N} x_i\right)^2}{N(N-1)}} = 2.966 \qquad (4\text{-}2)$$

$$\text{median} = 4$$

If these values are placed in the file INPUT4 and the program is run with that file as an input, the results are

```
C:\book\fortran\chap6>stats_4
Enter the file name containing the input data:
input4
  The mean of this data set is:        4.400000
  The median of this data set is:      4.000000
  The standard deviation is:           2.966479
  The number of data points is:              5
```

The program gives the correct answers for our test data set.

Note the use of names on loops and branches in the above program. These names help us to keep the loops and branches straight. This becomes more and more important as programs get larger. Even in this simple program, loops and branches are nested four deep at some points!

6.6
WHEN SHOULD YOU USE AN ARRAY?

We have now learned *how* to use arrays in our Fortran programs, but we have not yet learned *when* to use them. At this point in a typical Fortran course, many students are tempted to use arrays to solve problems whether they are needed or not, just because they know how to do so. How can we decide whether or not it makes sense to use an array in a particular problem?

In general, *if much or all of the input data must be in memory at the same time in order to solve a problem efficiently, then the use of arrays to hold that data is appropriate for that problem.* Otherwise, arrays are not needed. For example, let's contrast the statistics programs that we have written in Examples 4-1 and 6-4. Example 4-1 calculated the mean and standard deviation of a data set, while Example 6-4 calculated the mean, median, and standard deviation of a data set.

Recall that the equations for the mean and standard deviation of a data set are

$$\bar{x} = \frac{1}{N}\sum_{i=1}^{N} x_i = \frac{1}{5}(22) = 4.4 \tag{4-1}$$

and

$$s = \sqrt{\frac{N\sum_{i=1}^{N} x_i^2 - \left(\sum_{i=1}^{N} x_i\right)^2}{N(N-1)}} = 2.966 \tag{4-2}$$

The sums in Equations (4-1) and (4-2) that are required to find the mean and standard deviation can easily be formed as data values are read in one by one. There is no need to wait until all of the data is read before starting to build the sums. Therefore, a program to calculate the mean and standard deviation of a data set does not need to use arrays. You *could* use an array to hold all of the input values before calculating the mean and standard deviation, but since the array is not necessary, you should not do so. Example 4-1 works fine, and is built entirely without arrays.

On the other hand, finding the median of a data set requires that the data be sorted into ascending order. Since sorting requires all data to be in memory, a program that calculates the median must use an array to hold all of the input data before the calculations start. Therefore, Example 6-4 uses an array to hold its input data.

What's wrong with using an array within a program even if it is not needed? There are two major problems associated with using unnecessary arrays:

1. *Unnecessary arrays waste memory.* Unnecessary arrays can eat up a lot of memory, making a program larger than it needs to be. A large program requires more memory to run it, which makes the computer that it runs on more expensive. In some cases, the extra size may make it impossible to run on a particular computer at all.
2. *Unnecessary arrays restrict program capabilities.* To understand this point, let's consider an example program that calculates the mean and standard deviation of a data set. If the program is designed with a 1000-element static input array, then it will only work for data sets with up to 1000 elements. If we encounter a data set with more than 1000 elements, the program would have to be recompiled and relinked with a larger array size. On the other hand, a program that calculates the mean and standard deviation of a data set as the values are input has no upper limit on data set size.

Good Programming Practice
Do not use arrays to solve a problem unless they are actually needed.

6.7
SUMMARY

In this chapter, we presented an introduction to arrays and to their use in Fortran programs. An array is a group of variables, all of the same type, which are referred to by a single name. An individual variable within the array is called an array element. Individual array elements are addressed by means of one or more (up to 15) subscripts. Arrays with one subscript (rank 1 arrays) were discussed in this chapter. Arrays with more than one subscript will be discussed in Chapter 8.

An array is declared using a type declaration statement by naming the array and specifying the maximum (and, optionally, the minimum) subscript values with the `DIMENSION` attribute. The compiler uses the declared subscript ranges to reserve space in the computer's memory to hold the array.

As with any variable, an array must be initialized before use. An array may be initialized at compile time using array constructors in the type declaration statements, or at runtime using array constructors, `DO` loops, or Fortran `READ`s.

Individual array elements may be used freely in a Fortran program just like any other variable. They may appear in assignment statements on either side of the equal sign. Entire arrays and array sections may also be used in calculations and assignment statements as long as the arrays are conformable with each other. Arrays are conformable if they have the same number of dimensions (rank) and the same extent in each dimension. A scalar is also conformable with any array. An operation between two conformable arrays is performed on an element-by-element basis. Scalar values are also conformable with arrays.

Arrays are especially useful for storing data values that change as a function of some variable (time, location, etc.). Once the data values are stored in an array, they can be easily manipulated to derive statistics or other information that may be desired.

6.7.1 Summary of Good Programming Practice

The following guidelines should be adhered to when working with arrays.

1. Before writing a program that uses arrays, you should decide whether an array is really needed to solve the problem or not. If arrays are not needed, don't use them!
2. All array sizes should be declared using named constants. If the sizes are declared using named constants, and if those same named constants are used in any size tests within the program, then it will be very easy to modify the maximum capacity of the program at a later time.
3. All arrays should be initialized before use. The results of using an uninitialized array are unpredictable and vary from processor to processor.
4. The most common problem when programming with arrays is attempting to read from or write to locations outside the bounds of the array. To detect these problems, the bounds checking option of your compiler should always be turned

on during program testing and debugging. Because bounds checking slows down the execution of a program, the bounds checking option may be turned off once debugging is completed.

6.7.2 Summary of Fortran Statements and Constructs

Type Declaration Statements with Arrays:

```
type, DIMENSION( [i1:]i2 ) :: array1, ...
```

Examples:

```
REAL, DIMENSION(100) :: array
INTEGER, DIMENSION(-5:5) :: i
```

Description:
These type declaration statements declare both type and the size of an array.

Implied DO loop structure:

```
READ (unit,format) (arg1, arg2, ... , index = istart, iend, incr)
WRITE (unit,format) (arg1, arg2, ... , index = istart, iend, incr)
[ (arg1, arg2, ... , index = istart, iend, incr) ]
```

Examples:

```
WRITE (*,*) ( array(i), i = 1, 10 )
INTEGER, DIMENSION(100) :: values
values = [ (i, i=1,100) ]
```

Description:
The implied DO loop is used to repeat the values in an argument list a known number of times. The values in the argument list may be functions of the DO loop index variable. During the first iteration of the DO loop, the variable *index* is set to the value *istart*. *index* is incremented by *incr* in each successive loop until its value exceeds *iend*, at which time the loop terminates.

6.7.3 Exercises

6-1. How may arrays be declared?

6-2. What is the difference between an array and an array element?

6-3. Execute the following Fortran program on your computer with both bounds checking turned on and bounds checking turned off. What happens in each case?

```
PROGRAM bounds
IMPLICIT NONE
REAL, DIMENSION(5) :: test = [ 1., 2., 3., 4., 5. ]
REAL, DIMENSION(5) :: test1
INTEGER :: i
DO i = 1, 6
   test1(i) = SQRT(test(i))
   WRITE (*,100) 'SQRT(',test(i), ') = ', test1(i)
   100 FORMAT (A,F6.3,A,F14.4)
END DO
END PROGRAM bounds
```

6-4. Determine the shape and size of the arrays specified by the following declaration statements, and the valid subscript range for each dimension of each array.

(a) `CHARACTER(len=80), DIMENSION(60) :: line`

(b) `INTEGER, PARAMETER :: ISTART = 32`
`INTEGER, PARAMETER :: ISTOP = 256`
`INTEGER, DIMENSION(ISTART:ISTOP) :: char`

(c) `INTEGER, PARAMETER :: NUM_CLASS = 3`
`INTEGER, PARAMETER :: NUM_STUDENT = 35`
`LOGICAL, DIMENSION(NUM_STUDENT,NUM_CLASS) :: passfail`

6-5. Determine which of the following Fortran program fragments are valid. For each valid statement, specify what will happen in the program. (Assume default typing for any variables that are not explicitly typed within the program fragments.)

(a) `INTEGER, DIMENSION(100) :: icount, jcount`
...
`icount = [ (i, i=1, 100) ]`
`jcount = icount + 1`

(b) `REAL, DIMENSION(10) :: value`
`value(1:10:2) = [ 5., 4., 3., 2., 1. ]`
`value(2:11:2) = [ 10., 9., 8., 7., 6. ]`
`WRITE (*,100) value`
`100 FORMAT ('Value = ',/,(F10.2))`

(c) `INTEGER, DIMENSION(6) :: a`
`INTEGER, DIMENSION(6) :: b`
`a = [1,-3,0,-5,-9,3]`
`b = [-6,6,0,5,2,-1]`
`WRITE (*,*) a > b`

6-6. What is meant by each of the following array terms? (a) size, (b) shape, (c) extent, (d) rank, (e) conformable.

6-7. Given an array `my_array` defined as shown and containing the values shown below, determine whether each of the following array sections is valid. Specify the shape and contents of each valid array section.

`REAL,DIMENSION(-2:7) :: my_array = [-3 -2 -1 0 1 2 3 4 5 6]`

(a) `my_array(-3,3)`

(b) `my_array(-2:2)`

(c) `my_array(1:5:2)`

(d) `INTEGER, DIMENSION(5) :: list = [ -2, 1, 2, 4, 2 ]`
`my_array(list)`

6-8. What will be the output from each of the WRITE statements in the following program? Why is the output of the two statements different?

```
PROGRAM test_output
IMPLICIT NONE
INTEGER, DIMENSION(0:7) :: my_data
INTEGER :: i, j
my_data = [ 1, 2, 3, 4, 5, 6, 7, 8 ]

DO i = 0,1
   WRITE (*,100) (my_data(4*i+j), j=0,3)
   100 FORMAT (6(1X,I4))
END DO
WRITE (*,100) ((my_data(4*i+j), j=0,3), i=0,1)
END PROGRAM test_output
```

6-9. An input data file INPUT1 contains the following values:

```
 27   17   10    8    6
 11   13  -11   12  -21
 -1    0    0    6   14
-16   11   21   26  -16
 04   99  -99   17    2
```

Assume that file INPUT1 has been opened on i/o unit 8, and that array values is a 16-element integer array, all of whose elements have been initialized to zero. What will be the contents of array values after each of the following READ statements has been executed?

(a) `DO i = 1, 4`
`   READ (8,*) (values(4*(i-1)+j), j = 1, 4)`
`END DO`

(b) `READ (8,*) ((values(4*(i-1)+j), j = 1, 4), i = 1, 4)`

(c) `READ (8,'(4I6)') ((values(4*(i-1)+j), j = 1, 4), i = 1, 4)`

6-10. Polar to Rectangular Conversion A *scalar quantity* is a quantity that can be represented by a single number. For example, the temperature at a given location is a scalar. In contrast, a *vector* is a quantity that has both a magnitude and a direction associated with it. For example, the velocity of an automobile is a vector, since it has both a magnitude and a direction.

Vectors can be defined either by a magnitude and a direction, or by the components of the vector projected along the axes of a rectangular coordinate system. The two representations are equivalent. For two-dimensional vectors, we can convert back and forth between the representations using the following equations:

$$\mathbf{V} = V \angle \theta = V_x \mathbf{i} + V_y \mathbf{j} \tag{6-2}$$

$$V_x = V \cos\theta \tag{6-3}$$

$$V_x = V \sin\theta \tag{6-4}$$

$$V = \sqrt{V_x^2 + V_y^2} \tag{6-5}$$

$$\theta = \tan^{-1} \frac{V_y}{V_x} \text{ over all four quadrants} \tag{6-6}$$

where **i** and **j** are the unit vectors in the *x* and *y* directions, respectively. The representation of the vector in terms of magnitude and angle is known as *polar coordinates*, and the representation of the vector in terms of components along the axes is know as *rectangular coordinates* (Figure 6-19).

Write a program that reads the polar coordinates (magnitude and angle) of a 2D vector into a rank 1 array polar (polar(1) will contain the magnitude V and polar(2) will contain the angle θ in degrees), and converts the vector from polar to rectangular form, storing the result in a rank 1 array rect. The first element of rect should contain the *x*-component of the vector, and the second element should contain the *y*-component of the vector. After the conversion, display the contents of array rect. Test your program by converting the following polar vectors to rectangular form:

(a) 5∠ − 36.87°

(b) 10∠45°

(c) 25∠233.13°

6-11. Rectangular to Polar Conversion Write a program that reads the rectangular components of a 2D vector into a rank 1 array rect (rect(1) will contain the component V_x and rect(2) will contain the component V_y) and converts the vector from rectangular to polar form, storing the result in a rank 1 array polar. The first element of polar should contain the magnitude of the vector, and the second element should contain the angle of the vector in degrees. After the conversion, display the contents of array polar. (*Hint:* Look up function ATAN2D in Appendix B.) Test your program by converting the following rectangular vectors to polar form:

(a) 3**i** − 4**j**

(b) 5**i** + 5**j**

(c) −5**i** + 12**j**

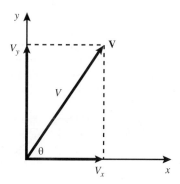

FIGURE 6-19
Representations of a vector.

6-12. Assume that `values` is a 101-element array containing a list of measurements from a scientific experiment, which has been declared by the statement

```
REAL, DIMENSION(-50:50) :: values
```

Write the Fortran statements that would count the number of positive values, negative values, and zero values in the array, and write out a message summarizing how many values of each type were found.

6-13. Write Fortran statements that would print out every fifth value in the array `values` described in Exercise 6-12. The output should take the form

```
values(-50) = xxx.xxxx
values(-45) = xxx.xxxx
...
values( 50) = xxx.xxxx
```

6-14. Dot Product A 3D vector can be represented in rectangular coordinates as

$$\mathbf{V} = V_x \mathbf{i} + V_y \mathbf{j} + V_z \mathbf{k} \tag{6-7}$$

where V_x is the component of vector $\mathbf{V}$ in the x direction, V_y is the component of vector $\mathbf{V}$ in the y direction, and V_z is the component of vector $\mathbf{V}$ in the z direction. Such a vector can be stored in a rank 1 array containing three elements, since there are three dimensions in the coordinate system. The same idea applies to an n-dimensional vector. An n-dimensional vector can be stored in a rank 1 array containing n elements. This is the reason why rank 1 arrays are sometimes called vectors.

One common mathematical operation between two vectors is the *dot product*. The dot product of two vectors $\mathbf{V}_1 = V_{x1} \mathbf{i} + V_{y1} \mathbf{j} + V_{z1} \mathbf{k}$ and $\mathbf{V}_2 = V_{x2} \mathbf{i} + V_{y2} \mathbf{j} + V_{z2} \mathbf{k}$ is a scalar quantity defined by the equation

$$\mathbf{V}_1 \bullet \mathbf{V}_2 = V_{x1} V_{x2} + V_{y1} V_{y2} + V_{z1} V_{z2} \tag{6-8}$$

Write a Fortran program that will read two vectors $\mathbf{V}_1$ and $\mathbf{V}_2$ into two 1D arrays in computer memory, and then calculate their dot product according to the equation given above. Test your program by calculating the dot product of vectors $\mathbf{V}_1 = 5\mathbf{i} - 3\mathbf{j} + 2\mathbf{k}$ and $\mathbf{V}_2 = 2\mathbf{i} + 3\mathbf{j} + 4\mathbf{k}$.

6-15. Power Supplied to an Object If an object is being pushed by a force $\mathbf{F}$ at a velocity $\mathbf{v}$ (Figure 6-20), then the power supplied to the object by the force is given by the equation

$$P = \mathbf{F} \bullet \mathbf{v} \tag{6-9}$$

where the force $\mathbf{F}$ is measured in newtons, the velocity $\mathbf{v}$ is measured in meters per second, and the power P is measured in watts. Use the Fortran program written in the Exercise 6-14 to calculate the power supplied by a force of $\mathbf{F} = 4\mathbf{i} + 3\mathbf{j} - 2\mathbf{k}$ newtons to an object moving with a velocity of $\mathbf{v} = 4\mathbf{i} - 2\mathbf{j} + 1\mathbf{k}$ meters per second.

FIGURE 6-20
A force $\mathbf{F}$ applied to an object moving with velocity $\mathbf{v}$.

6-16. Cross Product Another common mathematical operation between two vectors is the *cross product*. The cross product of two vectors $\mathbf{V}_1 = V_{x1}\mathbf{i} + V_{y1}\mathbf{j} + V_{z1}\mathbf{k}$ and $\mathbf{V}_2 = V_{x2}\mathbf{i} + V_{y2}\mathbf{j} + V_{z2}\mathbf{k}$ is a vector quantity defined by the equation

$$\mathbf{V}_1 \times \mathbf{V}_2 = (V_{y1}V_{z2} - V_{y2}V_{z1})\mathbf{i} + (V_{z1}V_{x2} - V_{z2}V_{x1})\mathbf{j} + (V_{x1}V_{y2} - V_{x2}V_{y1})\mathbf{k} \quad (6\text{-}10)$$

Write a Fortran program that will read two vectors $\mathbf{V}_1$ and $\mathbf{V}_2$ into arrays in computer memory, and then calculate their cross product according to the equation given above. Test your program by calculating the cross product of vectors $\mathbf{V}_1 = 5\mathbf{i} - 3\mathbf{j} + 2\mathbf{k}$ and $\mathbf{V}_2 = 2\mathbf{i} + 3\mathbf{j} + 4\mathbf{k}$.

6-17. Velocity of an Orbiting Object The vector angular velocity ω of an object moving with a velocity $\mathbf{v}$ at a distance $\mathbf{r}$ from the origin of the coordinate system (Figure 6-21) is given by the equation

$$\mathbf{v} = \mathbf{r} \times \omega \quad (6\text{-}11)$$

where $\mathbf{r}$ is the distance in meters, ω is the angular velocity in radians per second, and $\mathbf{v}$ is the velocity in meters per second. If the distance from the center of the earth to an orbiting satellite is $\mathbf{r} = 300{,}000\mathbf{i} + 400{,}000\mathbf{j} + 50{,}000\mathbf{k}$ meters, and the angular velocity of the satellite is $\omega = -6 \times 10^{-3}\mathbf{i} + 2 \times 10^{-3}\mathbf{j} - 9 \times 10^{-4}\mathbf{k}$ radians per second, what is the velocity of the satellite in meters per second? Use the program written in the previous exercise to calculate the answer.

6-18. Program `stats_4` in Example 6-4 will behave incorrectly if a user enters an invalid value in the input data set. For example, if the user enters the characters `1.o` instead of `1.0` on a line, then the READ statement will return a nonzero status for that line. This nonzero status will be misinterpreted as the end of the data set, and only a portion of the input data will be processed. Modify the program to protect against invalid values in the input data file. If a bad value is encountered in the input data file, the program should display the line number containing the bad value, and skip it. The program should process all of the good values in the file, even those after a bad value.

6-19. In Set Theory, the union of two sets is the list of all elements that appear in *either* (or both) of the sets, and the intersection of the two sets is the list of all elements that appear in *both* sets only. For example, if one set A consists of the elements

$$A \in \{1\ 3\ 7\ 6\ 2\ 5\}$$

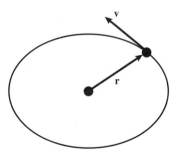

FIGURE 6-21
Calculating the velocity of an object in orbit.

and a second set B consists of the elements

$$B \in \{-1\ 2\ 0\ 5\ 8\ 9\}$$

then the union of the two sets would be

$$A \cup B \in \{-1\ 0\ 1\ 2\ 3\ 5\ 6\ 7\ 8\ 9\}$$

and the intersection of the two sets would be

$$A \cap B \in \{2\ 5\}$$

Write a program that will read in two arrays of integers representing the elements of two sets from two different user-specified input files, and calculate both the union and the intersection of the two sets. Use arrays to contain the input sets, and also to build both the union and the intersection. Note that the input sets may not be sorted in order, so your algorithm must work regardless of the order in which set elements are entered.

Test your program on two files named inputA.dat and inputB.dat, containing the following two sets:

File inputA.dat: 0, 1, −3, 5, −11, 6, 8, 11, 17, 15
File inputB.dat: 0, −1, 3, 7, −6, 16, 5, 12, 21

6-20. The location of any point P in a 3D space can be represented by a set of three values (x, y, z), where x is the distance along the x axis to the point, y is the distance along the y axis to the point, and z is the distance along the z axis to the point. Thus, a point can be represented by a three-element vector containing the values x, y, and z. If two points P_1 and P_2 are represented by the values (x_1, y_1, z_1) and (x_2, y_2, z_2), then the distance between the points P_1 and P_2 can be calculated from the equation

$$\text{distance} = \sqrt{(x_1 - x_2)^2 + (y_1 - y_2)^2 + (z_1 - z_2)^2} \qquad (6\text{-}12)$$

Write a Fortran program to read in two points (x_1, y_1, z_1) and (x_2, y_2, z_2), and to calculate the distance between them. Test your program by calculating the distance between the points $(-1, 4, 6)$ and $(1, 5, -2)$.

7

Introduction to Procedures

OBJECTIVES

- Learn how Fortran procedures help with good program design.
- Know the difference between a subroutine and a function.
- Be able to create and call subroutines.
- Understand and be able to use the INTENT attribute.
- Understand the pass-by-reference scheme for variable passing.
- Understand the differences among explicit-shape dummy arrays, assumed-shape dummy arrays, and assumed-size dummy arrays.
- Understand why assumed-size dummy arrays should never be used.
- Know how to share data between procedures using modules.
- Understand explicit interfaces, and why it is good to define procedures within modules.
- Be able to create and invoke user-defined functions.
- Know how to pass Fortran procedures as calling arguments to other procedures.

In Chapter 3, we learned the importance of good program design. The basic technique that we employed is **top-down design.** In top-down design, the programmer starts with a statement of the problem to be solved and the required inputs and outputs. Next, he or she describes the algorithm to be implemented by the program in broad outline, and applies *decomposition* to break the algorithm down into logical subdivisions called subtasks. Then, the programmer breaks down each subtask until he or she winds up with many small pieces, each of which does a simple, clearly understandable job. Finally, the individual pieces are turned into Fortran code.

Although we have followed this design process in our examples, the results have been somewhat restricted, because we have had to combine the final Fortran code generated for each subtask into a single large program. There has been no way to code, verify, and test each subtask independently before combining them into the final program.

Fortunately, Fortran has a special mechanism designed to make subtasks easy to develop and debug independently before building the final program. It is possible to code each subtask as a separate **program unit**[1] called an **external procedure,** and each external procedure can be compiled, tested, and debugged independently all of the other subtasks (procedures) in the program.[2]

There are two kinds of external procedures in Fortran: **subroutines** and **function subprograms** (or just **functions**). Subroutines are procedures that are invoked by naming them in a separate CALL statement, and that can return multiple results through calling arguments. Functions subprograms are procedures that are invoked by naming them in an expression, and whose result is a single value that is used in the evaluation of the expression. Both type of procedures will be described in this chapter.

Well-designed procedures enormously reduce the effort required on a large programming project. Their benefits include:

1. **Independent testing of subtasks.** Each subtask can be coded and compiled as an independent unit. The subtask can be tested separately to ensure that it performs properly by itself before combining it into the larger program. This step is known as **unit testing.** It eliminates a major source of problems before the final program is even built.
2. **Reusable code.** In many cases, the same basic subtask is needed in many parts of a program. For example, it may be necessary to sort a list of values into ascending order many different times within a program, or even in other programs. It is possible to design, code, test, and debug a *single* procedure to do the sorting, and then to reuse that procedure whenever sorting is required. This reusable code has two major advantages: It reduces the total programming effort required, and it simplifies debugging, since the sorting function only needs to be debugged once.
3. **Isolation from unintended side effects.** Subprograms communicate with the main programs that invoke them through a list of variables called an **argument list.** *The only variables in the main program that can be changed by the procedure are those in the argument list*. This is very important, since accidental programming mistakes can only affect the variables in the procedure in which the mistake occurred.

Once a large program is written and released, it has to be *maintained*. Program maintenance involves fixing bugs and modifying the program to handle new and unforeseen circumstances. The programmer who modifies a program during maintenance is often not the person who originally wrote it. In poorly written programs, it is common for the programmer modifying the program to make a change in one region of the code, and to have that change cause unintended side effects in a totally different part of the program. This happens because variable names are reused in different portions of the program. When the programmer changes the values left behind in some of the variables, those values are accidentally picked up and used in other portions of the code.

---

[1] A program unit is a *separately compiled* portion of a Fortran program. Main programs, subroutines, and function subprograms are all program units.
[2] Fortran also supports **internal procedures,** which are procedures entirely contained within another program unit. Internal procedures will be described in Chapter 9. Unless otherwise indicated, the references in this chapter to procedures, subroutines, and functions refer to external procedures, external subroutines, and external functions.

The use of well-designed procedures minimizes this problem by **data hiding.** All of the variables in the procedure except for those in the argument list are not visible to the main program, and therefore mistakes or changes in those variables cannot accidentally cause unintended side effects in the other parts of the program.

Good Programming Practice
Break large program tasks into procedures whenever practical to achieve the important benefits of independent component testing, reusability, and isolation from undesired side effects.

We will now examine the two different types of Fortran procedures: subroutines and functions.

7.1
SUBROUTINES

A **subroutine** is a Fortran procedure that is invoked by naming it in a CALL statement, and that receives its input values and returns its results through an **argument list.** The general form of a subroutine is

```
SUBROUTINE subroutine_name ( argument_list )
...
   (Declaration section)
...
   (Execution section)
...
RETURN
END SUBROUTINE [subroutine_name]
```

The SUBROUTINE statement marks the beginning of a subroutine. It specifies the name of the subroutine and the argument list associated with it. The subroutine name must follow standard Fortran conventions: It may be up to 63 characters long and contain both alphabetic characters and digits, but the first character must be alphabetic. The argument list contains a list of the variables and/or arrays that are being passed from the calling program to the subroutine. These variables are called **dummy arguments,** since the subroutine does not actually allocate any memory for them. They are just placeholders for actual arguments that will be passed from the calling program unit when the subroutine is invoked.

Note that like any Fortran program, a subroutine must have a declaration section and an execution section. When a program calls the subroutine, the execution of the calling program is suspended, and the execution section of the subroutine is run. When a RETURN or END SUBROUTINE statement is reached in the subroutine, the calling program starts running again at the line following the subroutine call.

Each subroutine is an independent program unit, beginning with a SUBROUTINE statement and terminated by an END SUBROUTINE statement. It is compiled separately from the main program and from any other procedures. Because each program unit in a program is compiled separately, local variable names and statement labels may be reused in different routines without causing an error.

Any executable program unit may call a subroutine, including another subroutine. (However, a subroutine may not call itself unless it is declared to be **recursive;** recursion will be explained in Chapter 13.) To call a subroutine, the calling program uses a CALL statement. The form of a CALL statement is

CALL subroutine_name (argument_list)

where the order and type of the **actual arguments** in the argument list must match the order and type of the dummy arguments declared in the subroutine.

A simple example subroutine is shown in Figure 7-1. This subroutine calculates the hypotenuse of a right triangle from the lengths of the other two sides.

FIGURE 7-1
A simple subroutine to calculate the hypotenuse of a right triangle.

```
SUBROUTINE calc_hypotenuse ( side_1, side_2, hypotenuse )
!
!   Purpose:
!     To calculate the hypotenuse of a right triangle from the two
!     other sides.
!
!   Record of revisions:
!       Date        Programmer          Description of change
!       ====        ==========          =====================
!     11/22/15     S. J. Chapman        Original code
!
IMPLICIT NONE

! Data dictionary: declare calling parameter types & definitions
REAL, INTENT(IN) :: side_1         ! Length of side 1
REAL, INTENT(IN) :: side_2         ! Length of side 2
REAL, INTENT(OUT) :: hypotenuse    ! Length of hypotenuse

! Data dictionary: declare local variable types & definitions
REAL :: temp                       ! Temporary variable

! Calculate hypotenuse
temp = side_1**2 + side_2**2
hypotenuse = SQRT ( temp )

END SUBROUTINE calc_hypotenuse
```

This subroutine has three arguments in its dummy argument list. Arguments side_1 and side_2 are placeholders for real values containing the lengths of sides 1 and 2 of the triangle. These dummy arguments are used to pass data to the subroutine but are not changed inside the subroutine, so they are declared to be input values with the "INTENT(IN)" attribute. Dummy argument hypotenuse is a placeholder for a real variable that will receive the length of the hypotenuse of the triangle. The value of

Introduction to Procedures

hypotenuse is set in the subroutine, so it is declared to be an output variable with the "INTENT(OUT)" attribute.

The variable temp is actually defined within the subroutine. It is used in the subroutine, but it is not accessible to any calling program. Variables that are used within a subroutine and that are not accessible by calling programs are called **local variables**.

Finally, the RETURN statement in the subroutine is optional. Execution automatically returns to the calling program when the END SUBROUTINE statement is reached. A RETURN statement is only necessary when we wish to return to the calling program before the end of the subroutine is reached. As a result, the RETURN statement is rarely used.

To test a subroutine, it is necessary to write a program called a **test driver program**. The test driver program is a small program that calls the subroutine with a sample data set for the specific purpose of testing it. A test driver program for subroutine calc_hypotenuse is shown in Figure 7-2:

FIGURE 7-2
A test driver program for subroutine calc_hypotenuse.

```
PROGRAM test_calc_hypotenuse
!
!   Purpose:
!     Program to test the operation of subroutine calc_hypotenuse.
!
!   Record of revisions:
!       Date         Programmer          Description of change
!       ====         ==========          =====================
!     11/22/15      S. J. Chapman        Original code
!
IMPLICIT NONE

! Data dictionary: declare variable types & definitions
REAL :: s1           ! Length of side 1
REAL :: s2           ! Length of side 2
REAL :: hypot        ! Hypotenuse

! Get the lengths of the two sides.
WRITE (*,*) 'Program to test subroutine calc_hypotenuse: '
WRITE (*,*) 'Enter the length of side 1: '
READ (*,*) s1
WRITE (*,*) 'Enter the length of side 2: '
READ (*,*) s2

! Call calc_hypotenuse.
CALL calc_hypotenuse ( s1, s2, hypot )

! Write out hypotenuse.
WRITE (*,1000) hypot
1000 FORMAT ('The length of the hypotenuse is: ', F10.4 )

END PROGRAM test_calc_hypotenuse
```

This program calls subroutine `calc_hypotenuse` with an actual argument list of variables `s1`, `s2`, and `hypot`. Therefore, wherever the dummy argument `side_1` appears in the subroutine, variable `s1` is really used instead. Similarly, the hypotenuse is really written into variable `hypot`.

7.1.1 Example Problem—Sorting

Let us now reexamine the sorting problem of Example 6-3, using subroutines where appropriate.

EXAMPLE 7-1

Sorting Data:

Develop a program to read in a data set from a file, sort it into ascending order, and display the sorted data set. Use subroutines where appropriate.

SOLUTION
The program in Example 6-3 read an arbitrary number of real input data values from a user-supplied file, sorted the data into ascending order, and wrote the sorted data to the standard output device. The sorting process would make a good candidate for a subroutine, since only the array `a` and its length `nvals` are in common between the sorting process and the rest of the program. The rewritten program using a sorting subroutine is shown in Figure 7-3:

FIGURE 7-3
Program to sort real data values into ascending order using a `sort` subroutine.

```
PROGRAM sort3
!
!  Purpose:
!    To read in a real input data set, sort it into ascending order
!    using the selection sort algorithm, and to write the sorted
!    data to the standard output device.  This program calls subroutine
!    "sort" to do the actual sorting.
!
!  Record of revisions:
!      Date       Programmer          Description of change
!      ====       ==========          =====================
!    11/22/15    S. J. Chapman        Original code
!
IMPLICIT NONE

! Data dictionary: declare constants
INTEGER, PARAMETER :: MAX_SIZE = 10   ! Max input data size

! Data dictionary: declare variable types & definitions
```

(*continued*)

(*continued*)

```
      REAL, DIMENSION(MAX_SIZE) :: a      ! Data array to sort
      LOGICAL :: exceed = .FALSE.         ! Logical indicating that array
                                          !    limits are exceeded.
      CHARACTER(len=20) :: filename       ! Input data file name
      INTEGER :: i                        ! Loop index
      CHARACTER(len=80) :: msg            ! Error message
      INTEGER :: nvals = 0                ! Number of data values to sort
      INTEGER :: status                   ! I/O status: 0 for success
      REAL :: temp                        ! Temporary variable for reading

      ! Get the name of the file containing the input data.
      WRITE (*,*) 'Enter the file name with the data to be sorted: '
      READ (*,1000) filename
      1000 FORMAT ( A20 )

      ! Open input data file.  Status is OLD because the input data must
      ! already exist.
      OPEN ( UNIT=9, FILE=filename, STATUS='OLD', ACTION='READ', &
             IOSTAT=status, IOMSG=msg )

      ! Was the OPEN successful?
      fileopen: IF ( status == 0 ) THEN       ! Open successful

         ! The file was opened successfully, so read the data to sort
         ! from it, sort the data, and write out the results.
         ! First read in data.
         DO
            READ (9, *, IOSTAT=status) temp   ! Get value
            IF ( status /= 0 ) EXIT           ! Exit on end of data
            nvals = nvals + 1                 ! Bump count
            size: IF ( nvals <= MAX_SIZE ) THEN  ! Too many values?
               a(nvals) = temp                ! No: Save value in array
            ELSE
               exceed = .TRUE.                ! Yes: Array overflow
            END IF size
         END DO

         ! Was the array size exceeded?  If so, tell user and quit.
         toobig: IF ( exceed ) THEN
            WRITE (*,1010) nvals, MAX_SIZE
            1010 FORMAT (' Maximum array size exceeded: ', I6, ' > ', I6 )
         ELSE

            ! Limit not exceeded: sort the data.
            CALL sort (a, nvals)

            ! Now write out the sorted data.
            WRITE (*,'(A)') ' The sorted output data values are: '
            WRITE (*,'(3X,F10.4)') ( a(i), i = 1, nvals )

         END IF toobig
```

(*continued*)

(concluded)

```fortran
    ELSE fileopen

       ! Else file open failed.  Tell user.
       WRITE (*,1050) TRIM(msg)
       1050 FORMAT ('File open failed--error = ', A)

    END IF fileopen

    END PROGRAM sort3

    !*****************************************************************
    !*****************************************************************

    SUBROUTINE sort (arr, n)
    !
    !  Purpose:
    !    To sort real array "arr" into ascending order using a selection
    !    sort.
    !
    IMPLICIT NONE

    ! Data dictionary: declare calling parameter types & definitions
    INTEGER, INTENT(IN) :: n                  ! Number of values
    REAL, DIMENSION(n), INTENT(INOUT) :: arr  ! Array to be sorted

    ! Data dictionary: declare local variable types & definitions
    INTEGER :: i             ! Loop index
    INTEGER :: iptr          ! Pointer to smallest value
    INTEGER :: j             ! Loop index
    REAL :: temp             ! Temp variable for swaps

    ! Sort the array
    outer: DO i = 1, n-1

           ! Find the minimum value in arr(I) through arr(N)
           iptr = i
           inner: DO j = i+1, n
              minval: IF ( arr(j) < arr(iptr) ) THEN
                 iptr = j
              END IF minval
           END DO inner

           ! iptr now points to the minimum value, so swap arr(iptr)
           ! with arr(i) if i /= iptr.
           swap: IF ( i /= iptr ) THEN
              temp      = arr(i)
              arr(i)    = arr(iptr)
              arr(iptr) = temp
           END IF swap

    END DO outer

    END SUBROUTINE sort
```

This new program can be tested just as the original program was, with identical results. If the following data set is placed in file `input2`:

```
13.3
12.
-3.0
0.
4.0
6.6
4.
-6.
```

then the results of the test run will be:

```
C:\book\fortran\chap7>sort3
Enter the file name containing the data to be sorted:
input2
The sorted output data values are:
  -6.0000
  -3.0000
    .0000
   4.0000
   4.0000
   6.6000
  12.0000
  13.3000
```

The program gives the correct answers for our test data set, as before.

Subroutine `sort` performs the same function as the sorting code in the original example, but now `sort` is an independent subroutine that we can reuse unchanged whenever we need to sort any array of real numbers.

Note that the array was declared in the sort subroutine as

```
REAL, DIMENSION(n), INTENT(INOUT) :: arr  ! Array to be sorted
```

The statement tells the Fortran compiler that dummy argument `arr` is an array whose length is n, where n is also a calling argument. The dummy argument `arr` is only a placeholder for whatever array is passed as an argument when the subroutine is called. The actual size of the array will be the size of the array that is passed from the calling program.

Also, note that n was declared to be an input parameter *before* it was used to define `arr`. Most compilers will require n to be declared first, so that its meaning is known before it is used in the array declaration. If the order of the declarations were reversed, most compilers will generate an error saying that n is undefined when `arr` is declared.

Finally, note that the dummy argument `arr` was used both to pass the data to subroutine `sort` and to return the sorted data to the calling program. Since it is used for both input and output, it is declared with the `INTENT(INOUT)` attribute.

7.1.2 The INTENT Attribute

Dummy subroutine arguments can have an INTENT attribute associated with them. The INTENT attribute is associated with the type declaration statement that declares each dummy argument. The attribute can take one of three forms:

INTENT(IN)	Dummy argument is used only to pass input data to the subroutine.
INTENT(OUT)	Dummy argument is used only to return results to the calling program.
INTENT(INOUT) or INTENT(IN OUT)	Dummy argument is used both to pass input data to the subroutine and to return results to the calling program.

The purpose of the INTENT attribute is to tell the compiler how the programmer intends each dummy argument to be used. Some arguments may be intended only to provide input data to the subroutine, and some may be intended only to return results from the subroutine. Finally, some may be intended to both provide data and return results. The appropriate INTENT attribute should *always* be declared for each argument.[3]

Once the compiler knows what we intend to do with each dummy argument, it can use that information to help catch programming errors at compile time. For example, suppose that a subroutine accidentally modifies an input argument. Changing that input argument will cause the value of the corresponding variable in the calling program to be changed, and the changed value will be used in all subsequent processing. This type of programming error can be very hard to locate, since it is caused by the interaction between procedures.

A simple example is shown below. Here subroutine sub1 calculates an output value, but also accidentally modifies its input value.

```
SUBROUTINE sub1(input,output)
IMPLICIT NONE
REAL, INTENT(IN) :: input
REAL, INTENT(OUT) :: output

output = 2. * input
input = -1.        ! This line is an error!
END SUBROUTINE sub1
```

By declaring our intent for each dummy argument, the compiler can spot this error for us at compilation time. When this subroutine is compiled with the Intel Fortran compiler, the results are

---

[3] The intent of a dummy argument may also be declared in a separate INTENT statement of the form

```
INTENT(IN) :: arg1, arg2, ...
```

```
C:\book\fortran\chap7>ifort sub1.f90
Intel(R) Visual Fortran Intel(R) 64 Compiler for applications running on
Intel(R) 64, Version 16.0.2.180 Build 20160204
Copyright (C) 1985-2016 Intel Corporation. All rights reserved.

sub1.f90(7): error #6780: A dummy argument with the INTENT(IN) attribute
shall not be defined nor become undefined.   [INPUT]
input = -1.
^
compilation aborted for sub1.f90 (code 1)
```

Similarly, a variable declared with INTENT(OUT) *must* be defined within the subroutine, or the compiler will produce an error.

The INTENT attribute is only valid for dummy procedure arguments. It is an error to declare the intent of local variables in a subroutine, or of variables in a main program.

As we will see later, declaring the intent of each dummy argument will also help us spot errors that occur in the calling sequence *between* procedures. You should always declare the intent of every dummy argument in every procedure.

Good Programming Practice

Always declare the intent of every dummy argument in every procedure.

7.1.3 Variable Passing in Fortran: The Pass-By-Reference Scheme

Fortran programs communicate with their subroutines using a **pass-by-reference** scheme. When a subroutine call occurs, the main program passes a pointer to the location in memory of each argument in the actual argument list. The subroutine looks at the memory locations pointed to by the calling program to get the values of the dummy arguments it needs. This process is illustrated in Figure 7-4.

The figure shows a main program test calling a subroutine sub1. There are three actual arguments being passed to the subroutine, a real variable a, a four-element real array b, and an integer variable next. These variables occupy memory addresses 001, 002–005, and 006 respectively in some computer. Three dummy arguments are declared in sub1: a real variable x, a real array y, and an integer variable i. When the main program calls sub1, *what is passed to the subroutine are the pointers to the memory locations containing the calling arguments*: 001, 002, and 006. Whenever variable x is referred to in the subroutine, the contents of memory location 001 are accessed, etc. This parameter passing scheme is called pass-by-reference, since only pointers to the values are passed to the subroutine, not the actual values themselves.

There are some possible pitfalls associated with the pass-by-reference scheme. *The programmer must ensure that the values in the calling argument list match the subroutine's calling parameters in number, type, and order*. If there is a mismatch, the Fortran program will not be able to recognize that fact, and it will misuse the parameters without informing you of the problem. *This is the most common error made by*

FIGURE 7-4
The pass-by-reference memory scheme. Note that only pointers to the memory addresses of the actual arguments are passed to the subroutine.

programmers when using Fortran subroutines. For example, consider the program shown in Figure 7-5:

FIGURE 7-5
Example illustrating the effects of a type mismatch when calling a subroutine.

```
PROGRAM bad_call
!
!  Purpose:
!    To illustrate misinterpreted calling arguments.
!
IMPLICIT NONE
REAL :: x = 1.                ! Declare real variable x.
CALL bad_argument ( x )       ! Call subroutine.
END PROGRAM bac_call

SUBROUTINE bad_argument ( i )
IMPLICIT NONE
INTEGER :: i                  ! Declare argument as integer.
WRITE (*,*) 'i = ', i         ! Write out i.
END SUBROUTINE bad_argument
```

The argument in the call to subroutine `bad_argument` is real, but the corresponding dummy argument is type integer. Fortran will pass the address of the real variable x to the subroutine, which will then treat it as an integer. The results are quite bad. When the program is compiled with the Intel Fortran compiler, we get:

```
C:\book\fortran\chap7>bad_call
I =    1065353216
```

Another serious problem can occur if a variable is placed in the calling argument list in a position at which an array is expected. The subroutine cannot tell the difference between a variable and an array, so it will treat the variable and the variables following it in memory as though they were all part of one big array! This behavior can produce a world of problems. A subroutine containing a variable named x in its calling sequence could wind up modifying another variable y that wasn't even passed to the subroutine, just because y happens to be allocated after x in the computer's memory. Problems like that can be *extremely* difficult to find and debug.

In Section 7.3, we will learn how to get a Fortran compiler to automatically check the number, type, intent, and order of each argument in each subroutine call, so that the compiler can catch these errors for us at compilation time.

Programming Pitfalls

Make sure that the values in the argument list of a subroutine call match the subroutine's declared parameters in number, type, and order. Very bad results may occur if you do not ensure that the arguments match properly.

7.1.4 Passing Arrays to Subroutines

A calling argument is passed to a subroutine by passing a pointer to the memory location of the argument. If the argument happens to be an array, then the pointer points to the first value in the array. However, the subroutine needs to know both the location and the size of the array to ensure that it stays within the boundaries of the array, and in order to perform array operations. How can we supply this information to the subroutine?

There are three possible approaches to specifying the length of a dummy array in a subroutine. One approach is to pass the bounds of each dimension of the array to the subroutine as arguments in the subroutine call, and to declare the corresponding dummy array to be that length. The dummy array is thus an **explicit-shape dummy array,** since each of its bounds is explicitly specified. If this is done, the subroutine will know the shape of each dummy array when it is executed. Since the shape of the array is known, the bounds checkers on most Fortran compilers will be able to detect and report out-of-bounds memory references. For example, the following code declares two arrays data1 and data2 to be of extent n, and then processes nvals values in the arrays. If an out-of-bounds reference occurs in this subroutine, it can be detected and reported.

```
SUBROUTINE process ( data1, data2, n, nvals )
INTEGER, INTENT(IN) :: n, nvals
REAL, INTENT(IN),  DIMENSION(n) :: data1 ! Explicit shape
REAL, INTENT(OUT), DIMENSION(n) :: data2 ! Explicit shape

DO i = 1, nvals
   data2(i) = 3. * data1(i)
END DO
END SUBROUTINE process
```

When explicit-shape dummy arrays are used, the size and shape of each dummy array is known to the compiler. Since the size and shape of each array is known, it is possible to use array operations and array sections with the dummy arrays. The following subroutine uses array sections; it will work because the dummy arrays are explicit-shape arrays.

```
SUBROUTINE process2 ( data1, data2, n, nvals )
INTEGER, INTENT(IN) :: nvals
REAL, INTENT(IN), DIMENSION(n)  :: data1 ! Explicit shape
REAL, INTENT(OUT), DIMENSION(n) :: data2 ! Explicit shape

data2(1:nvals) = 3. * data1(1:nvals)
END SUBROUTINE process2
```

A second approach is to declare all dummy arrays in a subroutine as **assumed-shape dummy arrays** and to create an explicit interface to the subroutine. This approach will be explained in Section 7.3.

The third (and oldest) approach is to declare the length of each dummy array with an asterisk as an **assumed-size dummy array.** In this case, the compiler knows nothing about the length of the actual array passed to the subroutine. Bounds checking, whole array operations, and array sections will not work for assumed-size dummy arrays, because the compiler does not know the actual size and shape of the array. For example, the following code declares two assumed-size dummy arrays `data1` and `data2`, and then processes `nvals` values in the arrays.

```
SUBROUTINE process3 ( data1, data2, nvals )
REAL, INTENT(IN),  DIMENSION(*) :: data1 ! Assumed size
REAL, INTENT(OUT), DIMENSION(*) :: data2 ! Assumed size
INTEGER, INTENT(IN) :: nvals

DO i = 1, nvals
   data2(i) = 3. * data1(i)
END DO
END SUBROUTINE process3
```

Arrays `data1` and `data2` must be at least `nvals` values long. If they are not, the Fortran code will either abort with an error at runtime or overwrite other locations in memory. Subroutines written like this are hard to debug, since the bounds checking option of most compilers will not work for unknown-length arrays. They also cannot use whole array operations or array sections.

Assumed-size dummy arrays are a holdover from earlier versions of Fortran. *They should never be used in any new programs.*

Introduction to Procedures

Good Programming Practice
Use explicit-shape or assumed-shape dummy arrays in all new procedures. This permits whole array operations to be used within the procedure. It also allows for easier debugging, since out-of-bounds references can be detected. *Assumed-size dummy arrays should never be used.* They are undesirable, and are likely to be eliminated from a future version of the Fortran language.

EXAMPLE 7-2   *Bounds Checking in Subroutines:*

Write a simple Fortran program containing a subroutine that oversteps the limits of an array in its argument list. Compile and execute the program both with bounds checking turned off and with bounds checking turned on.

SOLUTION
The program in Figure 7-6 allocates a 5-element array a. It initializes all the elements of a to zero, and then calls subroutine sub1. Subroutine sub1 modifies six elements of array a, despite the fact that a has only five elements.

FIGURE 7-6
A program illustrating the effect of exceeding the boundaries of an array in a subroutine.

```
PROGRAM array2
!
!   Purpose:
!     To illustrate the effect of accessing an out-of-bounds
!     array element.
!
!   Record of revisions:
!       Date         Programmer          Description of change
!       ====         ==========          =====================
!     11/22/15      S. J. Chapman        Original code
!
IMPLICIT NONE

! Declare the and initialize the variables used in this program.
INTEGER :: i                              ! Loop index
REAL, DIMENSION(5) :: a = 0.              ! Array

! Call subroutine sub1.
CALL sub1( a, 5, 6 )

! Write out the values of array a
DO i = 1, 6
   WRITE (*,100) i, a(i)
   100 FORMAT ( 'A(', I1, ') = ', F6.2 )
END DO
```

(continued)

(*concluded*)

```
!*************************************************************
!*************************************************************

END PROGRAM array2

SUBROUTINE sub1 ( a, ndim, n )
IMPLICIT NONE

INTEGER, INTENT(IN) :: ndim                ! size of array
REAL, INTENT(OUT), DIMENSION(ndim) :: a    ! Dummy argument
INTEGER, INTENT(IN) :: n                   ! # elements to process
INTEGER :: i                               ! Loop index

DO i = 1, n
   a(i) = i
END DO

END SUBROUTINE sub1
```

When this program is compiled with the Intel Fortran compiler with bounds checking turned *off*, the result is

```
C:\book\fortran\chap7>array2
a(1) =   1.00
a(2) =   2.00
a(3) =   3.00
a(4) =   4.00
a(5) =   5.00
a(6) =   6.00
```

In this case, the subroutine has written beyond the end of array a, into memory that was allocated for some other purpose. If this memory were allocated to another variable, then the contents of that variable would have been changed without the user knowing that anything can happen. This can produce a very subtle and hard to find bug!

If the program is recompiled with the Intel Fortran compiler with bounds checking turned *on*, the result is

```
C:\book\fortran\chap7>array2
forrtl: severe (408): fort: (10): Subscript #1 of the array A has value 6
which is greater than the upper bound of 5

Image              PC                 Routine     Line        Source
array2.exe         00007FF60DA3B81E   Unknown     Unknown     Unknown
array2.exe         00007FF60DA31383   Unknown     Unknown     Unknown
array2.exe         00007FF60DA31085   Unknown     Unknown     Unknown
array2.exe         00007FF60DA8132E   Unknown     Unknown     Unknown
array2.exe         00007FF60DA81BE8   Unknown     Unknown     Unknown
KERNEL32.DLL       00007FFA56B38102   Unknown     Unknown     Unknown
ntdll.dll          00007FFA594DC5B4   Unknown     Unknown     Unknown
```

Here the program detected the out-of-bounds reference and shut down after telling the user where the problem occurred.

Introduction to Procedures

7.1.5 Passing Character Variables to Subroutines

When a character variable is used as a dummy subroutine argument, the length of the character variable is declared with an asterisk. Since no memory is actually allocated for dummy arguments, it is not necessary to know the length of the character argument when the subroutine is compiled. A typical dummy character argument is shown below:

```
SUBROUTINE sample ( string )
CHARACTER(len=*), INTENT(IN) :: string
...
```

When the subroutine is called, the length of the dummy character argument will be the length of the actual argument passed from the calling program. If we need to know the length of the character string passed to the subroutine during execution, we can use the intrinsic function LEN( ) to determine it. For example, the following simple subroutine displays the length of any character argument passed to it.

```
SUBROUTINE sample ( string )
CHARACTER(len=*), INTENT(IN) :: string
WRITE (*,'(A,I3)') 'Length of variable = ', LEN(string)
END SUBROUTINE sample
```

7.1.6 Error Handling in Subroutines

What happens if a program calls a subroutine with insufficient or invalid data for proper processing? For example, suppose that we are writing a subroutine that subtracts two input variables and takes the square root of the result. What should we do if the difference of the two variables is a negative number?

```
SUBROUTINE process (a, b, result)
IMPLICIT NONE
REAL, INTENT(IN) :: a, b
REAL, INTENT(OUT) :: result
REAL :: temp
temp = a - b
result = SQRT ( temp )
END SUBROUTINE process
```

For example, suppose that a is 1 and b is 2. If we just process the values in the subroutine, a runtime error will occur when we attempt to take the square root of a negative number, and the program will abort. This is clearly not an acceptable result.

An alternative version of the subroutine is shown below. In this version, we test for a negative number, and if one is present, we print out an informative error message and stop.

```
SUBROUTINE process (a, b, result)
IMPLICIT NONE
REAL, INTENT(IN) :: a, b
REAL, INTENT(OUT) :: result
```

```
REAL :: temp
temp = a - b
IF ( temp >= 0. ) THEN
   result = SQRT ( temp )
ELSE
   WRITE (*,*) 'Square root of negative value in subroutine "process"!'
   STOP
END IF
END SUBROUTINE process
```

While better than the previous example, this design is also bad. If `temp` is ever negative, the program will just stop without ever returning from subroutine `process`. If this happens, the user will lose all of the data and processing that has occurred up to that point in the program.

A much better way to design the subroutine is to detect the possible error condition, and to report it to the calling program by setting a value into an **error flag.** The calling program can then take appropriate actions about the error. For example, it can be designed to recover from the error, if possible. If not, it can at least write out an informative error message, save the partial results calculated so far, and then shut down gracefully.

In the example shown below, a 0 returned in the error flag means successful completion, and a 1 means that the square-root-of-a-negative-number error occurred.

```
SUBROUTINE process (a, b, result, error)
IMPLICIT NONE
REAL, INTENT(IN) :: a, b
REAL, INTENT(OUT) :: result
INTEGER, INTENT(OUT) :: error
REAL :: temp
temp = a - b
IF ( temp >= 0. ) THEN
   result = SQRT ( temp )
   error = 0
ELSE
   result = 0
   error = 1
END IF
END SUBROUTINE process
```

Programming Pitfalls

Never include `STOP` statements in any of your subroutines. If you do, you might create a working program, and release it to users, only to find that it mysteriously halts from time to time on certain unusual data sets.

Good Programming Practice

If there are possible error conditions within a subroutine, you should test for them, and set an error flag to be returned to the calling program. The calling program should test for the error conditions after a subroutine call, and take appropriate actions.

Quiz 7-1

This quiz provides a quick check to see if you have understood the concepts introduced in Section 7.1. If you have trouble with the quiz, reread the section, ask your instructor, or discuss the material with a fellow student. The answers to this quiz are found in the back of the book.

For questions 1 through 3, determine whether the subroutine calls are correct or not. If they are in error, specify what is wrong with them.

1.  ```
    PROGRAM test1
    REAL, DIMENSION(120) :: a
    REAL :: average, sd
    INTEGER :: n
    ...
    CALL ave_sd ( a, 120, n, average, sd )
    ...
    END PROGRAM test1
    SUBROUTINE ave_sd( array, nvals, n, average, sd )
    REAL, INTENT(IN) :: nvals, n
    REAL, INTENT(IN), DIMENSION(nvals) :: array
    REAL, INTENT(OUT) :: average, sd
    ...
    END SUBROUTINE ave_sd
    ```
2.  ```
    PROGRAM test2
    CHARACTER(len=12) :: str1, str2
    str1 = 'ABCDEFGHIJ'
    CALL swap_str (str1, str2)
    WRITE (*,*) str1, str2
    END PROGRAM test2
    SUBROUTINE swap_str (string1, string2)
    CHARACTER(len=*),INTENT(IN) :: string1
    CHARACTER(len=*),INTENT(OUT) :: string2
    INTEGER :: i, length
    length = LEN(string1)
    DO i = 1, length
       string2(length-i+1:length-i+1) = string1(i:i)
    END DO
    END SUBROUTINE swap_str
    ```
3.  ```
    PROGRAM test3
    INTEGER, DIMENSION(25) :: idata
    REAL :: sum
    ...
    CALL sub3 ( idata, sum )
    ...
    END PROGRAM test3
    SUBROUTINE sub3( iarray, sum )
    INTEGER, INTENT(IN), DIMENSION(*) :: iarray
    REAL, INTENT(OUT) :: sum
    INTEGER :: i
    sum = 0.
    ```

(continued)
    ```

(*concluded*)

```
      DO i = 1, 30
         sum = sum + iarray(i)
      END DO
      END SUBROUTINE sub3
```

7.1.7 Examples

EXAMPLE 7-3

Statistics Subroutines:

Develop a set of reusable subroutines capable of determining the statistical properties of a data set of real numbers in an array. The set of subroutines should include:

1. A subroutine to determine the maximum value in a data set, and the sample number containing that value.
2. A subroutine to determine the minimum value in a data set, and the sample number containing that value.
3. A subroutine to determine the average (mean) and standard deviation of the data set.
4. A subroutine to determine the median of the data set.

SOLUTION
We will be generating four different subroutines, each of which works on a common input data set consisting of an array of real numbers.

1. **State the problem.**

    The problem is clearly stated above. We will write four different subroutines: `rmax` to find the maximum value and the location of that value in a real array, `rmin` to find the minimum value and the location of that value in a real array, `ave_sd` to find the average and standard deviation of a real array, and `median` to find the median of a real array.

2. **Define the inputs and outputs.**

    The input to each subroutine will be array of values, plus the number of values in the array. The outputs will be as follows:

    (*a*) The output of subroutine `rmax` will be a real variable containing the maximum value in the input array, and an integer variable containing the offset in the array at which the maximum value occurred.

    (*b*) The output of subroutine `rmin` will be a real variable containing the minimum value in the input array, and an integer variable containing the offset in the array at which the minimum value occurred.

    (*c*) The output of subroutine `ave_sd` will be two real variables containing the average and standard deviation of the input array.

    (*d*) The output of subroutine `median` will be a real variable containing the median value of the input array.

3. **Describe the algorithm.**
The pseudocode for the `rmax` routine is:

```
! Initialize "real_max" to the first value in the array
! and "imax" to 1.
real_max ← a(1)
imax ← 1

! Find the maximum value in a(1) through a(n)
DO for i = 2 to n
   IF a(i) > real_max THEN
      real_max ← a(i)
      imax ← i
   END of IF
END of DO
```

The pseudocode for the `rmin` routine is:

```
! Initialize "real_min" to the first value in the array
! and "imin" to 1.
real_min ← a(1)
imin ← 1

! Find the maximum value in a(1) through a(n)
DO for i = 2 to n
   IF a(i) < real_min THEN
      real_min ← a(i)
      imin ← i
   END of IF
END of DO
```

The pseudocode for the `ave_sd` routine is essentially the same as that in Example 6-4. It will not be repeated here. For the `median` calculation, we will be able to take advantage of the `sort` subroutine that we have already written. (Here is an example of reusable code saving us time and effort.) The pseudocode for the `median` subroutine is:

```
CALL sort ( n, a )
IF n is an even number THEN
   med ← (a(n/2) + a(n/2+1)) / 2.
ELSE
   med ← a(n/2+1)
END of IF
```

4. **Turn the algorithm into Fortran statements.**
The resulting Fortran subroutines are shown in Figure 7-7.

FIGURE 7-7
The subroutines `rmin`, `rmax`, `ave_sd`, and `median`.

```
SUBROUTINE rmax ( a, n, real_max, imax )
! Purpose:
!    To find the maximum value in an array, and the location
!    of that value in the array.
```

(*continued*)

(*continued*)

```
!
IMPLICIT NONE
! Data dictionary: declare calling parameter types & definitions
INTEGER, INTENT(IN) :: n                ! No. of vals in array a.
REAL, INTENT(IN), DIMENSION(n) :: a     ! Input data.
REAL, INTENT(OUT) :: real_max           ! Maximum value in a.
INTEGER, INTENT(OUT) :: imax            ! Location of max value.

! Data dictionary: declare local variable types & definitions
INTEGER :: i                            ! Index variable

! Initialize the maximum value to first value in array.
real_max = a(1)
imax = 1

! Find the maximum value.
DO i = 2, n
   IF ( a(i) > real_max ) THEN
      real_max = a(i)
      imax = i
   END IF
END DO

END SUBROUTINE rmax

!*****************************************************************
!*****************************************************************

SUBROUTINE rmin ( a, n, real_min, imin )
!
! Purpose:
!    To find the minimum value in an array, and the location
!    of that value in the array.
!
IMPLICIT NONE

! Data dictionary: declare calling parameter types & definitions
INTEGER, INTENT(IN) :: n                ! No. of vals in array a.
REAL, INTENT(IN), DIMENSION(n) :: a     ! Input data.
REAL, INTENT(OUT) :: real_min           ! Minimum value in a.
INTEGER, INTENT(OUT) :: imin            ! Location of min value.

! Data dictionary: declare local variable types & definitions
INTEGER :: i                            ! Index variable

! Initialize the minimum value to first value in array.
real_min = a(1)
imin = 1

! Find the minimum value.
DO I = 2, n
   IF ( a(i) < real_min ) THEN
      real_min = a(i)
      imin = i
```

(*continued*)

(*continued*)

```fortran
      END IF
END DO

END SUBROUTINE rmin

!*******************************************************************
!*******************************************************************

SUBROUTINE ave_sd ( a, n, ave, std_dev, error )
!
!  Purpose:
!    To calculate the average and standard deviation of an array.
!
IMPLICIT NONE

! Data dictionary: declare calling parameter types & definitions
INTEGER, INTENT(IN) :: n                ! No. of vals in array a.
REAL, INTENT(IN), DIMENSION(n) :: a     ! Input data.
REAL, INTENT(OUT) :: ave                ! Average of a.
REAL, INTENT(OUT) :: std_dev            ! Standard deviation.
INTEGER, INTENT(OUT) :: error           ! Flag: 0 — no error
                                        !       1 — sd invalid
                                        !       2 — ave & sd invalid

! Data dictionary: declare local variable types & definitions
INTEGER :: i                            ! Loop index
REAL :: sum_x                           ! Sum of input values
REAL :: sum_x2                          ! Sum of input values squared

! Initialize the sums to zero.
sum_x = 0.
sum_x2 = 0.

! Accumulate sums.
DO I = 1, n
   sum_x = sum_x + a(i)
   sum_x2 = sum_x2 + a(i)**2
END DO

! Check to see if we have enough input data.
IF ( n >= 2 ) THEN ! we have enough data

   ! Calculate the mean and standard deviation
   ave = sum_x / REAL(n)
   std_dev = SQRT( (REAL(n) * sum_x2 - sum_x**2) &
          / (REAL(n) * REAL(n - 1)) )
   error = 0

ELSE IF ( n == 1 ) THEN ! no valid std_dev

   ave = sum_x
   std_dev = 0.                         ! std_dev invalid
   error = 1
```

(*continued*)

(concluded)

```
      ELSE
          ave = 0.                              ! ave invalid
          std_dev = 0.                          ! std_dev invalid
          error = 2

      END IF
      END SUBROUTINE ave_sd

!*****************************************************************
!*****************************************************************

      SUBROUTINE median ( a, n, med )
      !
      ! Purpose:
      !   To calculate the median value of an array.
      !
      IMPLICIT NONE

      ! Data dictionary: declare calling parameter types & definitions
      INTEGER, INTENT(IN) :: n                ! No. of vals in array a.
      REAL, INTENT(IN), DIMENSION(n) :: a     ! Input data.
      REAL, INTENT(OUT) :: med                ! Median value of a.

      ! Sort the data into ascending order.
      CALL sort ( a, n )

      ! Get median.
      IF ( MOD(n,2) == 0 ) THEN
         med = ( a(n/2) + a(n/2+1) ) / 2.
      ELSE
         med = a(n/2+1)
      END IF
      END SUBROUTINE median
```

5. **Test the resulting Fortran programs.**

To test these subroutines, it is necessary to write a driver program to read the input data, call the subroutines, and write out the results. This test is left as an exercise to the student (see Exercise 7-13 at the end of the chapter).

7.2

SHARING DATA USING MODULES

We have seen that programs exchange data with the subroutines they call through an argument list. Each item in the argument list of the program's CALL statement must be matched by a dummy argument in the argument list of the subroutine being invoked. A pointer to the location of each argument is passed from the calling program to the subroutine for use in accessing the arguments.

Introduction to Procedures

In addition to the argument list, Fortran programs, subroutines, and functions can also exchange data through modules. A **module** is a separately-compiled program unit that contains the definitions and initial values of the data that we wish to share between program units.[4] If the module's name is included in a USE statement within a program unit, then the data values declared in the module may be used within that program unit. Each program unit that uses a module will have access to the same data values, so *modules provide a way to share data between program units*.

A module begins with a MODULE statement, which assigns a name to the module. The name may be up to 63 characters long, and must follow the standard Fortran naming conventions. The module ends with an END MODULE statement, which may optionally include the module's name. The declarations of the data to be shared are placed between these two statements. An example module is shown in Figure 7-8.

FIGURE 7-8
A simple module used to share data among program units.

```
MODULE shared_data
!
!  Purpose:
!    To declare data to share between two routines.
IMPLICIT NONE
SAVE

INTEGER, PARAMETER :: num_vals = 5         ! Max number of values in array
REAL, DIMENSION(num_vals) :: values        ! Data values
END MODULE shared_data
```

The SAVE statement guarantees that all data values declared in the module will be preserved between references in different procedures. It should always be included in any module that declares sharable data. SAVE statements will be discussed in detail in Chapter 9.

To use the values in this module, a program unit must declare the module name in a USE statement. The form of a USE statement is

USE *module_name*

USE statements must appear before any other statements in a program unit (except for the PROGRAM or SUBROUTINE statement, and except for comments, which may appear anywhere). The process of accessing information in a module with a USE statement is known as **USE association**.

An example that uses module shared_data to share data between a main program and a subroutine is shown in Figure 7-9.

---

[4] Modules also have other functions, as we shall see in Section 7.3 and in Chapter 13.

FIGURE 7-9
An example program using a module to share data between a main program and a subroutine.

```
PROGRAM test_module
!
!  Purpose:
!    To illustrate sharing data via a module.
!
USE shared_data                 ! Make data in module "test" visible
IMPLICIT NONE

REAL, PARAMETER :: PI = 3.141592 ! Pi

values = PI * [ 1., 2., 3., 4., 5. ]

CALL sub1                       ! Call subroutine

END PROGRAM test_module

!**********************************************************************
!**********************************************************************

SUBROUTINE sub1
!
!  Purpose:
!    To illustrate sharing data via a module.
!
USE shared_data                 ! Make data in module "test" visible
IMPLICIT NONE

WRITE (*,*) values

END SUBROUTINE sub1
```

The contents of module `shared_data` are being shared between the main program and subroutine `sub1`. Any other subroutines or functions within the program could also have access to the data by including the appropriate USE statements.

Note that the array `values` is defined in the module, and used in both program `test_module` and subroutine `sub1`. However, the array `values` does *not* have a type declaration in either the program or the subroutine; the definition is inherited through USE association. In fact, it is an error to declare a variable within a procedure that has the same name as one inherited through USE association.

Programming Pitfalls

Do not declare local variables with the same name as variables inherited through USE association. This redefinition of a variable name will produce a compilation error.

Modules are especially useful for sharing large volumes of data among many program units, and for sharing data among a group of related procedures while keeping it invisible from the invoking program unit.

Introduction to Procedures

Good Programming Practice
You may use modules to pass large amounts of data between procedures within a program. If you do so, always include the SAVE statement within the module to ensure that the contents of the module remain unchanged between uses. To access the data in a particular program unit, include a USE statement as the *first noncomment statement* after the PROGRAM, SUBROUTINE, or FUNCTION statement within the program unit.

EXAMPLE 7-4   *Random Number Generator:*

It is always impossible to make perfect measurements in the real world. There will always be some *measurement noise* associated with each measurement. This fact is an important consideration in the design of systems to control the operation of such real-world devices as airplanes, refineries, etc. A good engineering design must take these measurement errors into account, so that the noise in the measurements will not lead to unstable behavior (no plane crashes or refinery explosions!).

Most engineering designs are tested by running *simulations* of the operation of the system before it is ever built. These simulations involve creating mathematical models of the behavior of the system, and feeding the models a realistic string of input data. If the models respond correctly to the simulated input data, then we can have reasonable confidence that the real-world system will respond correctly to the real-world input data.

The simulated input data supplied to the models must be corrupted by a simulated measurement noise, which is just a string of random numbers added to the ideal input data. The simulated noise is usually produced by a *random number generator*.

A random number generator is a procedure that will return a different and apparently random number each time it is called. Since the numbers are in fact generated by a deterministic algorithm, they only appear to be random.[5] However, if the algorithm used to generate them is complex enough, the numbers will be random enough to use in the simulation.

One simple random number generator algorithm is shown below.[6] It relies on the unpredictability of the modulo function when applied to large numbers. Consider the following equation:

$$n_{i+1} = \mathrm{mod}(8121 n_i + 28411, 134456) \tag{7-1}$$

Assume that n_i is a nonnegative integer. Then because of the modulo function, n_{i+1} will be a number between 0 and 134,455 inclusive. Next, n_{i+1} can be fed into the equation to produce a number n_{i+2} that is also between 0 and 134,455. This process can be repeated forever to produce a series of numbers in the range [0, 134,455].

---

[5] For this reason, some people refer to these procedures as *pseudorandom number generators*.
[6] This algorithm is adapted from the discussion found in Chapter 7 of *Numerical Recipes: The Art of Scientific Programming*, by Press, Flannery, Teukolsky, and Vetterling, Cambridge University Press, 1986.

If we didn't know the numbers 8121, 28,411, and 134,456 in advance, it would be impossible to guess the order in which the values of n would be produced. Furthermore, it turns out that there is an equal (or uniform) probability that any given number will appear in the sequence. Because of these properties, Equation (7-1) can serve as the basis for a simple random number generator with a uniform distribution.

We will now use Equation (7-1) to design a random number generator whose output is a real number in the range [0.0, 1.0).[7]

SOLUTION

We will write a subroutine that generates one random number in the range $0 \leq ran < 1.0$ each time that it is called. The random number will be based on the equation

$$ran_i = \frac{n_i}{134{,}456} \tag{7-2}$$

where n_i is a number in the range 0 to 134,455 produced by Equation (7-1).

The particular sequence produced by Equations (7-1) and (7-2) will depend on the initial value of n_0 (called the *seed*) of the sequence. We must provide a way for the user to specify n_0 so that the sequence may be varied from run to run.

1. **State the problem.**

    Write a subroutine random0 that will generate and return a single number ran with a uniform probability distribution in the range $0 \leq ran < 1.0$, based on the sequence specified by Equations (7-1) and (7-2). The initial value of the seed n_0 will be specified by a call to a subroutine called seed.

2. **Define the inputs and outputs.**

    There are two subroutines in this problem: seed and random0. The input to subroutine seed is an integer to serve as the starting point of the sequence. There is no output from this subroutine. There is no input to subroutine random0, and the output from the subroutine is a single real value in the range [0.0, 1.0).

3. **Describe the algorithm.**

    The pseudocode for subroutine random0 is very simple:

    ```
    SUBROUTINE random0 ( ran )
    n ← MOD (8121 * n + 28411, 134456 )
    ran ← REAL(n) / 134456.
    END SUBROUTINE random0
    ```

    where the value of n is saved between calls to the subroutine. The pseudocode for subroutine seed is also trivial:

    ```
    SUBROUTINE seed ( iseed )
    n ← ABS ( iseed )
    END SUBROUTINE seed
    ```

---

[7] The notation [0.0,1.0) implies that the range of the random numbers is between 0.0 and 1.0, including the number 0.0, but excluding the number 1.0.

Introduction to Procedures

The absolute value function is used so that the user can enter any integer as the starting point. The user will not have to know in advance that only positive integers are legal seeds.

The variable n will be placed in a module so that it may be accessed by both subroutines. In addition, we will initialize n to a reasonable value so that we get good results even if subroutine seed is not called to set the seed before the first call to random0.

4. **Turn the algorithm into Fortran statements.**

The resulting Fortran subroutines are shown in Figure 7-10.

FIGURE 7-10
Subroutines to generate a random number sequence, and to set the seed of the sequence.

```fortran
MODULE ran001
!
!   Purpose:
!     To declare data shared between subs random0 and seed.
!
!   Record of revisions:
!       Date         Programmer        Description of change
!       ====         ==========        =====================
!     11/23/15     S. J. Chapman       Original code
!
IMPLICIT NONE
SAVE
INTEGER :: n = 9876
END MODULE ran001

!******************************************************************
!******************************************************************

SUBROUTINE random0 ( ran )
!
!   Purpose:
!     Subroutine to generate a pseudorandom number with a uniform
!     distribution in the range 0. <= ran < 1.0.
!
!   Record of revisions:
!       Date         Programmer        Description of change
!       ====         ==========        =====================
!     11/23/15     S. J. Chapman       Original code
!
USE ran001                          ! Shared seed
IMPLICIT NONE

! Data dictionary: declare calling parameter types & definitions
REAL, INTENT(OUT) :: ran      ! Random number

! Calculate next number
n = MOD (8121 * n + 28411, 134456 )

! Generate random value from this number
```

(continued)

(*concluded*)

```
ran = REAL(n) / 134456.

END SUBROUTINE random0

!***********************************************************
!***********************************************************

SUBROUTINE seed ( iseed )
!
!  Purpose:
!    To set the seed for random number generator random0.
!
!  Record of revisions:
!      Date        Programmer          Description of change
!      ====        ==========          =====================
!    11/23/15    S. J. Chapman         Original code
!
USE ran001                          ! Shared seed
IMPLICIT NONE

! Data dictionary: declare calling parameter types & definitions
INTEGER, INTENT(IN) :: iseed   ! Value to initialize sequence

! Set seed
n = ABS ( iseed )

END SUBROUTINE seed
```

5. **Test the resulting Fortran programs.**

If the numbers generated by these routines are truly uniformly distributed random numbers in the range $0 \leq \text{ran} < 1.0$, then the average of many numbers should be close to 0.5. To test the results, we will write a test program that prints out the first 10 values produced by random0 to see if they are indeed in the range $0 \leq \text{ran} < 1.0$. Then, the program will average five consecutive 1000-sample intervals to see how close the averages come to 0.5. The test code to call subroutines seed and random0 is shown in Figure 7-11:

FIGURE 7-11
Test driver program for subroutines seed and random0.

```
PROGRAM test_random0
!
!  Purpose:
!    Subroutine test the random number generator random0.
!
!  Record of revisions:
!      Date        Programmer          Description of change
!      ====        ==========          =====================
!    11/23/15    S. J. Chapman         Original code
!
IMPLICIT NONE

! Data dictionary: declare variable types & definitions
```

(*continued*)

(concluded)

```
REAL :: ave              ! Average of random numbers
INTEGER :: i             ! DO loop index
INTEGER :: iseed         ! Seed for random number sequence
INTEGER :: iseq          ! DO loop index
REAL :: ran              ! A random number
REAL :: sum              ! Sum of random numbers

! Get seed.
WRITE (*,*) 'Enter seed: '
READ (*,*) iseed

! Set seed.
CALL SEED ( iseed )

! Print out 10 random numbers.
WRITE (*,*) '10 random numbers: '
DO i = 1, 10
   CALL random0 ( ran )
   WRITE (*,'(3X,F16.6)') ran
END DO
! Average 5 consecutive 1000-value sequences.
WRITE (*,*) 'Averages of 5 consecutive 1000-sample sequences:'
DO iseq = 1, 5
   sum = 0.
   DO i = 1, 1000
      CALL random0 ( ran )
      sum = sum + ran
   END DO
   ave = sum / 1000.
   WRITE (*,'(3X,F16.6)') ave
END DO

END PROGRAM test_random0
```

The results of compiling and running the test program are shown below:

```
C:\book\fortran\chap7>test_random0
Enter seed:
12
10 random numbers:
         .936091
         .203204
         .431167
         .719105
         .064103
         .789775
         .974839
         .881686
         .384951
         .400086
   Averages of 5 consecutive 1000-sample sequences:
         .504282
         .512665
         .496927
         .491514
         .498117
```

The numbers do appear to be between 0.0 and 1.0, and the averages of long sets of these numbers are nearly 0.5, so these subroutines appear to be functioning correctly. You should try them again using different seeds to see if they behave consistently.

Fortran includes an intrinsic subroutine `RANDOM_NUMBER` to generate sequences of random numbers. That subroutine will typically produce more nearly random results than the simple subroutine developed in this example. The full details of how to use subroutine `RANDOM_NUMBER` are found in Appendix B.

7.3
MODULE PROCEDURES

In addition to data, modules may also contain complete subroutines and functions, which are known as **module procedures.** These procedures are compiled as a part of the module, and are made available to a program unit by including a USE statement containing the module name in the program unit. Procedures that are included within a module must follow any data objects declared in the module, and must be preceded by a CONTAINS statement. The CONTAINS statement tells the compiler that the following statements are included procedures.

A simple example of a module procedure is shown below. Subroutine sub1 is contained within module my_subs.

```
MODULE my_subs
IMPLICIT NONE

(Declare shared data here)

CONTAINS
   SUBROUTINE sub1 ( a, b, c, x, error )
   IMPLICIT NONE
   REAL, DIMENSION(3), INTENT(IN) :: a
   REAL, INTENT(IN) :: b, c
   REAL, INTENT(OUT) :: x
   LOGICAL, INTENT(OUT) :: error
   ...
   END SUBROUTINE sub1
END MODULE my_subs
```

Subroutine sub1 is made available for use in a calling program unit if the statement "USE my_subs" is included as the first noncomment statement within the program unit. The subroutine can be called with a standard CALL statement as shown below:

```
PROGRAM main_prog
USE my_subs
IMPLICIT NONE

...
CALL sub1 ( a, b, c, x, error )

...
END PROGRAM main_prog
```

7.3.1 Using Modules to Create Explicit Interfaces

Why would we bother to include a procedure in a module? We already know that it is possible to separately compile a subroutine and to call it from another program unit, so why go through the extra steps of including the subroutine in a module, compiling the module, declaring the module in a USE statement, and then calling the subroutine?

The answer is that *when a procedure is compiled within a module and the module is used by a calling program, all of the details of the procedure's interface are made available to the compiler.* When the calling program is compiled, the compiler can automatically check the number of arguments in the procedure call, the type of each argument, whether or not each argument is an array, and the INTENT of each argument. In short, the compiler can catch most of the common errors that a programmer might make when using procedures!

A procedure compiled within a module and accessed by USE association is said to have an **explicit interface,** since all of the details about every argument in the procedure are explicitly known to the Fortran compiler whenever the procedure is used, and the compiler checks the interface to ensure that it is being used properly.

In contrast, procedures not in a module are said to have an **implicit interface.** A Fortran compiler has no information about these procedures when it is compiling a program unit that invokes them, so it just *assumes* that the programmer got the number, type, intent, etc., of the arguments right. If the programmer actually got the calling sequence wrong, then the program will fail in strange and hard-to-find ways.

To illustrate this point, let's reexamine the program in Figure 7-5. In that program, there is an implicit interface between program bad_call and subroutine bad_argument. A real value is passed to the subroutine when an integer argument is expected and the number is misinterpreted by the subroutine. As we see from that example, the Fortran compiler did not catch the error in the calling arguments.

Figure 7-12 shows the program rewritten to include the subroutine within a module.

FIGURE 7-12
Example illustrating the effects of a type mismatch when calling a subroutine included within a module.

```
MODULE my_subs
CONTAINS
   SUBROUTINE bad_argument ( i )
   IMPLICIT NONE
   INTEGER, INTENT(IN) :: i    ! Declare argument as integer.
   WRITE (*,*) ' I = ', i      ! Write out i.
   END SUBROUTINE
END MODULE my_subs

!*****************************************************************
!*****************************************************************

PROGRAM bad_call2
!
```

(continued)

(*concluded*)

```
!  Purpose:
!    To illustrate misinterpreted calling arguments.
!
USE my_subs
IMPLICIT NONE
REAL :: x = 1.                   ! Declare real variable x.
CALL bad_argument ( x )          ! Call subroutine.
END PROGRAM bad_call2
```

When this program is compiled, the Fortran compiler will catch the argument mismatch for us.

```
C:\book\fortran\chap7>ifort bad_call2.f90
Intel(R) Visual Fortran Intel(R) 64 Compiler for applications running on
Intel(R) 64, Version 16.0.2.180 Build 20160204
Copyright (C) 1985-2016 Intel Corporation. All rights reserved.
bad_call2.f90(21): error #6633: The type of the actual argument differs
from the type of the dummy argument.   [X]
CALL bad_argument ( x )     ! Call subroutine.
compilation aborted for bad_call2.f90 (code 1)
```

There is also another way to allow a Fortran compiler to explicitly check procedure interfaces—the INTERFACE block. We will learn more about it in Chapter 13.

Good Programming Practice
Use either assumed-shape arrays or explicit-shape arrays as dummy array arguments in procedures. If assumed-shape arrays are used, an explicit interface is required. Whole array operations, array sections, and array intrinsic functions may be used with the dummy array arguments in either case. *Never use assumed-size arrays in any new program.*

Quiz 7-2

This quiz provides a quick check to see if you have understood the concepts introduced in Sections 7.2 through 7.3. If you have trouble with the quiz, reread the sections, ask your instructor, or discuss the material with a fellow student. The answers to this quiz are found in the back of the book.

1. How can we share data between two or more procedures without passing it through a calling interface? Why would we want to do this?
2. Why should you gather up the procedures in a program and place them into a module?

For questions 3 and 4, determine whether there are any errors in these programs. If possible, tell what the output from each program will be.

(*continued*)

Introduction to Procedures

(*concluded*)

3.
```
   MODULE mydata
   IMPLICIT NONE
   REAL, SAVE, DIMENSION(8) :: a
   REAL, SAVE :: b
   END MODULE mydata

   PROGRAM test1
   USE mydata
   IMPLICIT NONE
   a = [ 1.,2.,3.,4.,5.,6.,7.,8. ]
   b = 37.
   CALL sub2
   END PROGRAM test1

   SUBROUTINE sub1
   USE mydata
   IMPLICIT NONE
   WRITE (*,*) 'a(5) = ', a(5)
   END SUBROUTINE sub1
```

4.
```
   MODULE mysubs
   CONTAINS
      SUBROUTINE sub2(x,y)
      REAL, INTENT(IN) :: x
      REAL, INTENT(OUT) :: y
      y = 3. * x - 1.
      END SUBROUTINE sub2
   END MODULE

   PROGRAM test2
   USE mysubs
   IMPLICIT NONE
   REAL :: a = 5.
   CALL sub2 (a, -3.)
   END PROGRAM test2
```

7.4
FORTRAN FUNCTIONS

A Fortran **function** is a procedure whose result is a single number, logical value, character string, or array. The result of a function is a single value or single array that can be combined with variables and constants to form Fortran expressions. These expressions may appear on the right side of an assignment statement in the calling program. There are two different types of functions: **intrinsic functions** and **user-defined functions** (or **function subprograms**).

Intrinsic functions are those functions built into the Fortran language, such as SIN(X), LOG(X), etc. Some of these functions were described in Chapter 2; all of them are detailed in Appendix B. User-defined functions or function subprograms are functions defined by individual programmers to meet a specific need not addressed by the standard intrinsic functions. They are used just like intrinsic functions in expressions. The general form of a user-defined Fortran function is:

```
FUNCTION name ( argument_list )
...
(Declaration section must declare type of name)
...
(Execution section)
...
name = expr
RETURN
END FUNCTION [name]
```

The function must begin with a FUNCTION statement and end with an END FUNCTION statement. The name of the function may be up to 63 alphabetic, numeric, and underscore characters long, but the first letter must be alphabetic. The name must be specified in the FUNCTION statement, and is optional on the END FUNCTION statement.

A function is invoked by naming it in an expression. When a function is invoked, execution begins at the top of the function, and ends when either a RETURN statement or the END FUNCTION statement is reached. Because execution ends at the END FUNCTION statement anyway, the RETURN statement is not actually required in most functions, and is rarely used. When the function returns, the returned value is used to continue evaluating the Fortran expression that it was named in.

The name of the function must appear on the left side of a least one assignment statement in the function. The value assigned to *name* when the function returns to the invoking program unit will be the value of the function.

The argument list of the function may be blank if the function can perform all of its calculations with no input arguments. The parentheses around the argument list are required even if the list is blank.

Since a function returns a value, it is necessary to assign a type to the function. If IMPLICIT NONE is used, *the type of the function must be declared both in the function procedure and in the calling programs*. If IMPLICIT NONE is not used, the default type of the function will follow the standard rules of Fortran unless they are overridden by a type declaration statement. The type declaration of a user-defined Fortran function can take one of two equivalent forms:

```
INTEGER FUNCTION my_function ( i, j )
```

or

```
FUNCTION my_function ( i, j )
INTEGER :: my_function
```

An example of a user-defined function is shown in Figure 7-13. Function quadf evaluates a quadratic expression with user-specified coefficients at a user-specified value x.

FIGURE 7-13
A function to evaluate a quadratic polynomial of the form $f(x) = ax^2 + bx + c$.

```fortran
REAL FUNCTION quadf ( x, a, b, c )
!
!  Purpose:
!    To evaluate a quadratic polynomial of the form
!       quadf = a * x**2 + b * x + c
!
!  Record of revisions:
!      Date       Programmer         Description of change
!      ====       ==========         =====================
!    11/23/15   S. J. Chapman        Original code
!
IMPLICIT NONE

! Data dictionary: declare calling parameter types & definitions
REAL, INTENT(IN) :: x    ! Value to evaluate expression for
REAL, INTENT(IN) :: a    ! Coefficient of X**2 term
REAL, INTENT(IN) :: b    ! Coefficient of X term
REAL, INTENT(IN) :: c    ! Coefficient of constant term

! Evaluate expression.
quadf = a * x**2 + b * x + c

END FUNCTION quadf
```

This function produces a result of type real. Note that the INTENT attribute is not used with the declaration of the function name quadf, since it must always be used for output only. A simple test program using the function is shown in Figure 7-14.

FIGURE 7-14
A test driver program for function quadf.

```fortran
PROGRAM test_quadf
!
!  Purpose:
!    Program to test function quadf.
!
IMPLICIT NONE

! Data dictionary: declare variable types & definitions
REAL :: quadf                ! Declare function
REAL :: a, b, c, x           ! Declare local variables

! Get input data.
WRITE (*,*) 'Enter quadratic coefficients a, b, and c: '
READ (*,*) a, b, c
WRITE (*,*) 'Enter location at which to evaluate equation: '
READ (*,*) x

! Write out result.
WRITE (*,100) 'quadf(', x, ') = ', quadf(x,a,b,c)
100 FORMAT (A,F10.4,A,F12.4)
END PROGRAM test_quadf
```

Notice that function `quadf` is declared as type real both in the function itself and in the test program. In this example, function `quadf` was used in the argument list of a `WRITE` statement. It could also have been used in assignment statements or wherever a Fortran expression is permissible.

Good Programming Practice
Be sure to declare the type of any user-defined functions both in the function itself and in any routines that call the function.

7.4.1 Unintended Side Effects in Functions

Input values are passed to a function through its argument list. Functions use the same argument-passing scheme as subroutines. A function receives pointers to the locations of its arguments, and it can deliberately or accidentally modify the contents of those memory locations. Therefore, *it is possible for a function subprogram to modify its own input arguments*. If any of the function's dummy arguments appear on the left side of an assignment statement within the function, then the values of the input variables corresponding to those arguments will be changed. A function that modifies the values in its argument list is said to have **side effects.**

By definition, a function should produce a *single output value* using one or more input values, and it should have no side effects. The function should never modify its own input arguments. If a programmer needs to produce more than one output value from a procedure, then the procedure should be written as a subroutine and not as a function. To ensure that a function's arguments are not accidentally modified, they should always be declared with the `INTENT(IN)` attribute.

Good Programming Practice
A well-designed Fortran function should produce a single output value from one or more input values. It should never modify its own input arguments. To ensure that a function does not accidentally modify its input arguments, always declare the arguments with the `INTENT(IN)` attribute.

7.4.2 Using Functions with Deliberate Side Effects

Programmers who regularly work with C++ and some other languages are used to writing functions that work with a different calling convention. These functions accept input data through arguments and return output data through other arguments, just like a subroutine. In this design, the function return value is a *status* indicating the success or failure of the operation performed by the function. By convention, zero is usually returned from the function for a successful operation, and nonzero values are returned to indicate various error codes. People with this background often design their Fortran functions the same way. They deliberately write functions with side effects to return the data, and with the function returns indicating the status of the operation.

This is a perfectly acceptable programming style, but it is good practice to be consistent in writing functions. If you use this programming style, use it consistently.

Quiz 7-3

This quiz provides a quick check to see if you have understood the concepts introduced in Section 7.4. If you have trouble with the quiz, reread the section, ask your instructor, or discuss the material with a fellow student. The answers to this quiz are found in the back of the book.

Write a user-defined function to perform the following calculations:

1. $f(x) = \dfrac{x - 1}{x + 1}$
2. The hyperbolic tangent function $\tanh(x) = \dfrac{e^x - e^{-x}}{e^x + e^{-x}}$
3. The factorial function $n! = (n)(n - 1) \ldots (3)(2)(1)$
4. Write a logical function that has two input arguments, x and y. The function should return a true value if $x^2 + y^2 > 1$, and a false value otherwise.

For questions 5 to 7, determine whether there are any errors in these functions. If so, show how to correct them.

5.  ```
    REAL FUNCTION average ( x, n )
    IMPLICIT NONE
    INTEGER, INTENT(IN) :: n
    REAL, DIMENSION(n), INTENT(IN) :: x
    INTEGER :: j
    REAL :: sum
    DO j = 1, n
       sum = sum + x(j)
    END DO
    average = sum / n
    END FUNCTION average
    ```

6.  ```
    FUNCTION fun_2 ( a, b, c )
    IMPLICIT NONE
    REAL, INTENT(IN) :: a, b, c
    a = 3. * a
    fun_2 = a**2 - b + c
    END FUNCTION fun_2
    ```

7.  ```
    LOGICAL FUNCTION badval ( x, y )
    IMPLICIT NONE
    REAL, INTENT(IN) :: x, y
    badval = x > y
    END FUNCTION badval
    ```
    ```

EXAMPLE 7-5

The sinc function:

The sinc function is defined by the equation:

$$\text{sinc}(x) = \frac{\sin x}{x} \tag{7-3}$$

This function occurs in many different types of engineering analysis problems. For example, the sinc function describes the frequency spectrum of a rectangular time pulse. A plot of the function sinc(x) versus x is shown in Figure 7-15. Write a user-defined Fortran function to calculate the sinc function.

SOLUTION
The sinc function looks easy to implement, but there is a calculation problem when $x = 0$. The value of sinc(0) = 1, since

$$\text{sinc}(0) = \lim_{x \to 0} \left[\frac{\sin(x)}{x} \right] = 1$$

Unfortunately, a computer program would blow up on the division-by-zero. We must include a logical IF construct in the function to handle the special case where x is nearly 0.

1. **State the problem.**
   Write a Fortran function that calculates sinc(x).

2. **Define the inputs and outputs.**
   The input to the function is the real argument x. The function is of type real, and its output is the value of sinc(x).

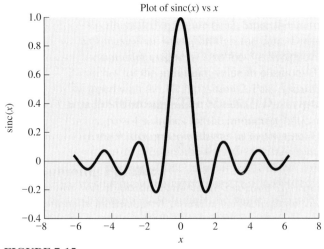

FIGURE 7-15
Plot of sinc(x) versus x.

Introduction to Procedures

3. **Describe the algorithm.**
   The pseudocode for this function is

   ```
   IF |x| > epsilon THEN
       sinc ← SIN(x) / x
   ELSE
       sinc ← 1.
   END IF
   ```

   where `epsilon` is chosen to ensure that the division does not cause divide-by-zero errors. For most computers, a good choice for `epsilon` might be 1.0E-30.

4. **Turn the algorithm into Fortran statements.**
   The resulting Fortran subroutines are shown in Figure 7-16.

FIGURE 7-16
The Fortran function sinc(x).

```
FUNCTION sinc ( x )
!
!  Purpose:
!     To calculate the sinc function
!        sinc(x) = sin(x) / x
!
! Record of revisions:
!     Date         Programmer         Description of change
!     ====         ==========         =====================
!  11/23/15    S. J. Chapman          Original code
!
IMPLICIT NONE

! Data dictionary: declare calling parameter types & definitions
REAL, INTENT(IN) :: x         ! Value for which to evaluate sinc
REAL :: sinc                  ! Output value sinc(x)

! Data dictionary: declare local constants
REAL, PARAMETER :: EPSILON = 1.0E-30  ! the smallest value for which
                                      ! to calculate SIN(x)/x

! Check to see of ABS(x) > EPSILON.
IF ( ABS(x) > EPSILON ) THEN
   sinc = SIN(x) / x
ELSE
   sinc = 1.
END IF

END FUNCTION sinc
```

5. **Test the resulting Fortran program.**
   To test this function, it is necessary to write a driver program to read an input value, call the function, and write out the results. We will calculate several values of sinc(x) on a hand calculator and compare them with the results of the test program. Note that we must verify the function of the program for input values both greater than and less than `epsilon`.

   A test driver program is shown in Figure 7-17:

FIGURE 7-17
A test driver program for the function sinc(x).

```
PROGRAM test_sinc
!
! Purpose:
!   To test the sinc function sinc(x)
!
IMPLICIT NONE

! Data dictionary: declare function types
REAL :: sinc              ! sinc function

! Data dictionary: declare variable types & definitions
REAL :: x                 ! Input value to evaluate

! Get value to evaluate
WRITE (*,*) 'Enter x: '
READ (*,*) x

! Write answer.
WRITE (*,'(A,F8.5)') 'sinc(x) = ', sinc(x)

END PROGRAM test_sinc
```

Hand calculations yield the following values for sinc(x):

x	sinc(x)
0	1.00000
10^{-29}	1.00000
$\dfrac{\pi}{2}$	0.63662
π	0.00000

The results from the test program for these input values are:

```
C:\book\fortran\chap7>test_sinc
Enter x:
0
sinc(x) = 1.0000

C:\book\fortran\chap7>test_sinc
Enter x:
1.E-29
sinc(x) = 1.0000

C:\book\fortran\chap7>test_sinc
Enter x:
1.570796
sinc(x) = 0.63662

C:\book\fortran\chap7>test_sinc
Enter x:
3.141593
sinc(x) = 0.0000
```

The function appears to be working correctly.

7.5
PASSING PROCEDURES AS ARGUMENTS TO OTHER PROCEDURES

When a procedure is invoked, the actual argument list is passed to the procedure as a series of pointers to specific memory locations. How the memory at each location is interpreted depends on the type and size of the dummy arguments declared in the procedure.

This pass-by-reference approach can be extended to permit us to pass a pointer to a *procedure* instead of a memory location. Both functions and subroutines can be passed as calling arguments. For simplicity, we will first discuss passing user-defined functions to procedures, and afterward discuss passing subroutines to procedures.

7.5.1 Passing User-Defined Functions as Arguments

If a user-defined function is named as an actual argument in a procedure call, then a *pointer to that function* is passed to the procedure. If the corresponding formal argument in the procedure is used as a function, then when the procedure is executed, the function in the calling argument list will be used in place of the dummy function name in the procedure. Consider the following example:

```
PROGRAM :: test
REAL, EXTERNAL :: fun_1, fun_2
REAL :: x, y, output
...
CALL evaluate ( fun_1, x, y, output )
CALL evaluate ( fun_2, x, y, output )
...
END PROGRAM test

SUBROUTINE evaluate ( fun, a, b, result )
REAL, EXTERNAL :: fun
REAL, INTENT(IN) :: a, b
REAL, INTENT(OUT) :: result
result = b * fun(a)
END SUBROUTINE evaluate
```

Assume that `fun_1` and `fun_2` are two user-supplied functions. Then a pointer to function `fun_1` is passed to subroutine `evaluate` on the first occasion that it is called, and function `fun_1` is used in place of the dummy formal argument `fun` in the subroutine. A pointer to function `fun_2` is passed to subroutine `evaluate` the second time that it is called, and function `fun_2` is used in place of the dummy formal argument `fun` in the subroutine.

User-supplied functions may only be passed as calling arguments if they are declared to be *external* in the calling and the called procedures. When a name in an argument list is declared to be external, this tells the compiler that a separately-compiled function is being passed in the argument list instead of a variable. A function may be declared to be external either with an EXTERNAL attribute or in an EXTERNAL statement. The `EXTERNAL` attribute is included in a type declaration statement, just like any other attribute. An example is

```
REAL, EXTERNAL :: fun_1, fun_2
```

The `EXTERNAL` statement is a specification statement of the form

```
EXTERNAL fun_1, fun_2
```

Either of the above forms state that `fun_1`, `fun_2`, etc., are names of procedures that are defined outside of the current routine. If used, the `EXTERNAL` statement must appear in the declaration section, before the first executable statement.[8]

EXAMPLE 7-6

Passing Functions to Procedures in an Argument List:

The function `ave_value` in Figure 7-18 determines the average amplitude of a function between user-specified limits `first_value` and `last_value` by sampling the function at n evenly-spaced points, and calculating the average amplitude between those points. The function to be evaluated is passed to function `ave_value` as the dummy argument `func`.

FIGURE 7-18

Function `ave_value` calculates the average amplitude of a function between two points `first_value` and `last_value`. The function is passed to function `ave_value` as a calling argument.

```
REAL FUNCTION ave_value ( func, first_value, last_value, n )
!
!  Purpose:
!    To calculate the average value of function "func" over the
!    range [first_value, last_value] by taking n evenly-spaced
!    samples over the range, and averaging the results. Function
!    "func" is passed to this routine via a dummy argument.
!
!  Record of revisions:
!     Date         Programmer          Description of change
!     ====         ==========          =====================
!   11/24/15     S. J. Chapman         Original code
!
IMPLICIT NONE

! Data dictionary: declare calling parameter types & definitions
REAL, EXTERNAL :: func                 ! Function to be evaluated
REAL, INTENT(IN) :: first_value        ! First value in range
REAL, INTENT(IN) :: last_value         ! Last value in rnage
INTEGER, INTENT(IN) :: n               ! Number of samples to average
```

(continued)

---

[8] There is also another way to pass functions to procedures using **function pointers.** Function pointers will be described in Chapter 15.

Introduction to Procedures

(concluded)

```
! Data dictionary: declare local variable types & definitions
REAL :: delta                  ! Step size between samples
INTEGER :: i                   ! Index variable
REAL :: sum                    ! Sum of values to average

! Get step size.
delta = ( last_value - first_value ) / REAL(n-1)

! Accumulate sum.
sum = 0.
DO i = 1, n
   sum = sum + func ( REAL(i-1) * delta )
END DO

! Get average.
ave_value = sum / REAL(n)

END FUNCTION ave_value
```

A test driver program to test function `ave_value` is shown in Figure 7-19. In that program, function `ave_value` is called with the user-defined function `my_function` as a calling argument. Note that `my_function` is declared as EXTERNAL in the test driver program `test_ave_value`. The function `my_function` is averaged over 101 samples in the interval [0,1], and the results are printed out.

FIGURE 7-19
Test driver program for function `ave_value`, illustrating how to pass a user-defined function as a calling argument.

```
PROGRAM test_ave_value
!
! Purpose:
!   To test function ave_value by calling it with a user-defined
!   function my_func.
!
! Record of revisions:
!   Date          Programmer        Description of change
!   ====          ==========        =====================
!   11/24/15      S. J. Chapman     Original code
!
IMPLICIT NONE

! Data dictionary: declare function types
REAL :: ave_value              ! Average value of function
REAL, EXTERNAL :: my_function  ! Function to evaluate

! Data dictionary: declare local variable types & definitions
REAL :: ave                    ! Average of my_function

! Call function with func=my_function.
ave = ave_value ( my_function, 0., 1., 101 )
WRITE (*,1000) 'my_function', ave
1000 FORMAT ('The average value of ',A,' between 0. and 1. is ', &
             F16.6,'.')

END PROGRAM test_ave_value
```

(continued)

(concluded)

```
REAL FUNCTION my_function( x )
IMPLICIT NONE
REAL, INTENT(IN) :: x
my_function = 3. * x
END FUNCTION my_function
```

When program test_ave_value is executed, the results are

C:\book\fortran\chap7>**test_ave_value**
The average value of my_function between 0. and 1. is 1.500000.

Since for this case my_function is a straight line between (0,0) and (1,3), it is obvious that the average value was correctly calculated as 1.5.

7.5.2 Passing Subroutines as Arguments

Subroutines may also be passed to procedures as calling arguments. If a subroutine is to be passed as a calling argument, it must be declared in an EXTERNAL statement. The corresponding dummy argument should appear in a CALL statement in the procedure.

EXAMPLE 7-7

Passing Subroutines to Procedures in an Argument List:

The function subs_as_arguments in Figure 7-20 accepts two input arguments x and y, and passes them to a subroutine for calculations. The name of the subroutine to execute is also passed as a command line argument.

FIGURE 7-20
Subroutine subs_as_arguments calls a subroutine to perform an operation on values x and y. The name of the subroutine to execute is also passed as a command line argument.

```
SUBROUTINE subs_as_arguments(x, y, sub, result )
!
!  Purpose:
!    To test passing subroutine names as arguments.
!
IMPLICIT NONE
! Data dictionary: declare calling parameter types & definitions
EXTERNAL :: sub                   ! Dummy subroutine name
REAL, INTENT(IN) :: x             ! First value
REAL, INTENT(IN) :: y             ! Last value
REAL, INTENT(OUT) :: result       ! Result
CALL sub(x, y, result)
END SUBROUTINE subs_as_arguments
```

A test driver program to test subroutine test_subs_as_arguments is shown in Figure 7-21. In that program, subroutine subs_as_arguments is called twice with the user-defined subroutines prod and sum passed as calling arguments. Note that the dummy argument sub is declared as EXTERNAL in subroutine subs_as_arguments, and that the actual subroutines prod and sum are declared external in the main program.

FIGURE 7-21
Test driver program for subroutine subs_as_arguments, illustrating how to pass a user-defined subroutine as a calling argument.

```
PROGRAM test_subs_as_arguments
!
!  Purpose:
!     To test passing subroutine names as arguments.
!
IMPLICIT NONE

! Data dictionary: declare calling parameter types & definitions
EXTERNAL :: sum, prod          ! Name of subroutines to call
REAL :: x                      ! First value
REAL :: y                      ! Last value
REAL :: result                 ! Result

! Get the x and y values
WRITE (*,*) 'Enter x:'
READ (*,*) x
WRITE (*,*) 'Enter y:'
READ (*,*) y

! Calculate product
CALL subs_as_arguments(x, y, prod, result)
WRITE (*,*) 'The product is ', result

! Calculate product and sum
CALL subs_as_arguments(x, y, sum, result)
WRITE (*,*) 'The sum is ', result

END PROGRAM test_subs_as_arguments
!*******************************************************************
!*******************************************************************
SUBROUTINE prod ( x, y, result )
!
!  Purpose:
!     To calculate product of two real numbers.
!
IMPLICIT NONE

! Data dictionary: declare calling parameter types & definitions
REAL, INTENT(IN) :: x          ! First value
REAL, INTENT(IN) :: y          ! Last value
REAL, INTENT(OUT) :: result    ! Result

! Calculate value.
```

(continued)

(concluded)

```
result = x * y

END SUBROUTINE prod
!***********************************************************************
!***********************************************************************
SUBROUTINE sum ( x, y, result )
!
! Purpose:
!   To calculate sum of two real numbers.
!
IMPLICIT NONE

! Data dictionary: declare calling parameter types & definitions
REAL, INTENT(IN)  :: x           ! First value
REAL, INTENT(IN)  :: y           ! Last value
REAL, INTENT(OUT) :: result      ! Result

! Calculate value.
result = x + y

END SUBROUTINE sum
```

When program `test_subs_as_arguments` is executed, the results are

```
C:\book\fortran\chap7>test_subs_as_arguments
Enter x:
4
  Enter y:
5
  The product is    20.00000
  The sum is     9.000000
```

Here subroutine `subs_as_arguments` is being executed twice, once with subroutine `prod` and once with subroutine `sum`.

7.6
SUMMARY

In this chapter, we presented an introduction to Fortran procedures. Procedures are independently compiled program units with their own declaration sections, execution sections, and termination sections. They are extremely important to the design, coding, and maintenance of large programs. Procedures permit the independent testing of subtasks as a project is being built, allow time savings through reusable code, and improve reliability through variable hiding.

There are two types of procedures: subroutines and functions. Subroutines are procedures whose results include one or more values. A subroutine is defined using a `SUBROUTINE` statement, and is executed using a `CALL` statement. Input data is passed to a subroutine and results are returned from the subroutine through argument lists on the `SUBROUTINE` statement and `CALL` statement. When a subroutine is called, pointers are passed to the subroutine pointing to the locations of each argument in the argument list. The subroutine reads from and writes to those locations.

The use of each argument in a subroutine's argument list can be controlled by specifying an INTENT attribute in the argument's type declaration statement. Each argument can be specified as either input only (IN), output only (OUT), or both input and output (INOUT). The Fortran compiler checks to see that each argument is used properly, and so can catch many programming errors at compile time.

Data can also be passed to subroutines through modules. A module is a separately compiled program unit that can contain data declarations, procedures, or both. The data and procedures declared in the module are available to any procedure that includes the module with a USE statement. Thus, two procedures can share data by placing the data and a module, and having both procedures USE the module.

If procedures are placed in a module and that module is used in a program, then the procedures have an explicit interface. The compiler will automatically check to ensure that number, type, and use of all arguments in each procedure call match the argument list specified for the procedure. This feature can catch many common errors.

Fortran functions are procedures whose results are a single number, logical value, character string, or array. There are two types of Fortran functions: intrinsic (built-in) functions and user-defined functions. Some intrinsic functions were discussed in Chapter 2, and all intrinsic functions are included in Appendix C. User-defined functions are declared using the FUNCTION statement and are executed by naming the function as a part of a Fortran expression. Data may be passed to a user-defined function through calling arguments or via modules. A properly-designed Fortran function should not change its input arguments. It should *only* change the single output value.

It is possible to pass a function or subroutine to a procedure via a calling argument, provided that the function or subroutine is declared EXTERNAL in the calling program.

7.6.1 Summary of Good Programming Practice

The following guidelines should be adhered to when working with subroutines and functions.

1. Break large program tasks into smaller, more understandable procedures whenever possible.
2. Always specify the INTENT of every dummy argument in every procedure to help catch programming errors.
3. Make sure that the actual argument list in each procedure invocation matches the dummy argument list in *number, type, intent,* and *order*. Placing procedures in a module and then accessing the procedures by USE association will create an explicit interface, which will allow the compiler to automatically check that the argument lists are correct.
4. Test for possible error conditions within a subroutine, and set an error flag to be returned to the calling program unit. The calling program unit should test for error conditions after the subroutine call, and take appropriate actions if an error occurs.
5. Always use either explicit-shape dummy arrays or assumed-shape dummy arrays for dummy array arguments. Never use assumed-size dummy arrays in any new program.
6. Modules may be used to pass large amounts of data between procedures within a program. The data values may be declared only once in the module, and all procedures needing access to that data use that module. Be sure to include a

SAVE statement in the module to guarantee that the data is preserved between accesses by different procedures.
7. Collect the procedures that you use in a program and place them in a module. When they are a module, the Fortran compiler will automatically verify the calling argument list each time that they are used.
8. Be sure to declare the type of any function both in the function itself and in any program units that invoke the function.
9. A well-designed Fortran function should produce a single output value from one or more input values. It should never modify its own input arguments. To ensure that a function does not accidentally modify its input arguments, always declare the arguments with the INTENT(IN) attribute.[9]

7.6.2 Summary of Fortran Statements and Structures

CALL Statement:

```
CALL subname( arg1, arg2, ... )
```

Example:

```
CALL sort ( number, data1 )
```

Description:
This statement transfers execution from the current program unit to the subroutine, passing pointers to the calling arguments. The subroutine executes until either a RETURN or an END SUBROUTINE statement is encountered, and then execution will continue in the calling program unit at the next executable statement following the CALL statement.

CONTAINS Statement:

```
CONTAINS
```

Examples:

```
MODULE test
...
CONTAINS
   SUBROUTINE sub1(x, y)
   ...
   END SUBROUTINE sub1
END MODULE test
```

Description:
The CONTAINS statement specifies that the following statements are separate procedure(s) within a module. The CONTAINS statement and the module procedures following it must appear after any type and data definitions within the module.

---

[9] However, certain programmers use a different style in which function return results as arguments and the function return is a status. If you program in that style, this Good Programming Practice does not apply to you.

Introduction to Procedures

END **Statements:**

```
END FUNCTION [name]
END MODULE [name]
END SUBROUTINE [name]
```

Example:

```
END FUNCTION my_function
END MODULE my_mod
END SUBROUTINE my_sub
```

Description:
These statements end user-defined Fortran functions, modules, and subroutines, respectively. The name of the function, module, or subroutine may optionally be included, but it is not required.

EXTERNAL **Attribute:**

```
type, EXTERNAL :: name1, name2, ...
```

Example:

```
REAL, EXTERNAL :: my_function
```

Description:
This attribute declares that a particular name is an externally defined function. It is equivalent to naming the function in an EXTERNAL statement.

EXTERNAL **Statement:**

```
EXTERNAL name1, name2, ...
```

Example:

```
EXTERNAL my_function
```

Description:
This statement declares that a particular name is an externally defined procedure. Either it or the EXTERNAL attribute must be used in the calling program unit and in the called procedure if the procedure specified in the EXTERNAL statement is to be passed as an actual argument.

FUNCTION Statement:

> [*type*] FUNCTION *name*(*arg1*, *arg2*, ...)

Examples:

> INTEGER FUNCTION max_value (num, iarray)
> FUNCTION gamma(x)

Description:
This statement declares a user-defined Fortran function. The type of the function may be declared in the FUNCTION statement, or it may be declared in a separate type declaration statement. The function is executed by naming it in an expression in the calling program. The dummy arguments are placeholders for the calling arguments passed when the function is executed. If a function has no arguments, then it must be declared with an empty pair of parentheses [*name*( )].

INTENT Attribute:

> *type*, INTENT(*intent_type*) :: *name1*, *name2*, ...

Example:

> REAL, INTENT(IN) :: value
> INTEGER, INTENT(OUT) :: count

Description:
This attribute declares the intended use of a particular dummy procedure argument. Possible values of *intent_type* are IN, OUT, and INOUT. The INTENT attribute allows the Fortran compiler to know the intended use of the argument and to check that it is used in the way intended. This attribute may only appear on dummy arguments in procedures.

INTENT Statement:

> INTENT(*intent_type*) :: *name1*, *name2*, ...

Example:

> INTENT(IN) :: a, b
> INTENT(OUT) :: result

Description:
This statement declares the intended use of a particular dummy procedure argument. Possible values of *intent_type* are IN, UT, and INOUT. The INTENT statement allows the Fortran compiler to know the intended use of the argument, and to check that it is used in the way intended. Only dummy arguments may appear in INTENT statements. **Do not use this statement; use the INTENT attribute instead.**

Introduction to Procedures

MODULE Statement:

```
MODULE name
```

Example:

```
MODULE my_data_and_subs
```

Description:
This statement declares a module. The module may contain data, procedures, or both. The data and procedures are made available for use in a program unit by declaring the module name in a USE statement (USE association).

RETURN Statement:

```
RETURN
```

Example:

```
RETURN
```

Description:
When this statement is executed in a procedure, control returns to the program unit that invoked the procedure. This statement is optional at the end of a subroutine or function, since execution will automatically return to the calling routine whenever an END SUBROUTINE or END FUNCTION statement is reached.

SUBROUTINE Statement:

```
SUBROUTINE name ( arg1, arg2, ... )
```

Example:

```
SUBROUTINE sort ( num, data1 )
```

Description:
This statement declares a Fortran subroutine. The subroutine is executed with a CALL statement. The dummy arguments are placeholders for the calling arguments passed when the subroutine is executed.

USE Statement:

```
USE module1, module2, ...
```

Example:

```
USE my_data
```

Description:
This statement makes the contents of one or more modules available for use in a program unit. USE statements must be the first noncomment statements within the program unit after the PROGRAM, SUBROUTINE, or FUNCTION statement.

7.6.3 Exercises

7-1. What is the difference between a subroutine and a function?

7-2. When a subroutine is called, how is data passed from the calling program to the subroutine, and how are the results of the subroutine returned to the calling program?

7-3. What are the advantages and disadvantages of the pass-by-reference scheme used in Fortran?

7-4. What are the advantages and disadvantages of using explicit-shape dummy arrays in procedures? What are the advantages and disadvantages of using assumed-shape dummy arrays? Why should assumed-size dummy arrays never be used?

7-5. Suppose that a 15-element array a is passed to a subroutine as a calling argument. What will happen if the subroutine attempts to write to element a(16)?

7-6. Suppose that a real value is passed to a subroutine in an argument that is declared to be an integer in the subroutine. Is there any way for the subroutine to tell that the argument type is mismatched? What happens on your computer when the following code is executed?

```
PROGRAM main
IMPLICIT NONE
REAL :: x
x = -5.
CALL sub1 ( x )
END PROGRAM main

SUBROUTINE sub1 ( i )
IMPLICIT NONE
INTEGER, INTENT(IN) :: i
WRITE (*,*) ' I = ', i
END SUBROUTINE sub1
```

7-7. How could the program in Exercise 7-6 be modified to ensure that the Fortran compiler catches the argument mismatch between the actual argument in the main program and the dummy argument in subroutine sub1?

7-8. What is the purpose of the INTENT attribute? Where can it be used? Why should it be used?

7-9. Determine whether the following subroutine calls are correct or not. If they are in error, specify what is wrong with them.

(a)
```
PROGRAM sum_sqrt
IMPLICIT NONE
INTEGER, PARAMETER :: LENGTH = 20
INTEGER :: result
REAL :: test(LENGTH) = &
     [ 1., 2., 3., 4., 5., 6., 7., 8., 9.,10., &
      11.,12.,13.,14.,15.,16.,17.,18.,19.,20. ]
...
CALL test_sub ( LENGTH, test, result )
...
END PROGRAM sum_sqrt

SUBROUTINE test_sub ( length, array, res )
```

```
      IMPLICIT NONE
      INTEGER, INTENT(IN) :: length
      REAL, INTENT(OUT) :: res
      INTEGER, DIEMNSION(length), INTENT(IN) :: array
      INTEGER, INTENT(INOUT) :: i
      DO i = 1, length
         res = res + SQRT(array(i))
      END DO
      END SUBROUTINE test_sub
```

(b)
```
   PROGRAM test
   IMPLICIT NONE
   CHARACTER(len=8) :: str = '1AbHz05Z'
   CHARACTER :: largest
   CALL max_char (str, largest)
   WRITE (*,100) str, largest
   100 FORMAT (' The largest character in ', A, ' is ', A)
   END PROGRAM test

   SUBROUTINE max_char(string, big)
   IMPLICIT NONE
   CHARACTER(len=10), INTENT(IN) :: string
   CHARACTER, INTENT(OUT) :: big
   INTEGER :: i
   big = string(1:1)
   DO i = 2, 10
      IF ( string(i:i) > big ) THEN
         big = string(i:i)
      END IF
   END DO
   END SUBROUTINE max_char
```

7-10. Is the following program correct or incorrect? If it is incorrect, what is wrong with it? If it is correct, what values will be printed out by the following program?

```
MODULE my_constants
IMPLICIT NONE
REAL, PARAMETER :: PI = 3.141593    ! Pi
REAL, PARAMETER :: G = 9.81         ! Accel. due to gravity
END MODULE my_constants

PROGRAM main
IMPLICIT NONE
USE my_constants
WRITE (*,*) 'SIN(2*PI) = ' SIN(2.*PI)
G = 17.
END PROGRAM main
```

7-11. Modify the selection sort subroutine developed in this chapter so that it sorts real values in *descending* order.

7-12. Write a subroutine ucase that accepts a character string, and converts any lowercase letter in the string to uppercase without affecting any nonalphabetic characters in the string.

7-13. Write a driver program to test the statistical subroutines developed in Example 7-3. Be sure to test the routines with a variety of input data sets. Did you discover any problems with the subroutines?

7-14. Write a subroutine that uses subroutine `random0` to generate a random number in the range [−1.0,1.0).

7-15. Dice Simulation It is often useful to be able to simulate the throw of a fair die. Write a Fortran function `dice()` that simulates the throw of a fair die by returning some random integer between 1 and 6 every time that it is called. (*Hint:* Call `random0` to generate a random number. Divide the possible values out of `random0` into six equal intervals and return the number of the interval that a given random number falls into.)

7-16. Road Traffic Density Subroutine `random0` produces a number with a *uniform* probability distribution in the range [0.0, 1.0). This subroutine is suitable for simulating random events if each outcome has an equal probability of occurring. However, in many events, the probability of occurrence is *not* equal for every event, and a uniform probability distribution is not suitable for simulating such events.

For example, when traffic engineers studied the number of cars passing a given location in a time interval of length t, they discovered that the probability of k cars passing during the interval is given by the equation

$$P(k, t) = e^{-\lambda t}\frac{(\lambda t)^k}{k!} \text{ for } t \geq 0, \lambda > 0, \text{ and } k = 0, 1, 2,\ldots \quad (7\text{-}4)$$

This probability distribution is known as the *Poisson distribution*; it occurs in many applications in science and engineering. For example, the number of calls k to a telephone switchboard in time interval t, the number of bacteria k in a specified volume t of liquid, and the number of failures k of a complicated system in time interval t all have Poisson distributions.

Write a function to evaluate the Poisson distribution for any k, t, and λ. Test your function by calculating the probability of 0, 1, 2, …, 5 cars passing a particular point on a highway in 1 minute, given that λ is 1.6 per minute for that highway.

7-17. What are two purposes of a module? What are the special advantages of placing procedures within modules?

7-18. Write three Fortran functions to calculate the hyperbolic sine, cosine, and tangent functions:

$$\sinh(x) = \frac{e^x - e^{-x}}{2} \quad \cosh(x) = \frac{e^x + e^{-x}}{2} \quad \tanh(x) = \frac{e^x - e^{-x}}{e^x + e^{-x}}$$

Use your functions to calculate the hyperbolic sines, cosines, and tangents of the following values: −2, −1.5, −1.0, −0.5, −0.25, 0.0, 0.25, 0.5, 1.0, 1.5, and 2.0. Sketch the shapes of the hyperbolic sine, cosine, and tangent functions.

7-19. Cross Product Write a function to calculate the cross product of two vectors $\mathbf{V}_1$ and $\mathbf{V}_2$:

$$\mathbf{V}_1 \times \mathbf{V}_2 = (V_{y1}V_{z2} - V_{y2}V_{z1})\mathbf{i} + (V_{z1}V_{x2} - V_{z2}V_{x1})\mathbf{j} + (V_{x1}V_{y2} - V_{x2}V_{y1})\mathbf{k}$$

where $\mathbf{V}_1 = V_{x1}\mathbf{i} + V_{y1}\mathbf{j} + V_{z1}\mathbf{k}$ and $\mathbf{V}_2 = V_{x2}\mathbf{i} + V_{y2}\mathbf{j} + V_{z2}\mathbf{k}$. Note that this function will return a real array as its result. Use the function to calculate the cross product of the two vectors $\mathbf{V}_1 = [-2, 4, 0.5]$ and $\mathbf{V}_2 = [0.5, 3, 2]$.

7-20. Sort with Carry It is often useful to sort an array `arr1` into ascending order, while simultaneously carrying along a second array `arr2`. In such a sort, each time an element of array `arr1` is exchanged with another element of `arr1`, the corresponding elements of array `arr2` are also swapped. When the sort is over, the elements of array `arr1` are in ascending order, while the elements of array `arr2` that were associated with particular elements of array `arr1` are still associated with them. For example, suppose we have the following two arrays:

Element	arr1	arr2
1.	6.	1.
2.	1.	0.
3.	2.	10.

After sorting array `arr1` while carrying along array `arr2`, the contents of the two arrays will be:

Element	arr1	arr2
1.	1.	0.
2.	2.	10.
3.	6.	1.

Write a subroutine to sort one real array into ascending order while carrying along a second one. Test the subroutine with the following two 9-element arrays:

```
REAL, DIMENSION(9) :: &
  a = [ 1., 11., -6., 17.,-23., 0., 5., 1., -1. ]
REAL, DIMENSION(9) :: &
  b = [ 31.,101., 36., -17., 0., 10., -8., -1., -1. ]
```

7-21. Minima and Maxima of a Function Write a subroutine that attempts to locate the maximum and minimum values of an arbitrary function $f(x)$ over a certain range. The function being evaluated should be passed to the subroutine as a calling argument. The subroutine should have the following input arguments:

first_value — The first value of x to search
last_value — The last value of x to search
num_steps — The number of steps to include in the search
func — The name of the function to search

The subroutine should have the following output arguments:

xmin — The value of x at which the minimum was found
min_value — The minimum value of $f(x)$ found
xmax — The value of x at which the maximum was found
max_value — The maximum value $f(x)$ found

7-22. Write a test driver program for the subroutine generated in the previous problem. The test driver program should pass to the subroutine the user-defined function $f(x) = x^3 - 5x^2 + 5x + 2$, and search for the minimum and maximum in 200 steps over the range $-1 \leq x \leq 3$. It should print out the resulting minimum and maximum values.

7-23. Derivative of a Function The *derivative* of a continuous function $f(x)$ is defined by the equation

$$\frac{d}{dx}f(x) = \lim_{\Delta x \to 0} \frac{f(x + \Delta x) - f(x)}{\Delta x} \tag{7-5}$$

In a sampled function, this definition becomes

$$f'(x_i) = \frac{f(x_{i+1}) - f(x_i)}{\Delta x} \qquad (7\text{-}6)$$

where $\Delta x = x_{i+1} - x_i$. Assume that a vector `vect` contains `nsamp` samples of a function taken at a spacing of `dx` per sample. Write a subroutine that will calculate the derivative of this vector from Equation (7-6). The subroutine should check to make sure that `dx` is greater than zero to prevent divide-by-zero errors in the subroutine.

To check you subroutine, you should generate a data set whose derivative is known, and compare the result of the subroutine with the known correct answer. A good choice for a test function is $\sin x$. From elementary calculus, we know that

$$\frac{d}{dx}(\sin x) = \cos x.$$

Generate an input vector containing 100 values of the function $\sin x$ starting at $x = 0$, and using a step size Δx of 0.05. Take the derivative of the vector with your subroutine, and then compare the resulting answers to the known correct answer. How close did your routine come to calculating the correct value for the derivative?

7-24. Derivative in the Presence of Noise We will now explore the effects of input noise on the quality of a numerical derivative (Figure 7-22). First, generate an input vector containing 100 values of the function $\sin x$ starting at $x = 0$, and using a step size Δx of 0.05, just as you did in the previous problem. Next, use subroutine `random0` to generate a small amount of random noise with a maximum amplitude of ± 0.02, and add that random noise to the samples in your input vector. Note that the peak amplitude of the noise is only 2% of the peak amplitude of your signal, since the maximum value of $\sin x$ is 1. Now take the derivative of the function using the derivative subroutine that you developed in the last problem. How close to the theoretical value of the derivative did you come?

7-25. Two's Complement Arithmetic As we learned in Chapter 1, an 8-bit integer in two's complement format can represent all the numbers between -128 and $+127$, including 0. The sidebar in Chapter 1 also showed us how to add and subtract binary numbers in two's complement format. Assume that a two's complement binary number is supplied in an eight-character variable containing 0s and 1s, and perform the following instructions:

(a) Write a subroutine or function that adds 2 two's complement binary numbers stored in character variables, and returns the result in a third character variable.

(b) Write a subroutine or function that subtracts 2 two's complement binary numbers stored in character variables, and returns the result in a third character variable.

(c) Write a subroutine or function that converts a two's complement binary number stored in a character variable into a decimal integer stored in an `INTEGER` variable, and returns the result.

(d) Write a subroutine or function that converts a decimal integer stored in an `INTEGER` variable into a two's complement binary number stored in a character variable, and returns the result.

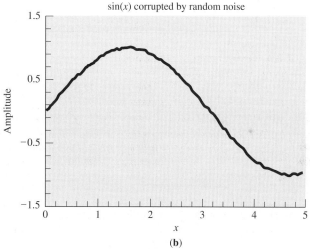

FIGURE 7-22
(*a*) A plot of sin *x* as a function of *x* with no noise added to the data. (*b*) A plot of sin *x* as a function of *x* with a 2% peak amplitude uniform random noise added to the data.

(e) Write a program that uses the four procedures created above to implement a two's complement calculator, in which the user can enter numbers in either decimal or binary form, and perform addition and subtraction on them. The results of any operation should be displayed in both decimal and binary form.

7-26. Linear Least Squares Fit Develop a subroutine that will calculate slope m and intercept b of the least-squares line that best fits an input data set. The input data points (x, y) will be passed to the subroutine in two input arrays, X and Y. The equations describing the slope and intercept of the least-squares line are

$$y = mx + b \tag{5-5}$$

$$m = \frac{(\Sigma xy) - (\Sigma x)\bar{y}}{(\Sigma x^2) - (\Sigma x)\bar{x}} \quad (5\text{-}6)$$

and

$$b = \bar{y} - m\bar{x} \quad (5\text{-}7)$$

where

Σx is the sum of the x values

Σx^2 is the sum of the squares of the x values

Σxy is the sum of the products of the corresponding x and y values

$\bar{x}$ is the mean (average) of the x values

$\bar{y}$ is the mean (average) of the y values

Test your routine using a test driver program and the following 20-point input data set:

Sample data to test least squares fit routine

No.	x	y	No.	x	y
1	−4.91	−8.18	11	−0.94	0.21
2	−3.84	−7.49	12	0.59	1.73
3	−2.41	−7.11	13	0.69	3.96
4	−2.62	−6.15	14	3.04	4.26
5	−3.78	−5.62	15	1.01	5.75
6	−0.52	−3.30	16	3.60	6.67
7	−1.83	−2.05	17	4.53	7.70
8	−2.01	−2.83	18	5.13	7.31
9	0.28	−1.16	19	4.43	9.05
10	1.08	0.52	20	4.12	10.95

7-27. Correlation Coefficient of Least Squares Fit Develop a subroutine that will calculate both the slope m and intercept b of the least-squares line that best fits an input data set, and also the correlation coefficient of the fit. The input data points (x, y) will be passed to the subroutine in two input arrays, X and Y. The equations describing the slope and intercept of the least-squares line are given in the previous problem, and the equation for the correlation coefficient is

$$r = \frac{n(\Sigma xy) - (\Sigma x)(\Sigma y)}{\sqrt{[(n\Sigma x^2) - (\Sigma x)^2][(n\Sigma y^2) - (\Sigma y)^2]}} \quad (7\text{-}7)$$

where

Σx is the sum of the x values

Σy is the sum of the y values

Σx^2 is the sum of the squares of the x values

Σy^2 is the sum of the squares of the y values

Σxy is the sum of the products of the corresponding x and y values

n is the number of points included in the fit

Test your routine using a test driver program and the 20-point input data set given in the previous problem.

7-28. The Birthday Problem The Birthday Problem is: if there is a group of n people in a room, what is the probability that two or more of them have the same birthday? It is possible to determine the answer to this question by simulation. Write a function that calculates the probability that two or more of n people will have the same birthday, where n is a calling argument. (*Hint:* To do this, the function should create an array of size n and generate n birthdays in the range 1 to 365 randomly. It should then check to see if any of the n birthdays are identical. The function should perform this experiment at least 10,000 times and calculate the fraction of those times in which two or more people had the same birthday.) Write a main program that calculates and prints out the probability that two or more of n people will have the same birthday for $n = 2, 3, \ldots, 40$.

7-29. Elapsed Time Measurement When testing the operation of procedures, it is very useful to have a set of *elapsed time subroutines*. By starting a timer running before a procedure executes, and then checking the time after the execution is completed, we can see how fast or slow the procedure is. In this manner, a programmer can identify the time-consuming portions of his or her program and rewrite them if necessary to make them faster.

Write a pair of subroutines named `set_timer` and `elapsed_time` to calculate the elapsed time in seconds between the last time that subroutine `set_timer` was called and the time that subroutine `elapsed_time` is being called. When subroutine `set_timer` is called, it should get the current time and store it into a variable in a module. When subroutine `elapsed_time` is called, it should get the current time and then calculate the difference between the current time and the stored time in the module. The elapsed time in seconds between the two calls should be returned to the calling program unit in an argument of subroutine `elapsed_time`. (*Note*: The intrinsic subroutine to read the current time is called `DATE_AND_TIME`; see Appendix B.)

7-30. Use subroutine `random0` to generate a set of three arrays of random numbers. The three arrays should be 100, 1000, and 10,000 elements long. Then, use your elapsed time subroutines to determine the time that it takes subroutine `sort` to sort each array. How does the elapsed time to sort increase as a function of the number of elements being sorted? (*Hint:* On a fast computer, you will need to sort each array many times and calculate the average sorting time in order to overcome the quantization error of the system clock.)

7-31. Evaluating Infinite Series The value of the exponential function can be calculated by evaluating the following infinite series:

$$e^x = \sum_{n=0}^{\infty} \frac{x}{n!} \qquad (7\text{-}8)$$

Write a Fortran function that calculates using the first 12 terms of the infinite series. Compare the result of your function with the result of the intrinsic function EXP(x) for $x = -10., -5., -1., 0., 1., 5., 10.,$ and 15.

7-32. Use subroutine random0 to generate an array containing 10,000 random numbers between 0.0 and 1.0. Then, use the statistics subroutines developed in this chapter to calculate the average and standard deviation of values in the array. The theoretical average of a uniform random distribution in the range [0,1) is 0.5, and the theoretical standard deviation of the uniform random distribution is $1/\sqrt{2}$. How close does the random array generated by random0 come to behaving like the theoretical distribution?

7-33. Gaussian (Normal) Distribution Subroutine random0 returns a uniformly-distributed random variable in the range [0,1), which means that there is an equal probability of any given number in the range occurring on a given call to the subroutine. Another type of random distribution is the Gaussian distribution, in which the random value takes on the classic bell-shaped curve shown in Figure 7-23. A Gaussian distribution with an average of 0.0 and a standard deviation of 1.0 is called a *standardized normal distribution*, and the probability of any given value occurring in the standardized normal distribution is given by the equation

$$p(x) = \frac{1}{\sqrt{2\pi}} e^{-x^2/2} \tag{7-9}$$

It is possible to generate a random variable with a standardized normal distribution starting from a random variable with a uniform distribution in the range [−1,1) as follows:

1. Select two uniform random variables x_1 and x_2 from the range [−1,1) such that $x_1^2 + x_2^2 < 1$. To do this, generate two uniform random variables in the range [−1,1), and see if the sum of their squares happens to be less than 1. If so, use them. If not, try again.

2. Then each of the values y_1 and y_2 in the equations below will be a normally distributed random variable.

$$y_1 = \sqrt{\frac{-2\log_e r}{r}} x_1 \tag{7-10}$$

$$y_2 = \sqrt{\frac{-2\log_e r}{r}} x_2 \tag{7-11}$$

where

$$r = x_1^2 + x_2^2 \tag{7-12}$$

Write a subroutine that returns a normally-distributed random value each time that it is called. Test your subroutine by getting 1000 random values and calculating the standard deviation. How close to 1.0 was the result?

7-34. Gravitational Force The gravitational force F between two bodies of masses m_1 and m_2 is given by the equation

FIGURE 7-23
A Normal probability distribution.

$$F = \frac{Gm_1 m_2}{r^2} \tag{7-13}$$

where G is the gravitation constant (6.672×10^{-11} N m²/kg²), m_1 and m_2 are the masses of the bodies in kilograms, and r is the distance between the two bodies. Write a function to calculate the gravitation force between two bodies given their masses and the distance between them. Test your function by determining the force on a 1000-kg satellite in orbit 38,000 km above the Earth. (The mass of the Earth is 5.98×10^{24} kg.)

7-35. Heapsort The selection sort subroutine that is introduced in this chapter is by no means the only type of sorting algorithms available. One alternate possibility is the *heapsort* algorithm, the description of which is beyond the scope of this book. However, an implementation of the *heapsort* algorithm is included in file `heapsort.f90`, which is available among the Chapter 7 files at the book's website.

If you have not done so previously, write a set of elapsed time subroutines for your computer, as described in Exercise 7-29. Generate an array containing 10,000 random values. Use the elapsed time subroutines to compare the time required to sort these 10,000 values using the selection sort and the heapsort algorithms. Which algorithm is faster? (*Note*: Be sure that you are sorting the same array each time. The best way to do this is to make a copy of the original array before sorting, and then sort the two arrays with the different subroutines.)

8

Additional Features of Arrays

OBJECTIVES

- Know how to declare and use 2D or rank 2 arrays.
- Know how to declare and use multidimensional or rank n arrays.
- Know how and when to use the WHERE construct.
- Know how and when to use the FORALL construct.
- Understand how to allocate, use, and deallocate allocatable arrays.

In Chapter 6, we learned how to use simple 1D (rank 1) arrays. This chapter picks up where Chapter 6 left off, covering advanced topics such as multidimensional arrays, array functions, and allocatable arrays.

8.1
2D OR RANK 2 ARRAYS

The arrays that we have worked with so far in Chapter 6 are **1D arrays** or **rank 1 arrays** (also known as **vectors**). These arrays can be visualized as a series of values laid out in a column, with a single subscript used to select the individual array elements (Figure 8-1a). Such arrays are useful to describe data that is a function of one independent variable, such as a series of temperature measurements made at fixed intervals of time.

Some types of data are functions of more than one independent variable. For example, we might wish to measure the temperature at five different locations at four different times. In this case, our 20 measurements could logically be grouped into five different columns of four measurements each, with a separate column for each location (Figure 8-1b). Fortran has a mechanism especially designed to hold this sort of data—a **2D** or **rank 2 array** (also called a **matrix**).

Rank 2 arrays are arrays whose elements are addressed with two subscripts, and any particular element in the array is selected by simultaneously choosing values for both of them. For example, Figure 8-2a shows a set of four generators whose power output has been measured at six different times. Figure 8-2b shows an array consisting of the six different power measurements for each of the four different generators.

Additional Features of Arrays

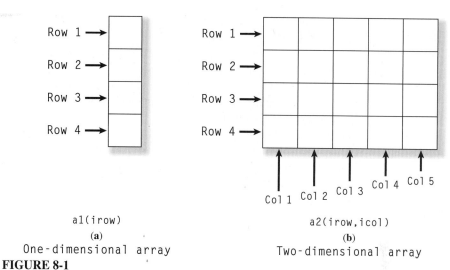

FIGURE 8-1
Representations of 1- and 2D arrays.

In this example, each row specifies a measurement time, and each column specifies a generator number. The array element containing the power supplied by generator 3 at time 4 would be power(4,3); its value is 41.1 MW.

8.1.1 Declaring Rank 2 Arrays

The type and size of a rank 2 array must be declared to the compiler using a type declaration statement. Some example array declarations are shown below:

1. `REAL, DIMENSION(3,6) :: sum`
   This type statement declares a real array consisting of 3 rows and 6 columns, for a total of 18 elements. The legal values of the first subscript are 1 to 3, and the

(a) Power measurements from 4 different generators at 6 different times.   (b) Two-dimensional matrix of power measurements.

FIGURE 8-2
(a) A power generating station consisting of four different generators. The power output of each generator is measured at six different times. (b) 2D matrix of power measurements.

legal values of the second subscript are 1 to 6. Any other subscript values are out of bounds.

2. `INTEGER, DIMENSION(0:100,0:20) :: hist`
   This type statement declares an integer array consisting of 101 rows and 21 columns, for a total of 2121 elements. The legal values of the first subscript are 0 to 100, and the legal values of the second subscript are 0 to 20. Any other subscript values are out of bounds.

3. `CHARACTER(len=6), DIMENSION(-3:3,10) :: counts`
   This type statement declares an array consisting of 7 rows and 10 columns, for a total of 70 elements. Its type is CHARACTER, with each array element capable of holding six characters. The legal values of the first subscript are −3 to 3, and the legal values of the second subscript are 1 to 10. Any other subscript values are out of bounds.

8.1.2 Rank 2 Array Storage

We have already learned that a rank 1 array of length N occupies N successive locations in the computer's memory. Similarly, a rank 2 array of size M by N occupies $M \times N$ successive locations in the computer's memory. How are the elements of the array arranged in the computer's memory? Fortran always allocates array elements in **column major order.** That is, Fortran allocates the first column in memory, then the second one, and then the third one, etc., until all columns have been allocated. Figure 8-3 illustrates this memory allocation scheme for a 3 × 2 array a. As we can see from the picture, the array element a(2,2) is really the fifth location reserved in

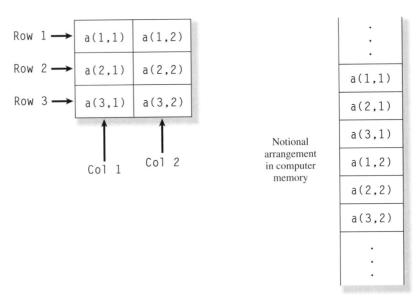

FIGURE 8-3
Notional memory allocation for a 3 × 2 rank 2 array a.

memory. The order of memory allocation will become important when we discuss data initialization and I/O statements later in this section.[1]

8.1.3 Initializing Rank 2 Arrays

Rank 2 arrays may be initialized with assignment statements, in type declaration statements, or Fortran READ statements.

Initializing rank 2 arrays with assignment statements

Initial values may be assigned to an array on an element-by-element basis using assignment statements in a nested DO loop or all at once with an array constructor. For example, suppose we have a 4 × 3 integer array istat that we wish to initialize with the values shown in Figure 8-4.

This array could be initialized at run time on an element-by-element basis with DO loops, as shown below:

```
INTEGER, DIMENSION(4,3) :: istat
DO i = 1, 4
   DO j = 1, 3
      istat(i,j) = j
   END DO
END DO
```

The array could also be initialized in a single statement with an array constructor. However, this is not as simple as it might seem. The notional data pattern in memory that would initialize the array is shown in Figure 8-4b. It consists of four 1s, followed by four 2s, followed by four 3s. The array constructor that would produce this pattern in memory is

[1,1,1,1,2,2,2,2,3,3,3,3]

so it would seem that the array could be initialized with the assignment statement

istat = [1,1,1,1,2,2,2,2,3,3,3,3]

Unfortunately, *this assignment statement will not work*. The array constructor produces a 1 × 12 array, while array istat is a 4 × 3 array. Although they both have the same number of elements, the two arrays are not **conformable** because they have different shapes, and so cannot be used in the same operation. This assignment statement will produce a compile-time error on a Fortran compiler.

---

[1] The Fortran standard does not actually *require* that the elements of an array occupy successive locations in memory. It only requires that they *appear* to be successive when addressed with appropriate subscripts or when used in operations such as I/O statements. To keep this distinction clear, we will refer to the *notional order* of the elements in memory, with the understanding that the actual order implemented by the processor could be anything. (As a practical matter, though, every Fortran compiler that the author has ever seen allocates the elements of an array in successive memory locations.) The allocation of array elements in memory was deliberately not constrained by the standard to make it easier to implement Fortran on massively parallel computers, where different memory models might be appropriate.

FIGURE 8-4
(a) Initial values for integer array istat.
(b) Notional layout of values in memory for array istat.

Array constructors always produce rank 1 arrays. So how can we overcome this limitation to use array constructors to initialize rank 2 arrays? Fortran provides a special intrinsic function, called RESHAPE, which changes the shape of an array without changing the number of elements in it. The form of the RESHAPE function is

output = RESHAPE (array1, array2)

where array1 contains the data to reshape, and array2 is a rank 1 array describing the new shape. The number of elements in array2 is the number of dimensions in the output array, and the value of each element in array2 is the extent of each dimension. *The number of elements in array1 must be the same as the number of elements in the shape specified in array2,* or the RESHAPE function will fail. The assignment statement to initialize array istat becomes:

istat = RESHAPE ([1,1,1,1,2,2,2,2,3,3,3,3], [4,3])

The RESHAPE function converts the 1 × 12 array constructor into a 4 × 3 array that can be assigned to istat.

Note that when RESHAPE changes the shape of an array, it maps the elements from the old shape to the new shape in column major order. Thus, the first element in the array constructor becomes istat(1,1), the second one becomes istat(2,1), etc.

Additional Features of Arrays

Good Programming Practice
Use the RESHAPE function to change the shape an array. This is especially useful when used with an array constructor to create array constants of any desired shape.

Initializing rank 2 arrays with type declaration statements

Initial values may also be loaded into the array at compilation time using type declaration statements. When a type declaration statement is used to initialize a rank 2 array, the data values are loaded into the array in the order in which memory is notionally allocated by the Fortran compiler. *Since arrays are allocated in column order, the values listed in the type declaration statement must be in column order.* That is, all of the elements in column 1 must be listed in the statement first, and then all of the elements in column 2, etc. Array istat contains four rows and three columns, so to initialize the array with a type declaration statement the four values of column 1 must be listed first, then the four values for column 2, and finally the four values for column 3.

The values used to initialize the array must have the same shape as the array, so the RESHAPE function must be used as well. Therefore, array istat could be initialized at compilation time with the following statement:

```
INTEGER, DIMENSION(4,3) :: istat(4,3) = &
         RESHAPE ( [ 1,1,1,1,2,2,2,2,3,3,3,3 ], [4,3] )
```

Initializing rank 2 arrays with READ statements

Arrays may be initialized with Fortran READ statements. If an array name appears without subscripts in the argument list of a READ statement, the program will attempt to read values for all of the elements in the array and the values will be assigned to the array elements in the order in which they are notionally stored in the computer's memory. Therefore, if file INITIAL.DAT contains the values

```
1  1  1  1  2  2  2  2  3  3  3  3
```

then the following code will initialize array istat to have the values shown in Figure 8-4.

```
INTEGER, DIMENSION(4,3) :: istat
OPEN (7, FILE='initial.dat', STATUS='OLD', ACTION='READ')
READ (7,*) istat
```

Implied DO loops may be used in READ statements to change the order in which array elements are initialized or to initialize only a portion of an array. For example, if file INITIAL1.DAT contains the values

```
1  2  3  1  2  3  1  2  3  1  2  3
```

then the following code will initialize array istat to have the values shown in Figure 8-4.

```
INTEGER :: i, j
INTEGER, DIMENSION(4,3) :: istat
OPEN (7, FILE='initial1.dat', STATUS='OLD', ACTION='READ')
READ (7,*) ((istat(i,j), j=1,3), i=1,4)
```

The values would have been read from file INITIAL1.DAT in a different order than in the previous example, but the implied DO loops would ensure that the proper input values went into the proper array elements.

8.1.4 Example Problem

EXAMPLE 8-1

Electric Power Generation:

Figure 8-2 shows a series of electrical output power measurements at six different times for four different generators at the Acme Electric Power generating station. Write a program to read these values from a disk file, and to calculate the average power supplied by each generator over the measurement period and the total power supplied by all of the generators at each time in the measurement period.

SOLUTION

1. **State the problem.**

    Calculate the average power supplied by each generator in the station over the measurement period and the total instantaneous power supplied by the generating station at each time within the measurement period. Write those values out on the standard output device.

2. **Define the inputs and outputs.**

    There are two types of inputs to this program:

    (a) A character string containing the file name of the input data file. This string will come from the standard input device.
    (b) The 24 real data values in the file, representing the power supplied by each of the 4 generators at each of 6 different times. The data in the input file must be organized so that the six values associated with generator G_1 appear first, followed by the six values associated with generator G_2, etc.

    The outputs from this program are the average power supplied by each generator in the station over the measurement period and the total instantaneous power supplied by the generating station at each time within the measurement period.

3. **Describe the algorithm.**

    This program can be broken down into six major steps

    ```
    Get the input file name
    Open the input file
    Read the input data into an array
    Calculate the total instantaneous output power at each time
    Calculate the average output power of each generator
    Write the output values
    ```

(continued)

Additional Features of Arrays

(*concluded*)

The detailed pseudocode for the problem is given below:

```
Prompt user for the input file name "filename"
Read file name "filename"
OPEN file "filename"
IF OPEN is successful THEN
   Read array power

   ! Calculate the instantaneous output power of the station
   DO for itime = 1 to 6
      DO for igen = 1 to 4
         power_sum(itime) ← power(itime,igen) + power_sum(itime)
      END of DO
   END of DO

   ! Calculate the average output power of each generator
   DO for igen = 1 to 4
      DO for itime = 1 to 6
         power_ave(igen) ← power(itime,igen) + power_ave(igen)
      END of DO
      power_ave(igen) ← power_ave(igen) / 6
   END of DO

   ! Write out the total instantaneous power at each time
   Write out power_sum for itime = 1 to 6

   ! Write out the average output power of each generator
   Write out power_ave for igen = 1 to 4

End of IF
```

4. **Turn the algorithm into Fortran statements.**
   The resulting Fortran program is shown in Figure 8-5.

FIGURE 8-5
Program to calculate the instantaneous power produced by a generating station and the average power produced by each generator within the station.

```
PROGRAM generate
!
!   Purpose:
!      To calculate total instantaneous power supplied by a generating
!      station at each instant of time, and to calculate the average
!      power supplied by each generator over the period of measurement.
!
!   Record of revisions:
!       Date          Programmer         Description of change
!       ====          ==========         =====================
!     11/23/15     S. J. Chapman         Original code
!
```

(*continued*)

(*continued*)

```
IMPLICIT NONE

! Data dictionary: declare constants
INTEGER, PARAMETER :: MAX_GEN = 4      ! Max number of generators
INTEGER, PARAMETER :: MAX_TIME = 6     ! Max number of times

! Data dictionary: declare variable types, definitions, & units
CHARACTER(len=20) :: filename          ! Input data file name
INTEGER :: igen                        ! Loop index: generators
INTEGER :: itime                       ! Loop index: time
CHARACTER(len=80) :: msg               ! Error message
REAL, DIMENSION(MAX_TIME,MAX_GEN) :: power
                                       ! Pwr of each gen at each time (MW)
REAL, DIMENSION(MAX_GEN) :: power_ave  ! Ave power of each gen (MW)
REAL, DIMENSION(MAX_TIME) :: power_sum ! Total power at each time (MW)
INTEGER :: status                      ! I/O status: 0 = success

! Initialize sums to zero.
power_ave = 0.
power_sum = 0.

! Get the name of the file containing the input data.
WRITE (*,1000)
1000 FORMAT ('Enter the file name containing the input data: ')
READ (*,'(A20)') filename

! Open input data file. Status is OLD because the input data must
! already exist.
OPEN ( UNIT=9, FILE=filename, STATUS='OLD', ACTION='READ', &
       IOSTAT=status, IOMSG=msg )

! Was the OPEN successful?
fileopen: IF ( status == 0 ) THEN

   ! The file was opened successfully, so read the data to process.
   READ (9, *, IOSTAT=status) power

   ! Calculate the instantaneous output power of the station at
   ! each time.
   sum1: DO itime = 1, MAX_TIME
      sum2: DO igen = 1, MAX_GEN
         power_sum(itime) = power(itime,igen) + power_sum(itime)
      END DO sum2
   END DO sum1

   ! Calculate the average output power of each generator over the
   ! time being measured.
   ave1: DO igen = 1, MAX_GEN
      ave2: DO itime = 1, MAX_TIME
         power_ave(igen) = power(itime,igen) + power_ave(igen)
      END DO ave2
```

(*continued*)

Additional Features of Arrays

(*concluded*)

```
         power_ave(igen) = power_ave(igen) / REAL(MAX_TIME)
      END DO ave1

      ! Tell user.
      out1: DO itime = 1, MAX_TIME
         WRITE (*,1010) itime, power_sum(itime)
         1010 FORMAT ('The instantaneous power at time ', I1, ' is ', &
                   F7.2, ' MW.')
      END DO out1

      out2: DO igen = 1, MAX_GEN
         WRITE (*,1020) igen, power_ave(igen)
         1020 FORMAT ('The average power of generator ', I1, ' is ', &
                   F7.2, ' MW.')
      END DO out2

   ELSE fileopen

      ! Else file open failed. Tell user.
      WRITE (*,1030) msg
      1030 FORMAT ('File open failed: ', A)

   END IF fileopen

END PROGRAM generate
```

5. **Test the program.**

To test this program, we will place the data from Figure 8-2 into a file called gendat. The contents of file gendat are shown below:

```
            20.0  19.8  20.1  20.0  20.0  19.9
            40.3  40.1  40.0  39.5  39.9  40.0
            42.0  41.5  41.3  41.1  39.8  41.0
            20.4  26.9  38.4  42.0  12.2   6.0
```

Note that each row of the file corresponds to a specific generator, and each column corresponds to a specific time. Next, we will calculate the answers by hand for one generator and one time, and compare the results with those from the program. At time 3, the total instantaneous power being supplied by all of the generators is

$$P_{TOT} = 20.1 \text{ MW} + 40.0 \text{ MW} + 41.3 \text{ MW} + 38.4 \text{ MW} = 139.8 \text{ MW}$$

The average power for Generator 1

$$P_{G1,AVE} = \frac{20.1 + 19.8 + 20.1 + 20.0 + 20.0 + 19.9}{6} = 19.98 \text{ MW}$$

The output from the program is

```
C:\book\fortran\chap8>generate
Enter the file name containing the input data:
gendat
The instantaneous power at time 1 is 122.70 MW.
```

```
The instantaneous power at time 2 is 128.30 MW.
The instantaneous power at time 3 is 139.80 MW.
The instantaneous power at time 4 is 142.60 MW.
The instantaneous power at time 5 is 111.90 MW.
The instantaneous power at time 6 is 106.90 MW.
The average power of generator  1 is  19.97 MW.
The average power of generator  2 is  39.97 MW.
The average power of generator  3 is  41.12 MW.
The average power of generator  4 is  24.32 MW.
```

so the numbers match and the program appears to be working correctly.

Note that in this problem the raw data array power was organized as a 6 × 4 matrix (6 times by 4 generators), but the input data file was organized as a 4 × 6 matrix (4 generators by 6 times)! This reversal is caused by the fact that Fortran stores array data in columns, but reads in data along lines. In order for the columns to be filled correctly in memory, the data had to be transposed in the input file! Needless to say, this can be very confusing for people having to work with the program and its input data.

It would be much better if we could eliminate this source of confusion by making the organization of the data in the input file match the organization of the data within the computer. How can we do this? With implied DO loops! If we were to replace the statement

```
READ (9,*,IOSTAT=status) power
```

with the statement

```
READ (9,*,IOSTAT=status) ((power(itime,igen), igen=1,max_gen), itime=1, max_time)
```

then the data along a row in the input file would go into the corresponding row of the matrix in the computer's memory. With the new READ statement, the input data file could be structured as follows

```
20.0  40.3  42.0  20.4
19.8  40.1  41.5  26.9
20.1  40.0  41.3  38.4
20.0  39.5  41.1  42.0
20.0  39.9  39.8  12.2
19.9  40.0  41.0   6.0
```

and after the READ statement, the contents of array power would be

$$\text{power} = \begin{bmatrix} 20.0 & 40.3 & 42.0 & 20.4 \\ 19.8 & 40.1 & 41.5 & 26.9 \\ 20.1 & 40.0 & 41.3 & 38.4 \\ 20.0 & 39.5 & 41.1 & 42.0 \\ 20.0 & 39.9 & 39.8 & 12.2 \\ 19.9 & 40.0 & 41.0 & 6.0 \end{bmatrix}$$

Additional Features of Arrays

Good Programming Practice
Use DO loops and/or implied DO loops when reading or writing rank 2 arrays in order to keep the structure of the matrix in the file the same as the structure of the matrix within the program. This correspondence makes the programs easier to understand.

8.1.5 Whole Array Operations and Array Subsets

Two arrays may be used together in arithmetic operations and assignment statements as long as they are conformable (i.e., as long as they either have the same shape one of them is a scalar). If they are conformable, then the corresponding operation will be performed on an element-by-element basis.

Array subsets may be selected from rank 2 arrays using subscript triplets or vectors subscripts. A separate subscript triplet or vector subscript is used for each dimension in the array. For example, consider the following 5×5 array.

$$a = \begin{bmatrix} 1 & 2 & 3 & 4 & 5 \\ 6 & 7 & 8 & 9 & 10 \\ 11 & 12 & 13 & 14 & 15 \\ 16 & 17 & 18 & 19 & 20 \\ 21 & 22 & 23 & 24 & 25 \end{bmatrix}$$

The array subset corresponding to the first column of this array is selected as a(:,1):

$$a(:,1) = \begin{bmatrix} 1 \\ 6 \\ 11 \\ 16 \\ 21 \end{bmatrix}$$

and the array subset corresponding to the first row is selected as a(1,:):

$$a(1,:) = \begin{bmatrix} 1 & 2 & 3 & 4 & 5 \end{bmatrix}$$

Array subscripts may be used independently in each dimension. For example, the array subset a(1:3,1:5:2) selects rows 1 through 3 and columns 1, 3, and 5 from array a. This array subset is:

$$a(1:3, 1:5:2) = \begin{bmatrix} 1 & 3 & 5 \\ 6 & 8 & 10 \\ 11 & 13 & 15 \end{bmatrix}$$

Similar combinations of subscripts can be used to select any rows or columns out of a rank 2 array.

8.2
MULTIDIMENSIONAL OR RANK *n* ARRAYS

Fortran supports more complex arrays with up to 15 different subscripts. These larger arrays are declared, initialized, and used in the same manner as the rank 2 arrays described in the previous section.

Rank *n* arrays are notionally allocated in memory in a manner that is an extension of the column order used for rank 2 arrays. Memory allocation for a $2 \times 2 \times 2$ rank 3 array is illustrated in Figure 8-6. Note that the first subscript runs through its complete range before the second subscript is incremented, and the second subscript runs through its complete range before the third subscript is incremented. This process repeats for whatever number of subscripts are declared for the array, with the first subscript always changing most rapidly and the last subscript always changing most slowly. We must keep this allocation structure in mind if we wish to initialize or perform I/O operations with rank *n* arrays.

Notional arrangement in computer memory:

```
a(1,1,1)
a(2,1,1)
a(1,2,1)
a(2,2,1)
a(1,1,2)
a(2,1,2)
a(1,2,2)
a(2,2,2)
```

FIGURE 8-6
Notional memory allocation for a $2 \times 2 \times 2$ array a. Array elements are allocated so that the first subscript changes most rapidly, the second subscript the next most rapidly, and the third subscript the least rapidly.

Quiz 8-1

This quiz provides a quick check to see if you have understood the concepts introduced in Sections 8.1 and 8.2. If you have trouble with the quiz, reread the sections, ask your instructor, or discuss the material with a fellow student. The answers to this quiz are found in the back of the book.

For questions 1 to 3, determine the number of elements in the array specified by the declaration statements and the valid subscript range(s) for each array.

1. `REAL, DIMENSION(-64:64,0:4) :: data_input`
2. `INTEGER, PARAMETER :: MIN_U = 0, MAX_U = 70`
   `INTEGER, PARAMETER :: MAXFIL = 3`
   `CHARACTER(len=24), DIMENSION(MAXFIL,MIN_U:MAX_U) :: filenm`
3. `INTEGER, DIMENSION(-3:3,-3:3,6) :: in`

Determine which of the following Fortran statements are valid. For each valid statement, specify what will happen in the program. Assume default typing for any variables that are not explicitly typed.

4. `REAL, DIMENSION(0:11,2) :: dist`
   `dist = [0.00,  0.25,  1.00,  2.25,  4.00,  6.25, &`
   `        9.00, 12.25, 16.00, 20.25, 25.00, 30.25, &`
   `       -0.00, -0.25, -1.00, -2.25, -4.00, -6.25, &`
   `       -9.00,-12.25,-16.00,-20.25,-25.00,-30.25]`

5. `REAL, DIMENSION(0:11,2) :: dist`
   `dist = RESHAPE([0.00, 0.25,  1.00,  2.25,  4.00,  6.25, &`
   `                9.00,12.25, 16.00, 20.25, 25.00,30.25, &`
   `                0.00, 0.25,  1.00,  2.25,  4.00,  6.25, &`
   `                9.00,12.25, 16.00, 20.25, 25.00,30.25], &`
   `               [12,2])`

6. `REAL, DIMENSION(-2:2,-1:0) :: data1 = &`
   `           RESHAPE ( [ 1.0, 2.0, 3.0, 4.0, 5.0, &`
   `                       6.0, 7.0, 8.0, 9.0, 0.0 ], &`
   `                     [ 5, 2 ] )`
   `REAL, DIMENSION(0:4,2) :: data2 = &`
   `           RESHAPE ( [ 0.0, 9.0, 8.0, 7.0, 6.0, &`
   `                       5.0, 4.0, 3.0, 2.0, 1.0 ], &`
   `                     [ 5, 2 ] )`
   `REAL, DIMENSION(5,2) :: data_out`
   `data_out = data1 + data2`
   `WRITE (*,*) data_out(:,1)`
   `WRITE (*,*) data_out(3,:)`

(continued)

(*concluded*)

7.  ```
    INTEGER, DIMENSION(4) :: list1 = [1,4,2,2]
    INTEGER, DIMENSION(3) :: list2 = [1,2,3]
    INTEGER, DIMENSION(5,5) :: array
    DO i = 1,5
      DO j = 1,5
        array(i,j) = i + 10 * j
      END DO
    END DO
    WRITE (*,*) array(list1, list2)
    ```

8.  ```
    INTEGER, DIMENSION(4) :: list = [2,3,2,1]
    INTEGER, DIMENSION(10) :: vector = [ (10*k, k = -4,5) ]
    vector(list) = [ 1, 2, 3, 4 ]
    WRITE (*,*) vector
    ```

Suppose that a file `input` is opened on i/o unit 2, and contains the following data:

11.2	16.5	31.3	3.1414	16.0	12.0
1.1	9.0	17.1	11.	15.0	-1.3
10.0	11.0	12.0	13.0	14.0	5.0
15.1	16.7	18.9	21.1	24.0	-22.2

What data would be read from file `input` by each of the following statements? What would the value of `mydata(2,4)` be in each case?

9.  ```
    REAL, DIMENSION(3,5) :: mydata
    READ (2,*) mydata
    ```

10. ```
    REAL, DIMENSION(0:2,2:6) :: mydata
    READ (2,*) mydata
    ```

11. ```
    REAL, DIMENSION(3,5) :: mydata
    READ (2,*) ((mydata(i,j), j=1,5), i=1,3)
    ```

12. ```
    REAL, DIMENSION(3,5) :: mydata
    DO i = 1, 3
      READ (2,*) (mydata(i,j), j=1,5)
    END DO
    ```

Answer the following questions:

13. What is the value of `dist(6,2)` in Question 5 of this quiz?
14. What is the rank of `mydata` in Question 10 of this quiz?
15. What is the shape of `mydata` in Question 10 of this quiz?
16. What is the extent of the first dimension of `data_input` in Question 1 of this quiz?
17. What is the maximum number of dimensions that an array can have in Fortran?

8.3

USING FORTRAN INTRINSIC FUNCTIONS WITH ARRAYS

There are three classes of Fortran intrinsic functions: **elemental functions, inquiry functions,** and **transformational functions.** Some of the functions from each of these classes are designed for use with array arguments. We will now examine a few of them. A more complete description of all Fortran intrinsic functions and subroutines is found in Appendix B.

8.3.1 Elemental Intrinsic Functions

Elemental intrinsic functions are ones that are specified for scalar arguments, but that may also be applied to array arguments. If the argument of an elemental function is a scalar, then the result of the function will be a scalar. If the argument of the function is an array, then the result of the function will be an array of the same shape as the input array. Note that if there is more than one input argument, all of the arguments must have the same shape. If an elemental function is applied to an array, the result will be the same as if the function were applied to each element of the array on an element-by-element basis. Thus, the following two sets of statements are equivalent:

```
REAL, DIMENSION(4) :: x = [ 0., 3.141592, 1., 2. ]
REAL, DIMENSION(4) :: y
INTEGER :: i

y = SIN(x)                ! Whole array at once

DO i = 1,4
   y(i) = SIN(x(i))       ! Element by element
END DO
```

Most of the Fortran intrinsic functions that accept scalar arguments are elemental, and so can be used with arrays. This includes common functions such as ABS, SIN, COS, TAN, EXP, LOG, LOG10, MOD, and SQRT.

8.3.2 Inquiry Intrinsic Functions

Inquiry intrinsic functions are functions whose value depends on the properties of an object being investigated. For example, the function UBOUND(arr) is an inquiry function that returns the largest subscript(s) of array arr. A list of some of the common array inquiry functions is shown in Table 8-1. Any function arguments shown in italics are optional; they may or may not be present when the function is invoked.

These functions are useful for determining the properties of an array, such as its size, shape, extent, and the legal subscript range in each extent. They will be especially important once we begin passing arrays to procedures in Chapter 9.

TABLE 8-1
Some common array inquiry functions

Function name and calling sequence	Purpose
ALLOCATED(ARRAY)	Determines allocation status of an allocatable array (see Section 8.6).
LBOUND(ARRAY, DIM)	Returns all of the lower bounds of ARRAY if *DIM* is absent, or a specified lower bound of ARRAY if *DIM* is present. The result is a rank 1 array if *DIM* is absent, or a scalar if *DIM* is present.
SHAPE(SOURCE)	Returns the shape of array SOURCE.
SIZE(ARRAY, DIM)	Returns either the extent of ARRAY along a particular dimension if *DIM* is present; otherwise, it returns the total number of elements in the array.
UBOUND(ARRAY, DIM)	Returns all of the upper bounds of ARRAY if *DIM* is absent, or a specified upper bound of ARRAY if *DIM* is present. The result is a rank 1 array if *DIM* is absent, or a scalar if *DIM* is present.

EXAMPLE 8-2

Determining the Properties of an Array:

To illustrate the use of the array inquiry functions, we will declare a rank 2 array a, and use the functions to determine its properties.

SOLUTION
The program in Figure 8-7 invokes the functions SHAPE, SIZE, LBOUND, and UBOUND to determine the properties of the array.

FIGURE 8-7
Program to determine the properties of an array.

```
PROGRAM check_array
!
!  Purpose:
!    To illustrate the use of array inquiry functions.
!
!  Record of revisions:
!      Date          Programmer          Description of change
!      ====          ==========          =====================
!    11/23/15       S. J. Chapman        Original code
!
IMPLICIT NONE

! List of variables:
REAL,DIMENSION(-5:5,0:3) :: a = 0.   ! Array to examine

! Get the shape, size, and bounds of the array.
WRITE (*,100) SHAPE(a)
100 FORMAT ('The shape of the array is:            ',7I6)
```

(continued)

Additional Features of Arrays

(concluded)

```
WRITE (*,110) SIZE(a)
110 FORMAT ('The size of the array is:         ',I6)

WRITE (*,120) LBOUND(a)
120 FORMAT ('The lower bounds of the array are:  ',7I6)

WRITE (*,130) UBOUND(a)
130 FORMAT ('The upper bounds of the array are:  ',7I6)

END PROGRAM check_array
```

When the program is executed, the results are:

```
C:\book\fortran\chap8>check_array
The shape of the array is:         11    4
The size of the array is:          44
The lower bounds of the array are: -5    0
The upper bounds of the array are:  5    3
```

These are obviously the correct answers for array a.

8.3.3 Transformational Intrinsic Functions

Transformational intrinsic functions are functions that have one or more array-valued arguments or an array-valued result. Unlike elemental functions, which operate on an element-by-element basis, transformational functions operate on arrays as a whole. The output of a transformational function will often not have the same shape as the input arguments. For example, the function DOT_PRODUCT has two vector input arguments of the same size and produces a scalar output.

There are *many* transformational intrinsic functions in Fortran. Some of the more common ones are summarized in Table 8-2. Some of the functions listed in Table 8-2 have additional optional arguments that are not mentioned. The complete details of

TABLE 8-2
Some common transformational functions

Function name and calling sequence	Purpose
ALL(MASK)	Logical function that returns TRUE if *all* of the values in array MASK are true.
ANY(MASK)	Logical function that returns TRUE if *any* of the values in array MASK are true.
COUNT(MASK)	Returns the number of TRUE elements in array MASK.
DOT_PRODUCT(VECTOR_A, VECTOR_B)	Calculates the dot product of two equal-sized vectors.
MATMUL(MATRIX_A, MATRIX_B)	Performs matrix multiplication on to conformable matrices.

(continued)

(*concluded*)

MAXLOC(ARRAY, *MASK*)	Returns the *location* of the maximum value in ARRAY among those elements for which *MASK* was true. The result is a rank 1 array with one element for each subscript in ARRAY. (*MASK* is optional.)
MAXVAL(ARRAY, *MASK*)[1]	Returns the maximum value in ARRAY among those elements for which *MASK* was true. (*MASK* is optional.)
MINLOC(ARRAY, *MASK*)	Returns the *location* of the minimum value in ARRAY among those elements for which *MASK* was true. The result is a rank 1 array with one element for each subscript in ARRAY. (*MASK* is optional.)
MINVAL(ARRAY, *MASK*)[1]	Returns the minimum value in ARRAY among those elements for which *MASK* was true. (*MASK* is optional.)
PRODUCT(ARRAY, *MASK*)[1]	Calculates the product of the elements in ARRAY for which the *MASK* is true. *MASK* is optional; if not present, it calculates the product of all of the elements in the array.
RESHAPE(SOURCE,SHAPE)	Constructs an array of the specified shape from the elements of array SOURCE. SHAPE is a rank 1 array containing the extents of each dimension in the array to be built.
SUM(ARRAY, *MASK*)[1]	Calculates the sum of the elements in ARRAY for which the *MASK* is true. *MASK* is optional; if not present, it calculates the sum of all of the elements in the array.
TRANSPOSE(MATRIX)	Returns the transpose of a rank 2 array.

[1] If a MASK is specified in these functions, it must be specified in the form MASK=*mask_expr*, where *mask_expr* is the logical array specifying the mask. The reason for this form is explained in Chapter 9 and Appendix B.

each function, including any additional arguments, are found in Appendix B. Any function arguments shown in italics are optional; they may or may not be present when the function is invoked.

We have already seen the RESHAPE function used to initialize arrays. A number of other transformational functions will appear in the exercises at the end of this chapter.

8.4

MASKED ARRAY ASSIGNMENT: THE WHERE CONSTRUCT

We have already seen that Fortran permits us to use either array elements or entire arrays in array assignment statements. For example, we could take the logarithm of the elements in a rank 2 array value in either of the following ways:

```
DO i = 1, ndim1
   DO j = 1, ndim2
      logval(i,j) = LOG(value(i,j))
   END DO
END DO
```

```
logval = LOG ( value )
```

Both of the above examples take the logarithm of all of the elements in array value and store the result in array logval.

Suppose that we would like to take the logarithm of *some* of the elements of array value, but not all of them. For example, suppose that we only want to take the logarithm of *positive* elements, since the logarithms of zero and negative numbers are not defined and produce runtime errors. How could we do this? One way would be to do it on an element-by-element basis using a combination of DO loops and an IF construct. For example,

```
DO i = 1, ndim1
  DO j = 1, ndim2
    IF ( value(i,j) > 0. ) THEN
       logval(i,j) = LOG(value(i,j))
    ELSE
       logval(i,j) = -99999.
    END IF
  END DO
END DO
```

We can also perform this calculation all at once using a special form of array assignment statement known as **masked array assignment.** A masked array assignment statement is an assignment statement whose operation is controlled by a logical array of the *same shape* as the array in the assignment. The assignment operation is *only* performed for the elements of the array that correspond to TRUE values in the mask. In Fortran, masked array assignments are implemented using the WHERE construct or statement.

8.4.1 The WHERE Construct

The general form of a WHERE construct is

```
[name:] WHERE (mask_expr1)
    Array Assignment Statement(s)    ! Block 1
ELSEWHERE (mask_expr2) [name]
    Array Assignment Statement(s)    ! Block 2
ELSEWHERE [name]
    Array Assignment Statement(s)    ! Block 3
END WHERE [name]
```

where each *mask_expr1* is a logical array of the same shape as the array(s) being manipulated in the array assignment statements. This construct applies the operation or set of operations in Block 1 to all of the elements of the array for which *mask_expr1* is TRUE. It applies the operation or set of operations in Block 2 to all of the elements of the array for which *mask_expr1* is FALSE and *mask_expr2* is TRUE. Finally, it applies the operation or set of operations in Block 3 to all the elements of the array for which both *mask_expr1* and *mask_expr2* are FALSE. There can be as many masked ELSEWHERE clauses as desired in a Fortran WHERE construct.

Note that *at most one block of statements will be executed for any given element in the array.*

A name may be assigned to a WHERE construct, if desired. If the WHERE statement at the beginning of a construct is named, then the associated END WHERE statement must also have the same name. The name is optional on an ELSEWHERE statement even if it is used on the corresponding WHERE and END WHERE statements.

The example given above could be implemented with a WHERE construct as:

```
WHERE ( value > 0. )
   logval = LOG(value)
ELSEWHERE
   logval = -99999.
END WHERE
```

The expression "value > 0." produces a logical array whose elements are TRUE where the corresponding elements of value are greater than zero, and FALSE where the corresponding elements of value are less than or equal zero. This logical array then serves as a mask to control the operation of the array assignment statement.

The WHERE construct is generally more elegant than element-by-element operations, especially for multidimensional arrays.

Good Programming Practice
Use WHERE constructs to modify and assign array elements when you want to modify and assign only those elements that pass some test.

8.4.2 The WHERE Statement

Fortran also includes a single-line WHERE statement:

    WHERE (*mask_expr*) Array Assignment Statement

The assignment statement is applied to those elements of the array for which the mask expression is true.

EXAMPLE 8-3 *Limiting the Maximum and Minimum Values in an Array:*

Suppose that we are writing a program to analyze an input data set whose values should be in the range [21000, 1000]. If numbers greater than 1000 or less than 1000 would cause problems with our processing algorithm, it might be desirable to put in a test limiting all data values to the acceptable range. Write such a test for a 10,000-element rank 1 real array input using both DO and IF constructs and a WHERE construct.

SOLUTION
The test using DO and IF constructs is

```
DO i = 1, 10000
   IF ( input(i) > 1000. ) THEN
      input(i) = 1000.
   ELSE IF ( input(i) < -1000. ) THEN
      input(i) = -1000.
   END IF
END DO
```

(*continued*)

(*concluded*)

The test using a Fortran WHERE construct is

```
WHERE ( input > 1000. )
   input = 1000.
ELSEWHERE ( input < -1000. )
   input = -1000.
END WHERE
```

The WHERE construct is simpler than the DO and IF constructs for this example.

8.5
THE FORALL CONSTRUCT

Fortran also includes a construct that is designed to permit a set of operations to be applied on an element-by-element basis to a subset of the elements in an array. The elements to be operated on may be chosen *both* by subscript index *and* by a logical condition. The operations will only be applied to those elements of the array that satisfy both the index constraints and the logical condition. This construct is called the FORALL construct.

8.5.1 The Form of the FORALL Construct

The general form of the FORALL construct is

```
[name:] FORALL (in1=triplet1[, in2=triplet2, ..., logical_expr])
   Statement 1
   Statement 2
   ...
   Statement n
END FORALL [name]
```

Each index in the FORALL statement is specified by a subscript triplet of the form

$$subscript\_1 : subscript\_2 : stride$$

where *subscript_1* is the starting value of the index, *subscript_2* is the ending value, and *stride* is index step. Statements 1 through *n* in the body of the construct are assignment statements that manipulate the elements of arrays having the selected indices and satisfying the logical expression on an element-by-element basis.

A name may be assigned to a FORALL construct, if desired. If the FORALL statement at the beginning of a construct is named, then the associated END FORALL statement must also have the same name.

A simple example of a FORALL construct is shown below. These statements create a 10 × 10 identity matrix, which has 1s along the diagonal and 0s everywhere else.

```
REAL, DIMENSION(10,10) :: i_matrix = 0.
...
FORALL ( i=1:10 )
   i_matrix(i,i) = 1.0
END FORALL
```

As a more complex example, let's suppose that we would like to take the reciprocal of all of the elements in an n × m array `work`. We might do this with the simple assignment statement

```
work = 1. / work
```

but this statement would cause a runtime error and abort the program if any of the elements of `work` happened to be zero. A FORALL construct that avoids this problem is

```
FORALL ( i=1:n, j=1:m, work(i,j) /= 0. )
   work(i,j) = 1. / work(i,j)
END FORALL
```

8.5.2 The Significance of the FORALL Construct

In general, any expression that can be written in a FORALL construct could also be written as a set of nested DO loops combined with a block IF construct. For example, the previous FORALL example could be written as

```
DO i = 1, n
   DO j = 1, m
      IF ( work(i,j) /= 0.
         work(i,j) = 1. / work(i,j)
      END IF
   END DO
END DO
```

What is the difference between these two sets of statements, and why is the FORALL construct included in the Fortran language at all?

The answer is that *the statements in the DO loop structure must be executed in a strict order, while the statements in the FORALL construct may be executed in any order*. In the DO loops, the elements of array work are processed in the following strict order:

```
work(1,1)
work(1,2)
  ...
work(1,m)
work(2,1)
work(2,2)
  ...
work(2,m)
  ...
work(n,m)
```

In contrast, the FORALL construct processes the same set of elements *in any order selected by the processor*. This freedom means that massively parallel computers can optimize the program for maximum speed by parceling out each element to a separate processor, and the processors can finish their work in any order without impacting the final answer.

If the body of a FORALL construct contains more than one statement, then the processor completely finishes all of the selected elements of the first statement before starting any of the elements of the second statement. In the example below, the values for a(i,j) that are calculated in the first statement are used to calculate b(i,j) in the second statement. All of the a values are calculated before the first b value is calculated.

```
FORALL (i=2:n-1, j=2:n-1)
    a(i,j) = SQRT(a(i,j))
    b(i,j) = 1.0 / a(i,j)
END FORALL
```

Because each element must be capable of being processed independently, the body of a FORALL construct cannot contain transformational functions whose results depend on the values in the entire array. However, the body can contain nested FORALL and WHERE constructs.[2]

8.5.3 The FORALL Statement

Fortran also includes a single-line FORALL statement:

```
FORALL (ind1=triplet1[, ..., logical_expr]) Assignment Statement
```

The assignment statement is executed for those indices and logical expressions that satisfy the FORALL control parameters. This simpler form is the same as a FORALL construct with only one statement.

8.6
ALLOCATABLE ARRAYS

In all of the examples that we have seen so far, the size of each array was declared in a type declaration statement at the beginning of the program. This type of array declaration is called **static memory allocation,** since the size of each array is set at compilation time and never changes. *The size of each array must be made large enough to hold the largest problem that a particular program will ever have to solve*, which can be a very serious limitation. If we declare the array sizes to be large enough to handle the largest problem that we will ever need to solve, then the program will waste memory 99% of the time that it is run. In addition, the program might not run at all on small computers that don't have enough memory to hold it. If the arrays are made small, then the program cannot solve large problems at all.

---

[2] The proposed Fortran 2015 Draft Standard (currently proposed for approval in 2018) declares FORALL to be obsolescent, indicating that it should not be used in new programs. It has been replaced by better mechanisms of allocating work amongst processors, as we shall see later.

What can a programmer do about this problem? If the program is well designed, then the array limitations could be modified by just changing one or two array size parameters in the source code and recompiling it. This process will work for in-house programs for which the source code is available, but it is not very elegant. It won't work at all for programs whose source code is unavailable, such as those programs that you buy from someone else.

A much better solution is to design a program that uses **dynamic memory allocation:** it dynamically sets the sizes of the arrays each time it is executed to be just large enough to solve the current problem. This approach does not waste computer memory and will allow the same program to run on both small and large computers.

8.6.1 Fortran Allocatable Arrays

A Fortran array using dynamic memory is declared using the ALLOCATABLE attribute in the type declaration statement, and is actually allocated with an ALLOCATE statement. When the program is through using the memory, it should free it up for other uses with a DEALLOCATE statement. The structure of a typical array declaration with the ALLOCATABLE attribute[3] is

```
REAL, ALLOCATABLE, DIMENSION(:,:) :: arr1
```

Note that colons are used as placeholders in the declaration since we do not know how big the array will actually be. The *rank* of the array is declared in the type declaration statement, but not the *size* of the array.

An array declared with colons for dimensions is known as a **deferred-shape array,** because the actual shape of the array is deferred until the memory for the array is allocated. (In contrast, an array whose size is explicitly declared in a type declaration statement is known as an **explicit-shape array.**)

When the program executes, the actual size of the array will be specified with an ALLOCATE statement. The forms of an ALLOCATE statement are

```
ALLOCATE (list of arrays, STAT=status, ERRMSG=err_msg)
ALLOCATE (array to allocate, SOURCE=source_expr, STAT=status, ERRMSG=string)
```

A typical example of the first form of the ALLOCATE statement is

```
ALLOCATE (arr1(100,0:10), STAT=status, ERR_MSG=msg)
```

---

[3] An array may also be declared to be allocatable in a separate ALLOCATABLE statement of the form

```
ALLOCATABLE :: arr1
```

It is preferable *not* to use this statement, since it is always possible to specify the ALLOCATABLE attribute in a type declaration statement and the array will appear in a type declaration statement anyway. The only time when a separate ALLOCATABLE statement is necessary is when default typing is used and there is no type declaration statement. Since we should *never* use default typing in any program, there is never a need for this statement.

This statement allocates a 100 × 11 array `arr1` at execution time. The STAT= and ERR_MSG= clauses are optional. If it is present, STAT= returns an integer status. If the allocation is successful, the integer value returned by the STAT= clause will be 0, and the character variable in the ERRMSG= clause will not be changed. If the allocation is unsuccessful, the integer value returned by the STAT= clause will be a nonzero code indicating the type of the error, and the character variable in the ERRMSG= clause will contain a descriptive message indicating what the problem is for display to the user.

In the second form of the ALLOCATE statement, the array allocated has the same shape as the source expression, and the data from the source expression is copied to the newly allocated array. For example, if array `source_array` is a 10 × 20 array, then array `myarray` will be allocated as a 10 × 20 array and the contents of the two arrays will be identical.

```
ALLOCATE (myarray, SOURCE=source_array, STAT=istat, ERRMSG=msg)
```

The most common source of failure for any allocate statement is not having enough free memory to allocate the array. If the allocation fails and the STAT= clause is not present, then the program will abort. You should always use the STAT= clause so that the program can terminate gracefully if there is not enough memory available to allocate the array.

Good Programming Practice
Always include the STAT= clause in any ALLOCATE statement and always check the returned status, so that a program can be shut down gracefully if there is insufficient memory to allocate the necessary arrays.

An allocatable array *may not be used in any way* in a program until memory is allocated for it. Any attempt to use an allocatable array that is not currently allocated will produce a runtime error and cause the program to abort. Fortran includes the logical intrinsic function ALLOCATED( ) to allow a program to test the allocation status of an array before attempting to use it. For example, the following code tests the status of allocatable array `input_data` before attempting to reference it:

```
REAL, ALLOCATABLE, DIMENSION(:) :: input_data
...
IF ( ALLOCATED(input_data) ) THEN
   READ (8,*) input_data
ELSE
   WRITE (*,*) 'Warning: Array not allocated!'
END IF
```

This function can be very helpful in large programs involving many procedures, in which memory is allocated in one procedure and used in a different one.

At the end of the program or procedure in which an allocatable array is used, you should *deallocate* the memory to make it available for reuse. This is done with a DEALLOCATE statement. The structure of a DEALLOCATE statement is

DEALLOCATE (*list of arrays to deallocate*, STAT=*status*)

A typical example is

DEALLOCATE (arr1, STAT=status)

where the status clause has the same meaning as in the ALLOCATE statement. After a DEALLOCATE statement is executed, the data in the deallocated arrays is no longer available for use.

You should always deallocate any allocatable arrays once you are finished with them. This frees up the memory to be used elsewhere in the program, or in other programs running on the same computer.

Good Programming Practice
Always deallocate allocatable arrays with a DEALLOCATE statement as soon as you are through using them.

EXAMPLE 8-4   *Using Allocatable Arrays:*

To illustrate the use of allocatable arrays, we will rewrite the statistical analysis program of Example 6-4 to dynamically allocate only the amount of memory needed to solve the problem. To determine how much memory to allocate, the program will read the input data file and count the number of values. It will then allocate the array, rewind the file, read in the values, and calculate the statistics.

SOLUTION
The modified program with allocatable arrays is shown in Figure 8-8.

FIGURE 8-8
A modified form of the statistics program that uses allocatable arrays.

```
PROGRAM stats_5
!
! Purpose:
!    To calculate mean, median, and standard deviation of an input
!    data set read from a file.  This program uses allocatable arrays
!    to use only the memory required to solve each problem.
!
```

(continued)

(*continued*)

```
!   Record of revisions:
!      Date         Programmer         Description of change
!      ====         ==========         =====================
!    11/18/15     S. J. Chapman        Original code
! 1. 11/23/15     S. J. Chapman        Modified for dynamic memory
!
IMPLICIT NONE

! Data dictionary: declare variable types & definitions
REAL,ALLOCATABLE,DIMENSION(:) :: a  ! Data array to sort
CHARACTER(len=20) :: filename   ! Input data file name
INTEGER :: i                    ! Loop index
INTEGER :: iptr                 ! Pointer to smallest value
INTEGER :: j                    ! Loop index
REAL :: median                  ! The median of the input samples
CHARACTER(len=80) :: msg        ! Error message
INTEGER :: nvals = 0            ! Number of values to process
INTEGER :: status               ! Status: 0 for success
REAL :: std_dev                 ! Standard deviation of input samples
REAL :: sum_x = 0.              ! Sum of input values
REAL :: sum_x2 = 0.             ! Sum of input values squared
REAL :: temp                    ! Temporary variable for swapping
REAL :: x_bar                   ! Average of input values

! Get the name of the file containing the input data.
WRITE (*,1000)
1000 FORMAT ('Enter the file name with the data to be sorted:')
READ (*,'(A20)') filename

! Open input data file.  Status is OLD because the input data must
! already exist.
OPEN ( UNIT=9, FILE=filename, STATUS='OLD', ACTION='READ', &
       IOSTAT=status, IOMSG=msg )

! Was the OPEN successful?
fileopen: IF ( status == 0 ) THEN         ! Open successful

   ! The file was opened successfully, so read the data to find
   ! out how many values are in the file, and allocate the
   ! required space.
   DO
      READ (9, *, IOSTAT=status) temp    ! Get value
      IF ( status /= 0 ) EXIT            ! Exit on end of data
      nvals = nvals + 1                  ! Bump count
   END DO

   ! Allocate memory
   WRITE (*,*) 'Allocating a: size = ', nvals
   ALLOCATE ( a(nvals), STAT=status)       ! Allocate memory

   ! Was allocation successful? If so, rewind file, read in
   ! data, and process it.
   allocate_ok: IF ( status == 0 ) THEN
```

(*continued*)

(*continued*)

```
      REWIND ( UNIT=9 )                      ! Rewind file

      ! Now read in the data.  We know that there are enough
      ! values to fill the array.
      READ (9, *) a                          ! Get value

      ! Sort the data.
      outer: DO i = 1, nvals-1

         ! Find the minimum value in a(i) through a(nvals)
         iptr = i
         inner: DO j = i+1, nvals
            minval: IF ( a(j) < a(iptr) ) THEN
               iptr = j
            END IF minval
         END DO inner

         ! iptr now points to the minimum value, so swap a(iptr)
         ! with a(i) if i /= iptr.
         swap: IF ( i /= iptr ) THEN
            temp     = a(i)
            a(i)     = a(iptr)
            a(iptr) = temp
         END IF swap

      END DO outer

      ! The data is now sorted.  Accumulate sums to calculate
      ! statistics.
      sums: DO i = 1, nvals
         sum_x  = sum_x + a(i)
         sum_x2 = sum_x2 + a(i)**2
      END DO sums

      ! Check to see if we have enough input data.
      enough: IF ( nvals < 2 ) THEN

         ! Insufficient data.
         WRITE (*,*) 'At least 2 values must be entered.'

      ELSE

         ! Calculate the mean, median, and standard deviation
         x_bar   = sum_x / real(nvals)
         std_dev = sqrt( (real(nvals) * sum_x2 - sum_x**2) &
                 / (real(nvals) * real(nvals-1)) )
         even: IF ( mod(nvals,2) == 0 ) THEN
            median = ( a(nvals/2) + a(nvals/2+1) ) / 2.
         ELSE
            median = a(nvals/2+1)
         END IF even
```

(*continued*)

```
            ! Tell user.
            WRITE (*,*) 'The mean of this data set is:   ', x_bar
            WRITE (*,*) 'The median of this data set is:', median
            WRITE (*,*) 'The standard deviation is:      ', std_dev
            WRITE (*,*) 'The number of data points is:   ', nvals

         END IF enough

         ! Deallocate the array now that we are done.
         DEALLOCATE ( a, STAT=status )

      END IF allocate_ok

   ELSE fileopen

      ! Else file open failed.  Tell user.
      WRITE (*,1050) TRIM(msg)
      1050 FORMAT ('File open failed--status = ', A)

   END IF fileopen

END PROGRAM stats_5
```

To test this program, we will run it with the same data set as Example 6-4.

```
C:\book\fortran\chap8>stats_5
Enter the file name containing the input data:
input4
Allocating a: size =          5
The mean of this data set is:        4.400000
The median of this data set is:      4.000000
The standard deviation is:           2.966479
The number of data points is:        5
```

The program gives the correct answers for our test data set.

8.6.2 Using Fortran Allocatable Arrays in Assignment Statements

We have already seen how to allocate and deallocate allocatable arrays using ALLOCATE and DEALLOCATE statements. In addition, Fortran 2003 and later allow allocatable arrays to be allocated and deallocated automatically by simply assigning data to them.

If an expression is assigned to an allocatable array of the same rank, then the array is *automatically allocated* to the correct shape if it is unallocated, or it is *automatically deallocated and reallocated* to the correct shape if it was previously allocated with an incompatible shape. No ALLOCATE and DEALLOCATE statements are required. If the shape of the data being assigned is the same as the shape already allocated, it is just reused without reallocating. This means that the arrays can be used seamlessly in calculations with data of different sizes.

For example, consider the following program.

```
PROGRAM test_allocatable_arrays
IMPLICIT NONE

! Declare data
REAL, DIMENSION(:), ALLOCATABLE :: arr1
REAL, DIMENSION(8) :: arr2 = [ 1., 2., 3., 4., 5., 6., 7., 8. ]
REAL, DIMENSION(3) :: arr3 = [ 1., -2., 3. ]

! Automatically allocate arr1 as a 3 element array
arr1 = 2. * arr3
WRITE (*,*) arr1

! Automatically allocate arr1 as a 4 element array
arr1 = arr2(1:8:2)
WRITE (*,*) arr1

! Reuse arr1 as a 4 element array without deallocating
arr1 = 2. * arr2(1:4)
WRITE (*,*) arr1
END PROGRAM test_allocatable_arrays
```

When this program is compiled and executed, the results are:

```
C:\book\fortran\chap8>ifort/standard-semantics test_allocatable_arrays.f90
Intel(R) Visual Fortran Intel(R) 64 Compiler for applications running on
Intel(R) 64, Version 16.0.2.180 Build 20160204
Copyright (C) 1985-2016 Intel Corporation. All rights reserved.

Microsoft (R) Incremental Linker Version 12.00.40629.0
Copyright (C) Microsoft Corporation. All rights reserved.

-out:test_allocatable_arrays.exe
-subsystem:console
test_allocatable_arrays.obj

C:\book\fortran\chap8>test_allocatable_arrays
 2.000000  -4.000000   6.000000
 1.000000   3.000000   5.000000   7.000000
 2.000000   4.000000   6.000000   8.000000
```

When the first assignment statement is executed, `arr1` is unallocated, so it is automatically allocated as a 3-element array and the values [2. –4. 6.] are stored in it. When the second assignment statement is executed, `arr1` is allocated as a 3-element array, which is the wrong size, so the array is automatically deallocated and reallocated with four elements and the values [1. 3. 5. 7.] are stored in it. When the third assignment statement is executed, arr1 is allocated as a 4-element array, which is the correct size, so the array is *not* reallocated and the values [2. 4. 6. 8.] are stored in the existing allocation.[4]

---

[4] Note that it is necessary to use the /standard-semantics option with the Intel Fortran compiler to enable Fortran 2003 allocatable array behaviors. Different options may be required for other compilers.

Note that this automatic allocation and deallocation works only if the allocatable variable is the same rank as the expression being assigned to it. If the ranks differ, the assignment will produce a compile-time error.

```
REAL, DIMENSION(:), ALLOCATABLE :: arr1
REAL, DIMENSION(2,2), :: arr2 = RESHAPE ( [ 1,2,3,4 ], [2,2] )
...
arr1 = arr2     ! Error
```

Good Programming Practice
When allocatable arrays are used in a Fortran 2003 or later program, they are automatically resized to match the size of the data assigned to them as long as that data has the same rank as the allocatable array.

Fortran 2003 allocatable arrays declared without a SAVE attribute[5] are automatically deallocated whenever the program unit containing them finishes. Thus, allocatable arrays in subroutines or functions do not need to be deallocated with a DEALLOCATE statement at the end of the subroutine or function.

Good Programming Practice
Allocatable arrays declared in a subroutine or function without a SAVE attribute will be automatically deallocated when the subroutine or function exits. No DEALLOCATE statements are required.

Quiz 8-2

This quiz provides a quick check to see if you have understood the concepts introduced in Sections 8.3 through 8.6. If you have trouble with the quiz, reread the sections, ask your instructor, or discuss the material with a fellow student. The answers to this quiz are found in the back of the book.

For questions 1 to 6, determine what will be printed out by the WRITE statements.

```
1. REAL, DIMENSION(-3:3,0:50) :: values
   WRITE (*,*) LBOUND(values,1)
   WRITE (*,*) UBOUND(values,2)
   WRITE (*,*) SIZE(values,1)
   WRITE (*,*) SIZE(values)
   WRITE (*,*) SHAPE(values)
```

(continued)

---

[5] The SAVE attribute is described in Chapter 9.

(*continued*)

2.  ```
    REAL, ALLOCATABLE, DIMENSION(:,:,:) :: values
    ...
    ALLOCATE( values(3,4,5), STAT=istat )
    WRITE (*,*) UBOUND(values,2)
    WRITE (*,*) SIZE(values)
    WRITE (*,*) SHAPE(values)
    ```

3.  ```
    REAL, DIMENSION(5,5) :: input1
    DO i = 1, 5
       DO j = 1, 5
          input1(i,j) = i+j-1
       END DO
    END DO
    WRITE (*,*) MAXVAL(input1)
    WRITE (*,*) MAXLOC(input1)
    ```

4.  ```
    REAL, DIMENSION(2,2) :: arr1
    arr1 = RESHAPE( [3.,0.,-3.,5.], [2,2] )
    WRITE (*,*) SUM( arr1 )
    WRITE (*,*) PRODUCT( arr1 )
    WRITE (*,*) PRODUCT( arr1, MASK=arr1 /= 0. )
    WRITE (*,*) ANY(arr1 > 0.)
    WRITE (*,*) ALL(arr1 > 0.)
    ```

5.  ```
    INTEGER, DIMENSION(2,3) :: arr2
    arr2 = RESHAPE( [3,0,-3,5,-8,2], [2,3] )
    WHERE ( arr2 > 0 )
       arr2 = 2 * arr2
    END WHERE
    WRITE (*,*) SUM( arr2, MASK=arr2 > 0. )
    ```

6.  ```
    REAL, ALLOCATABLE, DIMENSION(:) :: a, b, c
    a = [ 1., 2., 3. ]
    b = [ 6., 5., 4. ]
    c = a + b
    WRITE (*,*) c
    ```

Determine which of the following sets of Fortran statements are valid. For each set of valid statements, specify what will happen in the program. For each set of invalid statements, specify what is wrong. Assume default typing for any variables that are not explicitly typed.

7.  ```
    REAL, DIMENSION(6) :: dist1
    REAL, DIMENSION(5) :: time
    dist1 = [ 0.00, 0.25, 1.00, 2.25, 4.00, 6.25 ]
    ```

(*continued*)

> *(concluded)*
>
> ```
>       time = [ 0.0, 1.0, 2.0, 3.0, 4.0 ]
>       WHERE ( time > 0. )
>          dist1 = SQRT(dist1)
>       END WHERE
>    8. REAL, DIMENSION(:), ALLOCATABLE :: time
>       time = [ 0.00, 0.25, 1.00, 2.25, 4.00, 6.25, &
>                         9.00, 12.25, 16.00, 20.25]
>       WRITE (*,*) time
> 
>    9. REAL, DIMENSION(:,:), ALLOCATABLE :: test
>       WRITE (*,*) ALLOCATED(test)
> ```

8.7
SUMMARY

In this chapter, we presented 2D (rank 2) and multidimensional arrays (rank *n*). Fortran allows up to seven dimensions in an array.

A multidimensional array is declared using a type declaration statement by naming the array and specifying the maximum (and, optionally, the minimum) subscript values with the DIMENSION attribute. The compiler uses the declared subscript ranges to reserve space in the computer's memory to hold the array. The array elements are allocated in the computer's memory in an order such that the first subscript of the array changes most rapidly and the last subscript of the array changes most slowly.

As with any variable, an array must be initialized before use. An array may be initialized at compile time using array constructors in the type declaration statements or at run time using array constructors, DO loops, or Fortran READs.

Individual array elements may be used freely in a Fortran program just like any other variable. They may appear in assignment statements on either side of the equal sign. Entire arrays and array sections may also be used in calculations and assignment statements as long as the arrays are conformable with each other. Arrays are conformable if they have the same number of dimensions (rank) and the same extent in each dimension. A scalar is also conformable with any array. An operation between two conformable arrays is performed on an element-by-element basis. Scalar values are also conformable with arrays.

Fortran contains three basic types of intrinsic functions: elemental functions, inquiry functions, and transformational functions. Elemental functions are defined for a scalar input and produce a scalar output. When applied to an array, an elemental function produces an output that is the result of applying the operation separately to each element of the input array. Inquiry functions return information about an array, such as its size or bounds. Transformational functions operate on entire arrays and produce an output that is based on all of the elements of the array.

The `WHERE` construct permits an array assignment statement to be performed on only those elements of an array that meet specified criteria. It is useful for preventing errors caused by out-of-range data values in the array.

The `FORALL` construct is a method of applying an operation to many elements of an array without specifying the order in which the operation must be applied to the individual elements.

Arrays may either be static or allocatable. The size of static arrays is declared at compilation time, and they may only be modified by recompiling the program. The size of dynamic arrays may be declared at execution time, allowing a program to adjust its memory requirements to fit the size of the problem to be solved. Allocatable arrays are declared using the `ALLOCATABLE` attribute, are allocated during program execution using the `ALLOCATE` statement, and are deallocated using the `DEALLOCATE` statement. In Fortran 2003 and later versions, allocatable arrays can also be automatically allocated and deallocated using assignment statements, and allocatable arrays without a `SAVE` attribute are automatically deallocated at the end of the execution of a subroutine or function.

8.7.1 Summary of Good Programming Practice

The following guidelines should be adhered to when working with arrays.

1. Use the `RESHAPE` function to change the shape of an array. This is especially useful when used with an array constructor to create array constants of any desired shape.
2. Use implicit `DO` loops to read in or write out rank 2 arrays so that each row of the array appears as a row of the input or output file. This correspondence makes it easier for a programmer to relate the data in the file to the data present within the program.
3. Use `WHERE` constructs to modify and assign array elements when you want to modify and assign only those elements that pass some test.
4. Use allocatable arrays to produce programs that automatically adjust their memory requirements to the size of the problem being solved. Declare allocatable arrays with the `ALLOCATABLE` attribute, allocate memory to them with the `ALLOCATE` statement and deallocate memory with the `DEALLOCATE` statement.
5. Always include the `STAT=` clause in any `ALLOCATE` statement, and always check the returned status, so that a program can be shut down gracefully if there is insufficient memory to allocate the necessary arrays.
6. Always deallocate allocatable arrays with a `DEALLOCATE` statement as soon as you are through using them.
7. When allocatable arrays are used in a Fortran 2003 or later program, they are automatically resized to match the size of the data assigned to them as long as that data has the same rank as the allocatable array.
8. Allocatable arrays declared in a subroutine or function without a SAVE attribute will be automatically deallocated when the subroutine or function exits. No `DEALLOCATE` statements are required.

8.7.2 Summary of Fortran Statements and Constructs

ALLOCATABLE Attribute:

```
type, ALLOCATABLE, DIMENSION(:,[:, ...]) :: array1, ...
```

Examples:

```
REAL, ALLOCATABLE, DIMENSION(:) :: array1
INTEGER, ALLOCATABLE, DIMENSION(:,:,:) :: indices
```

Description:
The ALLOCATABLE attribute declares that the size of an array is dynamic. The size will be specified in an ALLOCATE statement at run time. The type declaration statement must specify the rank of the array but not the extent in each dimension. Each dimension is specified using a colon as a placeholder.

ALLOCATABLE Statement:

```
ALLOCATABLE :: array1, ...
```

Example:

```
ALLOCATABLE :: array1
```

Description:
The ALLOCATABLE statement declares that the size of an array is dynamic. It duplicates the function of the ALLOCATABLE attribute associated with a type declaration statement. **Do not use this statement.** Use the ALLOCATABLE attribute instead.

ALLOCATE Statement:

```
ALLOCATE (array1( [i1:]i2, [j1:]j2, ... ), ... , STAT=status, ERRMSG=msg)
ALLOCATE (array1, SOURCE=expr, STAT=status, ERRMSG=msg)
```

Examples:

```
ALLOCATE (array1(10000), STAT=istat)
ALLOCATE (indices(-10:10,-10:10,5), STAT=istat)
ALLOCATE (array1, SOURCE=array2, STAT=istat, ERRMSG=msg)
```

Description:
The ALLOCATE statement dynamically allocates memory to an array that was previously declared allocatable. In the first form of the ALLOCATE statement, the extent of each dimension is specified in the ALLOCATE statement. The returned status will be zero for successful completion and will be a machine-dependent positive number in the case of an error.

In the second form of the ALLOCATE statement, the size of the array is the same as the size of the source array, and the contents of the array are the same as the source array.

The second form of the ALLOCATE statement, the SOURCE= clause, and the ERRMSG= clause are only supported in Fortran 2003 and later.

DEALLOCATE Statement:

```
DEALLOCATE (array1, ... , STAT=status, ERRMSG=msg)
```

Example:

```
DEALLOCATE (array1, indices, STAT=status)
```

Description:
The DEALLOCATE statement dynamically deallocates the memory that was assigned by an ALLOCATE statement to one or more allocatable arrays. After the statement executes, the memory associated with those arrays is no longer accessible. The returned status will be zero for successful completion and will be a machine-dependent positive number in the case of an error.

FORALL Construct:

```
[name: ] FORALL (index1=triplet1[, ..., logical_expr])
   Assignment Statement(s)
END FORALL [name]
```

Example:

```
FORALL (i=1:3, j=1:3, i > j)
   arr1(i,j) = ABS(i-j) + 3
END FORALL
```

Description:
The FORALL construct permits assignment statements to be executed for those indices that meet the triplet specifications and the optional logical expression, but it does not specify the order in which they are executed. There may be as many indices as desired, and each index will be specified by a subscript triplet. The logical expression is applied as a mask to the indices, and those combinations of specified indices for which the logical expression is TRUE will be executed.

FORALL Statement:

```
FORALL (index1=triplet1[, ..., logical_expr]) Assignment Statement
```

Description:
The FORALL statement is a simplified version of the FORALL construct in which there is only one assignment statement.

WHERE Construct:

```
[name:] WHERE ( mask_expr1 )
   Block 1
ELSEWHERE ( mask_expr2 ) [name]
   Block 2
```

(continued)

Additional Features of Arrays

(*concluded*)

```
        ELSEWHERE [name]
           Block 3
        END WHERE [name]
```

Description:
The WHERE construct permits operations to be applied to the elements of an array that match a given criterion. A different set of operations may be applied to the elements that do not match. Each *mask_expr* must be a logical array of the same shape as the arrays being manipulated within the code blocks. If a given element of the *mask_expr1* is true, then the array assignment statements in Block 1 will be applied to the corresponding element in the arrays being operated on. If an element of the *mask_expr1* is false and the corresponding element of the *mask_expr2* is true, then the array assignment statements in Block 2 will be applied to the corresponding element in the arrays being operated on. If both mask expressions are false, then the array assignment statements in Block 3 will be applied to the corresponding element in the arrays being operated on.

The ELSEWHERE clauses are optional in this construct. There can be as many masked ELSEWHERE clauses are desired, and up to one plain ELSEWHERE.

WHERE Statement:

```
        WHERE ( mask expression ) array_assignment_statement
```

Description:
The WHERE statement is a simplified version of the WHERE construct in which there is only one array assignment statement and no ELSEWHERE clause.

8.7.3 Exercises

8-1. Determine the shape and size of the arrays specified by the following declaration statements, and the valid subscript range for each dimension of each array.

(*a*) `CHARACTER(len=80), DIMENSION(3,60) :: line`

(*b*) `INTEGER, DIMENSION(-10:10,0:20) :: char`

(*c*) `REAL, DIMENSION(-5:5,-5:5,-5:5,-5:5,-5:5) :: range`

8-2. Determine which of the following Fortran program fragments are valid. For each valid statement, specify what will happen in the program. (Assume default typing for any variables that are not explicitly typed within the program fragments.)

(*a*)
```
REAL, DIMENSION(6,4) :: b
...
DO i = 1, 6
   DO j = 1, 4
      temp   = b(i,j)
      b(i,j) = b(j,i)
      b(j,i) = temp
   END DO
END DO
```

(b)
```
INTEGER, DIMENSION(9) :: info
info = [1,-3,0,-5,-9,3,0,1,7]
WHERE ( info > 0 )
   info = -info
ELSEWHERE
   info = -3 * info
END WHERE
WRITE (*,*) info
```

(c)
```
INTEGER, DIMENSION(8) :: info
info = [1,-3,0,-5,-9,3,0,7]
WRITE (*,*) info <= 0
```

(d)
```
REAL, DIMENSION(4,4) :: z = 0.
...
FORALL ( i=1:4, j=1:4 )
   z(i,j) = ABS(i-j)
END FORALL
```

8-3. Given a 5 × 5 array my_array containing the values shown below, determine the shape and contents of each of the following array sections.

$$my\_array = \begin{bmatrix} 1 & 2 & 3 & 4 & 5 \\ 6 & 7 & 8 & 9 & 10 \\ 11 & 12 & 13 & 14 & 15 \\ 16 & 17 & 18 & 19 & 20 \\ 21 & 22 & 23 & 24 & 25 \end{bmatrix}$$

(a) my_array(3,:)
(b) my_array(:,2)
(c) my_array(1:5:2,:)
(d) my_array(:,2:5:2)
(e) my_array(1:5:2,1:5:2)
(f) `INTEGER, DIMENSION(3) :: list = [ 1, 2, 4 ]`
    my_array(:,list)

8-4. What will be the output from each of the WRITE statements in the following program? Why is the output of the two statements different?

```
PROGRAM test_output1
IMPLICIT NONE
INTEGER, DIMENSION(0:1,0:3) :: my_data
INTEGER :: i, j
my_data(0,:) = [ 1, 2, 3, 4 ]
my_data(1,:) = [ 5, 6, 7, 8 ]
!
DO i = 0,1
   WRITE (*,100) (my_data(i,j), j=0,3)
   100 FORMAT (6(1X,I4))
END DO
WRITE (*,100) ((my_data(i,j), j=0,3), i=0,1)
END PROGRAM test_output1
```

Additional Features of Arrays

8-5. An input data file `input1` contains the following values:

```
27   17   10    8    6
11   13  -11   12  -21
-1    0    0    6   14
-16  11   21   26  -16
04   99  -99   17    2
```

Assume that file `input1` has been opened on i/o unit 8, and that array `values` is a 4 × 4 integer array, all of whose elements have been initialized to zero. What will be the contents of array `values` after each of the following READ statements has been executed?

(a)
```
DO i = 1, 4
   READ (8,*) (values(i,j), j = 1, 4)
END DO
```

(b) `READ (8,*) ((values(i,j), j = 1, 4), i=1,4)`

(c)
```
DO i = 1, 4
   READ (8,*) values(i,:)
END DO
```

(d) `READ (8,*) values`

8-6. What will be printed out by the following program?

```
PROGRAM test
IMPLICIT NONE
INTEGER, PARAMETER :: N = 5, M = 10
INTEGER, DIMENSION(N:M,M-N:M+N) :: info

WRITE (*,100) SHAPE(info)
100 FORMAT ('The shape of the array is:          ',2I6)
WRITE (*,110) SIZE(info)
110 FORMAT ('The size of the array is:           ',I6)
WRITE (*,120) LBOUND(info)
120 FORMAT ('The lower bounds of the array are:  ',2I6)
WRITE (*,130) UBOUND(info)
130 FORMAT ('The upper bounds of the array are:  ',2I6)
END PROGRAM test
```

8-7. Assume that `values` is a 10,201-element array containing a list of measurements from a scientific experiment, which has been declared by the statement

`REAL, DIMENSION(-50:50,0:100) :: values`

(a) Create a set of Fortran statements that would count the number of positive values, negative values, and zero values in the array, and write out a message summarizing how many values of each type were found. Do not use any intrinsic functions in your code.

(b) Use the transformational intrinsic function COUNT to create a set of Fortran statements that would count the number of positive values, negative values, and zero values in the array, and write out a message summarizing how many values of each type were found. Compare the complexity of this code to the complexity of the code in (a).

8-8. Write a program that can read in a rank 2 array from an input disk file, and calculate the sums of all the data in each row and each column in the array. The size of the array to read in will be specified by two numbers on the first line in the input file, and the elements in each row of the array will be found on a single line of the input file. Size the program to handle arrays of up to 100 rows and 100 columns. An example of an input data file containing a 2 row × 4 column array is shown below:

```
      2         4
   -24.0    -1121.      812.1     11.1
    35.6     8.1E3      135.23   -17.3
```

Write out the results in the form:

```
Sum of row  1 =
Sum of row  2 =
 ...
Sum of col  1 =
 ...
```

8-9. Test the program that you wrote in Exercise 8-8 by running it on the following array:

$$\text{array} = \begin{bmatrix} 33. & -12. & 16. & 0.5 & -1.9 \\ -6. & -14. & 3.5 & 11. & 2.1 \\ 4.4 & 1.1 & -7.1 & 9.3 & -16.1 \\ 0.3 & 6.2 & -9.9 & -12. & 6.8 \end{bmatrix}$$

8-10. Modify the program you wrote in Exercise 8-8 to use allocatable arrays that are adjusted to match the number of rows and columns in the problem each time the program is run.

8-11. Write a set of Fortran statements that would search a rank 3 array `arr` and limit the maximum value of any array element to be less than or equal to 1000. If any element exceeds 1000, its value should be set to 1000. Assume that array `arr` has dimensions 1000 × 10 × 30. Write two sets of statements, one checking the array elements one at a time using DO loops and one using the WHERE construct. Which of the two approaches is easier?

8-12. Average Annual Temperature As a part of a meteorological experiment, average annual temperature measurements were collected at 36 locations specified by latitude and longitude as shown in the chart below.

	90.0° W long	90.5° W long	91.0° W long	91.5° W long	92.0° W long	92.5° W long
30.0° N lat	68.2	72.1	72.5	74.1	74.4	74.2
30.5° N lat	69.4	71.1	71.9	73.1	73.6	73.7
31.0° N lat	68.9	70.5	70.9	71.5	72.8	73.0
31.5° N lat	68.6	69.9	70.4	70.8	71.5	72.2
32.0° N lat	68.1	69.3	69.8	70.2	70.9	71.2
32.5° N lat	68.3	68.8	69.6	70.0	70.5	70.9

Write a Fortran program that calculates the average annual temperature along each latitude and each longitude included in the experiment. Finally, calculate the average annual temperature for all of the locations in the experiment. Take advantage for intrinsic functions where appropriate to make your program simpler.

Additional Features of Arrays

8-13. Matrix Multiplication Matrix multiplication is only defined for two matrices in which *the number of columns in the first matrix is equal to the number of rows in the second matrix.* If matrix A is an $N \times L$ matrix and matrix B is an $L \times M$ matrix, then the product $C = A \times B$ is an $N \times M$ matrix whose elements are given by the equation

$$c_{ik} = \sum_{j=1}^{L} a_{ij} b_{jk} \qquad (8\text{-}1)$$

For example, if matrices A and B are 2×2 matrices

$$A = \begin{bmatrix} 3.0 & -1.0 \\ 1.0 & 2.0 \end{bmatrix} \quad \text{and} \quad B = \begin{bmatrix} 1.0 & 4.0 \\ 2.0 & -3.0 \end{bmatrix}$$

then the elements of matrix C will be

$$\begin{aligned} c_{11} &= a_{11}b_{11} + a_{12}b_{21} = (3.0)(1.0) + (-1.0)(2.0) = 1.0 \\ c_{12} &= a_{11}b_{12} + a_{12}b_{22} = (3.0)(4.0) + (-1.0)(-3.0) = 15.0 \\ c_{21} &= a_{21}b_{11} + a_{22}b_{21} = (1.0)(1.0) + (2.0)(2.0) = 5.0 \\ c_{22} &= a_{21}b_{12} + a_{22}b_{22} = (1.0)(4.0) + (2.0)(-3.0) = -2.0 \end{aligned}$$

Write a program that can read two matrices of arbitrary size from two input disk files, and multiply them if they are of compatible sizes. If they are of incompatible sizes, an appropriate error message should be printed. The number of rows and columns in each matrix will be specified by two integers on the first line in each file, and the elements in each row of the matrix will be found on a single line of the input file. Use allocatable arrays to hold both the input matrices and the resulting output matrix. Verify your program by creating two input data files containing matrices of the compatible sizes, calculating the resulting values, and checking the answers by hand. Also, verify the proper behavior of the program if it is given that the two matrices are of incompatible sizes.

8-14. Use the program produced in Exercise 8-14 to calculate $C = A \times B$ where:

$$A = \begin{bmatrix} 1. & -5. & 4. & 2. \\ -6. & -4. & 2. & 2. \end{bmatrix}$$

$$B = \begin{bmatrix} 1. & -2. & -1. \\ 2. & 3. & 4. \\ 0. & -1. & 2. \\ 0. & -3. & 1. \end{bmatrix}$$

How many rows and how many columns are present in the resulting matrix C?

8-15. Fortran includes an intrinsic function MATMUL to perform matrix multiplication. Rewrite the program of Exercise 8-13 to use function MATMUL to multiply the matrices together.

8-16. Relative Maxima A point in a rank 2 array is said to be a *relative maximum* if it is higher than any of the eight points surrounding it. For example, the element at position

(2, 2) in the array shown below is a relative maximum, since it is larger than any of the surrounding points.

$$\begin{bmatrix} 11 & 7 & -2 \\ -7 & 14 & 3 \\ 2 & -3 & 5 \end{bmatrix}$$

Write a program to read a matrix A from an input disk file and to scan for all relative maxima within the matrix. The first line in the disk file should contain the number of rows and the number of columns in the matrix and then the next lines should contain the values in the matrix, with all of the values in a given row on a single line of the input disk file. (Be sure to use the proper form of implied DO statements to read in the data correctly.) Use allocatable arrays. The program should only consider interior points within the matrix, since any point along an edge of the matrix cannot be completely surrounded by points lower than itself. Test your program by finding all of the relative maxima in the following matrix, which can be found in file FINDPEAK:

$$A = \begin{bmatrix} 2. & -1. & -2. & 1. & 3. & -5. & 2. & 1. \\ -2. & 0. & -2.5 & 5. & -2. & 2. & 1. & 0. \\ -3. & -3. & -3. & 3. & 0. & 0. & -1. & -2. \\ -4.5 & -4. & -7. & 6. & 1. & -3. & 0. & 5. \\ -3.5 & -3. & -5. & 0. & 4. & 17. & 11. & 5. \\ -9. & -6. & -5. & -3. & 1. & 2. & 0. & 0.5 \\ -7. & -4. & -5. & -3. & 2. & 4. & 3. & -1. \\ -6. & -5. & -5. & -2. & 0. & 1. & 2. & 5. \end{bmatrix}$$

8-17. Temperature Distribution on a Metallic Plate Under steady-state conditions, the temperature at any point on the surface of a metallic plate will be the average of the temperatures of all points surrounding it. This fact can be used in an iterative procedure to calculate the temperature distribution at all points on the plate.

Figure 8-9 shows a square plate divided in 100 squares or nodes by a grid. The temperatures of the nodes form a 2D array T. The temperature in all nodes at the edges of the plate is constrained to be 20° C by a cooling system, and the temperature of the node (3, 8) is fixed at 100° C by exposure to boiling water.

A new estimate of the temperature in any given node can be calculated from the average of the temperatures in all segments surrounding it:

$$T_{ij,\text{new}} = \frac{1}{4}(T_{i+1,j} + T_{i-1,j} + T_{i,j+1} + T_{i,j-1}) \tag{8-2}$$

To determine the temperature distribution on the surface of a plate, an initial assumption must be made about the temperatures in each node. Then Equation 8-2 is applied to each node whose temperature is not fixed to calculate a new estimate of the temperature in that node. These updated temperature estimates are used to calculate newer estimates, and the process is repeated until the new temperature estimates in each node differ from the old ones by only a small amount. At that point, a steady-state solution has been found.

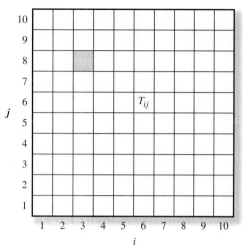

FIGURE 8-9
A metallic plate divided into 100 small segments.

Write a program to calculate the steady-state temperature distribution throughout the plate, making an initial assumption that all interior segments are at a temperature of 50° C. Remember that all outside segments are fixed at a temperature of 20° C and segment (3, 8) is fixed at a temperature of 100° C. The program should apply Equation 8-1 iteratively until the maximum temperature change between iterations in any node is less than 0.01 degree. What will the steady-state temperature of segment (5, 5) be?

9

Additional Features of Procedures

OBJECTIVES

- Learn how to use multidimensional arrays in Fortran procedures.
- Understand how and when to use the SAVE attribute or statement.
- Understand the difference between allocatable and automatic arrays, and when to use each in a procedure.
- Understand pure and elemental procedures.
- Learn how to declare and use internal subroutines and functions.
- Learn how to separate procedure interfaces and executable code using SUBMODULEs.

In Chapter 7, we learned the basics of using Fortran subroutines, function subprograms, and modules. This chapter describes more advanced features of procedures, including multidimensional arrays in procedures and the use of internal procedures.

9.1 PASSING MULTIDIMENSIONAL ARRAYS TO SUBROUTINES AND FUNCTIONS

Multidimensional arrays can be passed to subroutines or functions in a manner similar to 1D arrays. However, the subroutine or function will need to know *both the number of dimensions and the extent of each dimension* in order to use the array properly. There are three possible ways to pass this information to the subprogram.

9.1.1 Explicit Shape Dummy Arrays

The first approach is to use **explicit shape dummy arrays.** In this case, we pass the array and the extent of each dimension of the array to the subroutine. The extent values are used to declare the size of the array in the subroutine, and thus the subroutine knows all about the array. An example subroutine using explicit shape dummy arrays is shown below.

```
SUBROUTINE process1 (data1, data2, n, m)
INTEGER, INTENT(IN) :: n, m
REAL, INTENT(IN),  DIMENSION(n,m) :: data1   ! Explicit shape
REAL, INTENT(OUT), DIMENSION(n,m) :: data2   ! Explicit shape

data2 = 3. * data1

END SUBROUTINE process1
```

When explicit-shape dummy arrays are used, the size and shape of each dummy array in the subprogram are known to the compiler. Since the size and shape of each array are known, it is possible to use array operations and array sections with the dummy arrays.

9.1.2 Assumed-Shape Dummy Arrays

The second approach is to declare all dummy arrays in a subroutine as **assumed-shape dummy arrays.** Assumed-shape arrays are declared using colons as placeholders for each subscript of the array. These arrays work only if the subroutine or function has an explicit interface, so that the calling program knows everything about the subroutine interface. This is normally accomplished by placing the subprogram into a module, and then USEing the module in the calling program.

Whole array operations, array sections, and array intrinsic functions can all be used with assumed-shape dummy arrays, because the compiler can determine the size and shape of each array from the information in the interface. If needed, the actual size and extent of an assumed-shape array can be determined by using the array inquiry functions in Table 8-1. However, the upper and lower bounds of each dimension cannot be determined, since only the *shape* of the actual array but not the *bounds* are passed to the procedure. If the actual bounds are needed for some reason in a particular procedure, then an explicit-shape dummy array must be used.

Assumed-shape dummy arrays are generally better than explicit-shape dummy arrays in that we don't have to pass every bound from the calling program unit to a procedure. However, assumed-shape arrays only work if a procedure has an explicit interface.

An example subroutine using assumed-shape dummy arrays is shown below:

```
MODULE test_module
CONTAINS

   SUBROUTINE process2 ( data1, data2 )
   REAL, INTENT(IN),  DIMENSION(:,:) :: data1   ! Explicit shape
   REAL, INTENT(OUT), DIMENSION(:,:) :: data2   ! Explicit shape

   data2 = 3. * data1

   END SUBROUTINE process2

END MODULE test_module
```

9.1.3 Assumed-Size Dummy Arrays

The third (and oldest) approach is to use an **assumed-size dummy array.** These are arrays in which the length of one of the array dimensions is an asterisk. Assumed-size dummy arrays are a holdover from earlier versions of Fortran. *They should never be used in any new programs*, so we will not discuss them here.

Good Programming Practice
Use either assumed-shape arrays or explicit-shape arrays as dummy array arguments in procedures. If assumed-shape arrays are used, an explicit interface is required. Whole array operations, array sections, and array intrinsic functions may be used with the dummy array arguments in either case. *Never use assumed-size arrays in any new program.*

EXAMPLE 9-1

Gauss-Jordan Elimination:

Many important problems in science and engineering require the solution of a system of N simultaneous linear equations in N unknowns. Some of these problems require the solution of small systems of equations, say 3×3 or 4×4. Such problems are relatively easy to solve. Other problems might require the solution of really large sets of simultaneous equations, like 1000 equations in 1000 unknowns. Those problems are *much* harder to solve and the solution requires a variety of special iterative techniques. A whole branch of the science of numerical methods is devoted to different ways to solve systems of simultaneous linear equations.

We will now develop a subroutine to solve a system of simultaneous linear equations using the straightforward approach known as Gauss-Jordan elimination. The subroutine that we develop should work fine for systems of up to about 20 equations in 20 unknowns.

Gauss-Jordan elimination depends on the fact that you can multiply one equation in a system of equations by a constant and add it to another equation, and the new system of equations will still be equivalent to the original one. In fact, it works in exactly the same way that we solve systems of simultaneous equations by hand.

To understand the technique, consider the 3×3 system of equations shown below.

$$1.0x_1 + 1.0x_2 + 1.0x_3 = 1.0$$
$$2.0x_1 + 1.0x_2 + 1.0x_3 = 2.0 \qquad (9\text{-}1)$$
$$1.0x_1 + 3.0x_2 + 2.0x_3 = 4.0$$

We would like to manipulate this set of equations by multiplying one of the equations by a constant and adding it to another one until we eventually wind up with a set of equations of the form

Additional Features of Procedures

$$1.0x_1 + 0.0x_2 + 0.0x_3 = b_1$$
$$0.0x_1 + 1.0x_2 + 0.0x_3 = b_2 \quad (9\text{-}2)$$
$$0.0x_1 + 0.0x_2 + 1.0x_3 = b_3$$

When we get to this form, the solution to the system will be obvious: $x_1 = b_1$, $x_2 = b_2$, and $x_3 = b_3$.

To get from Equations (9-1) to (9-2) *equation reference goes here*, we must go through three steps:

1. Eliminate all coefficients of x_1 except in the first equation.
2. Eliminate all coefficients of x_2 except in the second equation.
3. Eliminate all coefficients of x_3 except in the third equation.

First, we will eliminate all coefficients of x_1 except that in the first equation. If we multiply the first equation by -2 and add it to the second equation, and multiply the first equation by -1 and add it to the third equation, the results are:

$$1.0x_1 + 1.0x_2 + 1.0x_3 = 1.0$$
$$0.0x_1 - 1.0x_2 - 1.0x_3 = 0.0 \quad (9\text{-}3)$$
$$0.0x_1 + 2.0x_2 + 1.0x_3 = 3.0$$

Next, we will eliminate all coefficients of x_2 except in the second equation. If we add the second equation as it is to the first equation, and multiply the second equation by 2 and add it to the third equation, the results are:

$$1.0x_1 + 0.0x_2 + 0.0x_3 = 1.0$$
$$0.0x_1 - 1.0x_2 - 1.0x_3 = 0.0 \quad (9\text{-}4)$$
$$0.0x_1 + 0.0x_2 - 1.0x_3 = 3.0$$

Finally, we will eliminate all coefficients of x_3 except in the third equation. In this case, there is no coefficient of x_3 in the first equation, so we don't have to do anything there. If we multiply the third equation by -1 and add it to the second equation, the results are:

$$1.0x_1 + 0.0x_2 + 0.0x_3 = 1.0$$
$$0.0x_1 - 1.0x_2 + 0.0x_3 = -3.0 \quad (9\text{-}5)$$
$$0.0x_1 + 0.0x_2 - 1.0x_3 = 3.0$$

The last step is almost trivial. If we divide the equation 1 by the coefficient of x_1, equation 2 by the coefficient of x_2, and equation 3 by the coefficient of x_3, then the solution to the equations will appear on the right hand side of the equations.

$$1.0x_1 + 0.0x_2 + 0.0x_3 = 1.0$$
$$0.0x_1 + 1.0x_2 + 0.0x_3 = 3.0 \quad (9\text{-}6)$$
$$0.0x_1 + 0.0x_2 + 1.0x_3 = -3.0$$

The final answer is $x_1 = 1$, $x_2 = 3$, and $x_3 = -3$!

Sometimes the technique shown above does not produce a solution. This happens when the set of equations being solved are not all *independent*. For example, consider the following 2×2 system of simultaneous equations:

$$2.0x_1 + 3.0x_2 = 4.0$$
$$4.0x_1 + 6.0x_2 = 8.0 \quad (9\text{-}7)$$

If equation 1 is multiplied by -2 and added to equation 1, we get

$$2.0x_1 + 3.0x_2 = 4.0$$
$$0.0x_1 + 0.0x_2 = 0.0 \quad (9\text{-}8)$$

There is no way to solve this system for a unique solution, since there are infinitely many values of x_1 and x_2 that satisfy Equations (9-8). These conditions can be recognized by the fact that the coefficient of x_2 in the second equation is 0. The solution to this system of equations is said to be nonunique. Our computer program will have to test for problems like this and report them with an error code.

SOLUTION

We will now write a subroutine to solve a system of N simultaneous equations in N unknowns. The computer program will work in exactly the manner shown above, except that at each step in the process, we will reorder the equations. In the first step, we will reorder the N equations such that the first equation is the one with the largest coefficient (absolute value) of the first variable. In the second step, we will reorder second equation through the Nth equation such that the second equation is the one with the largest coefficient (absolute value) of the second variable. This process is repeated for each step in the solution. Reordering the equations is important, because it reduces round-off errors in large systems of equations and also avoids divide-by-zero errors. (This reordering of equations is called the *maximum pivot* technique in the literature of numerical methods.)

1. **State the problem.**

   Write a subroutine to solve a system of N simultaneous equations in N unknowns using Gauss-Jordan elimination and the maximum pivot technique to avoid round-off errors. The subroutine must be able to detect singular sets of equations and set an error flag if they occur. Use explicit-shape dummy arrays in the subroutine.

2. **Define the inputs and outputs.**

   The input to the subroutine consists of an `ndim` x `ndim` matrix `a`, containing an `n` x `n` set of coefficients for the simultaneous equations and an `ndim` vector `b`, with the contents of the right-hand sides of the equations. The size of the matrix `ndim` must be greater than or equal to the size of the set of simultaneous equations `n`. Since the subroutine is to have explicit-shape dummy arrays, we will also have to pass `ndim` to the subroutine and use it to declare the dummy array sizes. The outputs from the subroutine are the solutions to the set of equations (in vector `b`) and an error flag. Note that the matrix of coefficients `a` will be destroyed during the solution process.

Additional Features of Procedures

3. **Describe the algorithm.**
The pseudocode for this subroutine is:

```
DO for irow = 1 to n

   ! Find peak pivot for column irow in rows i to n
   ipeak ← irow
   DO for jrow = irow+1 to n
      IF |a(jrow,irow)| > |a(ipeak,irow)| then
         ipeak ← jrow
      END of IF
   END of DO

   ! Check for singular equations
   IF |a(ipeak,irow)| < epsilon THEN
      Equations are singular; set error code & exit
   END of IF

   ! Otherwise, if ipeak /= irow, swap equations irow & ipeak
   IF ipeak <> irow
      DO for kcol = 1 to n
         temp ← a(ipeak,kcol)
         a(ipeak,kcol) ← a(irow,kcol)
         a(irow,kcol) ← temp
      END of DO
      temp ← b(ipeak)
      b(ipeak) ← b(irow)
      b(irow) ← temp
   END of IF

   ! Multiply equation irow by -a(jrow,irow)/a(irow,irow),
   ! and add it to Eqn jrow
   DO for jrow = 1 to n except for irow
      factor ← -a(jrow,irow)/a(irow,irow)
      DO for kcol = 1 to n
         a(jrow,kcol) ← a(irow,kcol) * factor + a(jrow,kcol)
      END of DO
      b(jrow) ← b(irow) * factor + b(jrow)
   END of DO
END of DO

! End of main loop over all equations. All off-diagonal
! terms are now zero. To get the final answer, we must
! divide each equation by the coefficient of its on-diagonal
! term.
DO for irow = 1 to n
   b(irow) ← b(irow) / a(irow,irow)
   a(irow,irow) ← 1.
END of DO
```

4. **Turn the algorithm into Fortran statements.**
The resulting Fortran subroutine is shown in Figure 9-1. Note that the sizes of arrays a and b are passed explicitly to the subroutine as a(ndim,ndim) and b(ndim). By doing so, we can use the compiler's bounds checker while we are debugging the

subroutine. Note also that the subroutine's large outer loops and `IF` structures are all named to make it easier for us to understand and keep track of them.

FIGURE 9-1
Subroutine `simul`.

```fortran
SUBROUTINE simul ( a, b, ndim, n, error )
!
!  Purpose:
!    Subroutine to solve a set of n linear equations in n
!    unknowns using Gaussian elimination and the maximum
!    pivot technique.
!
!  Record of revisions:
!      Date        Programmer          Description of change
!      ====        ==========          =====================
!    11/25/15     S. J. Chapman        Original code
!
IMPLICIT NONE

! Data dictionary: declare calling parameter types & definitions
INTEGER, INTENT(IN) :: ndim          ! Dimension of arrays a and b
REAL, INTENT(INOUT), DIMENSION(ndim,ndim) :: a
                                     ! Array of coefficients (n x n).
                                     ! This array is of size ndim x
                                     ! ndim, but only n x n of the
                                     ! coefficients are being used.
                                     ! The declared dimension ndim
                                     ! must be passed to the sub, or
                                     ! it won't be able to interpret
                                     ! subscripts correctly. (This
                                     ! array is destroyed during
                                     ! processing.)
REAL, INTENT(INOUT), DIMENSION(ndim) :: b
                                     ! Input: Right-hand side of eqns.
                                     ! Output: Solution vector.
INTEGER, INTENT(IN) :: n             ! Number of equations to solve.
INTEGER, INTENT(OUT) :: error        ! Error flag:
                                     !   0 -- No error
                                     !   1 -- Singular equations

! Data dictionary: declare constants
REAL, PARAMETER :: EPSILON = 1.0E-6  ! A "small" number for comparison
                                     ! when determining singular eqns

! Data dictionary: declare local variable types & definitions
REAL :: factor                       ! Factor to multiply eqn irow by
                                     ! before adding to eqn jrow
INTEGER :: irow                      ! Number of the equation currently
                                     ! being processed
INTEGER :: ipeak                     ! Pointer to equation containing
                                     ! maximum pivot value
INTEGER :: jrow                      ! Number of the equation compared
                                     ! to the current equation
```

(continued)

(*continued*)

```
      INTEGER :: kcol                        ! Index over all columns of eqn
      REAL :: temp                           ! Scratch value
      ! Process n times to get all equations...
      mainloop: DO irow = 1, n

         ! Find peak pivot for column irow in rows irow to n
         ipeak = irow
         max_pivot: DO jrow = irow+1, n
            IF (ABS(a(jrow,irow)) > ABS(a(ipeak,irow))) THEN
               ipeak = jrow
            END IF
         END DO max_pivot

         ! Check for singular equations.
         singular: IF ( ABS(a(ipeak,irow)) < EPSILON ) THEN
            error = 1
            RETURN
         END IF singular

         ! Otherwise, if ipeak /= irow, swap equations irow & ipeak
         swap_eqn: IF ( ipeak /= irow ) THEN
            DO kcol = 1, n
               temp            = a(ipeak,kcol)
               a(ipeak,kcol)   = a(irow,kcol)
               a(irow,kcol)    = temp
            END DO
            temp      = b(ipeak)
            b(ipeak)  = b(irow)
            b(irow)   = temp
         END IF swap_eqn

         ! Multiply equation irow by -a(jrow,irow)/a(irow,irow),
         ! and add it to Eqn jrow (for all eqns except irow itself).
         eliminate: DO jrow = 1, n
            IF ( jrow /= irow ) THEN
               factor = -a(jrow,irow)/a(irow,irow)
               DO kcol = 1, n
                  a(jrow,kcol) = a(irow,kcol)*factor + a(jrow,kcol)
               END DO
               b(jrow) = b(irow)*factor + b(jrow)
            END IF
         END DO eliminate
      END DO mainloop

      ! End of main loop over all equations. All off-diagonal
      ! terms are now zero. To get the final answer, we must
      ! divide each equation by the coefficient of its on-diagonal
      ! term.
      divide: DO irow = 1, n
         b(irow)        = b(irow) / a(irow,irow)
         a(irow,irow)   = 1.
      END DO divide
```

(*continued*)

(*concluded*)

```
! Set error flag to 0 and return.
error = 0
END SUBROUTINE simul
```

5. **Test the resulting Fortran programs.**

To test this subroutine, it is necessary to write a driver program. The driver program will open an input data file to read the equations to be solved. The first line of the file will contain the number of equations n in the system, and each of the next n lines will contain the coefficients of one of the equations. To show that the simultaneous equation subroutine is working correctly, we will display the contents of arrays a and b both before and after the call to simul.

The test driver program for subroutine simul is shown in Figure 9-2.

FIGURE 9-2
Test driver routine for subroutine simul.

```
PROGRAM test_simul
!
!  Purpose:
!    To test subroutine simul, which solves a set of N linear
!    equations in N unknowns.
!
!  Record of revisions:
!      Date         Programmer           Description of change
!      ====         ==========           =====================
!    11/25/15      S. J. Chapman         Original code
!
IMPLICIT NONE

! Data dictionary: declare constants
INTEGER, PARAMETER :: MAX_SIZE = 10      ! Max number of eqns

! Data dictionary: declare local variable types & definitions
REAL, DIMENSION(MAX_SIZE,MAX_SIZE) :: a
                                         ! Array of coefficients (n x n).
                                         ! This array is of size ndim x
                                         ! ndim, but only n x n of the
                                         ! coefficients are being used.
                                         ! The declared dimension ndim
                                         ! must be passed to the sub, or
                                         ! it won't be able to interpret
                                         ! subscripts correctly. (This
                                         ! array is destroyed during
                                         ! processing.)
REAL, DIMENSION(MAX_SIZE) :: b           ! Input: Right-hand side of eqns.
                                         ! Output: Solution vector.
INTEGER :: error                         ! Error flag:
                                         !   0 -- No error
                                         !   1 -- Singular equations
```

(*continued*)

Additional Features of Procedures

(*continued*)

```fortran
CHARACTER(len=20) :: file_name    ! Name of file with eqns
INTEGER :: i                       ! Loop index
INTEGER :: j                       ! Loop index
CHARACTER(len=80) :: msg           ! Error message
INTEGER :: n                       ! Number of simul eqns (<= MAX_SIZE)
INTEGER :: istat                   ! I/O status

! Get the name of the disk file containing the equations.
WRITE (*,"('Enter the file name containing the eqns: ')")
READ (*,'(A20)') file_name

! Open input data file. Status is OLD because the input data must
! already exist.
OPEN ( UNIT=1, FILE=file_name, STATUS='OLD', ACTION='READ', &
       IOSTAT=istat, IOMSG=msg )

! Was the OPEN successful?
fileopen: IF ( istat == 0 ) THEN
   ! The file was opened successfully, so read the number of
   ! equations in the system.
   READ (1,*) n

   ! If the number of equations is <= MAX_SIZE, read them in
   ! and process them.
   size_ok: IF ( n <= MAX_SIZE ) THEN
      DO i = 1, n
         READ (1,*) (a(i,j), j=1,n), b(i)
      END DO

      ! Display coefficients.
      WRITE (*,"(/,'Coefficients before call:')")
      DO i = 1, n
         WRITE (*,"(1X,7F11.4)") (a(i,j), j=1,n), b(i)
      END DO

      ! Solve equations.
      CALL simul (a, b, MAX_SIZE, n, error )

      ! Check for error.
      error_check: IF ( error /= 0 ) THEN

         WRITE (*,1010)
         1010 FORMAT (/'Zero pivot encountered!', &
                    //'There is no unique solution to this system.')

      ELSE error_check

         ! No errors. Display coefficients.
         WRITE (*,"(/,'Coefficients after call:')")
         DO  i = 1, n
             WRITE (*,"(1X,7F11.4)") (a(i,j), j=1,n), b(i)
         END DO
```

(*continued*)

(concluded)

```
        ! Write final answer.
        WRITE (*,"(/,'The solutions are:')")
        DO i = 1, n
           WRITE (*,"(2X,'X(',I2,') = ',F16.6)") i, b(i)
        END DO

      END IF error_check
   END IF size_ok
ELSE fileopen

   ! Else file open failed. Tell user.
   WRITE (*,1020) msg
   1020 FORMAT ('File open failed: ', A)

END IF fileopen
END PROGRAM test_simul
```

To test the subroutine, we need to call it with two different data sets. One of them should have a unique solution and the other one should be singular. We will test the system with two sets of equations. The original equations that we solved by hand will be placed in file `inputs1`

$$1.0x_1 + 1.0x_2 + 1.0x_3 = 1.0$$
$$2.0x_1 + 1.0x_2 + 1.0x_3 = 2.0 \quad (9\text{-}1)$$
$$1.0x_1 + 3.0x_2 + 2.0x_3 = 4.0$$

and the following set of equations will be placed in file `inputs2`.

$$1.0x_1 + 1.0x_2 + 1.0x_3 = 1.0$$
$$2.0x_1 + 6.0x_2 + 4.0x_3 = 8.0 \quad (9\text{-}9)$$
$$1.0x_1 + 3.0x_2 + 2.0x_3 = 4.0$$

The second equation of this set is a multiple of the third equation, so the second set of equations is singular. When we run program `test_simul` with these data sets, the results are:

```
C:\book\fortran\chap9>test_simul
Enter the file name containing the eqns:
inputs1

Coefficients before call:
    1.0000    1.0000    1.0000    1.0000
    2.0000    1.0000    1.0000    2.0000
    1.0000    3.0000    2.0000    4.0000

Coefficients after call:
    1.0000     .0000     .0000    1.0000
     .0000    1.0000     .0000    3.0000
     .0000     .0000    1.0000   -3.0000
```

Additional Features of Procedures

```
The solutions are:
  X( 1) =         1.000000
  X( 2) =         3.000000
  X( 3) =        -3.000000

C:\book\fortran\chap9>test_simul
Enter the file name containing the eqns:
inputs2
Coefficients before call:
    1.0000      1.0000      1.0000      1.0000
    2.0000      6.0000      4.0000      8.0000
    1.0000      3.0000      2.0000      4.0000
Zero pivot encountered!
There is no unique solution to this system.
```

The subroutine appears to be working correctly for both unique and singular sets of simultaneous equations.

Note that subroutine `simul` uses explicit-shape arrays. You will be asked to modify this subroutine to use assumed-shape dummy arrays in an end-of-chapter exercise.

EXAMPLE 9-2 *Using Assumed-Shape Dummy Arrays:*

A simple procedure using an assumed-shape dummy array is shown in Figure 9-3. This procedure declares an assumed-shape dummy array "array," and then determines its size, shape, and bounds using array intrinsic functions. Note that the subroutine is contained in a module, so it has an explicit interface.

FIGURE 9-3
Subroutine to illustrate the use of assumed-shape arrays.

```
MODULE test_module
! Purpose:
!   To illustrate the use of assumed-shape arrays.
!
CONTAINS
   SUBROUTINE test_array(array)
   IMPLICIT NONE
   REAL, DIMENSION(:,:) :: array          ! Assumed-shape array
   INTEGER :: i1, i2                      ! Bounds of first dimension
   INTEGER :: j1, j2                      ! Bounds of second dimension
```

(*continued*)

(*concluded*)

```fortran
      ! Get details about array.
      i1 = LBOUND(array,1)
      i2 = UBOUND(array,1)
      j1 = LBOUND(array,2)
      j2 = UBOUND(array,2)
      WRITE (*,100) i1, i2, j1, j2
  100 FORMAT ('The bounds are: (',I2,':',I2,',',I2,':',I2,')')
      WRITE (*,110) SHAPE(array)
  110 FORMAT ('The shape is:      ',2I4)
      WRITE (*,120) SIZE(array)
  120 FORMAT ('The size is:       ',I4)
      END SUBROUTINE test_array

END MODULE test_module

PROGRAM assumed_shape
!
!  Purpose:
!     To illustrate the use of assumed-shape arrays.
!
USE test_module
IMPLICIT NONE

! Declare local variables
REAL, DIMENSION(-5:5,-5:5) :: a = 0.   ! Array a
REAL, DIMENSION(10,2) :: b = 1.        ! Array b

! Call test_array with array a.
WRITE (*,*) 'Calling test_array with array a:'
CALL test_array(a)

! Call test_array with array b.
WRITE (*,*) 'Calling test_array with array b:'
CALL test_array(b)

END PROGRAM assumed_shape
```

When program `assumed_shape` is executed, the results are:

```
C:\book\fortran\chap9>assumed_shape
Calling test_array with array a:
The bounds are: ( 1:11, 1:11)
The shape is:        11  11
The size is:        121
Calling test_array with array b:
The bounds are: ( 1:10, 1: 2)
The shape is:        10   2
The size is:         20
```

Note that the subroutine has complete information about the rank, shape, and size of each array passed to it, but not about the bounds used for the array in the calling program.

9.2
THE SAVE ATTRIBUTE AND STATEMENT

According to the Fortran standard, *the values of all the local variables and arrays in a procedure become undefined whenever we exit the procedure.* Any local allocatable arrays will also be deleted when we exit the procedure. The next time that the procedure is invoked, the values of the local variables and arrays may or may not be the same as they were the last time we left it, depending on the behavior of the particular compiler being used. If we write a procedure that depends on having its local variables undisturbed between calls, the procedure will work fine on some computers and fail miserably on other ones!

Fortran provides a way to guarantee that local variables and arrays are saved unchanged between calls to a procedure: the SAVE attribute. The SAVE attribute appears in a type declaration statement like any other attribute. *Any local variables declared with the SAVE attribute will be saved unchanged between calls to the procedure.* For example, a local variable sums could be declared with the SAVE attribute as

```
REAL, SAVE :: sums
```

In addition, *any local variable that is initialized in a type declaration statement is automatically saved.* The SAVE attribute may be specified explicitly, if desired, but the value of the variable will be saved whether or not the attribute is explicitly included. Thus, the following two variables are both saved between invocations of the procedure containing them.

```
REAL, SAVE :: sum_x = 0.
REAL :: sum_x2 = 0.
```

Local allocatable arrays with a SAVE attribute will not be deallocated and will be saved unchanged between invocations of the procedure containing them.

Fortran also includes a SAVE statement. It is a nonexecutable statement that goes into the declaration portion of the procedure along with the type declaration statements. Any local variables listed in the SAVE statement will be saved unchanged between calls to the procedure. If no variables are listed in the SAVE statement, then *all* of the local variables will be saved unchanged. The format of the SAVE statement is

```
SAVE :: var1, var2, ...
```

or simply

```
SAVE
```

The SAVE attribute may not appear associated with dummy arguments or with data items declared with the PARAMETER attribute. Similarly, neither of these items may appear in a SAVE statement.

The SAVE statement should appear in any module used to share data, to ensure that the values in the module remain intact between calls to procedures that USE the module. Figure 7-8 showed a sample module that included a SAVE statement.

Good Programming Practice

If a procedure requires that the value of a local variable not change between successive invocations, include the SAVE attribute in the variable's type declaration statement or include the variable in a SAVE statement, or initialize the variable in its type declaration statement. If you do not do so, the subroutine will work correctly on some processors but will fail on other ones.

EXAMPLE 9-3   *Running Averages:*

It is sometimes desirable to keep running statistics on a data set as the values are being entered. The subroutine running_average shown in Figure 9-4 accumulates running averages and standard deviations for use in problems where we would like to keep statistics on data as it is coming in to the program. As each new data value is added, the running averages and standard deviations of all data up to that point are updated. The running sums used to derive the statistics are reset when the subroutine is called with the logical argument reset set to true. Note that the sums n, sum_x, and sum_x2 are being accumulated in local variables in this subroutine. To ensure that they remain unchanged between subroutine calls, *those local variables must appear in a* SAVE *statement or with a* SAVE *attribute.*

FIGURE 9-4
A subroutine to calculate the running mean and standard deviation of an input data set.

```
SUBROUTINE running_average ( x, ave, std_dev, nvals, reset )
!
!  Purpose:
!    To calculate the running average, standard deviation,
!    and number of data points as data values x are received.
!    If "reset" is .TRUE., clear running sums and exit.
!
!  Record of revisions:
!      Date        Programmer          Description of change
!      ====        ==========          =====================
!    11/25/15      S. J. Chapman       Original code
!
IMPLICIT NONE

! Data dictionary: declare calling parameter types & definitions
REAL, INTENT(IN) :: x              ! Input data value.
REAL, INTENT(OUT) :: ave           ! Running average.
REAL, INTENT(OUT) :: std_dev       ! Running standard deviation.
INTEGER, INTENT(OUT) :: nvals      ! Current number of points.
LOGICAL, INTENT(IN) :: reset       ! Reset flag: clear sums if true

! Data dictionary: declare local variable types & definitions
INTEGER, SAVE :: n                 ! Number of input values.
```

(continued)

(*concluded*)

```
REAL, SAVE :: sum_x            ! Sum of input values.
REAL, SAVE :: sum_x2           ! Sum of input values squared.

! If the reset flag is set, clear the running sums at this time.
calc_sums: IF ( reset ) THEN

   n       = 0
   sum_x   = 0.
   sum_x2  = 0.
   ave     = 0.
   std_dev = 0.
   nvals   = 0

ELSE

   ! Accumulate sums.
   n      = n + 1
   sum_x  = sum_x + x
   sum_x2 = sum_x2 + x**2

   ! Calculate average.
   ave = sum_x / REAL(n)

   ! Calculate standard deviation.
   IF ( n >= 2 ) THEN
      std_dev = SQRT( (REAL(n) * sum_x2 - sum_x**2) &
              / (REAL(n) * REAL(n-1)) )
   ELSE
      std_dev = 0.
   END IF

   ! Number of data points.
   nvals = n

END IF calc_sums

END SUBROUTINE running_average
```

A test driver for this subroutine is shown in Figure 9-5.

FIGURE 9-5
A test driver program to test subroutine `running_average`.

```
PROGRAM test_running_average
!
!  Purpose:
!     To test running average subroutine.
!
IMPLICIT NONE

! Declare variables:
INTEGER :: istat               ! I/O status
REAL :: ave                    ! Average
```

(*continued*)

(*concluded*)

```
REAL :: std_dev                       ! Standard deviation
CHARACTER(len=80) :: msg              ! Error message
INTEGER :: nvals                      ! Number of values
REAL :: x                             ! Input data value
CHARACTER(len=20) :: file_name        ! Input data file name

! Clear the running sums.
CALL running_average ( 0., ave, std_dev, nvals, .TRUE. )

! Get the name of the file containing the input data.
WRITE (*,*) 'Enter the file name containing the data: '
READ (*,'(A20)') file_name

! Open input data file. Status is OLD because the input data must
! already exist.
OPEN ( UNIT=21, FILE=file_name, STATUS='OLD', ACTION='READ', &
       IOSTAT=istat, IOMSG=msg )

! Was the OPEN successful?
openok: IF ( istat == 0 ) THEN

   ! The file was opened successfully, so read the data to calculate
   ! running averages for.
   calc: DO
      READ (21,*,IOSTAT=istat) x      ! Get next value
      IF ( istat /= 0 ) EXIT          ! EXIT if not valid.

      ! Get running average & standard deviation
      CALL running_average ( x, ave, std_dev, nvals, .FALSE. )

      ! Now write out the running statistics.
      WRITE (*,1020) 'Value = ', x, ' Ave = ', ave, &
                     ' Std_dev = ', std_dev, &
                     ' Nvals = ', nvals
      1020 FORMAT (3(A,F10.4),A,I6)
   END DO calc

ELSE openok

   ! Else file open failed. Tell user.
   WRITE (*,1030) msg
   1030 FORMAT ('File open failed: ', A)

END IF openok

END PROGRAM test_running_average
```

To test this subroutine, we will calculate running statistics by hand for a set of five numbers and compare the hand calculations to the results from the computer program. Recall that the average and standard deviation are defined as

$$\bar{x} = \frac{1}{N}\sum_{i=1}^{N} x_i \tag{4-1}$$

and

Additional Features of Procedures

$$s = \sqrt{\frac{N\sum_{i=1}^{N} x_i^2 - \left(\sum_{i=1}^{N} x_i\right)^2}{N(N-1)}} \quad (4\text{-}2)$$

where x_i is sample i out of N samples. If the five values are:

$$3., \quad 2., \quad 3., \quad 4., \quad 2.8$$

then the running statistics calculated by hand would be:

Value	n	Σx	Σx^2	Average	Std_dev
3.0	1	3.0	9.0	3.00	0.000
2.0	2	5.0	13.0	2.50	0.707
3.0	3	8.0	22.0	2.67	0.577
4.0	4	12.0	38.0	3.00	0.816
2.8	5	14.8	45.84	2.96	0.713

The output of the test program for the same data set is:

```
C:\book\fortran\chap9>test_running_average
Enter the file name containing the data:
input6
Value =    3.0000  Ave =    3.0000  Std_dev =   0.0000  Nvals =   1
Value =    2.0000  Ave =    2.5000  Std_dev =   0.7071  Nvals =   2
Value =    3.0000  Ave =    2.6667  Std_dev =   0.5774  Nvals =   3
Value =    4.0000  Ave =    3.0000  Std_dev =   0.8165  Nvals =   4
Value =    2.8000  Ave =    2.9600  Std_dev =   0.7127  Nvals =   5
```

so the results check to the accuracy shown in the hand calculations.

9.3
ALLOCATABLE ARRAYS IN PROCEDURES

In Chapter 7, we learned how to declare and allocate memory for **allocatable arrays.** Allocatable arrays could be adjusted to exactly the size required by the particular problem being solved.

An allocatable array that is used in a procedure must be declared as a local variable in that procedure. If the allocatable array is declared with the SAVE attribute or appears in a SAVE statement, then the array would be allocated once using an ALLOCATE statement the first time the procedure is called. That array would be used in the calculations and then its contents would be preserved intact between calls to the procedure.

If the allocatable array is declared *without* the SAVE attribute, then the array must be allocated using an ALLOCATE statement[1] *every time* the procedure is called. That array would be used in the calculations, and then its contents would be automatically deallocated when execution returns to the calling program.

---
[1] Or by direct assignment in the case of a Fortran 2003 or later program.

9.4
AUTOMATIC ARRAYS IN PROCEDURES

Fortran provides another, simpler way to *automatically* create temporary arrays while a procedure is executing and to automatically destroy them when execution returns from the procedure. These arrays are called **automatic arrays.** An automatic array is *a local explicit-shape array with nonconstant bounds.* (The bounds are specified either by dummy arguments or through data from modules.)

For example, array `temp` in the following code is an automatic array. Whenever subroutine `sub1` is executed, dummy arguments n and m are passed to the subroutine. Note that arrays x and y are explicit-shape dummy arrays of size n × m that have been *passed to* the subroutine, while array `temp` is an automatic array that is *created* within the subroutine. When the subroutine starts to execute, an array `temp` of size n × m is automatically created, and when the subroutine ends, the array is automatically destroyed.

```
SUBROUTINE sub1 ( x, y, n, m )
IMPLICIT NONE
INTEGER, INTENT(IN) :: n, m
REAL, INTENT(IN), DIMENSION(n,m) :: x    ! Dummy array
REAL, INTENT(OUT), DIMENSION(n,m) :: y   ! Dummy array
REAL, DIMENSION(n,m) :: temp             ! Automatic array
temp = 0.
...
END SUBROUTINE sub1
```

Automatic arrays may not be initialized in their type declaration statements, but they may be initialized by assignment statements at the beginning of the procedure in which they are created. They may be passed as calling arguments to other procedures invoked by the procedure in which they are created. However, they cease to exist when the procedure in which they are created executes a `RETURN` or `END` statement. It is illegal to specify the `SAVE` attribute for an automatic array.

9.4.1 Comparing Automatic Arrays and Allocatable Arrays

Both automatic arrays and allocatable arrays can be used to create temporary working arrays in a program. What is the difference between them, and when should we choose one type of array or another for a particular application? The major differences between the two types of arrays are:

1. *Automatic arrays are allocated automatically whenever a procedure containing them is entered, while allocatable arrays must be allocated manually (deallocation is still automatic).* This feature favors the use of automatic arrays when the temporary memory is only needed within a single procedure and any procedures that may be invoked by it.
2. *Allocatable arrays are more general and flexible*, since they may be created and destroyed in separate procedures. For example, in a large program, we might create a special subroutine to allocate all arrays to be just the proper size to solve

the current problem, and we might create a different subroutine to deallocate them after they have been used. Also, allocatable arrays may be used in a main program, while automatic arrays may not.

3. *Allocatable arrays can be resized during a calculation.* A programmer can change the size of an allocatable array during execution using `DEALLOCATE` and `ALLOCATE` statements,[2] so a single array can serve multiple purposes requiring different shapes within a single procedure. In contrast, an automatic array is automatically allocated to the specified size at the beginning of the procedure execution and the size cannot be changed during that particular execution.

Automatic arrays should normally be used to create temporary working arrays within a single procedure, while allocatable arrays should be used to create arrays in main programs, or arrays that will be created and destroyed in different procedures, or arrays that must be able to change size within a given procedure.

Good Programming Practice
Use automatic arrays to create local temporary working arrays in procedures. Use allocatable arrays to create arrays in main programs, or arrays that will be created and destroyed in different procedures, or arrays that must be able to change size within a given procedure.

9.4.2 Example Program

EXAMPLE 9-4

Using Automatic Arrays in a Procedure:

As an example using automatic arrays in a procedure, we will write a new version of subroutine `simul` that does not destroy its input data while calculating the solution.

To avoid destroying the data, it will be necessary to add a new dummy argument to return the solution to the system of equations. This argument will be called `soln` and will have `INTENT(OUT)`, since it will only be used for output. Dummy arguments `a` and `b` will now have `INTENT(IN)`, since they will not be modified at all in the subroutine. In addition, we will take advantages of array sections to simplify the nested `DO` loops found in the original subroutine `simul`.

The resulting subroutine is shown in Figure 9-6. Note that arrays `a1` and `temp1` are automatic arrays, since they are local to the subroutine but their bounds are passed to the subroutine as dummy arguments. Arrays `a`, `b`, and `soln` are explicit-shape dummy arrays, because they appear in the argument list of the subroutine.

---

[2] Or by direct assignment in the case of a Fortran 2003 program.

FIGURE 9-6
A rewritten version of subroutine `simul` using allocatable arrays. This version does not destroy its input arrays. The declarations of automatic arrays `a1` and `temp1` and the use of array sections are shown in bold face.

```
SUBROUTINE simul2 ( a, b, soln, ndim, n, error )
!
!  Purpose:
!    Subroutine to solve a set of N linear equations in N
!    unknowns using Gaussian elimination and the maximum
!    pivot technique. This version of simul has been
!    modified to use array sections and allocatable arrays
!    It DOES NOT DESTROY the original input values.
!
!  Record of revisions:
!      Date        Programmer          Description of change
!      ====        ==========          =====================
!    11/25/15    S. J. Chapman         Original code
! 1. 11/25/15    S. J. Chapman         Add automatic arrays
!
IMPLICIT NONE

! Data dictionary: declare calling parameter types & definitions
INTEGER, INTENT(IN) :: ndim          ! Dimension of arrays a and b
REAL, INTENT(IN), DIMENSION(ndim,ndim) :: a
                                     ! Array of coefficients (N x N).
                                     ! This array is of size ndim x
                                     ! ndim, but only N x N of the
                                     ! coefficients are being used.
REAL, INTENT(IN), DIMENSION(ndim) :: b
                                     ! Input: Right-hand side of eqns.
REAL, INTENT(OUT), DIMENSION(ndim) :: soln
                                     ! Output: Solution vector.
INTEGER, INTENT(IN) :: n             ! Number of equations to solve.
INTEGER, INTENT(OUT) :: error        ! Error flag:
                                     !   0 -- No error
                                     !   1 -- Singular equations

! Data dictionary: declare constants
REAL, PARAMETER :: EPSILON = 1.0E-6  ! A "small" number for comparison
                                     ! when determining singular eqns

! Data dictionary: declare local variable types & definitions
REAL, DIMENSION(n,n) :: a1           ! Copy of "a" which will be
                                     ! destroyed during the solution
REAL :: factor                       ! Factor to multiply eqn irow by
                                     ! before adding to eqn jrow
INTEGER :: irow                      ! Number of the equation currently
                                     ! being processed
INTEGER :: ipeak                     ! Pointer to equation containing
                                     ! maximum pivot value
INTEGER :: jrow                      ! Number of the equation compared
                                     ! to the current equation
REAL :: temp                         ! Scratch value
```

(continued)

Additional Features of Procedures

(*continued*)

```fortran
REAL, DIMENSION(n) :: temp1            ! Scratch array

! Make copies of arrays "a" and "b" for local use
a1 = a(1:n,1:n)
soln = b(1:n)

! Process N times to get all equations...
mainloop: DO irow = 1, n

   ! Find peak pivot for column irow in rows irow to N
   ipeak = irow
   max_pivot: DO jrow = irow+1, n
      IF (ABS(a1(jrow,irow)) > ABS(a1(ipeak,irow))) THEN
         ipeak = jrow
      END IF
   END DO max_pivot

   ! Check for singular equations.
   singular: IF ( ABS(a1(ipeak,irow)) < EPSILON ) THEN
      error = 1
      RETURN
   END IF singular

   ! Otherwise, if ipeak /= irow, swap equations irow & ipeak
   swap_eqn: IF ( ipeak /= irow ) THEN
      temp1 = a1(ipeak,1:n)
      a1(ipeak,1:n) = a1(irow,1:n)    ! Swap rows in a
      a1(irow,1:n) = temp1
      temp = soln(ipeak)
      soln(ipeak) = soln(irow)        ! Swap rows in b
      soln(irow)  = temp
   END IF swap_eqn

   ! Multiply equation irow by -a1(jrow,irow)/a1(irow,irow),
   ! and add it to Eqn jrow (for all eqns except irow itself).
   eliminate: DO jrow = 1, n
      IF ( jrow /= irow ) THEN
         factor = -a1(jrow,irow)/a1(irow,irow)
         a1(jrow,:) = a1(irow,1:n)*factor + a1(jrow,1:n)
         soln(jrow) = soln(irow)*factor + soln(jrow)
      END IF
   END DO eliminate
END DO mainloop

! End of main loop over all equations. All off-diagonal terms
! are now zero. To get the final answer, we must divide
! each equation by the coefficient of its on-diagonal term.
divide: DO irow = 1, n
   soln(irow) = soln(irow) / a1(irow,irow)
   a1(irow,irow) = 1.
END DO divide
```

(*continued*)

(*concluded*)

```
! Set error flag to 0 and return.
error = 0
END SUBROUTINE simul2
```

Testing this subroutine is left as an exercise to the student (see Exercise 9-9).

A PROFUSION (AND CONFUSION!) OF FORTRAN ARRAY TYPES

We have now seen many different types of Fortran arrays, and no doubt produced a little confusion along the way. Let's step back and review the different array types, seeing just where each type is used and how they relate to each other.

1. Explicit-Shape Arrays with Constant Bounds
Explicit-shape arrays with constant bounds are nondummy arrays whose shape is explicitly specified in their type declaration statements. They may be declared either in main programs or in procedures, but they *do not appear* in the dummy argument list of a procedure. Explicit-shape arrays with constant bounds allocate fixed, permanent arrays for use in a program. They may be initialized in their type declaration statements.

If an explicit-shape array with constant bounds is allocated in a procedure, the data stored in it is only guaranteed to be intact from invocation to invocation if the array is declared with the SAVE attribute, or if the array is initialized in the type declaration statement.

Two examples of explicit-shape arrays with constant bounds are

```
INTEGER, PARAMETER :: NDIM = 100
REAL, DIMENSION(NDIM,NDIM) :: input_data = 1.
REAL, DIMENSION(-3:3) :: scratch = 0.
```

2. Dummy Arrays
Dummy arrays are arrays that appear in the dummy argument list of procedures. They are placeholders for the actual arrays passed to the procedure when it is invoked. No actual memory is allocated for dummy arrays. There are three types of dummy arrays: *explicit-shape dummy arrays*, *assumed-shape dummy arrays*, and *assumed-size dummy arrays*.

a. Explicit-Shape Dummy Arrays
Explicit-shape dummy arrays are arrays that appear in the dummy argument list of a procedure, and whose dimensions are explicitly declared by arguments in the procedure's argument list. All of the advanced features of Fortran arrays can be used with explicit-shape dummy arrays, including whole array operations, array sections, and array intrinsic functions. An example of an explicit-shape dummy array is

```
SUBROUTINE test ( array, n, m1, m2 )
INTEGER, INTENT(IN) :: n, m1, m2
REAL, DIMENSION(n,m1:m2) :: array
```

b. Assumed-Shape Dummy Arrays

Assumed-shape dummy arrays are arrays that appear in the dummy argument list of a procedure, and whose dimensions are declared by colons. The type declaration statement specifies the type and rank of the array, but not the extent of each dimension. *An assumed-shape dummy array is only usable in a procedure with an explicit interface.* These arrays assume the shape of whatever actual array is passed to the procedure when it is invoked. All of the advanced features of Fortran arrays can be used with assumed-shape dummy arrays, including whole array operations, array sections, and array intrinsic functions. An example of an assumed-shape dummy array is

```
SUBROUTINE test ( array )
REAL, DIMENSION(:,:) :: array
```

c. Assumed-Size Dummy Arrays

Assumed-size dummy arrays are arrays that appear in the dummy argument list of a procedure, and whose last dimension is declared with an asterisk. The size of all dimensions except for the last must be explicitly specified so that the procedure can determine how to locate specific array elements in memory. An assumed-size dummy array cannot be used with whole array operations or with many of the array intrinsic functions, because the shape of the actual array is unknown. Assumed-size dummy arrays are a holdover from earlier versions of Fortran; **they should never be used in any new programs.** An example of an assumed-size dummy array is

```
SUBROUTINE test ( array )
REAL, DIMENSION(10,*) :: array
```

3. Automatic Arrays

Automatic arrays are explicit-shape arrays with nonconstant bounds that appear in procedures. They do *not* appear in the procedure's argument list, but the *bounds of the array* are either passed via the argument list or by shared data in a module.

When the procedure is invoked, an array of the shape and size specified by the nonconstant bounds is *automatically* created. When the procedure ends, the array is automatically destroyed. If the procedure is invoked again, a new array will be created that could be either the same shape as or a different shape from the previous one. Data is *not* preserved in automatic arrays between invocations of the procedure, and it is illegal to specify either a SAVE attribute or a default initialization for an automatic array. An example of an automatic array is:

```
SUBROUTINE test ( n, m )
INTEGER, INTENT(IN) :: n, m
REAL, DIMENSION(n,m) :: array    ! Bounds in argument list, but not array
```

4. Deferred-Shape Arrays

Deferred-shape arrays are allocatable arrays or pointer arrays (pointer arrays are covered in Chapter 15). A deferred-shape array is declared in a type declaration statement with an ALLOCATABLE (or POINTER) attribute, and with the dimensions declared by colons. It may appear in either main programs or procedures. The array may not be used in any fashion (except as an argument to the ALLOCATED function) until memory is actually allocated for it. Memory is allocated using an ALLOCATE statement and deallocated using a DEALLOCATE statement. (In Fortran 2003, memory can also be allocated automatically by an assignment statement.) A deferred-shape array may not be initialized in its type declaration statement.

If an allocatable array is declared and allocated in a procedure, and if it is desired to keep the array between invocations of the procedure, it must be declared with the SAVE attribute. If the array is not needed, it should *not* be declared with the SAVE attribute. In that case, the allocatable array will be automatically deallocated at the end of the procedure. An unneeded pointer array (defined later) should be explicitly deallocated to avoid possible problems with "memory leaks."

An example of a deferred-shape allocatable array is:

```
INTEGER, ALLOCATABLE :: array(:,:)
ALLOCATE ( array(1000,1000), STATUS=istat )
...
DEALLOCATE ( array, STATUS=istat )
```

Quiz 9-1

This quiz provides a quick check to see if you have understood the concepts introduced in Sections 9.1 through 9.3. If you have trouble with the quiz, reread the sections, ask your instructor, or discuss the material with a fellow student. The answers to this quiz are found in the back of the book.

1. When should a SAVE statement or attribute be used in a program or procedure? Why should it be used?
2. What is the difference between an automatic and an allocatable array? When should each of them be used?
3. What are the advantages and disadvantages of assumed-shape dummy arrays?

For questions 4 through 6, determine whether there are any errors in these programs. If possible, tell what the output from each program will be.

4. ```
   PROGRAM test1
   IMPLICIT NONE
   INTEGER, DIMENSION(10) :: i
   INTEGER :: j
   DO j = 1, 10
   ```

(continued)
   ```

(*concluded*)

```
      CALL sub1 ( i(j) )
      WRITE (*,*) ' i = ', i(j)
   END DO
   END PROGRAM test1

   SUBROUTINE sub1 ( ival )
   IMPLICIT NONE
   INTEGER, INTENT(INOUT) :: ival
   INTEGER :: isum
   isum = isum + 1
   ival = isum
   END SUBROUTINE sub1
```

5. ```
   PROGRAM test2
   IMPLICIT NONE
   REAL, DIMENSION(3,3) :: a
   a(1,:) = [ 1., 2., 3. ]
   a(2,:) = [ 4., 5., 6. ]
   a(3,:) = [ 7., 8., 9. ]
   CALL sub2 (a, b, 3)
   WRITE (*,*) b
   END PROGRAM test2

   SUBROUTINE sub2(x, y, nvals)
   IMPLICIT NONE
   REAL, DIMENSION(nvals), INTENT(IN) :: x
   REAL, DIMENSION(nvals), INTENT(OUT) :: y
   REAL, DIMENSION(nvals) :: temp
   temp = 2.0 * x**2
   y = SQRT(x)
   END SUBROUTINE sub2
   ```

6. ```
   PROGRAM test3
   IMPLICIT NONE
   REAL, DIMENSION(2,2) :: a = 1., b = 2.
   CALL sub3(a, b)
   WRITE (*,*) a
   END PROGRAM test3

   SUBROUTINE sub3(a,b)
   REAL, DIMENSION(:,:), INTENT(INOUT) :: a
   REAL, DIMENSION(:,:), INTENT(IN) :: b
   a = a + b
   END SUBROUTINE sub3
   ```

9.5
ALLOCATABLE ARRAYS AS DUMMY ARGUMENTS IN PROCEDURES

Allocatable arrays have been made more flexible in Fortran 2003 and later versions. Two of the changes in allocatable arrays affect procedures:

1. It is now possible to have allocatable dummy arguments.
2. It is now possible for a function to return an allocatable value.

9.5.1 Allocatable Dummy Arguments

If a subroutine has an explicit interface, it is possible for subroutine dummy arguments to be allocatable. If a dummy argument is declared to be allocatable, then the corresponding actual arguments used to call the subroutine must be allocatable as well.

Allocatable dummy arguments are allowed to have an INTENT attribute. The INTENT affects the operation of the subroutine as follows:

1. If an allocatable argument has the INTENT(IN) attribute, then the array is not permitted to be allocated or deallocated in the subroutine and the values in the array cannot be modified.
2. If the allocatable argument has the INTENT(INOUT) attribute, then the status (allocated or not) and the data of the corresponding actual argument will be passed to the subroutine when it is called. The array may be deallocated, reallocated, or modified in any way in the subroutine, and the final status (allocated or not) and the data of the dummy argument will be passed back to the calling program in the actual argument.
3. If the allocatable argument has the INTENT(OUT) attribute, *then the actual argument in the calling program will be automatically deallocated on entry*, so any data in the actual array will be lost. The subroutine can then use the unallocated argument in any way, and the final status (allocated or not) and the data of the dummy argument will be passed back to the calling program in the actual argument.

A program that illustrates the use of allocatable array dummy arguments is shown in Figure 9-7. This program allocates and initializes an allocatable array and passes it to subroutine test_alloc. The data in the array on entry to test_alloc is the same as the originally initialized values. The array is deallocated, reallocated, and initialized in the subroutine, and that data is present in the main program when the subroutine returns.

FIGURE 9-7
Program illustrating the use of allocatable array dummy arguments.

```
MODULE test_module
!  Purpose:
!    To illustrate the use of allocatable arguments
!    in a subroutine.
!
```

(continued)

(*continued*)

```fortran
CONTAINS

   SUBROUTINE test_alloc(array)
   IMPLICIT NONE
   REAL,DIMENSION(:),ALLOCATABLE,INTENT(INOUT) :: array
                                    ! Test array

   ! Local variables
   INTEGER :: i            ! Loop index
   INTEGER :: istat        ! Allocate status

   ! Get the status of this array
   IF ( ALLOCATED(array) ) THEN
      WRITE (*,'(A)') 'Sub:  the array is allocated'
      WRITE (*,'(A,6F4.1)') 'Sub:  Array on entry = ', array
   ELSE
      WRITE (*,*) 'Sub:  the array is not allocated'
   END IF

   ! Deallocate the array
   IF ( ALLOCATED(array) ) THEN
      DEALLOCATE( array, STAT=istat )
   END IF

   ! Reallocate as a 5 element vector
   ALLOCATE(array(5), STAT=istat )

   ! Save data
   DO i = 1, 5
      array(i) = 6 - i
   END DO

   ! Display contents of array a on exit
   WRITE (*,'(A,6F4.1)') 'Sub:  Array on exit = ', array

   ! Return to caller
   END SUBROUTINE test_alloc

END MODULE test_module


PROGRAM test_allocatable_arguments
!
!  Purpose:
!     To illustrate the use of allocatable arguments
!     in a subroutine.
!
USE test_module
IMPLICIT NONE

! Declare local variables
REAL,ALLOCATABLE,DIMENSION(:) :: a    ! Allocatable array
INTEGER :: istat                      ! Allocate status
```

(*continued*)

(*concluded*)

```
! Allocate the array initially
ALLOCATE( a(6), STAT=istat )

! Initialize array
a = [ 1., 2., 3., 4., 5., 6. ]

! Display a before call
WRITE (*,'(A,6F4.1)') 'Main: Array a before call = ', a

! Call subroutine
CALL test_alloc(a)

! Display a after call
WRITE (*,'(A,6F4.1)') 'Main: Array a after call = ', a

END PROGRAM test_allocatable_arguments
```

When this program executes, the results are as shown below:

```
C:\book\fortran\chap9>test_allocatable_arguments
Main: Array a before call =   1.0 2.0 3.0 4.0 5.0 6.0
Sub:  the array is allocated
Sub:  Array on entry =   1.0 2.0 3.0 4.0 5.0 6.0
Sub:  Array on exit  =   5.0 4.0 3.0 2.0 1.0
Main: Array a after call =   5.0 4.0 3.0 2.0 1.0
```

9.5.2 Allocatable Functions

A Fortran function result is permitted to return a value with an ALLOCATABLE attribute. The return variable will not be allocated on entry to the function. The variable can be allocated and deallocated as often as desired inside the function, but it must be allocated and contain a value before the function returns.

A program that illustrates the use of allocatable functions is shown in Figure 9-8. This program calls function test_alloc_fun with a parameter specifying the number of values to return in the allocatable array. The function allocates the result variable, saves data into it, and returns to the main program for display.

FIGURE 9-8
Program illustrating the use of allocatable functions.

```
MODULE test_module
!  Purpose:
!    To illustrate the use of allocatable function
!    return values.
!
CONTAINS

   FUNCTION test_alloc_fun(n)
   IMPLICIT NONE
   INTEGER,INTENT(IN) :: n   ! Number of elements to return
```

(*continued*)

(concluded)

```fortran
REAL,ALLOCATABLE,DIMENSION(:) :: test_alloc_fun

! Local variables
INTEGER :: i              ! Loop index
INTEGER :: istat          ! Allocate status

  ! Get the status of the array
IF ( ALLOCATED(test_alloc_fun) ) THEN
   WRITE (*,'(A)') 'Array is allocated'
ELSE
   WRITE (*,'(A)') 'Array is NOT allocated'
END IF

  ! Allocate as an n element vector
ALLOCATE(test_alloc_fun(n), STAT=istat )

  ! Initialize data
DO i = 1, n
   test_alloc_fun(i) = 6 - i
END DO

  ! Display contents of array a on exit
WRITE (*,'(A,20F4.1)') 'Array on exit = ', test_alloc_fun

  ! Return to caller

END FUNCTION test_alloc_fun

END MODULE test_module


PROGRAM test_allocatable_function

!! Purpose:
!   To illustrate the use of allocatable function
!   return values.

!USE test_module
IMPLICIT NONE

! Declare local variables
INTEGER :: n = 5                          ! Number of elements to allocate
REAL,DIMENSION(:),ALLOCATABLE :: res      ! Result

! Call function and display results
res = test_alloc_fun(n)
WRITE (*,'(A,20F4.1)') 'Function return = ', res

END PROGRAM test_allocatable_function
```

When this program executes, the results are as shown below:

```
C:\book\fortran\chap9>test_allocatable_function
Array is NOT allocated
Array on exit =  5.0 4.0 3.0 2.0 1.0
Function return =  5.0 4.0 3.0 2.0 1.0
```

9.6
PURE AND ELEMENTAL PROCEDURES

As we mentioned in previous chapters, the Fortran language has been evolving in ways to make it easier to execute on massively parallel processors. As a part of this evolution, Fortran 95 introduced two new classifications of procedures: **pure procedures** and **elemental procedures.**

9.6.1 Pure Procedures

Pure functions *are functions that do not have side effects.* That is, they do not modify their input arguments and any other data (such as data in modules) that is visible outside the function. In addition, local variables may not have the SAVE attribute, and may not be initialized in type declaration statements (since such initialization implies the SAVE attribute). Any procedures invoked by a pure function must also be pure.

Because pure functions do not have side effects, it is safe to invoke them in a FORALL construct, where they might be executed in any order. This is very helpful on massively parallel processors, because each processor can take one combination of control indices from the FORALL construct and execute it in parallel with all of the others.

Every argument in a pure function must be declared with INTENT(IN), and any subroutine or functions invoked by the function must itself be pure. In addition, the function must not do any external file I/O operations, and must not contain a STOP statement. These constraints are easy to abide by—all of the functions that we have created so far are pure.

A pure function is declared by adding a PURE prefix to the function statement. For example, the following function is pure:

```
PURE FUNCTION length(x, y)
IMPLICIT NONE
REAL, INTENT(IN) :: x, y
REAL :: length
length = SQRT(x**2 + y**2)
END FUNCTION length
```

Pure subroutines *are subroutines that do not have side effects.* Their constraints are exactly the same as those on pure functions, except that they are permitted to modify arguments declared with INTENT(OUT) or INTENT(INOUT). Pure subroutines are declared by adding the PURE prefix to the SUBROUTINE statement.

9.6.2 Elemental Procedures

Elemental functions are functions that are specified for scalar arguments, but that may also be applied to array arguments. If the argument(s) of an elemental function are scalars, then the result of the function will be a scalar. If the argument(s) of the function are arrays, then the result of the function will be an array of the same shape as the input argument(s).

User-defined elemental functions must normally be PURE functions, and must satisfy the following additional constraints:

1. All dummy arguments must be scalars, and must not have the POINTER attribute. (We will learn about pointers in Chapter 15.)
2. The function result must be a scalar, and must not have the POINTER attribute.
3. Dummy arguments must not be used in type declaration statements except as arguments of certain intrinsic functions. This constraint prohibits the use of automatic arrays in elemental functions.

A user-defined elemental function is declared by adding an ELEMENTAL prefix to the function statement. For example, the function sinc(x) from Figure 7-16 meets the requirements of an elemental function, so it could be declared as:

```
ELEMENTAL FUNCTION sinc( x )
```

If the sinc function is declared ELEMENTAL, then the function can also accept array arguments and return array results.

Elemental subroutines are subroutines that are specified for scalar arguments, but that may also be applied to array arguments. They must meet the same constraints as elemental functions. An elemental subroutine is declared by adding an ELEMENTAL prefix to the subroutine statement. For example,

```
ELEMENTAL SUBROUTINE convert(x, y, z)
```

9.6.3 Impure Elemental Procedures

Elemental procedures can also be designed to modify their calling arguments. If so, they are **impure elemental procedures**. Such procedures must be declared with an IMPURE keyword, and any arguments that are modified must be declared with INTENT(INOUT). When an impure elemental procedure is called on an array, the procedure is executed element-by-element *in array order*: a(1), a(2), a(3), ..., a(n). If it is a multidimensional array, the elements are executed in column major order: a(1,1), a(2,1), ..., etc.

For an example, consider the impure elemental function cum below. This function replaces each value in an array by the sum of all values up to that point in the array.

```
IMPURE ELEMENTAL REAL FUNCTION cum(a, sum)
IMPLICIT NONE
REAL, INTENT(IN) :: a
REAL, INTENT(INOUT) :: sum
sum = sum + a
cum = sum
END FUNCTION cum
```

A test program for this function is shown below:

```
PROGRAM test_cum
REAL,DIMENSION(5) :: a, b
```

```
          REAL :: sum
          INTEGER :: i

          sum = 0.
          a = [ 1., 2., 3., 4., 5.]
          b = cum(a,sum)

          WRITE (*,*) b

          END PROGRAM test_cum
```

When this program is executed, the value in each element of array b is the sum of all elements in array a up to and including the corresponding index:

     1.00000   3.000000   6.000000   10.000000   15.000000

9.7
INTERNAL PROCEDURES

In Chapter 7, we learned about **external procedures** and **module procedures.** There is also a third type of procedure—the **internal procedure.** An internal procedure is a procedure that is entirely contained within another program unit, called the **host program unit,** or just the **host.** The internal procedure is compiled together with the host and it can only be invoked from the host program unit. Like module procedures, internal procedures are introduced by a CONTAINS statement. *An internal procedure must follow all of the executable statements within the host procedure* and must be introduced by a CONTAINS statement.

Why would we want to use internal procedures? In some problems, there are low-level manipulations that must be performed repeatedly as a part of the solution. These low-level manipulations can be simplified by defining an internal procedure to perform them.

A simple example of an internal procedure is shown in Figure 9-9. This program accepts an input value in degrees and uses an internal procedure to calculate the secant of that value. Although the internal procedure secant is invoked only once in this simple example, it could have been invoked repeatedly in a larger problem to calculate secants of many different angles.

FIGURE 9-9
Program to calculate the secant of an angle in degrees using an internal procedure.

```
PROGRAM test_internal
!
!   Purpose:
!     To illustrate the use of an internal procedure.
!
!   Record of revisions:
!       Date         Programmer         Description of change
!       ====         ==========         =====================
!     11/25/15      S. J. Chapman       Original code
!
```

(continued)

(*concluded*)

```
IMPLICIT NONE

! Data dictionary: declare constants
REAL, PARAMETER :: PI = 3.141592        ! PI

! Data dictionary: declare variable types & definitions
REAL :: theta                           ! Angle in degrees

! Get desired angle
WRITE (*,*) 'Enter desired angle in degrees: '
READ (*,*) theta

! Calculate and display the result.
WRITE (*,'(A,F10.4)') 'The secant is ', secant(theta)

! Note that the WRITE above was the last executable statement.
! Now, declare internal procedure secant:
CONTAINS
   REAL FUNCTION secant(angle_in_degrees)
   !
   ! Purpose:
   !    To calculate the secant of an angle in degrees.
   !
   REAL :: angle_in_degrees

   ! Calculate secant
   secant = 1. / cos( angle_in_degrees * pi / 180. )

   END FUNCTION secant

END PROGRAM test_internal
```

Note that the internal function `secant` appears after the last executable statement in program `test`. It is not a part of the executable code of the host program. When program `test` is executed, the user is prompted for an angle and the internal function `secant` is called to calculate the secant of the angle as a part of the final `WRITE` statement. When this program is executed, the results are:

```
C:\book\fortran\chap9>test
Enter desired angle in degrees:
45
The secant is     1.4142
```

An internal procedure functions exactly like an external procedure, with the following three exceptions:

1. The internal procedure can *only* be invoked from the host procedure. No other procedure within the program can access it.
2. The name of an internal procedure may not be passed as a command line argument to another procedure.
3. An internal procedure inherits all of the data entities (parameters and variables) of its host program unit by **host association.**

The last point requires more explanation. When an internal procedure is defined within a host program unit, all of the parameters and variables within the host program unit are also usable within the internal procedure. Look at Figure 9-9 again. Note that there is no IMPLICIT NONE statement within the internal procedure, because the one in the host program applies to the internal procedure as well. Note also that the named constant PI, which is defined in the host program, is used in the internal procedure.

The only time when an internal procedure cannot access a data entity defined in its host is *when the internal procedure defines a different data entity with the same name*. In that case, the data entity defined in the host is not accessible in the procedure and the data entity in the host will be totally unaffected by any manipulations that occur within the internal procedure.

Good Programming Practice
Use internal procedures to perform low-level manipulations that must be performed repeatedly, but are only needed by one program unit.

9.8
SUBMODULES

In Chapter 7, we learned about **module procedures.** Procedures that are declared in a module have a full explicit interface, and these procedures can be used in any other procedures in a program by declaring the module in a USE statement. Modules can be used to store libraries of procedures, which can then be used by other parts of a program.

A procedure is placed into a module by including the whole procedure after the keyword CONTAINS. The Fortran compiler automatically generates an explicit interface for the procedure and also automatically compiles the code to execute from the procedure description.

```
MODULE test_module
IMPLICIT NONE

CONTAINS

    SUBROUTINE procedure1(a, b, c)
    IMPLICIT NONE
    REAL,INTENT(IN) :: a
    REAL,INTENT(IN) :: b
    REAL,INTENT(OUT) :: c
    ...
    END SUBROUTINE procedure1

    REAL FUNCTION func2(a, b)
    IMPLICIT NONE
    REAL,INTENT(IN) :: a
    REAL,INTENT(IN) :: b
    ...
```

```
        END FUNCTION func2

     END MODULE test_module
```

Unfortunately, if anything changes in a module, it has to be recompiled, and *any other parts of a program depending on it will also need to be recompiled*. This can result in a massive recompilation taking a very long time if even one line of a key module is changed. This long compile cycle can be very inefficient during the development of a large program.

Why should we have to recompile everything depending on some module? The only part of a module procedure that is visible by calling procedures is the **interface,** the list of calling and returning parameters from the procedure. Any executable code inside the procedure is *not* visible to the calling procedure, and so any changes inside it should not force us to completely recompile the calling program.

Fortran has a mechanism to do this, known as **submodules.** If a programmer uses submodules, he or she splits the procedures in a module into two pieces. The first piece is the module itself containing the *interface* (calling arguments) to each module procedure and the second piece is a submodule that contains the actual *executable code* for the procedures. If the interface for any procedure changes, all of the other procedures USEing the module must be recompiled. If only the implementation (executable code) of a procedure in the submodule is changed, then only the submodule needs to be recompiled. The interface to the procedures in the submodule has no changes, so rest of the program does not need to be modified or recompiled (see Figure 9-10).

A procedure is placed into a module/submodule combination by including the *interface* to the procedure in the module and the *executable code* in the submodule.

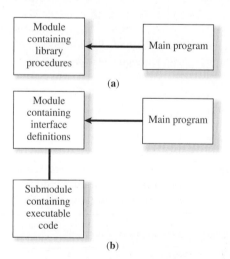

FIGURE 9-10
(*a*) A library in a module can be accessed by a main program using USE association. Any change in the library will force a recompilation of the main program. (*b*) A library in a module/submodule combination. The interface is in the module and the executable code is in the submodule. The module can be accessed by a main program using USE association. Any change in the executable code of the library that does not change the interface will not require a recompilation of the main program.

Note that the module contains an INTERFACE block, not a CONTAINS statement, and that the interface to each procedure is introduced by the keyword MODULE. The Fortran compiler automatically generates an explicit interface from the interface block.

```
MODULE test_module
IMPLICIT NONE

INTERFACE

    MODULE SUBROUTINE procedure1(a, b, c)
    IMPLICIT NONE
    REAL,INTENT(IN) :: a
    REAL,INTENT(IN) :: b
    REAL,INTENT(OUT) :: c
    END SUBROUTINE procedure1

    MODULE REAL FUNCTION func2(a, b)
    IMPLICIT NONE
    REAL,INTENT(IN) :: a
    REAL,INTENT(IN) :: b
    END FUNCTION func2

END INTERFACE

END MODULE test_module
```

Then the executable code is placed in a submodule, as shown below:

```
SUBMODULE (test_module) test_module_exec
IMPLICIT NONE
CONTAINS

    MODULE PROCEDURE procedure1
    ...
    END PROCEDURE procedure1

    MODULE PROCEDURE func2
    ...
    END PROCEDURE func2

END SUBMODULE test_module_exec
```

This submodule is declared to be a submodule of test_module by the SUBMODULE statement. Note that there is no definition for the input and output parameters of each module procedure—they are inherited from the interface definition in the module. If the code is written this way, then the contents of the submodule can be changed and recompiled without having to recompile the portions of the program that depend on it.

EXAMPLE 9-5 *Use of Submodules:*

Rewrite the simultaneous equations solving subroutine simul created in Example 9-1 so that it is in a module/submodule to create an explicit interface and to isolate the executable code from the interface.

Additional Features of Procedures

The Fortran module is shown in Figure 9-11 and the submodule is shown in Figure 9-12. Note that the interface definition for the subroutine is in the module and the executable code for the subroutine is in the submodule.

FIGURE 9-11
The interface for subroutine `simul` is placed in module `solver`.

```
MODULE solvers

! This module contains simultaneous equation solvers.

INTERFACE

    MODULE SUBROUTINE simul ( a, b, ndim, n, error )
    !
    ! Purpose:
    !   Subroutine to solve a set of n linear equations in n
    !   unknowns using Gaussian elimination and the maximum
    !   pivot technique.
    !
    ! Record of revisions:
    !     Date        Programmer          Description of change
    !     ====        ==========          =====================
    !   12/23/15     S. J. Chapman        Original code
    !
    IMPLICIT NONE

    ! Data dictionary: declare calling parameter types & definitions
    INTEGER, INTENT(IN) :: ndim          ! Dimension of arrays a and b
    REAL, INTENT(INOUT), DIMENSION(ndim,ndim) :: a
                    ! Array of coefficients (n x n).
                    ! This array is of size ndim x
                    ! ndim, but only n x n of the
                    ! coefficients are being used.
                    ! The declared dimension ndim
                    ! must be passed to the sub, or
                    ! it won't be able to interpret
                    ! subscripts correctly. (This
                    ! array is destroyed during
                    ! processing.)
    REAL, INTENT(INOUT), DIMENSION(ndim) :: b
                    ! Input: Right-hand side of eqns.
                    ! Output: Solution vector.
    INTEGER, INTENT(IN) :: n             ! Number of equations to solve.
    INTEGER, INTENT(OUT) :: error        ! Error flag:
                    !   0 -- No error
                    !   1 -- Singular equations
    END SUBROUTINE simul

END INTERFACE

END MODULE solvers
```

FIGURE 9-12
The executable code for subroutine `simul` is placed in submodule `solver_exec`.

```fortran
SUBMODULE (solvers) solvers_exec

! This submodule contains executable code for simultaneous
! equation solvers.

CONTAINS

   MODULE PROCEDURE simul

   ! Data dictionary: declare constants
   REAL, PARAMETER :: EPSILON = 1.0E-6   ! A "small" number for comparison
                                         ! when determining singular eqns

   ! Data dictionary: declare local variable types & definitions
   REAL :: factor               ! Factor to multiply eqn irow by
                                ! before adding to eqn jrow
   INTEGER :: irow              ! Number of the equation currently
                                ! being processed
   INTEGER :: ipeak             ! Pointer to equation containing
                                ! maximum pivot value
   INTEGER :: jrow              ! Number of the equation compared
                                ! to the current equation
   INTEGER :: kcol              ! Index over all columns of eqn
   REAL :: temp                 ! Scratch value

   ! Process n times to get all equations...
   mainloop: DO irow = 1, n

      ! Find peak pivot for column irow in rows irow to n
      ipeak = irow
      max_pivot: DO jrow = irow+1, n
         IF (ABS(a(jrow,irow)) > ABS(a(ipeak,irow))) THEN
            ipeak = jrow
         END IF
      END DO max_pivot

      ! Check for singular equations.
      singular: IF ( ABS(a(ipeak,irow)) < EPSILON ) THEN
         error = 1
         RETURN
      END IF singular

      ! Otherwise, if ipeak /= irow, swap equations irow & ipeak
      swap_eqn: IF ( ipeak /= irow ) THEN
         DO kcol = 1, n
            temp         = a(ipeak,kcol)
            a(ipeak,kcol) = a(irow,kcol)
            a(irow,kcol) = temp
         END DO
         temp     = b(ipeak)
         b(ipeak) = b(irow)
```

(*continued*)

Additional Features of Procedures

(*concluded*)

```
         b(irow)  = temp
      END IF swap_eqn

      ! Multiply equation irow by -a(jrow,irow)/a(irow,irow),
      ! and add it to Eqn jrow (for all eqns except irow itself).
      eliminate: DO jrow = 1, n
         IF ( jrow /= irow ) THEN
            factor = -a(jrow,irow)/a(irow,irow)
            DO kcol = 1, n
               a(jrow,kcol) = a(irow,kcol)*factor + a(jrow,kcol)
            END DO
            b(jrow) = b(irow)*factor + b(jrow)
         END IF
      END DO eliminate
   END DO mainloop

   ! End of main loop over all equations. All off-diagonal
   ! terms are now zero. To get the final answer, we must
   ! divide each equation by the coefficient of its on-diagonal
   ! term.
   divide: DO irow = 1, n
      b(irow)     = b(irow) / a(irow,irow)
      a(irow,irow) = 1.
   END DO divide

   ! Set error flag to 0 and return.
   error = 0

   END PROCEDURE simul

END SUBMODULE solvers_exec
```

The test driver program for subroutine `simul` is shown in Figure 9-13. Note that this test program USEs module solvers but *not* the submodule.

FIGURE 9-13
Test driver routine for subroutine `simul`.

```
PROGRAM test_simul_2
!
!  Purpose:
!    To test subroutine simul, which solves a set of N linear
!    equations in N unknowns.
!
!  Record of revisions:
!     Date        Programmer         Description of change
!     ====        ==========         =====================
!     12/23/15    S. J. Chapman      Original code
!
```

(*continued*)

(*continued*)

```
USE solvers
IMPLICIT NONE

! Data dictionary: declare constants
INTEGER, PARAMETER :: MAX_SIZE = 10     ! Max number of eqns

! Data dictionary: declare local variable types & definitions
REAL, DIMENSION(MAX_SIZE,MAX_SIZE) :: a
                                        ! Array of coefficients (n x n).
                                        ! This array is of size ndim x
                                        ! ndim, but only n x n of the
                                        ! coefficients are being used.
                                        ! The declared dimension ndim
                                        ! must be passed to the sub, or
                                        ! it won't be able to interpret
                                        ! subscripts correctly. (This
                                        ! array is destroyed during
                                        ! processing.)
REAL, DIMENSION(MAX_SIZE) :: b          ! Input: Right-hand side of eqns.
                                        ! Output: Solution vector.
INTEGER :: error                        ! Error flag:
                                        !   0 -- No error
                                        !   1 -- Singular equations
CHARACTER(len=20) :: file_name          ! Name of file with eqns
INTEGER :: i                            ! Loop index
INTEGER :: j                            ! Loop index
CHARACTER(len=80) :: msg                ! Error message
INTEGER :: n                            ! Number of simul eqns (<= MAX_SIZE)
INTEGER :: istat                        ! I/O status

! Get the name of the disk file containing the equations.
WRITE (*,"(' Enter the file name containing the eqns: ')")
READ (*,'(A20)') file_name

! Open input data file. Status is OLD because the input data must
! already exist.
OPEN ( UNIT=1, FILE=file_name, STATUS='OLD', ACTION='READ', &
       IOSTAT=istat, IOMSG=msg )

! Was the OPEN successful?
fileopen: IF ( istat == 0 ) THEN
   ! The file was opened successfully, so read the number of
   ! equations in the system.
   READ (1,*) n

   ! If the number of equations is <= MAX_SIZE, read them in
   ! and process them.
   size_ok: IF ( n <= MAX_SIZE ) THEN
      DO i = 1, n
         READ (1,*) (a(i,j), j=1,n), b(i)
      END DO
```

(*continued*)

(*concluded*)

```fortran
      ! Display coefficients.
      WRITE (*,"(/,'Coefficients before call:')")
      DO i = 1, n
         WRITE (*,"(1X,7F11.4)") (a(i,j), j=1,n), b(i)
      END DO

      ! Solve equations.
      CALL simul (a, b, MAX_SIZE, n, error )

      ! Check for error.
      error_check: IF ( error /= 0 ) THEN

         WRITE (*,1010)
         1010 FORMAT (/'Zero pivot encountered!', &
                     //'There is no unique solution to this system.')

      ELSE error_check

         ! No errors. Display coefficients.
         WRITE (*,"(/,'Coefficients after call:')")
         DO  i = 1, n
             WRITE (*,"(1X,7F11.4)") (a(i,j), j=1,n), b(i)
         END DO

         ! Write final answer.
         WRITE (*,"(/,'The solutions are:')")
         DO i = 1, n
            WRITE (*,"(2X,'X(',I2,') = ',F16.6)") i, b(i)
         END DO

      END IF error_check
   END IF size_ok
ELSE fileopen

   ! Else file open failed. Tell user.
   WRITE (*,1020) msg
   1020 FORMAT ('File open failed: ', A)

END IF fileopen
END PROGRAM test_simul_2
```

To test the subroutine, we need to call it with the same two data sets as before:

```
C:\book\fortran\chap9\solvers>test_simul_2
Enter the file name containing the eqns:
inputs1

Coefficients before call:
     1.0000     1.0000     1.0000     1.0000
     2.0000     1.0000     1.0000     2.0000
     1.0000     3.0000     2.0000     4.0000
```

```
Coefficients after call:
    1.0000      .0000      .0000     1.0000
     .0000     1.0000      .0000     3.0000
     .0000      .0000     1.0000    -3.0000

The solutions are:
   X( 1) =        1.000000
   X( 2) =        3.000000
   X( 3) =       -3.000000

C:\book\fortran\chap9\solvers>test_simul_2
Enter the file name containing the eqns:
inputs2
Coefficients before call:
    1.0000     1.0000     1.0000     1.0000
    2.0000     6.0000     4.0000     8.0000
    1.0000     3.0000     2.0000     4.0000

Zero pivot encountered!

There is no unique solution to this system.
```

The subroutine appears to be working correctly for both unique and singular sets of simultaneous equations.

Good Programming Practice
Use submodules to separate executable code from procedure interfaces, making it easier to modify internal code without forcing major recompilations.

There can be more than one submodule associated with a given module, and submodules can have submodules of their own. This flexibility can help us to organize our code in a structured way.

The procedures located in submodules are also called **separate procedures.**

9.9
SUMMARY

Multidimensional arrays can be passed to a subroutine or function subprogram either as explicit-shape dummy arrays or as assumed-shape dummy arrays. If multidimensional arrays are passed as explicit-shape dummy arrays, then the extent of each array dimension must also be passed to the subroutine as a calling argument and must be used to declare the array. If multidimensional arrays are passed as assumed-shape dummy arrays, then the procedure must have an explicit interface and the dimensions of the arrays are declared with colons as placeholders.

When a procedure finishes executing, the Fortran standard says that the local variables in the procedure become undefined. When the procedure is called again, the local variables might or might not have the same values as they did during the previous call, depending on the compiler and compiler options you are using. If a procedure needs for some local variables to be preserved between calls, the variables must be declared with the SAVE attribute or in a SAVE statement.

Automatic arrays are automatically created when a procedure starts executing and are automatically destroyed when the procedure finishes executing. Automatic arrays are local arrays whose dimensions are set by calling arguments, so they can have different sizes each time that the procedure is called. Automatic arrays are used as temporary work areas within a procedure.

Allocatable arrays may be used as dummy arguments and function return values in Fortran, as long as the subroutine or function has an explicit interface. If an allocatable array is declared with INTENT(IN), then the array cannot be deallocated or modified in the subroutine or function. If an allocatable array is declared with INTENT(OUT), then the array will be automatically deallocated before the subroutine or function starts to execute. If an allocatable array is declared with INTENT(INOUT), then the array will be unchanged at the start of the subroutine or function, but that subroutine or function is free to modify the data and/or the allocation of the array.

An internal procedure is a procedure defined entirely within another program unit, which is called the host program unit. It is only accessible from the host program unit. Internal procedures are included in the host program unit after all of the executable statements of the program unit and are preceded by a CONTAINS statement. An internal procedure has access to all of the data items defined in its host program unit by host association, unless the internal procedure contains a data item of the same name as a data item in the host. In that case, the data item in the host is not accessible to the internal procedure.

Submodules can be used to separate the interface definition of a procedure from the executable code of the procedure. If they are used, then a programmer can freely modify the executable code in the submodule without forcing all other code dependent on it to be recompiled, as long as the interface is not changed.

9.9.1 Summary of Good Programming Practice

The following guidelines should be adhered to when working with subroutines and functions.

1. Always use either explicit-shape dummy arrays or assumed-shape dummy arrays for dummy array arguments. Never use assumed-size dummy arrays in any new program.
2. If a procedure requires that the value of a local variable not change between successive invocations of the procedure, specify the SAVE attribute in the variable's type declaration statement, or include the variable in a SAVE statement, or initialize the variable in its type declaration statement.

3. Use automatic arrays to create local temporary working arrays in procedures. Use allocatable arrays to create arrays in main programs, or arrays that will be created and destroyed in different procedures, or arrays that must be able to change size within a given procedure.
4. Use internal procedures to perform low-level manipulations that must be performed repeatedly, but are only needed by one program unit.
5. Use submodules to separate executable code from procedure interfaces, making it easier to modify internal code without forcing major recompilations.

9.9.2 Summary of Fortran Statements and Structures

CONTAINS Statement:

```
            CONTAINS
```

Example:

```
            PROGRAM main
            ...
            CONTAINS
               SUBROUTINE sub1(x, y)
               ...
               END SUBROUTINE sub1
            END PROGRAM
```

Description:
The CONTAINS statement is a statement that specifies that the following statements are one or more separate procedures within the host unit. When used within a module, the CONTAINS statement marks the beginning of one or more module procedures. When used within a main program or an external procedure, the CONTAINS statement marks the beginning of one or more internal procedures. The CONTAINS statement must appear after any type, interface, and data definitions within a module, and must follow the last executable statement within a main program or an external procedure.

ELEMENTAL Prefix:

```
            ELEMENTAL FUNCTION name( arg1, ...)
            ELEMENTAL SUBROUTINE name( arg1, ...)
```

Example:

```
            ELEMENTAL FUNCTION my_fun (a, b, c)
```

Description:
This prefix declares that a procedure is ELEMENTAL, which means that it is defined with scalar inputs and outputs, but can be used with array inputs and outputs. When it is used with arrays, the operation defined by the elemental procedure is applied on an element-by-element basis to every element in the input array.

Additional Features of Procedures

END SUBMODULE Statement:

> END SUBMODULE [module_name]

Examples:

> END SUBMODULE solvers_exec

Description:
This statement marks the end of a submodule.

PURE Prefix:

> PURE FUNCTION name(arg1, ...)
> PURE SUBROUTINE name(arg1, ...)

Example:

> PURE FUNCTION my_fun (a, b, c)

Description:
This prefix declares that a procedure is PURE, which means that it has no side effects.

SAVE Attribute:

> type, SAVE :: name1, name2, ...

Example:

> REAL, SAVE :: sum

Description:
This attribute declares that the value of a *local variable* in a procedure must remain unchanged between successive invocations of the procedure. It is equivalent to the naming of the variable in a SAVE statement.

SAVE Statement:

> SAVE [var1, var2, ...]

Examples:

> SAVE count, index
> SAVE

Description:
This statement declares that the value of a *local variable* in a procedure must remain unchanged between successive invocations of the procedure. If a list of variables is included, only those variables will be saved. If no list is included, every local variable in the procedure or module will be saved.

SUBMODULE Statement:

SUBMODULE (parent_module) module_name

Examples:

SUBMODULE (solvers) solvers_exec

Description:
This statement declares a submodule, which can separate the executable code in a procedure from the interface (the calling sequence), which is declared in the parent module.

9.9.3 Exercises

9-1. What are the advantages and disadvantages of using explicit-shape dummy arrays in procedures? What are the advantages and disadvantages of using assumed-shape dummy arrays? Why should assumed-size dummy arrays never be used?

9-2. What are the differences between internal procedures and external procedures? When should an internal procedure be used instead of an external procedure?

9-3. What is the purpose of the SAVE statement and attribute? When should they be used?

9-4. Is the following program correct or not? If it is correct, what is printed out when it executes? If not, what is wrong with it?

```
PROGRAM junk
IMPLICIT NONE
REAL :: a = 3, b = 4, output
INTEGER :: i = 0
call sub1(a, i, output)
WRITE (*,*) 'The output is ', output

CONTAINS

    SUBROUTINE sub1(x, j, junk)
    REAL, INTENT(IN) :: x
    INTEGER, INTENT(IN) :: j
    REAL, INTENT(OUT) :: junk
    junk = (x - j) / b
    END SUBROUTINE sub1

END PROGRAM junk
```

9-5. What is printed out when the following code is executed? What are the values of x, y, i, and j at each point in the program? If a value changes during the course of execution, explain why it changes.

```
PROGRAM exercise9_5
IMPLICIT NONE
REAL :: x = 12., y = -3., result
```

```
         INTEGER :: i = 6, j = 4
         WRITE (*,100) 'Before call: x, y, i, j = ', x, y, i, j
   100   FORMAT (A,2F6.1,2I6)
         result = exec(y,i)
         WRITE (*,*) 'The result is ', result
         WRITE (*,100) 'After call: x, y, i, j = ', x, y, i, j
         CONTAINS

            REAL FUNCTION exec(x,i)
            REAL, INTENT(IN) :: x
            INTEGER, INTENT(IN) :: i
            WRITE (*,100) 'In exec:   x, y, i, j = ', x, y, i, j
      100   FORMAT (A,2F6.1,2I6)
            exec = ( x + y ) / REAL ( i + j )
            j = i
            END FUNCTION exec

         END PROGRAM exercise9_5
```

9-6. Matrix Multiplication Write a subroutine to calculate the product of two matrices if they are of compatible sizes and if the output array is large enough to hold the result. If the matrices are not of compatible sizes or if the output array is too small, set an error flag and return to the calling program. The dimensions of all three arrays a, b, and c should be passed to the subroutines from the calling program so that explicit-shape dummy arrays can be used and size checking can be done. (*Note:* The definition of matrix multiplication may be found in Exercise 8-13.) Check your subroutine by multiplying the following two pairs of arrays both with the subroutine and with the intrinsic subroutine MATMUL.

(a) $\quad a = \begin{bmatrix} 2 & -1 & 2 \\ -1 & -3 & 4 \\ 2 & 4 & 2 \end{bmatrix} \qquad b = \begin{bmatrix} 1 & 2 & 3 \\ 2 & 1 & 2 \\ 3 & 2 & 1 \end{bmatrix}$

(b) $\quad a = \begin{bmatrix} 1 & -1 & -2 \\ 2 & 2 & 0 \\ 3 & 3 & 3 \\ 5 & 4 & 4 \end{bmatrix} \qquad b = \begin{bmatrix} -2 \\ 5 \\ 2 \end{bmatrix}$

9-7. Write a new version of the matrix multiplication subroutine from Exercise 9-6 that uses an explicit interface and assumed-shape arrays. Before multiplying the matrices, this version should check to ensure that the input arrays are compatible and that the output array is large enough to hold the product of the two matrices. It can check for compatibility using the inquiry intrinsic functions found in Table 8-1. If these conditions are not satisfied, the subroutine should set an error flag and return.

9-8. Write a new version of the matrix multiplication subroutine from Exercise 9-6 that uses submodules to separate the explicit interface from the executable code.

9-9. Modify subroutine simul from Example 9-1 to use assumed-shape arrays. Use the two data sets in Example 9-1 to test the subroutine.

9-10. Write a test driver program to test subroutine simul2 in Figure 9-6. Use the two data sets in Example 9-1 to test the subroutine.

9-11. Why should the data in a module be declared with the SAVE attribute?

9-12. Modify program `test_alloc` in Figure 9-7 so that the allocatable dummy argument has an `INTENT(IN)` attribute. Does this program work now? If so, what does it do? If not, why not?

9-13. Modify program `test_alloc` in Figure 9-7 so that the allocatable dummy argument has an `INTENT(OUT)` attribute. Does this program work now? If so, what does it do? If not, why not?

9-14. Simulating Dice Throws Assume that a programmer is writing a game program. As a part of the program, it is necessary to simulate the throw of a pair of dice. Write a subroutine called `throw` to return two random values from 1 to 6 each time that it is called. The subroutine should contain an internal function called `die` to actually calculate the result of each toss of a die, and that function should be called twice by the subroutine to get the two results to return to the calling routine. (*Note:* It is possible to generate a random die result by using the intrinsic subroutine `RANDOM_NUMBER`.)

9-15. Create a set of ELEMENTAL functions to calculate the sine, cosine, and tangent of an angle θ, where θ is measured in degrees. Create a set of ELEMENTAL functions to calculate the arcsine, arccosine, and arctangent functions, returning the results in degrees. Test your functions by attempting to calculate the sine, cosine, and tangent of the 2×3 array `arr1`, and then inverting the calculations with the inverse functions. Array `arr1` is defined as follows:

$$\text{arr1} = \begin{bmatrix} 10.0 & 20.0 & 30.0 \\ 40.0 & 50.0 & 60.0 \end{bmatrix} \tag{9-10}$$

You should attempt to apply each function to the entire array in a single statement. Did your functions work properly with an array input?

9-16. Convert the ELEMENTAL functions of the previous exercise into PURE functions and try the problem again. What results do you get with PURE functions?

9-17. Second-Order Least-Squares Fits Sometimes, it makes no sense to fit a set of data points to a straight line. For example, consider a thrown ball. We know from basic physics that the height of the ball versus time will follow a parabolic shape, not a linear shape. How do we fit noisy data to a shape that is not a straight line?

It is possible to extend the idea of least-squares fits to find the best (in a least-squares sense) fit to a polynomial more complicated than a straight line. Any polynomial may be represented by an equation of the form

$$y(x) = c_0 + c_1 x + c_2 x^2 + c_3 x^3 + c_4 x^4 + \ldots \tag{9-11}$$

where the order of the polynomial corresponds to the highest power of x appearing in the polynomial. To perform a least-squares fit to a polynomial of order n, we must solve for the coefficients $c_0, c_1, \ldots, c_n$ that minimize the error between the polynomial and the data points being fit.

Additional Features of Procedures

The polynomial being fitted to the data may be of any order as long as there are at least as many distinct data points as there are coefficients to solve for. For example, the data may be fitted to a first order polynomial of the form

$$y(x) = c_0 + c_1 x \tag{9-12}$$

as long as there are at least two distinct data points in the fit. This is a straight line, where c_0 is the intercept of the line and c_1 is the slope of the line. Similarly, the data may be fitted to a second order polynomial of the form

$$y(x) = c_0 + c_1 x + c_2 x^2 \tag{9-13}$$

as long as there are at least three distinct data points in the fit. This is a quadratic expression whose shape is parabolic.

It can be shown[3] that the coefficients of a linear least squares fit to the polynomial $y(x) = c_0 + c_1 x$ are the solutions of the following system of equations

$$\begin{aligned} Nc_0 + (\Sigma x)c_1 &= \Sigma y \\ (\Sigma x)c_0 + (\Sigma x^2)c_1 &= \Sigma xy \end{aligned} \tag{9-14}$$

where

(x_i, y_i) is the ith sample measurement
N is the number of sample measurements included in the fit
Σx is the sum of the x_i values of all measurements
Σx^2 is the sum of the squares of the x_i values of all measurements
Σxy is the sum of the products of the corresponding x_i and y_i values

Any number of sample measurements (x_i, y_i) may be used in the fit, as long as the number of measurements is greater than or equal to 2.

The formulation shown above can be extended to fits of higher-order polynomials. For example, it can be shown that the coefficients of a least-squares fit to the second order polynomial $y(x) = c_0 + c_1 x + c_2 x^2$ are the solutions of the following system of equations

$$\begin{aligned} Nc_0 + (\Sigma x)c_1 + (\Sigma x^2)c_2 &= \Sigma y \\ (\Sigma x)c_0 + (\Sigma x^2)c_1 + (\Sigma x^3)c_2 &= \Sigma xy \\ (\Sigma x^2)c_0 + (\Sigma x^3)c_1 + (\Sigma x^4)c_2 &= \Sigma x^2 y \end{aligned} \tag{9-15}$$

where the various terms have meanings similar to the ones described above. Any number of sample measurements (x_i, y_i) may be used in the fit, as long as the number of distinct measurements is greater than or equal to 3. The least-squares fit of the data to a parabola can be found by solving Equations (9-15) for c_0, c_1, and c_2.

Create a subroutine to perform a least-squares fit to a second order polynomial (a parabola), and use that subroutine to fit a parabola to the position data contained in Table 9-1 below. Use an internal subroutine to solve the system of simultaneous equations given in Equations (9-15).

---

[3] *Probability and Statistics*, by Athanasios Papoulis, Prentice-Hall, 1990, pp. 392–393.

TABLE 9-1
Measured position and velocity of a ball versus time

Time (sec)	Position (m)	Velocity (m/s)
0.167	49.9	−5.1
0.333	52.2	−12.9
0.500	50.6	−15.1
0.667	47.0	−6.8
0.833	47.7	−12.3
1.000	42.3	−18.0
1.167	37.9	−5.7
1.333	38.2	−6.3
1.500	38.0	−12.7
1.667	33.8	−13.7
1.833	26.7	−26.7
2.000	24.8	−31.3
2.167	22.0	−22.9
2.333	16.5	−25.6
2.500	14.0	−25.7
2.667	5.6	−25.2
2.833	2.9	−35.0
3.000	0.8	−27.9

9-18. Create a test data set by calculating points (x_i, y_i) along the curve $y(x) = x^2 - 4x + 3$ for $x_i = 0, 0.1, 0.2, \ldots, 5.0$. Next, use the intrinsic subroutine RANDOM_NUMBER to add random noise to each of the y_i values. Then, use the subroutine created in Exercise 9-16 to try to estimate the coefficients of the original function that generated the data set. Try this when the added random noise has the range:

(a) 0.0 (No added noise)

(b) $[-0.1, 0.1]$

(c) $[-0.5, 0.5]$

(d) $[-1.0, 1.0]$

(e) $[-2.0, 2.0]$

(f) $[1.-, 3.0)$

How did the quality of the fit change as the amount of noise in the data increased?

9-19. Higher-Order Least-Squares Fits It can be shown that the coefficients of a least-squares fit to the nth order polynomial $y(x) = c_0 + c_1 x + c_2 x^2 + \ldots + c_n x^n$ are the solutions of the following system of n equations in n unknowns

$$\begin{aligned}
N c_0 + (\Sigma x) c_1 + (\Sigma x^2) c_2 + \ldots + (\Sigma x^n) c_n &= \Sigma y \\
(\Sigma x) c_0 + (\Sigma x^2) c_1 + (\Sigma x^3) c_2 + \ldots + (\Sigma x^{n+1}) c_n &= \Sigma xy \\
(\Sigma x^2) c_0 + (\Sigma x^3) c_1 + (\Sigma x^4) c_2 + \ldots + (\Sigma x^{n+2}) c_n &= \Sigma x^2 y \\
&\vdots \\
(\Sigma x^n) c_0 + (\Sigma x^{n+1}) c_1 + (\Sigma x^{n+2}) c_2 + \ldots + (\Sigma x^{2n}) c_n &= \Sigma x^n y
\end{aligned} \quad (9\text{-}16)$$

Write a subroutine that implements a least-squares fit to any polynomial of any order. (*Note:* Use dynamic memory allocation to create arrays of the proper size for the problem being solved.)

9-20. Create a test data set by calculating points (x_i, y_i) along the curve $y(x) = x^4 - 3x^3 - 4x^2 + 2x + 3$ for $x_i = 0, 0.1, 0.2, \ldots, 5.0$. Next, use the intrinsic subroutine RANDOM_NUMBER to add random noise to each of the y_i values. Then, use the higher-order least-squares fit subroutine created in Exercise 9-19 to try to estimate the coefficients of the original function that generated the data set. Try this when the added random noise has the range:

(a) 0.0 (No added noise)

(b) $[-0.1, 0.1]$

(c) $[-0.5, 0.5]$

(d) $[-1.0, 1.0]$

How did the quality of the fit change as the amount of noise in the data increased? How does the quality of the higher-order fit for a given amount of noise compare to the quality of a quadratic fit (Exercise 9-18) for the same amount of noise?

9-21. Place your second order least-squares fit subroutine and your higher-order least-squares fit subroutine into a common library that could be reused by other programs. Place them into a module and declare that the two subroutines have PUBLIC access. Rerun the test programs with the new module to show that the code works identically.

9-22. Interpolation A least-squares fit of order n calculates the nth order polynomial that "best fits" an (x, y) data set in a least-squares sense. Once this polynomial has been calculated, it can be used to estimate the expected value y_0 associated with any location x_0 within the data set. This process is called *interpolation*. Write a program that calculates a quadratic least-squares fit to the data set given below and then uses that fit to estimate the expected value y_0 at $x_0 = 3.5$.

Noisy Measurements	
x	y
0.00	-23.22
1.00	-13.54
2.00	-4.14
3.00	-0.04
4.00	3.92
5.00	4.97
6.00	3.96
7.00	-0.07
8.00	-5.67
9.00	-12.29
10.00	-20.25

9-23. Extrapolation Once a least-squares fit has been calculated, the resulting polynomial can also be used to estimate the values of the function *beyond the limits of the original input data set*. This process is called *extrapolation*. Write a program that calculates a linear least-squares fit to the data set given below, and then uses that fit to estimate the expected value y_0 at $x_0 = 14.0$.

Noisy Measurements

x	y
0.00	−14.22
1.00	−10.54
2.00	−5.09
3.00	−3.12
4.00	0.92
5.00	3.79
6.00	6.99
7.00	8.95
8.00	11.33
9.00	14.71
10.00	18.75

10

More about Character Variables

OBJECTIVES

- Understand the kinds of characters available in Fortran compilers, including possible Unicode support.
- Understand how relational operations work with character data.
- Understand the lexical functions `LLT`, `LLE`, `LGT`, and `LGE`, and why they are safer to use than the corresponding relational operators.
- Know how to use the character intrinsic functions `CHAR`, `ICHAR`, `ACHAR`, `IACHAR`, `LEN`, `LEN_TRIM`, `TRIM`, and `INDEX`.
- Know how to use internal files to convert numeric data to character form, and vice versa.

A **character variable** is a variable that contains character information. In this context, a "character" is any symbol found in a **character set.** There are two basic character sets in common use in the United States: ASCII (American Standard Code for Information Interchange, ISO/IEC 646:1991) and Unicode (ISO 10646).[1]

The ASCII character set is a system in which each character is stored in 1 byte (8 bits). Such a system allows for 256 possible characters, and the ASCII standard defines the first 128 of these possible values. The 8-bit codes corresponding to each letter and number in the ASCII coding system are given in Appendix A. The remaining 128 possible values that can be stored in a 1-byte character can have different definitions in different countries, depending on the "code page" used in that particular country, or sometimes on the operating system that the compiler runs on. These characters are defined in the ISO-8859 standard series.

The Unicode character set uses *multiple bytes* to represent each character, allowing a maximum of 1,112,064 possible characters. The Unicode character set includes the characters required to represent almost every language on Earth. The most common

---

[1] Previous versions of this book also discussed the EBCDIC character set, which was another 1-byte character set used in older IBM mainframes. The author has not seen an EBCDIC-coded computer in 32 years now, so all discussions of that character set have been dropped.

character encoding scheme is UTF-8, which uses a variable number of bytes to represent different characters. The 127 base ASCII characters are also the first 127 characters in Unicode and can be represented in a single byte. Characters higher up in the set may require 2, 3, or 4 bytes to encode.

Every Fortran compiler supports a 1-byte characters set called the *default character set*. The bottom 127 characters will be the ASCII character set. Fortran compilers are allowed to support other character sets such as Unicode as well, and many now do so.

10.1
CHARACTER COMPARISON OPERATIONS

Character strings may be compared to each other using either relational operators or special character comparison functions called lexical functions. Lexical functions have an advantage over the relational operators when program portability is considered.

10.1.1 The Relational Operators with Character Data

Character strings can be compared in logical expressions using the **relational operators** ==, /-, <, <=, >, and >=. The result of the comparison is a logical value that is either true or false. For instance, the expression '123' == '123' is true, while the expression '123' == '1234' is false.

How are two characters compared to determine if one is greater than the other? The comparison is based on the **collating sequence** of the characters. The collating sequence of the characters is the order in which they occur within a specific character set. For example, the character 'A' is character number 65 in the ASCII character set, while the character 'B' is character number 66 in the set (see Appendix A). Therefore, the logical expression 'A' < 'B' is true in the ASCII character set. On the other hand, the character 'a' is character number 97 in the ASCII set, so 'a' is greater than 'A'.

If a particular computer uses a different character set, then it is possible that the results of relational comparisons could differ because characters might occur in different order.[2]

We can make some comparisons safely regardless of character set. The letters 'A' to 'Z' are always in alphabetical order, the numbers '0' to '9' are always in numerical sequence, and the letters and numbers are not intermingled in the collating sequence. Beyond that, however, all bets are off. The relationships among the special symbols and the relationship between the uppercase and lowercase letters may differ for different character sets.

How are two strings compared to determine if one is greater than the other? The comparison begins with the first character in each string. If they are the same, then the second two characters are compared. This process continues until the first difference is found between the strings. For example, 'AAAAAB' > 'AAAAAA'.

---

[2] We will see later that there are special functions to allow comparisons to be done in a character-set independent manner.

What happens if the strings are different lengths? The comparison begins with the first letter in each string and progresses through each letter until a difference is found. If the two strings are the same all the way to the end of one of them, then the other string is considered the larger of the two. Therefore,

$$'AB' > 'AAAA' \quad \text{and} \quad 'AAAAA' > 'AAAA'$$

EXAMPLE 10-1

Alphabetizing Words:

It is often necessary to alphabetize lists of character strings (names, places, etc.). Write a subroutine that will accept a character array and alphabetize the data in the array.

SOLUTION

Since relational operators work for character strings the same way that they work for real values, it is easy to modify the sorting subroutine that we developed in Chapter 7 to alphabetize an array of character variables. All we have to do is to substitute character array declarations for the real declarations in the sorting routines. The rewritten program is shown in Figure 10-1:

FIGURE 10-1
A program to alphabetize character strings using a version of the selection sort algorithm adapted for character strings.

```
PROGRAM sort4
!
!  Purpose:
!    To read in a character input data set, sort it into ascending
!    order using the selection sort algorithm, and to write the
!    sorted data to the standard output device. This program calls
!    subroutine "sortc" to do the actual sorting.
!
!  Record of revisions:
!      Date        Programmer          Description of change
!      ====        ==========          =====================
!    11/28/15     S. J. Chapman        Original code
!
IMPLICIT NONE

! Data dictionary: declare constants
INTEGER, PARAMETER :: MAX_SIZE = 10    ! Max number to sort

! Data dictionary: declare variable types & definitions
CHARACTER(len=20), DIMENSION(MAX_SIZE) :: a
                                       ! Data array to sort
LOGICAL :: exceed = .FALSE.            ! Logical indicating that array
                                       !   limits are exceeded.
CHARACTER(len=20) :: filename          ! Input data file name
INTEGER :: i                           ! Loop index
CHARACTER(len=80) :: msg                ! Error message
```

(continued)

(*continued*)

```fortran
INTEGER :: nvals = 0              ! Number of data values to sort
INTEGER :: status                 ! I/O status: 0 for success
CHARACTER(len=20) :: temp         ! Temporary variable for reading

! Get the name of the file containing the input data.
WRITE (*,*) 'Enter the file name with the data to be sorted: '
READ (*,'(A20)') filename

! Open input data file. Status is OLD because the input data must
! already exist.
OPEN ( UNIT=9, FILE=filename, STATUS='OLD', ACTION='READ', &
       IOSTAT=status, IOMSG=msg )

! Was the OPEN successful?
fileopen: IF ( status == 0 ) THEN    ! Open successful

   ! The file was opened successfully, so read the data to sort
   ! from it, sort the data, and write out the results.
   ! First read in data.
   DO
      READ (9, *, IOSTAT=status) temp   ! Get value
      IF ( status /= 0 ) EXIT           ! Exit on end of data
      nvals = nvals + 1                 ! Bump count
      size: IF ( nvals <= MAX_SIZE ) THEN   ! Too many values?
         a(nvals) = temp                ! No: Save value in array
      ELSE
         exceed = .TRUE.                ! Yes: Array overflow
      END IF size
   END DO

   ! Was the array size exceeded? If so, tell user and quit.
   toobig: IF ( exceed ) THEN
      WRITE (*,1010) nvals, MAX_SIZE
      1010 FORMAT (' Maximum array size exceeded: ', I6, ' > ', I6 )
   ELSE

      ! Limit not exceeded: sort the data.
      CALL sortc (a, nvals)

      ! Now write out the sorted data.
      WRITE (*,*) 'The sorted output data values are: '
      WRITE (*,'(4X,A)') ( a(i), i = 1, nvals )

   END IF toobig

ELSE fileopen

   ! Else file open failed. Tell user.
   WRITE (*,1050) TRIM(msg)
   1050 FORMAT ('File open failed--error = ', A)

END IF fileopen

END PROGRAM sort4
```

(*continued*)

(*concluded*)

```
SUBROUTINE sortc (array, n)
!
!   Purpose:
!     To sort a character array into ascending order using a
!     selection sort.
!
!   Record of revisions:
!       Date          Programmer        Description of change
!       ====          ==========        =====================
!     11/28/15      S. J. Chapman       Original code
!
IMPLICIT NONE

! Data dictionary: declare calling parameter types & definitions
INTEGER, INTENT(IN) :: n                    ! Number of values
CHARACTER(len=20), DIMENSION(n), INTENT(INOUT) :: array
                                            ! Array to be sorted

! Data dictionary: declare local variable types & definitions
INTEGER :: i                ! Loop index
INTEGER :: iptr             ! Pointer to smallest value
INTEGER :: j                ! Loop index
CHARACTER(len=20) :: temp   ! Temp variable for swaps

! Sort the array
outer: DO i = 1, n-1

   ! Find the minimum value in array(i) through array(n)
   iptr = i
   inner: DO j = i-1, n
      minval: IF ( array(j) < array(iptr) ) THEN
         iptr = j
      END IF minval
   END DO inner

   ! iptr now points to the minimum value, so swap array(iptr)
   ! with array(i) if i /= iptr.
   swap: IF ( i /= iptr ) THEN
      temp        = array(i)
      array(i)    = array(iptr)
      array(iptr) = temp
   END IF swap

END DO outer

END SUBROUTINE sortc
```

To test this program, we will place the following character values in file `inputc`:

```
Fortran
fortran
ABCD
ABC
```

```
XYZZY
9.0
A9IDL
```

If we compile and execute the program on a computer with an ASCII collating sequence, the results of the test run will be:

```
C:\book\fortran\chap10>sort4
Enter the file name containing the data to be sorted:
inputc
The sorted output data values are:
   9.0
   A9IDL
   ABC
   ABCD
   Fortran
   XYZZY
   fortran
```

Note that the number 9 was placed before any of the letters and that the lowercase letters were placed after the uppercase letters. These locations are in accordance with the ASCII table in Appendix A.

10.1.2 The Lexical Functions LLT, LLE, LGT, and LGE

The result of the sort subroutine in the previous example could depend on the character set used by the processor on which it was executed. This dependence is bad, since it makes our Fortran program less portable between processors. We need some way to ensure that programs produce the *same answer* regardless of the computer on which they are compiled and executed.

Fortunately, the Fortran language includes a set of four logical intrinsic functions for just this purpose: LLT (lexically less than), LLE (lexically less than or equal to), LGT (lexically greater than), and LGE (lexically greater than or equal to). These functions are the exact equivalent of the relational operators <, <=, >, and >=, except that *they always compare characters according to the ASCII collating sequence, regardless of the computer they are running on*. If these **lexical functions** are used instead of the relational operators to compare character strings, the results will be the same on every computer.

A simple example using the LLT function is shown below. Here, character variables string1 and string2 are being compared using the relational operator < and the logical function LLT. The value of result1 will vary from processor to processor depending on the character set used, but the value of result2 will always be true on any processor.

```
LOGICAL :: result1, result2
CHARACTER(len=6) :: string1, string2
string1 = 'A1'
string2 = 'a1'
result1 = string1 < string2
result2 = LLT( string1, string2 )
```

Good Programming Practice

If there is any chance that your program will have to run on computers with different character sets, use the logical functions LLT, LLE, LGT, and LGE to test for inequality between two character strings. Do not use the relational operators <, <=, >, and >= with character strings, since their results may vary from computer to computer.

10.2
INTRINSIC CHARACTER FUNCTIONS

The Fortran language contains several additional intrinsic functions that are important for manipulating character data. Seven of these functions are CHAR, ICHAR, ACHAR, IACHAR, LEN, LEN_TRIM, TRIM, and INDEX. We will now discuss these functions and describe their use.

The CHAR function converts an input integer value into a corresponding output character. An example of the CHAR function is shown below:

```
CHARACTER :: out
INTEGER :: input = 65
out = CHAR(input)
```

The input to the CHAR function is a single integer argument, and the output from the function is *the character whose collating sequence number matches the input argument* for the particular processor. For example, if a processor uses the ASCII collating sequence, then CHAR(65) is the character 'A'.

The ICHAR function converts an input character into a corresponding output integer. An example of the ICHAR function is shown below:

```
CHARACTER :: input = 'A'
INTEGER :: out
out = ICHAR(input)
```

The input to the ICHAR function is a single character, and the output from the function is *the integer whose collating sequence number matches the input character* for the particular processor. For example, if a processor uses the ASCII collating sequence, then ICHAR('A') is the integer 65.

The functions ACHAR and IACHAR are exactly the same as the functions CHAR and ICHAR, except that they work with the ASCII collating sequence *regardless of the character set used by a particular processor*. Therefore, the results of the functions ACHAR and IACHAR will be the same on any computer. They should be used instead of the previous functions to improve the portability of the programs that you write.

Good Programming Practice

Use functions ACHAR and IACHAR instead of CHAR and ICHAR, since the result of the first set of functions is independent of the processor on which they are executed, while the result of the second set of functions varies depending on the collating sequence of the particular processor on which they are executed.

Function `LEN` returns the declared length of a character string. The input to `LEN` is a character string `str1`, and the output from it is an integer containing the number of characters in `str1`. An example of the `LEN` function is shown below:

```
CHARACTER(len=20) :: str1
INTEGER :: out
str1 = 'ABC XYZ'
out = LEN(str1)
```

The output from `LEN` is 20. Note that the output of `LEN` is the declared size of the string and not the number of nonblank characters in the string.

Function `LEN_TRIM` returns the length of a character string without trailing blanks. The input to `LEN_TRIM` is a character string `str1`, and the output from it is an integer containing the number of characters in `str1`, excluding trailing blanks. If `str1` is entirely blank, then function `LEN_TRIM` returns a 0. An example of the `LEN_TRIM` function is shown below:

```
CHARACTER(len=20) :: str1
INTEGER :: out
str1 = 'ABC XYZ'
out = LEN_TRIM(str1)
```

Function `TRIM` returns a character string without trailing blanks. The input to `TRIM` is a character string `str1`, and the output from it is the same string, excluding trailing blanks. If `str1` is entirely blank, then function `LEN_TRIM` returns a blank string. An example of the `LEN_TRIM` function is shown below:

```
CHARACTER(len=20) :: str1
str1 = 'ABC XYZ'
WRITE (*,*) '"', TRIM(str1), '"'
```

The output from `TRIM` is a seven-character string containing `'ABC XYZ'`.

The `INDEX` function searches for a pattern in a character string. The inputs to the function are two strings: `str1` containing the string to search and `str2` containing the pattern that we are looking for. The output from the function is an integer containing the position in the character string `str1` at which the pattern was found. If no match is found, `INDEX` returns a 0. An example of the `INDEX` function is shown below:

```
CHARACTER(len=20) :: str1 = 'THIS IS A TEST!'
CHARACTER(len=20) :: str2 = 'TEST'
INTEGER :: out
out = INDEX(str1,str2)
```

The output of this function is the integer 11, since `TEST` begins at character 11 in the input character string.

If `str2` were `'IS'`, then what would the value of `INDEX(str1,str2)` be? The answer is 3, since `'IS'` occurs within the word `'THIS'`. The `INDEX` function will never see the word `'IS'` because it stops searching at the first occurrence of the search pattern in the string.

The `INDEX` function can also have an optional third argument, *back*. If present, the argument *back* must be a logical value. If *back* is present and true, then the search

More about Character Variables

TABLE 10-1
Some common character intrinsic functions

Function name and argument(s)	Argument types	Result type	Comments
ACHAR(ival)	INT	CHAR	Returns the character corresponding to ival in the ASCII collating sequence
CHAR(ival)	INT	CHAR	Returns the character corresponding to ival in the processor's collating sequence
IACHAR(char)	CHAR	INT	Returns the integer corresponding to char in the ASCII collating sequence
ICHAR(char)	CHAR	INT	Returns the integer corresponding to char in the processor's collating sequence
INDEX(str1, str2,back)	CHAR, LOG	INT	Returns the character number of the first location in str1 to contain the pattern in str2 (0=no match) Argument *back* is optional; if present and true, then the search starts from the end of str1 instead of the beginning
LEN(str1)	CHAR	INT	Returns length of str1
LEN_TRIM(str1)	CHAR	INT	Returns length of str1, excluding any trailing blanks
LLT(str1,str2)	CHAR	LOG	True if str1 < str2 according to the ASCII collating sequence
LLE(str1,str2)	CHAR	LOG	True if str1 <= str2 according to the ASCII collating sequence
LGT(str1,str2)	CHAR	LOG	True if str1 > str2 according to the ASCII collating sequence
LGE(str1,str2)	CHAR	LOG	True if str1 >= str2 according to the ASCII collating sequence
TRIM(str1)	CHAR	CHAR	Returns str1 with trailing blanks removed

starts from the end of string str1 instead of from the beginning. An example of the INDEX function with the optional third argument is shown below:

```
CHARACTER(len=20) :: str1 = 'THIS IS A TEST!'
CHARACTER(len=20) :: str2 = 'IS'
INTEGER :: out
OUT = INDEX(str1,str2,.TRUE.)
```

The output of this function is the integer 6, since the last occurrence of IS begins at character 6 in the input character string.

10.3

PASSING CHARACTER VARIABLES TO SUBROUTINES AND FUNCTIONS

In Example 10-1, we created a subroutine to alphabetize an array of character variables. The character array in that subroutine was declared as

```
INTEGER, INTENT(IN) :: n
CHARACTER(len=20), DIMENSION(n), INTENT(INOUT) :: array
```

This subroutine will sort character in an array with any number of elements, but it will *only* sort the array when each element in the array is 20 characters long. If we wanted to sort data in an array whose elements were a different length, we would need a whole new subroutine to do it! This behavior is unreasonable. It should be possible to write a single subroutine to process character data in a given fashion regardless of the number of characters in each element.

Fortran contains a feature to support this. The language allows a special form of the character type declaration for dummy character arguments in procedures. This special declaration takes the form

CHARACTER(len=*) :: char_var

where char_var is the name of a dummy character argument. This declaration says that dummy argument char_var is a character variable, but the length of the character variable is not explicitly known at compilation time. If the procedure using char_var needs to know its length, it can call function LEN to get that information. The dummy arguments in subroutine sortc could have been declared as

```
INTEGER, INTENT(IN) :: n
CHARACTER(len=*), DIMENSION(n), INTENT(INOUT) :: array
```

If they were declared in this manner, the subroutine would work equally well for arrays of character variables containing elements of any length.

Good Programming Practice
Use the CHARACTER(len=*) type statement to declare dummy character arguments in procedures. This feature allows the procedure to work with strings of arbitrary lengths. If the procedure needs to know the actual length of a particular variable, it may call the LEN function with that variable as a calling argument.

Remember that dummy arguments are just placeholders for the variables that will be passed to the procedure when it is invoked. No actual memory is allocated for the dummy arguments. Since no memory is being allocated, the Fortran compiler does not need to know the length of the character variables that will be passed to the procedure in advance. Therefore, we can use the CHARACTER(len=*) type declaration statement for dummy character arguments in a procedure.

On the other hand, any character variables that are local to the procedure must be declared with explicit lengths. Memory will be allocated in the procedure for these local variables, and we must explicitly specify the length of each one for the compiler to know how much memory to allocate for it. This creates a problem for local variables that must be the same length as a dummy argument passed to the procedure. For example, in subroutine sortc, the variable temp was used for swapping, and must be the same length as an element of the dummy argument array.

How can we adjust the size of a temporary variable to fit the size of a dummy array whenever the subroutine is called? *If we declare the length of the variable to be the length of a dummy subroutine argument, then when the subroutine is executed, an* **automatic character variable** *of that size will be allocated.* (This is very similar to the behavior of automatic arrays described in the last chapter.) When the subroutine execution ends, that automatic variable will be destroyed. Like automatic arrays, this automatic character variable may not be initialized in its type declaration statement.

For example, the following statements create an automatic character variable temp of the same length as the dummy argument string:

More about Character Variables

```
SUBROUTINE sample ( string )
CHARACTER(len=*) :: string
CHARACTER(len=len(string)) :: temp
```

A version of the character sort subroutine that will work for character arrays of any length, with any number of elements, and on any processor is shown in Figure 10-2.

FIGURE 10-2
A modified version of subroutine `sortc` that will work for arrays of any size and array elements of any length.

```
SUBROUTINE sortc (array, n)
!
!  Purpose:
!    To sort character array "array" into ascending order using
!    a selection sort. This version of the subroutine sorts
!    according to the ASCII collating sequence. It works for
!    character arrays with any number of elements, with array
!    elements of any length, and on processors regardless of
!    character set.
!
!  Record of revisions:
!      Date         Programmer         Description of change
!      ====         ==========         =====================
!    11/28/15      S. J. Chapman       Original code
! 1. 11/28/15      S. J. Chapman       Modified to work with lexical
!                                      fns and arbitrary element
!                                      lengths
IMPLICIT NONE

! Declare calling parameters:
INTEGER, INTENT(IN) :: n                    ! Number of values
CHARACTER(len=*), DIMENSION(n), INTENT(INOUT) :: array
                                            ! Array to be sorted
! Declare local variables:
INTEGER :: i                   ! Loop index
INTEGER :: iptr                ! Pointer to smallest value
INTEGER :: j                   ! Loop index
CHARACTER(len=len(array)) :: temp ! Temp variable for swaps

! Sort the array
outer: DO i = 1, n-1

   ! Find the minimum value in array(i) through array(n)
   iptr = i
   inner: DO j = i+1, n
      minval: IF ( LLT(array(j),array(iptr)) ) THEN
         iptr = j
      END IF minval
   END DO inner

   ! iptr now points to the minimum value, so swap array(iptr)
   ! with array(i) if i /= iptr.
```

(continued)

(*concluded*)

```
  swap: IF ( i /= iptr ) THEN
     temp        = array(i)
     array(i)    = array(iptr)
     array(iptr) = temp
  END IF swap
END DO outer

END SUBROUTINE sortc
```

EXAMPLE 10-2

Shifting Strings to Uppercase:

We saw in Example 10-1 that lowercase character strings were not alphabetized properly with uppercase strings, since the collating sequence positions of the lowercase letters were different from the collating sequence numbers of the corresponding uppercase letters. The difference between upper- and lowercase letters also causes a problem when we are attempting to match a pattern within a character variable, since `'STRING'` is not the same as `'string'` or `'String'`. It is often desirable to shift all character variables to uppercase to make matching and sorting easier. Write a subroutine to convert all of the lowercase letters in a character string to uppercase, while leaving any other characters in the string unaffected.

SOLUTION

This problem is made more complicated by the fact that we don't know which collating sequence is used by the computer that the subroutine will be running on. In the vast majority of cases, assuming that the compiler uses the ASCII character set would be valid. However, we can bullet-proof the code by doing comparisons and conversions according to the ASCII character set regardless of the one actually used on the computer by taking advantage of the lexical functions and functions ACHAR and IACHAR.

Appendix A shows the ASCII collating sequence. If we look at Appendix A, we can see that *there is a fixed offset* between an uppercase letter and the corresponding lowercase letter in each collating sequence, so shifting from lowercase to uppercase is a matter of subtracting a fixed offset from each alphabetical letter in the string. If we use the lexical functions for comparisons and the ACHAR and IACHAR functions for conversions, then we can act as though the processor were ASCII and be assured of correct results regardless of the collating sequence of the actual machine.

1. **State the problem.**

   Write a subroutine to convert all of the lowercase letters in a character string to uppercase, while not affecting numeric and special characters. Design the subroutine to work properly on any processor by using functions that are independent of collating sequence.

2. **Define the inputs and outputs.**

   The input to the subroutine is the character argument `string`. The output from the subroutine is also in `string`. `string` can be of arbitrary length.

3. **Describe the algorithm.**

Looking at the ASCII table in Appendix A, we note that the uppercase letters begin at sequence number 65, while the lowercase letters begin at sequence number 97. There are exactly 32 numbers between each uppercase letter and its lowercase equivalent. Furthermore, there are no other symbols mixed into the middle of the alphabet.

These facts give us our basic algorithm for shifting strings to uppercase. We will determine if a character is lowercase by deciding if it is between 'a' and 'z' in the ASCII character set. If it is, then we will subtract 32 from its sequence number to convert it to uppercase using the ACHAR and IACHAR functions. The initial pseudocode for this algorithm is

```
Determine if character is lower case. If so,
  Convert to integer form
  Subtract 32 from the integer
  Convert back to character form
End of IF
```

The final pseudocode for this subroutine is

```
! Get length of string
length ← LEN(string)

DO for i = 1 to length
   IF LGE(string(i:i),'a') .AND. LLE(string(i:i),'z') THEN
       string(i:i) ← ACHAR ( IACHAR (string(i:i) - 32 ) )
   END of IF
END of DO
```

where length is the length of the input character string.

4. **Turn the algorithm into Fortran statements.**

The resulting Fortran subroutines are shown in Figure 10-3.

FIGURE 10-3
Subroutine ucase.

```
SUBROUTINE ucase ( string )
!
! Purpose:
!   To shift a character string to upper case on any processor,
!   regardless of collating sequence.
!
! Record of revisions:
!      Date          Programmer           Description of change
!      ====          ==========           =====================
!   11/28/15        S. J. Chapman         Original code
!
IMPLICIT NONE

! Declare calling parameters:
CHARACTER(len=*), INTENT(INOUT) :: string
```

(continued)

(*concluded*)

```
! Declare local variables:
INTEGER :: i              ! Loop index
INTEGER :: length         ! Length of input string

! Get length of string
length = LEN ( string )

! Now shift lower case letters to upper case.
DO i = 1, length
   IF ( LGE(string(i:i),'a') .AND. LLE(string(i:i),'z') ) THEN
      string(i:i) = ACHAR ( IACHAR ( string(i:i) ) - 32 )
   END IF
END DO

END SUBROUTINE ucase
```

5. **Test the resulting Fortran program.**
To test this subroutine, it is necessary to write a driver program to read a character string, call the subroutine, and write out the results. A test driver program is shown in Figure 10-4:

FIGURE 10-4
Test driver program for subroutine ucase.

```
PROGRAM test_ucase
!
!  Purpose:
!     To test subroutine ucase.
!
IMPLICIT NONE
CHARACTER(len=20) string
WRITE (*,*) 'Enter test string (up to 20 characters): '
READ (*,'(A20)') string
CALL ucase(string)
WRITE (*,*) 'The shifted string is: ', string
END PROGRAM test_ucase
```

The results from the test program for two input strings are:

```
C:\book\fortran\chap10>test_ucase
  Enter test string (up to 20 characters):
This is a test!...
  The shifted string is: THIS IS A TEST!...
C:\book\fortran\chap10>test_ucase
  Enter test string (up to 20 characters):
abcf1234^&*$po()-
  The shifted string is: ABCF1234^&*$PO()-
```

The subroutine is shifting all lowercase letters to uppercase, while leaving everything else alone. It appears to be working correctly.

10.4
VARIABLE-LENGTH CHARACTER FUNCTIONS

We have already seen that subroutines can work with strings of variable lengths by declaring them with the CHARACTER(len=*) declaration. Is there a way to write a character function that can return a string of arbitrary length?

The answer is yes. We can create an **automatic length character function,** where the length returned by the function is specified by a calling argument. Figure 10-5 shows a simple example. Function abc returns the first n characters of the alphabet, where n is specified in the call to the function.

FIGURE 10-5
A sample function that returns a variable-length character string.

```
MODULE character_subs

CONTAINS
   FUNCTION abc( n )
   !
   ! Purpose:
   !   To return a string containing the first N characters
   !   of the alphabet.
   !
   ! Record of revisions:
   !      Date         Programmer        Description of change
   !      ====         ==========        =====================
   !    11/28/15      S. J. Chapman       Original code
   !
   IMPLICIT NONE

   ! Declare calling parameters:
   INTEGER, INTENT(IN) :: n    ! Length of string to return
   CHARACTER(len=n) abc        ! Returned string

   ! Declare local variables:
   character(len=26) :: alphabet = 'abcdefghijklmnopqrstuvwxyz'

   ! Get string to return
   abc = alphabet(1:n)

   END FUNCTION abc
END MODULE character
```

A test driver program for this function is shown in Figure 10-6. The module containing the function must be named in a USE statement in the calling program.

FIGURE 10-6
Program to test function abc.

```
PROGRAM test_abc
!
```

(continued)

(*concluded*)

```
!  Purpose:
!    To test function abc.
!
USE character_subs
IMPLICIT NONE

INTEGER :: n                              ! String length

WRITE(*,*) 'Enter string length:'         ! Get string length
READ (*,*) n

WRITE (*,*) 'The string is: ', abc(n)     ! Tell user

END PROGRAM test_abc
```

When this program is executed, the results are:

```
C:\book\fortran\chap10>test_abc
Enter string length:
10
The string is: abcdefghij

C:\book\fortran\chap10>test_abc
Enter string length:
3
The string is: abc
```

Quiz 10-1

This quiz provides a quick check to see if you have understood the concepts introduced in Sections 10.1 through 10.4. If you have trouble with the quiz, reread the sections, ask your instructor, or discuss the material with a fellow student. The answers to this quiz are found in the back of the book.

For questions 1 to 3, state the result of the following expressions. Assume that the processor is using the ASCII character set.

1. 'abcde' < 'ABCDE'
2. LLT ('abcde','ABCDE')
3. '1234' == '1234 '

For questions 4 and 5, state whether each of the following statements is legal or not. If they are legal, tell what they do. If they are not legal, state why they are not legal.

4. ```
   FUNCTION day(iday)
   IMPLICIT NONE
   INTEGER, INTENT(IN) :: iday
   CHARACTER(len=3) :: day
   CHARACTER(len=3), DIMENSION(7) :: days = &
   ```

(*continued*)
   ```

More about Character Variables

(concluded)

```
            ['SUN', 'MON', 'TUE', 'WED', 'THU', 'FRI', 'SAT']
   IF ( ( iday >= 1 ) .AND. ( iday <= 7 ) ) THEN
      day = days(iday)
   END IF
   END FUNCTION day
```

5.
```
   FUNCTION swap_string(string)
   IMPLICIT NONE
   CHARACTER(len=*), INTENT(IN) :: string
   CHARACTER(len=len(string)) :: swap_string
   INTEGER :: length, i
   length = LEN(string)
   DO i = 1, length
      swap_string(length-i+1:length-i+1) = string(i:i)
   END DO
   END FUNCTION swap_string
```

For questions 6 to 8, state the contents of each variable after the code has been executed.

6.
```
   CHARACTER(len=20) :: last = 'JOHNSON'
   CHARACTER(len=20) :: first = 'JAMES'
   CHARACTER :: middle_initial = 'R'
   CHARACTER(len=42) name
   name = last // ',' // first // middle_initial
```

7.
```
   CHARACTER(len=4) :: a = '123'
   CHARACTER(len=12) :: b
   b = 'ABCDEFGHIJKLMNOPQRSTUVWXYZ'
   b(5:8) = a(2:3)
```

8.
```
   CHARACTER(len=80) :: line
   INTEGER :: ipos1, ipos2, ipos3, ipos4
   line = 'This is a test line containing some input data!'
   ipos1 = INDEX (LINE, 'in')
   ipos2 = INDEX (LINE, 'Test')
   ipos3 = INDEX (LINE, 't l')
   ipos4 = INDEX (LINE, 'in', .TRUE.)
```

10.5

INTERNAL FILES

We learned how to manipulate numeric data in the previous chapters of this book. In this chapter, we have learned how to manipulate character data. What we have *not* learned yet is how to convert numeric data into character data, and vice versa. There is a special mechanism in Fortran for such conversions, known as **internal files.**

Internal files are a special extension of the Fortran I/O system in which the READs and WRITEs occur to internal character buffers (internal files) instead of disk files (external files). Anything that can be written to an external file can also be written to an internal file, where it will be available for further manipulation. Likewise, anything that can be read from an external file can be read from an internal file.

The general form of a READ from an internal file is

READ (*buffer*,*format*) *arg1*, *arg2*, ...

where *buffer* is the input character buffer, *format* is the format for the READ, and *arg1*, *arg2*, etc., are the variables whose values are to be read from the buffer. The general form of a WRITE to an internal file is

WRITE (*buffer*,*format*) *arg1*, *arg2*, ...

where *buffer* is the output character buffer, *format* is the format for the WRITE, and *arg1*, *arg2*, etc., are the values to be written to the buffer.

A common use of internal files is to convert character data into numeric data, and vice versa. For example, if the character variable input contains the string '135.4', then the following code will convert the character data into a real value:

```
CHARACTER(len=5) :: input = '135.4'
REAL :: value
READ (input,*) value
```

Certain I/O features are not available with internal files. For example, the OPEN, CLOSE, BACKSPACE, and REWIND statements may not be used with them.

Good Programming Practice

Use internal files to convert data from character format to numeric format, and vice versa.

10.6
EXAMPLE PROBLEMS

EXAMPLE 10-3

Varying a format to match the data to be output:

So far, we have seen three format descriptors to write real data values. The F*w.d* format descriptor displays the data in a format with a fixed decimal point, and the E*w.d* and ES*w.d* format descriptors display the data in exponential notation. The F format descriptor displays data in a way that is easier for a person to understand quickly, but it will fail to display the number correctly if the absolute value of the number is either too small or too large. The E and ES format descriptors will display the number correctly regardless of size, but it is harder for a person to read at a glance.

Write a Fortran function that converts a real number into characters for display in a 12-character-wide field. The function should check the size of the number to be printed out and modify the format statement to display the data in F12.4 format for as long as

More about Character Variables   475

possible until the absolute value of the number either gets too big or too small. When the number is out of range for the F format, the function should switch to ES format.

SOLUTION
In the F12.4 format, the function displays four digits to the right of the decimal place. One additional digit is required for the decimal point, and another one is required for the minus sign, if the number is negative. After subtracting those characters, there are seven characters left over for positive numbers and six characters left over for negative numbers. Therefore, we must convert the number to exponential notation for any positive number larger than 9,999,999 and any negative number smaller than −999,999.

If the absolute value of the number to be displayed is smaller than 0.01, then the display should shift to ES format, because there will not be enough significant digits displayed by the F12.4 format. However, an exact zero value should be displayed in normal F format rather than exponential format.

When it is necessary to switch to exponential format, we will use the ES12.5 format, since the number appears in ordinary scientific notation.

1. **State the problem.**
   Write a function to convert a real number into 12 characters for display in a 12-character-wide field. Display the number in F12.4 format if possible, unless the number overflows the format descriptor or gets too small to display with enough precision in an F12.4 field. When it is not possible to display the number in F12.4 format, switch to the ES12.5 format. However, display an exact zero in F12.4 format.

2. **Define the inputs and outputs.**
   The input to the function is a real number passed through the argument list. The function returns a 12-character expression containing the number in a form suitable for displaying.

3. **Describe the algorithm.**
   The basic requirements for this function were discussed above. The pseudocode to implement these requirements is shown below:

```
IF value > 9999999. THEN
   Use ES12.5 format
ELSE IF value < -999999. THEN
   Use ES12.5 format
ELSE IF value == 0. THEN
   Use F12.4 format
ELSE IF ABS(value) < 0.01
   Use ES12.5 format
ELSE
   USE F12.4 format
END of IF
WRITE value to buffer using specified format
```

4. **Turn the algorithm into Fortran statements.**
   The resulting Fortran function is shown in Figure 10-7. Function real_to_char illustrates both how to use internal files and how to use a character variable to contain format descriptors. The proper format descriptor for the real-to-character conversion is

stored in variable `fmt`, and an internal `WRITE` operation is used to write the character string into buffer `string`.

FIGURE 10-7
Character function `real_to_char`.

```
FUNCTION real_to_char ( value )
!
! Purpose:
!   To convert a real value into a 12-character string, with the
!   number printed in as readable a format as possible considering
!   its range. This routine prints out the number according to the
!   following rules:
!      1. value > 9999999.                  ES12.5
!      2. value < -999999.                  ES12.5
!      3. 0. < ABS(value) < 0.01            ES12.5
!      4. value = 0.0                       F12.4
!      5. Otherwise                         F12.4
!
! Record of revisions:
!    Date         Programmer        Description of change
!    ====         ==========        =====================
!   11/28/15    S. J. Chapman       Original code
!
IMPLICIT NONE

! Data dictionary: declare calling parameter types & definitions
REAL, INTENT(IN) :: value              ! value to convert to char form
CHARACTER (len=12) :: real_to_char     ! Output character string

! Data dictionary: declare local variable types & definitions
CHARACTER(len=9) :: fmt                ! Format descriptor
CHARACTER(len=12) :: string            ! Output string

! Clear string before use
string = ' '

! Select proper format
IF ( value > 9999999. ) THEN
   fmt = '(ES12.5)'
ELSE IF ( value < -999999. ) THEN
   fmt = '(ES12.5)'
ELSE IF ( value == 0. ) THEN
   fmt = '(F12.4)'
ELSE IF ( ABS(value) < 0.01 ) THEN
   fmt = '(ES12.5)'
ELSE
   fmt = '(F12.4)'
END IF

! Convert to character form.
WRITE (string,fmt) value
real_to_char = string

END FUNCTION real_to_char
```

5. **Test the resulting Fortran program.**
To test this function, it is necessary to write a driver program to read a real number, call the subroutine, and write out the results. A test driver program is shown in Figure 10-8:

FIGURE 10-8
Test driver program for function `real_to_char`.

```
PROGRAM test_real_to_char
!
!  Purpose:
!    To test function real_to_char.
!
!  Record of revisions:
!      Date         Programmer          Description of change
!      ====         ==========          =====================
!    11/28/15     S. J. Chapman         Original code
!
!  External routines:
!    real_to_char -- Convert real to character string
!    ucase        -- Shift string to upper case
!
IMPLICIT NONE

! Declare external functions:
CHARACTER(len=12), EXTERNAL :: real_to_char

! Data dictionary: declare variable types & definitions
CHARACTER :: ch                ! Character to hold Y/N response.
CHARACTER(len=12) :: result    ! Character output
REAL :: value                  ! Value to be converted

while_loop: DO
   ! Prompt for input value.
   WRITE (*,'(A)') 'Enter value to convert:'
   READ (*,*) value

   ! Write converted value, and see if we want another.
   result = real_to_char(value)
   WRITE (*,'(A,A,A)') 'The result is ', result, &
                      ': Convert another one? (Y/N) [N]'
   ! Get answer.

   READ (*,'(A)') ch

   ! Convert answer to upper case to make match.
   CALL ucase ( ch )

   ! Do another?
      IF ( ch /= 'Y' ) EXIT
END DO while_loop

END PROGRAM test_real_to_char
```

To verify that this function is working correctly for all cases, we must supply test values that fall within each of the ranges that it is designed to work for. Therefore, we will test it with the following numbers:

```
         0.
         0.001234567
      1234.567
  12345678.
   -123456.7
  -1234567.
```

The results from the test program for the six input values are:

```
C:\book\fortran\chap10>test_real_to_char
Enter value to convert:
0.
The result is    .0000: Convert another one? (Y/N) [N]
y
Enter value to convert:
0.001234567
The result is 1.23457E-03: Convert another one? (Y/N) [N]
Y
Enter value to convert:
1234.567
The result is   1234.5670: Convert another one? (Y/N) [N]
Y
Enter value to convert:
12345678.
The result is 1.23457E+07: Convert another one? (Y/N) [N]
y
Enter value to convert:
-123456.7
The result is -123456.7000: Convert another one? (Y/N) [N]
y
Enter value to convert:
-1234567.
The result is -1.23457E+06: Convert another one? (Y/N) [N]
n
```

The function appears to be working correctly for all possible input values.

The test program `test_real_to_char` also contains a few interesting features. Since we would normally use the program to test more than one value, it is structured as a while loop. The user is prompted by the program to determine whether or not to repeat the loop. The first character of the user's response is stored in variable `ch` and is compared to the character `'Y'`. If the user responded with a `'Y'`, the loop is repeated; otherwise, it is terminated. Note that subroutine `ucase` is called to shift the contents of `ch` to uppercase, so that both `'y'` and `'Y'` will be interpreted as "yes" answers. This form of repetition control is very useful in interactive Fortran programs.

More about Character Variables

> ## Quiz 10-2
>
> This quiz provides a quick check to see if you have understood the concepts introduced in Sections 10.5 and 10.6. If you have trouble with the quiz, reread the sections, ask your instructor, or discuss the material with a fellow student. The answers to this quiz are found in the back of the book.
>
> For questions 1 to 3, state whether each of the following groups of statements is correct or not. If correct, describe the results of the statements.
>
> 1. ```
>    CHARACTER(len=12) :: buff
>    CHARACTER(len=12) :: buff1 = 'ABCDEFGHIJKL'
>    INTEGER :: i = -1234
>    IF ( buff1(10:10) == 'K' ) THEN
>       buff = "(1X,I10.8)"
>    ELSE
>       buff = "(1X,I10)"
>    END IF
>    WRITE (*,buff) i
>    ```
>
> 2. ```
>    CHARACTER(len=80) :: outbuf
>    INTEGER :: i = 123, j, k = -11
>    j = 1023 / 1024
>    WRITE (outbuf,*) i, j, k
>    ```
>
> 3. ```
>    CHARACTER(len=30) :: line = &
>             '123456789012345678901234567890'
>    CHARACTER(len=30) :: fmt = &
>             '(3X,I6,12X,I3,F6.2)'
>    INTEGER :: ival1, ival2
>    REAL :: rval3
>    READ (line,fmt) ival1, ival2, rval3
>    ```

10.7
SUMMARY

A character variable is a variable that contains character information. Two character strings may be compared using the relational operators. However, the result of the comparison may differ depending on the collating sequence of the characters on a particular processor. It is safer to test character strings for inequality using the lexical functions, which always return the same value on any computer regardless of collating sequence.

It is possible to declare automatic character variables in procedures. The length of an automatic character variable is specified by either a dummy argument or by a value passed in a module. Each time the procedure is run, a character variable of the specified length is automatically generated, and the variable is automatically destroyed when the execution of the procedure ends.

It is possible to generate character functions that can return character strings of variable length provided that there is an explicit interface between the function and any invoking program units. The easiest way to generate an explicit interface is to package the function within a module, and then to use that module in the calling procedure.

Internal files provide a means to convert data from character form to numeric form and vice versa within a Fortran program. They involve writes to and reads from a character variable within the program.

10.7.1 Summary of Good Programming Practice

The following guidelines should be adhered to when working with character variables:

1. If there is any chance that your program will have to run on computers with different character sets, use the logical functions LLT, LLE, LGT, and LGE to test for inequality between two character strings. Do not use the relational operators <, <=, >, and >= with character strings, since their results may vary from computer to computer.
2. Use functions ACHAR and IACHAR instead of functions CHAR and ICHAR, since the results of the first set of functions are independent of the processor on which they are executed, while the results of the second set of functions vary depending on the collating sequence of the particular processor that they are executed on.
3. Use the CHARACTER(len=*) type statement to declare dummy character arguments in procedures. This feature allows the procedure to work with strings of arbitrary lengths. If the subroutine or function needs to know the actual length of a particular variable, it may call the LEN function with that variable as a calling argument.
4. Use internal files to convert data from character format to numeric format, and vice versa.

10.7.2 Summary of Fortran Statements and Structures

Internal READ Statement:

> READ (*buffer,fmt*) *input_list*

Example:

> READ (line,'(1X, I10, F10.2)') i, slope

Description:
The internal READ statement reads the data in the input list according to the formats specified in *fmt*, which can be a character string, a character variable, the label of a FORMAT statement, or *. The data is read from the internal character variable *buffer*.

More about Character Variables

> **Internal WRITE Statement:**
>
> WRITE (*buffer,fmt*) *output_list*
>
> Example:
>
> WRITE (line,'(2I10,F10.2)') i, j, slope
>
> Description:
> The internal WRITE statement writes the data in the output list according to the formats specified in *fmt*, which can be a character string, a character variable, the label of a FORMAT statement, or *. The data is written to the internal character variable *buffer*.

10.7.3 Exercises

10-1. Determine the contents of each variable in the following code fragment after the code has been executed:

```
CHARACTER(len=16) :: a = '1234567890123456'
CHARACTER(len=16) :: b = 'ABCDEFGHIJKLMNOP', c
IF ( a > b ) THEN
   c = a(1:6) // b(7:12) // a(13:16)
ELSE
   c = b(7:12) // a(1:6) // a(13:16)
END IF
a(7:9) = '='
```

10-2. Determine the contents of each variable in the following code fragment after the code has been executed. How does the behavior of this code fragment differ from the behavior of the one in Exercise 10-1?

```
CHARACTER(len=16)  :: a = '1234567890123456'
CHARACTER(len=16)  :: b = 'ABCDEFGHIJKLMNOP', c
IF ( LGT(a,b) ) THEN
   c = a(1:6) // b(7:12) // a(13:16)
ELSE
   c = b(7:12) // a(1:6) // a(13:16)
END IF
a(7:9) = '='
```

10-3. Rewrite subroutine ucase as a character function. Note that this function must return a variable-length character string.

10-4. Write a subroutine lcase that properly converts a string to lowercase regardless of collating sequence.

10-5. Determine the order in which the following character strings will be sorted by the subroutine sortc of Example 10-1 according to the ASCII collating sequence.

```
'This is a test!'
'?well?'
```

```
'AbCd'
'aBcD'
'1DAY'
'2nite'
'/DATA/'
'quit'
```

10-6. Determine the contents of each variable in the following code fragment after the code has been executed:

```
CHARACTER(len=132) :: buffer
REAL :: a, b
INTEGER :: i = 1700, j = 2400
a = REAL(1700 / 2400)
b = REAL(1700) / 2400
WRITE (buffer,100) i, j, a, b
100 FORMAT (T11,I10,T31,I10,T51,F10.4,T28,F10.4)
```

10-7. Write a subroutine `caps` that searches for all of the words within a character variable and capitalizes the first letter of each word, while shifting the remainder of the word to lowercase. Assume that all nonalphabetic and nonnumeric characters can mark the boundaries of a word within the character variable (e.g., periods, commas, etc.). Nonalphabetic characters should be left unchanged. Test your routine on the following character variables:

```
CHARACTER(len=40) :: a = 'this is a test--does it work?'
CHARACTER(len=40) :: b = 'this iS the 2nd test!'
CHARACTER(len=40) :: c = '123 WHAT NOW?!? xxxooxxx.'
```

10-8. Rewrite subroutine `caps` as a variable-length character function, and test the function using the same data as in the previous exercise.

10-9. The intrinsic function `LEN` returns the number of characters that a character variable can store, *not* the number of characters actually stored in the variable. Write a function `len_used` that returns the number of characters actually used within a variable. The function should determine the number of characters actually used by determining the positions of the first and last nonblank characters in the variable and performing the appropriate math. Test your function with the following variables. Compare the results of function `len_used` with the results returned by `LEN` and `LEN_TRIM` for each of the values given.

```
CHARACTER(len=30) :: a(3)
a(1) = 'How many characters are used?'
a(2) = '   ...and how about this one?'
a(3) = '     !   !              '
```

10-10. When a relatively short character string is assigned to a longer character variable, the extra space in the variable is filled with blanks. In many circumstances, we would like to use a substring consisting of only the *nonblank* portions of the character variable. To do so, we need to know where the nonblank portions are within the variable. Write a subroutine that will accept a character string of arbitrary length, and return two integers containing the numbers of the first and last nonblank characters in the variable. Test your subroutine with several character variables of different lengths and with different contents.

10-11. Input Parameter File A common feature of large programs is an *input parameter file* in which the user can specify certain values to be used during the execution of the program. In simple programs, the values in the file must be listed in a specific order and none of them may be skipped. These values may be read with a series of consecutive READ statements. If a value is left out of the input file or an extra value is added to the input file, all subsequent READ statements are misaligned and the numbers will go into the wrong locations in the program.

In more sophisticated programs, default values are defined for the input parameters in the file. In such a system, *only the input parameters whose defaults need to be modified need to be included in the input file*. Furthermore, the values that do appear in the input file may occur in any order. Each parameter in the input file is recognized by a corresponding *keyword* indicating what that parameter is for.

For example, a numerical integration program might include default values for the starting time of the integration, the ending time of the integration, the step size to use, and whether or not to plot the output. These values could be overridden by lines in the input file. An input parameter file for this program might contain the following items:

```
start = 0.0
stop = 10.0
dt = 0.2
plot off
```

These values could be listed in any order, and some of them could be omitted if the default values are acceptable. In addition, the keywords might appear in uppercase, lowercase, or mixed case. The program will read this input file a line at a time and update the variable specified by the keyword with the value on the line.

Write a subroutine that accepts a character argument containing a line from the input parameter file, and has the following output arguments:

```
REAL :: start, stop, dt
LOGICAL :: plot
```

The subroutine should check for a keyword in the line and update the variable that matches that keyword. It should recognize the keywords 'START', 'STOP', 'DT', and 'PLOT'. If the keyword 'START' is recognized, the subroutine should check for an equal sign and use the value to the right of the equal sign update variable START. It should behave similarly for the other keywords with real values. If the keyword 'PLOT' is recognized, the subroutine should check for 'ON' or 'OFF' and update the logical accordingly. (*Hint:* Shift each line to uppercase for easy recognition. Then, use function INDEX to identify keywords.)

10-12. Histograms A *histogram* is a plot that shows how many times a particular measurement falls within a certain range of values. For example, consider the students in this class. Suppose that there are 30 students in the class and that their scores on the last exam fell within the following ranges:

Range	No. of Students
100–95	3
94–90	6
89–85	9
84–80	7
79–75	4
74–70	2
69–65	1

A plot of the number of students scoring in each range of numbers is a histogram.

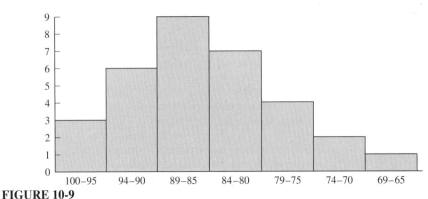

FIGURE 10-9
Histogram of student scores on last test.

To create this histogram, we started with a set of data consisting of 30 student grades. We divided the range of possible grades on the test (0 to 100) into 20 bins, and then counted how many student scores fell within each bin. Then we plotted the number of grades in each bin. (Since no one scored below 65 on the exam, we didn't bother to plot all of the empty bins between 0 and 64 in Figure 10-9.)

Write a subroutine that will accept an array of real input data values, divide them into a user-specified number of bins over a user-specified range, and accumulate the number of samples that fall within each bin. Create a simple plot of the histogram using asterisks to represent the levels in each bin.

10-13. Use the random-number subroutine `random0` that was developed in Chapter 7 to generate an array of 100,000 random numbers in the range [0,1). Use the histogram subroutine developed in the previous exercise to divide the range between 0 and 1 into 20 bins and to calculate a histogram of the 100,000 random numbers. How uniform was the distribution of the numbers generated by the random number generator?

10-14. Write a program that opens a user-specified disk file containing the source code for a Fortran program. The program should copy the source code from the input file to a user-specified output file, stripping out any comments during the copying process. Assume that the Fortran source file is in free format, so that the ! character marks the beginning of a comment.

11

Additional Intrinsic Data Types

OBJECTIVES

- Understand what is meant by different KINDs of a given data type.
- Understand how to select a specific kind of REAL, INTEGER, or CHARACTER data.
- Know how to select the precision and range of a real variable in a computer-independent manner.
- Know how to allocate and use variables of the COMPLEX data type.

In this chapter, we will examine alternate kinds of the REAL, INTEGER, and CHARACTER data types, and how to select the desired kind for a particular problem. Then, we will turn our attention to an additional data type that is built into the Fortran language: the COMPLEX data type. The COMPLEX data type is used to store and manipulate complex numbers, which have both real and imaginary components.

11.1
ALTERNATE KINDS OF THE REAL DATA TYPE

The REAL (or floating-point) data type is used to represent numbers containing decimal points. On most computers, a **default real** variable is 4 bytes (or 32 bits) long. It is divided into two parts, a **mantissa** and an **exponent.** Modern computers use the IEEE 754 Standard for floating-point variables to implement real numbers. In this implementation, 24 bits of the number are devoted to the mantissa and 8 bits are devoted to the exponent. The 24 bits devoted to the mantissa are enough to represent 6 to 7 significant decimal digits, so a real number can have up to about 7 significant digits.[1] Similarly, the 8 bits of the exponent are enough to represent numbers as large as 10^{38} and as small as 10^{-38}.

---

[1] One bit is used to represent the sign of the number, and 23 bits are used to represent the magnitude of the mantissa. Since $2^{23} = 8,388,608$, it is possible to represent between 6 and 7 significant digits with a real number.

There are times when a 4-byte floating-point number cannot adequately express a value that we need to solve a problem. Scientists and engineers sometimes need to express a number to more than seven significant digits of precision, or to work with numbers larger than 10^{38} or smaller than 10^{-38}. In either case, we cannot use a 32-bit variable to represent the number. Fortran includes at least one longer version of the real data type for use in these circumstances.

The longer version of the REAL data type is usually 8 bytes (or 64 bits) long. In a typical implementation,[2] 53 bits of the number are devoted to the mantissa and 11 bits are devoted to the exponent. The 53 bits devoted to the mantissa are enough to represent 15 to 16 significant decimal digits. Similarly, the 11 bits of the exponent are enough to represent numbers as large as 10^{308} and as small as 10^{-308}.

The Fortran Standard guarantees that a Fortran compiler will support *at least* two sizes of real numbers. However, they do *not* specify how many bits must be used for each size. For traditional reasons, the shorter version of the REAL data type on any particular computer is known as **single precision,** and the longer version of the REAL data type on any particular computer is known as **double precision.** On most computers, a single-precision real value is stored in 32 bits and a double-precision real value is stored in 64 bits. However, some 64-bit processors use 64 bit for single precision and 128 bits for double precision. There is no guarantee that a "single-precision" variable will be the same length on different processors. This variability makes the terms "single precision" and "double precision" poor choices for describing the accuracy of a floating-point value. We will introduce a better way to specify the accuracy of a floating-point value in Section 11.1.3 below.

Most Fortran compilers now also support a 16-byte (128-bit) REAL data type, which is usually known as **quadruple precision.** Quadruple precision can represent about 34 decimal digits and the exponent can cover numbers as large as 10^{4932} and as small as 10^{-4932}.

11.1.1 Kinds of REAL Constants and Variables

Since Fortran compilers have at least two different kinds of real variables, there must be some way to declare which of the types we want to use in a particular problem. This is done using a **kind type parameter.** Single-precision reals and double-precision reals are different **kinds** of the real data type, each with its own unique kind number. Examples of a real type declaration statement with a kind type parameter are shown below:

```
REAL(KIND=1) :: value_1
REAL(KIND=4) :: value_2
REAL(KIND=8), DIMENSION(20) :: array
REAL(4) :: temp
```

The kind of a real value is specified in parentheses after the REAL, either with or without the phrase KIND=. A variable declared with a kind type parameter is called a

---

[2] This statement refers to the IEEE Standard 754 for double-precision numbers. All new computer systems conform to this standard.

Additional Intrinsic Data Types

TABLE 11-1
KIND numbers for real values in some Fortran compilers

Computer/Compiler	32-bit real	64-bit real	128-bit real
PC/GNU Fortran	4*	8	16
PC/Intel Visual Fortran	4*	8	16
PC/NAGWare Fortran	1*	2	N/A

* denotes the **default real** type for a particular processor.

parameterized variable. If no kind is specified, then the default kind of real value is used. *The default kind may vary among different processors*, but is usually 32 bits long.

What do the kind numbers mean? Unfortunately, we do not know. Each compiler vendor is free to assign any kind number to any size of variable. For example, on some compilers, a 32-bit real value might be KIND=1 and a 64-bit real value might be KIND=2. On other compilers, a 32-bit real value might be KIND=4 and a 64-bit real value might be KIND=8. Table 11-1 shows examples of kind numbers for some representative computer/compiler combinations.

Therefore, to make your programs portable between computers, you should always assign kind numbers to a named constant and then use that named constant in all type declaration statements. It will then be possible to modify the program to run on different processors by changing only the value of the named constant. For example,

```
INTEGER, PARAMETER :: SGL = 4   ! Compiler dependent value
INTEGER, PARAMETER :: DBL = 8   ! Compiler dependent value
REAL(KIND=SGL) :: value_1
REAL(KIND=DBL), DIMENSION(20) :: array
REAL(SGL) :: temp
```

An even better approach for a large program would be to define the kind parameters within a module and to use that module in each procedure within the program. Then, it is possible to change the kind numbers for the entire program by editing a single file.

It is also possible to declare the kind of a real constant. The kind of a real constant is declared by appending an underscore and the kind number to the constant. The following are examples of valid real constants:

```
34.                ! Default kind
34._4              ! Only valid if 4 is a legal kind of real
34.E3              ! Single precision
1234.56789_DBL     ! Only valid if "DBL" is an integer named constant
```

The first example produces a constant of the *default kind* for the particular processor where the program is being executed. The second example is valid only if KIND=4 is a valid kind of real on the particular processor where the program is being executed. The third example produces a constant of the single-precision kind for the particular processor. The fourth example is only valid if DBL is a valid previously defined integer named constant, whose value is a valid kind number.

In addition to the above examples, a double-precision constant in exponential notation can be declared by using a D instead of an E to declare the exponent of the constant. For example,

        3.0E0   is a single-precision constant
        3.0D0   is a double-precision constant

Good Programming Practice
Always assign kind numbers to a named constant, and then use that named constant in all type declaration statements and constant declarations. This practice will make it easier to port the program to different computers that may use different kind numbers. For large programs, place the named constants containing the kind parameters in a single module, and then use that module in every procedure within the program.

11.1.2 Determining the KIND of a Variable

Fortran includes an intrinsic function KIND, which returns the kind number of a given constant or variable. This function can be used to determine the kind numbers in use by your compiler. For example, the program in Figure 11-1 determines the kind numbers associated with single- and double-precision variables on a particular processor.

FIGURE 11-1
Program to determine the kind numbers associated with single- and double-precision real variables on a particular computer system.

```
PROGRAM kinds
!
!  Purpose:
!    To determine the kinds of single and double precision real
!    values on a particular computer.
!
IMPLICIT NONE

! Write out the kinds of single & double precision values
WRITE (*,'("The KIND for single precision is",I2)') KIND(0.0)
WRITE (*,'("The KIND for double precision is",I2)') KIND(0.0D0)

END PROGRAM kinds
```

When this program is executed on a PC using the Intel Visual Fortran compiler, the results are:

```
C:\book\fortran\chap11>kinds
The KIND for single precision is 4
The KIND for double precision is 8
```

When the program is executed on a PC using the NAGWare Fortran compiler, the results are:

```
C:\book\fortran\chap11>kinds
The KIND for single precision is 1
The KIND for double precision is 2
```

As you can see, the kind numbers will vary from processor to processor. Try the program on your own computer/compiler and see what values you get.

11.1.3 Selecting Precision in a Processor-Independent Manner

A major problem encountered when porting a Fortran program from one computer to another one is the fact that the terms "single precision" and "double precision" are not precisely defined. Double-precision values have approximately twice the precision of single-precision values, but the number of bits associated with each kind of real is entirely up to the computer vendor. On most computers, a single-precision value is 32 bits long and a double-precision value is 64 bits long. However, on some computers such as Cray Supercomputers and those based on the 64-bit Intel® Itanium® chip, single precision is 64 bits long and double precision is 128 bits long. Thus, a program that runs properly in single precision on a Cray might need double precision to run properly when it is migrated to a 32-bit computer, and a program that requires double precision for proper operation on a 32-bit computer will only need single precision on a computer based on the 64-bit Itanium® chip.

How can we write programs so that they can be easily ported between processors with different word sizes and still function correctly? We can use an intrinsic function to *automatically select the proper kind of real value to use* as the program is moved between computers. This function is called SELECTED_REAL_KIND. When it is executed, it returns the kind number of the *smallest* type of real value that meets or exceeds the specified range and precision on that particular processor. The general form of this function is

```
kind_number = SELECTED_REAL_KIND(p=precision,r=range)
```

where *precision* is the number of decimal digits of precision required and *range* is the range of the exponent required in powers of 10. The two arguments *precision* and *range* are called optional arguments; either one or both may be supplied to specify the desired characteristics of the real value. The function returns the kind number of the smallest real kind satisfying the specified requirements. It returns a −1 if the specified precision is not available from any real data type on the processor, a −2 if the specified range is not available from any real data type on the processor, and a −3 if neither is available.

All of the following are legal uses of this function

```
kind_number = SELECTED_REAL_KIND(p=6,r=37)
kind_number = SELECTED_REAL_KIND(p=12)
kind_number = SELECTED_REAL_KIND(r=100)
kind_number = SELECTED_REAL_KIND(13,200)
kind_number = SELECTED_REAL_KIND(13)
kind_number = SELECTED_REAL_KIND(p=17)
```

On an Intel Core-i7-based computer using the Intel Visual Fortran compiler, the first of the functions will return a 4 (the kind number for single precision) and the next four will return an 8 (the kind number for double precision). The last function will return a 16, since Intel Visual Fortran supports a 16-byte real that supplies 17 decimal digits of precision.

Notice from the above example that the p= and r= are optional as long as *precision* and *range* are specified in that order, and the p= is optional if only the precision is specified. These are general characteristics of optional arguments, which we will learn more about in Chapter 13.

The function SELECTED_REAL_KIND should be used with a certain amount of caution, since over-specifying your program's requirements can increase the program's size and slow down execution. For example, 32-bit computers have between 6 and 7 decimal digits of precision in their single-precision variables. If you specify a real data type as SELECTED_REAL_KIND(6), then you will get single precision on those machines. However, if you specify a real data type as SELECTED_REAL_KIND(7), then you will get double precision and the program will be both larger and slower. Make sure that you really need that seventh decimal place before you ask for it![3]

Good Programming Practice
Use the function SELECTED_REAL_KIND to determine the kind numbers of the real variables needed to solve a problem. The function will return the proper kind numbers on any computer, making your programs more portable.

Three other intrinsic functions are available that can be used to determine kind of a real value and the precision and range of the real value on a particular computer. These functions are summarized in Table 11-2. The integer function KIND( ) returns the kind number of a specified value. The integer function PRECISION( ) returns the

TABLE 11-2
Common KIND-related intrinsic functions

Function	Description
SELECTED_REAL_KIND(p,r)	Return smallest kind of real value with a minimum of p decimal digits of precision and maximum range $\geq 10^r$.
SELECTED_INT_KIND(r)	Return smallest kind of integer value with a maximum range $\geq 10^r$.
KIND(X)	Return kind number of X, where X is a variable or constant of any intrinsic type.
PRECISION(X)	Return decimal precision of X, where X is a real or complex value.
RANGE(X)	Return the decimal exponent range for X, where X is an integer, real, or complex value.

---

[3] Fortran 2008 added a third optional parameter RADIX, which specifies the base of the numbering system desired (e.g., Base 2 versus Base 10). I know of no compiler that has yet implemented this feature.

number of decimal digits that can be stored in the real value, and the integer function RANGE( ) returns the exponent range that can be supported by the real value. The use of these functions is illustrated in the program in Figure 11-2.

FIGURE 11-2
Program to illustrate the use of function SELECTED_REAL_KIND( ) to select desired kinds of real variables in a processor-independent manner, and the use of functions KIND( ), PRECISION( ), and RANGE( ) to get information about real values.

```
PROGRAM select_kinds
!
!  Purpose:
!    To illustrate the use of SELECTED_REAL_KIND to select
!    desired kinds of real variables in a processor-independent
!    manner.
!
!  Record of revisions:
!      Date        Programmer          Description of change
!      ====        ==========          =====================
!    11/28/15      S. J. Chapman       Original code
!
IMPLICIT NONE

! Declare parameters:
INTEGER, PARAMETER :: SGL = SELECTED_REAL_KIND(p=6,r=37)
INTEGER, PARAMETER :: DBL = SELECTED_REAL_KIND(p=13,r=200)

! Declare variables of each type:
REAL(kind=SGL) :: var1 = 0.
REAL(kind=DBL) :: var2 = 0._DBL

! Write characteristics of selected variables.
WRITE (*,100) 'var1', KIND(var1), PRECISION(var1), RANGE(var1)
WRITE (*,100) 'var2', KIND(var2), PRECISION(var2), RANGE(var2)
100 FORMAT(A,': kind = ',I2,', Precision = ',I2,', Range = ',I3)

END PROGRAM select_kinds
```

When this program is executed on an Intel Core-i7-based PC using the Intel Visual Fortran compiler, the results are:

```
C:\book\fortran\chap11>select_kinds
var1: kind = 4, Precision =  6, Range =  37
var2: kind = 8, Precision = 15, Range = 307
```

Note that the program requested 13 decimal digits of precision and a range of 200 powers of 10 for the second variable, but the variable actually assigned by the processor has 15 digits of precision and a range of 308 powers of 10. This type of real variable was the smallest size available on the processor that met or exceeded the request. Try this program on your own computer and see what values you get.

11.1.4 Determining the KINDs of Data Types on a Particular Processor

Fortran includes an intrinsic module called iso_Fortran_env that contains information about the kinds of the data types that are available on a given processor and also standard names for constants describing the different types of data.[4] Some of the constants described in this intrinsic module are given in Table 11-3.

We can use the constants in module called iso_Fortran_env to select data sizes in a processor-independent manner. For example, we can use the following code to request a 16-bit integer and a 128-bit real variable on any computer in a processor-independent manner:

```
USE iso_Fortran_env
!
INTEGER(KIND=INT16) :: i
REAL(KIND=INT128) :: x
```

This is a very good way to specify data sizes in a processing-independent manner. However, it is relatively new and some compilers have not implemented this feature yet.

Good Programming Practice

Use the constants in module iso_Fortran_env to specify data sizes in a processor-independent manner.

11.1.5 Mixed-Mode Arithmetic

When an arithmetic operation is performed between a double-precision real value and another real or integer value, Fortran converts the other value to double precision and performs the operation in double precision with a double-precision result.

TABLE 11-3
Common KIND-constants in intrinsic module iso_Fortran_env

Function	Description
CHARACTER_KINDS	Returns a default integer array holding all the kind values supported for type CHARACTER.
INTEGER_KINDS	Returns a default integer array holding all the kind values supported for type INTEGER.
LOGICAL_KINDS	Returns a default integer array holding all the kind values supported for type LOGICAL.
REAL_KINDS	Returns a default integer array holding all the kind values supported for type REAL.
INT8, INT16, INT32, INT64	Standard constants to request 8-, 16-, 32-, and 64-bit integers on the current processor.
REAL32, REAL64, REAL128	Standard constants to request 32-, 64-, and 128-bit real or complex numbers on the current processor.

---

[4] This feature was added in Fortran 2008.

However, the automatic mode conversion *does not occur* until both the double-precision number and the other number appear in the same operation. Therefore, it is possible for a portion of an expression to be evaluated in integer or single-precision real arithmetic, followed by another portion evaluated in double-precision real arithmetic.

For example, suppose that a particular processor uses 32 bits to represent single-precision real values and 64 bits to represent double-precision real values. Then suppose that we want to add ⅓ to ⅓, and get the answer to 15 significant digits. We might try to calculate the answer with any of the following expressions:

	Expression	Result
1.	1.D0/3. + 1/3	3.333333333333333E-001
2.	1./3. + 1.D0/3.	6.666666333333333E-001
3.	1.D0/3. + 1./3.D0	6.666666666666666E-001

1. In the first expression, the single-precision constant 3. is converted to double precision before dividing into the double-precision constant 1.D0, producing the result 3.333333333333333E-001. Next, the integer constant 1 is divided by the integer constant 3, producing an integer 0. Finally, the integer 0 is converted into double precision and added to first number, producing the final value of 3.333333333333333E-001.

2. In the second expression, 1./3. is evaluated in single precision producing the result 3.333333E-01, and 1./3.D0 is evaluated in double precision, producing the result 3.333333333333333E-001. Then, the single-precision result is converted to double precision and added to the double-precision result to produce the final value of 6.666666333333333E-001.

3. In the third expression, both terms are evaluated in double precision, leading to a final value of 6.666666666666666E-001.

As we can see, adding ⅓ + ⅓ produces significantly different answers depending on the type of numbers used in each part of the expression. The third expression shown above yields the answer that we really wanted, while the first two are inaccurate to a greater or lesser degree. This result should serve as a warning: If you really need double-precision arithmetic, you should be very careful to ensure that *all* intermediate portions of a calculation are performed with double-precision arithmetic and that all intermediate results are stored in double-precision variables.

A special case of mixed-mode arithmetic occurs during the initialization of double-precision real variables in type declaration statements and DATA statements. *If the constant used to initialize the variable is written in single-precision form, then the variable will only be initialized to single-precision accuracy*, regardless of the number of significant digits written in the constant.[5] For example, the variable a1 in

---

[5] FORTRAN 77 behaved differently here—it would permit all of the digits of a constant to be used in an initialization statement, even if there were more digits than a single-precision value could support. This difference could cause problems when transporting a FORTRAN 77 program to modern Fortran.

the following program is only initialized to seven significant digits even though it is double precision:

```
PROGRAM test_initial
INTEGER, PARAMETER :: DBL = SELECTED_REAL_KIND(p=13)
REAL(KIND=DBL) :: a1 = 6.666666666666666
REAL(KIND=DBL) :: a2 = 6.666666666666666_DBL
WRITE (*,*) a1, a2
END PROGRAM test_initial
```

When this program is executed, the value of a1 is valid to only seven significant digits:

```
C:\book\fortran\chap11>test_initial
   6.666666507720947         6.666666666666666
```

Programming Pitfalls
Always be careful to initialize double-precision real variables with double-precision real constants so that the full precision of the constant is preserved.

11.1.6 Higher Precision Intrinsic Functions

All generic functions that support single-precision real values will also support double-precision real values. If the input value is single precision, then the function will be calculated with a single-precision result. If the input value is double precision, then the function will be calculated with a double-precision result.

One important intrinsic function is DBLE. This function converts any numeric input argument to double precision on the particular processor where it is executed.

11.1.7 When to Use High-Precision Real Values

We have seen that 64-bit real numbers are better than 32-bit real numbers, offering more precision and greater range. If they are so good, why bother with 32-bit real numbers at all? Why don't we just use 64-bit real numbers all the time?

There are a couple of good reasons for not using 64-bit real numbers all the time. For one thing, every 64-bit real number requires twice as much memory as a 32-bit real number. This extra size makes programs using them much larger, and computers with more memory are required to run the programs. Another important consideration is speed. Higher precision calculations are normally slower than lower precision calculations, so computer programs using higher precision calculations run more slowly than computer programs using lower precision calculations.[6] Because of these disadvantages, we should only use higher precision numbers when they are actually needed.

---

[6] Intel-based PCs are an exception to this general rule. The math processor performs hardware calculations with 80-bit accuracy regardless of the precision of the data being processed. As a result, there is little speed penalty for double precision operations on a PC.

When are 64-bit numbers actually needed? There are three general cases:

1. *When the dynamic range of the calculation requires numbers whose absolute values are smaller than 10^{-39} or larger than 10^{39}.* In this case, the problem must either be rescaled or 64-bit variables must be used.
2. *When the problem requires numbers of very different sizes to be added to or subtracted from one another.* If two numbers of very different sizes must be added or subtracted from one another, the resulting calculation will lose a great deal of precision. For example, suppose we wanted to add the number 3.25 to the number 1000000.0. With 32-bit numbers, the result would be 1000003.0. With 64-bit numbers, the result would be 1000003.25.
3. *When the problem requires two numbers of very nearly equal size to be subtracted from one another.* When two numbers of very nearly equal size must be subtracted from each other, small errors in the last digits of the two numbers become greatly exaggerated.

For example, consider two nearly equal numbers that are the result of a series of single-precision calculations. Because of the round-off error in the calculations, each of the numbers is accurate to 0.0001%. The first number a1 should be 1.0000000, but through round-off errors in previous calculations is actually 1.0000010, while the second number a2 should be 1.0000005, but through round-off errors in previous calculations is actually 1.0000000. The difference between these numbers should be

$$\text{true\_result} = \text{a1} - \text{a2} = -0.0000005$$

but the actual difference between them is

$$\text{actual\_result} = \text{a1} - \text{a2} = 0.0000010$$

Therefore, the error in the subtracted number is

$$\%\text{error} = \frac{\text{actual\_result} - \text{true\_result}}{\text{true\_result}} \times 100\%$$

$$\%\text{error} = \frac{0.0000010 - (-0.0000005)}{-0.0000005} \times 100\% = -300\%$$

The single-precision math created a 0.0001% error in a1 and a2, and then the subtraction blew that error up into a 300% error in the final answer! When two nearly equal numbers must be subtracted as a part of a calculation, then the entire calculation should be performed in higher precision to avoid round-off error problems.

EXAMPLE 11-1 *Numerical Calculation of Derivatives:*

The derivative of a function is defined mathematically as

$$\frac{d}{dx}f(x) = \lim_{\Delta x \to 0} \frac{f(x + \Delta x) - f(x)}{\Delta x} \tag{11-1}$$

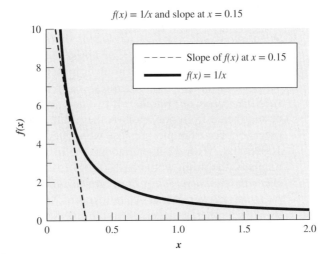

FIGURE 11-3
Plot of the function $f(x) = 1/x$, showing the slope at $x = 0.15$.

The derivative of a function is a measure of the instantaneous slope of the function at the point being examined. In theory, the smaller Δx, the better the estimate of the derivative is. However, the calculation can go bad if there is not enough precision to avoid round-off errors. Note that as Δx gets small, we will be subtracting two numbers that are very nearly equal, and the effects of round-off errors will be multiplied.

To test the effects of precision on our calculations, we will calculate the derivative of the function

$$f(x) = \frac{1}{x} \tag{11-2}$$

for the location $x = 0.15$. This function is shown in Figure 11-3.

SOLUTION
From elementary calculus, the derivative of $f(x)$ is

$$\frac{d}{dx} f(x) = \frac{d}{dx} \frac{1}{x} = -\frac{1}{x^2} \tag{11-3}$$

For $x = 0.15$,

$$\frac{d}{dx} f(x) = \frac{d}{dx} \frac{1}{x} = -\frac{1}{x^2} = -44.444444444\ldots \tag{11-4}$$

We will now attempt to evaluate the derivative of Equation (11-2) for sizes of Δx from 10^{-1} to 10^{-10} using both 32- and 64-bit mathematics on a computer that has a 32-bit single-precision and a 64-bit double-precision real data type. We will print out the results for each case, together with the true analytical solution and the resulting error.

A Fortran program to evaluate the derivative of Equation (11-2) is shown in Figure 11-4.

FIGURE 11-4
Program to evaluate the derivative of the function $f(x) = 1/x$ at $x = 0.15$ using both single-precision and double-precision arithmetic.

```fortran
PROGRAM diff
!
!  Purpose:
!    To test the effects of finite precision by differentiating
!    a function with 10 different step sizes, with both single
!    precision and double precision. The test will be based on
!    the function F(X) = 1./X.
!
!  Record of revisions:
!      Date        Programmer          Description of change
!      ====        ==========          =====================
!    12/01/15      S. J. Chapman       Original code
!
IMPLICIT NONE

! Data dictionary: declare constants
INTEGER, PARAMETER :: SGL = SELECTED_REAL_KIND(p=6,r=37)
INTEGER, PARAMETER :: DBL = SELECTED_REAL_KIND(p=13)

! List of local variables:
REAL(KIND=DBL) :: ans              ! True (analytic) answer
REAL(KIND=DBL) :: d_ans            ! Double precision answer
REAL(KIND=DBL) :: d_error          ! Double precision percent error
REAL(KIND=DBL) :: d_fx             ! Double precision F(x)
REAL(KIND=DBL) :: d_fxdx           ! Double precision F(x+dx)
REAL(KIND=DBL) :: d_dx             ! Step size
REAL(KIND=DBL) :: d_x = 0.15_DBL   ! Location to evaluate dF(x)/dx
INTEGER :: i                       ! Index variable
REAL(KIND=SGL) :: s_ans            ! Single precision answer
REAL(KIND=SGL) :: s_error          ! Single precision percent error
REAL(KIND=SGL) :: s_fx             ! Single precision F(x)
REAL(KIND=SGL) :: s_fxdx           ! Single precision F(x+dx)
REAL(KIND=SGL) :: s_dx             ! Step size
REAL(KIND=SGL) :: s_x = 0.15_SGL   ! Location to evaluate dF(x)/dx

! Print headings.
WRITE (*,1)
1 FORMAT ('      DX         TRUE ANS       SP ANS         DP ANS  ', &
          '     SP ERR    DP ERR   ')

! Calculate analytic solution at x=0.15.
ans = - ( 1.0_DBL / d_x**2 )

! Calculate answer from definition of differentiation
step_size: DO i = 1, 10

   ! Get delta x.
   s_dx = 1.0 / 10.0**i
   d_dx = 1.0_DBL / 10.0_DBL**i
```

(*continued*)

(*concluded*)

```
! Calculate single precision answer.
s_fxdx = 1. / ( s_x + s_dx )
s_fx   = 1. / s_x
s_ans  = ( s_fxdx - s_fx ) / s_dx

! Calculate single precision error, in percent.
s_error = ( s_ans - REAL(ans) ) / REAL(ans) * 100.

! Calculate double precision answer.
d_fxdx = 1.0_DBL / ( d_x + d_dx )
d_fx   = 1.0_DBL / d_x
d_ans  = ( d_fxdx - d_fx ) / d_dx

! Calculate double precision error, in percent.
d_error = ( d_ans - ans ) / ans * 100.

! Tell user.
WRITE (*,100) d_dx, ans, s_ans, d_ans, s_error, d_error
100 FORMAT (ES10.3, F12.7, F12.7, ES22.14, F9.3, F9.3)

END DO step_size

END PROGRAM diff
```

When this program is compiled and executed using Intel Visual Fortran Version 16 on a PC, the following results are obtained:

```
C:\book\fortran\chap11>diff
      DX       TRUE ANS     SP ANS       DP ANS                  SP ERR   DP ERR
   1.000E-01  -44.4444444 -26.6666641 -2.66666666666667E+01    -40.000  -40.000
   1.000E-02  -44.4444444 -41.6666527 -4.16666666666667E+01     -6.250   -6.250
   1.000E-03  -44.4444444 -44.1503487 -4.41501103752762E+01     -0.662   -0.662
   1.000E-04  -44.4444444 -44.4173813 -4.44148345547379E+01     -0.061   -0.067
   1.000E-05  -44.4444444 -44.4412231 -4.44414816790584E+01     -0.007   -0.007
   1.000E-06  -44.4444444 -44.3458557 -4.44441481501912E+01     -0.222   -0.001
   1.000E-07  -44.4444444 -47.6837158 -4.44444148151035E+01      7.288    0.000
   1.000E-08  -44.4444444 -47.6837158 -4.44444414604561E+01      7.288    0.000
   1.000E-09  -44.4444444   0.0000000 -4.44444445690806E+01   -100.000    0.000
   1.000E-10  -44.4444444   0.0000000 -4.44444481217943E+01   -100.000    0.000
```

When Δx was fairly large, both the single- and double-precision results give essentially the same answer. In that range, the accuracy of the result is only limited by the step size. As Δx gets smaller and smaller, the single-precision answer gets better and better until $\Delta x \approx 10^{-5}$. For step sizes smaller than 10^{-5}, round-off errors start to dominate the solution. The double-precision answer gets better and better until $\Delta x \approx 10^{-9}$. For step sizes smaller than 10^{-9}, double-precision round-off errors start to get progressively worse.

In this problem, the use of double precision allowed us to improve the quality of our answer from four correct significant digits to eight correct significant digits. The problem also points out the critical importance of a proper Δx size in producing a right answer. Such concerns occur in all computer programs performing scientific and engineering calculations. In all such programs, there are parameters that *must* be chosen correctly, or else round-off errors will result in bad answers. The design of proper algorithms for use on computers is a whole discipline in itself, known as *numerical analysis*.

11.1.8 Solving Large Systems of Simultaneous Linear Equations

In Chapter 9, we introduced the method of Gauss-Jordan elimination to solve systems of simultaneous linear equations of the form

$$a_{11}x_1 + a_{12}x_2 + \cdots + a_{1n}x_n = b_1$$
$$a_{21}x_1 + a_{22}x_2 + \cdots + a_{2n}x_n = b_2 \qquad (11\text{-}5)$$
$$\cdots$$
$$a_{n1}x_1 + a_{n2}x_2 + \cdots + a_{nn}x_n = b_n$$

In the Gauss-Jordan method, the first equation in the set is multiplied by a constant and added to all of the other equations in the set to eliminate x_1, and then the process is repeated with the second equation in the set multiplied by a constant and added to all of the other equations in the set to eliminate x_2, and so forth for all of the equations. *This type of solution is subject to cumulative round-off errors that eventually make the answers unusable.* Any round-off errors in eliminating the coefficients of x_1 are propagated into even bigger errors when eliminating the coefficients of x_2, which are propagated into even bigger errors when eliminating the coefficients of x_3, etc. For a large enough system of equations, the cumulative round-off errors will produce unacceptably bad solutions.

How big must a system of equations be before round-off error makes it impossible to solve them using Gauss-Jordan elimination? There is no easy answer to this question. Some systems of equations are more sensitive to slight round-off errors than others are. To understand why this is so, let's look at the two simple sets of simultaneous equations shown in Figure 11-5. Figure 11-5a shows a plot of the two simultaneous equations

$$3.0x - 2.0y = 3.0$$
$$5.0x + 3.0y = 5.0 \qquad (11\text{-}6)$$

The solution to this set of equations is $x = 1.0$ and $y = 0.0$. The point (1.0, 0.0) is the intersection of the two lines on the plot in Figure 11-5a. Figure 11-5b shows a plot of the two simultaneous equations

$$1.00x - 1.00y = -2.00$$
$$1.03x - 0.97y = -2.03 \qquad (11\text{-}7)$$

The solution to this set of equations is $x = -1.5$ and $y = 0.5$. The point (−1.5, 0.5) is the intersection of the two lines on the plot in Figure 11-5b.

Now let's compare the sensitivity of Equations (11-6) and (11-7) to slight errors in the coefficients of the equations. (A slight error in the coefficients of the equations is similar to the effect of round-off errors on the equations.) Assume that coefficient

FIGURE 11-5
(a) Plot of a well-conditioned 2 × 2 set of equations. (b) Plot of an ill-conditioned 2 × 2 set of equations.

a_{11} of Equations (11-6) is in error 1 percent, so that a_{11} is really 3.03 instead of 3.00. Then the solution to the equations becomes $x = 0.995$ and $y = 0.008$, which is almost the same as the solution to the original equations. Now, let's assume that coefficient a_{11} of Equations (11-7) is in error by 1 percent, so that a_{11} is really 1.01 instead of 1.00. Then the solution to the equations becomes $x = 1.789$ and $y = 0.193$, which is a major shift compared to the previous answer. Equations (11-6) are relatively insensitive to small coefficient errors, while Equations (11-7) are *very* sensitive to small coefficient errors.

Additional Intrinsic Data Types

If we examine Figure 11-5*b* closely, it will be obvious why Equations (11-7) are so sensitive to small changes in coefficients. The lines representing the two equations are almost parallel to each other, so a tiny change in one of the equations moves their intersection point by a very large distance. If the two lines had been exactly parallel to each other, then the system of equations would either have had no solutions or an infinite number of solutions. In the case where the lines are nearly parallel, there is a single unique solution, but its location is very sensitive to slight changes in the coefficients. Therefore, systems like Equations (11-7) will be very sensitive to accumulated round-off noise during Gauss-Jordan elimination.

Systems of simultaneous equations that behave well like Equations (11-6) are called **well-conditioned systems,** and systems of simultaneous equations that behave poorly like Equations (11-7) are called **ill-conditioned systems.** Well-conditioned systems of equations are relatively immune to round-off error, while ill-conditioned systems are very sensitive to round-off error.

When working with very large systems of equations or ill-conditioned systems of equations, it is helpful to work in double-precision arithmetic. Double precision arithmetic dramatically reduces round-off errors, allowing Gauss-Jordan elimination to produce correct answers even for difficult systems of equations.

EXAMPLE 11-2

Solving Large Systems of Linear Equations:

For large and/or ill-conditioned systems of equations, Gauss-Jordan elimination will only produce a correct answer if double-precision arithmetic is used to reduce round-off error. Write a subroutine that uses double-precision arithmetic to solve a system of simultaneous linear equations. Test your subroutine by comparing it to the single-precision subroutine `simul` created in Chapter 9. Compare the two subroutines on both well-defined and ill-defined systems of equations.

SOLUTION
The double-precision subroutine `dsimul` will be essentially the same as the single-precision subroutine `simul2` in Figure 9-6 that we developed in Chapter 9. Subroutine `simul2`, which is renamed `simul` here, is used as the starting point because that version includes both the use of array operations and automatic arrays for simplicity and flexibility, and because it does not destroy its input data.

1. **State the problem.**
   Write a subroutine to solve a system of *N* simultaneous equations in *N* unknowns using Gauss-Jordan elimination, double-precision arithmetic, and the maximum pivot technique to avoid round-off errors. The subroutine must be able to detect singular sets of equations and set an error flag if they occur.

2. **Define the inputs and outputs.**
   The input to the subroutine consists of an $N \times N$ double-precision matrix a with the coefficients of the variables in the simultaneous equations, and a double-precision vector b with the contents of the right-hand sides of the equations. The outputs from

the subroutine are the solutions to the set of equations (in vector `soln`) and an error flag.

3. **Describe the algorithm.**
   The pseudocode for this subroutine is the same as the pseudocode for subroutine `simul2` in Chapter 9, and is not repeated here.

4. **Turn the algorithm into Fortran statements.**
   The resulting Fortran subroutine is shown in Figure 11-6. Note that we are using the constants in intrinsic module `iso_Fortran_env` to specify that we want 64-bit real variables in this subroutine.

FIGURE 11-6
Subroutine to solve a system of simultaneous equations in double precision.

```
SUBROUTINE dsimul ( a, b, soln, ndim, n, error )
!
!  Purpose:
!    Subroutine to solve a set of N linear equations in N
!    unknowns using Gaussian elimination and the maximum
!    pivot technique. This version of simul has been
!    modified to use array sections and automatic arrays.
!    It uses double precision arithmetic to avoid
!    cumulative roundoff errors. It DOES NOT DESTROY the
!    original input values.
!
!  Record of revisions:
!      Date          Programmer        Description of change
!      ====          ==========        =====================
!    11/25/15      S. J. Chapman       Original code
! 1. 11/25/15      S. J. Chapman       Add automatic arrays
! 2. 12/01/15      S. J. Chapman       Double precision
!
USE iso_Fortran_env
IMPLICIT NONE

! Data dictionary: declare constants
REAL(KIND=REAL64), PARAMETER :: EPSILON = 1.0E-12
                                ! A "small" number for comparison
                                ! when determining singular eqns

! Data dictionary: declare calling parameter types & definitions
INTEGER, INTENT(IN) :: ndim              ! Dimension of arrays a and b
REAL(KIND=REAL64), INTENT(IN), DIMENSION(ndim,ndim) :: a
                                ! Array of coefficients (N x N).
                                ! This array is of size ndim x
                                ! ndim, but only N x N of the
                                ! coefficients are being used.
REAL(KIND=REAL64), INTENT(IN), DIMENSION(ndim) :: b
                                ! Input: Right-hand side of eqns.
REAL(KIND=REAL64), INTENT(OUT), DIMENSION(ndim) :: soln
                                ! Output: Solution vector.
```

(continued)

Additional Intrinsic Data Types

(*continued*)

```fortran
INTEGER, INTENT(IN) :: n              ! Number of equations to solve.
INTEGER, INTENT(OUT) :: error         ! Error flag:
                                      !   0 -- No error
                                      !   1 -- Singular equations

! Data dictionary: declare local variable types & definitions
REAL(KIND=REAL64), DIMENSION(n,n) :: a1 ! Copy of "a" which will be
                                      ! destroyed during the solution
REAL(KIND=REAL64) :: factor           ! Factor to multiply eqn irow by
                                      ! before adding to eqn jrow
INTEGER :: irow                       ! Number of the equation currently
                                      ! currently being processed
INTEGER :: ipeak                      ! Pointer to equation containing
                                      ! maximum pivot value
INTEGER :: jrow                       ! Number of the equation compared
                                      ! to the current equation
REAL(KIND=REAL64) :: temp             ! Scratch value
REAL(KIND=REAL64),DIMENSION(n) :: temp1 ! Scratch array

! Make copies of arrays "a" and "b" for local use
a1 = a(1:n,1:n)
soln = b(1:n)

! Process N times to get all equations...
mainloop: DO irow = 1, n

   ! Find peak pivot for column irow in rows irow to N
   ipeak = irow
   max_pivot: DO jrow = irow+1, n
      IF (ABS(a1(jrow,irow)) > ABS(a1(ipeak,irow))) THEN
         ipeak = jrow
      END IF
   END DO max_pivot

   ! Check for singular equations.
   singular: IF ( ABS(a1(ipeak,irow)) < EPSILON ) THEN
      error = 1
      RETURN
   END IF singular

   ! Otherwise, if ipeak /= irow, swap equations irow & ipeak
   swap_eqn: IF ( ipeak /= irow ) THEN
      temp1 = a1(ipeak,1:n)
      a1(ipeak,1:n) = a1(irow,1:n)   ! Swap rows in a
      a1(irow,1:n) = temp1
      temp = soln(ipeak)
      soln(ipeak) = soln(irow)        ! Swap rows in b
      soln(irow) = temp
   END IF swap_eqn

   ! Multiply equation irow by -a1(jrow,irow)/a1(irow,irow),
   ! and add it to Eqn jrow (for all eqns except irow itself).
   eliminate: DO jrow = 1, n
```

(*continued*)

(concluded)

```
      IF ( jrow /= irow ) THEN
         factor = -a1(jrow,irow)/a1(irow,irow)
         a1(jrow,1:n) = a1(irow,1:n)*factor + a1(jrow,1:n)
         soln(jrow) = soln(irow)*factor + soln(jrow)
      END IF
   END DO eliminate
END DO mainloop

! End of main loop over all equations. All off-diagonal
! terms are now zero. To get the final answer, we must
! divide each equation by the coefficient of its on-diagonal
! term.
divide: DO irow = 1, n
   soln(irow) = soln(irow) / a1(irow,irow)
END DO divide

! Set error flag to 0 and return.
error = 0

END SUBROUTINE dsimul
```

5. **Test the resulting Fortran programs.**

To test this subroutine, it is necessary to write a driver program. The driver program will open an input data file to read the equations to be solved. The first line of the file will contain the number of equations N in the system, and each of the next N lines will contain the coefficients of one of the equations. The coefficients will be stored in a single-precision array and sent to subroutine `simul` for solution, and will also be stored in a double-precision array and sent to subroutine `dsimul` for solution. To verify that the solutions are correct, they will be plugged back into the original equations and the resulting errors will be calculated. The solutions and errors for single- and double-precision arithmetic will be displayed in a summary table.

The test driver program for subroutine `dsimul` is shown in Figure 11-7. Note that it uses allocatable arrays throughout, so it will work with input data sets of any size.

FIGURE 11-7
Test driver program for subroutine `dsimul`.

```
PROGRAM test_dsimul
!
! Purpose:
!    To test subroutine dsimul, which solves a set of N linear
!    equations in N unknowns. This test driver calls subroutine
!    simul to solve the problem in single precision, and subrou-
!    tine dsimul to solve the problem in double precision. The
!    results of the two solutions together with their errors are
!    displayed in a summary table.
!
```

(continued)

(*continued*)

```fortran
!   Record of revisions:
!      Date        Programmer        Description of change
!      ====        ==========        =====================
!    12/01/15    S. J. Chapman       Original code
!
USE iso_Fortran_env
IMPLICIT NONE

! List of local variables
REAL(KIND=REAL32), ALLOCATABLE, DIMENSION(:,:) :: a
                                 ! Single-precision coefficients
REAL(KIND=REAL32), ALLOCATABLE, DIMENSION(:) :: b
                                 ! Single-precision constant values
REAL(KIND=REAL32), ALLOCATABLE, DIMENSION(:) :: soln
                                 ! Single-precision solution
REAL(KIND=REAL32), ALLOCATABLE, DIMENSION(:) :: serror
                                 ! Array of single-precision errors
REAL(KIND=REAL32) :: serror_max  ! Max single precision error
REAL(KIND=REAL64), ALLOCATABLE, DIMENSION(:,:) :: da
                                 ! Double-precision coefficients
REAL(KIND=REAL64), ALLOCATABLE, DIMENSION(:) :: db
                                 ! Double-precision constant values
REAL(KIND=REAL64), ALLOCATABLE, DIMENSION(:) :: dsoln
                                 ! Double-precision solution
REAL(KIND=REAL64), ALLOCATABLE, DIMENSION(:) :: derror
                                 ! Array of double-precision errors
REAL(KIND=REAL64) :: derror_max  ! Max double precision error
INTEGER :: error_flag            ! Error flag from subroutines
INTEGER :: i, j                  ! Loop index
INTEGER :: istat                 ! I/O status
CHARACTER(len=80) :: msg         ! Error message
INTEGER :: n                     ! Size of system of eqns to solve
CHARACTER(len=20) :: filename    ! Input data file name

! Get the name of the disk file containing the equations.
WRITE (*,*) 'Enter the file name containing the eqns: '
READ (*,'(A20)') filename

! Open input data file. Status is OLD because the input data must
! already exist.
OPEN ( UNIT=1, FILE=filename, STATUS='OLD', ACTION='READ', &
       IOSTAT=istat, IOMSG=msg )

! Was the OPEN successful?
open_ok: IF ( istat == 0 ) THEN

   ! The file was opened successfully, so read the number of
   ! equations in the system.
   READ (1,*) n

   ! Allocate memory for that number of equations
   ALLOCATE ( a(n,n), b(n), soln(n), serror(n), &
              da(n,n), db(n), dsoln(n), derror(n), STAT=istat )
```

(*continued*)

(*continued*)

```
      ! If the memory is available, read in equations and
      ! process them.
      solve: IF ( istat == 0 ) THEN

         DO i = 1, n
            READ (1,*) (da(i,j), j=1,n), db(i)
         END DO

         ! Copy the coefficients in single precision for the
         ! single precision solution.
         a = da
         b = db

      ! Display coefficients.
      WRITE (*,1010)
      1010 FORMAT (/,'Coefficients:')
      DO i = 1, n
        WRITE (*,'(7F11.4)') (a(i,j), j=1,n), b(i)
      END DO

      ! Solve equations.
      CALL simul  (a, b, soln, n, n, error_flag )
      CALL dsimul (da, db, dsoln, n, n, error_flag )

      ! Check for error.
      error_check: IF ( error_flag /= 0 ) THEN
         WRITE (*,1020)
         1020 FORMAT (/,'Zero pivot encountered!', &
             //,'There is no unique solution to this system.')

      ELSE error_check

         ! No errors. Check for roundoff by substituting into
         ! the original equations, and calculate the differences.
         serror_max = 0.
         derror_max = 0._REAL64
         serror = 0.
         derror = 0._REAL64
         DO i = 1, n
            serror(i) = SUM ( a(i,:)  * soln(:) ) - b(i)
            derror(i) = SUM ( da(i,:) * dsoln(:) ) - db(i)
         END DO
         serror_max = MAXVAL ( ABS ( serror ) )
         derror_max = MAXVAL ( ABS ( derror ) )

         ! Tell user about it.
         WRITE (*,1030)
         1030 FORMAT (/,'  i    SP x(i)     DP x(i)    ', &
              '     SP Err      DP Err ')
         WRITE (*,1040)
         1040 FORMAT ( ' ===  ==========    ==========     ', &
              '    ========    ======== ')
```

(*continued*)

Additional Intrinsic Data Types

(*concluded*)

```
      DO i = 1, n
         WRITE (*,1050) i, soln(i), dsoln(i), serror(i), derror(i)
         1050 FORMAT (I3, 2X, G15.6, G15.6, F15.8, F15.8)
      END DO

      ! Write maximum errors.
      WRITE (*,1060) serror_max, derror_max
      1060 FORMAT (/,'Max single-precision error:',F15.8, &
           /,'Max double-precision error:',F15.8)

   END IF error_check
END IF solve

! Deallocate dynamic memory
DEALLOCATE ( a, b, soln, serror, da, db, dsoln, derror )

ELSE open_ok
   ! Else file open failed. Tell user.
   WRITE (*,1070) filename
   1070 FORMAT ('ERROR: File ',A,' could not be opened!')
   WRITE (*,'(A)') TRIM(msg)
END IF open_ok

END PROGRAM test_dsimul
```

To test the subroutine, we will call it with three different data sets. The first of them should be a well-conditioned system of equations, the second one should be an ill-conditioned system of equations, and the third should have no unique solution. The first system of equations that we will use to test the subroutine is the 6×6 system of equations shown below:

$$\begin{aligned}
-2x_1 + 5x_2 + x_3 + 3x_4 + 4x_5 - x_6 &= 0 \\
2x_1 - x_2 - 5x_3 - 2x_4 + 6x_5 - 4x_6 &= 1 \\
-x_1 + 6x_2 - 4x_3 - 5x_4 + 3x_5 - x_6 &= -6 \\
4x_1 - 3x_2 - 6x_3 - 5x_4 - 2x_5 - 2x_6 &= 10 \\
-3x_1 + 6x_2 + 4x_3 + 2x_4 - 6x_5 + 4x_6 &= -6 \\
2x_1 + 4x_2 + 4x_3 + 4x_4 + 4x_5 - 4x_6 &= -2
\end{aligned} \quad (11\text{-}8)$$

If this system of equations is placed in a file called sys6.wel and program test_dsimul is run on this file, the results are:

```
C:\book\fortran\chap11>test_dsimul
Enter the file name containing the eqns:
sys6.wel

Coefficients:
    -2.0000    5.0000    1.0000    3.0000    4.0000   -1.0000    0.0000
     2.0000   -1.0000   -5.0000   -2.0000    6.0000    4.0000    1.0000
    -1.0000    6.0000   -4.0000   -5.0000    3.0000   -1.0000   -6.0000
     4.0000    3.0000   -6.0000   -5.0000   -2.0000   -2.0000   10.0000
    -3.0000    6.0000    4.0000    2.0000   -6.0000    4.0000   -6.0000
     2.0000    4.0000    4.0000    4.0000    5.0000   -4.0000   -2.0000
```

i	SP x(i)	DP x(i)	SP Err	DP Err
1	0.662556	0.662556	-0.00000048	0.00000000
2	-0.132567	-0.132567	0.00000060	-0.00000000
3	-3.01373	-3.01373	0.00000095	0.00000000
4	2.83548	2.83548	0.00000095	0.00000000
5	-1.08520	-1.08520	-0.00000048	0.00000000
6	-0.836043	-0.836043	-0.00000072	-0.00000000

```
Max single-precision error:  0.00000095
Max double-precision error:  0.00000000
```

For this well-conditioned system, the results of single- and double-precision calculations were essentially identical. The second system of equations that we will use to test the subroutine is the 6 × 6 system of equations shown below. Note that the second and sixth equations are almost identical, so this system is ill-conditioned.

$$\begin{aligned}
-2x_1 + 5x_2 + x_3 + 3x_4 + 4x_5 - x_6 &= 0 \\
2x_1 - x_2 - 5x_3 - 2x_4 + 6x_5 - 4x_6 &= 1 \\
-x_1 + 6x_2 - 4x_3 - 5x_4 + 3x_5 - x_6 &= -6 \\
4x_1 - 3x_2 - 6x_3 - 5x_4 - 2x_5 - 2x_6 &= 10 \\
-3x_1 + 6x_2 + 4x_3 + 2x_4 - 6x_5 + 4x_6 &= -6 \\
2x_1 - 1.00001x_2 - 5x_3 - 2x_4 + 6x_5 - 4x_6 &= 1.01
\end{aligned} \quad (11\text{-}9)$$

If this system of equations is placed in a file called sys6.ill and program test_dsimul is run on this file, the results are:

```
C:\book\fortran\chap11>test_dsimul
Enter the file name containing the eqns:
sys6.ill

Coefficients:
  -2.0000   5.0000   1.0000   3.0000   4.0000  -1.0000   0.0000
   2.0000  -1.0000  -5.0000  -2.0000   6.0000   4.0000   1.0000
  -1.0000   6.0000  -4.0000  -5.0000   3.0000  -1.0000  -6.0000
   4.0000   3.0000  -6.0000  -5.0000  -2.0000  -2.0000  10.0000
  -3.0000   6.0000   4.0000   2.0000  -6.0000   4.0000  -6.0000
   2.0000  -1.0000  -5.0000  -2.0000   6.0000   4.0000   1.0100
```

i	SP x(i)	DP x(i)	SP Err	DP Err
1	-3718.09	-3970.67	-0.00042725	0.00000000
2	-936.408	-1000.00	0.00073242	-0.00000000
3	-4191.41	-4475.89	0.00152588	0.00000000
4	2213.83	2364.00	0.00109863	-0.00000000
5	-1402.07	-1497.22	-0.00024414	0.00000000
6	-404.058	-431.444	0.00049806	-0.00000000

```
Max single-precision error:  0.00152588
Max double-precision error:  0.00000000
```

For this ill-conditioned system, the results of the single- and double-precision calculations were significantly different. The single-precision numbers x(i) differ from the true answers by 6 to 7 percent, while the double-precision answers are almost exactly correct. Double-precision calculations are essential for a correct answer to this problem! The third system of equations that we will use to test the subroutine is the 6×6 system of equations shown below:

$$\begin{aligned}
-2x_1 + 5x_2 + x_3 + 3x_4 + 4x_5 - x_6 &= 0 \\
2x_1 - x_2 - 5x_3 - 2x_4 + 6x_5 - 4x_6 &= 1 \\
-x_1 + 6x_2 - 4x_3 - 5x_4 + 3x_5 - x_6 &= -6 \\
4x_1 - 3x_2 - 6x_3 - 5x_4 - 2x_5 - 2x_6 &= 10 \\
-3x_1 + 6x_2 + 4x_3 + 2x_4 - 6x_5 + 4x_6 &= -6 \\
2x_1 - x_2 - 5x_3 - 2x_4 + 6x_5 - 4x_6 &= 1
\end{aligned} \qquad (11\text{-}8)$$

If this system of equations is placed in a file called sys6.sng and program test_dsimul is run on this file, the results are:

```
C:\book\fortran\chap11>test_dsimul
Enter the file name containing the eqns:
sys6.sng

Coefficients before calls:
   -2.0000    5.0000    1.0000    3.0000    4.0000   -1.0000     .0000
    2.0000   -1.0000   -5.0000   -2.0000    6.0000    4.0000    1.0000
   -1.0000    6.0000   -4.0000   -5.0000    3.0000   -1.0000   -6.0000
    4.0000    3.0000   -6.0000   -5.0000   -2.0000   -2.0000   10.0000
   -3.0000    6.0000    4.0000    2.0000   -6.0000    4.0000   -6.0000
    2.0000   -1.0000   -5.0000   -2.0000    6.0000    4.0000    1.0000

Zero pivot encountered!

There is no unique solution to this system.
```

Since the second and sixth equations of this set are identical, there is no unique solution to this system of equations. The subroutine correctly identified and flagged this situation.

Subroutine dsimul seems to be working correctly for all three cases: well-conditioned systems, ill-conditioned systems, and singular systems. Furthermore, these tests showed the clear advantage of the double-precision subroutine over the single-precision subroutine for ill-conditioned systems.

11.2

ALTERNATE LENGTHS OF THE INTEGER DATA TYPE

The Fortran standard also allows (but does not require) a Fortran compiler to support integers of multiple lengths. The idea of having integers of different lengths is that shorter integers could be used for variables that have a restricted range in order to reduce the size of a program, while longer integers could be used for variables that needed the extra range.

TABLE 11-4
KIND numbers for integer values in some Fortran compilers

Computer/Compiler	INT8	INT16	INT32	INT64
PC/GNU Fortran	1	2	4*	8
PC/Intel Visual Fortran	1	2	4*	8
PC/NAGWare Fortran	1	2	3*	4

* denotes the **default integer** type for a particular processor.

The lengths of supported integers will vary from processor to processor, and the kind type parameters associated with a given length will also vary. You will have to check with your particular compiler vendor to see what lengths are supported by your compiler. The lengths and kind type parameters of integers supported by several processors are shown in Table 11-4. (In the table, INT8 is an 8-bit integer, INT16 is a 16-bit integer, etc. These are the constant names in intrinsic module iso_Fortran_env.) Both the lengths of integers supported and the kind type parameters assigned to them differ from processor to processor. This variation creates a problem when we want to write programs that are portable across different types of processors.

How can we write programs so that they can be easily ported between processors with different kind numbers and still function correctly? The best approach is to use a Fortran intrinsic function to *automatically select the proper kind of integer to use* as the program is moved from processor to processor. This function is called SELECTED_INT_KIND. When it is executed, it returns the kind type parameter of the smallest kind of integer value that meets the specified range on that particular computer. The general form of this function is

```
kind_number = SELECTED_INT_KIND(range)
```

where *range* is the required range of the integer in powers of 10. The function returns the kind number of the smallest integer kind satisfying the specified requirements. It returns a −1 if the specified range is not available from any integer data type on the processor.

The following examples are legal uses of this function

```
kind_number = SELECTED_INT_KIND(3)
kind_number = SELECTED_INT_KIND(9)
kind_number = SELECTED_INT_KIND(12)
kind_number = SELECTED_INT_KIND(20)
```

On a Core-i7-based computer using the Intel Visual Fortran compiler, the first of the functions will return a 2 (the kind number for 2-byte integers), since the specified range is -10^3 to $+10^3$ and a 2-byte integer can hold any number in the range −32,768 to 32,767. Similarly, the next function will return a 4 (the kind number for 4-byte integers), since the specified range is -10^9 to $+10^9$ and a 4-byte integer can hold any number in the range −2,147,483,648 to 2,147,483,647. The third function will return an 8 (the kind number for 8-byte integers), since the specified range is -10^{12} to $+10^{12}$ and an 8-byte integer can hold any number in the range −9,223,372,036,854,775,808 to 9,223,372,036,854,775,807. The last function will return a −1, since no integer data

Additional Intrinsic Data Types

type has a range of -10^{20} to $+10^{20}$. Different results will be returned on other processors; try it on yours and see what you get.

The following code sample illustrates the use of integer kinds in a processor-independent fashion. It declares two integer variables i1 and i2. Integer i1 is guaranteed to be able to hold integer values between $-1,000$ and $1,000$, while integer i2 is guaranteed to be able to hold integer values between $-1,000,000,000$ and $1,000,000,000$. The actual capacity of each integer may vary from computer to computer, but it will always satisfy this minimum guarantee.

```
INTEGER, PARAMETER :: SHORT = SELECTED_INT_KIND(3)
INTEGER, PARAMETER :: LONG = SELECTED_INT_KIND(9)
INTEGER(KIND=SHORT) :: i1
INTEGER(KIND=LONG) :: i2
```

It is also possible to declare the kind of an integer constant. The kind of an integer constant is declared by appending an underscore and the kind number to the constant. The following are examples of valid integer constants:

```
34          ! Default integer kind
34_4        ! Only valid if 4 is a legal kind of integer
24_LONG     ! Only valid if "LONG" is an integer named constant
```

The first example produces an integer constant of the *default kind* for the particular processor where the program is being executed. The second example is only valid if KIND=4 is a valid kind of integer on the particular processor where the program is being executed. The third example is only valid if LONG is a valid previously defined integer named constant, whose value is a valid kind number.

Good Programming Practice
Use the function SELECTED_INT_KIND to determine the kind numbers of the integer variables needed to solve a problem. The function will return the proper kind numbers on any processor, making your programs more portable.

Alternately, if you know the size of integer that you want, you can specify it directly using the iso_Fortran_env constants from Table 11-3.

11.3

ALTERNATE KINDS OF THE CHARACTER DATA TYPE

Fortran includes a provision for supporting multiple kinds of character sets. Support for multiple character sets is optional and may not be implemented on your processor. If present, this feature allows the Fortran language to support different character sets for the many different languages found around the world, or even special "languages" such as musical notation.

The general form of a character declaration with a kind parameter is

```
CHARACTER(kind=kind_num,len=length) :: string
```

where *kind_num* is the kind number of the desired character set.

Fortran 2003 and later includes a function called SELECTED_CHAR_KIND to return the kind number for a specific character set. When it is executed, it returns the kind type parameter matching a particular character set. The general form of this function is

```
kind_number = SELECTED_CHAR_KIND(name)
```

where *name* is a character expression of the default type containing one of the following values: 'DEFAULT', 'ASCII', or 'ISO_10646' (Unicode). The function returns the kind number of the corresponding character set if it is supported and a –1 if it is not supported.

The following examples are legal uses of this function

```
kind_number = SELECTED_CHAR_KIND('DEFAULT')
kind_number = SELECTED_CHAR_KIND('ISO_10646')
```

The Fortran standard does not *require* a compiler to support the Unicode character set, but it provides the support functions required to use Unicode characters if they are present. At the time of this writing, GNU Fortran supports both the ASCII and ISO-10646 character sets. Intel Fortran supports only the ASCII character set.

11.4
THE COMPLEX DATA TYPE

Complex numbers occur in many problems in science and engineering. For example, complex numbers are used in electrical engineering to represent alternating current voltages, currents, and impedances. The differential equations that describe the behavior of most electrical and mechanical systems also give rise to complex numbers. Because they are so ubiquitous, it is impossible to work as an engineer without a good understanding of the use and manipulation of complex numbers.

A complex number has the general form

$$c = a + bi \qquad (11\text{-}11)$$

where c is a complex number, a and b are both real numbers, and i is $\sqrt{-1}$. The number a is called the *real part* and b is called the *imaginary part* of the complex number c. Since a complex number has two components, it can be plotted as a point on a plane (see Figure 11-8). The horizontal axis of the plane is the real axis and the vertical axis of the plane is the imaginary axis, so that any complex number $a + bi$ can be represented as a single point a units along the real axis and b units along the imaginary axis. A complex number represented this way is said to be in *rectangular coordinates*, since the real and imaginary axes define the sides of a rectangle.

Additional Intrinsic Data Types

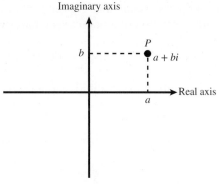

FIGURE 11-8
Representing a complex number in rectangular coordinates.

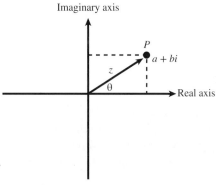

FIGURE 11-9
Representing a complex number in polar coordinates.

A complex number can also be represented as a vector of length z and angle θ pointing from the origin of the plane to the point P (see Figure 11-9). A complex number represented this way is said to be in *polar coordinates*.

$$c = a + bi = z\angle\theta \tag{11-12}$$

The relationships among the rectangular and polar coordinate terms a, b, z, and θ are:

$$a = z \cos \theta \tag{11-13}$$

$$b = z \sin \theta \tag{11-14}$$

$$z = \sqrt{a^2 + b^2} \tag{11-15}$$

$$\theta = \tan^{-1} \frac{b}{a} \tag{11-16}$$

Fortran uses rectangular coordinates to represent complex numbers. Each complex number consists of a pair of real numbers (a, b) occupying successive locations in memory. The first number (a) is the real part of the complex number and the second number (b) is the imaginary part of the complex number.

If complex numbers c_1 and c_2 are defined as $c_1 = a_1 + b_1 i$ and $c_2 = a_2 + b_2 i$, then the addition, subtraction, multiplication, and division of c_1 and c_2 are defined as:

$$c_1 + c_2 = (a_1 + a_2) + (b_1 + b_2)i \tag{11-17}$$

$$c_1 - c_2 = (a_1 - a_2) + (b_1 - b_2)i \tag{11-18}$$

$$c_1 \times c_2 = (a_1 a_2 - b_1 b_2) + (a_1 b_2 - b_1 a_2)i \tag{11-19}$$

$$\frac{c_1}{c_2} = \frac{a_1 a_2 + b_1 b_2}{a_2^2 + b_2^2} + \frac{b_1 a_2 - a_1 b_2}{a_2^2 + b_2^2} i \tag{11-20}$$

When two complex numbers appear in a binary operation, Fortran performs the required additions, subtractions, multiplications, or divisions between the two complex numbers using the above formulas.

11.4.1 Complex Constants and Variables

A **complex constant** consists of two numeric constants separated by commas and enclosed in parentheses. The first constant is the real part of the complex number and the second constant is the imaginary part of the complex number. For example, the following complex constants are equivalent to the complex numbers shown next to them:

```
(1., 0.)                1 + 0i
(0.7071,0.7071)         0.7071 + 0.7071i
(0, -1)                 -i
(1.01E6, 0.5E2)         1010000 + 50i
(1.12_DBL, 0.1_DBL)     1.12 + 0.1i   (Kind is DBL)
```

The last constant will only be valid if DBL is a named constant that has been set to a valid kind number for real data on the particular processor where the constant is used.

A named constant may be used to specify either the real or imaginary part of a complex constant. Thus, if PI is a named constant, then the following is a valid Fortran complex constant:

```
(PI,-PI)                π + πi
```

A complex variable is declared using a COMPLEX type declaration statement. The form of this statement is

```
COMPLEX(KIND=kind_num) :: :: variable_name1 [,variable_name2, ...]
```

The kind of the complex variable is optional; if it is left out, the default kind will be used. For example, the following statement declares a 256-element complex array.

Remember that we are actually allocating 512 default-length values, since two real values are required for each complex number.

```
COMPLEX, DIMENSION(256) :: array
```

There are at least two kinds of complex values on any processor, corresponding to the single precision and double precision kinds of real data. The single-precision version of the complex data type will have the same kind number as the single-precision version of the real data type and the double-precision version of the complex data type will have the same kind number as the double-precision version of the real data type. Therefore, the intrinsic function SELECTED_REAL_KIND can also be used to specify the size of complex data in a processor-independent manner.

The **default complex** kind will always be the same as the default real kind on any given processor.

11.4.2 Initializing Complex Variables

Like other variables, complex variables may be initialized by assignment statements, in type declaration statements, or by READ statements. The following code initializes all of the elements of array array1 to (0.,0.) using an assignment statement.

```
COMPLEX, DIMENSION(256) :: array1
array1 = (0.,0.)
```

A complex number may also be initialized in a type declaration statement using a complex constant. The following code declares and initializes variable a1 to (3.141592, -3.141592) using a type declaration statement.

```
COMPLEX :: a1 = (3.141592, -3.141592)
```

When a complex number is read or written with a formatted I/O statement, the first format descriptor encountered is used for the real part of the complex number and the second format descriptor encountered is used for the imaginary part of the complex number. The following code initializes variable a1 using a formatted READ statement.

```
COMPLEX :: a1
READ (*,'(2F10.2)') a1
```

The value in the first 10 characters of the input line will be placed in the real part of variable a1, and the value in the second 10 characters of the input line will be placed in the imaginary part of variable a1. Note that no parentheses are included on the input line when we read a complex number using formatted I/O. In contrast, when we read a complex number with a *list-directed* I/O statement, the complex number must be typed exactly like a complex constant, parentheses and all. The following READ statement

```
COMPLEX :: a1
READ (*,*) a1
```

requires that the input value be typed as shown: (1.0,0.25). When a complex number is written with a free format WRITE statement, it is printed as a complex value complete with parentheses. For example, the statements

```
COMPLEX :: a1 = (1.0,0.25)
WRITE (*,*) a1
```

produce the result:

```
(1.000000,2.500000E-01)
```

11.4.3 Mixed-Mode Arithmetic

When an arithmetic operation is performed between a complex number and another number (any kind of real or integer), Fortran converts the other number into a complex number and then performs the operation with a complex result. For example, the following code will produce an output of (300.,-300.):

```
COMPLEX :: c1 = (100.,-100.), c2
INTEGER :: i = 3
c2 = c1 * i
WRITE (*,*) c2
```

Initially, c1 is a complex variable containing the value (100.,-100.), and i is an integer containing the value 3. When the third line is executed, the integer i is converted into the complex number (3.,0.), and that number is multiplied by c1 to give the result (300., -300.).

When an arithmetic operation is performed between two complex or real numbers of different kinds, *both numbers are converted into the kind having the higher decimal precision* before the operation and the resulting value will have the higher precision.

If a real expression is assigned to a complex variable, the value of the expression is placed in the real part of the complex variable, and the imaginary part of the complex variable is set to zero. If two real values need to be assigned to the real and imaginary parts of a complex variable, then the CMPLX function (described below) must be used. When a complex value is assigned to a real or integer variable, *the real part of the complex number is placed in the variable and the imaginary part is discarded.*

11.4.4 Using Complex Numbers with Relational Operators

It is possible to compare two complex numbers with the == relational operator to see if they are equal to each other and to compare them with the /= operator to see if they are not equal to each other. However, they *cannot be compared with the >, <, >=, or <= operators*. The reason for this is that complex numbers consist of two separate parts. Suppose that we have two complex numbers $c_1 = a_1 + b_1 i$ and $c_2 = a_2 + b_2 i$, with $a_1 > a_2$ and $b_1 < b_2$. How can we possibly say which of these numbers is larger?

Additional Intrinsic Data Types

On the other hand, it is possible to compare the *magnitudes* of two complex numbers. The magnitude of a complex number can be calculated with the CABS intrinsic function (see below), or directly from Equation 11-21.

$$|c| = \sqrt{a^2 + b^2} \qquad (11-21)$$

Since the magnitude of a complex number is a real value, two magnitudes can be compared with any of the relational operators.

11.4.5 COMPLEX Intrinsic Functions

Fortran includes many specific and generic functions that support complex calculations. These functions fall into three general categories:

1. **Type conversion functions.** These functions convert data to and from the complex data type. Function CMPLX(a,b,kind) is a generic function that converts real or integer numbers a and b into a complex number whose real part has value a and whose imaginary part has value b. The kind parameter is optional; if it is specified, then the resulting complex number will be of the specified kind. Functions REAL() and INT() convert the *real part* of a complex number into the corresponding real or integer data type and throw away the imaginary part of the complex number. Function AIMAG() converts the *imaginary part* of a complex number into a real number.
2. **Absolute value function.** This function calculates the absolute value of a number. Function CABS(c) is a specific function that calculates the absolute value of a complex number using the equation

$$CBS\ (c) = \sqrt{a + b}$$

where $c = a + bi$

3. **Mathematical functions.** These functions include exponential functions, logarithms, trigonometric functions, and square roots. The generic functions SIN, COS, LOG10, SQRT, etc., will work as well with complex data as they will with real data.

Some of the intrinsic functions that support complex numbers are listed in Table 11-5.

It is important to be careful when converting a complex number to a real number. If we use the REAL() or DBLE() functions to do the conversion, only the *real* portion of the complex number is translated. In many cases, what we really want is the *magnitude* of the complex number. If so, we must use ABS() instead of REAL() to do the conversion.

Programming Pitfalls

Be careful when converting a complex number into a real number. Find out whether the real part of the number or the magnitude of the number is needed, and use the proper function to do the conversion.

TABLE 11-5
Some intrinsic functions that support complex numbers

Generic function	Specific function	Function value	Comments
ABS(c)	CABS(c)	$\sqrt{a^2 + b^2}$	Calculate magnitude of a complex number (result is a real value of the same kind as *c*).
CMPLX(a,b,*kind*)			Combines *a* and *b* into a complex number $a + bi$ (*a*, *b* may be integer, real, or double precision). *kind* is an optional integer. If present, it specifies the kind of the resulting complex number. If not specified, the kind will be default complex.
	CONJG(c)	c^*	Calculate the complex conjugate of *c*. If $c = a + bi$, then $c^* = a - bi$.
DBLE(c)			Convert real part of *c* into a double precision real number.
INT(c)			Convert real part of *c* into an integer.
REAL(c,*kind*)			Convert real part of *c* into a real number. *kind* is an optional integer. If present, it specifies the kind of the resulting real number.

Also, it is important to be careful when using double-precision variables with the function CMPLX. The Fortran standard states that *the function CMPLX returns a result of the default complex kind* regardless of its input arguments, unless another kind is explicitly specified. This can lead to a trap in which a programmer accidentally loses precision without being aware of it. Consider the following code as an example. In it we declare two double-precision real variables and a double-precision complex variable, and try to assign the contents of the two real variables to the complex variable. Because the kind is not specified in the CMPLX function, the accuracy of the information in the complex variable is reduced to single precision.

```
PROGRAM test_complex
INTEGER, PARAMETER :: DBL = SELECTED_REAL_KIND(p=13)
COMPLEX(KIND=DBL) :: c1 = (0.,0.)
REAL(KIND=DBL) :: a1 = 3.333333333333333_DBL
REAL(KIND=DBL) :: b1 = 6.666666666666666_DBL
c1 = CMPLX(a1,b1)
WRITE (*,*) c1
END PROGRAM test_complex
```

When this program is executed, the result is only accurate to single precision:

```
C:\book\fortran\chap11>test_complex
  (3.333333253860474,6.666666507720947)
```

Additional Intrinsic Data Types

To get the desired result, the CMPLX function must be rewritten as with the kind of the result specified:

```
c1 = CMPLX(a1,b1,DBL)
```

Programming Pitfalls

Be careful to specify the output kind type parameter when using the CMPLX function with double-precision values. Failure to do so can produce mysterious losses of precision within a program.

EXAMPLE 11-3

The Quadratic Equation (Revisited):

Write a general program to solve for the roots of a quadratic equation, regardless of type. Use complex variables so that no branches will be required based on the value of the discriminant.

SOLUTION

1. **State the problem.**

Write a program that will solve for the roots of a quadratic equation, whether they are distinct real roots, repeated real roots, or complex roots, without requiring tests on the value of the discriminant.

2. **Define the inputs and outputs.**

The inputs required by this program are the coefficients a, b, and c of the quadratic equation

$$ax^2 + bx + c = 0 \tag{3.1}$$

The output from the program will be the roots of the quadratic equation, whether they are real, repeated, or complex.

3. **Describe the algorithm.**

This task can be broken down into three major sections, whose functions are input, processing, and output:

```
Read the input data
Calculate the roots
Write out the roots
```

We will now break each of the above major sections into smaller, more detailed pieces. In this algorithm, the value of the discriminant is unimportant in determining how to proceed. The resulting pseudocode is:

```
Write 'Enter the coefficients A, B, and C: '
Read in a, b, c
```

```
discriminant ← CMPLX( b**2 - 4.*a*c, 0. )
x1 ← ( -b + SQRT(discriminant) ) / ( 2. * a )
x2 ← ( -b - SQRT(discriminant) ) / ( 2. * a )
Write 'The roots of this equation are: '
Write 'x1 = ', REAL(x1), ' +i ', AIMAG(x1)
Write 'x2 = ', REAL(x2), ' +i ', AIMAG(x2)
```

4. **Turn the algorithm into Fortran statements.**

The final Fortran code is shown in Figure 11-10.

FIGURE 11-10
A program to solve the quadratic equation using complex numbers.

```fortran
PROGRAM roots_2
!
! Purpose:
!   To find the roots of a quadratic equation
!      A * X**2 + B * X + C = 0.
!   using complex numbers to eliminate the need to branch
!   based on the value of the discriminant.
!
! Record of revisions:
!     Date        Programmer         Description of change
!     ====        ==========         =====================
!   12/01/15    S. J. Chapman        Original code
!
IMPLICIT NONE

! Data dictionary: declare variable types & definitions
REAL :: a               ! The coefficient of X**2
REAL :: b               ! The coefficient of X
REAL :: c               ! The constant coefficient
REAL :: discriminant    ! The discriminant of the quadratic eqn
COMPLEX :: x1           ! First solution to the equation
COMPLEX :: x2           ! Second solution to the equation

! Get the coefficients.
WRITE (*,1000)
1000 FORMAT ('Program to solve for the roots of a quadratic', &
         /,'equation of the form A * X**2 + B * X + C = 0. ' )
WRITE (*,1010)
1010 FORMAT ('Enter the coefficients A, B, and C: ')
READ (*,*) a, b, c

! Calculate the discriminant
discriminant = b**2 - 4. * a * c

! Calculate the roots of the equation
x1 = ( -b + SQRT( CMPLX(discriminant,0.) ) ) / (2. * a)
x2 = ( -b - SQRT( CMPLX(discriminant,0.) ) ) / (2. * a)

! Tell user.
WRITE (*,*) 'The roots are: '
WRITE (*,1020) ' x1 = ', REAL(x1), ' + i ', AIMAG(x1)
WRITE (*,1020) ' x2 = ', REAL(x2), ' + i ', AIMAG(x2)
1020 FORMAT (A,F10.4,A,F10.4)

END PROGRAM roots_2
```

5. **Test the program.**
Next, we must test the program using real input data. We will test cases in which the discriminant is greater than, less than, and equal to 0 to be certain that the program is working properly under all circumstances. From Equation (3-1), it is possible to verify the solutions to the equations given below:

$$x^2 + 5x + 6 = 0 \quad\quad x = -2 \text{ and } x = -3$$
$$x^2 + 4x + 4 = 0 \quad\quad x = -2$$
$$x^2 + 2x + 5 = 0 \quad\quad x = -1 \pm i2$$

When the above coefficients are fed into the program, the results are

```
C:\book\fortran\chap11>roots_2
Program to solve for the roots of a quadratic
equation of the form A * X**2 + B * X + C.
Enter the coefficients A, B, and C:
1,5,6
The roots are:
  X1 =    -2.0000 + i    .0000
  X2 =    -3.0000 + i    .0000

C:\book\fortran\chap11>roots_2
Program to solve for the roots of a quadratic
equation of the form A * X**2 + B * X + C.
Enter the coefficients A, B, and C:
1,4,4
The roots are:
  X1 =    -2.0000 + i    .0000
  X2 =    -2.0000 + i    .0000

C:\book\fortran\chap11>roots_2
Program to solve for the roots of a quadratic
equation of the form A * X**2 + B * X + C.
Enter the coefficients A, B, and C:
1,2,5
The roots are:
  X1 =    -1.0000 + i   2.0000
  X2 =    -1.0000 + i  -2.0000
```

The program gives the correct answers for our test data in all three possible cases. Note how much simpler this program is compared to the quadratic root solver found in Example 3-1. The use of the complex data type has greatly simplified our program.

Quiz 11-1

This quiz provides a quick check to see if you have understood the concepts introduced in Sections 11.1 through 11.4. If you have trouble with the quiz, reread the sections, ask your instructor, or discuss the material with a fellow student. The answers to this quiz are found in the back of the book.

(continued)

(*concluded*)

1. What kinds of real numbers and integers are supported by your compiler? What are the kind numbers associated with each one?
2. What will be written out by the code shown below?

   ```
   COMPLEX :: a, b, c, d
   a = ( 1., -1. )
   b = ( -1., -1. )
   c = ( 10., 1. )
   d = ( a + b ) / c
   WRITE (*,*) d
   ```

3. Use the definitions in Equations (11-17) through (11-20) to write a computer program that evaluates d in the problem above *without using complex numbers*. How much harder is it to evaluate this expression without the benefit of complex numbers?

11.5
SUMMARY

In this chapter, we introduced the concept of kinds and kind type parameters. Kinds are versions of the same basic data type, each differing in size, precision, range, etc.

All Fortran compilers support at least two kinds of real data, which are usually known as single precision and double precision. Double-precision data occupies twice the memory of single-precision data on most computers. Double-precision variables have both a greater range and more significant digits than single-precision variables.

The choice of precision for a particular real value is specified by the kind type parameter in the type declaration statement. Unfortunately, the numbers associated with each kind of real value vary among different processors. They can be determined by using the KIND intrinsic function on a particular processor, or the desired precision can be specified in a processor-independent manner using the SELECTED_REAL_KIND intrinsic function.

Double-precision real numbers take up more space and require more computer time to calculate than single-precision real numbers, so they should not be used indiscriminately. In general, they should be used when:

1. A problem requires many significant digits or a large range of numbers.
2. Numbers of dramatically different sizes must be added or subtracted.
3. Two nearly equal numbers must be subtracted, and the result used in further calculations.

Fortran permits (but does not require) a compiler to support multiple kinds of integers. Not all compilers will support multiple kinds of integers. The kind numbers associated with particular integer lengths vary from processor to processor. Fortran includes an intrinsic function `SELECTED_INT_KIND` to help programmers select the kind of integer required for a particular application in a processor-independent manner.

Fortran also allows a compiler to support multiple kinds of character sets. If your compiler implements this feature, you can use it to write out character data in different languages. Fortran 2003 and later also includes an intrinsic function `SELECTED_CHAR_KIND` to help programmers select the kind number of the ASCII or Unicode character set in a processor-independent manner.

Complex numbers consist of two real numbers in successive locations in memory. These two numbers are treated as though they were the real and imaginary parts of a complex number expressed in rectangular coordinates. They are processed according to the rules for complex addition, subtraction, multiplication, division, etc. There is a kind of complex number corresponding to each kind of real number available on a particular processor. The kind numbers are identical for real and complex data, so the desired precision of a complex value may be selected using the `SELECTED_REAL_KIND` intrinsic function.

Complex constants are written as two numbers in parentheses, separated by commas (e.g., `(1.,-1.)`). Complex variables are declared using a `COMPLEX` type declaration statement. They may be read and written using any type of real format descriptor (`E`, `ES`, `F`, etc.). When reading or writing complex numbers, the real and imaginary parts of the number are processed separately. The first value read will become the real part and the next value will become the imaginary part. If list-directed input is used with complex numbers, the input value must be typed as a complex constant, complete with parentheses.

In a binary operation involving a complex number and an integer or real number, the other number is first converted to complex and then the operation is performed using complex arithmetic. All arithmetic is performed at the highest precision of any number in the calculation.

11.5.1 Summary of Good Programming Practice

The following guidelines should be adhered to when working with parameterized variables, complex numbers, and derived data types:

1. Always assign kind numbers to a named constant, and then use that named constant in all type declaration statements and constant declarations. For large programs with many procedures, place the kind parameters in a single module, and then use that module in every procedure within the program.
2. Use the function `SELECTED_REAL_KIND` to determine the kind numbers of the real values needed to solve a problem. The function will return the proper kind numbers on any processor, making your programs more portable.
3. Use the constants in module `iso_Fortran_env` to specify data sizes in a processor-independent manner.
4. Use the function `SELECTED_INT_KIND` to determine the kind numbers of the integer variables needed to solve a problem.

5. Use double-precision real numbers instead of single-precision real numbers whenever:

    (a) A problem requires many significant digits or a large range of numbers.
    (b) Numbers of dramatically different sizes must be added or subtracted.
    (c) Two nearly equal numbers must be subtracted, and the result used in further calculations.

6. Be careful when you are converting a complex number to a real or double-precision number. If you use the REAL( ) or DBLE( ) functions, only the *real* portion of the complex number is translated. In many cases, what we really want is the *magnitude* of the complex number. If so, we must use CABS( ) instead of REAL( ) to do the conversion.

7. Be careful when you are converting a pair of double-precision real numbers into a complex number using function CMPLX. If you do not explicitly specify that the kind of the function result is double precision, the result will be of type default complex, and precision will be lost.

11.5.2 Summary of Fortran Statements and Structures

COMPLEX Statement:

    COMPLEX(KIND=*kind_no*) :: *variable_name1* [,*variable_name2*, ...]

Example:

    COMPLEX(KIND=single) :: volts, amps

Description:
The COMPLEX statement declares variables of the complex data type. The kind number is optional and machine dependent. If it is not present, the kind is the default complex kind for the particular machine (usually single precision).

REAL Statement with KIND parameter:

    REAL(KIND=*kind_no*) :: *variable_name1* [,*variable_name2*, ...]

Example:

    REAL(KIND=single), DIMENSION(100) :: points

Description:
The REAL statement is a type declaration statement that declares variables of the real data type. The kind number is optional and machine dependent. If it is not present, the kind is the default real kind for the particular machine (usually single precision).
To specify double-precision real values, the kind must be set to the appropriate number for the particular machine. The kind number may be found by using the function KIND(0.0D0) or by using the function SELECTED_REAL_KIND.

Additional Intrinsic Data Types

11.5.3 Exercises

11-1. What are kinds of the REAL data type? How many kinds of real data must be supported by a compiler according to the Fortran standard?

11-2. What kind numbers are associated with the different types of real variables available on your compiler/computer? Determine the precision and range associated with each type of real data.

11-3. What are the advantages and disadvantages of double-precision real numbers compared to single-precision real numbers? When should double precision real numbers be used instead of single-precision real numbers?

11-4. What is an ill-conditioned system of equations? Why is it hard to find the solution to an ill-conditioned set of equations?

11-5. State whether each of the following sets of Fortran statements are legal or illegal. If they are illegal, what is wrong with them? If they are legal, what do they do?

(a) Statements:

```
INTEGER, PARAMETER :: SGL = KIND(0.0)
INTEGER, PARAMETER :: DBL = KIND(0.0D0)
REAL(KIND=SGL) :: a
REAL(KIND=DBL) :: b
READ (*,'(F18.2)') a, b
WRITE (*,*) a, b
```

Input Data:

```
1111111111111111111111111111111111111111
2222222222222222222222222222222222222222
----|----|----|----|----|----|----|----|
    5   10   15   20   25   30   35   40
```

(b) Statements:

```
USE iso_Fortran_env
COMPLEX(kind=REAL32), DIMENSION(5) :: a1
INTEGER :: i
DO i = 1, 5
   a1(i) = CMPLX ( i, -2*i )
END DO
IF (a1(5) > a1(3)) THEN
   WRITE (*,100) (i, a1(i), i = 1, 5)
   100 FORMAT (3X,'a1(',I2,') = (',F10.4,',',F10.4,')')
END IF
```

11-6. Derivative of a Function Write a subroutine to calculate the derivative of a double-precision real function $f(x)$ at position $x = x_0$. The calling arguments to the subroutine should be the function $f(x)$, the location x_0 at which to evaluate the function, and the step size Δx to use in the evaluation. The output from the subroutine will be the derivative of the function at point $x = x_0$. To make your subroutine machine independent, define double precision as the kind of real value having at least 13 digits of precision. Note that the function to be evaluated should be passed to the subroutine as a calling argument! Test your subroutine by evaluating the function $f(x) = 10 \sin 20x$ at position $x_0 = 0$.

11-7. If you have not done so previously, write a set of elapsed time subroutines for your computer, as described in Exercise 7-29. Use the elapsed time subroutines to compare the time required to solve a 10×10 system of simultaneous equations in single precision and in double precision. To do this, you will need to write two test driver programs (one single precision and one double precision) that read the coefficients of the equations, start the timer running, solve the equations, and then calculate the elapsed time. How much slower is the double-precision solution than the single-precision solution on your computer? (*Hint*: You will need to create an inner loop and solve the system of equations 10 or more times in order to get a meaningful elapsed time.)

Test your program on the system of equations shown below (this set of equations is contained in file **sys10** in directory Chap11 at the book's website).:

$$\begin{aligned}
-2x_1 + 5x_2 + x_3 + 3x_4 + 4x_5 - x_6 + 2x_7 - x_8 - 5x_9 - 2x_{10} &= -5 \\
6x_1 + 4x_2 - x_3 + 6x_4 - 4x_5 - 5x_6 + 3x_7 - x_8 + 4x_9 + 3x_{10} &= -6 \\
-6x_1 - 5x_2 - 2x_3 - 2x_4 - 3x_5 + 6x_6 + 4x_7 + 2x_8 - 6x_9 + 4x_{10} &= -7 \\
2x_1 + 4x_2 + 4x_3 + 4x_4 + 5x_5 - 4x_6 + 0x_7 + 0x_8 - 4x_9 + 6x_{10} &= 0 \\
-4x_1 - x_2 + 3x_3 - 3x_4 - 4x_5 - 4x_6 - 4x_7 + 4x_8 + 3x_9 - 3x_{10} &= 5 \\
4x_1 + 3x_2 + 5x_3 + x_4 + x_5 + x_6 + 0x_7 + 3x_8 + 3x_9 + 6x_{10} &= -8 \\
x_1 + 2x_2 - 2x_3 + 0x_4 + 3x_5 - 5x_6 + 5x_7 + 0x_8 + x_9 - 4x_{10} &= 1 \\
-3x_1 - 4x_2 + 2x_3 - x_4 - 2x_5 + 5x_6 - x_7 - x_8 - 4x_9 + x_{10} &= -4 \\
5x_1 + 5x_2 - 2x_3 - 5x_4 + x_5 - 4x_6 - x_7 + 0x_8 - 2x_9 - 3x_{10} &= -7 \\
-5x_1 - 2x_2 - 5x_3 + 2x_4 + x_5 - 3x_6 + 4x_7 - x_8 - 4x_9 + 4x_{10} &= 6
\end{aligned}$$

11-8. Write a program to determine the kinds of integers supported by your particular compiler. The program should use the function SELECTED_INT_KIND with various input ranges to determine all legal kind numbers. What are the kind numbers and ranges associated with each kind of integer?

11-9. Simultaneous Equations with Complex Coefficients Create a subroutine csimul to solve for the unknowns in a system of simultaneous linear equations that have complex coefficients. Test your subroutine by solving the system of equations shown below:

$$\begin{aligned}
(-2 + 5i)x_1 + (1 + 3i)x_2 + (4 - i)x_3 &= (7 + 5i) \\
(2 - i)x_1 + (-5 - 2i)x_2 + (6 + 4i)x_3 &= (-10 - 8i) \\
(-1 + 6i)x_1 + (-4 - 5i)x_2 + (3 - i)x_3 &= (-3 - 3i)
\end{aligned} \qquad (11\text{-}22)$$

11-10. Amplitude and Phase of a Complex Number Write a subroutine that will accept a complex number $c = a + bi$ stored in a variable of type COMPLEX, and return the amplitude amp and the phase theta (in degrees) of the complex number in two real variables. (*Hint:* Use intrinsic function ATAN2D to help calculate the phase.)

11-11. Euler's Equation Euler's equation defines e raised to an imaginary power in terms of sinusoidal functions as follows:

$$e^{i\theta} = \cos\theta + i\sin\theta \qquad (11\text{-}23)$$

Write a function to evaluate $e^{i\theta}$ for any θ using Euler's equation. Also, evaluate $e^{i\theta}$ using the intrinsic complex exponential function CEXP. Compare the answers that you get by the two methods for the cases where $\theta = 0$, $\pi/2$, and π.

12

Derived Data Types

OBJECTIVES

- Learn how to declare a derived data type.
- Learn how to create and use variables of a derived data type.
- Learn how to create parameterized versions of a derived data type.
- Learn how to create derived data types that are extensions of other data types.
- Learn how to create and use type-bound procedures.
- Learn how to use the ASSOCIATE construct.

In this chapter, we will introduce derived data types. The derived data type is a mechanism for users to create special new data types to suit the needs of a particular problem that they may be trying to solve.

12.1
INTRODUCTION TO DERIVED DATA TYPES

So far, we have studied Fortran's **intrinsic data types:** integer, real, complex, logical, and character. In addition to these data types, the Fortran language permits us to create our own data types to add new features to the language, or to make it easier to solve specific types of problems. A user-defined data type may have any number and combination of components, but each component must be either an intrinsic data type or a user-defined data type that was previously defined. Because user-defined data types must be ultimately derived from intrinsic data types, they are called **derived data types.**

Basically, a derived data type is a convenient way to group together all of the information about a particular item. In some ways, it is like an array. Like an array, a single derived data type can have many components. Unlike an array, the components of a derived data type may have different types. One component may be an integer, while the next component is a real, the next a character string, and so forth. Furthermore, each component is known by a name instead of a number.

A derived data type is defined by a sequence of type declaration statements beginning with a TYPE statement and ending with an END TYPE statement. Between these two statements are the definitions of the components in the derived data type. The form of a derived data type is

```
TYPE [::] type_name
   component definitions
   ...
END TYPE [type_name]
```

where the double colons and the name on the END TYPE statement are optional. There may be as many component definitions in a derived data type as desired.

To illustrate the use of a derived data type, let's suppose that we were writing a grading program. The program would contain information about the students in a class such as name, social security number, age, sex, etc. We could define a special data type called person to contain all of the personal information about each person in the program:

```
TYPE :: person
   CHARACTER(len=14) :: first_name
   CHARACTER :: middle_initial
   CHARACTER(len=14) :: last_name
   CHARACTER(len=14) :: phone
   INTEGER :: age
   CHARACTER :: sex
   CHARACTER(len=11) :: ssn
END TYPE person
```

Once the derived type person is defined, variables of that type may be declared as shown:

```
TYPE (person) :: john, jane
TYPE (person), DIMENSION(100) :: people
```

The latter statement declares an array of 100 variables of type person. Each variable of a derived data type is known as a **structure.**

It is also possible to create unnamed constants of a derived data type. To do so, we use a **structure constructor.** A structure constructor consists of the name of the type followed by the components of the derived data type in parentheses. The components appear in the order in which they were declared in the definition of the derived type. For example, the variables john and jane could be initialized by constants of type person as follows:

```
john = person('John','R','Jones','323-6439',21,'M','123-45-6789')
jane = person('Jane','C','Bass','332-3060',17,'F','999-99-9999')
```

A derived data type may be used as a component within another derived data type. For example, a grading program could include a derived data type called grade_info containing a component of the type person defined above to contain personal information about the students in the class. The example below defines the derived type grade_info, and declares an array class to be 30 variables of this type.

```
TYPE :: grade_info
   TYPE (person) :: student
   INTEGER :: num_quizzes
```

```
         REAL, DIMENSION(10) :: quiz_grades
         INTEGER :: num_exams
         REAL, DIMENSION(10) :: exam_grades
         INTEGER :: final_exam_grade
         REAL :: average
      END TYPE
      TYPE (grade_info), DIMENSION(30) :: class
```

12.2
WORKING WITH DERIVED DATA TYPES

Each component in a variable of a derived data type can be addressed independently, and can be used just like any other variable of the same type. If the component is an integer, then it can be used just like any other integer, etc. A component is specified by a **component selector,** which consists of the name of the variable followed by a percent sign (%), and then followed by the component name. For example, the following statement sets the component age of variable john to 35:

```
john%age = 35
```

To address a component within an array of a derived data type, *place the array subscript after the array name and before the percent sign.* For example, to set the final exam grade for student 5 in array class above, we would write:

```
class(5)%final_exam_grade = 95
```

To address a component of a derived data type that is included within another derived data type, we simply concatenate their names separated by percent signs. Thus, we could set the age of student 5 within the class with the statement:

```
class(5)%student%age = 23
```

As you can see, it is easy to work with the components of a variable of a derived data type. However, it is *not* easy to work with variables of derived data types as a whole. It is legal to assign one variable of a given derived data type to another variable of the same type, but that is almost the only operation that is defined. Other intrinsic operations such as addition, subtraction, multiplication, division, comparison, etc., are not defined by default for these variables. We will learn how to extend these operations to work properly with derived data types in Chapter 13.

12.3
INPUT AND OUTPUT OF DERIVED DATA TYPES

If a variable of a derived data type is included in a WRITE statement, then by default each of the components of the variable is written out in the order in which they are declared in the type definition. If the WRITE statement uses formatted I/O, then the format descriptors must match the type and order of the components in the variable.[1]

---

[1] There is a way to modify this behavior, as we will see in Chapter 16.

Similarly, if a variable of a derived data type is included in a READ statement, then the input data must be supplied in the order in which each of the components are declared in the type definition. If the READ statement uses formatted I/O, then the format descriptors must match the type and order of the components in the variable.

The program shown in Figure 12-1 illustrates the output of a variable of type person using both formatted and free format I/O.

FIGURE 12-1
A program to illustrate output of variables of derived data types.

```
PROGRAM test_io
!
!  Purpose:
!    To illustrate I/O of variables of derived data types.
!
!  Record of revisions:
!      Date        Programmer          Description of change
!      ====        ==========          =====================
!    12/02/15    S. J. Chapman         Original code
!
IMPLICIT NONE

! Declare type person
TYPE :: person
   CHARACTER(len=14) :: first_name
   CHARACTER :: middle_initial
   CHARACTER(len=14) :: last_name
   CHARACTER(len=14) :: phone
   INTEGER :: age
   CHARACTER :: sex
   CHARACTER(len=11) :: ssn
END TYPE person

! Declare a variable of type person:
TYPE (person) :: john

! Initialize variable
john = person('John','R','Jones','323-6439',21,'M','123-45-6789')

! Output variable using free format I/O
WRITE (*,*) 'Free format: ', john

! Output variable using formatted I/O
WRITE (*,1000) john
1000 FORMAT (' Formatted I/O:',/,4(1X,A,/),1X,I4,/,1X,A,/,1X,A)

END PROGRAM test_io
```

When this program is executed, the results are:

```
C:\book\fortran\chap12>test_io
Free format: John          RJones          323-6439        21M123-45-6789

Formatted I/O:
John
R
Jones
```

```
323-6439
  21
M
123-45-6789
```

■ 12.4
DECLARING DERIVED DATA TYPES IN MODULES

As we have seen, the definition of a derived data type can be fairly bulky. This definition must be included in every procedure that uses variables or constants of the derived type, which can present a painful maintenance problem in large programs. To avoid this problem, it is customary to define all derived data types in a program in a single module, and then to use that module in all procedures needing to use the data type. This practice is illustrated in Example 12-1 below.

Good Programming Practice
For large programs using derived data types, declare the definitions of each data type in a module, and then use that module in each procedure of the program that needs to access the derived data type.

MEMORY ALLOCATION FOR DERIVED DATA TYPES

When a Fortran compiler allocates memory for a variable of a derived data type, the compiler is *not* required to allocate the elements of the derived data type in successive memory locations. Instead, it is free to place them anywhere it wants, as long as the proper element order is preserved during I/O operations. This freedom was deliberately built into the Fortran standard to allow compilers on massively parallel computers to optimize memory allocations for the fastest possible performance.

However, there are times when a strict order of memory allocations is important. For example, if we want to pass a variable of a derived data type to a procedure written in another language, it is necessary for the elements of that variable to be in strict order.

If the elements of a derived data type must be allocated in consecutive memory locations for some reason, a special SEQUENCE statement must be included in the type definition. An example of a derived data type whose elements will always be declared in consecutive locations in memory is:

```
TYPE :: vector
   SEQUENCE
   REAL :: a
   REAL :: b
   REAL :: c
END TYPE
```

EXAMPLE 12-1

Sorting Derived Data Types by Components:

To illustrate the use of derived data types, we will create a small customer database program that permits us to read in a database of customer names and addresses, and to sort and display the addresses by last name, by city, or by zip code.

SOLUTION
To solve this problem, we will create a simple derived data type containing the personal information about each customer in the database and initialize the customer database from a disk file. Once the database is initialized, we will prompt the user for the desired display order and sort the data into that order.

1. **State the problem.**
    Write a program to create a database of customers from a data file, and to sort and display that database in alphabetical order by last name, by city, or by zip code.

2. **Define the inputs and outputs.**
    The inputs to the program are the name of the customer database file, the customer database file itself, and an input value from the user specifying the order in which the data is to be sorted. The output from the program is the customer list sorted in alphabetical order by the selected field.

3. **Describe the algorithm.**
    The first step in writing this program will be to create a derived data type to hold all of the information about each customer. This data type will need to be placed in a module so that it can be used by each procedure in the program. An appropriate data type definition is shown below:

```
TYPE :: personal_info
   CHARACTER(len=12) :: first      ! First name
   CHARACTER         :: mi         ! Middle Initial
   CHARACTER(len=12) :: last       ! Last name
   CHARACTER(len=26) :: street     ! Street Address
   CHARACTER(len=12) :: city       ! City
   CHARACTER(len=2)  :: state      ! State
   INTEGER           :: zip        ! Zip code
END TYPE personal_info
```

The program can logically be broken up into two sections, a main program that reads and writes the customer database and a separate procedure that sorts the data into the selected order. The top-level pseudocode for the main program is

```
Get name of customer data file
Read customer data file
Prompt for sort order
Sort data in specified order
Write out sorted customer data
```

Now we must expand and refine the pseudocode for the main program. We must describe how the data will be read in, the sort order is selected, and the sorting is done in more detail. A detailed version of the pseudocode for the main program is shown below.

Derived Data Types

```
Prompt user for the input file name "filename"
Read the file name "filename"
OPEN file "filename"
IF OPEN is successful THEN
   WHILE
      Read value into temp
      IF read not successful EXIT
      nvals ← nvals + 1
      customers(nvals) ← temp
   End of WHILE

   Prompt user for type of sort (1=last name;2=city;3=zip)
   Read choice
   SELECT CASE (choice)
   CASE (1)
      Call sort_database with last_name comparison function
   CASE (2)
      Call sort_database with city comparison function
   CASE (3)
      Call sort_database with zip code comparison function
   CASE DEFAULT
      Tell user of illegal choice
   END of SELECT CASE

   Write out sorted customer data
END of IF
```

The sorting procedure will be a selection sort similar to any of the sorting routines that we have already encountered in Chapters 6, 7, and 10. The one tricky thing about this particular sorting process is that *we do not know in advance what component of the data type we will be sorting on*. Sometimes we will be sorting on the last name, while other times we will be sorting on the city or zip code. We must do something to make the sort procedure work properly regardless of the component of the data that we are sorting on.

The easiest way to get around this problem is to write a series of functions that compares individual components of two different variables of the data type to determine which is lesser of the two. One function will compare two last names to determine which is the lesser (lower in alphabetical order), while another function will compare two city names to determine which is the lesser (lower in alphabetical order), and a third will compare two zip codes to determine which is the lesser (lower in numerical sequence). Once the comparison functions are written, we will be able to sort the data in any order by passing the appropriate comparison function to the sorting subroutine as a command line argument.

The pseudocode for the last name comparison routine is:

```
LOGICAL FUNCTION lt_last (a, b)
lt_lastname ← LLT(a%last, b%last)
```

Note that the routine uses the LLT function to ensure that the sorting order is the same on all computers regardless of collating sequence. The pseudocode for the city comparison routine is

```
LOGICAL FUNCTION lt_city (a, b)
lt_city ← LLT(a%city, b%city)
```

Finally, the pseudocode for the zip code comparison routine is

```
LOGICAL FUNCTION lt_zip (a, b)
lt_zip ← a%zip < b%zip
```

The pseudocode for the sorting routine will be the same as the pseudocode for subroutine `sort` in Chapter 7, except that the comparison function will be passed as a command line argument. It is not reproduced here.

4. **Turn the algorithm into Fortran statements.**
The resulting Fortran subroutine is shown in Figure 12-2.

FIGURE 12-2
Program to sort a customer database according to a user-specified field.

```fortran
MODULE types
!
!  Purpose:
!    To define the derived data type used for the customer
!    database.
!
!  Record of revisions:
!      Date        Programmer          Description of change
!      ====        ==========          =====================
!    12/04/15    S. J. Chapman         Original code
!
IMPLICIT NONE

! Declare type personal_info
TYPE :: personal_info
    CHARACTER(len=12) :: first      ! First name
    CHARACTER         :: mi         ! Middle Initial
    CHARACTER(len=12) :: last       ! Last name
    CHARACTER(len=26) :: street     ! Street Address
    CHARACTER(len=12) :: city       ! City
    CHARACTER(len=2)  :: state      ! State
    INTEGER           :: zip        ! Zip code
END TYPE personal_info

END MODULE types

PROGRAM customer_database
!
!  Purpose:
!    To read in a character input data set, sort it into ascending
!    order using the selection sort algorithm, and to write the
!    sorted data to the standard output device. This program calls
!    subroutine "sort_database" to do the actual sorting.
!
!  Record of revisions:
!      Date        Programmer          Description of change
!      ====        ==========          =====================
!    12/04/15    S. J. Chapman         Original code
!
```

(continued)

Derived Data Types

(*continued*)

```
USE types                                 ! Declare the module types
IMPLICIT NONE

! Data dictionary: declare constants
INTEGER, PARAMETER :: MAX_SIZE = 100   ! Max addresses in database

! Data dictionary: declare external functions
LOGICAL, EXTERNAL :: lt_last       ! Comparison fn for last names
LOGICAL, EXTERNAL :: lt_city       ! Comparison fn for cities
LOGICAL, EXTERNAL :: lt_zip        ! Comparison fn for zip codes

! Data dictionary: declare variable types & definitions
TYPE(personal_info), DIMENSION(MAX_SIZE) :: customers
                                   ! Data array to sort
INTEGER :: choice                  ! Choice of how to sort database
LOGICAL :: exceed = .FALSE.        ! Logical indicating that array
                                   !    limits are exceeded.
CHARACTER(len=20) :: filename      ! Input data file name
INTEGER :: i                       ! Loop index
CHARACTER(len=80) :: msg           ! Error message
INTEGER :: nvals = 0               ! Number of data values to sort
INTEGER :: status                  ! I/O status: 0 for success
TYPE(personal_info) :: temp        ! Temporary variable for reading

! Get the name of the file containing the input data.
WRITE (*,*) 'Enter the file name with customer database: '
READ (*,'(A20)') filename

! Open input data file.  Status is OLD because the input data must
! already exist.
OPEN ( UNIT=9, FILE=filename, STATUS='OLD', IOSTAT=status, &
       IOMSG=msg )

! Was the OPEN successful?
fileopen: IF ( status == 0 ) THEN         ! Open successful

   ! The file was opened successfully, so read the customer
   ! database from it.
   DO
      READ (9, 1010, IOSTAT=status) temp   ! Get value
      1010 FORMAT (A12,1X,A1,1X,A12,1X,A26,1X,A12,1X,A2,1X,I5)
      IF ( status /= 0 ) EXIT              ! Exit on end of data
      nvals = nvals + 1                    ! Bump count
      size: IF ( nvals <= MAX_SIZE ) THEN  ! Too many values?
         customers(nvals) = temp           ! No: Save value in array
      ELSE
         exceed = .TRUE.                   ! Yes: Array overflow
      END IF size
   END DO

   ! Was the array size exceeded?  If so, tell user and quit.
   toobig: IF ( exceed ) THEN
      WRITE (*,1020) nvals, MAX_SIZE
      1020 FORMAT ('Maximum array size exceeded: ', I6, ' > ', I6 )
   ELSE
```

(*continued*)

(*continued*)

```fortran
      ! Limit not exceeded: find out how to sort data.
      WRITE (*,1030)
 1030 FORMAT ('Enter way to sort database:',/, &
              ' 1 -- By last name ',/, &
              ' 2 -- By city ',/, &
              ' 3 -- By zip code ')
      READ (*,*) choice

      ! Sort database
      SELECT CASE ( choice)
      CASE (1)
         CALL sort_database (customers, nvals, lt_last )
      CASE (2)
         CALL sort_database (customers, nvals, lt_city )
      CASE (3)
         CALL sort_database (customers, nvals, lt_zip )
      CASE DEFAULT
         WRITE (*,*) 'Invalid choice entered!'
      END SELECT

      ! Now write out the sorted data.
      WRITE (*,'(A)') 'The sorted database values are: '
      WRITE (*,1040) ( customers(i), i = 1, nvals )
 1040 FORMAT (A12,1X,A1,1X,A12,1X,A26,1X,A12,1X,A2,1X,I5)

   END IF toobig

ELSE fileopen

   ! Else file open failed. Tell user.
   WRITE (*,1050) msg
 1050 FORMAT ('File open failed: ', A)

END IF fileopen

END PROGRAM customer_database

SUBROUTINE sort_database (array, n, lt_fun )
!
!  Purpose:
!    To sort array "array" into ascending order using a selection
!    sort, where "array" is an array of the derived data type
!    "personal_info". The sort is based on the external
!    comparison function "lt_fun", which will differ depending on
!    which component of the derived type array is used for
!    comparison.
!
!  Record of revisions:
!      Date       Programmer          Description of change
!      ====       ==========          =====================
!    12/04/15    S. J. Chapman        Original code
!
USE types                          ! Declare the module types
```

(*continued*)

(*continued*)

```
IMPLICIT NONE

! Data dictionary: declare calling parameter types & definitions
INTEGER, INTENT(IN) :: n                   ! Number of values
TYPE(personal_info), DIMENSION(n), INTENT(INOUT) :: array
                                           ! Array to be sorted
LOGICAL, EXTERNAL :: lt_fun                ! Comparison function

! Data dictionary: declare local variable types & definitions
INTEGER :: i                  ! Loop index
INTEGER :: iptr               ! Pointer to smallest value
INTEGER :: j                  ! Loop index
TYPE(personal_info) :: temp   ! Temp variable for swaps

! Sort the array
outer: DO i = 1, n-1

   ! Find the minimum value in array(i) through array(n)
   iptr = i
   inner: DO j = i+1, n
      minval: IF ( lt_fun(array(J),array(iptr)) ) THEN
         iptr = j
      END IF minval
   END DO inner

   ! iptr now points to the minimum value, so swap array(iptr)
   ! with array(i) if i /= iptr.
   swap: IF ( i /= iptr ) THEN
      temp       = array(i)
      array(i)   = array(iptr)
      array(iptr) = temp
   END IF swap

END DO outer
END SUBROUTINE sort_database

LOGICAL FUNCTION lt_last (a, b)
!
!  Purpose:
!    To compare variables "a" and "b" and determine which
!    has the smaller last name (lower alphabetical order).
!
USE types                         ! Declare the module types
IMPLICIT NONE

! Data dictionary: declare calling parameter types & definitions
TYPE (personal_info), INTENT(IN) :: a, b

! Make comparison.
lt_last = LLT ( a%last, b%last )

END FUNCTION lt_last
```

(*continued*)

(concluded)

```fortran
LOGICAL FUNCTION lt_city (a, b)
!
!  Purpose:
!    To compare variables "a" and "b" and determine which
!    has the smaller city (lower alphabetical order).
!
USE types                     ! Declare the module types
IMPLICIT NONE

! Data dictionary: declare calling parameter types & definitions
TYPE (personal_info), INTENT(IN) :: a, b

! Make comparison.
lt_city = LLT ( a%city, b%city )

END FUNCTION lt_city

LOGICAL FUNCTION lt_zip (a, b)
!
!  Purpose:
!    To compare variables "a" and "b" and determine which
!    has the smaller zip code (lower numerical value).
!
USE types                     ! Declare the module types
IMPLICIT NONE

! Data dictionary: declare calling parameter types & definitions
TYPE (personal_info), INTENT(IN) :: a, b

! Make comparison.
lt_zip = a%zip < b%zip

END FUNCTION lt_zip
```

5. **Test the resulting Fortran programs.**

   To test this program, it is necessary to create a sample customer database. A simple customer database is shown in Figure 12-3; it is stored in the disk in a file called database.

FIGURE 12-3
Sample customer database used to test the program of Example 12-1.

```
John      Q Public      123 Sesame Street   Anywhere       NY 10035
James     R Johnson     Rt. 5 Box 207C      West Monroe    LA 71291
Joseph    P Ziskend     P. O. Box 433       APO            AP 96555
Andrew    D Jackson     Jackson Square      New Orleans    LA 70003
Jane      X Doe         12 Lakeside Drive   Glenview       IL 60025
Colin     A Jeffries    11 Main Street      Chicago        IL 60003
```

To test the program, we will execute it three times using this database, once with each possible sorting option.

```
C:\book\fortran\chap12>customer_database
Enter the file name with customer database:
```

```
database
Enter way to sort database:
  1 -- By last name
  2 -- By city
  3 -- By zip code
1
The sorted database values are:
Jane        X Doe         12 Lakeside Drive    Glenview       IL 60025
Andrew      D Jackson     Jackson Square       New Orleans    LA 70003
Colin       A Jeffries    11 Main Street       Chicago        IL 60003
James       R Johnson     Rt. 5 Box 207C       West Monroe    LA 71291
John        Q Public      123 Sesame Street    Anywhere       NY 10035
Joseph      P Ziskend     P. O. Box 433        APO            AP 96555

C:\book\fortran\chap12>customer_database
Enter the file name with customer database:
database
Enter way to sort database:
  1 -- By last name
  2 -- By city
  3 -- By zip code
2
The sorted database values are:
Joseph      P Ziskend     P. O. Box 433        APO            AP 96555
John        Q Public      123 Sesame Street    Anywhere       NY 10035
Colin       A Jeffries    11 Main Street       Chicago        IL 60003
Jane        X Doe         12 Lakeside Drive    Glenview       IL 60025
Andrew      D Jackson     Jackson Square       New Orleans    LA 70003
James       R Johnson     Rt. 5 Box 207C       West Monroe    LA 71291

C:\book\fortran\chap12>customer_database
Enter the file name with customer database:
database
Enter way to sort database:
  1 -- By last name
  2 -- By city
  3 -- By zip code
3
The sorted database values are:
John        Q Public      123 Sesame Street    Anywhere       NY 10035
Colin       A Jeffries    11 Main Street       Chicago        IL 60003
Jane        X Doe         12 Lakeside Drive    Glenview       IL 60025
Andrew      D Jackson     Jackson Square       New Orleans    LA 70003
James       R Johnson     Rt. 5 Box 207C       West Monroe    LA 71291
Joseph      P Ziskend     P. O. Box 433        APO            AP 96555
```

Note that the program is working correctly with one minor exception. When it sorted the data by city, it got "APO" and "Anywhere" out of order. Can you tell why this happened? You will be asked to rewrite this program to eliminate the problem in Exercise 12-1.

12.5
RETURNING DERIVED TYPES FROM FUNCTIONS

It is possible to create a function of a derived data type *if and only if the function has an explicit interface*. The easiest way to create such an interface is to place the function within a module and to access that module using a USE statement. Example 12-2 creates two sample functions that return a derived data type.

EXAMPLE 12-2 *Adding and Subtracting Vectors:*

To illustrate the use of functions with derived data types, create a derived data type containing a 2D vector, plus two functions to add and subtract them. Also, create a test driver program to test the vector functions.

SOLUTION

1. **State the problem.**
   Create a module containing a 2D vector data type, plus functions to add and subtract vectors. Create a test driver program that prompts the user for two input vectors, and then adds and subtracts them using the functions.

2. **Define the inputs and outputs.**
   The inputs to the program are two vectors v1 and v2. The outputs are the sum and differences of the two vectors.

3. **Describe the algorithm.**
   The first step in writing this program will be to create a derived data type to hold a 2D vector. This type can be defined as follows:

   ```
   TYPE :: vector
      REAL :: x              ! X value
      REAL :: y              ! Y value
   END TYPE vector
   ```

   We must also define two functions vector_add and vector_sub that add and subtract 2D vectors, respectively. The pseudocode for the vector_add function is

   ```
   TYPE(vector) FUNCTION vector_add (v1, v2)
   vector_add.x ← v1%x + v2%x
   vector_add.y ← v1%y + v2%y
   ```

   and the pseudocode for the vector_sub function is

   ```
   TYPE(vector) FUNCTION vector_sub (v1, v2)
   vector_sub.x ← v1%x - v2%x
   vector_sub.y ← v1%y - v2%y
   ```

   The top-level pseudocode for the main program is

   ```
   Prompt user for the vector v1
   Read v1
   ```

Derived Data Types

```
        Prompt user for the vector v2
        Read v2
        Write the sum of the two vectors
        Write the difference of the two vectors
```

4. **Turn the algorithm into Fortran statements.**
   The resulting Fortran vector module is shown in Figure 12-4.

FIGURE 12-4
2D vector module.

```
MODULE vector_module
!
!  Purpose:
!    To define the derived data type for 2D vectors,
!    plus addition and subtraction operations.
!
!  Record of revisions:
!      Date       Programmer          Description of change
!      ====       ==========          =====================
!    12/04/15    S. J. Chapman        Original code
!
IMPLICIT NONE

! Declare type vector
TYPE :: vector
   REAL :: x                          ! X value
   REAL :: y                          ! Y value
END TYPE vector

! Add procedures
CONTAINS

   TYPE (vector) FUNCTION vector_add ( v1, v2 )
   !
   !  Purpose:
   !    To add two vectors.
   !
   !  Record of revisions:
   !      Date       Programmer          Description of change
   !      ====       ==========          =====================
   !    12/04/15    S. J. Chapman        Original code
   !
   IMPLICIT NONE

   ! Data dictionary: declare calling parameter types & definitions
   TYPE (vector), INTENT(IN) :: v1          ! First vector
   TYPE (vector), INTENT(IN) :: v2          ! Second vector

   ! Add the vectors
   vector_add%x = v1%x + v2%x
   vector_add%y = v1%y + v2%y

   END FUNCTION vector_add
```

(continued)

(*concluded*)

```
    TYPE (vector) FUNCTION vector_sub ( v1, v2 )
    !
    !  Purpose:
    !    To subtract two vectors.
    !
    !  Record of revisions:
    !      Date       Programmer          Description of change
    !      ====       ==========          =====================
    !    12/04/15    S. J. Chapman        Original code
    !
    IMPLICIT NONE

    ! Data dictionary: declare calling parameter types & definitions
    TYPE (vector), INTENT(IN) :: v1        ! First point
    TYPE (vector), INTENT(IN) :: v2        ! Second point

    ! Add the points
    vector_sub%x = v1%x - v2%x
    vector_sub%y = v1%y - v2%y

    END FUNCTION vector_sub

END MODULE vector_module
```

The test driver program is shown in Figure 12-5.

FIGURE 12-5
Test driver program for the vector module.

```
PROGRAM test_vectors
!
!  Purpose:
!    To test adding and subtracting 2D vectors.
!
!  Record of revisions:
!      Date       Programmer          Description of change
!      ====       ==========          =====================
!    12/04/15    S. J. Chapman        Original code
!
USE vector_module
IMPLICIT NONE

! Enter first point
TYPE (vector) :: v1              ! First point
TYPE (vector) :: v2              ! Second point

! Get the first vector
WRITE (*,*) 'Enter the first vector (x,y):'
READ (*,*) v1.x, v1.y

! Get the second point
WRITE (*,*) 'Enter the second vector (x,y):'
READ (*,*) v2.x, v2.y
```

(*continued*)

Derived Data Types

(concluded)

```
! Add the points
WRITE (*,1000) vector_add(v1,v2)
1000 FORMAT('The sum of the points is (',F8.2,',',F8.2,')')

! Subtract the points
WRITE (*,1010) vector_sub(v1,v2)
1010 FORMAT('The difference of the points is (',F8.2,',',F8.2,')')

END PROGRAM test_vectors
```

5. **Test the resulting Fortran programs.**

We will test this program entering two vectors, and manually checking the resulting answer. If vector v1 is (−2, 2) and vector v2 is (4, 3), then the sum of the vectors will be v1 + v2 = (2, 5) and the difference of the vectors will be v1 - v2 = (−6, −1).

```
C:\book\fortran\chap12>test_vectors
Enter the first vector (x,y):
-2. 2.
Enter the second vector (x,y):
4. 3.
The sum of the points is (    2.00,    5.00)
The difference of the points is (  -6.00,   -1.00)
```

The functions appear to be working correctly.

Good Programming Practice

To create functions of a derived data type, declare them within a module, and access the module using a USE statement.

Quiz 12-1

This quiz provides a quick check to see if you have understood the concepts introduced in Sections 12.1 through 12.5. If you have trouble with the quiz, reread the sections, ask your instructor, or discuss the material with a fellow student. The answers to this quiz are found in the back of the book.

For questions 1 to 7, assume the derived data types defined below:

```
TYPE :: position
    REAL :: x
    REAL :: y
    REAL :: z
END TYPE position
TYPE :: time
```

(continued)

> (*concluded*)
>
> ```
>         INTEGER :: second
>         INTEGER :: minute
>         INTEGER :: hour
>         INTEGER :: day
>         INTEGER :: month
>         INTEGER :: year
>     END TYPE time
>     TYPE :: plot
>         TYPE (time) :: plot_time
>         TYPE (position) :: plot_position
>     END TYPE
>     TYPE (plot), DIMENSION(10) :: points
> ```
>
> 1. Write the Fortran statements to print out the date associated with the seventh plot point in format "DD/MM/YYYY HH:MM:SS".
> 2. Write the Fortran statements to print out the position associated with the seventh plot point.
> 3. Write the Fortran statements required to calculate the rate of motion between the second and third plot points. To do this, you will have to calculate the difference in position and the difference in time between the two points. The rate of motion will be Δ pos/Δ pos time.
>
> For questions 4 to 6, state whether each of the following statements is valid. If the statements are valid, describe what they do.
>
> 4. `WRITE (*,*) points(1)`
>
> 5. `WRITE (*,1000) points(4)`
>   `1000 FORMAT (1X, 3ES12.6, 6I6 )`
>
> 6. `dpos = points(2).plot_position - points(1).plot_position`

12.6

DYNAMIC ALLOCATION OF DERIVED DATA TYPES

A variable or array of a derived data type can be declared with the `ALLOCATABLE` attribute and can be dynamically allocated and deallocated.[2] For example, suppose that a derived data type is defined as follows:

```
TYPE :: personal_info
    CHARACTER(len=12) :: first        ! First name
    CHARACTER         :: mi           ! Middle Initial
```

---

[2] Fortran 2003 and later only.

```
         CHARACTER(len=12) :: last         ! Last name
         CHARACTER(len=26) :: street       ! Street Address
         CHARACTER(len=12) :: city         ! City
         CHARACTER(len=2)  :: state        ! State
         INTEGER           :: zip          ! Zip code
      END TYPE personal_info
```

Then an allocatable variable of this type can be declared as

```
      TYPE(personal_info),ALLOCATABLE :: person
```

and can be allocated with the statement

```
      ALLOCATE( person, STAT=istat )
```

Similarly, an allocatable array of this type can be declared as

```
      TYPE(personal_info),DIMENSION(:),ALLOCATABLE :: people
```

and can be allocated with the statement

```
      ALLOCATE( people(1000), STAT=istat )
```

12.7

PARAMETERIZED DERIVED DATA TYPES

Just as Fortran allows multiple KINDs of integer or real data types, Fortran allows a user to define a derived data type with parameters.[3] This is called a **parameterized derived data type.** Two types of parameters can be used with derived data type definitions: Ones that are known at compile time (called *kind* type parameters) and ones that are not known until runtime (called *length* type parameters). Dummy values representing the kind numbers and element lengths are specified in parentheses after the type name, and these dummy values are then used to define the actual kinds and lengths of the elements in the derived type. If no dummy values are specified, then the derived data type will be created using default values that are specified in the type definition.

For example, the following lines declare a vector data type with KIND and length parameters.

```
      TYPE :: vector(kind,n)
         INTEGER, KIND :: kind = KIND(0.)     ! Defaults to single precision
         INTEGER, n = 3                       ! Defaults to three elements
         REAL(kind),DIMENSION(n) :: v         ! Parameterized vector
      END TYPE vector
```

The following type declarations will produce a derived data type containing a three-element single-precision vector.

---

[3] Fortran 2003 and later only.

```
TYPE (vector(KIND(0.),3)) :: v1      ! Kind and length specified
TYPE (vector) :: v2                   ! Kind and length defaulted
```

Similarly, the following type declaration produces a derived data type containing a 20-element double-precision vector:

```
TYPE (vector(KIND(0.D0),20)) :: v3   ! Kind and length specified
```

The next type declaration produces an array of 100 items of a derived data type, each containing a 20-element double-precision vector:

```
TYPE (vector(KIND(0.D0),20)),DIMENSION(100) :: v4
```

Derived data types can be declared allocatable, with the length of the individual elements deferred until allocation time. The following type declaration creates an allocatable structure whose length is deferred until the actual ALLOCATE statement is executed:

```
TYPE (vector(KIND(0.),:)),ALLOCATABLE :: v5
```

12.8
TYPE EXTENSION

A derived type that does not have the SEQUENCE or BIND(C)[4] attribute is *extensible*. This means that an existing user-defined type can be used as the basis of a larger, more comprehensive type definition. For example, suppose that a 2D point data type is defined as:

```
TYPE :: point
    REAL :: x
    REAL :: y
END TYPE
```

Then a 3D point data can be defined as an extension of the existing 2D point data type as follows:

```
TYPE, EXTENDS(point) :: point3d
    REAL :: z
END TYPE
```

This new data type contains three elements x, y, and z. Elements x and y were defined in type point and *inherited* by type point3d, while element z is unique to type point3d. Data type point is referred to as the *parent* of data type point3d.

The components of an extended data type can be used just like the components of any other data type. For example, suppose that we declare a variable of type point3d as follows:

```
TYPE(point3d) :: p
```

---

[4] The BIND(C) attribute makes a Fortran 2003 type interoperable with C. It will be discussed in Appendix B.

Derived Data Types

Then p will contain three components, which are usually addressed as p%x, p%y, and p%z. These components can be used in any calculations where they are required.

The inherited components of a derived data type can also be addressed by reference to the parent data type. For example, the x and y components of the item can also be addressed as p%parent%x and p%parent%y. Here, parent refers to the data type from which point3d was derived. This alternate form of address is used when we want to pass only the inherited values to a procedure.

Type extension is a mechanism whereby a new derived data type can inherit and extend the data defined in a parent data type. As we shall see in Chapter 16, such inheritance is an important feature of object-oriented programming.

The program shown below illustrates the use of extended data types. It declares a point data type, and then extends it to a point3d data type.

```
PROGRAM test_type_extension
!
!   Purpose:
!     To illustrate type extension of derived data types.
!
!   Record of revisions:
!       Date        Programmer          Description of change
!       ====        ==========          =====================
!     12/04/15    S. J. Chapman         Original code
!
IMPLICIT NONE

! Declare type point
TYPE :: point
   REAL :: x
   REAL :: y
END TYPE

! Declare type point3d
TYPE, EXTENDS(point) :: point3d
   REAL :: z
END TYPE

! Declare a variable of type person:
TYPE (point3d) :: my_point

! Initialize variable
my_point%x = 1.
my_point%y = 2.
my_point%z = 3.

! Output variable using free format I/O
WRITE (*,*) 'my_point = ', my_point

END PROGRAM test_type_extension
```

When this program is executed, the results are:

```
C:\book\fortran\chap12>test_type_extension
my_point =    1.0000000       2.0000000       3.0000000
```

12.9
TYPE-BOUND PROCEDURES

Fortran also allows procedures to be specifically associated ("bound") with a derived data type.[5] These procedures can only be used with variables of the derived data type that they are defined in.[6] They are invoked using elements of the derived data type with a syntax similar to that used to access a data element of the type. For example, a data element x of a derived type might be accessed as name%x, and a bound procedure proc associated with the type might be accessed as name%proc(arg list).

Type-bound Fortran procedures are created by adding a CONTAINS statement to the type definition and declaring the bindings in that statement. For example, suppose that we wanted to include a function to add two items of type point together. Then, we would declare the type definition as follows:

```
TYPE :: point
   REAL :: x
   REAL :: y
CONTAINS
   PROCEDURE,PASS :: add
END TYPE
```

This definition would declare that a procedure called add is associated with (bound to) this data type. If p is a variable of type point, then the add procedure would be referenced as p%add(...), just as element x would be referenced as p%x. The attribute PASS indicates that the variable of type point used to invoke the procedure is automatically passed to add as the first calling argument whenever it is called.

The procedure add would then need to be defined *in the same module* as the type definition statement. An example of a module declaring type point and including a procedure add is shown below:

```
MODULE point_module
IMPLICIT NONE

! Type definition
TYPE :: point
   REAL :: x
   REAL :: y
CONTAINS
   PROCEDURE,PASS :: add
END TYPE

CONTAINS
```

---

[5] Fortran 2003 and later.
[6] This makes bound procedures similar to the methods in classes of object-oriented languages such as C++ and Java, and we shall use them to implement object-oriented Fortran in Chapter 16.

```
        TYPE(point) FUNCTION add(this, another_point)
        CLASS(point) :: this, another_point
        add%x = this%x + another_point%x
        add%y = this%y + another_point%y
        END FUNCTION add
    END MODULE point_module
```

The function `add` has two arguments, `this` and `another_point`. Argument `this` is the variable used to invoke the procedure. It is automatically passed to the procedure when it is invoked without it being explicit in the call, while argument `another_point` will show up in the list of calling arguments.

Note that the derived data types are declared in the bound procedure using the `CLASS` keyword. `CLASS` is a special version of the `TYPE` keyword with additional properties that support object-oriented programming; it will be discussed in Chapter 16.

Three objects of this type could be declared as follows:

```
    TYPE(point) :: a, b, c
    a%x = -10.
    a%y = 5.
    b%x = 4.
    b%y = 2.
```

With this definition, the following statement adds points a and b together and stores the result in point c.

```
    c = a%add(b)
```

This statement calls function `add`, automatically passing it a as its first argument and b as its second argument. The function returns a result of type `point`, which is stored in variable c. After the function call, `c%x` will contain the value −6 and `c%y` will contain the value 7.

If the procedure binding contains the attribute `NOPASS` instead of `PASS`, then the bound procedure will *not* automatically get the variable used to invoke it as a calling argument. If the data type were declared as follows:

```
    TYPE :: point
        REAL :: x
        REAL :: y
    CONTAINS
        PROCEDURE,NOPASS :: add
    END TYPE
```

then the bound function would have to be called with the first argument explicitly shown in the call:

```
    c = a%add(a,b)
```

If no attribute is given in a binding, the default attribute is `PASS`. As we shall see in Chapter 16, this feature is useful in object-oriented programming.

EXAMPLE 12-3 *Using Bound Procedures:*

Convert the vector module of Example 12-2 so that it uses bound procedures.

SOLUTION
If a derived data type uses bound procedures, then the procedures will be addressed by using variable name followed by the component selector (%), and the variable used to invoke the procedures will be automatically passed as the first calling argument. The modified vector module is shown in Figure 12-6.

FIGURE 12-6
2D vector module with bound procedures.

```
MODULE vector_module
!
!   Purpose:
!     To define the derived data type for 2D vectors,
!     plus addition and subtraction operations.
!
!   Record of revisions:
!       Date        Programmer          Description of change
!       ====        ==========          =====================
!     12/04/15    S. J. Chapman         Original code
! 1.  12/06/15    S. J. Chapman         Use bound procedures
!
IMPLICIT NONE

! Declare type vector
TYPE :: vector
    REAL :: x                       ! X value
    REAL :: y                       ! Y value
CONTAINS
    PROCEDURE,PASS :: vector_add
    PROCEDURE,PASS :: vector_sub
END TYPE vector

! Add procedures
CONTAINS

    TYPE (vector) FUNCTION vector_add ( this, v2 )
    !
    !   Purpose:
    !      To add two vectors.
    !
    !   Record of revisions:
    !       Date        Programmer          Description of change
    !       ====        ==========          =====================
    !     12/04/15    S. J. Chapman         Original code
    ! 1.  12/06/15    S. J. Chapman         Use bound procedures
    !
    IMPLICIT NONE
```

(continued)

Derived Data Types

(*concluded*)

```
    ! Data dictionary: declare calling parameter types & definitions
    CLASS(vector),INTENT(IN) :: this      ! First vector
    CLASS(vector),INTENT(IN) :: v2        ! Second vector

    ! Add the vectors
    vector_add%x = this%x + v2%x
    vector_add%y = this%y + v2%y

    END FUNCTION vector_add

    TYPE (vector) FUNCTION vector_sub ( this, v2 )
    !
    !  Purpose:
    !    To subtract two vectors.
    !
    !  Record of revisions:
    !      Date        Programmer          Description of change
    !      ====        ==========          =====================
    !    12/04/15    S. J. Chapman         Original code
    ! 1. 12/06/15    S. J. Chapman         Use bound procedures
    !
    IMPLICIT NONE

    ! Data dictionary: declare calling parameter types & definitions
    CLASS(vector),INTENT(IN) :: this      ! First vector
    CLASS(vector),INTENT(IN) :: v2        ! Second vector

    ! Add the points
    vector_sub%x = this%x - v2%x
    vector_sub%y = this%y - v2%y

    END FUNCTION vector_sub

END MODULE vector_module
```

The test driver program is shown in Figure 12-7.

FIGURE 12-7
Test driver program for the vector module with bound procedures.

```
PROGRAM test_vectors
!
!  Purpose:
!    To test adding and subtracting 2D vectors.
!
!  Record of revisions:
!      Date        Programmer          Description of change
!      ====        ==========          =====================
!    12/04/15    S. J. Chapman         Original code
! 1. 12/06/15    S. J. Chapman         Use bound procedures
!
USE vector_module
IMPLICIT NONE
```

(*continued*)

(*concluded*)

```
! Enter first point
TYPE(vector) :: v1              ! First point
TYPE(vector) :: v2              ! Second point

! Get the first vector
WRITE (*,*) 'Enter the first vector (x,y):'
READ (*,*) v1%x, v1%y

! Get the second point
WRITE (*,*) 'Enter the second vector (x,y):'
READ (*,*) v2%x, v2%y

! Add the points
WRITE (*,1000) v1%vector_add(v2)
1000 FORMAT('The sum of the points is (',F8.2,',',F8.2,')')

! Subtract the points
WRITE (*,1010) v1%vector_sub(v2)
1010 FORMAT('The difference of the points is (',F8.2,',',F8.2,')')

END PROGRAM test_vectors
```

We will test this program using the same data as in the previous example.

```
C:\book\fortran\chap12>test_vectors
Enter the first vector (x,y):
-2. 2.
Enter the second vector (x,y):
4. 3.
The sum of the points is (    2.00,    5.00)
The difference of the points is (   -6.00,   -1.00)
```

The functions appear to be working correctly.

12.10

THE ASSOCIATE CONSTRUCT

The ASSOCIATE construct allows a programmer to temporarily associate a name with a variable or expression during the execution of a code block. This construct is useful for simplifying multiple references to variables or expressions with long names and/or many subscripts.

The form of an associate construct is

```
[name:] ASSOCIATE (association_list)
    Statement 1
    Statement 2
    ...
    Statement n
END ASSOCIATE [name]
```

The *association_list* is a set of one or more associations of the form

```
assoc_name => variable, array element, or expression
```

If more than one association appears in the list, they are separated by commas.

Derived Data Types

To get a better understanding of the ASSOCIATE construct, let's examine a practical case. Suppose that a radar is tracking a series of objects, and each object's position is stored in a data structure of the form:

```
TYPE :: trackfile
    REAL :: x             ! X position (m)
    REAL :: y             ! Y position (m)
    REAL :: dist          ! Distance to target (m)
    REAL :: bearing       ! Bearing to target (rad)
END TYPE trackfile
TYPE(trackfile),DIMENSION(1000) :: active_tracks
```

Suppose that the location of the radar itself is stored in a data structure of the form:

```
TYPE :: radar_loc
    REAL :: x             ! X position (m)
    REAL :: y             ! Y position (m)
END TYPE radar_loc
TYPE(radar_loc) :: my_radar
```

We would like to calculate the range and bearing to all of the tracks. This can be done with the following statements:

```
DO itf = 1, n_tracks
   active_tracks(i)%dist    = SQRT( (my_radar%x - active_tracks(i)%x) ** 2 &
                                  + (my_radar%y - active_tracks(i)%y) ** 2 )
   active_tracks(i)%bearing = ATAN2D( (my_radar%y - active_tracks(i)%y), &
                                      (my_radar%x - active_tracks(i)%x) )
END DO
```

These statements are legal, but they are *not* very readable because of the long names involved. If instead we use the ASSOCIATE construct, the fundamental equations are much clearer:

```
DO itf = 1, n_tracks
   ASSOCIATE ( x => active_tracks(itf)%x, &
               y => active_tracks(itf)%y, &
               dist => active_tracks(itf)%dist, &
               bearing => active_tracks(itf)%bearing )
      dist = SQRT( (my_radar%x - x) ** 2 + (my_radar%y - y) ** 2 )
      bearing = ATAN2D( (my_radar%y - y), (my_radar%x - x) )
   END ASSOCIATE
END DO
```

The ASSOCIATE construct is never required, but it can be useful to simplify and emphasize the algorithm being used.

▪ 12.11
SUMMARY

Derived data types are data types defined by the programmer for use in solving a particular problem. They may contain any number of components, and each component may be of any intrinsic data type or any previously defined derived data type. Derived data types are defined using a TYPE ... END TYPE construct, and variables of that type are declared using a TYPE statement. Constants of a derived data type may be constructed using structure constructors. A variable or constant of a derived data type is called a structure.

The components of a variable of a derived data type may be used in a program just like any other variables of the same type. They are addressed by naming both the variable and the component separated by a percent sign (e.g., student%age). Variables of a derived data type may not be used with any Fortran intrinsic operations except for assignment. Addition, subtraction, multiplication, division, etc., are undefined for these variables. They may be used in I/O statements.

We will learn how to extend intrinsic operations to variables of a derived data type in Chapter 13.

12.11.1 Summary of Good Programming Practice

The following guideline should be adhered to when working with parameterized variables, complex numbers, and derived data types:

1. For large programs using derived data types, declare the definitions of each data type in a module, and then use that module in each procedure of the program that needs to access the derived data type.

12.11.2 Summary of Fortran Statements and Structures

ASSOCIATE Construct:

```
[name:] ASSOCIATE (association_list)
    Statement 1
    ...
    Statement n
END ASSOCIATE [name]
```

Example:

```
ASSOCIATE (x => target(i)%state_vector%x, &
           y => target(i)%state_vector%y )
    dist(i) = SQRT(x**2 + y**2)
END ASSOCIATE
```

Description:
The ASSOCIATE construct allows a programmer to address one or more variables with very long names by a shorter name within the body of the construct. The equations within the ASSOCIATE construct can be much more compact, because the individual variable names are not too cumbersome.

Derived Data Types

Derived Data Type:

```
TYPE [::] type_name
   component 1
   ...
   component n
CONTAINS
   PROCEDURE[,[NO]PASS] :: proc_name1[, proc_name2, ...]
END TYPE [type_name]
TYPE (type_name) :: var1 (, var2, ...)
```

Example:

```
TYPE :: state_vector
   LOGICAL :: valid                 ! Valid data flag
   REAL(kind=single) :: x           ! x position
   REAL(kind=single) :: y           ! y position
   REAL(kind=double) :: time        ! time of validity
   CHARACTER(len=12) :: id          ! Target ID
END TYPE state_vector
TYPE (state_vector), DIMENSION(50) :: objects
```

Description:
The derived data type is a structure containing a combination of intrinsic and previously defined derived data types. The type is defined by a TYPE ... END TYPE construct, and variables of that type are declared with a TYPE() statement.

Bound procedures in derived data types are only available in Fortran 2003 and later.

NOPASS Attribute:

```
TYPE :: name
   variable definitions
CONTAINS
   PROCEDURE,NOPASS :: proc_name
END TYPE
```

Example:

```
TYPE :: point
   REAL :: x
   REAL :: y
CONTAINS
   PROCEDURE,NOPASS :: add
END TYPE
```

Description:
The NOPASS attribute means that the variable used to invoke the procedure will *not* be automatically passed to the procedure as its first calling argument.

PASS Attribute:

```
TYPE :: name
    variable definitions
CONTAINS
    PROCEDURE,PASS :: proc_name
END TYPE
```

Example:

```
TYPE :: point
    REAL :: x
    REAL :: y
CONTAINS
    PROCEDURE,PASS :: add
END TYPE
```

Description:
The PASS attribute means that the variable used to invoke the procedure will be automatically passed to the procedure as its first calling argument. This is the default case for bound procedures.

12.11.3 Exercises

12-1. When the database was sorted by city in Example 12-1, "APO" was placed ahead of "Anywhere". Why did this happen? Rewrite the program in this example to eliminate this problem.

12-2. Create a derived data type called "polar" to hold a complex number expressed in polar (z, θ) format as shown in Figure 12-8. The derived data type will contain two components, a magnitude z and an angle θ, with the angle expressed in degrees. Write two functions that convert an ordinary complex number into a polar number, and that convert a polar number into an ordinary complex number.

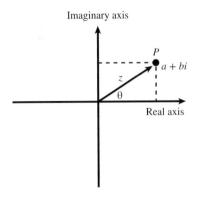

FIGURE 12-8
Representing a complex number in polar coordinates.

Derived Data Types

12-3. If two complex numbers are expressed in polar form, the two numbers may be multiplied by multiplying their magnitudes and adding their angles. That is, if $P_1 = z_1 \angle \theta_1$ and $P_2 = z_2 \angle \theta_2$, then $P_1 \cdot P_2 = z_1 z_2 \angle \theta_1 + \theta_2$. Write a function that multiplies two variables of type "polar" together using this expression and returns a result in polar form. Note that the resulting angle θ should be in the range $-180° \leq \theta \leq 180°$.

12-4. If two complex numbers are expressed in polar form, the two numbers may be divided by dividing their magnitudes and subtracting their angles. That is, if $P_1 = z_1 \angle \theta_1$ and $P_2 = z_2 \angle \theta_2$, then $\dfrac{P_1}{P_2} = \dfrac{z_1}{z_2} \angle \theta_1 - \theta_2$. Write a function that divides two variables of type "polar" together using this expression and returns a result in polar form. Note that the resulting angle θ should be in the range $-180° \leq \theta \leq 180°$.

12-5. Create a version of the `polar` data type with the functions defined in Exercises 12-2 through 12-4 as bound procedures. Write a test driver program to illustrate the operation of the data type.

12-6. A point can be located in a Cartesian plane by two coordinates (x, y), where x is the displacement of the point along the x axis from the origin and y is the displacement of the point along the y axis from the origin. Create a derived data type called "point" whose components are x and y. A line can be represented in a Cartesian plane by the equation

$$uy = mx + b \tag{12-1}$$

where m is the slope of the line and b is the y-axis intercept of the line. Create a derived data type called "line" whose components are m and b.

12-7. The distance between two points (x_1, y_1) and (x_2, y_2) is given by the equation

$$distance = \sqrt{(x_2 - x_1)^2 + (y_2 - y_1)^2} \tag{12-2}$$

Write a function that calculates the distance between two values of type "point" as defined in Exercise 12-5 above. The inputs should be two points and the output should be the distance between the two points expressed as a real number.

12-8. From elementary geometry, we know that two points uniquely determine a line as long as they are not coincident. Write a function that accepts two values of type "point", and returns a value of type "line" containing the slope and y-intercept of the line. If the two points are identical, the function should return zeros for both the slope and the intercept. From Figure 12-9, we can see that the slope of the line can be calculated from the equation

$$m = \frac{y_2 - y_1}{x_2 - x_1} \tag{12-3}$$

and the intercept can be calculated from the equation

$$b = y_1 - mx_1 \tag{12-4}$$

12-9. Tracking Radar Targets Many surveillance radars have antennas that rotate at a fixed rate, scanning the surrounding airspace. The targets detected by such radars are usually

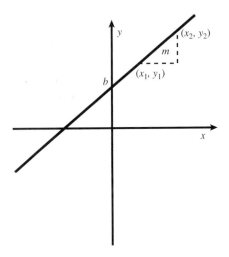

FIGURE 12-9
The slope and intercept of a line can be determined from two points and that lie along the line.

displayed on *Plan Position Indicator (PPI)* displays, such as the one shown in Figure 12-10. As the antenna sweeps around the circle, a bright line sweeps around the PPI display. Each target detected shows up on the display as a bright spot at a particular range r and angle θ, where θ is measured in compass degrees relative to North.

Each target will be detected at a different position every time that the radar sweeps around the circle, both because the target moves and because of inherent noise in the range and angle measurement process. The radar system needs to track detected targets through successive sweeps, and to estimate target position and velocity from the

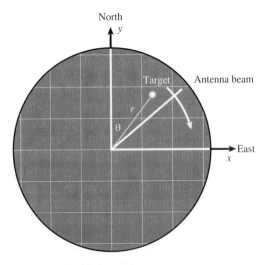

FIGURE 12-10
The PPI display of a track-while-scan radar. Target detections show up as bright spots on the display. Each detection is characterized by a range, compass azimuth, and detection time (r, θ, T_n).

successive detected positions. Radar systems that accomplish such tracking automatically are known as *track-while-scan* (TWS) radars. They work by measuring the position of the target each time it is detected, and passing that position to a *tracking algorithm*.

One of the simplest tracking algorithms is known as the α–β tracker. The α–β tracker works in Cartesian coordinates, so the first step in using the tracker is to convert each target detection from polar coordinates (r, θ) into rectangular coordinates (x, y). The tracker then computes a smoothed target position $(\bar{x}_n, \bar{y}_n)$ and velocity $(\bar{\dot{x}}_n, \bar{\dot{y}}_n)$ from the equations:

Updated position
$$\bar{x}_n = x_{pn} + \alpha(x_n - x_{pn})$$
$$\bar{y}_n = y_{pn} + \alpha(y_n - y_{pn})$$
(12-5)

Updated velocity
$$\bar{\dot{x}}_n = \bar{\dot{x}}_{n-1} + \frac{\beta}{T_S}(x_n - x_{pn})$$
$$\bar{\dot{y}}_n = \bar{\dot{y}}_{n-1} + \frac{\beta}{T_S}(y_n - y_{pn})$$
(12-6)

Predicted position
$$x_{pn} = \bar{x}_{n-1} + \bar{\dot{x}}_{n-1} T_S$$
$$y_{pn} = \bar{y}_{n-1} + \bar{\dot{y}}_{n-1} T_S$$
(12-7)

where (x_n, y_n) is the *measured* target position at time n, (x_{pn}, y_{pn}) is the *predicted* target position at time n, $(\bar{\dot{x}}_n, \bar{\dot{y}}_n)$ is the smoothed target velocity at time n, $(\bar{x}_{n-1}, \bar{y}_{n-1})$ and $(\bar{\dot{x}}_{n-1}, \bar{\dot{y}}_{n-1})$ are the smoothed positions and velocity from time $n-1$, α is the position smoothing parameter, β is the velocity smoothing parameter, and T_S is the time between observations.

Design a Fortran program that acts as a radar tracker. The input to the program will be a series of radar target detections (r, θ, T), where r is range in meters, θ is azimuth in compass degrees, and T is the time of the detection in seconds. The program should convert the observations to rectangular coordinates on an East-North grid, and use them to update the tracker as follows:

1. Calculate the time difference T_S since the last detection.
2. Predict the position of the target at the time of the new detection using Equations (12-7).
3. Update the smoothed position of the target using Equations (12-5). Assume that the position smoothing parameter $\alpha = 0.7$.
4. Update the smoothed velocity of the target using Equations (12-6). Assume that the velocity smoothing parameter $\beta = 0.38$.

A block diagram illustrating the operation of the tracker is shown in Figure 12-11.

The program should print out a table containing the observed position of the target, predicted position of the target, and smoothed position of the target each time that the target is measured. Finally, it should produce line printer plots of the estimated x and y velocity components of the target.

The program should include separate derived data types to hold the detections in polar coordinates (r_n, θ_n, T_n), the detections in rectangular coordinates (x_n, y_n, T_n), and the smoothed state vectors $(\bar{x}_n, \bar{y}_n, \bar{\dot{x}}_n, \bar{\dot{y}}_n, T_n)$. It should include separate procedures to perform the polar-to-rectangular conversion, target predictions, and target updates. (Be careful of the polar-to-rectangular conversion—since it uses compass angles, the equations to convert to rectangular coordinates will be different than what we saw earlier!)

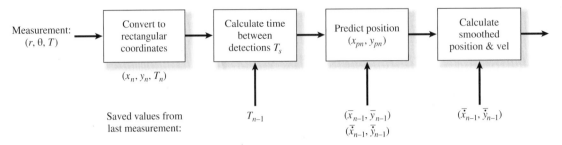

FIGURE 12-11
Block diagram of the operation of an α–β tracker. Note that the smoothed position, velocity, and time from the last update must be saved for use in the current tracker cycle.

Test your program by supplying it with both a noise-free and a noisy input data set. Both data sets are from a plane flying in a straight line, making a turn, and flying in a straight line again. The noisy data is corrupted by a Gaussian noise with a standard deviation of 200 m in range and 1.1° in azimuth. (The noise-free data can be found in file track1.dat and the noisy data can be found in file track2.dat on the disk accompanying the Instructor's Manual, or at the website for this book.) How well does the tracker work at smoothing out errors? How well does the tracker handle the turn?

13

Advanced Features of Procedures and Modules

OBJECTIVES

- Understand the four types of scope available in Fortran, and when each one applies.
- Learn about the `BLOCK` construct.
- Learn how to create recursive subroutines and functions.
- Learn how to create and use keyword arguments.
- Learn how to create and use optional arguments.
- Learn how to create explicit interfaces with Interface Blocks.
- Learn how to create user-defined generic procedures.
- Learn how to create bound generic procedures.
- Learn how to create user-defined operators.
- Learn how to create assignments and operators that are bound to a specific derived data type.
- Learn how to restrict access to entities defined within a Fortran module.
- Learn how to create and use type-bound procedures.
- Learn about the standard Fortran intrinsic modules.
- Learn the standard procedures for accessing command line arguments and environment variables.

This chapter introduces some more advanced features of Fortran procedures and modules. These features permit us to have better control over access to the information contained in procedures and modules, allow us to write more flexible procedures that support optional arguments and varying data types, and allow us to extend the Fortran language to support new operations on both intrinsic and derived data types.

13.1
SCOPE AND SCOPING UNITS

In Chapter 7, we learned that the main program and each external subroutine and function in a program are compiled independently, and are then associated together by a linker. Because they were compiled independently, variable names, constant names, loop names, statement labels, etc., could be reused in the different procedures without interfering with each other. For example, the name `my_data` could be declared and used as a character variable in one procedure, and declared and used as an integer array in another procedure without causing a conflict. There was no conflict because the **scope** of each name or label was restricted to a single procedure.

The **scope** of an object (a variable, named constant, procedure name, statement label, etc.) is the portion of a Fortran program over which it is defined. There are four levels of scope in a Fortran program. They are:

1. **Global Scope.** Global objects are objects that are defined throughout an entire program. The names of these objects must be unique within a program. The only global objects that we have encountered so far are the names of programs, external procedures, and modules. Each of these names must be unique within the entire program.[1]
2. **Local Scope.** Local objects are objects that are defined and must be unique within a single **scoping unit.** Examples of scoping units are programs, external procedures, and modules. A local object within a scoping unit must be unique within that unit, but the object name, statement label, etc., may be reused in another scoping unit without causing a conflict.
3. **Block Scope.** Blocks are constructs within a program or procedure that can define their own local variables that are independent of the variables in the containing procedure. Blocks will be described in the next section.
4. **Statement Scope.** The scope of certain objects may be restricted to a single statement within a program unit. The only examples that we have seen of objects whose scope is restricted to a single statement are the implied `DO` variable in an array constructor and the index variables in a `FORALL` statement. An example array constructor is

$$\text{array} = [\ (2*i, \ i=1,10,2) \ ]$$

Here the variable `i` is used to define the array values using an implied `DO` loop. This use of variable `i` should not interfere with the use of `i` in the surrounding program, because the scope of this variable is limited to this single statement.

---

[1] In some circumstances, there can be local objects with the same names as some global objects. For example, if a program contains an external subroutine called `sort`, then no other global object in the program can have the name `sort`. However, a *different subroutine* within the program could contain a local variable called `sort` without causing a conflict. Since the local variable is not visible outside the subroutine, it does not conflict with the global object of the same name.

Just what is a scoping unit? It is the portion of a Fortran program over which a local object is defined. The scoping units in a Fortran program are:

1. A main program, internal or external procedure, or module, excluding any derived type definitions or procedures contained within it.
2. A derived type definition.
3. An interface, which we will meet later in this chapter.
4. A code block, which we will meet later in this chapter.

Local objects within each of these scoping units must be unique, but they may be reused between scoping units. The fact that a derived type definition is a scoping unit means that we can have a variable named x as a component of the derived type definition, and also have a variable named x within the program containing the derived type definition, without the two variables conflicting with each other.

If one scoping unit completely surrounds another scoping unit, then it is called the **host scoping unit,** or just the **host,** of the inner scoping unit. The inner scoping unit automatically inherits the object definitions declared in the host scoping unit, unless the inner scoping unit explicitly redefines the objects. This inheritance is called **host association.** Thus, an internal procedure inherits all of the variable names and values defined in the host procedure unless the internal procedure explicitly redefines a variable name for its own use. If the internal procedure uses a variable name defined in the host unit without redefining it, then changes to that variable in the internal procedure will also change the variable in the host unit. In contrast, if the internal procedure redefines a variable name used in the host unit, then modifications to that local variable will not affect the value of the variable with the same name in the host unit.

Finally, objects defined in a module normally have the scope of that module, but their scope may be extended by **USE association.** If the module name appears in a USE statement in a program unit, then all of the objects defined in the module become objects defined in the program unit using the module, and the names of those objects must be unique. If an object named x is declared within a module and that module is used in a procedure, then no other object may be named x within the procedure.

EXAMPLE 13-1

Scope and Scoping Units:

When dealing with a subject as complex as scope and scoping units, it is helpful to look at an example. Figure 13-1 shows a Fortran program written specifically to explore the concept of scope. If we can answer the following questions about that program, then we will have a pretty good understanding of scope.

1. What are the scoping units within this program?
2. Which scoping units are hosts to other units?
3. Which objects in this program have global scope?
4. Which objects in this program have statement scope?
5. Which objects in this program have local scope?
6. Which objects in this program are inherited by host association?
7. Which objects in this program are made available by USE association?
8. Explain what will happen in this program as it is executed.

FIGURE 13-1
Program to illustrate the concept of scope and scoping units.

```
MODULE module_example
IMPLICIT NONE
REAL :: x = 100.
REAL :: y = 200.
END MODULE

PROGRAM scoping_test
USE module_example
IMPLICIT NONE
INTEGER :: i = 1, j = 2
WRITE (*,'(A25,2I7,2F7.1)') ' Beginning:', i, j, x, y
CALL sub1 ( i, j )
WRITE (*,'(A25,2I7,2F7.1)') ' After sub1:', i, j, x, y
CALL sub2
WRITE (*,'(A25,2I7,2F7.1)') ' After sub2:', i, j, x, y
CONTAINS
   SUBROUTINE sub2
   REAL :: x
   x = 1000.
   y = 2000.
   WRITE (*,'(A25,2F7.1)') ' In sub2:', x, y
   END SUBROUTINE sub2
END PROGRAM scoping_test

SUBROUTINE sub1 (i,j)
IMPLICIT NONE
INTEGER, INTENT(INOUT) :: i, j
INTEGER, DIMENSION(5) :: array
WRITE (*,'(A25,2I7)') 'In sub1 before sub2:', i, j
CALL sub2
WRITE (*,'(A25,2I7)') 'In sub1 after sub2:', i, j
array = [ (1000*i, i=1,5) ]
WRITE (*,'(A25,7I7)') 'After array def in sub2:', i, j, array
CONTAINS
   SUBROUTINE sub2
   INTEGER :: i
   i = 1000
   j = 2000
   WRITE (*,'(A25,2I7)') 'In sub1 in sub2:', i, j
   END SUBROUTINE sub2
END SUBROUTINE sub1
```

SOLUTION
The answers to the questions are given below.

1. *What are the scoping units within this program?*

    Each module, main program, and internal and external procedure is a scoping unit, so the scoping units are module `module_example`, main program `scoping_test`, external subroutine `sub1`, and the two internal subroutines `sub2`. If there had been any derived data types within the program, their definitions would also have been scoping units. Figure 13-2 illustrates the relationships among the five scoping units in this program.

Advanced Features of Procedures and Modules

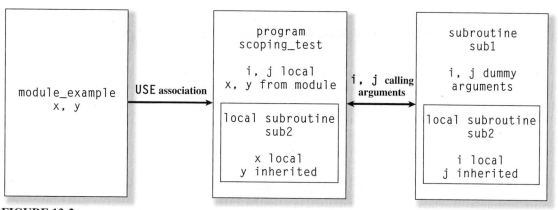

FIGURE 13-2
A block diagram illustrating the relationships among the various scoping units in program scoping_test.

2. *Which scoping units are hosts to other units?*
    The main program scoping_test is the host scoping unit for the internal subroutine sub2 contained within it, and the external subroutine sub1 is the host scoping unit for the internal subroutine sub2 contained within it. Note that the two internal subroutines are different, even though they have the same name!

3. *Which objects in this program have global scope?*
    The objects within this program that have global scope are the names of the module module_example, the main program scoping_test, and the external subroutine sub1. These names must be unique throughout the program. For example, there cannot be two external subroutines both named sub1 in a single program. In contrast, the names of the internal subroutines sub2 have local scope only, so it is legal to have two different local subroutines of the same name in two different scoping units.

4. *Which objects in this program have statement scope?*
    The only object within this program that has statement scope is the variable i within the array definition in subroutine sub1. Because that variable has statement scope, the value of variable i in subroutine sub1 will be unchanged by the use of i to define the array.

5. *Which objects in this program have local scope?*
    All other objects within this program have local scope, including the names of the internal subroutines sub2. Because each internal subroutine is local to its host scoping unit, there is no conflict involved in having two subroutine with the same name. Each of the internal subroutines is only defined within and callable from its host scoping unit.

6. *Which objects in this program are inherited by host association?*
    All objects in the two internal subroutines are inherited from their host scoping units by host association with the exception of those objects explicitly redefined within the internal subroutines. Thus, variable x is local to the first internal subroutine, while variable y is inherited from the main program, which is

the host scoping unit. Similarly, variable i is local to the second internal subroutine, while variable j is inherited from the subroutine sub1, which is the host scoping unit.

7. *Which objects in this program are made available by USE association?*

   Variables x and y are made available to the main program by USE association.

8. *Explain what will happen in this program as it is executed.*

   When this program begins execution, variables x and y are initialized to 100. and 200. respectively in module module_example, and variables i and j are initialized to 1 and 2 respectively in the main program. Variables x and y are visible in the main program by USE association.

   When subroutine sub1 is called, variables i and j are passed to sub1 as calling arguments. Subroutine sub1 then calls its local subroutine sub2, which sets i to 1000 and j to 2000. However, variable i is local to sub2, so changing it has no effect on variable i in sub1. Variable j is the same variable in sub1 and sub2 through host association, so when sub2 sets a new value for j, the value of j in sub1 is changed to 2000.

   Next a value is assigned to the array using variable i as an array constructor. Variable i takes on values from 1 to 5 as a part of the implied DO loop, but the scope of that variable is statement only, so in the next line of the subroutine the value of variable i remains 1 as it was before the array assignment.

   When execution returns from sub1 to the main program, i is still 1 and j is 2000. Next, the main program calls its own local subroutine sub2. Subroutine sub2 sets x to 1000. and y to 2000. However, variable x is local to sub2, so changing it has no effect on variable x in the main program. Variable y is the same variable in the main program and in sub2 through host association, so when sub2 sets a new value for y, the value of y in the main program is changed to 2000.

   After the call to sub2, the values of i, j, x, and y in the main program are 1, 2000, 100., and 2000., respectively.

   We can verify our analysis of the operation of this program by executing it and examining the results:

```
C:\book\fortran\chap13>scoping_test
              Beginning:      1       2    100.0   200.0
     In sub1 before sub2:     1       2
        In sub1 in sub2:   1000    2000
     In sub1 after sub2:      1    2000
  After array def in sub2:    1    2000    1000    2000    3000    4000    5000
              After sub1:     1    2000    100.0   200.0
              In sub2:    1000.0  2000.0
             After sub2:     1    2000    100.0  2000.0
```

The output of this program matches our analysis.

It is possible to reuse a local object name for different purposes in nested scoping units. For example, the integer i was defined in subroutine sub1 and would normally have been available to internal subroutine sub2 by host association. However, sub2 defined its own integer i, so in fact the integer i is different in the two scoping units. This sort of double definition is a recipe for confusion, and should be avoided in your code. Instead, just create a new variable name in the internal subroutine that does not conflict with any in the host.

Good Programming Practice
When working with nested scoping units, avoid redefining the meaning of objects that have the same name in both the inner and outer scoping units. This applies especially to internal procedures. You can avoid confusion about the behavior of variables in the internal procedure by simply giving them different names from the variables in the host procedure.

13.2
BLOCKS

Blocks are a new type of construct that were introduced with Fortran 2008. A block is an arbitrary block of code that appears within a host program or procedure. It begins with a BLOCK statement and ends with an END BLOCK statement. The block can contain any desired code, and it is possible to define local variables that are unique to the block.

The structure of a block construct is:

```
[name:] BLOCK
   Type definitions ...
   ...
   Executable code
   IF (...) EXIT [name]
   ...
END BLOCK [name]
```

Note that it is possible to exit a code block at any point in the block using EXIT statement. If the block exits, code execution continues at the first executable statement after the end of the block.

Each block can define local variables before the executable code in the block. When execution of the code block ends, all of the variables defined in the block become undefined. If allocatable arrays were defined in the block without a SAVE attribute, they will be automatically deallocated when block execution ends.

A block also has access to the local variables of its host by host association, unless the block defines a local variable with the same name.

Figure 13-3 shows a sample program containing a block construct. The program defines three variables i, j, and k, and writes out their values before the start of the block. The block construct defines a new local variable j, and has access to the variables i and k by host association. The DO loop in the block executes three times, and then the execution exits the code block.

FIGURE 13-3
Program illustrating a block construct.

```
PROGRAM test_blocks
IMPLICIT NONE

INTEGER :: i, j, k
i = 1
j = 2
k = 3

! Variables before the block
WRITE (*,*) 'Before block: i, j, k = ', i, j, k

! Declare block
test_block: BLOCK
   INTEGER :: j

   WRITE (*,*) 'In block before DO loop.'

   DO j = 1, 10

      ! Variables in the block
      WRITE (*,*) 'In block:     i, j, k = ', i, j, k

      IF ( j > 2 ) EXIT test_block

   END DO

   WRITE (*,*) 'In block after DO loop.'

END BLOCK test_block

! Variables after the block
WRITE (*,*) 'After block: i, j, k = ', i, j, k

END PROGRAM test_blocks
```

When this program is executed, the results are:

```
C:\book\fortran\chap13>test_blocks
 Before block: i, j, k =         1     2     3
 In block before DO loop.
 In block:     i, j, k =         1     1     3
 In block:     i, j, k =         1     2     3
 In block:     i, j, k =         1     3     3
 After block: i, j, k =          1     2     3
```

Note that the WRITE statement after the end of the DO loop was never executed, because program execution jumped to the first statement after the block when the EXIT statement was executed.

13.3
RECURSIVE PROCEDURES

An ordinary Fortran procedure may not invoke itself either directly or indirectly (i.e., by either invoking itself or by invoking another procedure that then invokes the original procedure). In other words, ordinary Fortran procedures are not **recursive.** However, there are certain classes of problems that are most easily solved recursively. For example, the factorial function can be defined as

$$N! = \begin{cases} N(N-1)! & N \geq 1 \\ 1 & N = 0 \end{cases} \qquad (13\text{-}1)$$

This definition can most easily be implemented recursively, with the procedure that calculates $N!$ calling itself to calculate $(N-1)!$, and that procedure calling itself to calculate $(N-2)!$, etc., until finally the procedure is called to calculate 0!.

To accommodate such problems, Fortran allows subroutines and functions to be declared recursive. If a procedure is declared recursive, then the Fortran compiler will implement it in such a way that it can invoke itself either directly or indirectly as often as desired.

A subroutine is declared recursive by adding the keyword `RECURSIVE` to the `SUBROUTINE` statement. Figure 13-4 shows an example subroutine that calculates the factorial function directly from Equation (13-1). It looks just like any other subroutine except that it is declared to be recursive. You will be asked to verify the proper operation of this subroutine in Exercise 13-2.

FIGURE 13-4
A subroutine to recursively implement the factorial function.

```
RECURSIVE SUBROUTINE factorial ( n, result )
!
!  Purpose:
!    To calculate the factorial function
!           | n(n-1)!   n >= 1
!    n ! = |
!           | 1         n = 0
!
!  Record of revisions:
!      Date        Programmer         Description of change
!      ====        ==========         =====================
!    12/17/15    S. J. Chapman        Original code
!
IMPLICIT NONE

! Data dictionary: declare calling parameter types & definitions
INTEGER, INTENT(IN)  :: n          ! Value to calculate
INTEGER, INTENT(OUT) :: result     ! Result

! Data dictionary: declare local variable types & definitions
INTEGER :: temp                    ! Temporary variable

IF ( n >= 1 ) THEN
   CALL factorial ( n-1, temp )
   result = n * temp
ELSE
   result = 1
END IF

END SUBROUTINE factorial
```

It is also possible to define recursive Fortran functions. However, there is an extra complication when working with recursive functions. Remember that a function is invoked by naming the function in an expression, while the value to be returned from the function is specifying by assigning it to the function name. Thus, if a function were to invoke itself, the function's name *would appear on the left-hand side of an assignment statement when its return value is being set, and on the right-hand side of an assignment statement when it is invoking itself recursively*. This double use of the function name could certainly cause confusion.

To avoid confusion between the two uses of the function name in a recursive function, *Fortran allows us to specify two different names for invoking the function recursively and for returning its result*. The actual name of the function is used whenever we want the function to invoke itself, and a special dummy argument is used whenever we want to specify a value to return. The name of this special dummy argument is specified in a RESULT clause in the FUNCTION statement. For example, the following line declares a recursive function fact that uses the dummy argument answer for the value returned to the invoking program unit:

```
RECURSIVE FUNCTION fact(n) RESULT(answer)
```

If a RESULT clause is included in a function, then the function name *may not* appear in a type declaration statement in the function. The name of the dummy result variable is declared instead. For example, Figure 13-5 shows a recursive function that calculates the factorial function directly from Equation (13-1). Note that the type of the result variable answer is declared, not the type of the function name fact. You will be asked to verify the proper operation of this function in Exercise 13-2.

FIGURE 13-5
A function to recursively implement the factorial function.

```
RECURSIVE FUNCTION fact(n) RESULT(answer)
!
!  Purpose:
!    To calculate the factorial function
!            | n(n-1)!    n >= 1
!    n ! =  |
!            | 1          n = 0
!
!  Record of revisions:
!      Date        Programmer          Description of change
!      ====        ==========          =====================
!    12/17/15    S. J. Chapman         Original code
!
IMPLICIT NONE

! Data dictionary: declare calling parameter types & definitions
INTEGER, INTENT(IN) :: n           ! Value to calculate
INTEGER :: answer                  ! Result variable

IF ( n >= 1 ) THEN
   answer = n * fact(n-1)
ELSE
   answer = 1
END IF

END FUNCTION fact
```

Fortran 2015 is the next standard of Fortran to be developed, possibly for adoption around 2018. In that standard, all subroutines and functions will be recursive by default. If you specifically want a procedure to *not* be recursive, the procedure is declared with a new NON_RECURSIVE keyword. Don't count on seeing this feature in the near future; it can take years for compiler vendors to catch up with changes in the standards.

13.4
KEYWORD ARGUMENTS AND OPTIONAL ARGUMENTS

In Chapter 7, we stated that when invoking a procedure, *the actual argument list used to invoke the procedure must match the dummy argument list exactly in number, type, and order.* If the first dummy argument is a real array, then the first actual argument must also be a real array, etc. If the procedure has four dummy arguments, then the procedure invocation must have four actual arguments.

This statement is usually true in Fortran. However, it is possible to change the order of the calling arguments in the list, or to specify actual arguments for only some of the procedure's dummy arguments *provided that the interface to the procedure is explicit*. A procedure interface can be made explicit by placing the procedure in a module and accessing that module in the invoking program by USE association. (A procedure interface can also be made explicit by using an interface block, as we will explain in the next section.)

If a procedure's interface is explicit, then it is possible to use **keyword arguments** in the calling program to provide increased flexibility. A keyword argument is an argument of the form

```
keyword = actual_argument
```

where keyword is the name of the dummy argument that is being associated with the actual argument. If the procedure invocation uses keyword arguments, then the calling arguments can be arranged in any order, because the keywords allow the compiler to sort out which actual argument goes with which dummy argument.

Let's illustrate this idea with an example. Figure 13-6 shows a function calc that takes three real arguments first, second, and third. The function is contained inside a module to make its interface explicit. The main program invokes this function in four different ways using the same arguments. The first time that the function is invoked, it is done the conventional way, in which the actual arguments match the dummy arguments in type, number, and order.

```
WRITE (*,*) calc ( 3., 1., 2. )
```

The next two times that the function is invoked, we use keyword arguments.

```
WRITE (*,*) calc ( first=3., second=1., third=2. )
WRITE (*,*) calc ( second=1., third=2., first=3. )
```

The final time that the function is called, we use a mixture of conventional arguments and keyword arguments. The first argument is conventional and so it is associated with the first dummy argument. The later arguments are keyword arguments, so they are associated with dummy arguments by their keywords. In general, it is legal to mix conventional calling arguments and keyword arguments, but *once a keyword argument appears in the list, all of the remaining arguments must also be keyword arguments.*

FIGURE 13-6
Program to illustrate the use of keyword arguments.

```
WRITE (*,*) calc ( 3., third=2., second=1. )

MODULE procs
CONTAINS
   REAL FUNCTION calc ( first, second, third )
   IMPLICIT NONE
   REAL, INTENT(IN) :: first, second, third
   calc = ( first - second ) / third
   END FUNCTION calc
END MODULE procs

PROGRAM test_keywords

USE procs
IMPLICIT NONE

WRITE (*,*) calc ( 3., 1., 2. )
WRITE (*,*) calc ( first=3., second=1., third=2. )
WRITE (*,*) calc ( second=1., third=2., first=3. )
WRITE (*,*) calc ( 3., third=2., second=1.)

END PROGRAM test_keywords
```

When the program in Figure 13-6 is executed, the results are

```
C:\book\fortran\chap13>test_keywords
    1.000000
    1.000000
    1.000000
    1.000000
```

The function calculated the same value every time regardless of the order in which the arguments were presented.

Keyword arguments allow us to change the order in which actual arguments are presented to a procedure, but by itself that is not very useful. It appears that all we are doing here is creating extra typing to accomplish the same goal! However, keyword arguments are useful when used with optional arguments.

An **optional argument** is a dummy procedure argument that does not always have to be present when the procedure is invoked. If it is present, then the procedure will use it. If not, then the procedure will function without it. Optional arguments are only possible in procedures with explicit interfaces. They are specified by including the OPTIONAL attribute in the declaration of a dummy argument:

```
INTEGER, INTENT(IN), OPTIONAL :: upper_limit
```

Advanced Features of Procedures and Modules

The procedure containing an optional argument must have some way to determine if the optional argument is present when the procedure is executed. This is accomplished by a logical intrinsic function PRESENT, which returns a true value if the optional argument is present and a false value if it is not present. For example, a procedure could take some action based on the presence or absence of an optional argument upper_limit as follows:

```
IF ( PRESENT(upper_limit) ) THEN
   ...
ELSE
   ...
END IF
```

Keywords are very useful for procedures with optional arguments. If the optional arguments are present and in order in the calling sequence, then no keywords are required. If only some of the optional arguments are present, but the ones that are present are in order, then no keywords are required. However, if optional arguments are out of order, or if some of the earlier optional arguments are missing while later ones are supplied, then keywords must be supplied, and the compiler will use the keywords to sort out which optional arguments are present and which ones are absent.

Incidentally, we have already met an intrinsic function that uses keywords and optional arguments. Recall that the function SELECTED_REAL_KIND accepts two arguments for the desired precision p and the desired range r of the real number. The default order for the two arguments is (p, r), so if the arguments are specified in that order no keywords are necessary. If they are specified out of order or if only the range is specified, then the keywords must be used. Examples of legal uses of the function include:

```
kind_num = SELECTED_REAL_KIND(13,100)
kind_num = SELECTED_REAL_KIND(13)
kind_num = SELECTED_REAL_KIND(r=100,p=13)
kind_num = SELECTED_REAL_KIND(r=100)
```

EXAMPLE 13-2

Finding the Extreme Values in a Data Set:

Suppose that we would like to write a subroutine that searches through a real array to locate the minimum and/or maximum values in the array, and the locations where the minimum and/or maximum values occur. This subroutine could be used in many different applications. On some occasions, we might be looking for only the maximum value in the array. At other times, we might only care about the minimum value. On still other occasions, we might be interested in both values (e.g., if we were setting the limits on a plotting program). Sometimes we will care where the extreme values occur within an array, and other times it will not matter.

To accommodate all of these possibilities in a single subroutine, we will write a subroutine that has four optional output arguments: The maximum value, the location of the maximum value, the minimum value, and the location of the minimum value. The values returned will depend on the arguments specified by the user in the subroutine call.

SOLUTION

The subroutine is shown in Figure 13-7. The subroutine that can return from one to four optional results in any possible combination. Note that the subroutine must have an explicit interface in order to support optional arguments, so it is placed inside a module.

FIGURE 13-7
A subroutine to locate the extreme values in a real array. The subroutine is embedded in a module to make its interface explicit.

```
MODULE procs

CONTAINS
   SUBROUTINE extremes(a, n, maxval, pos_maxval, minval, pos_minval)
   !
   !  Purpose:
   !    To find the maximum and minimum values in an array, and
   !    the location of those values in the array. This subroutine
   !    returns its output values in optional arguments.
   !
   !  Record of revisions:
   !      Date       Programmer          Description of change
   !      ====       ==========          =====================
   !    12/18/15    S. J. Chapman        Original code
   !
   IMPLICIT NONE

   ! Data dictionary: declare calling parameter types & definitions
   INTEGER, INTENT(IN) :: n                      ! # vals in array a
   REAL, INTENT(IN), DIMENSION(n) :: a           ! Input data.
   REAL, INTENT(OUT), OPTIONAL :: maxval         ! Maximum value.
   INTEGER, INTENT(OUT), OPTIONAL :: pos_maxval  ! Pos of maxval
   REAL, INTENT(OUT), OPTIONAL :: minval         ! Minimum value.
   INTEGER, INTENT(OUT), OPTIONAL :: pos_minval  ! Pos of minval

   ! Data dictionary: declare local variable types & definitions
   INTEGER :: i              ! Index
   REAL :: real_max          ! Max value
   INTEGER :: pos_max        ! Pos of max value
   REAL :: real_min          ! Min value
   INTEGER :: pos_min        ! Pos of min value

   ! Initialize the values to first value in array.
   real_max = a(1)
   pos_max = 1
   real_min = a(1)
   pos_min = 1

   ! Find the extreme values in a(2) through a(n).
   DO i = 2, n
      max: IF ( a(i) > real_max ) THEN
         real_max = a(i)
         pos_max = i
      END IF max
```

(continued)

(*concluded*)

```
      min: IF ( a(i) < real_min ) THEN
         real_min = a(i)
         pos_min = i
      END IF min
   END DO

   ! Report the results
   IF ( PRESENT(maxval) ) THEN
      maxval = real_max
   END IF
   IF ( PRESENT(pos_maxval) ) THEN
      pos_maxval = pos_max
   END IF
   IF ( PRESENT(minval) ) THEN
      minval = real_min
   END IF
   IF ( PRESENT(pos_minval) ) THEN
      pos_minval = pos_min
   END IF

   END SUBROUTINE extremes
END MODULE procs
```

You will be asked to verify the proper operation of this subroutine in Exercise 13-3 at the end of this chapter.

Quiz 13-1

This quiz provides a quick check to see if you have understood the concepts introduced in Sections 13.1 through 13.3. If you have trouble with the quiz, reread the sections, ask your instructor, or discuss the material with a fellow student. The answers to this quiz are found in the back of the book.

1. What is the scope of a variable in Fortran? What are the three levels of scope in Fortran?
2. What is host association? Explain how variables and constants are inherited by host association.
3. What is the value of z that is written out after the following code is executed? Explain how the value is produced.

```
PROGRAM x
REAL :: z = 10.
TYPE position
   REAL :: x
   REAL :: y
   REAL :: z
```

(*continued*)

(*concluded*)

```
            END TYPE position
            TYPE (position) :: xyz
            xyz = position(1., 2., 3.)
            z = fun1( z )
            WRITE (*,*) z
            CONTAINS
                REAL FUNCTION fun1(x)
                REAL, INTENT(IN) :: x
                fun1 = (x + xyz%x) / xyz%z
                END FUNCTION fun1
            END PROGRAM
```

4. What is the value of i after the following code is executed?

```
            PROGRAM xyz
            INTEGER :: i = 0
            INTEGER, DIMENSION(6) :: count
            i = i + 27
            count = (/ (2*i, i=6,1,-1) /)
            i = i - 7
            WRITE (*,*) i
            END PROGRAM xyz
```

5. Is the following program legal or illegal? Why or why not?

```
            PROGRAM abc
            REAL :: abc = 10.
            WRITE (*,*) abc
            END PROGRAM
```

6. What are recursive procedures? How are they declared?

7. Is the following function legal or illegal? Why or why not?

```
            RECURSIVE FUNCTION sum_1_n(n) RESULT(sum)
            IMPLICIT NONE
            INTEGER, INTENT(IN) :: n
            INTEGER :: sum_1_n
            IF ( n > 1 ) THEN
               sum = n + sum_1_n(n-1)
            ELSE
               sum = 1
            END IF
            END FUNCTION
```

8. What are keyword arguments? What requirement(s) must be met before they can be used? Why would you want to use a keyword argument?

9. What are optional arguments? What requirement(s) must be met before they can be used? Why would you want to use an optional argument?

13.5
PROCEDURE INTERFACES AND INTERFACE BLOCKS

As we have seen, a calling program unit must have an *explicit interface* to a procedure if it is to use advanced Fortran features such as keyword arguments and optional arguments. In addition, an explicit interface allows the compiler to catch many errors that occur in the calling sequences between procedures. These errors might otherwise produce subtle and hard-to-find bugs.

The easiest way to create an explicit interface is to place procedures in a module, and then use that module in the calling program unit. Any procedures placed in a module will always have an explicit interface.

Unfortunately, it is sometimes inconvenient or even impossible to place the procedures in a module. For example, suppose that a technical organization has a large library containing hundreds of subroutines and functions written in an earlier version of Fortran that are used both in old existing programs and in new programs. This is a very common occurrence because various versions of Fortran have been in general use since the late 1950s. Rewriting all of these subroutines and functions to place them into modules and add explicit interface characteristics such as the INTENT attribute would create a major problem. If the procedures were modified in this way, then the older programs would no longer be able to use them. Most organizations would not want to make two versions of each procedure, one with an explicit interface and one without, because this would create a significant configuration control problem whenever one of the library procedures is modified. Both versions of the procedure would have to be modified separately, and each one would have to be independently verified to be working properly.

The problem can be even worse, since the external library of procedures could be written in another language such as C++. In that case, it is completely impossible to place the procedures in a module.

13.5.1 Creating Interface Blocks

How do we take advantage of the features of an explicit interface when it is impossible or impractical to place procedures into a module? In these cases, Fortran allows us to define an **interface block** in the invoking program unit. The interface block specifies all of the interface characteristics of an external procedure, and the Fortran compiler uses the information in the interface block to perform its consistency checks and to apply such advanced features as keyword arguments.[2]

An interface block is created by duplicating the calling argument information of a procedure within the interface. The general form of an interface is

```
INTERFACE
    interface_body_1
    interface_body_2
    ...
END INTERFACE
```

---

[2] Fortran interface blocks are essentially equivalent to prototypes in the C language.

Each *interface_body* consists of the initial `SUBROUTINE` or `FUNCTION` statement of the corresponding external procedure, the type specification statements associated with its arguments, and an `END SUBROUTINE` or `END FUNCTION` statement. These statements provide enough information for the compiler to check the consistency of the interface between the calling program and the external procedure.

When an interface is used, it is placed in the header section of the invoking program unit along with all of the type declaration statements.

EXAMPLE 13-3

Creating an Interface to an External Subroutine:

In Example 7-1, we created a subroutine `sort` to sort an array of real values into ascending order. Assume that it is impossible to place that subroutine into a module and create an interface block to explicitly define the interface between the subroutine and a calling program unit. Use that interface to allow a program to call subroutine `sort` while using keyword arguments.

SOLUTION
First, we must create an interface for subroutine `sort`. The interface will consist of the `SUBROUTINE` statement, the type declaration statements of the subroutine's dummy arguments, and the `END SUBROUTINE` statement. It is

```
INTERFACE
    SUBROUTINE sort (array, n)
    IMPLICIT NONE
    REAL, DIMENSION(:), INTENT(INOUT) :: array
    INTEGER, INTENT(IN) :: n
    END SUBROUTINE sort
END INTERFACE
```

Next, we will use this interface in the calling program's header to explicitly define the interface to subroutine `sort`. Figure 13-8 shows a calling program that uses the interface block to create an explicit interface to subroutine `sort`.

FIGURE 13-8
A simple program illustrating the use of interface blocks.

```
PROGRAM interface_example
!
!  Purpose:
!    To illustrate the use of interface blocks to create explicit
!    interfaces. This program uses an interface block to create
!    an explicit interface to subroutine "sort", and then takes
!    advantage of that interface to use keyword arguments.
!
!  Record of revisions:
!     Date        Programmer         Description of change
!     ====        ==========         =====================
!     12/18/15    S. J. Chapman      Original code
!
IMPLICIT NONE
```

(continued)

Advanced Features of Procedures and Modules

(*concluded*)

```
! Declare interface to subroutine "sort"
INTERFACE
   SUBROUTINE sort(a,n)
   IMPLICIT NONE
   REAL, DIMENSION(:), INTENT(INOUT) :: a
   INTEGER, INTENT(IN) :: n
   END SUBROUTINE sort
END INTERFACE

! Data dictionary: declare local variable types & definitions
REAL, DIMENSION(6) :: array = [ 1., 5., 3., 2., 6., 4. ]
INTEGER :: nvals = 6

! Call "sort" to sort data into ascending order.
CALL sort ( N=nvals, A=array)

! Write out sorted array.
WRITE (*,*) array

END PROGRAM interface_example
```

When this program is compiled together with subroutine `sort` and executed, the results are:

```
C:\book\fortran\chap13>interface_example
    1.000000        2.000000        3.000000        4.000000
    5.000000        6.000000
```

The compiler used the interface block to correctly sort out the keyword arguments in the call to subroutine `sort`, and the program produced the correct answer.

13.5.2 Notes on the Use of Interface Blocks

How and when should interface blocks be used to best advantage in a program? When we look at the structure of an interface block, it seems that we are just creating extra work for ourselves by duplicating some of the statements from the original procedure in the interface block. When should we create an interface block, and why? The following notes provide guidance on the use of interface blocks in Fortran.

1. Whenever possible, *avoid interface blocks by simply placing all of your procedures in modules* and access the appropriate modules by USE association.

Good Programming Practice

Avoid interface blocks by placing your procedures in modules whenever possible.

2. An interface block must not specify the interface of a procedure already in a module available by USE association. This constitutes a double definition of the explicit interface, which is illegal and will cause a compiler error.

3. A common use of interface blocks is to provide explicit interfaces to separately-compiled procedures written in earlier versions of Fortran or in other languages such as C++. In this case, writing an interface block allows modern Fortran programs to have an explicit interface with full argument checking, while allowing older or non-Fortran programs to continue to use the procedures unchanged.
4. An easy way to make the interfaces for a large library of old subroutines or functions available to all calling program units is to place them in a module, and then to USE that module in each calling program unit. For example, the interface to subroutine sort could be placed in a module as follows:

```
MODULE interface_definitions
   INTERFACE
      SUBROUTINE sort (array, n)
      IMPLICIT NONE
      REAL, DIMENSION(:), INTENT(INOUT) :: array
      INTEGER, INTENT(IN) :: n
      END SUBROUTINE sort
      ...
      (insert other procedure interfaces here)
      ...
   END INTERFACE
END MODULE interface_definitions
```

Unlike module procedures, there is no CONTAINS statement when interfaces are included in a module.

Good Programming Practice
If you must create interfaces to many procedures, place all of the interfaces in a module so that they will be easily accessible to many program units by USE association.

5. Each interface is a separate scoping unit, so the same variable name may appear in an interface and in a program including that interface without causing a conflict.
6. The names used for dummy arguments in an interface block do not have to be the same as the names used for the dummy arguments in the corresponding procedures. The dummy arguments in the interface block must match the dummy arguments in the corresponding procedures in type, intent, array size, etc., but the names themselves do not have to match. However, there is no reason for you to ever rename the arguments in an interface. Even though it is legal to do so, it adds extra confusion and increases to possibility for error.
7. An interface block is an independent scoping unit, so any dummy variables used in the interface block *must be declared separately within the block*, even if they were declared in the surrounding scoping unit.

```
PROGRAM test_interface

! Declare variables
REAL,DIMENSION(10) :: x, y           ! x, y declared in main
INTEGER :: n                         ! n declared in main
...
INTERFACE
   SUBROUTINE proc (x, y, n)
   IMPLICIT NONE
   REAL, DIMENSION(:), INTENT(INOUT) :: x  ! Declared in interface block
   REAL, DIMENSION(:), INTENT(INOUT) :: y  ! Declared in interface block
   INTEGER, INTENT(IN) :: n                ! Declared in interface block
   END SUBROUTINE proc
END INTERFACE
...
CALL proc(x,y,n)
...
END PROGRAM test_interface
```

Fortran 2003 and later includes an IMPORT statement that can modify this behavior. If an IMPORT statement appears in an interface definition, then the variables specified in the IMPORT statement will be imported from the host scoping unit. If the IMPORT statement appears without a list of variables, then all of the variables in the host scoping unit will be imported. Examples of IMPORT statements are shown below:

```
IMPORT :: a, b    ! Import variables a and b only
IMPORT            ! Import all variables in host scoping unit
```

13.6
GENERIC PROCEDURES

The Fortran language includes both generic and specific intrinsic functions. A **generic function** is a function that can operate properly with many different types of input data, while a **specific function** is a function that requires one specific type of input data. For example, Fortran includes a generic function ABS( ) to take the absolute value of a number. It can function with integer data, single-precision real data, double-precision real data, or complex data. The language also includes the specific functions IABS( ) that requires an integer input value, ABS( ) that requires a single-precision real input value, DABS( ) that requires a double-precision real input value, and CABS( ) that requires a complex input value.

Now for a little secret: The generic function ABS( ) does not actually exist anywhere within a Fortran compiler. Instead, whenever the compiler encounters the generic function, it examines the arguments of the function and invokes the appropriate specific function for those arguments. For example, if the compiler detects the generic function ABS(-34) in a program, it will generate a call to the specific function IABS( ) because the calling argument of the function is an integer. When we use generic functions, we are allowing the compiler to do some of the detail work for us.

13.6.1 User-Defined Generic Procedures

Fortran allows us to define our own generic procedures in addition to the standard ones built into the compiler. For example, we might wish to define a generic subroutine `sort` that is capable of sorting integer data, single-precision real data, double-precision real data, or character data depending on the arguments supplied to it. We could use that generic subroutine in our programs instead of worrying about the specific details of the calling arguments each time that we want to sort a data set.

How is this accomplished? It is done with a special version of the interface block called a **generic interface block.** If we add a generic name to the INTERFACE statement, then every procedure interface defined within the interface block will be assumed to be a specific version of that generic procedure. The general form of an interface block used to declare a generic procedure is

```
INTERFACE generic_name
   specific_interface_body_1
   specific_interface_body_2
   ...
END INTERFACE
```

When the compiler encounters the generic procedure name in a program unit containing this generic interface block, it will examine the arguments associated with the call to the generic procedure to decide which of the specific procedures it should use.

In order for the compiler to determine which specific procedure to use, each of the specific procedures in the block must be *unambiguously* distinguished from the others. For example, one specific procedure might have real input data, while another one has integer input data, etc. The compiler can then compare the generic procedure's calling sequence to the calling sequences of each specific procedure to decide which one to use. The following rules apply to the specific procedures in a generic interface block:

1. Either all of the procedures in a generic interface block must be subroutines, or all of the procedures in the block must be functions. They cannot be mixed, because the generic procedure being defined must either be a subroutine or a function—it cannot be both.
2. Every procedure in the block must be distinguishable from all of the other procedures in the block by the type, number, and position of its nonoptional arguments. As long as each procedure is distinguishable from all of the other procedures in the block, the compiler will be able to decide which procedure to use by comparing the type, number, and position of the generic procedure's calling arguments with the type, number, and position of each specific procedure's dummy arguments.

Generic interface blocks may either be placed in the header of a program unit that invokes the generic procedure, or they may be placed in a module and that module may be used in the program unit that invokes the generic procedure.

Advanced Features of Procedures and Modules

Good Programming Practice
Use generic interface blocks to define procedures that can function with different types of input data. Generic procedures will add to the flexibility of your programs, making it easier for them to handle different types of data.

As an example, suppose that a programmer has written the following four subroutines to sort data into ascending order.

Subroutine	Function
SUBROUTINE sorti (array, nvals)	Sorts integer data
SUBROUTINE sortr (array, nvals)	Sorts single-precision real data
SUBROUTINE sortd (array, nvals)	Sorts double-precision real data
SUBROUTINE sortc (array, nvals)	Sorts character data

Now he or she wishes to create a generic subroutine `sort` to sort any of these types of data into ascending order. This can be done with the following generic interface block (parameters `single` and `double` will have to be previously defined):

```
INTERFACE sort
   SUBROUTINE sorti (array, nvals)
   IMPLICIT NONE
   INTEGER, INTENT(IN) :: nvals
   INTEGER, INTENT(INOUT), DIMENSION(nvals) :: array
   END SUBROUTINE sorti

   SUBROUTINE sortr (array, nvals)
   IMPLICIT NONE
   INTEGER, INTENT(IN) :: nvals
   REAL(KIND=single), INTENT(INOUT), DIMENSION(nvals) :: array
   END SUBROUTINE sortr

   SUBROUTINE sortd (array, nvals)
   IMPLICIT NONE
   INTEGER, INTENT(IN) :: nvals
   REAL(KIND=double), INTENT(INOUT), DIMENSION(nvals) :: array
   END SUBROUTINE sortd

   SUBROUTINE sortc (array, nvals)
   IMPLICIT NONE
   INTEGER, INTENT(IN) :: nvals
   CHARACTER(len=*), INTENT(INOUT), DIMENSION(nvals) :: array
   END SUBROUTINE sortc
END INTERFACE sort
```

This generic interface block satisfies the requirements stated above because all of the procedures are subroutines, and they can be distinguished from one another by the type of the array in their calling sequences.

13.6.2 Generic Interfaces for Procedures in Modules

In the above example, an explicit interface was given for each specific subroutine in the generic interface block defining the generic subroutine `sort`. This arrangement would be appropriate if each of the specific subroutines were separately compiled and did not have an explicit interface. But what happens if the individual subroutines are in a module, and so they already have explicit interfaces?

We learned in Section 13.4.2 that it is illegal to explicitly declare an interface for a procedure that already has an explicit interface by being in a module. If that is so, then how can we include procedures defined in modules in a generic interface block? To get around this problem, Fortran includes a special MODULE PROCEDURE statement that can be used in a generic interface block. The form of this statement is

```
MODULE PROCEDURE module_procedure_1 (, module_procedure_2, ...)
```

where `module_procedure_1`, etc., are the names of procedures whose interfaces are defined in a module that is available by USE association.

If the four sorting subroutines had been defined in a module instead of being separately compiled, then the generic interface for subroutine `sort` would become:

```
INTERFACE sort
    MODULE PROCEDURE sorti
    MODULE PROCEDURE sortr
    MODULE PROCEDURE sortd
    MODULE PROCEDURE sortc
END INTERFACE sort
```

This interface block should be placed in the module in which the procedures are defined.

EXAMPLE 13-4   *Creating a Generic Subroutine:*

Create a subroutine `maxval` that returns the maximum value in an array, and optionally the location of that maximum value. This subroutine should work correctly for integer, single-precision real, double-precision real, single-precision complex, or double-precision complex data. Since relational comparisons of complex data values are meaningless, the complex versions of the subroutine should look for the maximum absolute value in the array.

SOLUTION
We will be producing a generic subroutine that can work with five different types of input data, so in fact we create five different subroutines and relate them together using a generic interface block. Note that the subroutines must have an explicit interface in order to support optional arguments, so they will all be placed in a module.

1. **State the problem.**
   Write a generic subroutine to find the maximum value in an array and optionally the location of that maximum value. The subroutine should work for integer, single-precision real, double-precision real, single-precision complex, or double-precision

complex data. For complex data, the comparisons should be based on the magnitude of the values in the array.

2. **Define the inputs and outputs.**

There are five different subroutines in this problem. The input to each subroutine will be an array of values of the appropriate type, plus the number of values in the array. The outputs will be as follows:

(a) A variable containing the maximum value in the input array.
(b) An optional integer variable containing the offset in the array at which the maximum value occurred.

The types of the input and output arguments for each of the five subroutines are specified in Table 13-1.

3. **Describe the algorithm.**

The pseudocode for the first three specific subroutines is identical. It is:

```
! Initialize "value_max" to a(1) and "pos_max" to 1.
value_max ← a(1)
pos_max ← 1

! Find the maximum values in a(2) through a(nvals)
DO for i = 2 to nvals
   IF a(i) > value_max THEN
      value_max ← a(i)
      pos_max ← i
   END of IF
END of DO

! Report results
IF argument pos_maxval is present THEN
   pos_maxval ← pos_max
END of IF
```

The pseudocode for the two complex subroutines is slightly different, because comparisons must be with the absolute values. It is:

```
! Initialize "value_max" to ABS(a(1)) and "pos_max" to 1.
value_max ← ABS(a(1))
pos_max ← 1
```

TABLE 13-1
Arguments for the subroutines

Specific name	Input array type	Array length type	Output maximum value	Optional location of max value
maxval_i	Integer	Integer	Integer	Integer
maxval_r	Single-precision real	Integer	Single-precision real	Integer
maxval_d	Double-precision real	Integer	Double-precision real	Integer
maxval_c	Single-precision complex	Integer	Single-precision real	Integer
maxval_dc	Double-precision complex	Integer	Double-precision real	Integer

```
! Find the maximum values in a(2) through a(nvals)
DO for i = 2 to nvals
   IF ABS(a(i)) > value_max THEN
      value_max ← ABS(a(i))
      pos_max ← i
   END of IF
END of DO

! Report results
IF argument pos_maxval is present THEN
   pos_maxval ← pos_max
END of IF
```

4. **Turn the algorithm into Fortran statements.**
   The resulting Fortran subroutine is shown in Figure 13-9.

FIGURE 13-9
A generic subroutine `maxval` that finds the maximum value in an array and optionally the location of that maximum value.

```
MODULE generic_maxval
!
!  Purpose:
!    To produce a generic procedure maxval that returns the
!    maximum value in an array and optionally the location
!    of that maximum value for the following input data types:
!    integer, single precision real, double precision real,
!    single precision complex, and double precision complex.
!    Complex comparisons are done on the absolute values of
!    values in the input array.
!
!  Record of revisions:
!      Date        Programmer          Description of change
!      ====        ==========          =====================
!    12/18/15     S. J. Chapman        Original code
!
IMPLICIT NONE

! Declare parameters:
INTEGER, PARAMETER :: SGL = SELECTED_REAL_KIND(p=6)
INTEGER, PARAMETER :: DBL = SELECTED_REAL_KIND(p=13)

! Declare generic interface.
INTERFACE maxval
   MODULE PROCEDURE maxval_i
   MODULE PROCEDURE maxval_r
   MODULE PROCEDURE maxval_d
   MODULE PROCEDURE maxval_c
   MODULE PROCEDURE maxval_dc
END INTERFACE

CONTAINS
   SUBROUTINE maxval_i ( array, nvals, value_max, pos_maxval )
   IMPLICIT NONE
```

(continued)

(continued)

```fortran
! List of calling arguments:
INTEGER, INTENT(IN) :: nvals                          ! # vals
INTEGER, INTENT(IN), DIMENSION(nvals) :: array        ! Input data
INTEGER, INTENT(OUT) :: value_max                     ! Max value
INTEGER, INTENT(OUT), OPTIONAL :: pos_maxval          ! Position

! List of local variables:
INTEGER :: i                          ! Index
INTEGER :: pos_max                    ! Pos of max value

! Initialize the values to first value in array.
value_max = array(1)
pos_max = 1

! Find the extreme values in array(2) through array(nvals).
DO i = 2, nvals
   IF ( array(i) > value_max ) THEN
      value_max = array(i)
      pos_max = i
   END IF
END DO

! Report the results
IF ( PRESENT(pos_maxval) ) THEN
   pos_maxval = pos_max
END IF

END SUBROUTINE maxval_i

SUBROUTINE maxval_r ( array, nvals, value_max, pos_maxval )
IMPLICIT NONE

! List of calling arguments:
INTEGER, INTENT(IN) :: nvals
REAL(KIND=SGL), INTENT(IN), DIMENSION(nvals) :: array
REAL(KIND=SGL), INTENT(OUT) :: value_max
INTEGER, INTENT(OUT), OPTIONAL :: pos_maxval

! List of local variables:
INTEGER :: i                          ! Index
INTEGER :: pos_max                    ! Pos of max value

! Initialize the values to first value in array.
value_max = array(1)
pos_max = 1

! Find the extreme values in array(2) through array(nvals).
DO i = 2, nvals
   IF ( array(i) > value_max ) THEN
      value_max = array(i)
      pos_max = i
   END IF
END DO

! Report the results
```

(continued)

(*continued*)

```fortran
      IF ( PRESENT(pos_maxval) ) THEN
         pos_maxval = pos_max
      END IF

      END SUBROUTINE maxval_r

      SUBROUTINE maxval_d ( array, nvals, value_max, pos_maxval )
      IMPLICIT NONE

      ! List of calling arguments:
      INTEGER, INTENT(IN) :: nvals
      REAL(KIND=DBL), INTENT(IN), DIMENSION(nvals) :: array
      REAL(KIND=DBL), INTENT(OUT) :: value_max
      INTEGER, INTENT(OUT), OPTIONAL :: pos_maxval

      ! List of local variables:
      INTEGER :: i                    ! Index
      INTEGER :: pos_max              ! Pos of max value

      ! Initialize the values to first value in array.
      value_max = array(1)
      pos_max = 1

      ! Find the extreme values in array(2) through array(nvals).
      DO i = 2, nvals
         IF ( array(i) > value_max ) THEN
            value_max = array(i)
            pos_max = i
         END IF
      END DO

      ! Report the results
      IF ( PRESENT(pos_maxval) ) THEN
         pos_maxval = pos_max
      END IF

      END SUBROUTINE maxval_d

      SUBROUTINE maxval_c ( array, nvals, value_max, pos_maxval )
      IMPLICIT NONE

      ! List of calling arguments:
      INTEGER, INTENT(IN) :: nvals
      COMPLEX(KIND=SGL), INTENT(IN), DIMENSION(nvals) :: array
      REAL(KIND=SGL), INTENT(OUT) :: value_max
      INTEGER, INTENT(OUT), OPTIONAL :: pos_maxval

      ! List of local variables:
      INTEGER :: i                    ! Index
      INTEGER :: pos_max              ! Pos of max value

      ! Initialize the values to first value in array.
      value_max = ABS(array(1))
      pos_max = 1

      ! Find the extreme values in array(2) through array(nvals).
      DO i = 2, nvals
```

(*continued*)

(*concluded*)

```
         IF ( ABS(array(i)) > value_max ) THEN
            value_max = ABS(array(i))
            pos_max = i
         END IF
      END DO

      ! Report the results
      IF ( PRESENT(pos_maxval) ) THEN
         pos_maxval = pos_max
      END IF

   END SUBROUTINE maxval_c

   SUBROUTINE maxval_dc ( array, nvals, value_max, pos_maxval )
   IMPLICIT NONE

   ! List of calling arguments:
   INTEGER, INTENT(IN) :: nvals
   COMPLEX(KIND=DBL), INTENT(IN), DIMENSION(nvals) :: array
   REAL(KIND=DBL), INTENT(OUT) :: value_max
   INTEGER, INTENT(OUT), OPTIONAL :: pos_maxval

   ! List of local variables:
   INTEGER :: i                    ! Index
   INTEGER :: pos_max              ! Pos of max value

   ! Initialize the values to first value in array.
   value_max = ABS(array(1))
   pos_max = 1

   ! Find the extreme values in array(2) through array(nvals).
   DO i = 2, nvals
      IF ( ABS(array(i)) > value_max ) THEN
         value_max = ABS(array(i))
         pos_max = i
      END IF
   END DO

   ! Report the results
   IF ( PRESENT(pos_maxval) ) THEN
      pos_maxval = pos_max
   END IF

   END SUBROUTINE maxval_dc

END MODULE generic_maxval
```

5. **Test the resulting Fortran programs.**

To test this generic subroutine, it is necessary to write a test driver program to call the subroutine with the five different types of data that it supports, and display the results. The test driver program will also illustrate the use of keyword and optional arguments by calling the subroutine with different combinations and orders of arguments. Figure 13-10 shows an appropriate test driver program.

FIGURE 13-10
Test driver program for generic subroutine `maxval`.

```fortran
PROGRAM test_maxval
!
!  Purpose:
!    To test the generic subroutine maxval with five different types
!    of input data sets.
!
!  Record of revisions:
!     Date        Programmer        Description of change
!     ====        ==========        =====================
!     12/18/15    S. J. Chapman     Original code
!
USE generic_maxval
IMPLICIT NONE

! Data dictionary: declare variable types & definitions
INTEGER, DIMENSION(6) :: array_i              ! Integer array
REAL(KIND=SGL), DIMENSION(6) :: array_r       ! Sng prec real arr
REAL(KIND=DBL), DIMENSION(6) :: array_d       ! Dbl prec real arr
COMPLEX(KIND=SGL), DIMENSION(6) :: array_c    ! Sing. prec. cx arr
COMPLEX(KIND=DBL), DIMENSION(6) :: array_dc   ! Sing. prec. cx arr
INTEGER :: value_max_i                        ! Max value
REAL(KIND=SGL) :: value_max_r                 ! Max value
REAL(KIND=DBL) :: value_max_d                 ! Max value
INTEGER :: pos_maxval                         ! Pos of max value

! Initialize arrays
array_i  = [ -13, 3, 2, 0, 25, -2 ]
array_r  = [ -13., 3., 2., 0., 25., -2. ]
array_d  = [ -13._DBL, 3._DBL, 2._DBL, 0._DBL, &
             25._DBL, -2._DBL ]
array_c  = [ (1.,2.), (-4.,-6.), (4.,-7), (3.,4.), &
             (0.,1.), (6.,-8.) ]
array_dc = [ (1._DBL,2._DBL), (-4._DBL,-6._DBL), &
             (4._DBL,-7._DBL), (3._DBL,4._DBL), &
             (0._DBL,1._DBL), (6._DBL,-8._DBL) ]

! Test integer subroutine. Include optional argument.
CALL maxval ( array_i, 6, value_max_i, pos_maxval )
WRITE (*,1000) value_max_i, pos_maxval
1000 FORMAT ('Integer args: max value = ',I3, &
             '; position = ', I3 )

! Test single prec real subroutine. Leave out optional arg.
CALL maxval ( array_r, 6, value_max_r )
WRITE (*,1010) value_max_r
1010 FORMAT ('Single precision real args: max value = ',F7.3)

! Test double prec real subroutine. Use keywords.
CALL maxval ( ARRAY=array_d, NVALS=6, VALUE_MAX=value_max_d )
WRITE (*,1020) value_max_d
1020 FORMAT ('Double precision real args: max value = ',F7.3)
```

(continued)

(*concluded*)

```
! Test single prec cmplx subroutine. Use scrambled keywords.
CALL maxval ( NVALS=6, ARRAY=array_c, VALUE_MAX=value_max_r, &
              POS_MAXVAL=pos_maxval )
WRITE (*,1030) value_max_r, pos_maxval
1030 FORMAT (' Single precision complex args:' &
             ' max abs value = ',F7.3, &
             '; position = ', I3 )

! Test double prec cmplx subroutine. Leave out optional arg.
CALL maxval ( array_dc, 6, value_max_d )
WRITE (*,1040) value_max_r
1040 FORMAT (' Double precision complex args:' &
             ' max abs value = ',F7.3 )

END PROGRAM test_maxval
```

When the test driver program is executed, the results are:

```
C:\book\fortran\chap13>test_maxval
Integer arguments: max value = 25; position =  5
Single precision real arguments: max value = 25.000
Double precision real arguments: max value = 25.000
Single precision complex arguments: max abs value = 10.000; position =  6
Double precision complex arguments: max abs value = 10.000
```

It is obvious from inspection that the subroutine picked out the proper maximum values and locations for each data type.

13.6.3 Generic Bound Procedures

Fortran procedures bound to derived data types can also be generic. These procedures are declared using the GENERIC statement, as shown below:

```
TYPE :: point
   REAL :: x
   REAL :: y
CONTAINS
   GENERIC :: add => point_plus_point, point_plus_scalar
END TYPE point
```

This binding declares that the two procedures point_plus_point and point_plus_scalar will both be known by the generic procedure add, and will both be accessed using the component operator: p%add().

As with other generic interfaces, every procedure in the generic binding must be distinguishable from all of the other procedures in the binding by the type, number, and position of its nonoptional arguments. As long as each procedure is distinguishable from all of the other procedures in the binding, the compiler will be able to decide which procedure to use by comparing the type, number, and position of the generic procedure's calling arguments with the type, number, and position of each specific procedure's dummy arguments.

EXAMPLE 13-5

Using Generic Bound Procedures:

Create a vector data types with a bound generic procedure add. There should be two specific procedures associated with the generic procedure: one to add two vectors and one to add a vector to a scalar.

SOLUTION
A module using bound generic procedures to add either a vector or a scalar to another vector is shown in Figure 13-11.

FIGURE 13-11
2D vector module with bound generic procedures.

```
MODULE generic_procedure_module
!
!  Purpose:
!    To define the derived data type for 2D vectors,
!    plus two generic bound procedures.
!
!  Record of revisions:
!      Date        Programmer         Description of change
!      ====        ==========         =====================
!    12/20/15    S. J. Chapman        Original code
!
IMPLICIT NONE

! Declare type vector
TYPE :: vector
   REAL :: x                          ! X value
   REAL :: y                          ! Y value
CONTAINS
   GENERIC :: add => vector_plus_vector, vector_plus_scalar
   PROCEDURE,PASS :: vector_plus_vector
   PROCEDURE,PASS :: vector_plus_scalar
END TYPE vector

! Add procedures
CONTAINS

   TYPE (vector) FUNCTION vector_plus_vector ( this, v2 )
   !
   !  Purpose:
   !    To add two vectors.
   !
   !  Record of revisions:
   !      Date        Programmer         Description of change
   !      ====        ==========         =====================
   !    12/20/15    S. J. Chapman        Original code
   !
   IMPLICIT NONE
```

(continued)

(*concluded*)

```
    ! Data dictionary: declare calling parameter types & definitions
    CLASS(vector),INTENT(IN) :: this    ! First vector
    CLASS(vector),INTENT(IN) :: v2      ! Second vector

    ! Add the vectors
    vector_plus_vector%x = this%x + v2%x
    vector_plus_vector%y = this%y + v2%y

    END FUNCTION vector_plus_vector

    TYPE (vector) FUNCTION vector_plus_scalar ( this, s )
    !
    ! Purpose:
    !    To add a vector and a scalar.
    !
    ! Record of revisions:
    !     Date        Programmer        Description of change
    !     ====        ==========        =====================
    !   12/20/15    S. J. Chapman       Original code
    !
    IMPLICIT NONE

    ! Data dictionary: declare calling parameter types & definitions
    CLASS(vector),INTENT(IN) :: this    ! First vector
    REAL,INTENT(IN) :: s                ! Scalar

    ! Add the points
    vector_plus_scalar%x = this%x + s
    vector_plus_scalar%y = this%y + s

    END FUNCTION vector_plus_scalar
END MODULE generic_procedure_module
```

The test driver program is shown in Figure 13-12.

FIGURE 13-12
Test driver program for the vector module with bound procedures.

```
PROGRAM test_generic_procedures
!
! Purpose:
!   To test generic bound procedures.
!
! Record of revisions:
!     Date        Programmer        Description of change
!     ====        ==========        =====================
!   12/20/15    S. J. Chapman       Original code
!
USE generic_procedure_module
IMPLICIT NONE

! Enter first point
TYPE(vector) :: v1                      ! First vector
```

(*continued*)

(concluded)

```
TYPE(vector) :: v2              ! Second vector
REAL :: s                       ! Scalar

! Get the first vector
WRITE (*,*) 'Enter the first vector (x,y):'
READ (*,*) v1%x, v1%y

! Get the second vector
WRITE (*,*) 'Enter the second vector (x,y):'
READ (*,*) v2%x, v2%y

! Get a scalar
WRITE (*,*) 'Enter a scalar:'
READ (*,*) s

! Add the vectors
WRITE (*,1000) v1%add(v2)
1000 FORMAT('The sum of the vectors is (',F8.2,',',F8.2,')')

! Subtract the points
WRITE (*,1010) v1%add(s)
1010 FORMAT('The sum of the vector and scalar is (',F8.2,',',F8.2,')')

END PROGRAM test_generic_procedures
```

We will test this program using the same data as in the previous example.

```
C:\book\fortran\chap12>test_generic_procedures
Enter the first vector (x,y):
-2, 2.
Enter the second vector (x,y):
4., 3.
Enter a scalar:
2
The sum of the vectors is (    2.00,    5.00)
The sum of the vector and scalar is (    0.00,    4.00)
```

The functions appear to be working correctly.

13.7

EXTENDING FORTRAN WITH USER-DEFINED OPERATORS AND ASSIGNMENTS

When we were introduced to derived data types in Chapter 12, we learned that none of the intrinsic unary and binary operators are defined for derived data types. In fact, the only operation that was defined for derived data types was the assignment of one item of a derived data type to another variable of the same type. We were able to work freely with the *components* of derived data types, but not with the derived data types themselves. This is a serious limitation that reduces the usefulness of derived data types.

Fortunately, there is a way around this limitation. Fortran is an *extensible* language, which means that an individual programmer can add new features to it to accommodate special types of problems. The first examples of this extensibility were derived data types themselves. In addition, Fortran permits the programmer to define new unary and binary operators for both intrinsic and derived data types, and to define new extensions to standard operators for derived data types. With appropriate definitions, the Fortran language can be made to add, subtract, multiply, divide, compare, etc., two operands of a derived data type.

How can we define new operators or extend existing ones? The first step is to write a function that performs the desired task and place it into a module. For example, if we wanted to add two values of a derived data type, we would first create a function whose arguments are the two values to be added and whose result is the sum of the two values. The function will implement the instructions required to perform the addition. The next step is to associate the function with a user-defined or intrinsic operator using an **interface operator block.** The form of an interface operator block is

```
INTERFACE OPERATOR (operator_symbol)
    MODULE PROCEDURE function_1
    ...
END INTERFACE
```

where *operator_symbol* is any standard intrinsic operator (+, −, *, /, >, <, etc.) or any user-defined operator. A user-defined operator is a sequence of up to 63 letters surrounded by periods (numbers and underscore characters are not allowed in an operator name). For example, a user-defined operator might be named .INVERSE.. Each interface body can either be a complete description of the interface to the function if the function is not in a module or a MODULE PROCEDURE statement if the function is in a module. In either case, the function *must* have an explicit interface.

More than one function can be associated with the same operator symbol, but the functions must be distinguishable from one another by having different types of dummy arguments. When the compiler encounters the operator symbol in a program, it invokes the function whose dummy arguments match the operands associated with the operator symbol. If no associated function has dummy arguments that match the operands, then a compilation error results.

If the function associated with an operator has two dummy arguments, then the resulting operator will be a binary operator. If the function has only one dummy argument, then the operator will be a unary operator. Once defined, the operator will be treated as a reference to the function. For binary operations, the left-hand operand will become the first argument of the function and the right-hand operand will become the second argument of the function. The function must not modify its calling arguments. To ensure this, it is customary to declare all function arguments with INTENT(IN).

If the operator being defined by the interface is one of Fortran's intrinsic operators (+, −, *, /, >, etc.), then there are three additional constraints to consider:

1. It is not possible to change the meaning of an intrinsic operator for pre-defined intrinsic data types. For example, it is not possible to change the meaning of the addition operator (+) when it is applied to two integers. It is only possible to

extend the meaning of the operator by defining the actions to perform when the operator is applied to derived data types, or combinations of derived data types and intrinsic data types.
2. The number of arguments in a function must be consistent with the normal use of the operator. For example, multiplication (*) is a binary operator, so any function extending its meaning must have two arguments.
3. If a relational operator is extended, then the same extension applies regardless of which way the operator is written. For example, if the relational operator "greater than" is given an additional meaning, then the extension applies whether "greater than" is written as > or .GT.

It is possible to extend the meaning of the assignment operator (=) in a similar fashion. To define extended meanings for the assignment operator, we use an **interface assignment block**:

```
INTERFACE ASSIGNMENT (=)
    MODULE PROCEDURE subroutine_1
    ...
END INTERFACE
```

For an assignment operator, the interface body must refer to a *subroutine* instead of a function. The subroutine must have two arguments. The first argument is the output of the assignment statement and must have INTENT(OUT). The second dummy argument is the input to the assignment statement and must have INTENT(IN). The first argument corresponds to the left-hand side of the assignment statement, and the second argument corresponds to the right-hand side of the assignment statement.

More than one subroutine can be associated with the assignment symbol, but the subroutines must be distinguishable from one another by having different types of dummy arguments. When the compiler encounters the assignment symbol in a program, it invokes the subroutine whose dummy arguments match the types of the values on either side of the equal sign. If no associated subroutine has dummy arguments that match the values, then a compilation error results.

Good Programming Practice
Use interface operator blocks and interface assignment blocks to create new operators and to extend the meanings of existing operators to work with derived data types. Once proper operators are defined, working with derived data types can be very easy.

The best way to explain the use of user-defined operators and assignments is by an example. We will now define a new derived data type and create appropriate user-defined operations and assignments for it.

EXAMPLE 13-6 Vectors:

The study of the dynamics of objects in motion in 3D is an important area of engineering. In the study of dynamics, the position and velocity of objects, forces, torques, and so forth are usually represented by three-component vectors $\mathbf{v} = x\hat{\mathbf{i}} + y\hat{\mathbf{j}} + z\hat{\mathbf{k}}$, where the three components (x, y, z) represent the projection of the vector $\mathbf{v}$ along the x, y, and z axes respectively, and $\hat{\mathbf{i}}$, $\hat{\mathbf{j}}$, and $\hat{\mathbf{k}}$ are the unit vectors along the x, y, and z axes (see Figure 13-13). The solutions of many mechanical problems involve manipulating these vectors in specific ways.

The most common operations performed on these vectors are:

1. **Addition.** Two vectors are added together by separately adding their x, y, and z components. If $\mathbf{v}_1 = x_1\hat{\mathbf{i}} + y_1\hat{\mathbf{j}} + z_1\hat{\mathbf{k}}$ and $\mathbf{v}_2 = x_2\hat{\mathbf{i}} + y_2\hat{\mathbf{j}} + z_2\hat{\mathbf{k}}$, then $\mathbf{v}_1 + \mathbf{v}_2 = (x_1 + x_2)\hat{\mathbf{i}} + (y_1 + y_2)\hat{\mathbf{j}} + (z_1 + z_2)\hat{\mathbf{k}}$.
2. **Subtraction.** Two vectors are subtracted by separately subtracting their x, y, and z components. If $\mathbf{v}_1 = x_1\hat{\mathbf{i}} + y_1\hat{\mathbf{j}} + z_1\hat{\mathbf{k}}$ and $\mathbf{v}_2 = x_2\hat{\mathbf{i}} + y_2\hat{\mathbf{j}} + z_2\hat{\mathbf{k}}$, then $\mathbf{v}_1 - \mathbf{v}_2 = (x_1 - x_2)\hat{\mathbf{i}} + (y_1 - y_2)\hat{\mathbf{j}} + (z_1 - z_2)\hat{\mathbf{k}}$.
3. **Multiplication by a Scalar.** A vector is multiplied by a scalar by separately multiplying each component by the scalar. If $\mathbf{v} = x\hat{\mathbf{i}} + y\hat{\mathbf{j}} + z\hat{\mathbf{k}}$, then $a\mathbf{v} = ax\hat{\mathbf{i}} + ay\hat{\mathbf{j}} + az\hat{\mathbf{k}}$.

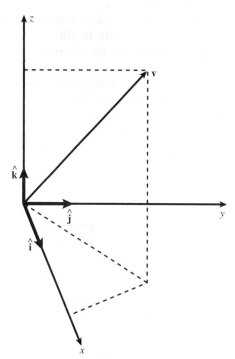

FIGURE 13-13
A 3D vector.

4. **Division by a Scalar.** A vector is divided by a scalar by separately dividing each component by the scalar. If $\mathbf{v} = x\hat{\mathbf{i}} + y\hat{\mathbf{j}} + z\hat{\mathbf{k}}$, then $\dfrac{\mathbf{v}}{a} = \dfrac{x}{a}\hat{\mathbf{i}} + \dfrac{y}{a}\hat{\mathbf{j}} + \dfrac{z}{a}\hat{\mathbf{k}}$.
5. **The Dot Product.** The dot product of two vectors is one form of multiplication operation performed on vectors. It produces a scalar that is the sum of the products of the vector's components. If $\mathbf{v}_1 = x_1\hat{\mathbf{i}} + y_1\hat{\mathbf{j}} + z_1\hat{\mathbf{k}}$ and $\mathbf{v}_2 = x_2\hat{\mathbf{i}} + y_2\hat{\mathbf{j}} + z_2\hat{\mathbf{k}}$, then the dot product of the vectors is $\mathbf{v}_1 \cdot \mathbf{v}_2 = x_1 x_2 + y_1 y_2 + z_1 z_2$.
6. **The Cross Product.** The cross product is another multiplication operation that appears frequently between vectors. The cross product of two vectors is another vector whose direction is perpendicular to the plane formed by the two input vectors. If $\mathbf{v}_1 = x_1\hat{\mathbf{i}} + y_1\hat{\mathbf{j}} + z_1\hat{\mathbf{k}}$ and $\mathbf{v}_2 = x_2\hat{\mathbf{i}} + y_2\hat{\mathbf{j}} + z_2\hat{\mathbf{k}}$, then the cross product of the two vectors is defined as $\mathbf{v}_1 \times \mathbf{v}_2 = (y_1 z_2 - y_2 z_1)\hat{\mathbf{i}} + (z_1 x_2 - z_2 x_1)\hat{\mathbf{j}} + (x_1 y_2 - x_2 y_1)\hat{\mathbf{k}}$.

Create a derived data type called `vector`, having three components x, y, and z. Define functions to create vectors from arrays, to convert vectors to arrays, and to perform the six vector operations defined above. Extend the intrinsic operators +, −, *, and / to have valid meanings when working with vectors, and create a new operator .DOT. for the dot product of two vectors. Finally, extend the assignment operator (=) to allow 3-element arrays to be assigned to vectors, and vectors to 3-element arrays.

SOLUTION

To make it easy to work with vectors, we should place the definition of the data type, the manipulating functions, and the operator definitions all in a single module. That one module can then be used by any programs wanting to manipulate vectors.

Note that six operations were defined for vectors, but *more than six functions must be written to implement them*. For example, the multiplication of a vector by a scalar could occur in either order: vector times scalar or scalar times vector. Both orders produce the same result, but the order of command line arguments for an implementing function is different in either case. Also, a scalar could be either an integer or a single-precision real number. To allow for all four possibilities, either order and either type of scalar, we actually have to write four functions!

1. **State the problem.**

   Create a derived data type called `vector`, having three single-precision real components x, y, and z. Write the following functions and subroutines for manipulating vectors:

   (*a*) Create a vector from a 3-element single-precision real array.
   (*b*) Convert a vector into a 3-element single-precision real array.
   (*c*) Add two vectors.
   (*d*) Subtract two vectors.
   (*e*) Multiply a single-precision real scalar by a vector.
   (*f*) Multiply a vector by a single-precision real scalar.
   (*g*) Multiply an integer scalar by a vector.
   (*h*) Multiply a vector by an integer scalar.

(i) Divide a vector by a single-precision real scalar.
(j) Divide a vector by an integer scalar.
(k) Calculate the dot product of two vectors.
(l) Calculate the cross product of two vectors.

Associate these functions and subroutines with the appropriate operators using the interface operator constructs and interface assignment constructs.

2. **Define the inputs and outputs.**

    Each of the procedures described above has its own inputs and outputs. The types of the input and output arguments for each function are specified in Table 13-2.

3. **Describe the algorithm.**

    The following definitions apply in the pseudocode for all of the above routines:

    (a) `vec_1`   First input argument (vector)
    (b) `vec_2`   Second input argument (vector)
    (c) `real_1`   First input argument (single-precision real)
    (d) `real_2`   Second input argument (single-precision real)
    (e) `int_1`   First input argument (integer)
    (f) `int_2`   Second input argument (integer)
    (g) `array`   Input argument (single-precision real array)
    (h) `vec_result`   Function result (vector)
    (i) `real_result`   Function result (single-precision real)
    (j) `array_result`   Function result (single-precision real array)

TABLE 13-2
Subroutines for manipulating vectors

Specific function / subroutine name	Input argument 1 type	Input argument 2 type	Output type
`array_to_vector` (subroutine)	3-element single-precision real array	N/A	Vector
`vector_to_array` (subroutine)	Vector	N/A	3-element single-precision real array
`vector_add`	Vector	Vector	Vector
`vector_subtract`	Vector	Vector	Vector
`vector_times_real`	Vector	Single-precision real	Vector
`real_times_vector`	Single-precision real	Vector	Vector
`vector_times_int`	Vector	Integer	Vector
`int_times_vector`	Integer	Vector	Vector
`vector_div_real`	Vector	Single-precision real	Vector
`vector_div_int`	Vector	Integer	Vector
`dot_product`	Vector	Vector	Single-precision real
`cross_product`	Vector	Vector	Vector

Given these definitions, the pseudocode for the `array_to_vector` subroutine is:

```
vec_result%x ← array(1)
vec_result%y ← array(2)
vec_result%z ← array(3)
```

The pseudocode for the `vector_to_array` subroutine is:

```
array_result(1) ← vec_1%x
array_result(2) ← vec_1%y
array_result(3) ← vec_1%z
```

The pseudocode for the `vector_add` function is:

```
vec_result%x ← vec_1%x + vec_2%x
vec_result%y ← vec_1%y + vec_2%y
vec_result%z ← vec_1%z + vec_2%z
```

The pseudocode for the `vector_subtract` function is:

```
vec_result%x ← vec_1%x - vec_2%x
vec_result%y ← vec_1%y - vec_2%y
vec_result%z ← vec_1%z - vec_2%z
```

The pseudocode for the `vector_times_real` function is:

```
vec_result%x ← vec_1%x * real_2
vec_result%y ← vec_1%y * real_2
vec_result%z ← vec_1%z * real_2
```

The pseudocode for the `real_times_vector` function is:

```
vec_result%x ← real_1 * vec_2%x
vec_result%y ← real_1 * vec_2%y
vec_result%z ← real_1 * vec_2%z
```

The pseudocode for the `vector_times_int` function is:

```
vec_result%x ← vec_1%x * REAL(int_2)
vec_result%y ← vec_1%y * REAL(int_2)
vec_result%z ← vec_1%z * REAL(int_2)
```

The pseudocode for the `int_times_vector` function is:

```
vec_result%x ← REAL(int_1) * vec_2%x
vec_result%y ← REAL(int_1) * vec_2%y
vec_result%z ← REAL(int_1) * vec_2%z
```

The pseudocode for the `vector_div_real` function is:

```
vec_result%x ← vec_1%x / real_2
vec_result%y ← vec_1%y / real_2
vec_result%z ← vec_1%z / real_2
```

The pseudocode for the `vector_div_int` function is:

```
vec_result%x ← vec_1%x / REAL(int_2)
vec_result%y ← vec_1%y / REAL(int_2)
vec_result%z ← vec_1%z / REAL(int_2)
```

The pseudocode for the `dot_product` function is:

real_result ← vec_1%x*vec_2%x + vec_1%y*vec_2%y + vec_1%z*vec_2%z

The pseudocode for the `cross_product` function is:

vec_result%x ← vec_1%y*vec_2%z - vec_1%z*vec_2%y
vec_result%y ← vec_1%z*vec_2%x - vec_1%x*vec_2%z
vec_result%z ← vec_1%x*vec_2%y - vec_1%y*vec_2%x

These twelve functions will be assigned to operators in interface operator and interface assignment blocks as follows:

Function	Operator
array_to_vector	=
vector_to_array	=
vector_add	+
vector_subtract	-
vector_times_real	*
real_times_vector	*
vector_times_int	*
int_times_vector	*
vector_div_real	/
vector_div_int	/
dot_product	.DOT.
cross_product	*

4. **Turn the algorithm into Fortran statements.**
   The resulting Fortran module is shown in Figure 13-14.

FIGURE 13-14
A module to create a derived data type `vector`, and to define mathematical operations that can be performed on values of type `vector`.

```
MODULE vectors
!
! Purpose:
!   To define a derived data type called vector, and the
!   operations which can be performed on it. The module
!   defines 8 operations which can be performed on vectors:
!
!                   Operation                        Operator
!                   =========                        ========
!       1.  Creation from a real array                  =
!       2.  Conversion to real array                    =
!       3.  Vector addition                             +
!       4.  Vector subtraction                          -
!       5.  Vector-scalar multiplication (4 cases)      *
```

(*continued*)

(*continued*)

```
!      6.  Vector-scalar division (2 cases)          /
!      7.  Dot product                             .DOT.
!      8.  Cross product                             *
!
!   It contains a total of 12 procedures to implement those
!   operations:  array_to_vector, vector_to_array, vector_add,
!   vector_subtract, vector_times_real, real_times_vector,
!   vector_times_int, int_times_vector, vector_div_real,
!   vector_div_int, dot_product, and cross_product.
!
!   Record of revisions:
!       Date        Programmer         Description of change
!       ====        ==========         =====================
!     12/21/15      S. J. Chapman      Original code
!
IMPLICIT NONE

! Declare vector data type:
TYPE :: vector
   REAL :: x
   REAL :: y
   REAL :: z
END TYPE

! Declare interface operators
INTERFACE ASSIGNMENT (=)
   MODULE PROCEDURE array_to_vector
   MODULE PROCEDURE vector_to_array
END INTERFACE

INTERFACE OPERATOR (+)
   MODULE PROCEDURE vector_add
END INTERFACE

INTERFACE OPERATOR (-)
   MODULE PROCEDURE vector_subtract
END INTERFACE

INTERFACE OPERATOR (*)
   MODULE PROCEDURE vector_times_real
   MODULE PROCEDURE real_times_vector
   MODULE PROCEDURE vector_times_int
   MODULE PROCEDURE int_times_vector
   MODULE PROCEDURE cross_product
END INTERFACE

INTERFACE OPERATOR (/)
   MODULE PROCEDURE vector_div_real
   MODULE PROCEDURE vector_div_int
END INTERFACE

INTERFACE OPERATOR (.DOT.)
   MODULE PROCEDURE dot_product
END INTERFACE
```

(*continued*)

(*continued*)

```fortran
! Now define the implementing functions.
CONTAINS
   SUBROUTINE array_to_vector(vec_result, array)
      TYPE (vector), INTENT(OUT) :: vec_result
      REAL, DIMENSION(3), INTENT(IN) :: array
      vec_result%x = array(1)
      vec_result%y = array(2)
      vec_result%z = array(3)
   END SUBROUTINE array_to_vector

   SUBROUTINE vector_to_array(array_result, vec_1)
      REAL, DIMENSION(3), INTENT(OUT) :: array_result
      TYPE (vector), INTENT(IN) :: vec_1
      array_result(1) = vec_1%x
      array_result(2) = vec_1%y
      array_result(3) = vec_1%z
   END SUBROUTINE vector_to_array

   FUNCTION vector_add(vec_1, vec_2)
      TYPE (vector) :: vector_add
      TYPE (vector), INTENT(IN) :: vec_1, vec_2
      vector_add%x = vec_1%x + vec_2%x
      vector_add%y = vec_1%y + vec_2%y
      vector_add%z = vec_1%z + vec_2%z
   END FUNCTION vector_add

   FUNCTION vector_subtract(vec_1, vec_2)
      TYPE (vector) :: vector_subtract
      TYPE (vector), INTENT(IN) :: vec_1, vec_2
      vector_subtract%x = vec_1%x - vec_2%x
      vector_subtract%y = vec_1%y - vec_2%y
      vector_subtract%z = vec_1%z - vec_2%z
   END FUNCTION vector_subtract

   FUNCTION vector_times_real(vec_1, real_2)
      TYPE (vector) :: vector_times_real
      TYPE (vector), INTENT(IN) :: vec_1
      REAL, INTENT(IN) :: real_2
      vector_times_real%x = vec_1%x * real_2
      vector_times_real%y = vec_1%y * real_2
      vector_times_real%z = vec_1%z * real_2
   END FUNCTION vector_times_real

   FUNCTION real_times_vector(real_1, vec_2)
      TYPE (vector) :: real_times_vector
      REAL, INTENT(IN) :: real_1
      TYPE (vector), INTENT(IN) :: vec_2
      real_times_vector%x = real_1 * vec_2%x
      real_times_vector%y = real_1 * vec_2%y
      real_times_vector%z = real_1 * vec_2%z
   END FUNCTION real_times_vector
```

(*continued*)

(*concluded*)

```
   FUNCTION vector_times_int(vec_1, int_2)
      TYPE (vector) :: vector_times_int
      TYPE (vector), INTENT(IN) :: vec_1
      INTEGER, INTENT(IN) :: int_2
      vector_times_int%x = vec_1%x * REAL(int_2)
      vector_times_int%y = vec_1%y * REAL(int_2)
      vector_times_int%z = vec_1%z * REAL(int_2)
   END FUNCTION vector_times_int

   FUNCTION int_times_vector(int_1, vec_2)
      TYPE (vector) :: int_times_vector
      INTEGER, INTENT(IN) :: int_1
      TYPE (vector), INTENT(IN) :: vec_2
      int_times_vector%x = REAL(int_1) * vec_2%x
      int_times_vector%y = REAL(int_1) * vec_2%y
      int_times_vector%z = REAL(int_1) * vec_2%z
   END FUNCTION int_times_vector

   FUNCTION vector_div_real(vec_1, real_2)
      TYPE (vector) :: vector_div_real
      TYPE (vector), INTENT(IN) :: vec_1
      REAL, INTENT(IN) :: real_2
      vector_div_real%x = vec_1%x / real_2
      vector_div_real%y = vec_1%y / real_2
      vector_div_real%z = vec_1%z / real_2
   END FUNCTION vector_div_real

   FUNCTION vector_div_int(vec_1, int_2)
      TYPE (vector) :: vector_div_int
      TYPE (vector), INTENT(IN) :: vec_1
      INTEGER, INTENT(IN) :: int_2
      vector_div_int%x = vec_1%x / REAL(int_2)
      vector_div_int%y = vec_1%y / REAL(int_2)
      vector_div_int%z = vec_1%z / REAL(int_2)
   END FUNCTION vector_div_int

   FUNCTION dot_product(vec_1, vec_2)
      REAL :: dot_product
      TYPE (vector), INTENT(IN) :: vec_1, vec_2
      dot_product = vec_1%x*vec_2%x + vec_1%y*vec_2%y &
                  + vec_1%z*vec_2%z
   END FUNCTION dot_product

   FUNCTION cross_product(vec_1, vec_2)
      TYPE (vector) :: cross_product
      TYPE (vector), INTENT(IN) :: vec_1, vec_2
      cross_product%x = vec_1%y*vec_2%z - vec_1%z*vec_2%y
      cross_product%y = vec_1%z*vec_2%x - vec_1%x*vec_2%z
      cross_product%z = vec_1%x*vec_2%y - vec_1%y*vec_2%x
   END FUNCTION cross_product

END MODULE vectors
```

5. Test the resulting Fortran programs.

To test this data type and its associated operations, it is necessary to write a test driver program that defines and manipulates vectors, and prints out the results. The program should exercise every operation defined for vectors in the module. Figure 13-15 shows an appropriate test driver program.

FIGURE 13-15
Test driver program to test the `vector` data type and associated operations.

```
PROGRAM test_vectors
!
!  Purpose:
!    To test the definitions, operations, and assignments
!    associated with the vector data type.
!
!  Record of revisions:
!      Date         Programmer          Description of change
!      ====         ==========          =====================
!    12/21/15     S. J. Chapman         Original code
!
USE vectors
IMPLICIT NONE

! Data dictionary: declare variable types & definitions
REAL, DIMENSION(3) :: array_out          ! Output array
TYPE (vector) :: vec_1, vec_2            ! Test vectors

! Test assignments by assigning an array to vec_1 and
! assigning vec_1 to array_out.
vec_1 = (/ 1., 2., 3. /)
array_out = vec_1
WRITE (*,1000) vec_1, array_out
1000 FORMAT (' Test assignments: ',/, &
             ' vec_1 =     ', 3F8.2,/, &
             ' array_out = ', 3F8.2)

! Test addition and subtraction.
vec_1 = (/ 10., 20., 30. /)
vec_2 = (/ 1., 2., 3. /)
WRITE (*,1010) vec_1, vec_2, vec_1 + vec_2, vec_1 - vec_2
1010 FORMAT (/' Test addition and subtraction: ',/, &
             ' vec_1 =         ', 3F8.2,/, &
             ' vec_2 =         ', 3F8.2,/, &
             ' vec_1 + vec_2 = ', 3F8.2,/, &
             ' vec_1 - vec_2 = ', 3F8.2)

! Test multiplication by a scalar.
vec_1 = (/ 1., 2., 3. /)
WRITE (*,1020) vec_1, 2.*vec_1, vec_1*2., 2*vec_1, vec_1*2
1020 FORMAT (/' Test multiplication by a scalar: ',/, &
             ' vec_1 =       ', 3F8.2,/, &
             ' 2. * vec_1 =  ', 3F8.2,/, &
```

(continued)

(*concluded*)

```
                          ' vec_1 * 2. =      ', 3F8.2,/, &
                          ' 2 * vec_1   =     ', 3F8.2,/, &
                          ' vec_1 * 2   =     ', 3F8.2)

! Test division by a scalar.
vec_1 = (/ 10., 20., 30. /)
WRITE (*,1030) vec_1, vec_1/5., vec_1/5
1030 FORMAT (/' Test division by a scalar: ',/, &
                          ' vec_1 =            ', 3F8.2,/, &
                          ' vec_1 / 5. =       ', 3F8.2,/, &
                          ' vec_1 / 5 =        ', 3F8.2)

! Test dot product.
vec_1 = (/ 1., 2., 3. /)
vec_2 = (/ 1., 2., 3. /)
WRITE (*,1040) vec_1, vec_2, vec_1 .DOT. vec_2
1040 FORMAT (/' Test dot product: ',/, &
                          ' vec_1 =                ', 3F8.2,/, &
                          ' vec_2 =                ', 3F8.2,/, &
                          ' vec_1 .DOT. vec_2 = ', 3F8.2)

! Test cross product.
vec_1 = (/ 1., -1., 1. /)
vec_2 = (/ -1., 1., 1. /)
WRITE (*,1050) vec_1, vec_2, vec_1*vec_2
1050 FORMAT (/' Test cross product: ',/, &
                          ' vec_1 =            ', 3F8.2,/, &
                          ' vec_2 =            ', 3F8.2,/, &
                          ' vec_1 * vec_2 = ', 3F8.2)
END PROGRAM test_vectors
```

When the test driver program is executed, the results are:

```
C:\book\fortran\chap13>test_vectors
Test assignments:
vec_1 =             1.00    2.00    3.00
array_out =         1.00    2.00    3.00

Test addition and subtraction:
vec_1 =            10.00   20.00   30.00
vec_2 =             1.00    2.00    3.00
vec_1 + vec_2 =    11.00   22.00   33.00
vec_1 - vec_2 =     9.00   18.00   27.00

Test multiplication by a scalar:
vec_1 =             1.00    2.00    3.00
2. * vec_1 =        2.00    4.00    6.00
vec_1 * 2. =        2.00    4.00    6.00
2 * vec_1 =         2.00    4.00    6.00
vec_1 * 2 =         2.00    4.00    6.00

Test division by a scalar:
vec_1 =            10.00   20.00   30.00
vec_1 / 5. =        2.00    4.00    6.00
vec_1 / 5 =         2.00    4.00    6.00
```

Advanced Features of Procedures and Modules

```
Test dot product:
vec_1 =                        1.00    2.00    3.00
vec_2 =                        1.00    2.00    3.00
vec_1 .DOT. vec_2 =           14.00

Test cross product:
vec_1 =                        1.00   -1.00    1.00
vec_2 =                       -1.00    1.00    1.00
vec_1 * vec_2 =               -2.00   -2.00     .00
```

The results of the program are correct, and we can verify them by calculating the answers from the definitions of the operations.

What would happen in a program if we tried to perform an operation with vectors that was not defined in the module? For example, what would happen if we tried to multiply a vector by a double-precision real scalar? A compilation error would result, because the compiler does not know how to perform the operation. When defining a new data type and its operations, be careful to define *every* combination of operations that you might wish to use.

13.8
BOUND ASSIGNMENTS AND OPERATORS

Assignments and operators can be bound to derived data types using the GENERIC statement. These procedures are declared using the GENERIC statement, as shown below.

```
TYPE :: point
   REAL :: x
   REAL :: y
CONTAINS
   GENERIC :: ASSIGNMENT(=) => assign1
   GENERIC :: OPERATOR(+) => plus1, plus2, plus3
END TYPE point
```

The bodies of the procedures implementing the operators must be declared in the same way as the generic assignments and operators defined in the previous section.

13.9
RESTRICTING ACCESS TO THE CONTENTS OF A MODULE

When a module is accessed by USE association, by default all of the entities defined within that module become available for use in the program unit containing the USE statement. In the past, we have used this fact to share data between program units, to make procedures with explicit interfaces available to program units, to create new operators, and to extend the meanings of existing operators.

In Example 13-6, we created a module called `vectors` to extend the Fortran language. Any program unit that accesses module `vectors` can define its own vectors, and can manipulate them using the binary operators +, −, *, /, and .DOT.. Unfortunately, the program will also be able to invoke such functions as `vector_add`, `vector_subtract`, etc., even though it should only be using them indirectly through the use of the defined operators. These procedure names are not needed in any program unit, but they are declared, and they might conflict with a procedure name defined in the program. A similar problem could occur when many data items are defined within a module, but only a few of them are needed by a particular program unit. All of the unnecessary data items will also be available in the program unit, making it possible for a programmer to modify them by mistake.

In general, *it is a good idea to restrict access to any procedures or data entities in a module to only those program units that must know about them.* This process is known as **data hiding**. The more access is restricted, the less chance there is of a programmer using or modifying an item by mistake. Restricting access makes programs more modular and easier to understand and maintain.

How can we restrict access to the entities in a module? Fortran provides a way to control the access to a particular item in a module by program units *outside* that module: the PUBLIC, PRIVATE, and PROTECTED attributes and statements. If the PUBLIC attribute or statement is specified for an item, then the item will be available to program units outside the module. If the PRIVATE attribute or statement is specified, then the item will not be available to program units outside the module, although procedures inside the module still have access to it. If the PROTECTED attribute or statement is specified, then the item will be available on a *read-only* basis to program units outside the module. Any attempt to modify the value of a PROTECTED variable outside the module in which it is defined will produce a compile-time error. The default attribute for all data and procedures in a module is PUBLIC, so by default any program unit that uses a module can have access to every data item and procedure within it.

The PUBLIC, PRIVATE, or PROTECTED status of a data item or procedure can be declared in one of two ways. It is possible to specify the status as an attribute in a type definition statement, or in an independent Fortran statement. Examples in which the attributes are declared as a part of a type definition statement are:

```
INTEGER, PRIVATE :: count
REAL, PUBLIC :: voltage
REAL, PROTECTED :: my_data
TYPE (vector), PRIVATE :: scratch_vector
```

This type of declaration can be used for data items and for functions, but not for subroutines. A PUBLIC, PRIVATE, or PROTECTED statement can also be used to specify the status of data items, functions, and subroutines. The form of a PUBLIC, PRIVATE, or PROTECTED statement is:

```
PUBLIC :: list of public items
PRIVATE :: list of private items
PROTECTED :: list of private items
```

If a module contains a PRIVATE statement without a list of private items, *then by default every data item and procedure in the module is private*. Any items that should be public must be explicitly listed in a separate `PUBLIC` statement. This is the preferred way to design modules, since only the items that are actually required by programs are exposed to them.

Good Programming Practice
It is good programming practice to hide any module data items or procedures that do not need to be directly accessed by external program units. The best way to do this is to include a `PRIVATE` statement in each module, and then list the specific items that you wish to expose in a separate `PUBLIC` statement.

As an example of the proper use of data hiding, let's reexamine module `vectors` from Example 13-6. Programs accessing this module need to define variables of type `vector`, and need to perform operations involving vectors. However, the programs do *not* need direct access to any of the subroutines or functions in the module. The proper declarations for this circumstance are shown in Figure 13-16.

FIGURE 13-16
The first part of module `vector`, modified to hide all nonessential items from external program units. Changes to the module are shown in bold type.

```
MODULE vectors
!
!   Purpose:
!     To define a derived data type called vector, and the
!     operations which can be performed on it. The module
!     defines 8 operations which can be performed on vectors:
!
!                       Operation                    Operator
!                       =========                    ========
!       1.  Creation from a real array                  =
!       2.  Conversion to real array                    =
!       3.  Vector addition                             +
!       4.  Vector subtraction                          -
!       5.  Vector-scalar multiplication (4 cases)      *
!       6.  Vector-scalar division (2 cases)            /
!       7.  Dot product                                 .DOT.
!       8.  Cross product                               *
!
!     It contains a total of 12 procedures to implement those
!     operations:  array_to_vector, vector_to_array, vector_add,
!     vector_subtract, vector_times_real, real_times_vector,
!     vector_times_int, int_times_vector, vector_div_real,
!     vector_div_int, dot_product, and cross_product. These
!     procedures are private to the module; they can only be
!     accessed from the outside via the defined operators.
!
```

(continued)

(*concluded*)

```
!   Record of revisions:
!       Date        Programmer          Description of change
!       ====        ==========          =====================
!    12/21/15     S. J. Chapman         Original code
! 1. 12/22/15     S. J. Chapman         Modified to hide non-
!                                       essential items.
!
IMPLICIT NONE

! Declare all items to be private except for type vector and
! the operators defined for it.
PRIVATE
PUBLIC :: vector, assignment(=), operator(+), operator(-), &
         operator(*), operator(/), operator(.DOT.)

! Declare vector data type:
TYPE :: vector
    REAL :: x
    REAL :: y
    REAL :: z
END TYPE
```

The following notes apply to PUBLIC and PRIVATE declarations for derived data types in modules.

1. The components of a derived data type declared in a module can be made inaccessible to program units outside of the module by including a PRIVATE statement *within* the derived data type. Note that *the derived data type as a whole is still available to outside program units, but its components cannot be accessed separately*. Outside program units may freely declare variables of the derived data type, but they may not work with individual components of those variables. An example of a derived data type with private components is:

   ```
   TYPE vector
       PRIVATE
       REAL :: x
       REAL :: y
   END TYPE
   ```

2. In contrast to the situation above, an entire derived data type can be declared to be private. An example is:

   ```
   TYPE, PRIVATE :: vector
       REAL :: x
       REAL :: y
   END TYPE
   ```

   In this case, the data type vector is not accessible by any program units that use the module. This differs from the previous case, in which the data type was available but its components could not be accessed separately. Such a derived data type can only be used for internal calculations within the module.

3. In Fortran 2003 and later, *individual components* of a derived data type can be declared to be public or private. An example is:

```
TYPE :: vector
   REAL,PUBLIC :: x
   REAL,PRIVATE :: y
END TYPE
```

In this case, outside program units may freely declare variables of type `vector`, and may freely access component x, but component y cannot be accessed outside the module in which the derived data type is defined. This feature supports object-oriented programming, as we shall see in Chapter 16.

4. Finally, it is possible to declare private variables of a derived data type even though the type itself is public. For example,

```
TYPE :: vector
   REAL :: x
   REAL :: y
END TYPE
TYPE (vector), PRIVATE :: vec_1
```

In this case, the derived data type `vector` is public and available in program units that use the module, but the variable `vec_1` may only be used within the module. This type of declaration might be used for variables used in internal calculations within the module.

13.10

ADVANCED OPTIONS OF THE USE STATEMENT

When a program unit accesses a module by USE association, by default it gets access to every data item, interface, and procedure in the module. It is possible for the module to restrict access to some items by declaring them to be PRIVATE. In addition to this control, it is possible for a program unit using the module to further restrict the list of items being used, and to modify the names of those items.

Why would we want to further restrict the list of items from a module that is accessed by USE association in a program unit? If a data item from a module is not needed in the program unit, then it is good defensive programming to make that item unavailable. This action will prevent the program unit from using or modifying the item *by mistake*, and will reduce the chance of developing hard-to-find bugs. A common problem of this sort would be to make a typographical error in a local variable name and not know it because the new name just accidentally happens to be declared in the module. Most typographical errors are caught by the compiler because the IMPLICIT NONE statement makes undeclared variables illegal. However, if the new name happens to be defined in the module, then using it will not be an error. Furthermore, since the contents of the module do not appear in the program unit listing, the programmer may not realize that a variable of that name was defined in the module! Problems like this can be hard to find.

To restrict access to certain specific items in a module, an `ONLY` clause may be added to the `USE` statement. The form of the statement is

```
USE module_name, ONLY: only_list
```

where `module_name` is the module name and `only_list` is the list of items from the module to be used, with items in the list separated by commas. As an example, we could further restrict access to operations in module `vectors` by using the statement

```
USE vectors, ONLY: vector, assignment(=)
```

In a procedure containing this statement, it would be legal to declare a variable of type `vector` and to assign a 3-element array to it, but it would not be legal to add two vectors together.

It is also possible to rename a data item or procedure in the `USE` statement. There are two reasons why we might wish to rename a data item or procedure when it is used by a program unit. One reason is that the item might have a name that is the same as a local data item or an item from another module also used by the program unit. In this case, renaming the item avoids a clash between the two definitions of the name.

The second reason to rename a module data item or procedure is that we might wish to shorten a name declared in a module when it is used very frequently in a program unit. For example, a module called `data_fit` might contain a procedure with the name `sp_real_least_squares_fit` to distinguish it from a double-precision version `dp_real_least_squares_fit`. When this module is used in a program unit, the programmer might wish to refer to the procedure by a less unwieldy name. He or she might wish to call the procedure simply `lsqfit` or something similar.

The forms of the `USE` statement that permit a programmer to rename a data item or procedure are

```
USE module_name, rename_list
USE module_name, ONLY: rename_list
```

where each item in the *rename_list* takes the form

```
local_name => module_name
```

In the first case, all public items in the module will be available to the program unit, but the ones in the rename list will be renamed. In the second case, only the items listed would be available, and they would be renamed. For example, the `USE` statement to rename the least-squares fit routine mentioned above while simultaneously restricting access to all other items in module `data_fits` would be

```
USE data_fit, ONLY: lsqfit => sp_real_least_squares_fit
```

A few complications can arise when multiple `USE` statements in a single program unit refer to the same module. It makes no sense to use more than one `USE` statement in a single routine to refer to a given module, so you should never have this problem in well-written code. However, if you do have more than one `USE` statement referring to the same module, the following rules apply:

1. If none of the `USE` statements have rename lists or `ONLY` clauses, then the statements are just duplicates of each other, which is legal but has no effect on the program.

2. If all of the USE statements include rename lists but no ONLY clauses, then the effect is the same as if all of the renamed items were listed in a single USE statement.
3. If all of the USE statements include ONLY clauses, then the effect is the same as if all of the lists were listed in a single USE statement.
4. If some USE statements have an ONLY clause and some do not, then the ONLY clauses have no effect on the program at all! This happens because the USE statements without ONLY clauses allow all public items in the module to be visible in the program unit.

Quiz 13-2

This quiz provides a quick check to see if you have understood the concepts introduced in Sections 13.4 to 13.8. If you have trouble with the quiz, reread the sections, ask your instructor, or discuss the material with a fellow student. The answers to this quiz are found in the back of the book.

1. What is an interface block? What are the two possible locations for interface blocks in a Fortran program?
2. Why would a programmer choose to create an interface block to a procedure instead of including the procedure in a module?
3. What items must appear in the interface body of an interface block?
4. Is the following program valid? Why or why not? If it is legal, what does it do?

```
PROGRAM test
IMPLICIT NONE
TYPE :: data
   REAL :: x1
   REAL :: x2
END TYPE
CHARACTER(len=20) :: x1 = 'This is a test.'
TYPE (data) :: x2
x2%x1 = 613.
x2%x2 = 248.
WRITE (*,*) x1, x2
END PROGRAM test
```

5. How is a generic procedure defined?
6. How is a generic bound procedure defined?
7. Is the following code valid? Why or why not? If it is legal, what does it do?

```
INTERFACE fit
   SUBROUTINE least_squares_fit (array, nvals, slope, intercept)
   IMPLICIT NONE
   INTEGER, INTENT(IN) :: nvals
   REAL, INTENT(IN), DIMENSION(nvals) :: array
```

(continued)

(concluded)
```
          REAL, INTENT(OUT) :: slope
          REAL, INTENT(OUT) :: intercept
       END SUBROUTINE least_squares_fit

       SUBROUTINE median_fit (data1, n, slope, intercept)
       IMPLICIT NONE
       INTEGER, INTENT(IN) :: n
       REAL, INTENT(IN), DIMENSION(n) :: data1
       REAL, INTENT(OUT) :: slope
       REAL, INTENT(OUT) :: intercept
       END SUBROUTINE median_fit
   END INTERFACE fit
```

8. What is a MODULE PROCEDURE statement? What is its purpose?
9. What is the difference in structure between a user-defined operator and a user-defined assignment? How are they implemented?
10. How can access to the contents of a module be controlled? Why would we wish to limit the access to some data items or procedures in a module?
11. What is the default type of access for items in a module?
12. How can a program unit accessing a module by USE association control that items in the module it sees? Why would a programmer wish to do this?
13. How can a program unit accessing a module by USE association rename data items or procedures in the module? Why would a programmer wish to do this?
14. Is the following code valid? Why or why not? If it is legal, what does it do?

```
MODULE test_module
TYPE :: test_type
   REAL :: x, y, z
   PROTECTED :: z
END TYPE test_type
END MODULE test_module

PROGRAM test
USE test_module
TYPE(test_type) :: t1, t2
t1%x = 10.
t1%y = -5.
t2%x = -2.
t2%y = 7.
t1%z = t1%x * t2%y
END PROGRAM test
```

13.11
INTRINSIC MODULES

Fortran also includes a concept called an **intrinsic module.** An intrinsic module is just like an ordinary Fortran module, except that it is pre-defined and coded by the creator of the Fortran compiler. Like ordinary modules, we access procedures and data in intrinsic modules via a USE statement.

There are a number of standard intrinsic modules in Fortran. The three most important ones are:

1. Module ISO_FORTRAN_ENV, which contains constants describing the characteristics of storage in a particular computer (how many bits in a standard integer, how many bits in a standard character, etc.), and also constants defining I/O units for the particular computer. (We will use this module in Chapter 14.)
2. Module ISO_C_BINDING, which contains data necessary for a Fortran compiler to interoperate with C on a given processor. (We will use this module in Appendix B.)
3. The IEEE modules, which describe the characteristics of IEEE 754 floating-point calculations on a particular processor. The standard IEEE modules are IEEE_EXCEPTIONS, IEEE_ARITHMETIC, and IEEE_FEATURES.

The Fortran standard requires compiler vendors to implement certain procedures in these intrinsic modules, but it allows them to add additional procedures, and also to define their own intrinsic modules. In the future, this should be a common way to ship special features with a compiler.

13.12
ACCESS TO COMMAND LINE ARGUMENTS AND ENVIRONMENT VARIABLES

Fortran includes standard procedures to allow a Fortran program to retrieve the command line that started the program, and to recover data from the program's environment. These mechanisms allow the user to pass parameters to the program at startup by typing them on the command line after the program name, or by including them as environment variables.

Fortran compiler vendors have allowed Fortran programs get access command line arguments and environment variables for many years, but since there was no *standard* way to do this, each vendor created its own special subroutines and functions. Since these procedures differed from vendor to vendor, Fortran programs tended to be less portable. Fortran has solved this problem by creating standard intrinsic procedures to retrieve command line parameters.

13.12.1 Access to Command Line Arguments

There are three standard intrinsic procedures for getting variables from the command line.

1. **Function** COMMAND_ARGUMENT_COUNT( ). This function returns the number of command line arguments present when the program started in an integer of the default type. It has no arguments.
2. **Subroutine** GET_COMMAND(COMMAND,LENGTH,STATUS). This subroutine returns the entire set of command line arguments in the character variable COMMAND, the length of the argument string in integer LENGTH, and the success or failure of the operation in integer STATUS. If the retrieval is successful, the STATUS will be zero. If the character variable COMMAND is too short to hold the argument, the STATUS will be −1. Any other error will cause a nonzero number to be returned. Note that all of these arguments are optional, so a user can include only some of them, using keyword syntax to specify which ones are present.
3. **Subroutine** GET_COMMAND_ARGUMENT(NUMBER,VALUE,LENGTH,STATUS). This subroutine returns a specified command argument. The integer value NUMBER specified which argument to return. The number must be in the range 0 to COMMAND_ARGUMENT_COUNT( ). The argument returned will be the program name if the number is zero, or the one corresponding to the specified number for a number greater than zero. The argument is returned in character variable VALUE, the length of the argument string in integer LENGTH, and the success or failure of the operation in integer STATUS. If the retrieval is successful, the STATUS will be zero. If the character variable VALUE is too short to hold the argument, the STATUS will be −1. Any other error will cause a nonzero number to be returned. Note that all of these arguments except NUMBER are optional, so a user can include only some of them, using keyword syntax to specify which ones are present.

A sample program that illustrates the use of these procedures is shown in Figure 13-17. This program recovers and displays the command line arguments used to start the program.

FIGURE 13-17
Program illustrating the use of intrinsic procedures to get command line arguments.

```
PROGRAM get_command_line

! Declare local variables
INTEGER :: i                          ! Loop index
CHARACTER(len=128) :: command         ! Command line
CHARACTER(len=80) :: arg              ! Single argument

! Get the program name
CALL get_command_argument(0, command)
WRITE (*,'(A,A)') 'Program name is: ', TRIM(command)

! Now get the individual arguments
DO i = 1, command_argument_count()
   CALL get_command_argument(i, arg)
   WRITE (*,'(A,I2,A,A)') 'Argument ', i, ' is ', TRIM(arg)
END DO

END PROGRAM get_command_line
```

When this program is executed, the results are:

```
C:\book\fortran\chap13>get_command_line 1 sdf 4 er4
Program name is: get_command_line
Argument   1 is 1
Argument   2 is sdf
Argument   3 is 4
Argument   4 is er4
```

13.12.2 Retrieving Environment Variables

The value of an environment variable can be retrieved using subroutine GET_ENVIRONMENT_VARIABLE. The arguments for this subroutine are:

```
CALL GET_ENVIRONMENT_VARIABLE(NAME,VALUE,LENGTH,STATUS,TRIM_NAME)
```

The argument NAME is a character expression supplied by the user, containing the name of the environment variable whose value is desired. The environment variable is returned in character variable VALUE, the length of the environment variable in integer LENGTH, and the success or failure of the operation in integer STATUS. If the retrieval is successful, the STATUS will be zero. If the character variable VALUE is too short to hold the argument, the STATUS will be –1. If the environment variable does not exist, the STATUS will be 1. If the processor does not support environment variables, the STATUS will be 2. If another error occurs, the status will be greater than 2. TRIM_NAME is a logical input argument. If it is true, then the command will ignore trailing blanks when matching the environment variable. If it is false, it will include the trailing blanks in the comparison.

Note that VALUE, LENGTH, STATUS, and TRIM_NAME are all optional arguments, so they can be included or left out, as desired.

A sample program that illustrates the use of GET_ENVIRONMENT_VARIABLE is shown in Figure 13-18. This program recovers and displays the value of the "windir" environment variable, which is defined on the computer where this text is being written.

FIGURE 13-18
Program illustrating the use of GET_ENVIRONMENT_VARIABLE.

```
PROGRAM get_env

! Declare local variables
INTEGER :: length                  ! Length
INTEGER :: status                  ! Status
CHARACTER(len=80) :: value         ! Environment variable value

! Get the value of the "windir" environment variable
CALL get_environment_variable('windir',value,length,status)

! Tell user
WRITE (*,*) 'Get "windir" environment variable:'
WRITE (*,'(A,I6)') 'Status = ', status
IF ( status <= 0 ) THEN
   WRITE (*,'(A,A)') 'Value = ', TRIM(value)
END IF

END PROGRAM get_env
```

When this program is executed, the results are:

```
C:\book\fortran\chap13>get_env
 Get 'windir' environment variable:
 Status =       0
 Value  = C:\WINDOWS
```

Good Programming Practice

Use the standard Fortran intrinsic procedures to retrieve the command line arguments used to start a program and the values of environment variables instead of the nonstandard procedures supplied by individual vendors.

13.13

THE VOLATILE ATTRIBUTE AND STATEMENT

When a Fortran compiler compiles a program for release, it usually runs an optimizer to increase the program's speed. The optimizer performs many techniques to increase the program's speed, but one very common approach is to hold the value of a variable in a CPU register between uses, since the access to registers is much faster than the access to main memory. This is commonly done for variables that are modified a lot in DO loops, provided that there are free registers to hold the data.

This optimization can cause serious problems if the variable being used is also accessed or modified by other processes outside the Fortran program. In that case, the external process might modify the value of the variable, while the Fortran program is using a different value that was previously stored in a register.

To avoid incompatible values, there must always be one and only one location where the data is stored. The Fortran compiler must know not to hold a copy of the variable in a register, and must know to update main memory as soon as any change happens to the value of the variable. This is accomplished by declaring a variable to be volatile. If a variable is volatile, the compiler does not apply any optimizations to it, and the program works directly with the location of the variable in main memory.

A variable is declared to be volatile with a VOLATILE attribute or statement. A volatile attribute takes the form

```
REAL,VOLATILE :: x      ! Volatile variable
REAL,VOLATILE :: y      ! Volatile variable
```

and a volatile statement takes the form

```
REAL :: x, y            ! Declarations
VOLATILE :: x, y        ! Volatile declaration
```

The VOLATILE attribute or statement is commonly used with massively parallel processing packages, which have methods to asynchronously transfer data between processes.

13.14
SUMMARY

This chapter introduced several advanced features of procedures and modules in Fortran. None of these features were available in earlier versions of Fortran.

Fortran supports four levels of scope: global, local, block, and statement. Global-scope objects include program, external procedure, and module names. The only statement-scope objects that we have seen so far are the variables in an implied `DO` loop in an array constructor, and the index variables in a `FORALL` statement. Local scope objects have a scope restricted to a single scoping unit, and block scope objects have a scope restricted to block in which they are defined. A scoping unit is a main program, a procedure, a module, a derived data type, or an interface. If one scoping unit is defined entirely inside another scoping unit, then the inner scoping unit inherits all of the data items defined in the host scoping unit by host association.

Ordinarily, Fortran subroutines and functions are not recursive—they cannot call themselves either directly or indirectly. However, they can be made recursive if they are declared to be recursive in the corresponding `SUBROUTINE` or `FUNCTION` statement. A recursive function declaration includes a `RESULT` clause specifying the name to be used to return the function result.

If a procedure has an explicit interface, then keyword arguments may be used to change the order in which calling arguments are specified. A keyword argument consists of the dummy argument's name followed by an equal sign and the value of the argument. Keyword arguments are very useful in supporting optional arguments.

If a procedure has an explicit interface, then optional arguments may be declared and used. An optional argument is an argument that may or may not be present in the procedure's calling sequence. An intrinsic function `PRESENT( )` is provided to determine whether or not a particular optional argument is present when the procedure gets called. Keyword arguments are commonly used with optional arguments because optional arguments often appear out of sequence in the calling procedure.

Interface blocks are used to provide an explicit interface for procedures that are not contained in a module. They are often used to provide Fortran interfaces to older pre-Fortran 90 code without rewriting all of the code. The body of an interface block must contain either a complete description of the calling sequence to a procedure, including the type and position of every argument in the calling sequence, or a `MODULE PROCEDURE` statement to refer to a procedure already defined in a module.

Generic procedures are procedures that can function properly with different types of input data. A generic procedure is declared using a generic interface block, which looks like an ordinary interface block with the addition of a generic procedure name. One or more specific procedures may be declared within the body of the generic interface block. Each specific procedure must be distinguishable from all other specific procedures by the type and sequence of its nonoptional dummy arguments. When a generic procedure is referenced in a program, the compiler uses the sequence of calling arguments associated with the reference to decide which of the specific procedures to execute.

Generic bound procedures can be declared using the `GENERIC` statement in a derived data type.

New operators may be defined and intrinsic operators may be extended to have new meanings in Fortran. A new operator may have a name consisting of up to 63 characters surrounded by periods. New operators and extended meanings of intrinsic operators are defined using an interface operator block. The first line of the interface operator block specifies the name of the operator to be defined or extended, and its body specifies the Fortran functions that are invoked to define the extended meaning. For binary operators, each function must have two input arguments; for unary operators, each function must have a single input argument. If several functions are present in the interface body, then they must be distinguishable from one another by the type and/or order of their dummy arguments. When the Fortran compiler encounters a new or extended operator, it uses the type and order of the operands to decide which of the functions to execute. This feature is commonly used to extend operators to support derived data types.

Generic bound operators can be declared using the GENERIC statement in a derived data type.

The assignment statement (=) may also be extended to work with derived data types. This extension is done using an interface assignment block. The body of the interface assignment block must refer to one or more subroutines. Each subroutine must have exactly two dummy arguments, with the first argument having INTENT(OUT) and the second argument having INTENT(IN). The first argument corresponds to the left-hand side of the equal sign, and the second argument corresponds to the right-hand side of the equal sign. All subroutines in the body of an interface assignment block must be distinguishable from one another by the type and order of their dummy arguments.

It is possible to control access to the data items, operators, and procedures in a module by using the PUBLIC, PRIVATE, and PROTECTED statements or attributes. If an entity in a module is declared PUBLIC, then it will be available to any program unit that accesses the module by USE association. If an entity is declared PRIVATE, then it will not be available to any program unit that accesses the module by USE association. However, it will remain available to any procedures defined within the module. If an entity is declared PROTECTED, then it will be read-only in any program unit that accesses the module by USE association.

The contents of a derived data type may be declared PRIVATE. If they are declared PRIVATE, then the components of the derived data type will not be separately accessible in any program unit that accesses the type by USE association. The data type as a whole will be available to the program unit, but its components will not be separately addressable. In addition, an entire derived data type may be declared PRIVATE. In that case, neither the data type nor its components are accessible.

The USE statement has two options. The statement may be used to rename specific data items or procedures accessed from a module, which can prevent name conflicts or provide simplified names for local use. Alternately, the ONLY clause may be used to restrict a program unit's access to only those items that appear in the list. Both options may be combined in a single USE statement.

Fortran includes intrinsic procedures to retrieve the command line arguments used to start a program and the values of environment variables. These new procedures replace nonstandard procedures that have varied from vendor to vendor. Use the new procedures instead of the nonstandard ones as soon as they become available to you.

13.14.1 Summary of Good Programming Practice

The following guidelines should be adhered to when working with the advanced features of procedures and modules:

1. When working with nested scoping units, avoid redefining the meaning of objects that have the same name in both the inner and outer scoping units. This applies especially to internal procedures. You can avoid confusion about the behavior of variables in the internal procedure by simply giving them different names from the variables in the host procedure.
2. Avoid interface blocks by placing your procedures in modules whenever possible.
3. If you must create interfaces to many procedures, place all of the interfaces in a module so that they will be easily accessible to program units by USE association.
4. Use user-defined generic procedures to define procedures that can function with different types of input data.
5. Use interface operator blocks and interface assignment blocks to create new operators and to extend the meanings of existing operators to work with derived data types. Once proper operators are defined, working with derived data types can be very easy.
6. It is good programming practice to hide any module data items or procedures that do not need to be directly accessed by external program units. This best way to do this is to include a PRIVATE statement in each module, and then list the specific items that you wish to expose in a separate PUBLIC statement.
7. Use the standard Fortran intrinsic procedures to retrieve the command line arguments used to start a program and the values of environment variables instead of the nonstandard procedures supplied by individual vendors.

13.14.2 Summary of Fortran Statements and Structures

BLOCK Construct:

```
                BLOCK
```

Example:

```
                [name:] BLOCK
                ...variable declarations
                   ...
                   Executable statements
                   ...
                   IF ( ) EXIT [name]
                   ...
                END BLOCK [name]
```

(continued)

(concluded)

Description:
The BLOCK construct is a block of code that can be located in any main program or procedure. It can define its own local variables, and also access the variables of the parent by host association. These variables become undefined when execution leaves the block.

CONTAINS Statement:

```
CONTAINS
```

Example:

```
PROGRAM main
   ...
CONTAINS
   SUBROUTINE sub1(x, y)
      ...
   END SUBROUTINE sub1
END PROGRAM
```

Description:
The CONTAINS statement is a statement that specifies that the following statements are one or more separate procedures within the host unit. When used within a module, the CONTAINS statement marks the beginning of one or more module procedures. When used within a main program or an external procedure, the CONTAINS statement marks the beginning of one or more internal procedures. The CONTAINS statement must appear after any type, interface, and data definitions within a module, and must follow the last executable statement within a main program or an external procedure.

GENERIC Statement:

```
TYPE [::] type_name
   component 1
   ...
   component n
CONTAINS
   GENERIC :: generic_name => proc_name1[, proc_name2, ...]
END TYPE [type_name]
```

Example:

```
TYPE :: point
   REAL :: x
   REAL :: y
CONTAINS
   GENERIC :: add => point_plus_point, point_plus_scalar
END TYPE point
```

Description:
The GENERIC statement defines a generic binding to a derived data type. The specific procedures associated with the generic procedure are listed after the => operator.

Advanced Features of Procedures and Modules

Generic Interface Block:

```
INTERFACE generic_name
    interface_body_1
    interface_body_2
    ...
END INTERFACE
```

Examples:

```
INTERFACE sort
    MODULE PROCEDURE sorti
    MODULE PROCEDURE sortr
END INTERFACE
```

Description:
A generic procedure is declared using a generic interface block. A generic interface block declares the name of the generic procedure on the first line, and then lists the explicit interfaces of the specific procedures associated with the generic procedure in the interface body. The explicit interface must be fully defined for any specific procedures not appearing in a module. Procedures appearing in a module are referred to with a MODULE PROCEDURE statement, since their interfaces are already known.

IMPORT Statement:

```
IMPORT :: var_name1 [, var_name2, ...]
```

Example:

```
IMPORT :: x, y
```

Description:
The IMPORT statement imports type definitions into an interface definition from the encompassing procedure.

Interface Assignment Block:

```
INTERFACE Assignment (=)
    interface_body
END INTERFACE
```

Example:

```
INTERFACE ASSIGNMENT (=)
    MODULE PROCEDURE vector_to_array
    MODULE PROCEDURE array_to_vector
END INTERFACE
```

Description:
An interface assignment block is used to extend the meaning of the assignment statement to support assignment operations between two different derived data types or between derived data types and intrinsic data types. Each procedure in the interface body must be a subroutine with two arguments. The first argument must have INTENT(OUT) and the second one must have INTENT(IN). All subroutines in the interface body must be distinguishable from each other by the order and type of their arguments.

Interface Block:

```
            INTERFACE
                interface_body_1
                ...
            END INTERFACE
```

Examples:

```
            INTERFACE
                SUBROUTINE sort(array,n)
                INTEGER, INTENT(IN) :: n
                REAL, INTENT(INOUT), DIMENSION(n) :: array
                END SUBROUTINE
            END INTERFACE
```

Description:
An interface block is used to declare an explicit interface for a separately-compiled procedure. It may appear in the header of a procedure that wishes to invoke the separately-compiled procedure, or it may appear in a module, and the module may be used by the procedure that wishes to invoke the separately-compiled procedure.

Interface Operator Block:

```
            INTERFACE OPERATOR (operator_symbol)
                interface_body
            END INTERFACE
```

Example:

```
            INTERFACE OPERATOR (*)
                MODULE PROCEDURE real_times_vector
                MODULE PROCEDURE vector_times_real
            END INTERFACE
```

Description:
An interface operator block is used to define a new operator, or to extend the meaning of an intrinsic operator to support derived data types. Each procedure in the interface must be a function whose arguments are INTENT(IN). If the operator is a binary operator, then the function must have two arguments. If the operator is a unary operator, then the function must have only one argument. All functions in the interface body must be distinguishable from each other by the order and type of their arguments.

MODULE PROCEDURE Statement:

         MODULE PROCEDURE module_procedure_1 [, module_procedure_2, ...]

Examples:

         INTERFACE sort
             MODULE PROCEDURE sorti
             MODULE PROCEDURE sortr
         END INTERFACE

Description:
The MODULE PROCEDURE statement is used in interface blocks to specify that a procedure contained in a module is to be associated with the generic procedure, operator, or assignment defined by the interface.

PROTECTED Attribute:

         type, PROTECTED :: name1[, name2, ...]

Examples:

         INTEGER,PROTECTED :: i_count
         REAL,PROTECTED :: result

Description:
The PROTECTED attribute declares that the value of a variable is "read-only" outside of the module in which it is declared. The value may be used but not modified in any procedure that accesses the defining module by USE access.

PROTECTED Statement:

         PROTECTED :: name1[, name2, ...]

Examples:

         PROTECTED :: i_count

Description:
The PROTECTED statement declares that the value of a variable is "read-only" outside of the module in which it is declared. The value may be used but not modified in any procedure that accesses the defining module by USE access.

Recursive FUNCTION Statement:

        RECURSIVE [type] FUNCTION name(arg1[, arg2, ...]) RESULT (res)

Example:

        RECURSIVE FUNCTION fact(n) RESULT (answer)
        INTEGER :: answer

Description:
This statement declares a recursive Fortran function. A recursive function is one that can invoke itself. The type of the function may either be declared in the FUNCTION statement or in a separate type declaration statement. (The type of the result variable *res* is declared, not the type of the function name.) The value returned by the function call is the value assigned to *res* within the body of the function.

USE Statement:

        USE module_name (, rename_list, ONLY: only_list)

Examples:

        USE my_procs
        USE my_procs, process_vector_input => input
        USE my_procs, ONLY: input => process_vector_input

Description:
The USE statement makes the contents of the named module available to the program unit in which the statement appears. In addition to its basic function, the USE statement permits module objects to be renamed as they are made available. The ONLY clause permits the programmer to specify that only certain objects from the module will be made available to the program unit.

VOLATILE Attribute:

        type, VOLATILE :: name1[, name2, ...]

Examples:

        INTEGER,VOLATILE :: i_count
        REAL,VOLATILE :: result

Description:
The VOLATILE attribute declares that the value of a variable might be changed at any time by some source external to the program, so all reads of the value in the variable must come directly from main memory, and all writes to the variable must go directly to main memory, not to a cached copy.

VOLATILE Statement:

> VOLATILE :: name1[, name2, ...]

Examples:

> VOLATILE :: x, y

Description:
The VOLATILE statement declares that the value of a variable might be changed at any time by some source external to the program, so all reads of the value in the variable must come directly from main memory, and all writes to the variable must go directly to main memory, not to a cached copy.

13.14.3 Exercises

13-1. In Example 12-1, the logical function lt_city failed to sort "APO" and "Anywhere" in proper order because all capital letters appear before all lowercase letters in the ASCII collating sequence. Add an internal procedure to function lt_city to avoid this problem by shifting both city names to uppercase before the comparison. Note that this procedure should *not* shift the names in the database to uppercase. It should only shift the names to uppercase temporarily as they are being used for the comparison.

13-2. Write test driver programs for the recursive subroutine factorial and the recursive function fact that were introduced in Section 13.3. Test both procedures by calculating 5! and 10! with each one.

13-3. Write a test driver program to verify the proper operation of subroutine extremes in Example 13-2.

13-4. What is printed out when the following code is executed? What are the values of x, y, i, and j at each point in the program? If a value changes during the course of execution, explain why it changes.

```
PROGRAM exercise13_4
IMPLICIT NONE
REAL :: x = 12., y = -3., result
INTEGER :: i = 6, j = 4
WRITE (*,100) 'Before call: x, y, i, j = ', x, y, i, j
100 FORMAT (A,2F6.1,2I6)
result = exec(y,i)
WRITE (*,*) 'The result is ', result
WRITE (*,100) 'After call:  x, y, i, j = ', x, y, i, j
CONTAINS
   REAL FUNCTION exec(x,i)
   REAL, INTENT(IN) :: x
   INTEGER, INTENT(IN) :: i
   WRITE (*,100) ' In exec:    x, y, i, j = ', x, y, i, j
   100 FORMAT (A,2F6.1,2I6)
   exec = ( x + y ) / REAL ( i + j )
```

```
      j = i
   END FUNCTION exec
END PROGRAM exercise13_4
```

13-5. Is the following program correct or not? If it is correct, what is printed out when it executes? If not, what is wrong with it?

```
PROGRAM exercise13_5
IMPLICIT NONE
REAL :: a = 3, b = 4, output
INTEGER :: i = 0
call sub1(a, i, output)
WRITE (*,*) 'The output is ', output

CONTAINS
   SUBROUTINE sub1(x, j, junk)
   REAL, INTENT(IN) :: x
   INTEGER, INTENT(IN) :: j
   REAL, INTENT(OUT) :: junk
   junk = (x - j) / b
   END SUBROUTINE sub1
END PROGRAM exercise13_5
```

13-6. What are the four levels of scope in Fortran? Give examples of objects of each type.

13-7. What are scoping units in Fortran? Name the different types of scoping units.

13-8. What is a keyword argument? Under what circumstances can keyword arguments be used?

13-9. Assuming the subroutine definition shown below, are the following calls legal or illegal? Assume that all calling arguments are of type real, and assume that the subroutine interface is explicit. Explain why each illegal call is illegal.

```
SUBROUTINE my_sub (a, b, c, d, e )
REAL, INTENT(IN) :: a, d
REAL, INTENT(OUT) :: b
REAL, INTENT(IN), OPTIONAL :: c, e
IF ( PRESENT(c) ) THEN
   b = (a - c) / d
ELSE
   b = a / d
END IF
IF ( PRESENT(e) ) b = b - e
END SUBROUTINE my_sub
```

(a) CALL my_sub (1., x, y, 2., z)
(b) CALL my_sub (10., 21., x, y, z)
(c) CALL my_sub (x, y, 25.)
(d) CALL my_sub (p, q, d=r)
(e) CALL my_sub (a=p, q, d=r, e=s)
(f) CALL my_sub (b=q, a=p, c=t, d=r, e=s)

13-10. What is an interface block? When would interface blocks be needed in a Fortran program?

Advanced Features of Procedures and Modules

13-11. In Example 9-1, we created a subroutine `simul` to solve a system of N simultaneous equations in N unknowns. Assuming that the subroutine is independently compiled, it will not have an explicit interface. Write an interface block to define an explicit interface for this subroutine.

13-12. What is a generic procedure? How can a generic procedure be defined?

13-13. How are generic procedures defined for bound procedures?

13-14. In Example 9-4, we created an improved version of the single-precision subroutine `simul2` to solve a system of N simultaneous equations in N unknowns. In Example 11-2, we created a double-precision subroutine `dsimul` to solve a double-precision system of N simultaneous equations in N unknowns. In Exercise 11-9, we created a complex subroutine `csimul` to solve a complex system of N simultaneous equations in N unknowns. Write a generic interface block for these three procedures.

13-15. Are the following generic interface blocks legal or illegal? Why?

(a)
```
INTERFACE my_procedure
    SUBROUTINE proc_1 (a, b, c)
    REAL, INTENT(IN) ::a
    REAL, INTENT(IN) ::b
    REAL, INTENT(OUT) ::c
    END SUBROUTINE proc_1
    SUBROUTINE proc_2 (x, y, out1, out2)
    REAL, INTENT(IN) ::x
    REAL, INTENT(IN) ::y
    REAL, INTENT(OUT) ::out1
    REAL, INTENT(OUT), OPTIONAL ::out2
    END SUBROUTINE proc_2
END INTERFACE my_procedure
```

(b)
```
INTERFACE my_procedure
    SUBROUTINE proc_1 (a, b, c)
    REAL, INTENT(IN) ::a
    REAL, INTENT(IN) ::b
    REAL, INTENT(OUT) ::c
    END SUBROUTINE proc_1
    SUBROUTINE proc_2 (x, y, z)
    INTEGER, INTENT(IN) ::x
    INTEGER, INTENT(IN) ::y
    INTEGER, INTENT(OUT) :: z
    END SUBROUTINE proc_2
END INTERFACE my_procedure
```

13-16. How can a new Fortran operator be defined? What rules apply to the procedures in the body of an interface operator block?

13-17. How can an intrinsic Fortran operator be extended to have new meanings? What special rules apply to procedures in an interface operator block if an intrinsic operator is being extended?

13-18. How can the assignment operator be extended? What rules apply to the procedures in the body of an interface assignment block?

13-19. Polar Complex Numbers A complex number may be represented in one of two ways: rectangular or polar. The rectangular representation takes the form $c = a + bi$, where a is the real component and b is the imaginary component of the complex number. The polar representation is of the form $z\angle\theta$, where z is the magnitude of the complex number, and θ is the angle of the number (Figure 13-19). The relationship between these two representations of complex numbers is:

$$a = z\cos\theta \qquad (11\text{-}13)$$

$$b = z\sin\theta \qquad (11\text{-}14)$$

$$z = \sqrt{a^2 + b^2} \qquad (11\text{-}15)$$

$$\theta = \tan^{-1}\frac{b}{a} \qquad (11\text{-}16)$$

The COMPLEX data type represents a complex number in rectangular form. Define a new data type called POLAR that represents a complex number in polar form. Then, write a module containing an interface assignment block and the supporting procedures to allow complex numbers to be assigned to polar numbers, and vice versa.

13-20. If two complex numbers $P_1 = z_1\angle\theta_1$ and $P_2 = z_2\angle\theta_2$ are expressed in polar form, then the product of the numbers is $P_1 \cdot P_2 = z_1 z_2 \angle \theta_1 + \theta_2$. Similarly P_1 divided by P_2 is $\dfrac{P_1}{P_2} = \dfrac{z_1}{z_2}\angle\theta_1 - \theta_2$. Extend the module created in Exercise 13-19 to add an interface operator block and the supporting procedures to allow two POLAR numbers to be multiplied and divided.

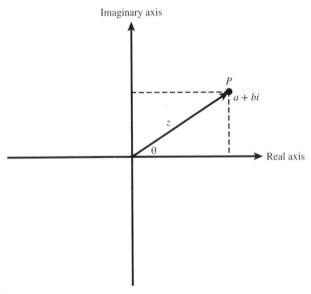

FIGURE 13-19
Representing a complex number in both rectangular and polar coordinates.

13-21. How can the access to data items and procedures in a module be controlled?

13-22. Are the following programs legal or illegal? Why?

(a)
```
MODULE my_module
IMPLICIT NONE
PRIVATE
REAL, PARAMETER :: PI = 3.141592
REAL, PARAMETER :: TWO_PI = 2 * PI
END MODULE my_module

PROGRAM test
USE my_module
IMPLICIT NONE
WRITE (*,*) 'Pi/2 =', PI / 2.
END PROGRAM test
```

(b)
```
MODULE my_module
IMPLICIT NONE
PUBLIC
REAL, PARAMETER :: PI = 3.141592
REAL, PARAMETER :: TWO_PI = 2 * PI
END MODULE my_module

PROGRAM test
USE my_module
IMPLICIT NONE
REAL :: TWO_PI
WRITE (*,*) 'Pi/2 =', PI / 2.
TWO_PI = 2. * PI
END PROGRAM test
```

13-23. Modify the module in Exercise 13-19 to only allow access to the definition of the POLAR type, the assignment operator, and to the multiplication and division operators. Restrict access to the functions that implement the operator definitions.

13-24. In each of the cases shown below, indicate which of the items defined in the module will be available in the program that accesses it.

(a)
```
MODULE module_1
  IMPLICIT NONE
  PRIVATE
  PUBLIC pi, two_pi, name
  REAL, PARAMETER :: PI = 3.141592
  REAL, PARAMETER :: TWO_PI = 2 * PI
  TYPE :: name
     CHARACTER(len=12) :: first
     CHARACTER :: mi
     CHARACTER(len=12) :: last
  END TYPE name
  TYPE (name), PUBLIC :: name1 = name("John","Q","Doe")
  TYPE (name) :: name2 = name("Jane","R","Public")
END MODULE module_1
```

```
        PROGRAM test
        USE module_1, sample_name => name1
        ...
        END PROGRAM test
(b) MODULE module_2
    IMPLICIT NONE
    REAL, PARAMETER :: PI = 3.141592
    REAL, PARAMETER :: TWO_PI = 2 * PI
    TYPE, PRIVATE :: name
       CHARACTER(len=12) :: first
       CHARACTER :: mi
       CHARACTER(len=12) :: last
    END TYPE name
    TYPE (name), PRIVATE :: name1 = name("John","Q","Doe")
    TYPE (name), PRIVATE :: name2 = name("Jane","R","Public")
    END MODULE module_2

    PROGRAM test
    USE module_2, ONLY: PI
    ...
    END PROGRAM test
```

14

Advanced I/O Concepts

OBJECTIVES

- Learn about all types of format descriptors available in Fortran.
- Learn additional options available for the OPEN, CLOSE, READ, and WRITE statements.
- Understand how to maneuver through a file using the REWIND, BACKSPACE, and ENDFILE statements.
- Understand how to check on file parameters using the INQUIRE statement.
- Know how to flush the output data to be written to disk using the FLUSH statement.
- Understand the differences between formatted and unformatted files, and between sequential and random access files. Learn when you should use each type of file.
- Learn about asynchronous I/O.

Chapter 5 introduced the basics of Fortran input and output statements. We learned how to read data using the formatted READ statement, and to write data using the formatted WRITE statement. We also learned about the most common format descriptors: A, E, ES, F, I, L, T, X, and /. Finally, we learned how to open, close, read, write, and position sequential disk files.

This chapter deals with the more advanced features of the Fortran I/O system. It includes a description of the additional format descriptors not yet mentioned, and provides more details about the operation of list-directed I/O statements. Next, it provides more details about the proper use of the various Fortran I/O statements, and introduces namelist I/O. Finally, the chapter explains the difference between formatted and unformatted disk files, and between sequential access and direct access disk files. We will learn when and how to properly use each type of file.

14.1
ADDITIONAL FORMAT DESCRIPTORS

A complete list of all Fortran format descriptors is shown in Table 14-1. Twelve of the format descriptors describe input/output data types: E, ES, EN, F, and D for single- and double-precision real values; I for integer values; B, O, and Z for either integer or real values; L for logical values; A for character values; and finally G for any type of value. Finally, there is a DT format descriptor for specifying the output format of derived data types. Five of the format descriptors control the horizontal and vertical positions of data: X, /, T, TL, and TR. The ':' character controls the way that formats associated with WRITE statements are scanned after the last variable in the WRITE statement has been output. Six of the format descriptors control the rounding of floating-point data: RU, RD, RN, RZ, RC, and RP. Two of them control the type of separator used between the integer and fractional parts of a number: DC and DP. Finally, a number of undesirable and/or obsolete format descriptors are briefly mentioned. The undesirable and/or obsolete format descriptors appear in shaded background in Table 14-1.

We will now discuss those format descriptors not previously described.

TABLE 14-1
Complete list of Fortran format descriptors

FORMAT Descriptors		Usage
Real data I/O descriptors		
D$w.d$		Double-precision data in exponential notation
E$w.d$	E$w.d$Ee	Real data in exponential notation
EN$w.d$	EN$w.d$Ee	Real data in engineering notation
ES$w.d$	ES$w.d$Ee	Real data in scientific notation
F$w.d$		Real data in decimal notation
Integer data I/O descriptor		
Iw	I$w.m$	Integer data in decimal format
Real or Integer data I/O descriptors		
Bw	B$w.m$	Data in binary format
Ow	O$w.m$	Data in octal format
Zw	Z$w.m$	Data in hexadecimal format
Logical data I/O descriptor		
Lw		Logical data
Character data I/O descriptors		
A	Aw	Character data
'$x...x$' "$x...x$"	nH$x...x$	Character constants (the nH$x...x$ form is *deleted* as of Fortran 95)
Generalized I/O descriptor		
G$w.d$	G$w.d$Ee	Generalized edit descriptor for any type of data
G0		Generalized edit descriptor with adjustable width for any type of data

(continued)

Advanced I/O Concepts

(concluded)

Derived type I/O descriptor	
`DT 'string' (vals)`	Derived type edit descriptor
Rounding Descriptors	
RU	Specify rounding up values for all descriptors following this descriptor in the current I/O statement
RD	Specify rounding down values for all descriptors following this descriptor in the current I/O statement
RZ	Specify rounding toward zero for all descriptors following this descriptor in the current I/O statement
RN	Specify rounding to nearest values for all descriptors following this descriptor in the current I/O statement
RC	Specify compatible rounding for all descriptors following this descriptor in the current I/O statement
RP	Specify processor-dependent rounding for all descriptors following this descriptor in the current I/O statement
Decimal descriptors	
DC	Use a comma as the character that separates the parts of a decimal for all descriptors following this descriptor in the current I/O statement
DP	Use a point as the character that separates the parts of a decimal for all descriptors following this descriptor in the current I/O statement
Positioning descriptors	
nX	Horizontal spacing: skip n spaces
/	Vertical spacing: move down one line
Tc	TAB: move to column c of current line
TLn	TAB: move left n columns in current line
TRn	TAB: move right n columns in current line
Scanning control descriptor	
:	Format scanning control character
Miscellaneous descriptors (undesirable)	
kP	Scale factor for display of real data
BN	Blank Null: ignore blanks in numeric input fields
BZ	Blank Zero: interpret blanks in a numeric input field as zeros
S	Sign control: Use default system convention
SP	Sign control: Display "+" before positive numbers
SS	Sign control: Suppress "+" before pos numbers

Where:
c column number
d number of digits to right of decimal place
e number of digits in exponent
k scale factor (number of places to shift decimal point)
m minimum number of digits to be displayed
r repetition count
w field width in characters

14.1.1 Additional Forms of the E and ES Format Descriptors

The E, ES, and F format descriptors were described in Chapter 5. In addition to the information presented there, there are optional forms of the E and ES descriptors that allow a programmer to specify the number of digits to display in the exponent of the real number. These forms are

$$rEw.dEe \quad \text{or} \quad rESw.dEe$$

where w, d, e, and r have the meanings given in Table 14-1. They function exactly as described in Chapter 5 except that the number of digits in the exponent is specified.

14.1.2 Engineering Notation—The EN Descriptor

Engineering notation is a modified version of scientific notation in which a real number is expressed as a value between 1.0 and 1000.0 times a power of 10, where the power of 10 is always a multiple of three. This form of notation is very convenient in the engineering world, because 10^{-6}, 10^{-3}, 10^3, 10^6, etc., all have standard, universally recognized prefixes. For example, 10^{-6} is known by the prefix *micro*, 10^{-3} is known by the prefix *milli*, and so forth. Engineers will commonly speak of 250 KΩ resistors and 50 nF capacitors instead of 2.5×10^5 Ω resistors and 5×10^{-8} F capacitors.

Fortran can print out numbers in engineering notation with the EN descriptor. When writing data, the EN descriptor displays a floating-point number with a mantissa in the range between 1 and 1000, while the exponent is always a power of 10 divisible by 3. The EN format descriptor has the form

$$rENw.d \quad \text{or} \quad rENw.dEe$$

where w, d, e, and r have the meanings given in Table 14-1.

For example, the following statements

```
a = 1.2346E7; b = 0.0001; c = -77.7E10
WRITE (*,'(3EN15.4)') a, b, c
```

will produce the output

```
     12.3460E+06     100.000E-06    -777.0000E+09
----|----|----|----|----|----|----|----|----|
    5   10   15   20   25   30   35   40   45
```

Note that all of the exponents are powers of 3. When reading data, the EN descriptor behaves exactly like the E, ES, and F descriptors.

14.1.3 Double-Precision Data—The D Descriptor

There is an obsolete format descriptor for use with double-precision data: the D format descriptor. The D format descriptor has the form

$$rDw.d$$

It is functionally identical to the E format descriptor, except that the exponent indicator is sometimes a D instead of an E. This descriptor is only preserved for backward compatibility with earlier versions of Fortran. *You should never use the D format descriptor in any new program.*

Advanced I/O Concepts

14.1.4 The Generalized (G) Format Descriptor

The F format descriptor is used to display real values in a fixed format. For example, the descriptor F7.3 will display a real value in the format ddd.ddd for positive numbers, or -dd.ddd for negative numbers. The F descriptor produces output data in a very easy-to-read format. Unfortunately, if the number to be displayed with an F7.3 descriptor is ≥ 1000 or ≤ -100, then the output data will be replaced by a field of asterisks: *******. In contrast, the E format descriptor will display a number regardless of its range. However, numbers displayed in the E format are not as easy to interpret as numbers displayed in the F format. Although the following two numbers are identical, the one displayed in the F format is easier to understand:

$$225.671 \qquad 0.225671E+03$$

Because the F format is easier to read, it would be really nice to have a format descriptor that displays numbers in the F format whenever possible, but then switches to the E format when they become too big or too small. The G (generalized) format descriptor behaves in just this fashion when used with real data.

The G format descriptor has the form

$$rGw.d \quad \text{or} \quad rGw.dEe$$

where w, d, e, and r have the meanings given in Table 14-1. A real value displayed with a G format descriptor will either be displayed in F or E format, depending on the exponent of the number. If the real value to be displayed is represented as $\pm 0.dddddd \times 10^k$ and the format descriptor to be used for the display is Gw.d, then the relationship between d and k will determine how the data is to be displayed. If $0 \leq k \leq d$, the value will be output in F format with a field width of $w - 4$ characters followed by four blanks. *The decimal point will be adjusted* (within the $w - 4$ characters) *as necessary to display as many significant digits as possible*. If the exponent is negative or is greater than d, the value will be output in E format. In either case, a total of d significant digits will be displayed.

The operation of the G format descriptor with real data is illustrated below. In the first example, k is -1, so the output comes out in E format. For the last example, k is 6 and d is 5, so the output again comes out in E format. For all of the examples in between, $0 \leq k \leq d$, so the output comes out in F format with the decimal point adjusted to display as many significant digits as possible.

Value	Exponent	G Descriptor	Output
0.012345	-1	G11.5	0.12345E-01
0.123450	0	G11.5	0.12345bbbb
1.234500	1	G11.5	1.23450bbbb
12.34500	2	G11.5	12.3450bbbb
123.4500	3	G11.5	123.450bbbb
1234.5600	4	G11.5	1234.50bbbb
12345.600	5	G11.5	12345.0bbbb
123456.00	6	G11.5	0.12345E+06

The generalized format descriptor can also be used with integer, logical, and character data. When it is used with integer data, it behaves like the I format descriptor. When it is used with logical data, it behaves like the L format descriptor. When it is used with character data, it behaves like the A format descriptor.

14.1.5 The G0 Format Descriptor

The G0 format descriptor is a generalized version of the G format descriptor that automatically adjusts its field to fit the type of data being displayed. For integer data, it will behave like an I0 descriptor. For logical data, it behaves like an L1 descriptor. For real data, it behaves like an $rESw.dEe$ descriptor. For character data, it behaves like an A descriptor.

14.1.6 The Binary, Octal, and Hexadecimal (B, O, and Z) Descriptors

The binary (B), octal (O), and hexadecimal (Z) descriptors can be used to read or write data in binary, octal, or hexadecimal formats. They work for both integer and real data. The general forms of these descriptors are

rBw or $rBw.m$
rOw or $rOw.m$
rZw or $rZw.m$

where w, m, and r have the meanings given in Table 14-1. The format descriptors must be large enough to display all of the digits in the appropriate notation, or the field will be filled with asterisks. For example, the statements

```
a = 16
b = -1
WRITE (*,'(A,B16,1X,B16)')       'Binary: ', a, b
WRITE (*,'(A,O11.4,1X,O11.4)') 'Octal:  ', a, b
WRITE (*,'(A,Z8,1X,Z8)')         'Hex:    ', a, b
```

will produce the output

```
Binary:              10000 ****************
Octal:            0020 37777777777
Hex:              10 FFFFFFFF
----|----|----|----|----|----|----|----|----|
    5   10   15   20   25   30   35   40   45
```

Since numbers are stored in two's complement format on this computer, a -1 will be 32 bits set to one. Therefore, the binary representation of b will consist of 32 ones. Since the B16 field is too small to display this number, it is filled with asterisks.

14.1.7 The TAB Descriptors

There are three TAB format descriptors: Tc, TLn, and TRn. We met the Tc descriptor in Chapter 5. In a formatted WRITE statement, it makes the output of the following descriptor begin at column c in the output buffer. In a formatted READ statement, it makes the field of the following descriptor begin at column c in the input buffer. For example, the following code will print the letter 'Z' in column 30 of the output line (remember that column 1 is used for carriage control and is not printed).

```
WRITE (*,'(T30,A)') 'Z'
```

The Tc descriptor performs an *absolute* tab function, in the sense that the output moves to column c regardless of where the previous output was. By contrast, the TLn and TRn descriptors are *relative* tab functions. TLn moves the output left by n columns, and TRn moves the output right by n columns. Where the next output will occur depends on the location of the previous output on the line. For example, the following code prints a 100 in columns 10–12 and a 200 in columns 17–19:

```
WRITE (*,'(T10,I3,TR4,I3)') 100, 200
```

14.1.8 The Colon (:) Descriptor

We have learned that if a WRITE statement runs out of variables before the end of its corresponding format, the use of the format continues until the first format descriptor without a corresponding variable, or until the end of the format, whichever comes first. For example, consider the statements

```
m = 1
voltage = 13800.
WRITE (*,40) m
40 FORMAT ('M = ', I3, ' N = ', I4, ' O = ', F7.2)
WRITE (*,50) voltage / 1000.
50 FORMAT ('Voltage = ', F8.1, ' kV')
```

These statements will produce the output

```
      M =    1 N =
Voltage =     13.8 kV
----|----|----|----|----|
    5   10   15   20   25
```

The use of the first FORMAT statement stops at I4, which is the first unmatched format descriptor. The use of the second FORMAT statement stops at the end of the statement, since there are no unmatched descriptors before that.

The colon descriptor (:) permits a user to modify the normal behavior of format descriptors during writes. The colon descriptor serves as a *conditional stopping point* for the WRITE statement. If there are more values to print out, the colon is ignored and the execution of the formatted WRITE statement continues according to the normal rules for using formats. However, if a colon is encountered in the format and there are no more values to write out, execution of the WRITE statement stops at the colon.

To help understand the use of the colon, let's examine the simple program shown in Figure 14-1.

FIGURE 14-1
Program illustrating the use of the colon format descriptor.

```
PROGRAM test_colon
IMPLICIT NONE
REAL, DIMENSION(8) :: x
INTEGER :: i
x = [ 1.1, 2.2, 3.3, 4.4, 5.5, 6.6, 7.7, 8.8 ]

WRITE (*,100) (i, x(i), i = 1, 8)
100 FORMAT (/'The output values are: '/, &
        3(5X,'X(',I2,') = ',F10.4))

WRITE (*,200) (i, x(i), i = 1, 8)
200 FORMAT (/'The output values are: '/, &
        3(:,5X,'X(',I2,') = ',F10.4))

END PROGRAM test_colon
```

This program contains an 8-element array whose values we wish to print out three-abreast across the page. Note that the portion of the format descriptors inside the parentheses has a repeat count of three, so each line will contain three values printed in identical format before the program advances to the next line. If the program is compiled and executed, the result is

```
C:\book\fortran\chap14>test
The output values are:
     X( 1) =       1.1000    X( 2) =     2.2000    X( 3) =     3.3000
     X( 4) =       4.4000    X( 5) =     5.5000    X( 6) =     6.6000
     X( 7) =       7.7000    X( 8) =     8.8000    X(

The output values are:
     X( 1) =       1.1000    X( 2) =     2.2000    X( 3) =     3.3000
     X( 4) =       4.4000    X( 5) =     5.5000    X( 6) =     6.6000
     X( 7) =       7.7000    X( 8) =     8.8000
```

The first WRITE statement and FORMAT statement run out of values to output after x(8) is written, but since it is in the middle of a format, the WRITE continues to execute until it comes to the first output descriptor without a corresponding variable. As a result, an extra 'X(' is printed out. The second WRITE statement and FORMAT are identical to the first pair, except that there is a colon at the beginning of the repeated portion of the FORMAT statement. This pair also runs out of values to output after x(8) is written. Since it is in the middle of a format, the WRITE continues to execute, but immediately bumps into the colon and stops. In this case, the extra 'X(' is not printed out.

The colon descriptor is most commonly used to terminate output cleanly in the middle of a line, as it was in the example above.

14.1.9 Scale Factors—The P Descriptor

The P descriptor adds a scale factor to any real values printed out with the E and F format descriptors. A scale factor has the form

$$nP$$

where n is the number of places by which to shift the decimal point. The P scale factor may precede either E or F format descriptors. The general forms of the descriptors with a scale factor are

$$nPrFw.d \quad \text{and} \quad nPrEw.d$$

With the F format descriptor, the P scale factor causes the displayed number to be multiplied by 10^n. With the E format descriptor, the P scale factor causes the fractional part of the displayed number to be multiplied by 10^n, and the exponent to be decreased by n.

The P scale factor has been made redundant by the introduction of the ES and EN format descriptors in Fortran 90. *It should never be used in any new program.*

14.1.10 The SIGN Descriptors

The SIGN format descriptors control the display of positive signs before positive numbers in an output line. There are three SIGN format descriptors: S, SP, and SS. The SP descriptor causes positive signs to be displayed before all positive numerical values following it in the same format statement, while the SS descriptor suppresses positive signs before all positive numerical values following it in the same format statement. The S descriptor restores the system default behavior for all positive numerical values following it. These format descriptors are almost never needed, and so are little used.

14.1.11 Blank Interpretation: The BN and BZ Descriptors

The BN (blank null) and BZ (blank zero) descriptors control the way in which blanks are interpreted in input data fields. If the BN descriptor is in effect, then blanks are ignored. If the BZ descriptor is in effect, then blanks are treated as zeros. In either case, if an entire input data field is blank, then the field is interpreted as 0. *The BN and BZ descriptors are never needed in any modern program.* They are only present for backward compatibility with the I/O behavior of FORTRAN 66.

14.1.12 Rounding Control: The RU, RD, RZ, RN, RC, and RP Descriptors

The RU (round up), RD (round down), RZ (round toward zero), RN (round nearest), RC (round compatible), and RP (round processor defined) descriptors control the way that data is rounded as it is read in or written out. Values such as 0.1 have no exact representation in the binary floating-point arithmetic used on IEEE 754 processors, so a number such as this must be *rounded* as it is saved into memory. Similarly, the binary representation of numbers inside the computer will not exactly match the decimal data written out in formatted files, so rounding must occur on output too. These descriptors control how the rounding works for a given input statement or output statement.

The RU descriptor specifies that all numeric values following it in the same READ or WRITE statement will be rounded up (toward positive infinity) during the conversion process. The RD descriptor specifies that all numeric values following it in the same READ or WRITE statement will be rounded down (toward negative infinity) during the conversion process. The RZ descriptor specifies that all numeric values following it in the same READ or WRITE statement will be rounded toward zero during the conversion process. The RN descriptor specifies that all numeric values following it in the same READ or WRITE statement will be rounded to the nearest representable value during the conversion process. If two representable values are equally distant, then the direction of rounding is not defined. The RC descriptor specifies that all numeric values following it in the same READ or WRITE statement will be rounded to the nearest representable value during the conversion process. If two representable values are equally distant, then the direction of rounding is away from zero. The RP descriptor specifies that all floating-point values following it in the same WRITE statement will be rounded in a processor-dependent manner.

14.1.13 Decimal Specifier: The DC and DP Descriptors

The DC (decimal comma) and DP (decimal point) descriptors control the character used to divide the integer part of an expression from the fractional part. If the DC descriptor is used, then all floating-point values following it in the same READ or WRITE statement will use a comma as the separator. If the DP descriptor is used, then all floating-point values following it in the same READ or WRITE statement will use a decimal point as the separator. Note that the default separator behavior for a given file is set by the DECIMAL= clause in the OPEN statement. The DC and DP descriptors are only used if we wish to temporarily override the choice made when the file was opened.

14.2
DEFAULTING VALUES IN LIST-DIRECTED INPUT

List-directed input has the advantage of being very simple to use, since no FORMAT statements need be written for it. A list-directed READ statement is very useful for getting input information from a user at a keyboard. The user may type the input data in any column, and the READ statement will still interpret it properly.

In addition, list-directed READ statements support *null values*. If an input data line contains two consecutive commas, then the corresponding variable in the input list will be left unchanged. This behavior permits a user to default one or more input data values to their previously defined values. Consider the following example

```
PROGRAM test_read
INTEGER :: i = 1, j = 2, k = 3
WRITE (*,*) 'Enter i, j, and k: '
READ (*,*) i, j, k
WRITE (*,*) 'i, j, k = ', i, j, k
END PROGRAM test_read
```

When this program is compiled and executed, the results are

```
C:\book\fortran\chap14>test_read
Enter i, j, and k:
1000,,-2002
i, j, k =         1000         2        -2002
```

Note that the value of j was defaulted to 2, while new values were assigned to i and k. It is also possible to default all of the remaining variables on a line by concluding it with a slash.

```
C:\book\fortran\chap14>test_read
Enter i, j, and k:
1000 /
i, j, k =         1000         2         3
```

Quiz 14-1

This quiz provides a quick check to see if you have understood the concepts introduced in Sections 14.1 and 14.2. If you have trouble with the quiz, reread the sections, ask your instructor, or discuss the material with a fellow student. The answers to this quiz are found in the back of the book.

For questions 1 to 4, determine what will be written out when the statements are executed.

1.  ```
    REAL :: a = 4096.07
    WRITE (*,1) a, a, a, a, a
    1 FORMAT (F10.1, F9.2, E12.5, G12.5, G11.4)
    ```

2.  ```
    INTEGER :: i
    REAL, DIMENSION(5) :: data1 = [ -17.2,4.,4.,.3,-2.22 ]
    WRITE (*,1) (i, data1(i), i=1, 5)
    1 FORMAT (2(5X,'Data1(',I3,') = ',F8.4,:,',',))
    ```

3.  ```
    REAL :: x = 0.0000122, y = 123456.E2
    WRITE (*,'(2EN14.6,/,1X,2ES14.6)') x, y, x, y
    ```

4.  ```
    INTEGER :: i = -2002, j = 1776, k = -3
    WRITE (*,*) 'Enter i, j, and k: '
    ```

(continued)

> (*concluded*)
> ```
>       READ (*,*) i, j, k
>       WRITE (*,1) i, j, k
>     1 FORMAT ('i = ',I10,' j = ',I10,' k = ',I10)
> ```
> where the input line is
> ```
>     ,         -1001/
>   ---------|---------|
>       10        20
> ```

14.3
DETAILED DESCRIPTION OF FORTRAN I/O STATEMENTS

A summary of Fortran I/O statements is shown in Table 14-2. These statements permit us to open and close files, check the status of files, go to a specific position within a file, and to read from or write to a file. In this section, we will learn about all of the statements found in the table. Some of them were introduced in simplified form in Chapter 5, but even the statements that we are already familiar with have many additional options to learn about.

The discussion of each i/o statement includes a table listing all of the possible clauses that can be used with the statement. Those clauses that should not be used in modern Fortran programs are shown with a shaded background.

14.3.1 The OPEN Statement

A disk file must be connected to an i/o unit before data can be read from or written to the file. Depending on the particular implementation of your compiler, a few files may

TABLE 14-2
Fortran I/O statements

Statement	Function
OPEN	Open a file (connect it to an i/o unit)
CLOSE	Close a file (disconnect it from an i/o unit)
INQUIRE	Check on properties of a file
READ	Read data from a file (via an i/o unit)
PRINT	Write data to the standard output device
WRITE	Write data to a file (via an i/o unit)
REWIND	Rewind a sequential file to the beginning
BACKSPACE	Move back one record in a sequential file
ENDFILE	Move to the end of a sequential file
FLUSH	Flush output buffers to disk
WAIT	Wait for asynchronous I/O to complete

be pre-connected to some of the i/o units when execution begins. If pre-connected files exist, it is possible to write data to them without opening them first. For example, Intel Visual Fortran automatically pre-connects a file called 'fort.21' to i/o unit 21, and so forth. The pre-connected file is automatically created the first time that a Fortran program writes to it.

Unfortunately, the number and the names of pre-connected files (if any) differ from processor to processor, so if you use this feature in your programs, they will be much less portable. You should always explicitly open any file that you use to improve the portability of your programs, and to allow you to choose your own name for each file.

Good Programming Practice
Do not rely on pre-connected files in your Fortran programs (except for the standard input and output devices). The number and the names of pre-connected files vary from processor to processor, so using them will reduce the portability of your programs.

An i/o unit is explicitly connected to a disk file using the OPEN statement. Once we are through using a file, the file should be disconnected from the i/o unit using the CLOSE statement. After the CLOSE statement has been executed, the i/o unit will no longer be connected to the file, and it may be connected to some other file using another OPEN statement.

The OPEN statement has the general form

OPEN (*open_list*)

where *open_list* consists of two or more clauses separated by commas. The possible clauses in an OPEN statement are summarized in Table 14-3. These clauses may be included in the OPEN statement in any order. Not all of the clauses will be included in every statement. Some of them are only meaningful for specific types of files. For example, the RECL= clause is only meaningful for direct access files. Also, some combinations of clauses have contradictory meanings, and will produce errors at compile time. We will point out some examples of these contradictions as we discuss the details of the clauses below.

The UNIT= clause
This clause specifies the number to be associated with the file. *Either the* UNIT= *clause or the* NEWUNIT= *clause must be present in any* OPEN *statement*. The i/o unit number specified here will be used in later READ and WRITE statements to access the file. The UNIT=io_unit clause may be abbreviated to just the io_unit number if it appears as the first clause in an OPEN statement. This feature is included in Fortran for backward compatibility with earlier versions of Fortran. Therefore, the following two statements are equivalent:

```
OPEN ( UNIT=10, ... )
OPEN ( 10, ... )
```

TABLE 14-3
Clauses allowed in the `OPEN` statement

Clause	Input or output	Purpose	Possible values
`[UNIT=]`*int_expr*	Input	Denotes i/o unit to attach file to. The "`UNIT=`" phrase is optional.	Processor-dependent integer.
`FILE=`*char_expr*	Input	Name of file to open.[1]	Character string.
`STATUS=`*char_expr*	Input	Specifies status for file to be opened.	`'OLD'`, `'NEW'`, `'SCRATCH'`, `'REPLACE'`, `'UNKNOWN'`
`NEWUNIT=`*int_var*	Output	Automatically select an i/o unit that does not clash with any existing open ones, and return the unit number used.	Processor-dependent integer *int_var* containing the i/o unit number.
`IOSTAT=`*int_var*	Output	I/O status at end of operation.	Processor-dependent integer *int_var*. 0 = success; positive = open failure.
`IOMSG=`*char_var*	Output	Character string describing any error that occurred during operation.	Character string.
`ACCESS=`*char_expr*	Input	Specified sequential, direct, or stream access.	`'SEQUENTIAL'`, `'DIRECT'`, `'STREAM'`
`ASYNCHRONOUS=`*char_expr*	Input	Specifies whether or not to use asynchronous I/O.[2]	`'YES'`, `'NO'`
`DECIMAL=`*char_expr*	Input	Specifies the separator to use between the integer and fractional parts of a number. (Default `'POINT'`)	`'COMMA'`, `'POINT'`
`ENCODING=`*char_expr*	Input	Specifies the type of character data to read/write from a file. `'UTF-8'` specifies a Unicode file.[3]	`'UTF-8'`, `'DEFAULT'`
`ROUND=`*char_expr*	Input	Specifies the type of rounding to perform during formatted I/O operations. (Default `'PROCESSOR DEFINED'`)	`'UP'`, `'DOWN'`, `'ZERO'`, `'NEAREST'`, `'COMPATIBLE'`, `'PROCESSOR DEFINED'`
`SIGN=`*char_expr*	Input	Specifies whether to display plus signs on positive output values during formatted write operations.	`'PLUS'`, `'SUPPRESS'`, `'PROCESSOR DEFINED'`
`FORM=`*char_expr*	Input	Specified formatted or unformatted data.	`'FORMATTED'`, `'UNFORMATTED'`
`ACTION=`*char_expr*	Input	Specifies whether file is read only, write only, or read/write.	`'READ'`, `'WRITE'`, `'READWRITE'`
`RECL=`*int_expr*	Input	For a formatted direct access file, the number of characters in each record. For an unformatted direct access file, the number of processor-dependent units in each record.[4]	Processor-dependent positive integer.
`POSITION=`*char_expr*	Input	Specifies the position of the file pointer after the file is opened.	`'REWIND'`, `'APPEND'`, `'ASIS'`
`DELIM=`*char_expr*	Input	Specifies whether list-directed character output is to be delimited by apostrophes, by quotation marks, or by nothing.[5] (Default `'NONE'`)	`'APOSTROPHE'`, `'QUOTE'`, `'NONE'`

(continued)

Advanced I/O Concepts

(*concluded*)

PAD=*variable*	Input	Specifies whether formatted input records are padded with blanks. (Default 'YES')	'YES', 'NO'
BLANK=*char_expr*	Input	Specifies whether blanks are to be treated as nulls or zeros. Nulls are the default case.[6]	'NULL', 'ZERO'
ERR=*label*	Input	Statement label to transfer control to if open fails.[7]	Statement labels in current scoping unit.

[1] The FILE= clause is not allowed for scratch files.
[2] The ASYNCHRONOUS= clause allows asynchronous I/O statements for this file. The default value is 'NO'.
[3] The ENCODING= clause is only defined for files connected for formatted I/O. The default value is 'DEFAULT', which is processor dependent, but normally 1-byte characters.
[4] The RECL= clause is only defined for files connected for direct access.
[5] The DELIM= clause is only defined for files connected for formatted I/O.
[6] The BLANK= clause is only defined for files connected for formatted I/O. This clause is never needed in a modern Fortran program.
[7] The ERR= clause is never needed in a modern Fortran program. Use the IOSTAT= and IOMSG= clauses instead.

The NEWUNIT= clause

This clause specifies that the file should be opened on an i/o unit that does not conflict with any other units currently in use. If the UNIT= clause is specified, Fortran will select an unused unit number, open the file on that unit, and return the unit number *in an output variable*. A typical usage is:

```
INTEGER :: lu
OPEN ( NEWUNIT=lu, ... )
```

After the file is opened, variable lu will contain the unit to use for reading and writing.

The FILE= clause

This clause specifies the name of the file to connect to the specified i/o unit. *A file name must be supplied for all files* except for scratch files.

The STATUS= clause

This clause specifies the status of the file to connect to the specified i/o unit. There are five possible file statuses: 'OLD', 'NEW', 'REPLACE', 'SCRATCH', and 'UNKNOWN'.

If the file status is 'OLD', then the file must already exist on the system when the OPEN statement is executed, or the OPEN will fail with an error. If the file status is 'NEW', then the file must *not* already exist on the system when the OPEN statement is executed, or the OPEN will fail with an error. If the file status is STATUS='REPLACE', then a new file will be opened whether it exists or not. If the file already exists, the program will delete it, create a new file, and then open it for output. The old contents of the file will be lost. If it does not exist, the program will create a new file by that name and open it.

If the file status is `'SCRATCH'`, then a **scratch file** will be created on the computer and attached to the i/o unit. A scratch file is a temporary file that is created by the computer that the program can use for temporary data storage while it is running. When a scratch file is closed or when the program ends, the file is automatically deleted from the system. Note that *the* `FILE=` *clause is not used with a scratch file*, since no permanent file is created. It is an error to specify a file name for a scratch file.

If the file status is `'UNKNOWN'`, then the behavior of the program will vary from processor to processor—the Fortran standard does not specify the behavior of this option. The most common behavior is for the program to first look for an existing file with the specified name, and open it if it exists. The contents of the file are not destroyed by the act of opening it with unknown status, but the original contents can be destroyed if we later write to the file. If the file does not exist, then the computer creates a new file with that name and opens it. Unknown status should be avoided in a program because the behavior of the `OPEN` statement is processor dependent, which could reduce the portability of the program.

If there is no `STATUS=` clause in an `OPEN` statement, then the default status is `'UNKNOWN'`.

The `IOSTAT=` clause

This clause specifies an integer variable that will contain the i/o status after the `OPEN` statement is executed. If the file is opened successfully, then the status variable will contain a zero. If the open failed, then the status variable will contain a processor-dependent positive value corresponding to the type of error that occurred.

The `IOMSG=` clause

This clause specifies a character variable that will contain the i/o status after the `OPEN` statement is executed. If the file is opened successfully, then the contents of this variable will be unchanged. If the open failed, then this variable will contain a message describing the problem that occurred.

The `ACCESS=` clause

This clause specifies the access method to be used with the file. There are three types of access methods: `'SEQUENTIAL'`, `'DIRECT'`, and `'STREAM'`. **Sequential access** involves opening a file and reading or writing its records in order from beginning to end. Sequential access is the default access mode in Fortran, and all files that we have seen so far have been sequential files. The records in a file opened with sequential access do not have to be of any particular length.

If a file is opened with **direct access,** it is possible to jump directly from one record to another within the file at any time without having to read any of the records in between. Every record in a file opened with direct access must be of the same length.

If a file is opened with **stream access,** data is written to the file or read from the file in "file storage units" (normally bytes). This mode differs from sequential access in that sequential access is record oriented, with end-of-record (newline) characters automatically inserted at the end of each record. In contrast, stream access just writes

The ASYNCHRONOUS= clause

This clause specifies whether or not asynchronous I/O is possible to or from this file. The default is `'NO'`.

The DECIMAL= clause

This clause specifies whether the separator between the integer and fraction values in a real number is a decimal point or a comma. The default is a **decimal** point.

The value in this clause can be overridden for a particular READ or WRITE statement by the DC and DP format descriptors

The ENCODING= clause

This clause specifies whether the character encoding in this file is standard ASCII or Unicode. If this value is `'UTF-8'`, then the character encoding is 2-byte Unicode. If this value is `'DEFAULT'`, then the character encoding is processor dependent, which for practical purposes means that it will be 1-byte ASCII characters.

The ROUND= clause

This clause specifies how rounding occurs when data is written to or read from formatted files. The options are `'UP'`, `'DOWN'`, `'ZERO'`, `'NEAREST'`, `'COMPATIBLE'`, `'PROCESSOR DEFINED'`. Values such as 0.1 have no exact representation in the binary floating-point arithmetic used on IEEE 754 processors, so a number such as this must be *rounded* as it is saved into memory. Similarly, the binary representation of numbers inside the computer will not exactly match the decimal data written out in formatted files, so rounding must occur on output too. This clause controls how the rounding works for a given file.

The `'UP'` option specifies that all numeric values will be rounded up (toward positive infinity) during the conversion process. The `'DOWN'` option specifies that all numeric values will be rounded down (toward negative infinity) during the conversion process. The `'ZERO'` option specifies that all numeric values will be rounded toward zero during the conversion process. The `'NEAREST'` option specifies that all numeric values will be rounded to the nearest representable value during the conversion process. If two representable values are equally distant, then the direction of rounding is not defined. The `'COMPATIBLE'` option is the same as the `'NEAREST'` option, except that if two representable values are equally distant, then the direction of rounding is away from zero. The PROCESSOR DEFINED specifies that all floating-point values will be rounded in a processor-dependent manner.

The value in this clause can be overridden for a particular READ or WRITE statement by the RU, RD, RZ, RN, RC, and RP format descriptors

The SIGN= clause

This clause controls the display of positive signs before positive numbers in an output line. The options are `'PLUS'`, `'SUPPRESS'`, and `'PROCESSOR DEFINED'`. The `'PLUS'` option causes positive signs to be displayed before all positive numerical values, while the `'SUPPRESS'` option suppresses positive signs before all positive numerical values. The `'PROCESSOR DEFINED'` option allows the computer to use the system default behavior for all positive numerical values. This is the default behavior.

The value in this clause can be overridden for a particular READ or WRITE statement by the S, SP, and SS format descriptors

The FORM= clause

This clause specifies the format status of the file. There are two file formats: `'FORMATTED'` and `'UNFORMATTED'`. The data in **formatted files** consists of recognizable characters, numbers, etc. These files are called formatted because we use format descriptors (or list-directed I/O statements) to convert their data into a form usable by the computer whenever we read or write them. When we write to a formatted file, the bit patterns stored in the computer's memory are translated into a series of characters that humans can read, and those characters are written to the file. The instructions for the translation process are included in the format descriptors. All of the disk files that we have used so far have been formatted files.

In contrast, **unformatted files** contain data that is an exact copy of the data stored in the computer's memory. When we write to an unformatted file, the exact bit patterns in the computer's memory are copied into the file. Unformatted files are much smaller than the corresponding formatted files, but the information in an unformatted file is coded in bit patterns that cannot be easily examined or used by people. Furthermore, the bit patterns corresponding to particular values vary among different types of computer systems, so unformatted files cannot easily be moved from one type of computer to another one.

If a file uses sequential access, the default file format is `'FORMATTED'`. If the file uses direct access, the default file format is `'UNFORMATTED'`.

The ACTION= clause

This clause specifies whether a file is to be opened for reading only, for writing only, or for both reading and writing. Possible values are `'READ'`, `'WRITE'`, or `'READWRITE'`. The default action is `'READWRITE'`.

The RECL= clause

This clause specifies the length of each record in a direct access file. For formatted files opened with direct access, this clause contains the length of each record in characters. For unformatted files, this clause contains the length of each record in processor-dependent units.

The POSITION= clause

This clause specifies the position of the file pointer after the file is opened. The possible values are `'REWIND'`, `'APPEND'`, or `'ASIS'`. If the expression is `'REWIND'`, then the file pointer points to the first record in the file. If the expression is `'APPEND'`, then the file pointer points just after the last record in the file and just before the end-of-file marker. If the expression is `'ASIS'`, then the position of the file pointer is unspecified and processor dependent. The default position is `'ASIS'`.

The DELIM= clause

This clause specifies which characters are to be used to delimit character strings in list-directed output and namelist output statements. The possible values are 'QUOTE', 'APOSTROPHE', or 'NONE'. If the expression is 'QUOTE', then the character strings will be delimited by quotation marks, and any quotation marks in the string will be doubled. If the expression is 'APOSTROPHE', then the character strings will be delimited by apostrophes, and any apostrophes in the string will be doubled. If the expression is 'NONE', then the character strings have no delimiters.

The PAD= clause

This clause has the possible values 'YES' or 'NO'. If this clause is 'YES', then the processor will pad out input data lines with blanks as required to match the length of the record specified in a READ format descriptor. If it is 'NO', then the input data line must be at least as long as the record specified in the format descriptor, or an error will occur. The default value is 'YES'.

The BLANK= clause

This clause specifies whether blank columns in numeric fields are to be treated as blanks or zeros. The possible values are 'ZERO' or 'NULL'. It is the equivalent of the BN and BZ format descriptors, except that the value specified here applies to the entire file. This clause provides backward compatibility with FORTRAN 66; it should never be needed in any new Fortran program.

The ERR= clause

This clause specifies the label of a statement to jump to if the file open fails. The ERR= clause provides a way to add special code to handle file open errors. (This clause should not be used in new programs; use the IOSTAT= and IOMSG= clauses instead.)

The importance of using the IOSTAT= and IOMSG= clauses

If a file open fails and there is no IOSTAT= clause or ERR= clause in the OPEN statement, then the Fortran program will print out an error message and abort. This behavior is very inconvenient in a large program that runs for a long period of time, since large amounts of work can be lost if the program aborts. It is *much* better to trap such errors, and let the user tell the program what to do about the problem. The user could specify a new disk file, or he or she could let the program shut down gracefully saving all the work done so far.

If either the IOSTAT= clause or ERR= clause is present in the OPEN statement, then the Fortran program will not abort when an open error occurs. If an error occurs and the IOSTAT= clause is present, then a positive i/o status will be returned specifying the type of error that occurred. If the IOMSG= clause is also present, then a user-readable character string describing the problem is also returned. The program can check for an error, and provide the user with options for continuing or shutting down gracefully. For example,

```
OPEN ( UNIT=8, FILE='test.dat', STATUS='OLD', IOSTAT=istat, IOMSG=msg)

! Check for OPEN error
in_ok: IF ( istat /= 0 ) THEN
    WRITE (*,*) 'Input file OPEN failed: istat = ', istat
    WRITE (*,*) 'Error message = ', msg
    WRITE (*,*) 'Shutting down...'
    ...
ELSE
    normal processing
    ...
END IF in_ok
```

In general, the IOSTAT= clause should be used instead of the ERR= clause in all new programs, since the IOSTAT= clause allows more flexibility and is better suited to modern structured programming. The use of the ERR= clause encourages "spaghetti code", in which execution jumps around in a fashion that is hard to follow and hard to maintain.

Good Programming Practice
Always use the IOSTAT= clause in OPEN statements to trap file open errors. When an error is detected, tell the user all about the problem before shutting down gracefully or requesting an alternate file.

Examples

Some example OPEN statements are shown below:

1. OPEN (UNIT=9, FILE='x.dat', STATUS='OLD', POSITION='APPEND', &
   ACTION='WRITE')

   This statement opens a file named x.dat and attaches it to i/o unit 9. The status of the file is 'OLD', so the file must already exist. The position is 'APPEND', so the file pointer will be positioned after the last record in the file, and just before the end-of-file marker. The file is a formatted file opened for sequential access, and is write-only. Since there is no IOSTAT= or ERR= clause, an open error would abort the program containing this statement.

2. OPEN (22, STATUS='SCRATCH')

   This statement creates a scratch file and attaches it to i/o unit 22. The scratch file is automatically given some unique name by the system, and is automatically deleted when the file is closed or the program ends. It is a formatted file and is opened for sequential access. Since there is no IOSTAT= or ERR= clause, an open error would abort the program containing this statement.

3. OPEN (FILE='input',UNIT=lu,STATUS='OLD',ACTION='READ',IOSTAT=istat)

   This statement opens an existing file named input, and attaches it to the i/o unit corresponding to the value of variable lu. The status of the file is 'OLD', so this

OPEN statement will fail if the file does not already exist. The file is a formatted file opened for sequential access, and is opened for reading only. A status code is returned in variable istat. It will be 0 for a successful file open, and positive for an unsuccessful file open. Since the IOSTAT= clause is present in this statement, an open error would not abort the program containing this statement.

4. OPEN (FILE='input',NEWUNIT=lu,ACTION='READ',IOSTAT=istat,IOMSG=msg)

This statement opens an existing file named input and attaches it to a program-defined i/o unit, and returns the corresponding value in variable lu. The status of the file is defaulted to 'UNKNOWN', so the behavior of the program is processor dependent. The file is a formatted file opened for sequential access, and is opened for reading only. A status code is returned in variable istat. It will be 0 for a successful file open, and positive for an unsuccessful file open. Since the IOSTAT= clause is present in this statement, an open error would not abort the program containing this statement. If an error does occur, a descriptive error message will be returned in the character variable msg.

14.3.2 The CLOSE Statement

Once a file is no longer needed, it should be disconnected from its i/o unit using the CLOSE statement. After the CLOSE statement has been executed, the i/o unit will no longer be connected to the file, and it may be connected to some other file using another OPEN statement.

A Fortran program will automatically update and close any open files whenever the program ends. Therefore, a CLOSE statement is not actually required unless we want to attach more than one file to the same i/o unit. However, it is good practice to close any file with a CLOSE statement just as soon as the program is finished using it. When a file has been opened by one program, no other program may have access to it at the same time. By closing the file as soon as possible, the file is made available for other programs to use. This is especially important for files that are shared by many people.

Good Programming Practice
Always explicitly close each disk file with a CLOSE statement as soon as possible after a program is finished using it, so that it may be available for use by others.

The CLOSE statement has the general form

    CLOSE (close_list)

where *close_list* consists of one or more clauses separated by commas. The possible clauses in the CLOSE statement are summarized in Table 14-4. They may be included in the CLOSE statement in any order.

TABLE 14-4
Clauses allowed in the `CLOSE` statement

Clause	Input or output	Purpose	Possible values
[UNIT=]*int_expr*	Input	I/o unit to close. The "UNIT=" phrase is optional.	Processor-dependent integer.
STATUS=*char_expr*	Input	Specifies whether file is to be kept or deleted after closing.	'KEEP', 'DELETE'
IOSTAT=*int_var*	Output	I/O status at end of operation.	Processor-dependent integer *int_var*. 0 = success; positive = close failure.
IOMSG=*char_var*	Output	Character string describing any error that occurred during operation.	Character string.
ERR=*label*	Input	Statement label to transfer control to if open fails.[1]	Statement labels in current scoping unit.

[1] The ERR= clause is never needed in a modern Fortran program. Use the IOSTAT= and IOMSG= clauses instead.

The UNIT= clause

This clause is exactly the same as the UNIT= clause in the OPEN statement. *The UNIT= clause must be present in any CLOSE statement.*

The STATUS= clause

This clause specifies the status of the file connected to the specified i/o unit. There are two possible file status: 'KEEP' and 'DELETE'. If the file status is 'KEEP', then the file is kept on the file system after it is closed. If the file status is 'DELETE', then the file is deleted after it is closed. A scratch file is always deleted when it is closed; it is not legal to specify keep status for a scratch file. For any other type of file, the default status is 'KEEP'.

The IOSTAT= clause

This clause specifies an integer variable that will contain the i/o status after the CLOSE statement is executed. If the file is closed successfully, then the status variable will contain a zero. If the close failed, then the status variable will contain a processor-dependent positive value corresponding to the type of error that occurred.

The IOMSG= clause

This clause specifies a character variable that will contain the i/o status after the CLOSE statement is executed. If the file is closed successfully, then the contents of this variable will be unchanged. If the close failed, then this variable will contain a message describing the problem that occurred.

The ERR= clause

This clause specifies the label of a statement to jump to if the file close fails. The ERR= clause provides a way to add special code to handle file close errors. (This clause should not be used in new programs; use the IOSTAT= clause instead.)

Advanced I/O Concepts

Examples

Some example `CLOSE` statements are shown below:

1. `CLOSE ( 9 )`
   This statement closes the file attached to i/o unit 9. If the file is a scratch file, it will be deleted; otherwise, it will be kept. Since there is no `IOSTAT=` or `ERR=` clause, an error would abort the program containing this statement.
2. `CLOSE ( UNIT=22, STATUS='DELETE', IOSTAT=istat, IOMSG=err_str )`
   This statement closes and deletes the file attached to i/o unit 22. An operation status code is returned in variable `istat`. It will be 0 for success, and positive for failure. Since the `IOSTAT=` clause is present in this statement, a close error will not abort the program containing this statement. If an error occurs, character variable `err_str` will contain a descriptive error message.

14.3.3 The INQUIRE Statement

It is often necessary to check on the status or properties of a file that we want to use in a Fortran program. The `INQUIRE` statement is used for this purpose. It is designed to provide detailed information about a file, either before or after the file has been opened.

There are three different versions of the `INQUIRE` statement. The first two versions of the statement are similar, except for the manner in which the file is looked up. The file can be found by either specifying the `FILE=` clause or the `UNIT=` clause (but not both simultaneously!). If a file has not yet been opened, it must be identified by name. If the file is already open, it may be identified by either name or i/o unit. There are many possible output clauses in the `INQUIRE` statement. To find out a particular piece of information about a file, just include the appropriate clause in the statement. A complete list of all clauses is given in Table 14-5.

TABLE 14-5
Clauses allowed in the INQUIRE statement

Clause	Input or output	Purpose	Possible values
[UNIT=]*int_expr*	Input	I/o unit of file to check.[1]	Processor-dependent integer.
FILE=*char_expr*	Input	Name of file to check.[1]	Processor-dependent character string.
IOSTAT=*int_var*	Output	I/O status.	Returns 0 for success; processor-dependent positive number for failure.
IOMSG=*char_var*	Output	I/O error message.	If a failure occurs, this variable will contain a descriptive error message
EXIST=*log_var*	Output	Does the file exist?	.TRUE., .FALSE.
OPENED=*log_var*	Output	Is the file opened?	.TRUE., .FALSE.

(continued)

(continued)

NUMBER=*int_var*	Output	I/o unit number of file, if opened. If file is not opened, this value is undefined.	Processor-dependent positive number.
NAMED=*log_var*	Output	Does the file have a name? (Scratch files are unnamed.)	.TRUE., .FALSE.
NAME=*char_var*	Output	Name of file if file is named; undefined otherwise.	File name
ACCESS=*char_var*	Output	Specifies type of access if the file is currently open.[2]	'SEQUENTIAL', 'DIRECT', 'STREAM'
SEQUENTIAL=*char_var*	Output	Specifies if file *can be opened* for sequential access.[2]	'YES', 'NO', 'UNKNOWN'
DIRECT=*char_var*	Output	Specifies if file *can be opened* for direct access.[2]	'YES', 'NO', 'UNKNOWN'
STREAM=*char_var*	Output	Specifies if file *can be opened* for stream access.[2]	'YES', 'NO', 'UNKNOWN'
FORM=*char_var*	Output	Specifies type of formatting for a file if the file is open.[3]	'FORMATTED', 'UNFORMATTED'
FORMATTED=*char_var*	Output	Specifies if file *can be* connected for formatted I/O.[3]	'YES', 'NO', 'UNKNOWN'
UNFORMATTED=*char_var*	Output	Specifies if file *can be* connected for unformatted I/O.[3]	'YES', 'NO', 'UNKNOWN'
RECL=*int_var*	Output	Specifies the record length of a direct access file; undefined for sequential files.	Record length is in processor-dependent units.
NEXTREC=*int_var*	Output	For a direct access file, one more than the number of the last record read from or written to the file; undefined for sequential files.	
BLANK=*char_var*	Output	Specifies whether blanks in numeric fields are treated as nulls or zeros.[4]	'ZERO', 'NULL'
POSITION=*char_var*	Output	Specifies location of file pointer when the file is first opened. This value is undefined for unopened files, or for files opened for direct access.	'REWIND', 'APPEND', 'ASIS', 'UNDEFINED'
ACTION=*char_var*	Output	Specifies read, write, or read-write status for opened files. This value is undefined for unopened files.[5]	'READ', 'WRITE', 'READWRITE', 'UNDEFINED'
READ=*char_var*	Output	Specifies whether file *can be* opened for read-only access.[5]	'YES', 'NO', 'UNKNOWN'
WRITE=*char_var*	Output	Specifies whether file *can be* opened for write-only access.[5]	'YES', 'NO', 'UNKNOWN'

(continued)

Advanced I/O Concepts

(concluded)

Clause	Type	Description	Values
READWRITE=*char_var*	Output	Specifies whether file *can be* opened for readwrite access.[5]	'YES', 'NO', 'UNKNOWN'
DELIM=*char_var*	Output	Specifies type of character delimiter used with list-directed and namelist I/O to this file.	'APOSTROPHE', 'QUOTE', 'NONE', 'UNKNOWN'
PAD=*char_var*	Output	Specifies whether or not input lines are to be padded with blanks. This value is always yes unless a file is explicitly opened with PAD='NO'.	'YES', 'NO'
IOLENGTH=*int_var*	Output	Returns the length of an unformatted record, in processor-dependent units. This clause is special to the third type of INQUIRE statement (see text).	
ASYNCHRONOUS=*char_var*	Output	Specifies whether or not asynchronous I/O is permitted for this file.	'YES', 'NO'
ENCODING=*char_var*	Output	Specifies type of character encoding for the file.[6]	'UTF-8', 'UNDEFINED', 'UNKNOWN'
ID=*int_expr*	Input	The ID number of a pending asynchronous data transfer. Results are returned in the ID= clause.	
PENDING=*log_var*	Output	Returns the status of the asynchronous I/O operation specified in the ID= clause.	.TRUE., .FALSE.
POS=*int_var*	Output	Returns the position in the file for the next read or write.	
ROUND=*char_var*	Output	Returns the type of rounding in use.	'UP', 'DOWN', 'ZERO', 'NEAREST', 'COMPATIBLE', 'PROCESSOR DEFINED'
SIGN=*char_var*	Output	Returns the option for printing + sign	'PLUS', 'SUPPRESS', 'PROCESSOR DEFINED'
ERR=*statement label*	Input	Statement to branch to if statement fails.[7]	Statement label in current program unit.

[1] One and only one of the FILE= and UNIT= clauses may be included in any INQUIRE statement.
[2] The difference between the ACCESS= clause and the SEQUENTIAL=, DIRECT=, and STREAM= clauses is that the ACCESS= clause tells what sort of access *is being used*, while the other three clauses tell what sort of access *can be used*.
[3] The difference between the FORM= clause and the FORMATTED= and UNFORMATTED= clauses is that the FORM= clause tells what sort of I/O *is being used*, while the other two clauses tell what sort of I/O *can be used*.
[4] The BLANK= clause is only defined for files connected for formatted I/O.
[5] The difference between the ACTION= clause and the READ=, WRITE=, and READWRITE= clauses is that the ACTION= clause specifies the action for which the file *is* opened, while the other clauses specify the action for which the file *can be* opened.
[6] The value 'UTF-8' is returned for Unicode files; the value 'UNDEFINED' is returned for unformatted files.
[7] The ERR= clause is never needed in a modern Fortran program. Use the IOSTAT= and IOMSG= clauses instead.

The third form of the `INQUIRE` statement is the inquire-by-output-list statement. This statement takes the form

```
INQUIRE (IOLENGTH=int_var) output_list
```

where *int_var* is an integer variable and *output_list* is a list of variables, constants, and expressions like the ones that would appear in a `WRITE` statement. The purpose of this statement is to return the length of the unformatted record that can contain the entities in the output list. As we will see later in this chapter, unformatted direct access files have a fixed record length that is measured in processor-dependent units, and so the length changes from processor to processor. Furthermore, this record length must be specified when the file is opened. This form of the `INQUIRE` statement provides us with a processor-independent way to specify the length of records in direct access files. An example of this form of `INQUIRE` statement will be shown when we introduce direct access files in Section 14.6.

EXAMPLE 14-1

Preventing Output Files from Overwriting Existing Data:

In many programs, the user is asked to specify an output file into which the results of the program will be written. It is good programming practice to check to see if the output file already exists before opening it and writing into it. If it already exists, the user should be asked if he or she *really* wants to destroy the data in the file before the program overwrites it. If so, the program can open the file and write into it. If not, the program should get a new output file name and try again. Write a program that demonstrates a technique for protection against overwriting existing files.

SOLUTION
The resulting Fortran program is shown in Figure 14-2.

FIGURE 14-2
Program illustrating how to prevent an output file from accidentally overwriting data.

```
PROGRAM open_file
!
!  Purpose:
!    To illustrate the process of checking before overwriting an
!    output file.
!
IMPLICIT NONE

! Data dictionary: declare variable types & definitions
INTEGER :: istat                    ! I/o status
LOGICAL :: lexist                   ! True if file exists
LOGICAL :: lopen = .FALSE.          ! True if file is open
CHARACTER(len=20) :: name           ! File name
CHARACTER :: yn                     ! Yes / No flag

! Do until file is open
openfile: DO
```

(continued)

Advanced I/O Concepts

(*concluded*)

```
      ! Get output file name.
      WRITE (*,*) 'Enter output file name: '
      READ (*,'(A)') file_name

      ! Does file already exist?
      INQUIRE ( FILE=file_name, EXIST=lexist )
      exists: IF ( .NOT. lexist ) THEN
         ! It's OK, the file didn't already exist. Open file.
         OPEN (UNIT=9, FILE=name, STATUS='NEW', ACTION='WRITE' ,IOSTAT=istat)
         lopen = .TRUE.
      ELSE
         ! File exists. Should we replace it?
         WRITE (*,*) 'Output file exists. Overwrite it? (Y/N) '
         READ (*,'(A)') yn
         CALL ucase ( yn )                ! Shift to upper case

         replace: IF ( yn == 'Y' ) THEN
            ! It's OK. Open file.
            OPEN (UNIT=9, FILE=name, STATUS='REPLACE', ACTION='WRITE',
                  IOSTAT=istat)
            lopen = .TRUE.
         END IF replace

      END IF exists
      IF ( lopen ) EXIT
END DO openfile

! Now write output data, and close and save file.
WRITE (9,*) 'This is the output file!'
CLOSE (9,STATUS='KEEP')

END PROGRAM open_file
```

Test this program for yourself. Can you suggest additional improvements to make this program work better? (*Hint*: What about the OPEN statements?)

Good Programming Practice

Check to see if your output file is overwriting an existing data file. If it is, make sure that the user really wants to do that before destroying the data in the file.

14.3.4 The READ Statement

The READ statement reads data from the file associated with a specified i/o unit, converts its format according to the specified FORMAT descriptors, and stores it into the variables in the I/O list. A READ statement keeps reading input lines until all of the variables in *io_list* have been filled, the end of the input file is reached, or an error occurs. A READ statement has the general form

    READ (*control_list*) *io_list*

where `control_list` consists of one or more clauses separated by commas. The possible clauses in a READ statement are summarized in Table 14-6. The clauses may be included in the READ statement in any order. Not all of the clauses will be included in any given READ statement.

TABLE 14-6
Clauses allowed in the READ statement

Clause	Input or output	Purpose	Possible Values
[UNIT=]*int_expr*	Input	I/o unit to read from.	Processor-dependent integer.
[FMT=]*statement_label* [FMT=]*char_expr* [FMT=]*	Input	Specifies the format to use when reading formatted data.	
IOSTAT=*int_var*	Output	I/O status at end of operation.	Processor-dependent integer *int_var*: 0 = success positive = READ failure −1 = End of file −2 = End of record
IOMSG=*char_var*	Output	I/O error message.	If a failure occurs, this variable will contain a descriptive error message.
REC=*int_expr*	Input	Specifies the record number to read in a direct access file.	
NML=*namelist*	Input	Specifies namelist of I/O entities to read.	Namelists defined in the current scoping unit, or accessed through use or host association.
ADVANCE=*char_expr*	Input	Specifies whether to perform advancing or nonadvancing I/O. Valid for sequential files only.	'YES', 'NO'
SIZE=*int_var*	Output	Specifies number of characters read during nonadvancing I/O. Valid for nonadvancing I/O only.	
EOR=*label*	Input	Statement label to transfer control to if end of record is reached during nonadvancing I/O. Valid for nonadvancing I/O only.	Statement labels in current scoping unit.
ASYNCHRONOUS=*char_expr*	Input	Specifies whether or not asynchronous I/O is used for this statement.[1] (Default = 'NO')	'YES', 'NO'
DECIMAL=*char_expr*	Input	Temporarily overrides the separator specification specified in the OPEN statement.	'COMMA', 'POINT'

(continued)

Advanced I/O Concepts

(concluded)

DELIM=*char_expr*	Input	Temporarily overrides the delimiter specification specified in the OPEN statement.	'APOSTROPHE', 'QUOTE', 'NONE'
ID=*int_var*	Output	Returns a unique ID associated with an asynchronous I/O transfer.[2]	
POS=*int_var*	Input	Specifies the read position in a file opened for STREAM access.[3]	
ROUND=*char_var*	Input	Temporarily overrides the rounding specification specified in the OPEN statement.	'UP', 'DOWN', 'ZERO', 'NEAREST', 'COMPATIBLE', 'PROCESSOR DEFINED'
SIGN=*char_var*	Input	Temporarily overrides the sign specification specified in the OPEN statement.	'PLUS', 'SUPPRESS', 'PROCESSOR DEFINED'
END=*statement_label*	Input	Statement label to transfer control to if end of file is reached.[4]	Statement labels in current scoping unit.
ERR=*statement_label*	Input	Statement label to transfer control to if an error occurs.[4]	Statement labels in current scoping unit.

[1] The ASYNCHRONOUS= clause can only be 'YES' if the file was opened to allow asynchronous I/O.
[2] The ID= clause can only be used if an asynchronous data transfer is specified.
[3] The POS= clause can only be used with a file opened for stream access.
[4] The END=, ERR= and EOR= clauses are never needed in a modern Fortran program. Use the IOSTAT= and IOMSG= clauses instead.

The UNIT= clause

This clause specifies the i/o unit number from which to read the data. An * indicates reading data from the standard input device. *The UNIT= clause must be present in any READ statement.*

The i/o unit may also be specified by just naming it in the READ statement without the UNIT= keyword. This feature is included in Fortran for backward compatibility with earlier versions of Fortran. If the i/o unit is specified in this alternate form, then it must be the first clause in the READ statement. The following two statements are equivalent:

```
READ ( UNIT=10, ... )
READ ( 10, ... )
```

The FMT= clause

This clause has the form

[FMT=] *statement_label*   or   [FMT=] *char_expr*   or   [FMT=] *

where *statement_label* is the label of a FORMAT statement, *char_expr* is a character string containing the format information, or * indicates list-directed I/O. An FMT= clause must be supplied for all formatted READ statements.

If the FMT= clause is *second* clause in a READ statement, and if the first clause is an abbreviated unit number without the UNIT= keyword, then the format clause

may be abbreviated by just naming the statement number, character variable, or *
containing the format. This feature is included in Fortran for backward compatibility with earlier versions of Fortran. Therefore, the following two statements are equivalent:

```
READ ( UNIT=10, FMT=100 ) data1
READ ( 10, 100 ) data1
```

The IOSTAT= clause

This clause specifies an integer variable that will contain the status after the READ statement is executed. If the read is successful, then the status variable will contain a zero. If an end-of-file condition is detected, then the status variable will contain a -1. If an end-of-record condition is encountered during nonadvancing i/o, the status variable will contain a -2. If the read fails, then the status variable will contain a positive value corresponding to the type of error that occurred.

The IOMSG= clause

This clause specifies a character variable that will contain the i/o status after the READ statement is executed. If the read is successful, then the contents of this variable will be unchanged. If the read failed, then this variable will contain a message describing the problem that occurred.

The REC= clause

This clause specifies the number of the record to read in a direct access file. It is only valid for direct access files.

The NML= clause

This clause specifies a named list of values to read in. The details of namelist I/O will be described in the Section 14.4.

The ADVANCE= clause

This clause specifies whether or not the current input buffer should be discarded at the end of the READ. The possible values are 'YES' or 'NO'. If the value is 'YES', then any remaining data in the current input buffer will be discarded when the READ statement is completed. If the value is 'NO', then the remaining data in the current input buffer will be saved and used to satisfy the next READ statement. The default value is 'YES'. This clause is only valid for sequential files.

The SIZE= clause

This clause specifies the name of an integer variable to contain the number of characters that have been read from the input buffer during a nonadvancing I/O operation. It may only be specified if the ADVANCE='NO' clause is specified.

The EOR= clause

This clause specifies the label of an executable statement to jump to if the end of the current record is detected during a nonadvancing READ operation. If the end of the

Advanced I/O Concepts

input record is reached during a nonadvancing I/O operation, then the program will jump to the statement specified and execute it. This clause may only be specified if the `ADVANCE='NO'` clause is specified. If the `ADVANCE='YES'` clause is specified, then the read will continue on successive input lines until all of the input data is read.

The `ASYNCHRONOUS=` clause

This clause specifies whether or not a particular read is to be asynchronous. This value can only be `'YES'` if the file was opened for asynchronous I/O.

The `DECIMAL=` clause

This clause temporarily overrides the specification of the decimal separator in the `OPEN` statement.

The value in this clause can be overridden for a particular `READ` or `WRITE` statement by the `DC` and `DP` format descriptors

The `DELIM=` clause

This clause temporarily overrides the specification of the delimiter in the `OPEN` statement.

The `ID=` clause

This clause returns a unique ID associated with an asynchronous I/O transfer. This ID can later be used in the `INQUIRE` statement to determine if the I/O transfer has completed.

The `POS=` clause

This clause specifies the position for the read from a stream file.

The `ROUND=` clause

This clause temporarily overrides the value of the `ROUND` clause specified in the `OPEN` statement. The value in this clause can be overridden for a particular value by the `RU`, `RD`, `RZ`, `RN`, `RC`, and `RP` format descriptors.

The `SIGN=` clause

This clause temporarily overrides the value of the `SIGN` clause specified in the `OPEN` statement. The value in this clause can be overridden for a particular value by the `S`, `SP`, and `SS` format descriptors.

The `END=` clause

This clause specifies the label of an executable statement to jump to if the end of the input file is detected. The `END=` clause provides a way to handle unexpected end-of-file conditions. This clause should not be used in modern programs; use the more general and flexible `IOSTAT=` clause instead.

The `ERR=` clause

This clause specifies the label of an executable statement to jump to if a read error occurs. The most common read error is a mismatch between the type of the input data in a field and the format descriptors used to read it. For example, if the characters 'A123' appeared by mistake in a field read with the I4 descriptor, an error would be generated. This clause should not be used in modern programs; use the more general and flexible IOSTAT= clause instead.

The importance of using IOSTAT= and IOMSG= clauses

If a read fails and there is no IOSTAT= clause or ERR= clause in the READ statement, the Fortran program will print out an error message and abort. If the end of the input file is reached and there is no IOSTAT= clause or END= clause, the Fortran program will abort. Finally, if the end of an input record is reached during nonadvancing i/o and there is no IOSTAT= clause or EOR= clause, the Fortran program will abort. If either the IOSTAT= clause or the ERR=, END=, and EOR= clauses are present in the READ statement, then the Fortran program will not abort when read errors, end-of-file, or end-of-record conditions occur. If the IOMSG= clause is also present, then a user-readable character string describing the problem is also returned. The programmer can do something to handle those conditions and allow the program to continue running.

The following code fragment shows how to use the IOSTAT= message to read an unknown number of input values without aborting when the end of the input file is reached. It uses a while loop to read data until the end of the input file is reached.

```
OPEN ( UNIT=8, FILE='test.dat', STATUS='OLD' )

! Read input data
nvals = 0
DO
    READ (8,100,IOSTAT=istat) temp
    ! Check for end of data
    IF ( istat < 0 ) EXIT
    nvals = nvals + 1
    array(nvals) = temp
END DO
```

The IOSTAT= clause should be used instead of the END=, ERR=, and EOR= clauses in all new programs, since the IOSTAT= clause allows more flexibility and is better suited to modern structured programming. The use of the other clauses encourages "spaghetti code", in which execution jumps around in a fashion that is hard to follow and hard to maintain.

Good Programming Practice

Use the IOSTAT= and IOMSG= clauses in READ statements to prevent programs from aborting on errors, end-of-file conditions, or end-of-record conditions. When one of these conditions is detected, the program can take appropriate actions to continue processing or to shut down gracefully.

14.3.5 Alternate Form of the READ Statement

There is an alternate form of the READ statement that only works for formatted reads or list-directed reads from the standard input device. This statement has the form

> READ fmt, io_list

where *fmt* is the format specification to use when reading the list of variables in the *io_list*. The format may be the number of a FORMAT statement, the name of a character variable containing the formatting information, a character string containing the formatting information, or an asterisk. Examples of this version of the read statement include:

> READ 100, x, y
> 100 FORMAT (2F10.2)
>
> READ '(2F10.2)', x, y

This version of the READ statement is much less flexible than the standard READ statement, since it can only work with the standard input device and cannot support any of the optional clauses. It is a holdover from an earlier version of FORTRAN. There is no need to ever use it in a modern program.

14.3.6 The WRITE Statement

The WRITE statement takes data from the variables in the I/O list, converts it according to the specified FORMAT descriptors, and writes it out to the file associated with the specified i/o unit. The WRITE statement has the general form

> WRITE (control_list) io_list

where *control_list* consists of one or more clauses separated by commas. The possible clauses in a WRITE statement are the same as those in the READ statement, except that there are no END=, SIZE=, or EOR= clauses.

14.3.7 The PRINT Statement

There is an alternate output statement called the PRINT statement that only works for formatted writes or list-directed writes to the standard output device. This statement has the form

> PRINT fmt, io_list

where *fmt* is the format specification to use when reading the list of variables in the *io_list*. The format may be the number of a FORMAT statement, the name of a character variable containing the formatting information, a character string containing the formatting information, or an asterisk. Examples of PRINT statement include:

> PRINT 100, x, y
> 100 FORMAT (2F10.2)

```
string = '(2F10.2)'
PRINT string, x, y
```

The `PRINT` statement is much less flexible than the standard `WRITE` statement, since it can only work with the standard output device and cannot support any of the optional clauses. It is a holdover from an earlier version of FORTRAN. There is no need to ever use it in a modern program. However, many Fortran programmers are stylistically committed to using this statement through long years of habit. It does work, and programs using the `PRINT` statement will continue to be supported indefinitely in the future. You should recognize the statement when you see it, but in the opinion of this author it is better not to use it in your own programs.

14.3.8 File Positioning Statements

There are two file positioning statements in Fortran: `REWIND` and `BACKSPACE`. The `REWIND` statement positions the file so that the next `READ` statement will read the first line in the file. The `BACKSPACE` statement moves the file back by one line. These statements are only valid for sequential files. The statements have the general form

```
REWIND (control_list)
BACKSPACE (control_list)
```

where `control_list` consists of one or more clauses separated by commas. The possible clauses in a file positioning statement are summarized in Table 14-7. The meanings of these clauses are the same as in the other I/O statements described above.

The i/o unit may be specified without the `UNIT=` keyword if it is in the first position of the control list. The following statements are examples of legal file positioning statements:

```
REWIND (unit_in)
BACKSPACE (UNIT=12, IOSTAT=istat)
```

TABLE 14-7
Clauses allowed in the REWIND, BACKSPACE, or ENDFILE statements

Clause	Input or output	Purpose	Possible values
[UNIT=]*int_expr*	Input	I/o unit to operate on. The "UNIT=" phrase is optional.	Processor-dependent integer.
IOSTAT=*int_var*	Output	I/O status at end of operation.	Processor-dependent integer *int_var*. 0 = success positive = failure
IOMSG=*char_var*	Output	Character string containing an error message if an error occurs.	
ERR=*statement_label*	Input	Statement label to transfer control to if an error occurs.[1]	Statement labels in current scoping unit.

[1] The ERR= clause is never needed in a modern Fortran program. Use the IOSTAT= and IOMSG= clauses instead.

Advanced I/O Concepts

For compatibility with earlier versions of FORTRAN, a file positioning statement containing only an i/o unit number can also be specified without parentheses:

```
REWIND 6
BACKSPACE unit_in
```

The `IOSTAT=` clause should be used instead of the `ERR=` clause in modern Fortran programs. It is better suited to modern structured programming techniques.

14.3.9 The ENDFILE Statement

The `ENDFILE` statement writes an end-of-file record at the current position in a sequential file, and then positions the file after the end-of-file record. After executing an `ENDFILE` statement on a file, no further `READ`s or `WRITE`s are possible until either a `BACKSPACE` or a `REWIND` statement is executed. Until then, any further `READ` or `WRITE` statements will produce an error. This statement has the general form

```
ENDFILE (control_list)
```

where *control_list* consists of one or more clauses separated by commas. The possible clauses in an `ENDFILE` statement are summarized in Table 14-7. The meanings of these clauses are the same as in the other I/O statements described above. The i/o unit may be specified without the `UNIT=` keyword if it is in the first position of the control list.

For compatibility with earlier version of Fortran, an `ENDFILE` statement containing only an i/o unit number can also be specified without parentheses. The following statements are examples of legal `ENDFILE` statements:

```
ENDFILE (UNIT=12, IOSTAT=istat)
ENDFILE 6
```

The `IOSTAT=` clause should be used instead of the `ERR=` clause in modern Fortran programs. It is better suited to modern structured programming techniques.

14.3.10 The WAIT Statement

When an asynchronous I/O transfer starts, execution returns to the program immediately before the I/O operation is completed. This allows the program to continue running in parallel with the I/O operation. It is possible that at some later point the program may need to guarantee that the operation is complete before progressing further. For example, the program may need to read back data that was being written during an asynchronous write.

If this is so, the program can use the `WAIT` statement to guarantee that the operation is complete before continuing. The form of this statement is

```
WAIT (unit)
```

where *unit* is the I/O unit to wait for. Control will only return from this statement when all pending I/O operations to that unit are complete.

14.3.11 The FLUSH Statement

The FLUSH statement causes all data being written to a file to be posted or otherwise available for use before the statement returns. It has the effect of forceably writing any data stored in temporary output buffers to disk. The form of this statement is

```
FLUSH (unit)
```

where *unit* is the I/O unit to flush. Control will only return from this statement when all data has been written to disk.

14.4
NAMELIST I/O

Namelist I/O is a convenient way to write out a fixed list of variable names and values, or to read in a fixed list of variable names and values. A **namelist** is just a list of variable names that are always read or written as a group. The form of a namelist is

```
NAMELIST / nl_group_name / var1 [, var2, ...]
```

where *nl_group_name* is the name of the namelist, and *var1*, *var2*, etc., are the variables in the list. The NAMELIST is a specification statement, and must appear before the first executable statement in a program. If there is more than one NAMELIST statement with the same name, then the variables in all statements are concatenated and treated as though they were in a single large statement. The variables listed in a NAMELIST may be read or written as a unit using namelist-directed I/O statements.

A NAMELIST I/O statement looks like a formatted I/O statement, except that the FMT= clause is replaced by an NML= clause. The form of a namelist-directed WRITE statement is

```
WRITE (UNIT=unit, NML=nl_group_name, [...])
```

where *unit* is the i/o unit to which the data will be written, and *nl_group_name* is the name of the namelist to be written. (Unlike most other clauses in I/O statements, the *nl_group_name* is not enclosed in apostrophes or quotes.) When a namelist-directed WRITE statement is executed, the names of all of the variables in the namelist are printed out together with their values in a special order. The first item to be printed is an ampersand (&) followed by the namelist name. Next comes a series of output values in the form "NAME=value". These output values may either appear on a single line separated by commas or appear on separate lines, depending on the way a particular processor implements the namelist. Finally, the list is terminated by a slash (/).

For example, consider the program shown in Figure 14-3.

FIGURE 14-3
A simple program using a NAMELIST-directed WRITE statement.

```
PROGRAM write_namelist
!  Purpose:
!    To illustrate a NAMELIST-directed WRITE statement.
!
IMPLICIT NONE

! Data dictionary: declare variable types & definitions
INTEGER :: i = 1, j = 2                      ! Integer variables
REAL :: a = -999., b = 0.                    ! Real variables
CHARACTER(len=12) :: string = 'Test string.' ! Char variables
NAMELIST / mylist / i, j, string, a, b       ! Declare namelist

OPEN (8,FILE='output.nml',DELIM='APOSTROPHE') ! Open output file
WRITE (UNIT=8, NML=mylist)                    ! Write namelist
CLOSE (8)                                     ! Close file

END PROGRAM write_namelist
```

After this program is executed, the file `output.nml` contains the lines:

```
&MYLIST
I =            1
J =            2
STRING = 'Test string.'
A =       -999.000000
B =       0.000000E+00
/
```

The namelist output begins with an ampersand and the list name, and concludes with a slash. Note that the character string is surrounded by apostrophes, because the file was opened with the clause `DELIM='APOSTROPHE'`.

The general form of a namelist-directed READ statement is

```
READ (UNIT=unit, NML=nl_group_name, [...])
```

where *unit* is the i/o unit from which the data will be read, and *nl_group_name* is the name of the namelist to be read. When a namelist-directed READ statement is executed, the program searches the input file for the marker &*nl_group_name*, which indicates the beginning of the namelist. It then reads all of the values in the namelist until a slash character (/) is encountered to terminate the READ. The values in the input list may appear on any line within the input file, as long as they are between the markers &*nl_group_name* and /. The values are assigned to the namelist variables according to the names given in the input list. The namelist READ statement does not have to set a value for every variable in the namelist. If some namelist variables are not included in the input file list, then their values will remain unchanged after the namelist READ executes.

Namelist-directed READ statements are *very* useful. Suppose that you are writing a program containing 100 input variables. The variables will be initialized to their usual values by default in the program. During any particular run of the program, anywhere from 1 to 10 of these values may need to be changed, but the others would remain at their default values. In this case, you could include all 100 values in a namelist and include a namelist-directed READ statement in the program. Whenever a user runs the program, he or she can just list the few values to be changed in the namelist input file, and all of the other input variables will remain unchanged. This approach is much better than using an ordinary READ statement, since all 100 values would need to be listed in the ordinary READ's input file, even if they were not being changed during a particular run.

Consider the example in Figure 14-4, which illustrates how a namelist READ can update selected values in the namelist.

FIGURE 14-4
A simple program using a NAMELIST-directed READ statement.

```
PROGRAM read_namelist
!  Purpose:
!    To illustrate a NAMELIST-directed READ statement.
!
IMPLICIT NONE
! Data dictionary: declare variable types & definitions
INTEGER :: i = 1, j = 2                          ! Integer variables
REAL :: a = -999., b = 0.                        ! Real variables
CHARACTER(len=12) :: string = 'Test string.'     ! Char variables
NAMELIST / mylist / i, j, string, a, b           ! Declare namelist

OPEN (7,FILE='input.nml',DELIM='APOSTROPHE')     ! Open input file.

! Write NAMELIST before update
WRITE (*,'(A)') 'Namelist file before update: '
WRITE (UNIT=*, NML=mylist)

READ (UNIT=7,NML=mylist)                         ! Read namelist file.

! Write NAMELIST after update
WRITE (*,'(A)') 'Namelist file after update: '
WRITE (UNIT=*, NML=mylist)

END PROGRAM read_namelist
```

If the file `input.nml` contains the following data:

```
&MYLIST
I =           -111
STRING = 'Test 1.'
STRING = 'Different!'
B =         123456.
/
```

then variable b will be assigned the value 123456., variable i will be assigned the value -111, and variable `string` will be assigned a value of `'Different!'`. Note that if more than one input value exists for the same variable, the last one in the namelist is the one that is used. The values of all variables other than b, i, and `string` will not be changed. The result of executing this program will be:

```
C:\book\fortran\chap14>namelist_read
Namelist file before update:
&MYLIST
I =           1
J =           2
```

Advanced I/O Concepts

```
STRING =  Test string.
A     =      -999.000000
B     =         0.000000E+00
/
Namelist file after update:
&MYLIST
I     =         -111
J     =            2
STRING =  Different!
A     =      -999.000000
B     =    123456.000000
/
```

If a namelist output file is opened with the character delimiter set to `'APOSTROPHE'` or `'QUOTE'`, then the output file written by a namelist WRITE statement is in a form that can be directly read by a namelist READ statement. This fact makes the namelist a great way to exchange a lot of data between separate programs or between different runs of the same program.

Good Programming Practice
Use NAMELIST I/O to save data to be exchanged between programs or between different runs of a single program. Also, you may use NAMELIST READ statements to update selected input parameters when a program begins executing.

Array names, array sections, and array elements may all appear in a NAMELIST statement. If an array name appears in a namelist, then when a namelist WRITE is executed, every element of the array is printed out in the output namelist one at a time, such as a(1) = 3., a(2) = -1., etc. When a namelist READ is executed, each element of the array may be set separately, and only the elements whose values are to be changed need to be supplied in the input file.

Dummy arguments and variables that are created dynamically may not appear in a NAMELIST. This includes array dummy arguments with nonconstant bounds, character variables with nonconstant lengths, automatic variables, and pointers.

14.5
UNFORMATTED FILES

All of the files that we have seen so far in this book have been **formatted files**. A formatted file contains recognizable characters, numbers, etc., stored in a standard coding scheme such as ASCII or Unicode. These files are easy to distinguish, because we can see the characters and numbers in the file when we display them on the screen or print them on a printer. However, to use data in a formatted file, a program must translate the characters in the file into the internal integer or real format used by the particular processor that the program is running on. The instructions for this translation are provided by format descriptors.

Formatted files have the advantage that we can readily see what sort of data they contain. However, they also have disadvantages. A processor must do a good deal of work to convert a number between the processor's internal representation and the characters contained in the file. All of this work is just wasted effort if we are going to be reading the data back into another program on the same processor. Also, the internal representation of a number usually requires much less space than the corresponding ASCII or Unicode representation of the number found in a formatted file. For example, the internal representation of a 32-bit real value requires 4 bytes of space. The ASCII representation of the same value would be $\pm.dddddddE\pm ee$, which requires 13 bytes of space (1 byte per character). Thus, storing data in ASCII or Unicode format is inefficient and wasteful of disk space.

Unformatted files overcome these disadvantages by copying the information from the processor's memory directly to the disk file with no conversions at all. Since no conversions occur, no processor time is wasted formatting the data. Furthermore, the data occupies a much smaller amount of disk space. On the other hand, unformatted data cannot be examined and interpreted directly by humans. In addition, it usually cannot be moved between different types of processors, because those types of processors have different internal ways to represent integers and real values.

Formatted and unformatted files are compared in Table 14-8. In general, formatted files are best for data that people must examine, or data that may have to be moved between different types of processors. Unformatted files are best for storing information that will not need to be examined by human beings, and that will be created and used on the same type of processor. Under those circumstances, unformatted files are both faster and occupy less disk space.

Unformatted I/O statements look just like formatted I/O statements, except that the `FMT=` clause is left out of the control list in the `READ` and `WRITE` statements. For example, the following two statements perform formatted and unformatted writes of array `arr`:

```
WRITE (UNIT=10,FMT=100,IOSTAT=istat) (arr(i), i = 1, 1000)
100 FORMAT ( 5E13.6 )

WRITE (UNIT=10,IOSTAT=istat) (arr(i), i = 1, 1000)
```

A file may be either `FORMATTED` or `UNFORMATTED`, but not both. Therefore, we cannot mix formatted and unformatted I/O statements within a single file. The `INQUIRE` statement can be used to determine the formatting status of a file.

Good Programming Practice
Use formatted files to create data that must be readable by humans, or that must be transferable between processors of different types. Use unformatted files to efficiently store large quantities of data that do not have to be directly examined, and that will remain on only one type of processor. Also, use unformatted files when I/O speed is critical.

TABLE 14-8
Comparison of formatted and unformatted files

Formatted files	Unformatted files
Can display data on output devices.	Cannot display data on output devices.
Can easily transport data between different computers.	Cannot easily transport data between computers with different internal data representations.
Requires a relatively large amount of disk space.	Requires relatively little disk space.
Slow: requires a lot of computer time.	Fast: requires little computer time.
Truncation or rounding errors possible in formatting.	No truncation or rounding errors.

14.6
DIRECT ACCESS FILES

Direct access files are files that are written and read using the direct access mode. The records in a sequential access file must be read in order from beginning to end. By contrast, the records in a direct access file may be read in arbitrary order. Direct access files are especially useful for information that may need to be accessed in any order, such as database files.

The key to the operation of a direct access file is that *every record in a direct access file must be of the same length*. If each record is the same length, then it is a simple matter to calculate exactly how far the *i*th record is into the disk file, and to read the disk sector containing that record directly without reading all of the sectors before it in the file. For example, suppose that we want to read the 120th record in a direct access file with 100-byte records. The 120th record will be located between bytes 11,901 and 12,000 of the file. The computer can calculate the disk sector containing those bytes, and read it directly.

A direct access file is opened by specifying ACCESS='DIRECT' in the OPEN statement. The length of each record in a direct access file must be specified in the OPEN statement using the RECL= clause. A typical OPEN statement for a direct access formatted file is shown below.

```
OPEN ( UNIT=8, FILE='dirio.fmt', ACCESS='DIRECT', FORM='FORMATTED', &
       RECL=40 )
```

The FORM= clause had to be specified here, because the default form for direct access is 'UNFORMATTED'.

For formatted files, the length of each record in the RECL= clause is specified in units of characters. Therefore, each record in file dirio.fmt above is 40 characters long. For unformatted files, the length specified in the RECL= clause may be in units of bytes, words, or some other machine-dependent quantity. You can use the INQUIRE statement to determine the record length required for an unformatted direct access file in a processor-independent fashion.

READ and WRITE statements for direct access files look like ones for sequential access files, except that the REC= clause may be included to specify the particular record to read or write (if the REC= clause is left out, then the next record in the file will be read or written). A typical READ statement for a direct access formatted file is shown below.

```
READ ( 8, '(I6)', REC=irec ) ival
```

Direct access, unformatted files whose record length is a multiple of the sector size of a particular computer are the most efficient Fortran files possible on that computer. Because they are direct access, it is possible to read any record in such a file directly. Because they are unformatted, no computer time is wasted in format conversions during reads or writes. Finally, because each record is exactly one disk sector long, only one disk sector will need to be read or written for each record. (Shorter records that are not multiples of the disk sector size might stretch across two disk sectors, forcing the computer to read both sectors in order to recover the information in the record.) Because these files are so efficient, many large programs written in Fortran are designed to use them.

A simple example program using a direct access, formatted file is shown in Figure 14-5. This program creates a direct access, formatted file named dirio.fmt with 40 characters per record. It fills the first 100 records with information, and then directly recovers whichever record the user specifies.

FIGURE 14-5
An example program using a direct access, formatted file.

```
PROGRAM direct_access_formatted
!
!   Purpose:
!     To illustrate the use of direct access Fortran files.
!
!   Record of revisions:
!       Date        Programmer          Description of change
!       ====        ==========          =====================
!    12/27/15     S. J. Chapman         Original code
!
IMPLICIT NONE

! Data dictionary: declare variable types & definitions
INTEGER :: i                    ! Index variable
INTEGER :: irec                 ! Number of record in file
CHARACTER(len=40) :: line       ! String containing current line.

! Open a direct access formatted file with 40 characters per record.
OPEN ( UNIT=8, FILE='dirio.fmt', ACCESS='DIRECT', &
       FORM='FORMATTED', STATUS='REPLACE', RECL=40 )

! Insert 100 records into this file.
DO i = 1, 100
   WRITE ( 8, '(A,I3,A)', REC=i ) 'This is record ', i, '.'
END DO

! Find out which record the user wants to retrieve.
WRITE (*,'(A)',ADVANCE='NO') ' Which record would you like to see? '
```

(continued)

Advanced I/O Concepts

(*concluded*)

```
READ (*,'(I3)') irec

! Retrieve the desired record.
READ ( 8, '(A)', REC=irec ) line

! Display the record.
WRITE (*, '(A,/,5X,A)' ) ' The record is: ', line

END PROGRAM direct_access_formatted
```

When the program is compiled and executed, the results are:

```
C:\book\fortran\chap14>direct_access_formatted
Which record would you like to see? 34
The record is:
    This is record 34.
```

This program also illustrates the use of the ADVANCE='NO' clause in a WRITE statement to allow a response to be entered on the same line that the prompt is printed on. The cursor did not advance to a new line when the WRITE statement was executed.

EXAMPLE 14-2

Comparing Direct Access Formatted and Unformatted Files:

To compare the operation of formatted and unformatted direct access files, create two files containing 50,000 records, each with 4 double-precision real values per line. One file should be formatted and the other one should be unformatted. Compare the sizes to the two files, and then compare the time that it takes to recover 50,000 records in random order from each file. Use subroutine random0 from Chapter 7 to generate the values placed in the files, and also the order in which the values are to be recovered. Use subroutine elapsed_time from Exercise 7-29 to determine how long it takes to read each of the files.

SOLUTION
A program to generate the files and then the access to them is shown in Figure 14-6. Note that the program uses the INQUIRE statement to determine how long each record in the unformatted file should be.

FIGURE 14-6
An example program comparing direct access, unformatted files to direct access, formatted files.

```
PROGRAM direct_access
!
!  Purpose:
!    To compare direct access formatted and unformatted files.
!
```

(*continued*)

(*continued*)

```
!   Record of revisions:
!      Date        Programmer          Description of change
!      ====        ==========          =====================
!    12/27/15    S. J. Chapman         Original code
!
USE timer                                   ! Timer module
IMPLICIT NONE

! List of parameters:
INTEGER, PARAMETER :: SINGLE = SELECTED_REAL_KIND(p=6)
INTEGER, PARAMETER :: DOUBLE = SELECTED_REAL_KIND(p=14)
INTEGER, PARAMETER :: MAX_RECORDS = 50000     ! Max # of records
INTEGER, PARAMETER :: NUMBER_OF_READS = 50000 ! # of reads

! Data dictionary: declare variable types & definitions
INTEGER :: i, j                        ! Index variable
INTEGER :: length_fmt = 84             ! Length of each record in
                                       ! formatted file
INTEGER :: length_unf                  ! Length of each record in
                                       ! unformatted file
INTEGER :: irec                        ! Number of record in file
REAL(KIND=SINGLE) :: time_fmt          ! Time for formatted reads
REAL(KIND=SINGLE) :: time_unf          ! Time for unformatted reads
REAL(KIND=SINGLE) :: value             ! Value returned from random0
REAL(KIND=DOUBLE), DIMENSION(4) :: values ! Values in record

! Get the length of each record in the unformatted file.
INQUIRE (IOLENGTH=length_unf) values
WRITE (*,'(A,I2)') ' The unformatted record length is ', &
                   length_unf
WRITE (*,'(A,I2)') ' The formatted record length is ', &
                   length_fmt

! Open a direct access unformatted file.
OPEN ( UNIT=8, FILE='dirio.unf', ACCESS='DIRECT', &
       FORM='UNFORMATTED', STATUS='REPLACE', RECL=length_unf )

! Open a direct access formatted file.
OPEN ( UNIT=9, FILE='dirio.fmt', ACCESS='DIRECT', &
       FORM='FORMATTED', STATUS='REPLACE', RECL=length_fmt )

! Generate records and insert into each file.
DO i = 1, MAX_RECORDS
   DO j = 1, 4
      CALL random0(value)                  ! Generate records
      values(j) = 30._double * value
   END DO
   WRITE (8,REC=i) values                  ! Write unformatted
   WRITE (9,'(4ES21.14)',REC=i) values     ! Write formatted
END DO

! Measure the time to recover random records from the
! unformatted file.
CALL set_timer
```

(*continued*)

Advanced I/O Concepts

(concluded)

```
DO i = 1, NUMBER_OF_READS
   CALL random0(value)
   irec = (MAX_RECORDS-1) * value + 1
   READ (8,REC=irec) values
END DO
CALL elapsed_time (time_unf)

! Measure the time to recover random records from the
! formatted file.
CALL set_timer
DO i = 1, NUMBER_OF_READS
   CALL random0(value)
   irec = (MAX_RECORDS-1) * value + 1
   READ (9,'(4ES21.14)',REC=irec) values
END DO
CALL elapsed_time (time_fmt)

! Tell user.
WRITE (*,'(A,F6.2)') ' Time for unformatted file = ', time_unf
WRITE (*,'(A,F6.2)') ' Time for formatted file =   ', time_fmt

END PROGRAM direct_access
```

When the program is compiled with the Intel Visual Fortran compiler and executed on a personal computer with an i7 chipset, the results are:

```
C:\book\fortran\chap14>direct_access
   The unformatted record length is 8
   The formatted record length is 80
   Time for unformatted file =  0.19
   Time for formatted file =    0.33
```

The length of each record in the unformatted file is 32 bytes, since each record contains four double-precision (64-bit or 8-byte) values. Since the Intel Visual Fortran compiler happens to measure record lengths in 4-byte units, the record length is reported as 8. On other processors or with other compilers, the length might come out in different, processor-dependent units. If we examine the files after the program executes, we see that the formatted file is much larger than the unformatted file, even though they both store the same information.

```
C:\book\fortran\chap14>dir dirio.*

Volume in drive C is SYSTEM
Volume Serial Number is 6462-A133

Directory of C:\book\fortran\chap14

12/27/2015  01:58 PM         4,200,000 dirio.fmt
12/27/2015  01:58 PM         1,600,000 dirio.unf
              2 File(s)      5,800,000 bytes
              0 Dir(s) 117,824,688,128 bytes free
```

Unformatted direct access files are both smaller and faster than formatted direct access files, but they are not portable between different kinds of processors.

14.7
STREAM ACCESS MODE

The **stream access** mode reads or writes a file byte by byte, without processing special characters such as carriage returns, line feeds, and so forth. This differs from sequential access in that sequential access reads data a record at a time, using the carriage return and/or line feed data to mark the end of the record to process. Stream access mode is similar to the C language I/O functions getc and putc, which can read or write data a byte at a time, and which treat control characters just like any others in the file.

A file is opened in stream access mode by specifying ACCESS='STREAM' in the OPEN statement. A typical OPEN statement for a stream access is shown below.

```
OPEN ( UNIT=8, FILE='infile.dat', ACCESS='STREAM', FORM='FORMATTED', &
       IOSTAT=istat )
```

Data can be written out to the file in a series of WRITE statements. When the programmer wishes to complete a line he or she should output a "newline" character (similar to outputting \n in C). Fortran includes an intrinsic function new_line(a) that returns a newline character of the same KIND as the input character a. For example, the following statements would open a file and write two lines to it.

```
OPEN (UNIT=8, FILE='x.dat', ACCESS='STREAM', FORM='FORMATTED', IOSTAT=istat)
WRITE (8, '(A)') 'Text on first line'
WRITE (8, '(A)') new_line(' ')
WRITE (8, '(A)') 'Text on second line'
WRITE (8, '(A)') new_line(' ')
CLOSE (8, IOSTAT=istat)
```

Good Programming Practice
Use sequential access files for data that is normally read and processed sequentially. Use direct access files for data that must be read and written in any arbitrary order.

Good Programming Practice
Use direct access, unformatted files for applications where large quantities of data must be manipulated quickly. If possible, make the record length of the files a multiple of the basic disk sector size for your computer.

14.8
NONDEFAULT I/O FOR DERIVED TYPES

We learned in Chapter 12 that, by default, derived data types are read in and written out in the order in which they are defined in the type definition statement, and the sequence of Fortran descriptors must match the order of the individual elements in the derived data type.

It is possible to create a nondefault *user-defined* way to read or write data for derived data types. This is done by binding procedures to the data type to handle the input and output. There can be four types of procedures, for formatted input, formatted output, unformatted input, and unformatted output, respectively. One or more of them can be declared and bound to the data type as shown below:

```
TYPE :: point
   REAL :: x
   REAL :: y
CONTAINS
   GENERIC :: READ(FORMATTED) => read_fmt
   GENERIC :: READ(UNFORMATTED) => read_unfmt
   GENERIC :: WRITE(FORMATTED) => write_fmt
   GENERIC :: WRITE(UNFORMATTED) => write_unfmt
END TYPE
```

The procedure name(s) specified on the generic READ(FORMATTED) line are called to perform formatted read output, and so forth for the other types of I/O.

The bound procedures are accessed by specifying the DT format descriptor in an I/O statement. The format of this descriptor is:

```
DT 'string' (10, -4, 2)
```

where the character string and the list of parameters are passed to the procedure that will perform the I/O function. The character string is optional, and may be deleted if it is not needed for a particular user-defined I/O operation.

The procedures that perform the I/O function must have the following interfaces:

```
SUBROUTINE formatted_io (dtv,unit,iotype,v_list,iostat,iomsg)
SUBROUTINE unformatted_io(dtv,unit,               iostat,iomsg)
```

where the calling arguments are as follows:

1. dtv is the derived data type to read or write. For WRITE statements, this value must be declared with INTENT(IN) and not modified. For READ statements, this value must be declared with INTENT(INOUT) and the data read in must be stored in it.
2. unit is the I/O unit number to read from or write to. It must be declared as an integer with INTENT(IN).
3. iotype is a CHARACTER(len=*) variable with INTENT(IN). It will contain one of three possible strings: 'LISTDIRECTED' if this is a list-directed I/O operation, 'NAMELIST' if this is a namelist I/O operation, 'DT' // string (where string is the string in the DT format descriptor) if this is ordinary formatted I/O.
4. v_list is an array of integers with INTENT(IN) that contains the set of integers in parentheses in the DT format descriptor.
5. iostat is the I/O status variable, set by the procedures when they complete their operations.v
6. iomsg is a CHARACTER(len=*) variable with INTENT(OUT). If iostat is non-zero, a message must be placed in this variable. Otherwise, it must not be changed.

Each subroutine will perform the specified type and direction of I/O in any way that the programmer desires. As long as the interface is honored, the nondefault I/O will function seamlessly with other Fortran I/O features.

> ## Quiz 14-2
>
> This quiz provides a quick check to see if you have understood the concepts introduced in Sections 14.3 to 14.6. If you have trouble with the quiz, reread the sections, ask your instructor, or discuss the material with a fellow student. The answers to this quiz are found in the back of the book.
>
> 1. What is the difference between a formatted and an unformatted file? What are the advantages and disadvantages of each type of file?
> 2. What is the difference between a direct access file and a sequential file? What are the advantages and disadvantages of each type of file?
> 3. What is the purpose of the INQUIRE statement? In what three ways can it be used?
>
> For questions 4 to 9, determine whether the following statements are valid. If not, specify what is wrong with them. If they are, what do they do?
>
> 4. ```
>    INTEGER :: i = 29
>    OPEN (UNIT=i,FILE='temp.dat',STATUS='SCRATCH')
>    WRITE (FMT="('The unit is ',I3)",UNIT=i) i
>    ```
> 5. ```
>    INTEGER :: i = 7
>    OPEN (i,STATUS='SCRATCH',ACCESS='DIRECT')
>    WRITE (FMT=''('The unit is ',I3)'',UNIT=i) i
>    ```
> 6. ```
>    INTEGER :: i = 7, j = 0
>    OPEN (UNIT=i,STATUS='SCRATCH',ACCESS='DIRECT',RECL=80)
>    WRITE (FMT='(I10)', UNIT=i) j
>    CLOSE (i)
>    ```
> 7. ```
>    INTEGER :: i
>    REAL,DIMENSION(9) :: a = [ (-100,i=1,5), (100,i=6,9) ]
>    OPEN (8,FILE='mydata',STATUS='REPLACE',IOSTAT=istat)
>    WRITE (8,'(3EN14.7)') (a(i), i = 1, 3)
>    WRITE (8,*) (a(i), i = 4, 6)
>    WRITE (UNIT=8) (a(i), I = 7, 9)
>    CLOSE (8)
>    ```
> 8. ```
>    LOGICAL :: exists
>    INTEGER :: lu = 11, istat
>    INQUIRE (FILE='mydata.dat',EXIST=exists,UNIT=lu,IOSTAT=istat)
>    ```
>
> *(continued)*

Advanced I/O Concepts

> (*concluded*)
>
> 9. What is the data file out.dat after the following statements are executed?
>
> ```
> INTEGER :: i, istat
> REAL, DIMENSION(5) :: a = [ (100.*i, i=-2,2) ]
> REAL :: b = -37, c = 0
> NAMELIST / local_data / a, b, c
> OPEN(UNIT=3,FILE='in.dat',ACTION='READ',STATUS='OLD',IOSTAT=istat)
> OPEN(UNIT=4,FILE='out.dat',ACTION='WRITE',IOSTAT=istat)
> READ(3,NML=local_data,IOSTAT=istat)
> WRITE(4,NML=local_data,IOSTAT=istat)
> ```
>
> Assume that the file in.dat contains the following information:
>
> &local_data A(2) = -17., A(5) = 30. /

EXAMPLE 14-3

Spare Parts Inventory:

Any engineering organization that maintains computers or test equipment needs to keep a supply of spare parts and consumable supplies on hand for use when equipment breaks, printers run out of paper, etc. They need to keep track of these supplies to determine how many of each type are being used in a given period of time, how many are in stock, and when to order more of a particular item. In actual practice, these functions are usually implemented with a database program. Here, we will write a simple Fortran program to keep track of stockroom supplies.

SOLUTION

A program to keep track of stockroom supplies needs to maintain a database of all available supplies, their descriptions, and their quantities. A typical database record might consist of:

1. **Stock Number** A unique number by which the item is known. Stock numbers start at 1 and go up to the number of items carried in the stockroom (six characters on disk; one integer in memory)
2. **Description** Description of item (30 characters)
3. **Vendor** The company that makes or sells the item (10 characters)
3. **Vendor Number** The number by which the item is known to the vendor (20 characters)
4. **Number in stock** (six characters on disk; one integer in memory)
5. **Minimum quantity** If less than this number of the item is in stock, it should be reordered. (six characters on disk; one integer in memory)

We will create a database file on disk in which the number of each record corresponds to the stock number of the item in the record. There will be as many records as there are items in stock, and each record will be 78 bytes long to hold the 78 characters of a database record. Furthermore, it may be necessary to withdraw items from stock in any order, so we should have direct access to any record in the database. We will implement the database using a *direct access, formatted* Fortran file with a record length of 78 bytes.

In addition, we will need a file containing information about the withdrawals from stock of various parts and supplies, and their replenishment by purchases from vendors. This *transaction file* will consist of stock numbers and quantities purchased or withdrawn (purchases of supplies are indicated by positive numbers, and withdrawals from stock are indicated by negative numbers). Since the transactions in the transaction file will be read in chronological sequence, it will be ok to use a sequential file for the transaction file.

Finally, we will need a file for reorders and error messages. This output file will contain reordering messages whenever the quantity of a stock item falls below the minimum quantity. It will also contain error messages if someone tries to withdraw an item that is not currently in stock.

1. **State the problem.**

   Write a program to maintain a database of stockroom supplies for a small company. The program will accept inputs describing the issues from the stockroom and replenishments of the stock, and will constantly update the database of stockroom supplies. It will also generate reorder messages whenever the supply of an item gets too low.

2. **Define the inputs and outputs.**

   The input to the program will be a sequential transaction file describing the issues from the stockroom and replenishments of the stocks. Each purchase or issue will be a separate line in the transaction file. Each record will consist of a *stock number* and *quantity* in free format.

   There are two outputs from the program. One will be the database itself, and the other will be a message file containing reordering and error messages. The database file will consist of 78-byte records structured as described above.

3. **Describe the algorithm.**

   When the program starts, it will open the database file, transaction file, and message file. It will then process each transaction in the transaction file, updating the database as necessary, and generating required messages. The high-level pseudocode for this program is

   ```
   Open the three files
   WHILE transactions file is not at end-of-file DO
     Read transaction
     Apply to database
     IF error or limit exceeded THEN
        Generate error / reorder message
     END of IF
   End of WHILE
   Close the three files
   ```

Advanced I/O Concepts

The detailed pseudocode for this program is

```
! Open files
Open database file for DIRECT access
Open transaction file for SEQUENTIAL access
Open message file for SEQUENTIAL access

! Process transactions
WHILE
   Read transaction
   IF end-of-file EXIT
   Add / subtract quantities from database
   IF quantity < 0 THEN
      Generate error message
   END of IF
   IF quantity < minimum THEN
      Generate reorder message
   END of IF
End of WHILE

! Close files
Close database file
Close transaction file
Close message file
```

4. **Turn the algorithm into Fortran statements.**
   The resulting Fortran subroutines are shown in Figure 14-7.

FIGURE 14-7
Program stock.

```
PROGRAM stock
!
!  Purpose:
!     To maintain an inventory of stockroom supplies, and generate
!     warning messages when supplies get low.
!
!  Record of revisions:
!      Date          Programmer           Description of change
!      ====          ==========           =====================
!    12/27/15      S. J. Chapman          Original code
!
IMPLICIT NONE

! Data dictionary: declare constants
INTEGER, PARAMETER :: LU_DB = 7       ! Unit for db file
INTEGER, PARAMETER :: LU_M = 8        ! Unit for message file
INTEGER, PARAMETER :: LU_T = 9        ! Unit for trans file

! Declare derived data type for a database item
TYPE :: database_record
   INTEGER :: stock_number             ! Item number
   CHARACTER(len=30) :: description    ! Description of item
```

(*continued*)

(*continued*)

```
       CHARACTER(len=10) :: vendor              ! Vendor of item
       CHARACTER(len=20) :: vendor_number       ! Vendor stock number
       INTEGER :: number_in_stock               ! Number in stock
       INTEGER :: minimum_quantity              ! Minimum quantity
    END TYPE

    ! Declare derived data type for transaction
    TYPE :: transaction_record
       INTEGER :: stock_number                  ! Item number
       INTEGER :: number_in_transaction         ! Number in transaction
    END TYPE

    ! Data dictionary: declare variable types & definitions
    TYPE (database_record) :: item              ! Database item
    TYPE (transaction_record) :: trans          ! Transaction item
    CHARACTER(len=3) :: file_stat               ! File status
    INTEGER :: istat                            ! I/O status
    LOGICAL :: exist                            ! True if file exists
    CHARACTER(len=120) :: msg                   ! Error message

    CHARACTER(len=24) :: db_file = 'stock.db'   ! Database file
    CHARACTER(len=24) :: msg_file = 'stock.msg' ! Message file
    CHARACTER(len=24) :: trn_file = 'stock.trn' ! Trans. file

    ! Begin execution: open database file, and check for error.
    OPEN (LU_DB, FILE=db_file, STATUS='OLD', ACCESS='DIRECT', &
          FORM='FORMATTED', RECL=78, IOSTAT=istat, IOMSG=msg )
    IF ( istat /= 0 ) THEN
       WRITE (*,100) db_file, istat
       100 FORMAT (' Open failed on file ',A,'. IOSTAT = ',I6)
       WRITE (*,'(A)') msg
       ERROR STOP 'Database file bad'
    END IF

    ! Open transaction file, and check for error.
    OPEN (LU_T, FILE=trn_file, STATUS='OLD', ACCESS='SEQUENTIAL', &
          IOSTAT=istat, IOMSG=msg )
    IF ( istat /= 0 ) THEN
       WRITE (*,100) trn_file, istat
       WRITE (*,'(A)') msg
       ERROR STOP 'Transaction file bad'
    END IF

    ! Open message file, and position file pointer at end of file.
    ! Check for error.
    INQUIRE (FILE=msg_file,EXIST=exist) ! Does the msg file exist?
    IF ( exist ) THEN
       file_stat = 'OLD'                        ! Yes, append to it.
    ELSE
       file_stat = 'NEW'                        ! No, create it.
    END IF
    OPEN (LU_M, FILE=msg_file, STATUS=file_stat, POSITION='APPEND', &
          ACCESS='SEQUENTIAL', IOSTAT=istat, IOMSG=msg )
    IF ( istat /= 0 ) THEN
       WRITE (*,100) msg_file, istat
```

(*continued*)

(*concluded*)

```
      WRITE (*,'(A)') msg
      ERROR STOP 'Message file bad'
   END IF

   ! Now begin processing loop for as long as transactions exist.
   process: DO
      ! Read transaction.
      READ (LU_T,*,IOSTAT=istat) trans

      ! If we are at the end of the data, exit now.
      IF ( istat /= 0 ) EXIT

      ! Get database record, and check for error.
      READ (LU_DB,'(I6,A30,A10,A20,I6,I6)',REC=trans%stock_number, &
            IOSTAT=istat) item

      IF ( istat /= 0 ) THEN
         WRITE (*,'(A,I6,A,I6)') &
            ' Read failed on database file record ', &
            trans%stock_number, ' IOSTAT = ', istat
      ERROR STOP 'Database read failed'
      END IF

      ! Read ok, so update record.
      item%number_in_stock = item%number_in_stock &
                           + trans%number_in_transaction
      ! Check for errors.
      IF ( item%number_in_stock < 0 ) THEN
         ! Write error message & reset quantity to zero.
         WRITE (LU_M,'(A,I6,A)') ' ERROR: Stock number ', &
               trans%stock_number, ' has quantity < 0! '
         item%number_in_stock = 0
      END IF

      ! Check for quantities < minimum.
      IF ( item%number_in_stock < item%minimum_quantity ) THEN
         ! Write reorder message to message file.
         WRITE (LU_M,110) ' Reorder stock number ', &
               trans%stock_number, ' from vendor ', &
               item%vendor, ' Description: ', &
               item%description
         110 FORMAT (A,I6,A,A,/,A,A)
      END IF

      ! Update database record
      WRITE (LU_DB,'(I6,A30,A10,A20,I6,I6)',REC=trans%stock_number, &
            IOSTAT=istat) item
   END DO process

   ! End of updates. Close files and exit.
   CLOSE ( LU_DB )
   CLOSE ( LU_T )
   CLOSE ( LU_M )

END PROGRAM stock
```

5. **Test the resulting Fortran program.**

To test this subroutine, it is necessary to create a sample database file and transaction file. The following sample database file has only four stock items:

```
1Paper,   8.5 x 11", 500 sheets   MYNEWCO   111-345                   12          5
2Toner,   Laserjet IIP            HP        92275A                     2          2
3Disks,   DVD-ROM, 50 ea          MYNEWCO   54242                     10         10
4Cable,   USB Printer             MYNEWCO   11-32-J6                   1          1
----|----|----|----|----|----|----|----|----|----|----|----|----|----|----|----|
    10        20        30        40        50        60        70        80
```

The following transaction file contains records of the dispensing of three reams of paper and five floppy disks. In addition, two new toner cartridges arrive and are placed in stock.

```
                            1  -3
                            3  -5
                            2   2
```

If the program is run against this transaction file, the new database becomes:

```
1Paper,   8.5 x 11", 500 sheets   MYNEWCO   111-345                    9          5
2Toner,   Laserjet IIP            HP        92275A                     2          2
3Disks,   DVD-ROM, 50 ea          MYNEWCO   54242                     10         10
4Cable,   USB Printer             MYNEWCO   11-32-J6                   1          1
----|----|----|----|----|----|----|----|----|----|----|----|----|----|----|----|
    10        20        30        40        50        60        70        80
```

and the message file contains the following lines:

```
            Reorder stock number    3 from vendor MYNEWCO
                Description: Disks, DVD-ROM, 50 ea
```

By comparing the before and after values in the database, we can see that the program is functioning correctly.

This example illustrated several advanced I/O features. The files that must exist for the program to work are opened with the 'OLD' status. The output message file may or may not previously exist, so is opened with the proper status 'OLD' or 'NEW' depending on the results of an INQUIRE statement. The example uses both direct access and sequential access files. The direct access file was used in the database, where it is necessary to be able to access any record in any order. The sequential files were used for simple input and output lists that were processed in sequential order. The message file was opened with the 'APPEND' option so that new messages could be written at the end of any existing messages.

The program also exhibits a few undesirable features. The principal one is the use of STOP statements whenever an error occurs. This was done here to keep the example simple for classroom purposes. However, in a real program, we should either close all files and shut down gracefully when an error occurs or offer the chance for the user to fix whatever problem is detected.

A real database would have probably used direct access *unformatted* files, instead of formatted files. We used formatted files here to make it easy to see the before-and-after effects on the database.

14.9
ASYNCHRONOUS I/O

Fortran 2003 and later has defined a new I/O mode called **asynchronous I/O.** In normal Fortran I/O operations, if a program writes data to a file with a `WRITE` statement, program execution halts at the `WRITE` statement until the data is completely written out, and then the program continues to run. Similarly, if a program reads data to a file with a `READ` statement, program execution halts at the `READ` statement until the data is completely read, and then the program continues to run. This is referred to a **synchronous I/O,** since the I/O operations are synchronized with the execution of the program.

In contrast, asynchronous I/O operations occur *in parallel* with the running of the program. If an asynchronous `WRITE` statement is executed, the data to be written is copied into some internal buffer, the write process is started, and control returns instantly to the calling program. In that fashion, the calling program can continue running a full speed while the write operation is going on.

The situation is a little more complex for an asynchronous read operation. If an asynchronous `READ` statement is executed, the read process is started, and control returns instantly to the calling program. At the time execution is returned to the calling program, *the variables being read are undefined*. They may have the old values, they may have the new values, or they may be in the middle of being updated, so the values must not be used until the read operation completes. The computer can go ahead and perform other calculations, but it must not use the variables in the asynchronous `READ` statement until the operation is complete.

How can a program using asynchronous reads know when the operation is complete? When it starts the I/O operation, it can get an ID for the operation using the `ID=` clause, and it can query the status of the operation using the `INQUIRE` statement. Alternately, the program can execute a `WAIT` or a file positioning statement (`REWIND`, `BACKSPACE`) on the I/O unit. In either case, control will not return to the calling program until all I/O operations on that unit are complete, so the program can safely use the new data after the execution resumes.

A typical way to use asynchronous I/O would be to start a read operation, do some other calculations in the meantime, and then call `WAIT` to ensure that the I/O operation has completed before using the data from the read. If programs are structured properly, it should be possible to keep running most of the time instead of being blocked by I/O operations.

Note that Fortran compilers are *allowed but not required* to implement asynchronous I/O. It is most likely to be found on systems designed to support many CPUs, where the I/O operations could proceed independently of the calculations on different CPUs. Massively parallel computers should always support asynchronous I/O operations.

14.9.1 Performing Asynchronous I/O

To use asynchronous I/O operations, a file must first be opened with the option to allow asynchronous I/O, and then each individual `READ` or `WRITE` statement must select the asynchronous I/O option. If an asynchronous `WRITE` is performed, no special actions need to be taken by the program. If an asynchronous `READ` is performed, then the program must wait for the `READ` to complete before using the variable.

An asynchronous `WRITE` operation is set up as shown below. Note that the `ASYNCHRONOUS=` clause must be in both the `OPEN` and the `WRITE` statement.

```
REAL,DIMENSION(5000,5000) :: data1
...
OPEN( UNIT=8, FILE='x.dat', ASYNCHRONOUS='yes', STATUS='NEW', &
      ACTION='WRITE', IOSTAT=istat)
...
! Write data to file
WRITE(8, 1000, ASYNCHRONOUS='yes', IOSTAT=istat) data1
1000 FORMAT( 10F10.6 )

(continue processing ...)
```

An asynchronous `READ` operation is set up as shown below. Note that the `ASYNCHRONOUS=` clause must be in both the `OPEN` and the `READ` statement.

```
REAL,DIMENSION(5000,5000) :: data2
...
OPEN( UNIT=8, FILE='y.dat', ASYNCHRONOUS='yes', STATUS='OLD', &
      ACTION='READ', IOSTAT=istat)
...
! Read data from file
READ(8, 1000, ASYNCHRONOUS='yes', IOSTAT=istat) data2
1000 FORMAT( 10F10.6 )

(continue processing but DO NOT USE data2 ...)

! Now wait for I/O completion
WAIT(8)

(Now it is safe to use data2 ...)
```

14.9.2. Problems with Asynchronous I/O

A major problem with asynchronous I/O operations can occur when Fortran compilers try to optimize execution speed. Modern optimizing compilers often move the order of actions around and do things in parallel to increase the overall speed of a program. This usually works fine, but it could cause a real problem if the compiler moved a statement using the data in an asynchronous `READ` from a point after to a point before a `WAIT` statement on that unit. In that case, the data being used might be the old information, the new information, of some combination of the two!

Fortran has defined an attribute to warn a compiler of this sort of problem with asynchronous I/O. The `ASYNCHRONOUS` attribute or statement provide this warning. For example, the following array is declared with the `ASYNCHRONOUS` attribute:

```
REAL,DIMENSION(1000),ASYNCHRONOUS :: data1
```

Advanced I/O Concepts

And the following statement declares that several variables have the ASYNCHRONOUS attribute:

```
ASYNCHRONOUS :: x, y, z
```

The ASYNCHRONOUS attribute is automatically assigned to a variable if it (or a component of it) appears in an input/output list or a namelist associated with an asynchronous I/O statement. There is no need to declare the variable ASYNCHRONOUS in that case, so as a practical matter you may not see this attribute explicitly declared very often.

14.10
ACCESS TO PROCESSOR-SPECIFIC I/O SYSTEM INFORMATION

Fortran includes an intrinsic module that provides a processor-independent way to get information about the I/O system for that processor. This module is called ISO_FORTRAN_ENV. It defines the constants shown in Table 14-9.

If you use these constants in a Fortran program instead of hard-coding the corresponding values, your program will be more portable. If the program is moved to another processor, the implementation of ISO_FORTRAN_ENV on that processor will contain the correct values for the new environment, and the code itself will not need to be modified.

To access the constants stored in this module, just include a USE statement in the corresponding program unit, and then access the constants by name:

```
USE ISO_FORTRAN_ENV
...
WRITE (OUTPUT_UNIT,*) 'This is a test'
```

TABLE 14-9
Constants defined in Module ISO_FORTRAN_ENV

Constant	Value/Description
INPUT_UNIT	This is an integer containing the unit number of the **standard input stream,** which is the unit accessed by a READ(*,*) statement.
OUTPUT_UNIT	This is an integer containing the unit number of the **standard output stream,** which is the unit accessed by a WRITE(*,*) statement.
ERROR_UNIT	This is an integer containing the unit number of the **standard error stream.**
IOSTAT_END	This is an integer containing the value returned by a READ statement in the IOSTAT= clause if the end of file is reached.
IOSTAT_EOR	This is an integer containing the value returned by a READ statement in the IOSTAT= clause if the end of record is reached.
NUMERIC_STORAGE_SIZE	This is an integer containing the number of bits in a default numeric value.
CHARACTER_STORAGE_SIZE	This is an integer containing the number of bits in a default character value.
FILE_STORAGE_SIZE	This is an integer containing the number of bits in a default file storage unit.

14.11
SUMMARY

In this chapter, we introduced the additional Fortran format descriptors EN, D, G, G0, B, O, Z, P, TL, TR, S, SP, SN, BN, BZ, RU, RD, RN, RZ, RC, RP, DC, and DP, and :. The EN descriptor provides a way to display data in engineering notation. The G and G0 descriptors provide a way to display any form of data. The B, O, and Z descriptors display integer or real data in binary, octal, and hexadecimal format, respectively. The TLn and TRn descriptors shift the position of data in the current line left and right by n characters. The colon descriptor (:) serves as a conditional stopping point for a WRITE statement. The D, P, S, SP, SN, BN, and BZ descriptors should not be used in new programs.

Then, we covered advanced features of Fortran I/O statements. The INQUIRE, PRINT, and ENDFILE statements were introduced, and all options were explained for all Fortran I/O statements. We introduced NAMELIST I/O, and explained the advantages of namelists for exchanging data between two programs or between two runs of the same program.

Fortran includes two file forms: *formatted* and *unformatted*. Formatted files contain data in the form of ASCII or Unicode characters, while unformatted files contain data that is a direct copy of the bits stored in the computer's memory. Formatted I/O requires a relatively large amount of processor time, since the data must be translated every time a read or write occurs. However, formatted files can be easily moved between processors of different types. Unformatted I/O is very quick, since no translation occurs. However, unformatted files cannot be easily inspected by humans, and cannot be easily moved between processors of different types.

Fortran includes three access methods: *sequential*, *direct*, and *stream access*. Sequential access files are files intended to be read or written in sequential order. There is a limited ability to move around within a sequential file using the REWIND and BACKSPACE commands, but the records in these files must basically be read one after another. Direct access files are files intended to be read or written in any arbitrary order. To make this possible, each record in a direct access file must be of a fixed length. If the length of each record is known, then it is possible to directly calculate where to find any specific record in the disk file, and to read or write only that record. Direct access files are especially useful for large blocks of identical records that might need to be accessed in any order. A common application for them is in databases.

The stream access mode reads or writes a file byte by byte, without processing special characters such as carriage returns, line feeds, and so forth. This differs from sequential access in that sequential access reads data a record at a time, using the carriage return and/or line feed data to mark the end of the record to process. Stream access mode is similar to the C language I/O functions getc and putc, which can read or write data a byte at a time, and which treat control characters just like any others in the file.

14.11.1 Summary of Good Programming Practice

The following guidelines should be adhered to when working with Fortran I/O:

1. Never use the D, P, BN, BZ, S, SP, or SS format descriptors in new programs.
2. Do not rely on pre-connected files in your Fortran programs (except for the standard input and output files). The number and the names of pre-connected files

vary from processor to processor, so using them will reduce the portability of your programs. Instead, always explicitly open each file that you use with an OPEN statement.
3. Always use the IOSTAT= and IOMSG= clauses in OPEN statements to trap errors. When an error is detected, tell the user all about the problem before shutting down gracefully or requesting an alternate file.
4. Always explicitly close each disk file with a CLOSE statement as soon as possible after a program is finished using it, so that it may be available for use by others in a multitasking environment.
5. Check to see if your output file is overwriting an existing data file. If it is, make sure that the user really wants to do that before destroying the data in the file.
6. Use the IOSTAT= and IOMSG= clauses in READ statements to prevent programs from aborting on errors, end-of-file, or end-of-record conditions. When an error or end-of-file condition is detected, the program can take appropriate actions to continue processing or to shut down gracefully.
7. Use NAMELIST I/O to save data to be exchanged between programs or between different runs of a single program. Also, you may use NAMELIST READ statements to update selected input parameters when a program begins executing.
8. Use formatted files to create data that must be readable by humans, or that must be transferable between different types of computers. Use unformatted files to efficiently store large quantities of data that do not have to be directly examined, and that will remain on only one type of computer. Also, use unformatted files when I/O speed is critical.
9. Use sequential access files for data that is normally read and processed sequentially. Use direct access files for data that must be read and written in any arbitrary order.
10. Use direct access, unformatted files for applications where large quantities of data must be manipulated quickly. If possible, make the record length of the files a multiple of the basic disk sector size for your computer.

14.11.2 Summary of Fortran Statements and Structures

BACKSPACE Statement:

```
              BACKSPACE (control_list)
or            BACKSPACE (unit)
or            BACKSPACE unit
```

Example:

```
              BACKSPACE (lu,IOSTAT=istat)
              BACKSPACE (8)
```

Description:
The BACKSPACE statement moves the current position of a file back by one record. Possible clauses in the control list are UNIT=, IOSTAT=, and ERR=.

ENDFILE Statement:

                    ENDFILE (*control_list*)
or   ENDFILE (*unit*)
or   ENDFILE *unit*

Examples:

                    ENDFILE (UNIT=lu, IOSTAT=istat)
                    ENDFILE (8)

Description:
The ENDFILE statement writes an end-of-file record to a file, and positions the file pointer beyond the end-of-file record. Possible clauses in the control list are UNIT=, IOSTAT=, and ERR=.

FLUSH Statement:

                    FLUSH (*control_list*)

Examples:

                    FLUSH (8)

Description:
The FLUSH statement forces any output data still in memory buffers to be written to the disk.

INQUIRE Statement:

                    INQUIRE (*control_list*)

Example:

                    LOGICAL :: lnamed
                    CHARACTER(len=12) :: filename, access
                    INQUIRE (UNIT=22, NAMED=lnamed, NAME=filename,
                    ACCESS=access)

Description:
The INQUIRE statement permits a user to determine the properties of a file. The file may be specified either by its file name or (after the file is opened) by its i/o unit number. The possible clauses in the INQUIRE statement are described in Table 14-5.

NAMELIST Statement:

                    NAMELIST / *nl_group_name* / *var1* [, *var2*, ...]

Examples:

                    NAMELIST / control_data / page_size, rows, columns
                    WRITE (8,NML=control_data)

Description:
The NAMELIST statement is a specification statement that associates a group of variables in a namelist. All of the variables in the namelist may be written or read as a unit using the namelist version of the WRITE and READ statements. When a namelist is read, only the values that appear in the input list will be modified by the READ. The values appear in the input list in a keyword format, and individual values may appear in any order.

Advanced I/O Concepts

PRINT Statement:

                PRINT *fmt, output_list*

Examples:

                PRINT *, intercept
                PRINT '(2I6)', i, j

Description:
The PRINT statement outputs the data in the output list *to the standard output device* according to the formats specified in the format descriptors. The format descriptors may be in a FORMAT statement or a character string, or the format might be defaulted to list-directed I/O with an asterisk.

REWIND Statement:

                REWIND (*control_list*)
or   REWIND (*lu*)
or   REWIND *lu*

Example:

                REWIND (8)
                REWIND (lu,IOSTAT=istat)
                REWIND 12

Description:
The REWIND statement moves the current position of a file back to the beginning of the file. Possible clauses in the control list are UNIT=, IOSTAT=, and ERR=.

WAIT Statement:

                WAIT (*control_list*)

Examples:

                WAIT (8)

Description:
The WAIT statement waits for any pending asynchronous I/O operations to complete before returning to the calling program.

14.11.3 Exercises

14-1. What is the difference between the ES and the EN format descriptor? How would the number 123,45.67 be displayed by each of these descriptors?

14-2. What types of data may be displayed with the B, O, Z descriptors? What do these descriptors do?

14-3. Write the form of the G format descriptor that will display seven significant digits of a number. What is the minimum width of this descriptor?

14-4. Write the following integers with the I8 and I8.8 format descriptors. How do the outputs compare? (*a*) 1024 (*b*) −128 (*c*) 30,000

14-5. Write the integers from the previous exercise with the B16 (binary), O11 (octal), and Z8 (hexadecimal) format descriptors.

14-6. Use subroutine `random0` developed in Chapter 7 to generate nine random numbers in the range [−100,000, 100,000). Display the numbers with the G11.5 and G0 format descriptors.

14-7. Suppose that you wanted to display the nine random numbers generated in the previous exercise in the following format:

```
VALUE(1) = ±xxxxxx.xx   VALUE(2) = ±xxxxxx.xx
VALUE(3) = ±xxxxxx.xx   VALUE(4) = ±xxxxxx.xx
VALUE(5) = ±xxxxxx.xx   VALUE(5) = ±xxxxxx.xx
VALUE(7) = ±xxxxxx.xx   VALUE(8) = ±xxxxxx.xx
VALUE(9) = ±xxxxxx.xx
----|----|----|----|----|----|----|----|----|----|----|----|
    10        20       30        40       50        60
```

Write a single format descriptor that would generate this output. Use the colon descriptor appropriately in the format statement.

14-8. Suppose that the following values were to be displayed with a G11.4 format descriptor. What would each output look like?

(*a*) -6.38765×10^{10}

(*b*) -6.38765×10^{2}

(*c*) -6.38765×10^{-1}

(*d*) 2345.6

(*e*) .TRUE.

(*f*) 'String!'

14-9. Suppose that the first four values from the previous exercise were to be displayed with an EN15.6 format descriptor. What would each output look like?

14-10. Explain the operation of NAMELIST I/O. Why is it especially suitable for initializing a program or sharing data between programs?

14-11. What will be written out by the statements shown below?

```
INTEGER :: i, j
REAL, DIMENSION(3,3) :: array
NAMELIST / io / array
array = RESHAPE( [ ((10.*i*j, j=1,3), i=0,2) ], [3,3] )
WRITE (*,NML=io)
```

14-12. What will be written out by the statements shown below?

```
INTEGER :: i, j
REAL, DIMENSION(3,3) :: a
NAMELIST / io / a
```

Advanced I/O Concepts

```
a = RESHAPE( [ ((10.*i*j, j=1,3), i=0,2) ], [3,3] )
READ (8,NML=io)
WRITE (*,NML=io)
```

Input data on unit 8:

```
&io a(1,1) = -100.
a(3,1) = 6., a(1,3) = -6. /
a(2,2) = 1000. /
```

14-13. What is the difference between using the TRn format descriptor and the nX format descriptor to move 10 characters to the right in an output format statement?

14-14. What is printed out by the following sets of Fortran statements?

(a)
```
REAL:: value = 356.248
INTEGER :: i
WRITE (*,200) 'Value = ', (value, i=1,5)
200 FORMAT (A,F10.4,G10.2,G11.5,G11.6,ES10.3)
```

(b)
```
INTEGER, DIMENSION(5) :: i
INTEGER :: j
DO j = 1, 5
   i(j) = j**2
END DO
READ (*,*) i
WRITE (*,500) i
500 FORMAT (3(10X,I5))
```

Input data:
```
       -101       ,,     17                    /
        20               71             ,,
```

14-15. Assume that a file is opened with the following statement:

```
OPEN ( UNIT=71, FILE='myfile' )
```

What is the status of the file when it is opened this way? Will the file be opened for sequential or direct access? Where will the file pointer be? Will it be formatted or unformatted? Will the file be opened for reading, writing, or both? How long will each record be? How will list-directed character strings that are written to the file be delimited? What will happen if the file is not found? What will happen if an error occurs during the open process?

14-16. Answer the questions of the previous exercise for the following files.

(a) `OPEN (UNIT=21, FILE='myfile', ACCESS='DIRECT', &`
      `FORM='FORMATTED', RECL=80, IOSTAT=istat, IOMSG=msg)`

(b) `OPEN (NEWUNIT=i, FILE='yourfile', ACCESS='DIRECT', ACTION='WRITE', &`
      `STATUS='REPLACE', RECL=80, IOSTAT=istat, IOMSG=msg)`

(c) `OPEN (5, FILE='file_5', ACCESS='SEQUENTIAL', &`
      `STATUS='OLD', DELIM='QUOTE', ACTION='READWRITE', &`
      `POSITION='APPEND', IOSTAT=istat )`

(d) `OPEN ( UNIT=1, STATUS='SCRATCH', IOSTAT=istat, IOMSG=msg )`

14-17. The `IOSTAT=` clause in a `READ` statement can return positive, negative, or zero values. What do positive values mean? Negative values? Zero values?

14-18. File Copy while Trimming Trailing Blanks Write a Fortran program that prompts the user for an input file name and an output file name, and then copies the input file to the output file, trimming trailing blanks off of the end of each line before writing it out. The program should use the `STATUS=` and `IOSTAT=` clauses in the `OPEN` statement to confirm that the input file already exists, and use the `STATUS=` and `IOSTAT=` clauses in the `OPEN` statement to confirm that the output file does not already exist. Be sure to use the proper `ACTION=` clause for each file. If the output file is already present, then prompt the user to see if it should be overwritten. If so, overwrite it, and if not, stop the program. After the copy process is completed, the program should ask the user whether or not to delete the original file. The program should set the proper status in the input file's `CLOSE` statement if the file is to be deleted.

14-19. Determine whether or not each of the following sets of Fortran statements is valid. If not, explain why not. If so, describe the output from the statements.

(a) Statements:
```
CHARACTER(len=10) :: acc, fmt, act, delim
INTEGER :: unit = 35
LOGICAL :: lexist, lnamed, lopen
INQUIRE (FILE='input',EXIST=lexist)
IF ( lexist ) THEN
   OPEN (unit, FILE='input', STATUS='OLD')
   INQUIRE (UNIT=unit,OPENED=lopen,EXIST=lexist, &
            NAMED=lnamed,ACCESS=acc,FORM=fmt, &
            ACTION=act, DELIM=delim)
   WRITE (*,100) lexist, lopen, lnamed, acc, fmt, &
            act, delim
   100 FORMAT ('File status: Exists = ',L1, &
              ' Opened = ', L1, ' Named = ',L1, &
              ' Access = ', A,/,' Format = ',A, &
              ' Action = ', A,/,' Delims = ',A)
END IF
```

(b) Statements:
```
INTEGER :: i1 = 10
OPEN (9, FILE='file1', ACCESS='DIRECT', FORM='FORMATTED', &
      STATUS='NEW')
WRITE (9,'(I6)') i1
```

14-20. Copying a File in Reversed Order Write a Fortran program that prompts the user for an input file name and an output file name, and then copies the input file to the output file *in reversed order*. That is, the last record of the input file is the first record of the output file. The program should use the `INQUIRE` statement to confirm that the input file already exists, and that the output file does not already exist. If the output file is already present, then prompt the user to see if it should be overwritten before proceeding. (*Hint:* Read all of the lines in the input file to count them, and then use `BACKSPACE` statements to work backward through the file. Be careful of the `IOSTAT` values!)

Advanced I/O Concepts

14-21. Comparing Formatted and Unformatted Files Write a Fortran program containing a real array with 100,000 random values in the range $[-10^6, 10^6)$. Then perform the following actions:

(a) Open a formatted sequential file and write the values to the file preserving the full seven significant digits of the numbers. (Use the ES format so that numbers of any size will be properly represented.) Write 10 values per line to the file, so that there are 100 lines in the file. How big is the resulting file?

(b) Open an unformatted sequential file and write the values to the file. Write 10 values per line to the file, so that there are 100 lines in the file. How big is the resulting file?

(c) Which file was smaller, the formatted file or the unformatted file?

(d) Use the subroutines set_timer and elapsed_time created in Exercise 7-29 to time the formatted and unformatted writes. Which one is faster?

14-22. Comparing Sequential and Direct Access Files Write a Fortran program containing a real array with 1,000 random values in the range $[-10^5, 10^5)$. Then perform the following actions:

(a) Open a *formatted sequential file*, and write the values to the file preserving the full seven significant digits of the numbers. (Use the ES14.7 format so that numbers of any size will be properly represented.) How big is the resulting file?

(b) Open a *formatted direct access file* with 14 characters per record, and write the values to the file preserving the full seven significant digits of the numbers. (Again, use the ES14.7 format.) How big is the resulting file?

(c) Open an *unformatted direct access file* and write the values to the file. Make the length of each record large enough to hold one number. (This parameter is computer dependent; use the INQUIRE statement to determine the length to use for the RECL= clause.) How big is the resulting file?

(d) Which file was smaller, the formatted direct access file or the unformatted direct access file?

(e) Now, retrieve 100 records from each of the three files in the following order: Record 1, Record 1000, Record 2, Record 999, Record 3, Record 998, etc. Use the subroutines set_timer and elapsed_time created in Exercise 7-29 to time the reads from each of the files. Which one is fastest?

(f) How did the sequential access file compare to the random access files when reading data in this order?

15

Pointers and Dynamic Data Structures

OBJECTIVES

- Understand dynamic memory allocation using pointers.
- Be able to explain what a target is, and why targets must be declared explicitly in Fortran.
- Understand the difference between a pointer assignment statement and a conventional assignment statement.
- Understand how to use pointers with array subsets.
- Know how to dynamically allocate and deallocate memory using pointers.
- Now how to create dynamic data structures such as linked lists using pointers.

In earlier chapters, we have created and used variables of the five intrinsic Fortran data types and of derived data types. These variables all had two characteristics in common: They all stored some form of data, and they were almost all **static,** meaning that the number and types of variables in a program were declared before program execution, and remained the same throughout program execution.[1]

Fortran includes another type of variable that contains no data at all. Instead, it contains the *address in memory* of another variable where the data is actually stored. Because this type of variable points to another variable, it is called a **pointer.** The difference between a pointer and an ordinary variable is illustrated in Figure 15-1.

FIGURE 15-1
The difference between a pointer and an ordinary variable: (*a*) A pointer stores the address of an ordinary variable in its memory location. (*b*) An ordinary variable stores a data value.

---

[1] Allocatable arrays, automatic arrays, and automatic character variables were the limited exceptions to this rule.

Both pointers and ordinary variables have names, but pointers store the addresses of ordinary variables, while ordinary variables store data values.

Pointers are primarily used in situations where variables and arrays must be created and destroyed dynamically during the execution of a program, and where it is not known before the program executes just how many of any given type of variable will be needed during a run. For example, suppose that a mailing list program must read in an unknown number of names and addresses, sort them into a user-specified order, and then print mailing labels in that order. The names and addresses will be stored in variables of a derived data type. If this program is implemented with static arrays, then the arrays must be as large as the largest possible mailing list ever to be processed. Most of the time the mailing lists will be much smaller, and this will produce a terrible waste of computer memory. If the program is implemented with allocatable arrays, then we can allocate just the required amount of memory, but we must still know in advance how many addresses there will be before the first one is read. By contrast, we will now learn how to *dynamically allocate a variable for each address as it is read in*, and how to use pointers to manipulate those addresses in any desired fashion. This flexibility will produce a much more efficient program.

We will first learn the basics of creating and using pointers, and then see several examples of how they can be used to write flexible and powerful programs.

15.1
POINTERS AND TARGETS

A Fortran variable is declared to be a pointer by either including the `POINTER` attribute in its type definition statement (the preferred choice) or by listing it in a separate `POINTER` statement. For example, each of the following statements declares a pointer `p1` that must point to a real variable.

```
REAL, POINTER :: p1
```

or

```
REAL :: p1
POINTER :: p1
```

Note that the *type* of a pointer must be declared, even though the pointer does not contain any data of that type. Instead, it contains the *address* of a variable of the declared type. A pointer is only allowed to point to variables of its declared type. Any attempt to point to a variable of a different type will produce a compilation error.

Pointers to variables of derived data types may also be declared. For example,

```
TYPE (vector), POINTER :: vector_pointer
```

declares a pointer to a variable of derived data type `vector`. Pointers may also point to an array. A pointer to an array is declared with a **deferred-shape array specification,** meaning that the rank of the array is specified, but the actual extent of the array in each dimension is indicated by colons. Two pointers to arrays are:

```
INTEGER, DIMENSION(:), POINTER :: ptr1
REAL, DIMENSION(:,:), POINTER :: ptr2
```

The first pointer can point to any 1D integer array, while the second pointer can point to any 2D real array.

A pointer can point to any variable or array of the pointer's type as long as the variable or array has been declared to be a **target.** A target is a data object whose address has been made available for use with pointers. A Fortran variable or array is declared to be a target by either including the TARGET attribute in its type definition statement (the preferred choice) or by listing it in a separate TARGET statement. For example, each of the following sets of statements declares two targets to which pointers may point.

```
REAL, TARGET :: a1 = 7
INTEGER, DIMENSION(10), TARGET :: int_array
```

or

```
REAL :: a1 = 7
INTEGER, DIMENSION(10) :: int_array
TARGET :: a1, int_array
```

They declare a real scalar value a1 and a rank 1 integer array int_array. Variable a1 may be pointed to by any real scalar pointer (such as the pointer p1 declared above), and int_array may be pointed to by any integer rank 1 pointer (such as pointer ptr1 above).

THE SIGNIFICANCE OF THE TARGET ATTRIBUTE

A pointer is a variable that contains the memory location of another variable, which is called the target. The target itself is just an ordinary variable of the same type as the pointer. Given that the target is just an ordinary variable, why is it necessary to attach a special TARGET attribute to the variable before a pointer can point to it? Other computer languages such as C have no such requirement.

The reason that the TARGET attribute is required has to do with the way Fortran compilers work. Fortran is normally used for large, numerically intensive mathematical problems, and most Fortran compilers are designed to produce output programs that are as fast as possible. These compilers include an *optimizer* as a part of the compilation process. The optimizer examines the code and rearranges it, unwraps loops, eliminates common subexpressions, etc., in order to increase the final execution speed. As a part of this optimization process, some of the variables in the original program can actually disappear, having been combined out of existence or replaced by temporary values in registers. So, what would happen if the variable that we wish to point to is optimized out of existence? There would be a problem pointing to it!

It is possible for a compiler to analyze a program and determine whether or not each individual variable is ever used as the target of a pointer, but that process is tedious. The TARGET attribute was added to the language to make it easier for the compiler writers. The attribute tells a compiler that a particular variable *could* be pointed to by a pointer, and therefore it must not be optimized out of existence.

Pointers and Dynamic Data Structures

15.1.1 Pointer Assignment Statements

A pointer can be **associated** with a given target by means of a **pointer assignment statement.** A pointer assignment statement takes the form

```
pointer => target
```

where *pointer* is the name of a pointer, and *target* is the name of a variable or array of the same type as the pointer. The pointer assignment operator consists of an equal sign followed by a greater than sign with no space in between.[2] When this statement is executed, the memory address of the target is stored in the pointer. After the pointer assignment statement, any reference to the pointer will actually be a reference to the data stored in the target.

If a pointer is already associated with a target, and another pointer assignment statement is executed using the same pointer, then the association with the first target is lost and the pointer now points to the second target. Any reference to the pointer after the second pointer assignment statement will actually be a reference to the data stored in the second target.

For example, the program in Figure 15-2 defines a real pointer p and two target variables t1 and t2. The pointer is first associated with variable t1 by a pointer assignment statement, and p is written out by a WRITE statement. Then the pointer is associated with variable t2 by another pointer assignment statement, and p is written out by a second WRITE statement.

FIGURE 15-2
Program to illustrate pointer assignment statements.

```
PROGRAM test_ptr
IMPLICIT NONE
REAL, POINTER :: p
REAL, TARGET :: t1 = 10., t2 = -17.
p => t1
WRITE (*,*) 'p, t1, t2 = ', p, t1, t2
p => t2
WRITE (*,*) 'p, t1, t2 = ', p, t1, t2
END PROGRAM test_ptr
```

When this program is executed, the results are:

```
C:\book\fortran\chap15>test_ptr
 p, t1, t2 =      10.000000       10.000000      -17.000000
 p, t1, t2 =     -17.000000       10.000000      -17.000000
```

It is important to note that p never contains either 10. or −17. Instead, it contains the addresses of the variables in which those values were stored, and the Fortran compiler treats a reference to the pointer as a reference to those addresses. Also, note that

---

[2] This sign is identical in form to the rename sign in the USE statement (see Chapter 13), but it has a different meaning.

a value could be accessed either through a pointer to a variable or through the variable's name, and the two forms of access can be mixed even within a single statement (Figure 15-3).

It is also possible to assign the value of one pointer to another pointer in a pointer assignment statement.

pointer1 => *pointer2*

After such a statement, *both pointers point directly and independently to the same target.* If either pointer is changed in a later assignment, the other one will be unaffected and will continue to point to the original target. If *pointer2* is disassociated (does not point to a target) at the time the statement is executed, then *pointer1* also becomes disassociated. For example, the program in Figure 15-4 defines two real pointers p1 and p2, and two target variables t1 and t2. The pointer p1 is first associated with variable t1 by a pointer assignment statement, and then pointer p2 is assigned the value of pointer p1 by another pointer assignment statement. After these statements, both pointers p1 and p2 are independently associated with variable t1. When pointer p1 is later associated with variable t2, pointer p2 remains associated with t1.

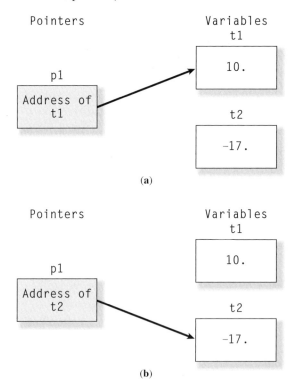

FIGURE 15-3
The relationship between the pointer and the variables in program test_ptr. (*a*) The situation after the first executable statement: p contains the address of variable t1, and a reference to p is the same as a reference to t1. (*b*) The situation after the third executable statement: p contains the address of variable t2, and a reference to p is the same as a reference to t2.

Pointers and Dynamic Data Structures

FIGURE 15-4
Program to illustrate pointer assignment between two pointers.

```
PROGRAM test_ptr2
IMPLICIT NONE
REAL, POINTER :: p1, p2
REAL, TARGET :: t1 = 10., t2 = -17.
p1 => t1
p2 => p1
WRITE (*,'(A,4F8.2)') ' p1, p2, t1, t2 = ', p1, p2, t1, t2
p1 => t2
WRITE (*,'(A,4F8.2)') ' p1, p2, t1, t2 = ', p1, p2, t1, t2
END PROGRAM test_ptr2
```

When this program is executed, the results are (Figure 15-5):

```
C:\book\fortran\chap15>test_ptr2
 p1, p2, t1, t2 =    10.00   10.00   10.00  -17.00
 p1, p2, t1, t2 =   -17.00   10.00   10.00  -17.00
```

(a)

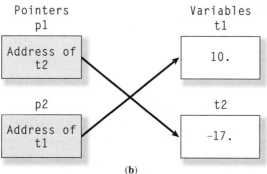

(b)

FIGURE 15-5
The relationship between the pointer and the variables in program `test_ptr2`. (*a*) The situation after the second executable statement: `p1` and `p2` both contain the address of variable `t1`, and a reference to either one is the same as a reference to `t1`. (*b*) The situation after the fourth executable statement: `p1` contains the address of variable `t2`, and `p2` contains the address of variable `t1`. Note that `p2` was unaffected by the reassignment of pointer `p1`.

15.1.2 Pointer Association Status

The **association status** of a pointer indicates whether or not the pointer currently points to a valid target. There are three possible statuses: **undefined, associated,** and **disassociated.** When a pointer is first declared in a type declaration statement, its pointer association status is *undefined*. Once a pointer has been associated with a target by a pointer assignment statement, its association status becomes *associated*. If a pointer is later disassociated from its target and is not associated with any new target, then its association status becomes *disassociated*.

How can a pointer be disassociated from its target? It can be disassociated from one target and simultaneously associated with another target by executing a pointer assignment statement. In addition, a pointer can be disassociated from all targets by executing a NULLIFY statement. A NULLIFY statement has the form

```
NULLIFY ( ptr1 [,ptr2, ...] )
```

where *ptr1*, *ptr2*, etc., are pointers. After the statement is executed, the pointers listed in the statement are disassociated from all targets.

A pointer can only be used to reference a target when it is associated with that target. Any attempt to use a pointer when it is not associated with a target will result in an error, and the program containing the error will abort. Therefore, we must be able to tell whether or not a particular pointer is associated with a particular target, or with any target at all. This can be done using the logical intrinsic function ASSOCIATED. The function comes in two forms, one containing a pointer as its only argument and one containing both a pointer and a target. The first form is

```
status = ASSOCIATED ( pointer )
```

This function returns a true value *if the pointer is associated with any target*, and a false value if it is not associated with any target. The second form is

```
status = ASSOCIATED ( pointer, target )
```

This function returns a true value *if the pointer is associated with the particular target included in the function*, and a false value otherwise.

A pointer's association status can only be undefined from the time that it is declared until it is first used. Thereafter, the pointer's status will always be either associated or disassociated. Because the undefined status is ambiguous, it is recommended that every pointer's status be clarified as soon as it is created by either assigning it to a target or nullifying it. For example, pointers could be declared and nullified in a program as follows:

```
REAL, POINTER :: p1, p2
INTEGER, POINTER :: i1
...
(additional specification statements)
...
NULLIFY (p1, p2, i1)
```

Good Programming Practice

Always nullify or assign all pointers in a program unit as soon as they are created. This eliminates any possible ambiguities associated with the undefined state.

Fortran also provides an intrinsic function NULL() that can be used to nullify a pointer at the time it is declared (or at any time during the execution of a program). Thus, pointers can be declared and nullified as follows:

```
REAL, POINTER :: p1 => NULL(), p2 => NULL()
INTEGER, POINTER :: i1 => NULL()
...
(additional specification statements)
```

The details of the NULL() function are described in Appendix B.

The simple program shown in Figure 15-6 illustrates the use of the NULL() function and the ASSOCIATED intrinsic function.

FIGURE 15-6
Program to illustrate the use of the NULLIFY statement and the ASSOCIATED function.

```
PROGRAM test_ptr3
IMPLICIT NONE
REAL, POINTER :: p1 => null(), p2 => null(), p3 => null()
REAL, TARGET :: a = 11., b = 12.5, c = 3.141592
WRITE (*,*) ASSOCIATED(p1)
p1 => a                    ! p1 points to a
p2 => b                    ! p2 points to b
p3 => c                    ! p3 points to c
WRITE (*,*) ASSOCIATED(p1)
WRITE (*,*) ASSOCIATED(p1,b)
END PROGRAM test_ptr3
```

The pointers p1, p2, and p3 will be nullified as soon as program execution begins. Thus, the result of the first ASSOCIATED(p1) function will be false. Then the pointers are associated with targets a, b, and c. When the second ASSOCIATED(p1) function is executed, the pointer will be associated, so the result of the function will be true. The third ASSOCIATED(p1,b) function checks to see if pointer p1 points to variable b. It doesn't, so the function returns false.

▪ 15.2
USING POINTERS IN ASSIGNMENT STATEMENTS

Whenever a pointer appears in a Fortran expression where a value is expected, *the value of the target pointed to is used* instead of the pointer itself. This process is known as **dereferencing** the pointer. We have already seen an example of dereferencing in the previous section: Whenever a pointer appeared in a WRITE statement, the value of the target pointed to was printed out instead. As another example, consider two pointers p1 and p2 that are associated with variables a and b, respectively. In the ordinary assignment statement

```
p2 = p1
```

both p1 and p2 appear in places where variables are expected, so they are dereferenced, and this statement is exactly identical to the statement

```
b = a
```

By contrast, in the pointer assignment statement

```
p2 => p1
```

p2 appears in a place where a pointer is expected, while p1 appears in a place where a target (an ordinary variable) is expected. As a result, p1 is dereferenced, while p2 refers to the pointer itself. The result is that the target pointed to by p1 is assigned to the pointer p2.

The program shown in Figure 15-7 provides another example of using pointers in place of variables:

FIGURE 15-7
Program to illustrate the use of pointers in place of variables in assignment statements.

```
PROGRAM test_ptr4
IMPLICIT NONE
REAL, POINTER :: p1 => null(), p2 => null(), p3 => null()
REAL, TARGET :: a = 11., b = 12.5, c
p1 => a                          ! p1 points to a
p2 => b                          ! p2 points to b
p3 => c                          ! p3 points to c
p3 = p1 + p2                     ! Same as c = a + b
WRITE (*,*) 'p3 = ', p3
p2 => p1                         ! p2 points to a
p3 = p1 + p2                     ! Same as c = a + a
WRITE (*,*) 'p3 = ', p3
p3 = p1                          ! Same as c = a
p3 => p1                         ! p3 points to a
WRITE (*,*) 'p3 = ', p3
WRITE (*,*) 'a, b, c = ', a, b, c
END PROGRAM test_ptr4
```

In this example, the first assignment statement p3 = p1 + p2 is equivalent to the statement c = a + b, since the pointers p1, p2, and p3 point to variables a, b, and c respectively, and since ordinary variables are expected in the assignment statement. The pointer assignment statement p2 => p1 causes pointer p1 to point to a, so the second assignment statement p3 = p1 + p2 is equivalent to the statement c = a + a. Finally, the assignment statement p3 = p1 is equivalent to the statement c = a, while the pointer assignment statement p3 => p1 causes pointer p3 to point to a. The output of this program is:

```
C:\book\fortran\chap15>test_ptr4
p3 =        23.500000
```

```
p3 =          22.000000
p3 =          11.000000
a, b, c =     11.000000      12.500000      11.000000
```

We will now show one way that pointers can improve the efficiency of a program. Suppose that it is necessary to swap two 100 × 100 element real arrays `array1` and `array2` in a program. To swap these arrays, we would normally use the following code:

```
REAL, DIMENSION(100,100) :: array1, array2, temp
...
temp = array1
array1 = array2
array2 = temp
```

The code is simple enough, but note that we are moving 10,000 real values in each assignment statement! All of that moving requires a lot of time. By contrast, we could perform the same manipulation with pointers and only exchange the *addresses* of the target arrays:

```
REAL, DIMENSION(100,100), TARGET :: array1, array2
REAL, DIMENSION(:,:), POINTER :: p1, p2, temp
p1 => array1
p2 => array2
...
temp => p1
p1 => p2
p2 => temp
```

In the latter case, we have only swapped the addresses, and not the entire 10,000-element arrays! This is enormously more efficient than the previous example.

Good Programming Practice
When sorting or swapping large arrays or derived data types, it is more efficient to exchange pointers to the data than it is to manipulate the data itself.

15.3
USING POINTERS WITH ARRAYS

A pointer can point to an array as well as a scalar. A pointer to an array must declare the type and the rank of the array that it will point to, but does *not* declare the extent in each dimension. Thus, the following statements are legal:

```
REAL, DIMENSION(100,1000), TARGET :: mydata
REAL, DIMENSION(:,:), POINTER :: pointer
pointer => array
```

A pointer can point not only to an array but also to a *subset* of an array (an array section). Any array section that can be defined by a subscript triplet can be used as the target of a pointer. For example, the program in Figure 15-8 declares a 16-element integer array `info`, and fills the array with the values 1 through 16. This array serves as the target for a series of pointers. The first pointer `ptr1` points to the entire array, while the second one points to the array section defined by the subscript triplet `ptr1(2::2)`. This will consist of the even subscripts 2, 4, 6, 8, 10, 12, 14, and 16 from the original array. The third pointer also uses the subscript triplet 2::2, and it points the even elements from the list pointed to by second pointer. This will consist of the subscripts 4, 8, 12, and 16 from the original array. This process of selection continues with the remaining pointers.

FIGURE 15-8
Program to illustrate the use of pointers with array sections defined by subscript triplets.

```
PROGRAM array_ptr
IMPLICIT NONE
INTEGER :: i
INTEGER, DIMENSION(16), TARGET :: info = [ (i, i=1,16) ]
INTEGER, DIMENSION(:), POINTER :: ptr1, ptr2, ptr3, ptr4, ptr5
ptr1 => info
ptr2 => ptr1(2::2)
ptr3 => ptr2(2::2)
ptr4 => ptr3(2::2)
ptr5 => ptr4(2::2)
WRITE (*,'(A,16I3)') ' ptr1 = ', ptr1
WRITE (*,'(A,16I3)') ' ptr2 = ', ptr2
WRITE (*,'(A,16I3)') ' ptr3 = ', ptr3
WRITE (*,'(A,16I3)') ' ptr4 = ', ptr4
WRITE (*,'(A,16I3)') ' ptr5 = ', ptr5
END PROGRAM array_ptr
```

When this program is executed, the results are:

```
C:\book\fortran\chap15>array_ptr
ptr1 =   1  2  3  4  5  6  7  8  9 10 11 12 13 14 15 16
ptr2 =   2  4  6  8 10 12 14 16
ptr3 =   4  8 12 16
ptr4 =   8 16
ptr5 =  16
```

Although pointers work with array sections defined by subscript triplets, *they do not work with array sections defined by vector subscripts*. Thus, the code in Figure 15-9 is illegal and will produce a compilation error.

FIGURE 15-9
Program to illustrate invalid pointer assignments to array sections defined with vector subscripts.

```
PROGRAM bad
IMPLICIT NONE
INTEGER :: i
INTEGER, DIMENSION(3) :: subs = [ 1, 8, 11 ]
INTEGER, DIMENSION(16), TARGET :: info = [ (i, i=1,16) ]
INTEGER, DIMENSION(:), POINTER :: ptr1
ptr1 => info(subs)
WRITE (*,'(A,16I3)') ' ptr1 = ', ptr1
END PROGRAM bad
```

15.4
DYNAMIC MEMORY ALLOCATION WITH POINTERS

One of the most powerful features of pointers is that they can be used to dynamically create variables or arrays whenever required, and then to release the space used by the dynamic variables or arrays once they are no longer needed. The procedure for doing this is similar to that used to create allocatable arrays. Memory is allocated using an ALLOCATE statement, and it is deallocated using a DEALLOCATE statement. The ALLOCATE statement has the same form as the ALLOCATE statement for an allocatable array. The statement takes the form

```
ALLOCATE (pointer(size),[ ...,] STAT=status)
```

where *pointer* is the name of a pointer to the variable or array being created, *size* is the dimension specification if the object being created is an array, and *status* is the result of the operation. If the allocation is successful, then the status will be 0. If it fails, a processor-dependent positive integer will be returned in the status variable. The STAT= clause is optional but should always be used, since a failed allocation statement without a STAT= clause will cause a program to abort.

This statement creates an unnamed data object of the specified size and the pointer's type, and sets the pointer to point to the object. Because the new data object is unnamed, it can only be accessed by using the pointer. After the statement is executed, the association status of the pointer will become *associated*. If the pointer was associated with another data object before the ALLOCATE statement is executed, then that association is lost.

The data object created by using the pointer ALLOCATE statement is unnamed, and so can only be accessed by the pointer. *If all pointers to that memory are either nullified or reassociated with other targets, then the data object will no longer be accessible by the program.* The object will still be present in memory, but it will no longer be possible to use it. Thus, careless programming with pointers can result in memory being filled with unusable space. This unusable memory is commonly referred to as a "memory leak". One symptom of this problem is that a program seems to grow larger and larger as it continues to execute, until it either fills the entire computer or uses all available memory. An example of a program with a memory leak is shown in Figure 15-10. In this program, 10-element arrays are allocated using both ptr1 and ptr2. The two arrays are initialized to different values, and those values are printed out. Then ptr2 is assigned to point to the same memory as ptr1 in a pointer assignment statement. *After that statement, the memory that was assigned to ptr2 is no longer accessible to the program.* That memory has been "lost", and will not be recovered until the program stops executing.

FIGURE 15-10
Program to illustrate "memory leaks" in a program.

```
PROGRAM mem_leak
IMPLICIT NONE
INTEGER :: i, istat
INTEGER, DIMENSION(:), POINTER :: ptr1, ptr2
```

(continued)

(*concluded*)

```
! Check associated status of ptrs.
WRITE (*,'(A,2L5)') ' Are ptr1, ptr2 associated? ', &
      ASSOCIATED(ptr1), ASSOCIATED(ptr2)

! Allocate and initialize memory
ALLOCATE (ptr1(1:10), STAT=istat)
ALLOCATE (ptr2(1:10), STAT=istat)
ptr1 = [ (i, i = 1,10 ) ]
ptr2 = [ (i, i = 11,20 ) ]

! Check associated status of ptrs.
WRITE (*,'(A,2L5)') ' Are ptr1, ptr2 associated? ', &
      ASSOCIATED(ptr1), ASSOCIATED(ptr2)

WRITE (*,'(A,10I3)') ' ptr1 = ', ptr1     ! Write out data
WRITE (*,'(A,10I3)') ' ptr2 = ', ptr2

ptr2 => ptr1                              ! Reassign ptr2

WRITE (*,'(A,10I3)') ' ptr1 = ', ptr1     ! Write out data
WRITE (*,'(A,10I3)') ' ptr2 = ', ptr2

NULLIFY(ptr1)                             ! Nullify pointer
DEALLOCATE(ptr2, STAT=istat)              ! Deallocate memory

END PROGRAM mem_leak
```

When program mem_leak executes, the results are:

```
C:\book\fortran\chap15>mem_leak
Are ptr1, ptr2 associated?     F    F
Are ptr1, ptr2 associated?     T    T
ptr1 =   1  2  3  4  5  6  7  8  9 10
ptr2 =  11 12 13 14 15 16 17 18 19 20
ptr1 =   1  2  3  4  5  6  7  8  9 10
ptr2 =   1  2  3  4  5  6  7  8  9 10
```

Memory that has been allocated with an ALLOCATE statement should be deallocated with a DEALLOCATE statement when the program is finished using it. If it is not deallocated, then that memory will be unavailable for any other use until the program finishes executing. When memory is deallocated in a pointer DEALLOCATE statement, the pointer to that memory is nullified at the same time. Thus, the statement

```
DEALLOCATE(ptr2, STAT=istat)
```

both deallocates the memory pointed to and nullifies the pointer ptr2.

The pointer DEALLOCATE statement can only deallocate memory that was created by an ALLOCATE statement. It is important to remember this fact. If the pointer in the statement happens to point to a target that was not created with an ALLOCATE statement, then the DEALLOCATE statement will fail and the program will abort unless the STAT= clause was specified. The association between such pointers and their targets can be broken by the use of the NULLIFY statement.

Pointers and Dynamic Data Structures

A potentially serious problem can occur when deallocating memory. Suppose that two pointers ptr1 and ptr2 both point to the same allocated array. If pointer ptr1 is used in a DEALLOCATE statement to deallocate the array, then that pointer is nullified. However, ptr2 will *not* be nullified. *It will continue to point to the memory location where the array used to be*, even if that memory location is reused for some other purpose by the program. If that pointer is used to either read data from or write data to the memory location, it will be either reading unpredictable values or overwriting memory used for some other purpose. In either case, using that pointer is a recipe for disaster! If a piece of allocated memory is deallocated, then *all* of the pointers to that memory should be nullified or reassigned. One of them will be automatically nullified by the DEALLOCATE statement, and any others should be nullified in NULLIFY statement(s).

Good Programming Practice
Always nullify or reassign *all* pointers to a memory location when that memory is deallocated. One of them will be automatically nullified by the DEALLOCATE statement, and any others should be manually nullified in NULLIFY statement(s) or reassigned in pointer assignment statements.

Figure 15-11 illustrates the effect of using a pointer after the memory to which it points has been deallocated. In this example, two pointers ptr1 and ptr2 both point to the same 10-element allocatable array. When that array is deallocated with ptr1, that pointer becomes disassociated. Pointer ptr2 remains associated, but now points to a piece of memory that can be freely reused by the program for other purposes. When ptr2 is accessed in the next WRITE statement, it points to an unallocated part of memory that could contain anything. Then, a new 2-element array is allocated using ptr1. Depending on the behavior of the compiler, this array could be allocated over the freed memory from the previous array, or it could be allocated somewhere else in memory.

FIGURE 15-11
Program to illustrate the effect of using a pointer after the memory to which it points has been deallocated.

```
PROGRAM bad_ptr
IMPLICIT NONE
INTEGER :: i, istat
INTEGER, DIMENSION(:), POINTER :: ptr1, ptr2

! Allocate and initialize memory
ALLOCATE (ptr1(1:10), STAT=istat)       ! Allocate ptr1
ptr1 = [ (i, i = 1,10 ) ]               ! Initialize ptr1
ptr2 => ptr1                            ! Assign ptr2

! Check associated status of ptrs.
WRITE (*,'(A,2L5)') ' Are ptr1, ptr2 associated? ', &
      ASSOCIATED(ptr1), ASSOCIATED(ptr2)
```

(*continued*)

(*concluded*)

```
WRITE (*,'(A,10I3)') ' ptr1 = ', ptr1    ! Write out data
WRITE (*,'(A,10I3)') ' ptr2 = ', ptr2

! Now deallocate memory associated with ptr1
DEALLOCATE(ptr1, STAT=istat)             ! Deallocate memory

! Check associated status of ptrs.
WRITE (*,'(A,2L5)') ' Are ptr1, ptr2 associated? ', &
      ASSOCIATED(ptr1), ASSOCIATED(ptr2)

! Write out memory associated with ptr2
WRITE (*,'(A,10I3)') ' ptr2 = ', ptr2

ALLOCATE (ptr1(1:2), STAT=istat)         ! Reallocate ptr1
ptr1 = [ 21, 22 ]

WRITE (*,'(A,10I3)') ' ptr1 = ', ptr1    ! Write out data
WRITE (*,'(A,10I3)') ' ptr2 = ', ptr2

END PROGRAM bad_ptr
```

These results of this program will vary from compiler to compiler, since deallocated memory may be treated differently on different processors. When this program is executed on the Lahey Fortran Compiler, the results are:

```
C:\book\fortran\chap15>bad_ptr
Are ptr1, ptr2 associated?     T    T
ptr1 =   1  2  3  4  5  6  7  8  9 10
ptr2 =   1  2  3  4  5  6  7  8  9 10
Are ptr1, ptr2 associated?     F    T
ptr2 =   1  2  3  4  5  6  7  8  9 10
ptr1 =  21 22
ptr2 =  21 22  3  4  5  6  7  8  9 10
```

After ptr1 was used to deallocate the memory, its pointer status changed to *disassociated*, while the status of ptr2 remained *associated*. When ptr2 was then used to examine memory, it pointed to the memory location *where the array used to be*, and saw the old values because the memory had not yet been reused. Finally, when ptr1 was used to allocate a new 2-element array, some of the freed-up memory was reused.

It is possible to mix pointers and allocatable arrays in a single ALLOCATE statement or DEALLOCATE statement, if desired.

15.5

USING POINTERS AS COMPONENTS OF DERIVED DATA TYPES

Pointers may appear as components of derived data types. Pointers in derived data types may even point to the derived data type being defined. This feature is very useful, since it permits us to construct various types of dynamic data structures linked together by successive pointers during the execution of a program. The simplest such structure is a **linked list,** which is a list of values linked together in a linear fashion by pointers. For example, the following derived data type contains a real number and a pointer to another variable of the same type:

```
TYPE :: real_value
   REAL :: value
   TYPE (real_value), POINTER :: p
END TYPE
```

A linked list is a series of variables of a derived data type, with the pointer from each variable pointing to the next variable in the list. The pointer in the last variable is nullified, since there is no variable after it in the list. Two pointers (say, head and tail) are also defined to point to the first and last variables in the list. Figure 15-12 illustrates this structure for variables of type real_value.

Linked lists are much more flexible than arrays. Recall that a static array must be declared with a fixed size when a program is compiled. As a result, we must size each such array to be large enough to handle the *largest problem* that a program will ever be required to solve. This large memory requirement can result in a program being too large to run on some computers, and also results in a waste of memory most of the time that the program is executed. Even allocatable arrays don't completely solve the problem. Allocatable arrays prevent memory waste by allowing us to allocate only the amount of memory needed for a specific problem, but we must know before we allocate the memory just how many values will be present during a particular run. In contrast, *linked lists permit us to add elements one at a time*, and we do not have to know in advance how many elements will ultimately be in the list.

When a program containing a linked list first starts to execute, there are no values in the list. In that case, the head and tail pointers have nothing to point to, so they are both nullified (see Figure 15-13a). When the first value is read, a variable of the derived data type is created, and the value is stored in that variable. The head and

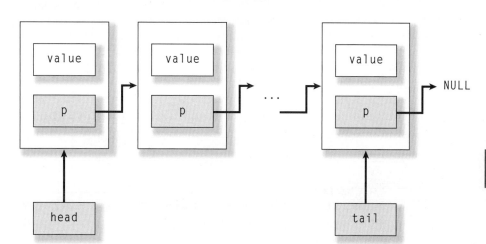

FIGURE 15-12
A typical linked list. Note that pointer in each variable points to the next variable in the list.

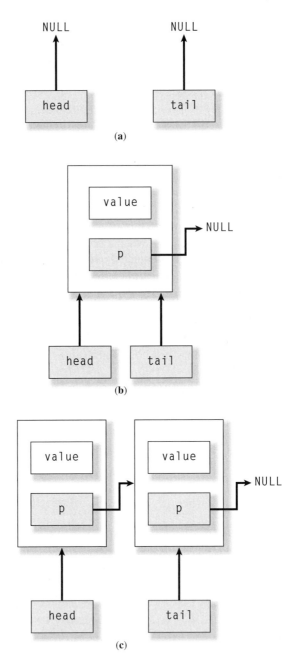

FIGURE 15-13
Building a linked list: (*a*) The initial situation with an empty list. (*b*) After adding one value to the list. (*c*) After adding a second value to the list.

`tail` pointers are set to point to the variable, and the pointer in the variable is nullified (Figure 15-13*b*).

When the next value is read, a new variable of the derived data type is created, the value is stored in that variable, and the pointer in the variable is nullified. The pointer in the previous variable and the `tail` pointer are set to point to the new variable. The `head` pointer does not change (Figure 15-13c). This process is repeated as each new value is added to the list.

Once all of the values are read, the program can process them by starting at the `head` pointer and following the pointers in the list until the `tail` pointer is reached.

EXAMPLE 15-1

Creating a Linked List:

In this example, we will write a simple program that reads in a list of real numbers, and then writes them out again. The number of values that the program can handle should only be limited by the amount of memory in the computer. This program doesn't do anything interesting by itself, but building a linked list in memory is a necessary first step in many practical problems. We will learn how to create the list in this example, and then start using lists to do useful work in later examples.

SOLUTION
We will use a linked list to hold the input values, since the size of a linked list can keep growing as long as additional memory can be allocated for new values. Each input value will be stored in a variable of the following derived data type, where the element p points to the next item in the list and the element `value` stores the input real value.

```
TYPE :: real_value
   REAL :: value
   TYPE (real_value), POINTER :: p
END TYPE
```

1. **State the problem.**

   Write a program to read an arbitrary number of real values from a file and to store them in a linked list. After all of the values have been read, the program should write them to the standard output device.

2. **Define the inputs and outputs.**

   The input to the program will be a file name, and a list of real values arranged one value per line in that file. The output from the program will be the real values in the file listed to the standard output device.

3. **Describe the algorithm.**

   This program can be broken down into four major steps:

   ```
   Get the input file name
   Open the input file
   Read the input data into a linked list
   Write the data to the standard output device
   ```

   The first three major steps of the program are to get the name of the input file, to open the file, and to read in the data. We must prompt the user for the input file name, read in the name, and open the file. If the file open is successful, we must read in the data, keeping track of the number of values read. Since we don't know how many data values to expect, a while loop is appropriate for the READ. The pseudocode for these steps is shown below:

```
Prompt user for the input file name "filename"
Read the file name "filename"
OPEN file "filename"
IF OPEN is successful THEN
   WHILE
      Read value into temp
      IF read not successful EXIT
      nvals ← nvals + 1
      (ALLOCATE new list item & store value)
   End of WHILE
   ...                    (Insert writing step here)
End of IF
```

The step of adding a new item to the linked list needs to be examined more carefully. When a new variable is added to the list, there are two possibilities: Either there is nothing in the list yet or there are already values in the list. If there is nothing in the list yet, then the head and tail pointers are nullified, so we will allocate the new variable using the head pointer, and point the tail pointer to the same place. The pointer p within the new variable must be nullified because there is nothing to point to yet, and the real value will be stored in the element value of the variable.

If there are already values in the list, then the tail pointer points to the last variable in the list. In that case, we will allocate the new variable using the pointer p within the last variable in the list, and then point the tail pointer to the new variable. The pointer p within the new variable must be nullified because there is nothing to point to, and the real value will be stored in the element value of the new variable. The pseudocode for steps is:

```
Read value into temp
IF read not successful EXIT
nvals ← nvals + 1
IF  head is not associated THEN
   ! The list is empty
   ALLOCATE head
   tail => head           ! Tail points to first value
   NULLIFY tail%p         ! Nullify p within 1st value
   tail%value ← temp      ! Store new number
ELSE
   ! The list already has values
   ALLOCATE tail%p
   tail => tail%p         ! Tail now points to new last value
   NULLIFY tail%p         ! Nullify p within new last value
   tail%value ← temp      ! Store new number
END of IF
```

The final step is to write the values in the linked list. To do this, we must go back to the head of the list and follow the pointers in it to the end of the list. We will define

Pointers and Dynamic Data Structures

a local pointer `ptr` to point to the value currently being printed out. The pseudocode for steps is:

```
ptr => head
WHILE ptr is associated
    WRITE ptr%value
    ptr => ptr%p
END of WHILE
```

4. **Turn the algorithm into Fortran statements.**
   The resulting Fortran subroutine is shown in Figure 15-14.

FIGURE 15-14
Program to read in a series of real values and store them in a linked list.

```
PROGRAM linked_list
!
!  Purpose:
!    To read in a series of real values from an input data file
!    and store them in a linked list. After the list is read in
!    it will be written back to the standard output device.
!
!  Record of revisions:
!      Date         Programmer        Description of change
!      ====         ==========        =====================
!    01/02/16      S. J. Chapman      Original code
!
IMPLICIT NONE

! Derived data type to store real values in
TYPE :: real_value
   REAL :: value
   TYPE (real_value), POINTER :: p
END TYPE

! Data dictionary: declare variable types & definitions
TYPE (real_value), POINTER :: head      ! Pointer to head of list
CHARACTER(len=20) :: filename           ! Input data file name
INTEGER :: nvals = 0                    ! Number of data read
TYPE (real_value), POINTER :: ptr       ! Temporary pointer
TYPE (real_value), POINTER :: tail      ! Pointer to tail of list
INTEGER :: istat                        ! Status: 0 for success
CHARACTER(len=80) :: msg                ! I/O Message
REAL :: temp                            ! Temporary variable

! Get the name of the file containing the input data.
WRITE (*,*) 'Enter the file name with the data to be read: '
READ (*,'(A20)') filename

! Open input data file.
OPEN ( UNIT=9, FILE=filename, STATUS='OLD', ACTION='READ', &
       IOSTAT=istat, IOMSG=msg )

! Was the OPEN successful?
```

(*continued*)

(*concluded*)

```
fileopen: IF ( istat == 0 ) THEN          ! Open successful

   ! The file was opened successfully, so read the data from
   ! it, and store it in the linked list.
   input: DO
      READ (9, *, IOSTAT=istat) temp       ! Get value
      IF ( istat /= 0 ) EXIT               ! Exit on end of data
      nvals = nvals + 1                    ! Bump count

      IF (.NOT. ASSOCIATED(head)) THEN     ! No values in list
         ALLOCATE (head,STAT=istat)        ! Allocate new value
         tail => head                      ! Tail pts to new value
         NULLIFY (tail%p)                  ! Nullify p in new value
         tail%value = temp                 ! Store number
      ELSE                                 ! Values already in list
         ALLOCATE (tail%p,STAT=istat)      ! Allocate new value
         tail => tail%p                    ! Tail pts to new value
         NULLIFY (tail%p)                  ! Nullify p in new value
         tail%value = temp                 ! Store number
      END IF
   END DO input

   ! Now, write out the data.
   ptr => head
   output: DO
      IF ( .NOT. ASSOCIATED(ptr) ) EXIT    ! Pointer valid?
      WRITE (*,'(F10.4)') ptr%value        ! Yes: Write value
      ptr => ptr%p                         ! Get next pointer
   END DO output

ELSE fileopen

   ! Else file open failed. Tell user.
   WRITE (*,'(A,I6)') 'File open failed--status = ', istat
   WRITE (*,*) msg

END IF fileopen

END PROGRAM linked_list
```

5. **Test the resulting Fortran programs.**

To test this program, we must generate a file of input data. If the following 10 real values are placed in a file called input.dat, then we can use that file to test the program: 1.0, 3.0, −4.4, 5., 2., 9.0, 10.1, −111.1, 0.0, −111.1. When the program is executed with this file, the results are:

```
C:\book\fortran\chap15>linked_list
Enter the file name with the data to be read:
input.dat
    1.0000
    3.0000
   -4.4000
    5.0000
    2.0000
```

Pointers and Dynamic Data Structures

```
     9.0000
    10.1000
  -111.1000
      .0000
  -111.1000
```

The program appears to be working properly. Note that the program does not check the status of the ALLOCATE statements. This was done deliberately to make the manipulations of the linked list as clear as possible. In any real program, these statuses should be checked to detect memory problems so that the program can shut down gracefully.

EXAMPLE 15-2 *The Insertion Sort:*

We introduced the selection sort in Chapter 6. That algorithm sorted a list by searching for the smallest value in the list and placing it at the top. Then it searched for the smallest value in the remaining portion of the list, and placed it in the second position, and so forth until all of the values were sorted.

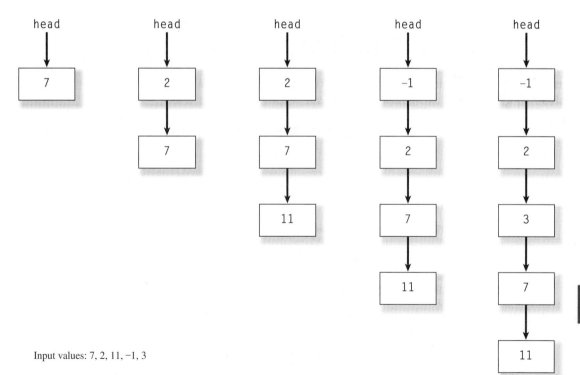

Input values: 7, 2, 11, −1, 3

FIGURE 15-15
Sorting the values 7, 2, 11, −1, and 3 with the insertion sort.

Another possible sorting algorithm is the *insertion sort*. The insertion sort works by placing each value in its proper position in the list as it is read in. If the value is smaller than any previous value in the list, then it is placed at the top. If the value is larger than any previous value in the list, then it is placed at the bottom. If the value is in between, then the number is inserted at the appropriate place in the middle of the list.

An insertion sort of the values 7, 2, 11, −1, and 3 is shown in Figure 15-15. The first value read is a 7. Since there are no other values in the list, it is placed at the top. The next value read is a 2. Since it is smaller than the 7, it is placed above the 7 in the list. The third value read is an 11. Since it is larger than any other value in the list, it is placed at the bottom. The fourth value read is a −1. Since it is smaller than any other value in the list, it is placed at the top. The fifth value read is a 3. Since it is larger than 2 and smaller than 7, it is placed between them in the list. In the insertion sort, the list is always kept sorted as each value is read.

Linked lists are ideally suited for implementing an insertion sort, since new values can be added at the front, at the end, or anywhere in the middle of the list by simply changing pointers. Use a linked list to implement an insertion sort algorithm to sort an arbitrary number of integer values.

SOLUTION

We will use a linked list to hold the input values, since it is easy to insert new values anywhere in the linked list by simply changing pointers. Each input value will be read and stored in a variable of the following derived data type, where the pointer `next_value` points to the next item in the list and the element `value` stores the input integer value.

```
TYPE :: int_value
   INTEGER :: value
   TYPE (int_value), POINTER :: next_value
END TYPE
```

Each value will be read, compared to all previous values, and inserted at the proper point in the list.

1. **State the problem.**

    Write a program to read an arbitrary number of integer values from a file and to sort them using an insertion sort. After all of the values have been read and sorted, the program should write the sorted list out to the standard output device.

2. **Define the inputs and outputs.**

    The input to the program will be a file name, and a list of integer values arranged one value per line in that file. The output from the program will be the sorted integer values listed to the standard output device.

3. **Describe the algorithm.**

    The pseudocode for this program is shown below:

    ```
    Prompt user for the input file name "filename"
    Read the file name "filename"
    OPEN file "filename"
    IF OPEN is successful THEN
       WHILE
    ```

```
         Read value into temp
         IF read not successful EXIT
         nvals ← nvals + 1
         ALLOCATE new data item & store value
         Insert item at proper point in list
      End of WHILE
      Write the data to the standard output device
   End of IF
```

The step of adding a new item to the linked list needs to be examined in more detail. When we add a new variable to the list, there are two possibilities: Either there is nothing in the list yet or there are already values in the list. If there is nothing in the list yet, then the head and tail pointers are nullified, so we will allocate the new variable using the head pointer, and point the tail pointer to the same place. The pointer next_value within the new variable must be nullified because there is nothing to point to yet, and the integer will be stored in the element value of the variable.

If there are already values in the list, then we must search to find the proper place to insert the new value into the list. There are three possibilities here. If the number is smaller than the first number in the list (pointed to by the head pointer), then we will add the value at the front of the list. If the number is greater than or equal to the last number in the list (pointed to by the tail pointer), then we will add the value at the end of the list. If the number is between those values, we will search until we locate the two values that it lies between, and insert the new value there. Note that we must allow for the possibility that the new value is equal to one of numbers already in the list. The pseudocode for these steps is:

```
   Read value into temp
   IF read not successful EXIT
   nvals ← nvals + 1
   ALLOCATE ptr
   ptr%value ← temp
   IF head is not associated THEN
      ! The list is empty
      head => ptr
      tail => head
      NULLIFY tail%next_value
   ELSE
      ! The list already has values. Check for
      ! location for new value.
      IF ptr%value < head%value THEN
         ! Add at front
         ptr%next_value => head
         head => ptr
      ELSE IF ptr%value >= tail%value THEN
         ! Add at rear
         tail%next_value => ptr
         tail => ptr
         NULLIFY tail%next_value
      ELSE
         ! Find place to add value
         ptr1 => head
         ptr2 => ptr1%next_value
```

```
            DO
               IF ptr%value >= ptr1%value AND
                  ptr%value < ptr2%value THEN
                  ! Insert value here
                  ptr%next_value => ptr2
                  ptr1%next_value => ptr
                  EXIT
               END of IF
               ptr1 => ptr2
               ptr2 => ptr2%next_value
            END of DO
         END of IF
      END of IF
```

The final step is to write the values in the linked list. To do this, we must go back to the head of the list and follow the pointers to the end of the list. We will use pointer ptr to point to the value currently being printed out. The pseudocode for steps is:

```
ptr => head
WHILE ptr is associated
   WRITE ptr%value
   ptr => ptr%next_value
END of WHILE
```

4. **Turn the algorithm into Fortran statements.**

The resulting Fortran subroutine is shown in Figure 15-16.

FIGURE 15-16
Program to read in a series of integer values and sort them using the insertion sort.

```
PROGRAM insertion_sort
!
!  Purpose:
!    To read a series of integer values from an input data file
!    and sort them using an insertion sort. After the values
!    are sorted, they will be written back to the standard
!    output device.
!
!  Record of revisions:
!      Date          Programmer         Description of change
!      ====          ==========         =====================
!    01/02/16      S. J. Chapman        Original code
!
IMPLICIT NONE

! Derived data type to store integer values in
TYPE :: int_value
   INTEGER :: value
   TYPE (int_value), POINTER :: next_value
END TYPE

! Data dictionary: declare variable types & definitions
TYPE (int_value), POINTER :: head    ! Pointer to head of list
```

(*continued*)

(continued)

```
CHARACTER(len=20) :: filename    ! Input data file name
INTEGER :: istat                 ! Status: ø for success
INTEGER :: nvals = 0             ! Number of data read
TYPE (int_value), POINTER :: ptr    ! Ptr to new value
TYPE (int_value), POINTER :: ptr1   ! Temp ptr for search
TYPE (int_value), POINTER :: ptr2   ! Temp ptr for search
TYPE (int_value), POINTER :: tail   ! Pointer to tail of list
INTEGER :: temp                  ! Temporary variable

! Get the name of the file containing the input data.
WRITE (*,*) 'Enter the file name with the data to be sorted: '
READ (*,'(A20)') filename

! Open input data file.
OPEN ( UNIT=9, FILE=filename, STATUS='OLD', ACTION='READ', &
       IOSTAT=istat )

! Was the OPEN successful?
fileopen: IF ( istat == 0 ) THEN             ! Open successful

   ! The file was opened successfully, so read the data value
   ! to sort, allocate a variable for it, and locate the proper
   ! point to insert the new value into the list.
   input: DO
      READ (9, *, IOSTAT=istat) temp         ! Get value
      IF ( istat /= 0 ) EXIT input           ! Exit on end of data
      nvals = nvals + 1                      ! Bump count

      ALLOCATE (ptr,STAT=istat)              ! Allocate space
      ptr%value = temp                       ! Store number

      ! Now find out where to put it in the list.
      new: IF (.NOT. ASSOCIATED(head)) THEN  ! No values in list
         head => ptr                         ! Place at front
         tail => head                        ! Tail pts to new value
         NULLIFY (ptr%next_value)            ! Nullify next ptr
      ELSE
         ! Values already in list. Check for location.
         front: IF ( ptr%value < head%value ) THEN
            ! Add at front of list
            ptr%next_value => head
            head => ptr
         ELSE IF ( ptr%value >= tail%value ) THEN
            ! Add at end of list
            tail%next_value => ptr
            tail => ptr
            NULLIFY ( tail%next_value )
         ELSE
            ! Find place to add value
            ptr1 => head
            ptr2 => ptr1%next_value
            search: DO
               IF ( (ptr%value >= ptr1%value) .AND. &
```

(continued)

(*concluded*)

```
                    (ptr%value < ptr2%value) ) THEN
                ! Insert value here
                ptr%next_value => ptr2
                ptr1%next_value => ptr
                EXIT search
             END IF
             ptr1 => ptr2
             ptr2 => ptr2%next_value
          END DO search
       END IF front
    END IF new
 END DO input

 ! Now, write out the data.
 ptr => head
 output: DO
    IF ( .NOT. ASSOCIATED(ptr) ) EXIT   ! Pointer valid?
    WRITE (*,'(I10)') ptr%value         ! Yes: Write value
    ptr => ptr%next_value               ! Get next pointer
 END DO output

ELSE fileopen

 ! Else file open failed. Tell user.
 WRITE (*,'(A,I6)') 'File open failed--status = ', istat

END IF fileopen

END PROGRAM insertion_sort
```

5. **Test the resulting Fortran programs.**

To test this program, we must generate a file of input data. If the following seven integer values are placed in a file called 'input1.dat', then we can use that file to test the program: 7, 2, 11, −1, 3, 2, and 0. When the program is executed with this file, the results are:

```
C:\book\fortran\chap15>insertion_sort
Enter the file name with the data to be sorted:
input1.dat
        -1
         0
         2
         2
         3
         7
        11
```

The program appears to be working properly. Note that this program also does not check the status of the ALLOCATE statements. This was done deliberately to make the manipulations as clear as possible. (At one point in the program, the DO and IF structures are nested 6 deep!) In any real program, these statuses should be checked to detect memory problems so that the program can shut down gracefully.

15.6
ARRAYS OF POINTERS

It is not possible to declare an array of pointers in Fortran. In a pointer declaration, the DIMENSION attribute *refers to the dimension of the pointer's target*, not to the dimension of the pointer itself. The dimension must be declared with a deferred-shape specification, and the actual size will be the size of the target with which the pointer is associated. In the example shown below, the subscript on the pointer refers the corresponding position in the target array, so the value of ptr(4) is 6.

```
REAL, DIMENSION(:), POINTER :: ptr
REAL, DIMENSION(5), TARGET :: tgt = [ -2, 5., 0., 6., 1 ]
ptr => tgt
WRITE (*,*) ptr(4)
```

There are many applications in which *arrays of pointers* are useful. Fortunately, we can create an array of pointers for those applications by using derived data types. It is illegal to have an array of pointers in Fortran, but it is perfectly legal to have an array of any derived data type. Therefore, we can declare a derived data type containing only a pointer, and then create an array of that data type! For example, the program in Figure 15-17 declares an array of a derived data type containing real pointers, each of which points to a real array.

FIGURE 15-17
Program illustrating how to create an array of pointers using a derived data type.

```
PROGRAM ptr_array
IMPLICIT NONE
TYPE :: ptr
   REAL, DIMENSION(:), POINTER :: p
END TYPE
TYPE (ptr), DIMENSION(3) :: p1
REAL, DIMENSION(4), TARGET :: a = [ 1.,   2.,   3.,   4. ]
REAL, DIMENSION(4), TARGET :: b = [ 5.,   6.,   7.,   8. ]
REAL, DIMENSION(4), TARGET :: c = [ 9.,  10.,  11.,  12. ]
p1(1)%p => a
p1(2)%p => b
p1(3)%p => c
WRITE (*,*) p1(3)%p
WRITE (*,*) p1(2)%p(3)
END PROGRAM ptr_array
```

With the declarations in program ptr_array, the expression p1(3)%p refers to the *third* array (array c), so the first WRITE statement should print out 9., 10., 11., and 12. The expression p1(2)%p(3) refers to the *third* value of the *second* array (array b), so the second WRITE statement prints out the value 7. When this program is compiled and executed with the Compaq Visual Fortran compiler, the results are:

```
C:\book\fortran\chap15>ptr_array
    9.000000      10.000000       11.000000       12.000000
    7.000000
```

Quiz 15-1

This quiz provides a quick check to see if you have understood the concepts introduced in Sections 15.1 through 15.6. If you have trouble with the quiz, reread the sections, ask your instructor, or discuss the material with a fellow student. The answers to this quiz are found in the back of the book.

1. What is a pointer? What is a target? What is the difference between a pointer and an ordinary variable?
2. What is a pointer assignment statement? What is the difference between a pointer assignment statement and an ordinary assignment statement?
3. What are the possible association statuses of a pointer? How can the association status be changed?
4. What is dereferencing?
5. How can memory be dynamically allocated with pointers? How can it be deallocated?

Are each of the following code segments valid or invalid? If a code segment is valid, explain what it does. If it is invalid, explain why.

6.
```
REAL, TARGET :: value = 35.2
REAL, POINTER :: ptr2
ptr2 = value
```

7.
```
REAL, TARGET :: value = 35.2
REAL, POINTER :: ptr2
ptr2 => value
```

8.
```
INTEGER, DIMENSION(10,10), TARGET :: array
REAL, DIMENSION(:,:), POINTER :: ptr3
ptr3 => array
```

9.
```
REAL, DIMENSION(10,10) :: array
REAL, DIMENSION(:,:) :: ptr4
POINTER :: ptr4
TARGET :: array
ptr4 => array
```

10.
```
INTEGER, POINTER :: ptr
WRITE (*,*) ASSOCIATED(ptr)
ALLOCATE (ptr)
ptr = 137
WRITE (*,*) ASSOCIATED(ptr), ptr
NULLIFY (ptr)
```

11.
```
INTEGER, DIMENSION(:), POINTER :: ptr1, ptr2
INTEGER :: istat
ALLOCATE (ptr1(10), STAT=istat)
```

(continued)

Pointers and Dynamic Data Structures

```
        (concluded)

            ptr1 = 0
            ptr1(3) = 17
            ptr2 => ptr1
            DEALLOCATE (ptr1)
            WRITE (*,*) ptr2
    12. TYPE mytype
            INTEGER, DIMENSION(:), POINTER :: array
        END TYPE
        TYPE (mytype), DIMENSION(10) :: p
        INTEGER :: i, istat
        DO i = 1, 10
            ALLOCATE (p(i).array(10), STAT=istat)
            DO j = 1, 10
                p(i)%array(j) = 10*(i-1) + j
            END DO
        END DO
        WRITE (*,'(10I4)') p(4).array
        WRITE (*,'(10I4)') p(7).array(1)
```

15.7
USING POINTERS IN PROCEDURES

Pointers may be used as dummy arguments in procedures and may be passed as actual arguments to procedures. In addition, a function result can be a pointer. The following restrictions apply if pointers are used in procedures:

1. If a procedure has dummy arguments with either the POINTER or TARGET attributes, then the procedure must have an explicit interface.
2. If a dummy argument is a pointer, then the actual argument passed to the procedure must be a pointer of the same type, kind, and rank.
3. A pointer dummy argument cannot appear in an ELEMENTAL procedure.

It is important to be careful when passing pointers to procedures. As programs get larger and more flexible, we will often get to a situation where pointers are allocated in one procedure, used in others, and finally deallocated and nullified in yet another. In such a complex program, it is *very* easy to make errors such as attempting to work with disassociated pointers, or allocating new arrays with pointers that are already in use. It is very important that the status results be checked for all ALLOCATE and DEALLOCATE statements, and that the status of pointers be checked using the ASSOCIATED function.

When a pointer is used to pass data to a procedure, we automatically know the type of the data associated with the pointer from the type of the pointer itself. If the pointer points to an array, we will know the rank of the array, but not its extent or size. If we need to know the extent or size of the array, then we can use the intrinsic functions LBOUND and UBOUND to determine the bounds of each dimension of the array.

EXAMPLE 15-3

Extracting the Diagonal Elements from a Matrix:

To illustrate the proper use of pointers, we will write a subroutine that accepts a pointer to a square matrix, and returns a pointer to an array containing the diagonal elements of the matrix.

SOLUTION

A subroutine with appropriate error checking is shown in Figure 15-18. This example subroutine accepts a pointer to a 2D square array, and returns the diagonal elements of the array in a 1D array that it allocates on a separate pointer. The subroutine checks the association status of the input pointer to ensure that it is currently associated, checks the array to make sure that it is square, and checks the association status of the output pointer to ensure that it is *not* currently associated. (The last test ensures that we don't accidentally reuse a pointer that is currently in use. Reusing the pointer might leave the original data inaccessible if there were no other pointer to it.) If any of the conditions fail, then an appropriate error flag is set and the subroutine returns to the calling program unit.

FIGURE 15-18
Subroutine to extract the diagonal elements from a square array. This subroutine illustrates the proper technique for working with pointers passed as calling arguments.

```
SUBROUTINE get_diagonal ( ptr_a, ptr_b, error )
!
!  Purpose:
!    To extract the diagonal elements from the rank two
!    square array pointed to by ptr_a, and store them in
!    a rank one array allocated on ptr_b. The following
!    error conditions are defined:
!       0 -- No error.
!       1 -- ptr_a not associated on input
!       2 -- ptr_b already associated on input
!       3 -- Array on ptr_a not sqare
!       4 -- Unable to allocate memory for ptr_b
!
!  Record of revisions:
!      Date        Programmer         Description of change
!      ====        ==========         =====================
!    01/03/16      S. J. Chapman      Original code
!
IMPLICIT NONE

! Data dictionary: declare calling parameter types & definitions
INTEGER, DIMENSION(:,:), POINTER :: ptr_a  ! Ptr to square array
INTEGER, DIMENSION(:),   POINTER :: ptr_b  ! Ptr to output array
INTEGER, INTENT(OUT) :: error              ! Errors flag

! Data dictionary: declare variable types & definitions
INTEGER :: i                               ! Loop counter
```

(continued)

(*concluded*)

```fortran
      INTEGER :: istat                      ! Allocate status
      INTEGER, DIMENSION(2) :: l_bound      ! Lower bounds on ptr_a
      INTEGER, DIMENSION(2) :: u_bound      ! Upper bounds on ptr_a
      INTEGER, DIMENSION(2) :: extent       ! Extent of array on ptr_a

      ! Check error conditions
      error_1: IF ( .NOT. ASSOCIATED ( ptr_a ) ) THEN
         error = 1
      ELSE IF ( ASSOCIATED ( ptr_b ) ) THEN
         error = 2
      ELSE
         ! Check for square array
         l_bound = LBOUND ( ptr_a )
         u_bound = UBOUND ( ptr_a )
         extent = u_bound - l_bound + 1
         error_3: IF ( extent(1) /= extent(2) ) THEN
            error = 3
         ELSE
            ! Everything is ok so far, allocate ptr_b.
            ALLOCATE ( ptr_b(extent(1)), STAT=istat )
            error_4: IF ( istat /= 0 ) THEN
               error = 4
            ELSE
               ! Everything is ok, extract diagonal.
               ok: DO i = 1, extent(1)
                  ptr_b(i) = ptr_a(l_bound(1)+i-1,l_bound(2)+i-1)
               END DO ok

               ! Reset error flag.
               error = 0
            END IF error_4
         END IF error_3
      END IF error_1

      END SUBROUTINE get_diagonal
```

A test driver program for this subroutine is shown in Figure 15-19. This program tests the first three possible error conditions, and also the proper operation of the subroutine when no error occurs. There is no easy way to get the memory allocation of `ptr_b` to fail, so there is no explicit test in the driver for that.

FIGURE 15-19
Test driver program for subroutine `get_diagonal`.

```fortran
PROGRAM test_diagonal
!
!  Purpose:
!     To test the diagonal extraction subroutine.
!
!  Record of revisions:
```

(*continued*)

(*continued*)

```fortran
!       Date        Programmer          Description of change
!       ====        ==========          =====================
!     01/03/16     S. J. Chapman         Original code
!
IMPLICIT NONE

! Declare interface to subroutine diagonal:
INTERFACE
   SUBROUTINE get_diagonal ( ptr_a, ptr_b, error )
   INTEGER, DIMENSION(:,:), POINTER :: ptr_a
   INTEGER, DIMENSION(:), POINTER :: ptr_b
   INTEGER, INTENT(OUT) :: error
   END SUBROUTINE get_diagonal
END INTERFACE

! Data dictionary: declare variable types & definitions
INTEGER :: i, j, k                       ! Loop counter
INTEGER :: istat                         ! Allocate status
INTEGER, DIMENSION(:,:), POINTER :: ptr_a ! Ptr to square array
INTEGER, DIMENSION(:), POINTER :: ptr_b   ! Ptr to output array
INTEGER :: error                         ! Errors flag

! Call diagonal with nothing defined to see what happens.
CALL get_diagonal ( ptr_a, ptr_b, error )
WRITE (*,*) 'No pointers allocated: '
WRITE (*,*) ' Error = ', error

! Allocate both pointers, and call the subroutine.
ALLOCATE (ptr_a(10,10), STAT=istat )
ALLOCATE (ptr_b(10), STAT=istat )
CALL get_diagonal ( ptr_a, ptr_b, error )
WRITE (*,*) 'Both pointers allocated: '
WRITE (*,*) ' Error = ', error

! Allocate ptr_a only, but with unequal extents.
DEALLOCATE (ptr_a, STAT=istat)
DEALLOCATE (ptr_b, STAT=istat)
ALLOCATE (ptr_a(-5:5,10), STAT=istat )
CALL get_diagonal ( ptr_a, ptr_b, error )
WRITE (*,*) 'Array on ptr_a not square: '
WRITE (*,*) ' Error = ', error

! Allocate ptr_a only, initialize, and get results.
DEALLOCATE (ptr_a, STAT=istat)
ALLOCATE (ptr_a(-2:2,0:4), STAT=istat )
k = 0
DO j = 0, 4
   DO i = -2, 2
      k = k + 1
      ptr_a(i,j) = k                    ! Store the numbers 1 .. 25
   END DO                               ! in row order in the array
END DO
```

(*continued*)

(*concluded*)

```
CALL get_diagonal ( ptr_a, ptr_b, error )
WRITE (*,*) 'ptr_a allocated & square; ptr_b not allocated: '
WRITE (*,*) ' Error = ', error
WRITE (*,*) ' Diag  = ', ptr_b

END PROGRAM test_diagonal
```

When the test driver program is executed, the results are:

```
C:\book\fortran\chap15>test_diagonal
No pointers allocated:
  Error =         1
Both pointers allocated:
  Error =         2
Array on ptr_a not square:
  Error =         3
ptr_a allocated & square; ptr_b not allocated:
  Error =         0
  Diag  =         1       7      13      19      25
```

All error were flagged properly, and the diagonal values are correct, so the subroutine appears to be working properly.

Good Programming Practice

Always test the association status of any pointers passed to a procedure as calling arguments. It is easy to make mistakes in a large program that result in an attempt to use an unassociated pointer, or an attempt to reallocate an already associated pointer (the latter case will produce a memory leak).

15.7.1 Using the INTENT Attribute with Pointers

If the INTENT attribute appears on a pointer dummy argument, it refers to the *pointer* and not to its target. Thus, if a subroutine has the following declaration

```
SUBROUTINE test(xval)
REAL,POINTER,DIMENSION(:),INTENT(IN) :: xval
...
```

then the pointer xval cannot be allocated, deallocated, or reassigned within the subroutine. However, the contents of the pointer's *target* can be changed. Therefore, the statement

```
xval(90:100) = -2.
```

would be legal within this subroutine if the target of the pointer has at least 100 elements.

15.7.2 Pointer-Valued Functions

It is also possible for a function to return a pointer value. If a function is to return a pointer, then the RESULT clause must be used in the function definition, and the RESULT variable must be declared to be a pointer. For example, the function in Figure 15-20 accepts a pointer to a rank 1 array, and returns a pointer to every fifth value in the array.

FIGURE 15-20
A pointer-valued function.

```
FUNCTION every_fifth (ptr_array) RESULT (ptr_fifth)
!
!   Purpose:
!     To produce a pointer to every fifth element in an
!     input rank one array.
!
!   Record of revisions:
!       Date        Programmer          Description of change
!       ====        ==========          =====================
!     01/03/16     S. J. Chapman        Original code
!
IMPLICIT NONE

! Data dictionary: declare calling parameter types & definitions
INTEGER, DIMENSION(:), POINTER :: ptr_array
INTEGER, DIMENSION(:), POINTER :: ptr_fifth

! Data dictionary: declare local variable types & definitions
INTEGER :: low              ! Array lower bound
INTEGER :: high             ! Array upper bound

low = LBOUND(ptr_array,1)
high = UBOUND(ptr_array,1)
ptr_fifth => ptr_array(low:high:5)

END FUNCTION every_fifth
```

A pointer-valued function must always have an explicit interface in any procedure that uses it. The explicit interface may be specified by an interface or by placing the function in a module and then using the module in the procedure. Once the function is defined, it can be used any place that a pointer expression can be used. For example, it can be used on the right-hand side of a pointer assignment statement as follows:

```
ptr_2 => every_fifth( ptr_1 )
```

The function can also be used in a location where an integer array is expected. In that case, the pointer returned by the function will automatically be dereferenced, and the values pointed to will be used. Thus, the following statement is legal, and will print out the values pointed to by the pointer returned from the function.

```
WRITE (*,*) every_fifth( ptr_1 )
```

As with any function, a pointer-valued function cannot be used on the left-hand side of an assignment statement.

15.8
PROCEDURE POINTERS

It is also possible for a Fortran pointer to refer to a *procedure* instead of a variable or array. A procedure pointer is declared by the statement:

```
PROCEDURE (proc), POINTER :: p => NULL()
```

This statement declares a pointer to a procedure that has the *same calling sequence* as procedure `proc`, which must have an explicit interface.

Once a procedure pointer is declared, a procedure can be assigned to it in the same fashion as for variables or arrays. For example, suppose that subroutine `sub1` has an explicit interface. Then a pointer to `sub1` could be declared as

```
PROCEDURE (sub1), POINTER :: p => NULL()
```

and the following assignment would be legal

```
p => sub1
```

After such an assignment, the following two subroutine calls are identical, producing exactly the same results.

```
CALL sub1(a, b, c)
CALL p(a, b, c)
```

Note that this pointer will work for *any* subroutine that has the same interface as `sub1`. For example, suppose that subroutines `sub1` and `sub2` both have the same interface (number, sequence, type, and intent of calling parameters). Then the first call to p below would call `sub1` and the second one would call `sub2`.

```
p => sub1
CALL p(a, b, c)
p => sub2
CALL p(a, b, c)
```

An example using function pointers is shown in Figure 15-21. This program declares three functions with the same signature in a module so that they have an explicit interface. The main program declares a procedure pointer of type `func1`, and it is useable with any function having the same signature as `func1`. The program assigns a function to the pointer based on user selection, and then evaluates the function using the pointer.

FIGURE 15-21
A program to store a database of names and phone numbers in a binary tree structure, and to retrieve a selected item from that tree.

```
MODULE test_functions
!
! Purpose:
!    Module containing test functions. The module creates
!    an explicit interface for the functions.
```

(continued)

(continued)

```
!
!   Record of revisions:
!      Date       Programmer          Description of change
!      ====       ==========          =====================
!    01/08/16    S. J. Chapman        Original code
!
IMPLICIT NONE

CONTAINS

   ! All of the following functions have the same signature,
   ! and they have an explicit interface because they are
   ! contained in a module.

   REAL FUNCTION func1(x)
   IMPLICIT NONE
   REAL,INTENT(IN) :: x
   func1 = x**2 - 2*x + 4
   END FUNCTION func1

   REAL FUNCTION func2(x)
   IMPLICIT NONE
   REAL,INTENT(IN) :: x
   func2 = exp(-x/5) * sin(2*x)
   END FUNCTION func2

   REAL FUNCTION func3(x)
   IMPLICIT NONE
   REAL,INTENT(IN) :: x
   func3 = cos(x)
   END FUNCTION func3

END MODULE test_functions

PROGRAM test_function_pointers
!
!   Purpose:
!     To test Fortran procedure pointers. The function
!     point will work with any procedure with an explicit
!     interface that has same signature as "func1".
!
!   Record of revisions:
!      Date       Programmer          Description of change
!      ====       ==========          =====================
!    01/08/16    S. J. Chapman        Original code
!
USE test_functions
IMPLICIT NONE

! Declare variables
INTEGER :: index                      ! Selection index
PROCEDURE(func1), POINTER :: p        ! Function pointer
REAL :: x                             ! Calling argument
```

(continued)

(*concluded*)

```
! Get the name of the file containing the input data.
WRITE (*,*) 'Select a function to associate with the pointer:'
WRITE (*,*) '   1: func1'
WRITE (*,*) '   2: func2'
WRITE (*,*) '   3: func3'
READ (*,*) index

! Is it valid?
IF ( (index < 1) .OR. (index > 3) ) THEN

   WRITE (*,*) 'Invalid selection made!'
   ERROR STOP 'Bad index'

ELSE

   ! Associate the pointer
   SELECT CASE (index)
      CASE (1)
         WRITE (*,*) 'func1 selected...'
         p => func1
      CASE (2)
         WRITE (*,*) 'func2 selected...'
         p => func2
      CASE (3)
         WRITE (*,*) 'func3 selected...'
         p => func3
   END SELECT

   ! Execute the function
   WRITE (*,'(A)',ADVANCE='NO') 'Enter x: '
   READ (*,*) x
   WRITE (*,'(A,F13.6)') 'f(x) = ', p(x)

END IF

END PROGRAM test_function_pointers
```

When this program is executed, the results are:

```
C:\book\fortran\chap15>test_function_pointers
Select a function to associate with the pointer:
   1: func1
   2: func2
   3: func3
3
 func3 selected...
Enter x: 3.14159
f(x) =     -1.000000
```

Since $\cos(\pi) = -1$, this is the correct answer.

Procedure pointers are very useful in Fortran programs, because a user can associate a specific procedure with a defined data type. For example, the following type declaration includes a pointer to a procedure that can invert the matrix declared in the derived data type.

```
TYPE matrix(m,n)
   INTEGER, LEN :: m,n
   REAL :: element(m,n)
   PROCEDURE (lu), POINTER :: invert
END TYPE
  :
TYPE(m=10,n=10) :: a
  :
CALL a%invert(...)
```

Note that this is different from binding the procedure to the data type in that binding is permanent, while the procedure pointed to by the function pointer can change during the course of program execution.

■ 15.9
BINARY TREE STRUCTURES

We have already seen one example of a dynamic data structure: the linked list. Another very important dynamic data structure is the **binary tree.** A binary tree consists of repeated components (or **nodes**) arranged in an inverted tree structure. Each component or node is a variable of a derived data type that stores some sort of data plus *two* pointers to other variables of the same data type. A sample derived data type might be:

```
TYPE :: person
   CHARACTER(len=10) :: last
   CHARACTER(len=10) :: first
   CHARACTER :: mi
   TYPE (person), POINTER :: before
   TYPE (person), POINTER :: after
END TYPE
```

This data type is illustrated in Figure 15-22. It could be extended to include further information about each person such as address, phone number, social security number, etc.

An important requirement for binary trees is that *the components must be sortable according to some known criterion.* For our example, the components may be sortable alphabetically by last name, first name, and middle initial. If the pointers in a component are associated, then the pointer before must point to another

FIGURE 15-22
A typical component of a binary tree.

component that falls before the current component in the sorting order, and the pointer `after` must point to another component that falls after the current component in the sorting order.

Binary trees start from a single node (the *root node*) that is the first value read into the program. When the first value is read, a variable is created to hold it, and the two pointers in the variable are nullified. When the next value is read, a new node is created to hold it, and it is compared to the value in the root node. If the new value is less than the value in the root node, then the `before` pointer of the root node is set to point to the new variable. If the new value is greater than the value in the root node, then the `after` pointer of the root node is set to point to the new variable. If a value is greater than the value in the root node but the `after` pointer is already in use, then we compare the new value to the value in the node pointed to by the `after` pointer, and insert the new node in the proper position below that node. This process is repeated as new values are added, producing nodes arranged in an inverted tree structure, with their values in order.

This process is best illustrated by an example. Let's add the following names to a binary tree structure consisting of variables of the type defined above.

```
Jackson, Andrew D
Johnson, James R
Johnson, Jessie R
Johnson, Andrew C
Chapman, Stephen J
Gomez, Jose A
Chapman, Rosa P
```

The first name read in is "Jackson, Andrew D". Since there is no other data yet, this name is stored in node 1 that becomes the root node of the tree, and both of the pointers in the variable are nullified (see Figure 15-23*a*). The next name read in is "Johnson, James R". This name is stored in node 2, and both pointers in the new variable are nullified. Next, the new value is compared to the root node. Since it is greater than the value in the root node, the pointer `after` of the root node is set to point to the new variable (see Figure 15-23*b*).

The third name read in is "Johnson, Jessie R". This name is stored in node 3, and both pointers in the new variable are nullified. Next, the new value is compared to the root node. It is greater than the value in the root node, but the `after` point of the root node already points to node 2, so we compare the new variable with the value in node 2. That value is "Johnson, James R". Since the new value is greater than that value, the new variable is attached below node 2, and the `after` pointer of node 2 is set to point to it (see Figure 15-23*c*).

The fourth name read in is "Johnson, Andrew C". This name is stored in node 4, and both pointers in the new variable are nullified. Next, the new value is compared to the root node. It is greater than the value in the root node, but the `after` point of the root node already points to node 2, so we compare the new variable with the value in node 2. That value is "Johnson, James R". Since the new value is less than that value, the new variable is attached below node 2, and the `before` pointer of node 2 is set to point to it (see Figure 15-23*d*).

The fifth name read in is "Chapman, Stephen J". This name is stored in node 5, and both pointers in the new variable are nullified. Next, the new value is compared to the root node. Since the new value is less than that value, the new variable is attached below the root node, and the before pointer of the root node is set to point to it (see Figure 15-23*e*).

The sixth name read in is "Gomez, Jose A". This name is stored in node 6, and both pointers in the new variable are nullified. Next, the new value is compared to the root node. It is less than the value in the root node, but the before point of the root node already points to node 5, so we compare the new variable with the value in node 5. That value is "Chapman, Stephen J". Since the new value is greater than that

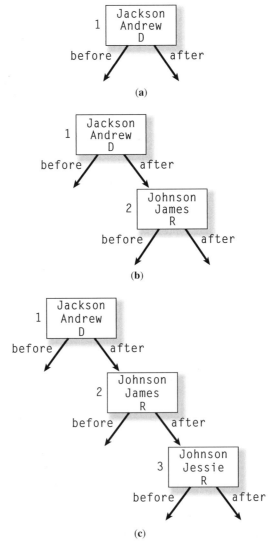

FIGURE 15-23
The development of a binary tree structure.

Pointers and Dynamic Data Structures

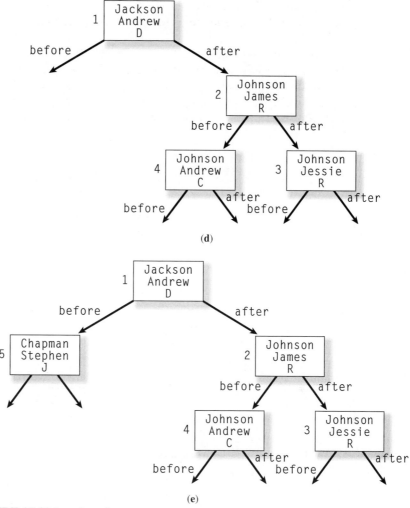

FIGURE 15-23 (*continued*)
The development of a binary tree structure.

value, the new variable is attached below node 5, and the after pointer of node 5 is set to point to it (see Figure 15-23f).

The seventh name read in is "Chapman, Rosa P". This name is stored in node 7, and both pointers in the new variable are nullified. Next, the new value is compared to the root node. It is less than the value in the root node, but the before point of the root node already points to node 5, so we compare the new variable with the value in node 5. That value is "Chapman, Stephen J". Since the new value is less than that value, the new variable is attached below node 5, and the before pointer of node 5 is set to point to it (see Figure 15-23g).

This process can be repeated indefinitely as more data values are added to the tree.

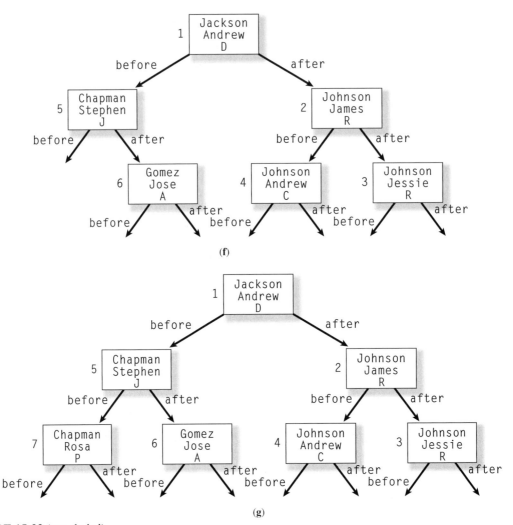

FIGURE 15-23 (*concluded*)
The development of a binary tree structure.

15.9.1 The Significance of Binary Tree Structures

Now let's examine the completed structure in Figure 15-23g. Notice that when the tree is finished, the values are arranged in *sorted order from left to right across the structure*. This fact means that the binary tree can be used as a way to sort a data set (Figure 15-24). (In this application, it is similar to the insertion sort described earlier in the chapter.)

However, there is something far more important about this data structure than the fact that it is sorted. Suppose that we wanted to search for a particular name in the original list of names. Depending on where the name appears in the list, we would

Pointers and Dynamic Data Structures 741

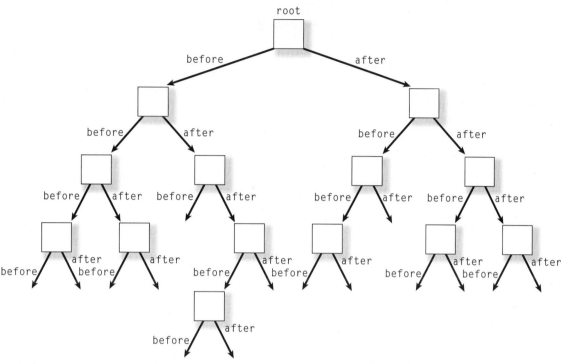

FIGURE 15-24
A binary tree structure whose lowest branches are not completely filled in.

have to check from one to seven names before locating the one we wanted. On the average, we would have to search 3½ names before spotting the desired one. In contrast, if the names are arranged in a binary tree structure, then starting from the root node *no more than three checks would be required to locate any particular name*. A binary tree is a very efficient way to search for and retrieve data values.

This advantage increases rapidly as the size of the database to be searched increases. For example, suppose that we have 32,767 values in a database. If we search through the linear list to try to find a particular value, from 1 to 32,767 values would have to be searched, and the average search length would be 16,384. In contrast, 32,767 values can be stored in a binary tree structure consisting of only 15 layers, so the maximum number of values to search to find any particular value would be 15! Binary trees are a *very* efficient way to store data for easy retrieval.

In practice, binary trees may not be quite this efficient. Since the arrangement of the nodes in a binary tree depends on the order in which data was read in, it is possible that there may be more layers of nodes in some parts of the tree than in others. In that case, there may be a few extra layers to search to find some of the values. However, the efficiency of a binary tree is so much greater than that of a linear list that binary trees are still better for data storage and retrieval.

The worst sort of data to store in a binary tree is sorted data. If sorted data is read, then each value is larger than the previous one, and so each new node is placed after

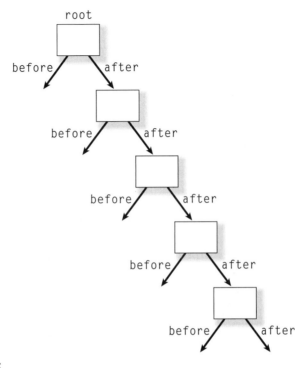

FIGURE 15-25
A binary tree resulting from sorted input data. Note that the tree has just become a list, and all of the advantages of the binary tree structure have been lost.

the previous one. In the end, we wind up with a binary tree consisting of only one branch, which just reproduces the structure of the original list (see Figure 15-25). The best sort of data to store in a binary tree is random data, since random values will fill in all branches of the tree roughly equally.

Many databases are structured as binary trees. These databases often include special techniques called *hashing techniques* to partially randomize the order of the data stored in the database, and so avoid the situation shown in Figure 15-25. They also often include special procedures to even out the bottom branches of the binary tree in order to make searching for data in the tree faster.

15.9.2 Building a Binary Tree Structure

Because each node of a binary tree looks and behaves just like any other node, binary trees are perfectly suited to recursive procedures. For example, suppose that we would like to add a value to a binary tree. A program could read the new value, create a new node for it, and call a subroutine named insert_node to insert the node into the tree. The subroutine will first be called with a pointer to the root node. The root node becomes the "current node" for the subroutine. If the current node doesn't exist, then

Pointers and Dynamic Data Structures

it will add the new node at that location. If the current node does exist, then it will compare the value in the current node to the value in the new node. If the value in the new node is less than the value in the current node, then the subroutine will call itself recursively using the `before` pointer from the current node. If the value in the new node is greater than the value in the current node, then the subroutine will call itself recursively using the `after` pointer from the current node. Subroutine `insert_node` will continue to call itself recursively until it reaches the bottom of the tree and locates the proper place to insert the new node.

Similar recursive subroutines can be written to retrieve specific values from the binary tree, or to write out all of the values in the tree in sorted order. The following example will illustrate the construction of a binary tree.

EXAMPLE 15-4

Storing and Retrieving Data in a Binary Tree:

Suppose that we would like to create a database containing the names and telephone numbers of a group of people. (This structure could easily accommodate more information about each person, but we will keep it simple for the purposes of this example.) Write a program to read the names and phone numbers, and store them in a binary tree. After reading all of the names, the program should be able to print out all of the names and phone numbers in alphabetical order. In addition, it should be able to recover the phone number of any individual given his or her name. Use recursive subroutines to implement the binary tree functions.

SOLUTION

The information about each person will be stored in a binary tree. We must create a derived data type to hold the information contained in each node: name, telephone number, and pointers to two other nodes. An appropriate derived data type is:

```
TYPE :: node
   CHARACTER(len=10) :: last
   CHARACTER(len=10) :: first
   CHARACTER :: mi
   CHARACTER(len=16) :: phone
   TYPE (node), POINTER :: before
   TYPE (node), POINTER :: after
END TYPE
```

The main program will read names and phone numbers from an input data file, and create nodes to hold them. When each node is created, it will call a recursive subroutine `insert_node` to locate the proper place in the tree to put the new node. Once all of the names and phone numbers are read in, the main program will call recursive subroutine `write_node` to list out all names and phone numbers in alphabetical order. Finally, the program will prompt the user to provide a name, and it will call recursive subroutine `find_node` to get the phone number associated with that name.

Note that for a binary tree to work, *there must be a way to compare two values of the derived data type representing each node*. In our case, we wish to sort and compare the data by last name, first name, and middle initial. Therefore, we will create extended definitions for the operators >, <, and ==, so that they can work with the derived data type.

1. **State the problem.**

   Write a program that reads a list of names and phone numbers from an input file, and stores them in a binary tree structure. After reading in all of the names, the program will print out all of the names and phone numbers in alphabetical order. Then, it will prompt the user for a specific name, and retrieve the phone number associated with that name. It will use recursive subroutines to implement the binary tree functions.

2. **Define the inputs and outputs.**

   The inputs to the program are a file name, and a list of names and phone numbers within the file. The names and phone numbers will be in the order: last, first, middle initial, phone number.

   The outputs from the program will be:

   (a) A list of all names and phone numbers in alphabetical order.
   (b) The phone number associated with a user-specified name.

3. **Describe the algorithm.**

   The basic pseudocode for the main program is

   ```
   Get input file name
   Read input data and store in binary tree
   Write out data in alphabetical order
   Get specific name from user
   Recover and display phone number associated with that name
   ```

   The data to be stored in the binary tree will be read from the input file using a while loop, and stored using recursive subroutine `add_node`. Once all of the data has been read, the sorted data will be written out to the standard output device using subroutine `write_node`, and then the user will be prompted to input the name of the record to find. Subroutine `find_node` will be used to search for the record. If the record is found, it will be displayed. The detailed pseudocode for the main program is:

   ```
   Prompt user for the input file name "filename"
   Read the file name "filename"
   OPEN file "filename"
   IF OPEN is successful THEN
      WHILE
         Create new node using pointer "temp"
         Read value into temp
         IF read not successful EXIT
         CALL add_node(root, temp) to put item in tree
      End of WHILE
      Call write_node(root) to write out sorted data
      Prompt user for name to recover; store in "temp"
      CALL find_node(root, temp, error)
      Write the data to the standard output device
   End of IF
   ```

   It is necessary to create a module containing the definition of the derived data type and the three recursive subroutines required to manipulate the binary tree structure. To add a node to the tree, we should start by looking at the root node. If the root node does not exist, then the new node will become the root node. If the root node exists, then we should compare the name in the new node to the name in the root node to determine if the new node is alphabetically less than or greater than the root node. If it is less, then we should check the `before` pointer of the root node. If that pointer is null, then we will add the new node there. Otherwise, we will check the node pointed to by the `before` pointer, and repeat the process. If the new node is alphabetically greater than or equal to the root node, then we should check the `after` pointer of the root node. If that pointer is null, then we will add the new node there. Otherwise, we will check the node pointed to by the `after` pointer, and repeat the process.

For each node we examine, we perform the same steps:

(a) Determine whether the new node is < or >= the current node.
(b) If it is less than the current node and the `before` pointer is null, add the new node there.
(c) If it is less than the current node and the `before` pointer is not null, examine the node pointed to.
(d) If the new node is greater than or equal to the current node and the `after` pointer is null, add the new node there.
(e) If it is greater than or equal to the current node and the `after` pointer is not null, examine the node pointed to.

Since the same pattern repeats over and over again, we can implement `add_node` as a recursive subroutine.

```
IF ptr is not associated THEN
   ! There is no tree yet. Add the node right here.
   ptr => new_node
ELSE IF new_node < ptr THEN
   ! Check to see if we can attach new node here.
   IF ptr%before is associated THEN
      ! Node in use, so call add_node recursively
      CALL add_node ( ptr%before, new_node )
   ELSE
      ! Pointer not in use. Add node here.
      ptr%before => new_node
   END of IF
ELSE
   ! Check to see if we can attach new node to after ptr.
   IF ptr%after is associated THEN
      ! Node in use, so call add_node recursively
      CALL add_node ( ptr%after, new_node )
   ELSE
      ! Pointer not in use. Add node here.
      ptr%after => new_node
   END of IF
END of IF
```

Subroutine `write_node` is a recursive subroutine to write out the values in the tree in alphabetical order. To do this, it starts at the root node and works its way down to the leftmost branch in the tree. Then, it works its way along from left to right through the structure. The pseudocode is shown below:

```
IF pointer "before" is associated THEN
   CALL write_node ( ptr%before )
END of IF
WRITE contents of current node
IF pointer "after" is associated THEN
   CALL write_node ( ptr%after )
END of IF
```

Subroutine `find_node` is a recursive subroutine to locate a particular node in the tree. To find a node in the tree, we start by looking at the root node. We should compare the name we are searching for to the name in the root node to determine if the name we want is alphabetically less than or greater than the root node. If it is less, then we should check the `before` pointer of the root node. If that pointer is null, then the desired node does not exist. Otherwise, we will check the node pointed to by the `before` pointer, and repeat the process. If the name we are searching for is alphabetically greater than or equal to the root node, then we should check the `after` pointer of the root node. If that pointer is null, then the desired node does not exist. Otherwise, we will check the node pointed to by the `after` pointer, and repeat the process. If the name we are searching for is equal to the root node, then the root node contains the data we want, and we will return it. This process is repeated recursively for each node called until either the desired data is found or a null pointer is reached. The pseudocode is shown below:

```
IF search_value < ptr THEN
   IF ptr%before is associated THEN
      CALL find_node ( ptr%before, search_value, error )
   ELSE ! not found
      error ← 1
   END of IF
ELSE IF search_value == ptr THEN
   search_value = ptr
   error ← 0
ELSE
   IF ptr%after is associated THEN
      CALL find_node ( ptr%after, search_value, error )
   ELSE ! not found
      error ← 1
   END of IF
END of IF
```

It is necessary to include in the module the definition of the derived data type and the definitions of the >, <, and == operators for that data type. To do this, we will include three INTERFACE OPERATOR blocks in the module. In addition, we must write the three private functions that implement the operators. The first function is called `greater_than`, the second one is called `less_than`, and the third one is called `equal_to`. These functions must compare the two last names to decide whether the first is greater, less, or the same as the second. If they are the same, then the functions must compare the two first names and middle initials. Note that all names should be

Pointers and Dynamic Data Structures

shifted to uppercase to avoid mixing upper- and lowercase during the comparisons. This will be done using a subroutine called `ushift`, which in turn calls the subroutine `ucase` that we developed in Chapter 10. The pseudocode for function `greater_than` is:

```
IF last1 > last2 THEN
   greater_than = .TRUE.
ELSE IF last1 < last2 THEN
   greater_than = .FALSE.
ELSE ! Last names match
   IF first1 > first2 THEN
      greater_than = .TRUE.
   ELSE IF first1 < first2 THEN
      greater_than = .FALSE.
   ELSE ! First names match
      IF mi1 > mi2 THEN
         greater_than = .TRUE.
      ELSE
         greater_than = .FALSE.
      END of IF
   END of IF
END of IF
```

The pseudocode for function `less_than` is:

```
IF last1 < last2 THEN
   less_than = .TRUE.
ELSE IF last1 > last2 THEN
   less_than = .FALSE.
ELSE ! Last names match
   IF first1 < first2 THEN
      less_than = .TRUE.
   ELSE IF first1 > first2 THEN
      less_than = .FALSE.
   ELSE ! First names match
      IF mi1 < mi2 THEN
         less_than = .TRUE.
      ELSE
         less_than = .FALSE.
      END of IF
   END of IF
END of IF
```

The pseudocode for function `equal_to` is:

```
IF last1 == last2 .AND. first1 == first2 .AND. mi1 == mi2 THEN
   equal_to = .TRUE.
ELSE
   equal_to = .FALSE.
END of IF
```

4. **Turn the algorithm into Fortran statements.**

The resulting Fortran program is shown in Figure 15-26. Module `btree` contains the definition of the derived data type and all of the supporting subroutines and functions, as well as defining the operators >, <, and == for the derived data type. Note that only the essential procedures in the module are PUBLIC. The main program accesses the procedures in the module by USE association, so the procedures have an explicit interface.

FIGURE 15-26
A program to store a database of names and phone numbers in a binary tree structure, and to retrieve a selected item from that tree.

```
MODULE btree
!
!  Purpose:
!    To define the derived data type used as a node in the
!    binary tree, and to define the operations >, <. and ==
!    for this data type. This module also contains the
!    subroutines to add a node to the tree, write out the
!    values in the tree, and find a value in the tree.
!
!  Record of revisions:
!      Date       Programmer          Description of change
!      ====       ==========          =====================
!    01/04/16    S. J. Chapman         Original code
!
IMPLICIT NONE

! Restrict access to module contents.
PRIVATE
PUBLIC :: node, OPERATOR(>), OPERATOR(<), OPERATOR(==)
PUBLIC :: add_node, write_node, find_node

! Declare type for a node of the binary tree.
TYPE :: node
   CHARACTER(len=10) :: last
   CHARACTER(len=10) :: first
   CHARACTER :: mi
   CHARACTER(len=16) :: phone
   TYPE (node), POINTER :: before
   TYPE (node), POINTER :: after
END TYPE

INTERFACE OPERATOR (>)
   MODULE PROCEDURE greater_than
END INTERFACE

INTERFACE OPERATOR (<)
   MODULE PROCEDURE less_than
END INTERFACE

INTERFACE OPERATOR (==)
   MODULE PROCEDURE equal_to
END INTERFACE
```

(continued)

(continued)

```
CONTAINS
   RECURSIVE SUBROUTINE add_node (ptr, new_node)
   !
   ! Purpose:
   !    To add a new node to the binary tree structure.
   !
   TYPE (node), POINTER :: ptr       ! Pointer to current pos. in tree
   TYPE (node), POINTER :: new_node  ! Pointer to new node

   IF ( .NOT. ASSOCIATED(ptr) ) THEN
      ! There is no tree yet. Add the node right here.
      ptr => new_node
   ELSE IF ( new_node < ptr ) THEN
      IF ( ASSOCIATED(ptr%before) ) THEN
         CALL add_node ( ptr%before, new_node )
      ELSE
         ptr%before => new_node
      END IF
   ELSE
      IF ( ASSOCIATED(ptr%after) ) THEN
         CALL add_node ( ptr%after, new_node )
      ELSE
         ptr%after => new_node
      END IF
   END IF
   END SUBROUTINE add_node

   RECURSIVE SUBROUTINE write_node (ptr)
   !
   ! Purpose:
   !    To write out the contents of the binary tree
   !    structure in order.
   !
   TYPE (node), POINTER :: ptr   ! Pointer to current pos. in tree

   ! Write contents of previous node.
   IF ( ASSOCIATED(ptr%before) ) THEN
      CALL write_node ( ptr%before )
   END IF

   ! Write contents of current node.
   WRITE (*,"(A,', ',A,1X,A)") ptr%last, ptr%first, ptr%mi

   ! Write contents of next node.
   IF ( ASSOCIATED(ptr%after) ) THEN
      CALL write_node ( ptr%after )
   END IF
   END SUBROUTINE write_node

   RECURSIVE SUBROUTINE find_node (ptr, search, error)
   !
   ! Purpose:
   !    To find a particular node in the binary tree structure.
```

(continued)

(*continued*)

```
!    "Search" is a pointer to the name to find, and will
!    also contain the results when the subroutine finishes
!    if the node is found.
!
TYPE (node), POINTER :: ptr      ! Pointer to curr pos. in tree
TYPE (node), POINTER :: search   ! Pointer to value to find.
INTEGER :: error                 ! Error: 0 = ok, 1 = not found

IF ( search < ptr ) THEN
   IF ( ASSOCIATED(ptr%before) ) THEN
      CALL find_node (ptr%before, search, error)
   ELSE
      error = 1
   END IF
ELSE IF ( search == ptr ) THEN
   search = ptr
   error = 0
ELSE
   IF ( ASSOCIATED(ptr%after) ) THEN
      CALL find_node (ptr%after, search, error)
   ELSE
      error = 1
   END IF
END IF
END SUBROUTINE find_node

LOGICAL FUNCTION greater_than (op1, op2)
!
!  Purpose:
!    To test to see if operand 1 is > operand 2
!    in alphabetical order.
!
TYPE (node), INTENT(IN) :: op1, op2
CHARACTER(len=10) :: last1, last2, first1, first2
CHARACTER :: mi1, mi2

CALL ushift (op1, last1, first1, mi1 )
CALL ushift (op2, last2, first2, mi2 )

IF (last1 > last2) THEN
   greater_than = .TRUE.
ELSE IF (last1 < last2) THEN
   greater_than = .FALSE.
ELSE ! Last names match
   IF (first1 > first2) THEN
      greater_than = .TRUE.
   ELSE IF (first1 < first2) THEN
      greater_than = .FALSE.
   ELSE ! First names match
      IF (mi1 > mi2) THEN
         greater_than = .TRUE.
      ELSE
```

(*continued*)

(continued)

```
                    greater_than = .FALSE.
               END IF
          END IF
     END IF
  END FUNCTION greater_than
  LOGICAL FUNCTION less_than (op1, op2)
  !
  !  Purpose:
  !     To test to see if operand 1 is < operand 2
  !     in alphabetical order.
  !
  TYPE (node), INTENT(IN) :: op1, op2
  CHARACTER(len=10) :: last1, last2, first1, first2
  CHARACTER :: mi1, mi2

  CALL ushift (op1, last1, first1, mi1 )
  CALL ushift (op2, last2, first2, mi2 )

  IF (last1 < last2) THEN
     less_than = .TRUE.
  ELSE IF (last1 > last2) THEN
     less_than = .FALSE.
  ELSE ! Last names match
     IF (first1 < first2) THEN
        less_than = .TRUE.
     ELSE IF (first1 > first2) THEN
        less_than = .FALSE.
     ELSE ! First names match
        IF (mi1 < mi2) THEN
           less_than = .TRUE.
        ELSE
           less_than = .FALSE.
        END IF
     END IF
  END IF
  END FUNCTION less_than

  LOGICAL FUNCTION equal_to (op1, op2)
  !
  !  Purpose:
  !     To test to see if operand 1 is equal to operand 2
  !     alphabetically.
  !
  TYPE (node), INTENT(IN) :: op1, op2

  CHARACTER(len=10) :: last1, last2, first1, first2
  CHARACTER :: mi1, mi2

  CALL ushift (op1, last1, first1, mi1 )
  CALL ushift (op2, last2, first2, mi2 )

  IF ( (last1 == last2) .AND. (first1 == first2) .AND. &
       (mi1 == mi2 ) ) THEN
```

(continued)

(*continued*)

```fortran
      equal_to = .TRUE.
   ELSE
      equal_to = .FALSE.
   END IF
   END FUNCTION equal_to

   SUBROUTINE ushift( op, last, first, mi )
   !
   !  Purpose:
   !    To create upshifted versions of all strings for
   !    comparison.
   !
   TYPE (node), INTENT(IN) :: op
   CHARACTER(len=10), INTENT(INOUT) :: last, first
   CHARACTER, INTENT(INOUT) :: mi

   last = op%last
   first = op%first
   mi = op%mi
   CALL ucase (last)
   CALL ucase (first)
   CALL ucase (mi)
   END SUBROUTINE ushift

   SUBROUTINE ucase ( string )
   !
   !  Purpose:
   !    To shift a character string to upper case on any processor,
   !    regardless of collating sequence.
   !
   !  Record of revisions:
   !      Date        Programmer          Description of change
   !      ====        ==========          =====================
   !    11/28/15     S. J. Chapman        Original code
   !
   IMPLICIT NONE

   ! Declare calling parameters:
   CHARACTER(len=*), INTENT(INOUT) :: string

   ! Declare local variables:
   INTEGER :: i                ! Loop index
   INTEGER :: length           ! Length of input string

   ! Get length of string
   length = LEN ( string )

   ! Now shift lower case letters to upper case.
   DO i = 1, length
      IF ( LGE(string(i:i),'a') .AND. LLE(string(i:i),'z') ) THEN
         string(i:i) = ACHAR ( IACHAR ( string(i:i) ) - 32 )
      END IF
   END DO

   END SUBROUTINE ucase

END MODULE btree
```

(*continued*)

(*continued*)

```fortran
PROGRAM binary_tree
!
!  Purpose:
!    To read in a series of random names and phone numbers
!    and store them in a binary tree. After the values are
!    stored, they are written out in sorted order. Then the
!    user is prompted for a name to retrieve, and the program
!    recovers the data associated with that name.
!
!  Record of revisions:
!      Date         Programmer         Description of change
!      ====         ==========         =====================
!    01/04/16      S. J. Chapman       Original code
!
USE btree
IMPLICIT NONE

! Data dictionary: declare variable types & definitions
INTEGER :: error                     ! Error flag: 0=success
CHARACTER(len=20) :: filename        ! Input data file name
INTEGER :: istat                     ! Status: 0 for success
CHARACTER(len=120) :: msg            ! Error message
TYPE (node), POINTER :: root         ! Pointer to root node
TYPE (node), POINTER :: temp         ! Temp pointer to node

! Nullify new pointers
NULLIFY ( root, temp )

! Get the name of the file containing the input data.
WRITE (*,*) 'Enter the file name with the input data: '
READ (*,'(A20)') filename

! Open input data file. Status is OLD because the input data must
! already exist.
OPEN ( UNIT=9, FILE=filename, STATUS='OLD', ACTION='READ', &
       IOSTAT=istat, IOMSG=msg )

! Was the OPEN successful?
fileopen: IF ( istat == 0 ) THEN           ! Open successful

   ! The file was opened successfully, allocate space for each
   ! node, read the data into that node, and insert it into the
   ! binary tree.
   input: DO
      ALLOCATE (temp,STAT=istat)                ! Allocate node
      NULLIFY ( temp%before, temp%after)        ! Nullify pointers

      READ (9, 100, IOSTAT=istat) temp%last, temp%first, &
         temp%mi, temp%phone                    ! Read data
      100 FORMAT (A10,1X,A10,1X,A1,1X,A16)
      IF ( istat /= 0 ) EXIT input              ! Exit on end of data
      CALL add_node(root, temp)                 ! Add to binary tree
   END DO input
```

(*continued*)

(*concluded*)

```fortran
   ! Now, write out the sorted data.
   WRITE (*,'(/,A)') 'The sorted data list is: '
   CALL write_node(root)

   ! Prompt for a name to search for in the tree.
   WRITE (*,'(/,A)') 'Enter name to recover from tree:'
   WRITE (*,'(A)',ADVANCE='NO') 'Last Name:      '
   READ (*,'(A)') temp%last
   WRITE (*,'(A)',ADVANCE='NO') 'First Name:     '
   READ (*,'(A)') temp%first
   WRITE (*,'(A)',ADVANCE='NO') 'Middle Initial: '
   READ (*,'(A)') temp%mi

   ! Locate record
   CALL find_node ( root, temp, error )
   check: IF ( error == 0 ) THEN
      WRITE (*,'(/,A)') 'The record is:'
      WRITE (*,'(7A)') temp%last, ', ', temp%first, ' ', &
                       temp%mi, '  ', temp%phone
   ELSE
      WRITE (*,'(/,A)') 'Specified node not found!'
   END IF check

ELSE fileopen

   ! Else file open failed. Tell user.
   WRITE (*,'(A,I6)') 'File open failed--status = ', istat
   WRITE (*,'(A)') msg

END IF fileopen

END PROGRAM binary_tree
```

5. **Test the resulting Fortran programs.**

To test this program, we will create an input data file containing names and telephone numbers, and we will execute the program with that data. The file "tree_in.dat" will be created containing the following data:

```
Leroux      Hector     A  (608) 555-1212
Johnson     James      R  (800) 800-1111
Jackson     Andrew     D  (713) 723-7777
Romanoff    Alexi      N  (212) 338-3030
Johnson     Jessie     R  (800) 800-1111
Chapman     Stephen    J  (713) 721-0901
Nachshon    Bini       M  (618) 813-1234
Ziskend     Joseph     J  (805) 238-7999
Johnson     Andrew     C  (504) 388-3000
Chi         Shuchung   F  (504) 388-3123
deBerry     Jonathan   S  (703) 765-4321
Chapman     Rosa       P  (713) 721-0901
Gomez       Jose       A  (415) 555-1212
Rosenberg   Fred       R  (617) 123-4567
```

We will execute the program twice. Once we will specify a valid name to look up and once we will specify an invalid one to test that the program is working properly in both cases. When the program is executed, the results are:

```
C:\book\fortran\chap15>binary_tree
Enter the file name with the input data:
tree_in.dat

The sorted data list is:
Chapman    , Rosa       P
Chapman    , Stephen    J
Chi        , Shuchung   F
deBerry    , Jonathan   S
Gomez      , Jose       A
Jackson    , Andrew     D
Johnson    , Andrew     C
Johnson    , James      R
Johnson    , Jessie     R
Leroux     , Hector     A
Nachshon   , Bini       M
Romanoff   , Alexi      N
Rosenberg  , Fred       R
Ziskend    , Joseph     J

Enter name to recover from tree:
Last Name:      Nachshon
First Name:     Bini
Middle Initial: M

The record is:
Nachshon   , Bini       M   (618) 813-1234

C:\book\fortran\chap15>binary_tree
Enter the file name with the input data:
tree_in.dat

The sorted data list is:
Chapman    , Rosa       P
Chapman    , Stephen    J
Chi        , Shuchung   F
deBerry    , Jonathan   S
Gomez      , Jose       A
Jackson    , Andrew     D
Johnson    , Andrew     C
Johnson    , James      R
Johnson    , Jessie     R
Leroux     , Hector     A
Nachshon   , Bini       M
Romanoff   , Alexi      N
Rosenberg  , Fred       R
Ziskend    , Joseph     J

Enter name to recover from tree:
Last Name:      Johnson
First Name:     James
Middle Initial: A

Specified node not found!
```

The program appears to be working. Please note that it properly stores the data into the binary tree regardless of capitalization (deBerry is in the proper place).

Can you determine what the tree structure that the program created looked like? What is the maximum number of layers that the program must search through to find any particular data item in this tree?

15.10
SUMMARY

A pointer is special type of variable that contains the *address* of another variable instead of containing a value. A pointer has a specified data type and (if it points to an array) rank, and it can *only* point to data items of that particular type and rank. Pointers are declared with the POINTER attribute in a type declaration statement or in a separate POINTER statement. The data item pointed to by a pointer is called a target. Only data items declared with the TARGET attribute in a type declaration statement or in a separate TARGET statement can be pointed to by pointers.

A pointer assignment statement places the address of a target in a pointer. The form of the statement is

```
pointer => target
pointer1 => pointer2
```

In the latter case, the address currently contained in `pointer2` is placed in `pointer1`, and both pointers independently point to the same target.

A pointer can have one of three possible association statuses: undefined, associated, or disassociated. When a pointer is first declared in a type declaration statement, its pointer association status is undefined. Once a pointer has been associated with a target by a pointer assignment statement, its association status becomes associated. If a pointer is later disassociated from its target and is not associated with any new target, then its association status becomes disassociated. A pointer should always be nullified or associated as soon as it is created. The function ASSOCIATED( ) can be used to determine the association status of a pointer.

Pointers can be used to dynamically create and destroy variables or arrays. Memory is allocated for data items in an ALLOCATE statement, and deallocated in a DEALLOCATE statement. The pointer in the ALLOCATE statement points to the data item that is created, and is the *only* way to access that data item. If that pointer is disassociated or is associated with another target before another pointer is set to point to the allocated memory, then the memory becomes inaccessible to the program. This is called a "memory leak".

When dynamic memory is deallocated in a DEALLOCATE statement, the pointer to the memory is automatically nullified. However, if there are other pointers pointing to that same memory, they must be manually nullified or reassigned. If not, the program might attempt to use them to read or write to the deallocated memory location, with potentially disastrous results.

Pointers may be used as components of derived data types, including the data type being defined. This feature permits us to create dynamic data structures such as linked lists and binary trees, where the pointers in one dynamically allocated data item point to the next item in the chain. This flexibility is extraordinarily useful in many problems.

It is not possible to declare an array of pointers, since the DIMENSION attribute in a pointer declaration refers to the dimension of the target, not the dimension of the pointer. When array of pointers are needed, they can be created by defining a derived data type containing only a pointer, and then creating an array of that derived data type.

Pointers may be passed to procedures as calling arguments provided that the procedure has an explicit interface in the calling program. A dummy pointer argument must not have an INTENT attribute. It is also possible for a function to return a pointer value if the RESULT clause is used and the result variable is declared to be a pointer.

15.10.1 Summary of Good Programming Practice

The following guidelines should be adhered to when working with the pointers:

1. Always nullify or assign all pointers in a program unit as soon as they are created. This eliminates any possible ambiguities associated with the undefined allocation status.
2. When sorting or swapping large arrays or derived data types, it is more efficient to exchange pointers to the data than it is to manipulate the data itself.
3. Always nullify or reassign *all* pointers to a memory location when that memory is deallocated. One of them will be automatically nullified by the DEALLOCATE statement, and any others must be manually nullified in NULLIFY statement(s) or reassigned in pointer assignment statements.
4. Always test the association status of any pointers passed to procedures as calling arguments. It is easy to make mistakes in a large program that result in an attempt to use an unassociated pointer, or an attempt to reallocate an already associated pointer (the latter case will produce a memory leak).

15.10.2 Summary of Fortran Statements and Structures

POINTER Attribute:

```
type, POINTER :: ptr1 [, ptr2, ... ]
```

Examples:

```
INTEGER, POINTER :: next_value
REAL, DIMENSION(:), POINTER :: array
```

Description:
The POINTER attribute declares the variables in the type definition statement to be pointers.

POINTER Statement:

        POINTER :: ptr1 [, ptr2, ...]

Example:

        POINTER :: p1, p2, p3

Description:
The POINTER statement declares the variables in its list to be pointers. It is generally preferable to use the pointer attribute in a type declaration statement to declare a pointer instead of this statement.

TARGET Attribute:

        type, TARGET :: var1 [, var2, ...]

Examples:

        INTEGER, TARGET :: num_values
        REAL, DIMENSION(100), TARGET :: array

Description:
The TARGET attribute declares the variables in the type definition statement to be legal targets for pointers.

TARGET Statement:

        TARGET :: var1 [, var2, ...]

Examples:

        TARGET :: my_data

Description:
The TARGET statement declares the variables in its list to be legal targets for pointers. It is generally preferable to use the target attribute in a type declaration statement to declare a target instead of this statement.

15.10.3 Exercises

15-1. What is the difference between a pointer variable and an ordinary variable?

15-2. How does a pointer assignment statement differ from an ordinary assignment statement? What happens in each of the two statements a = z and a => z below?

```
INTEGER :: x = 6, z = 8
INTEGER, POINTER :: a
a => x
a = z
a => z
```

Pointers and Dynamic Data Structures

15-3. Is the program fragment shown below correct or incorrect? If it is incorrect, explain what is wrong with it. If it is correct, what does it do?

```
PROGRAM ex15_3
REAL, POINTER :: p1
REAL:: x1 = 11.
INTEGER, POINTER :: p2
INTEGER :: x2 = 12
p1 => x1
p2 => x2
WRITE (*,'(A,4G8.2)') ' p1, p2, x1, x2 = ', p1, p2, x1, x2
p1 => p2
p2 => x1
WRITE (*,'(A,4G8.2)') ' p1, p2, x1, x2 = ', p1, p2, x1, x2
END PROGRAM ex15_3
```

15-4. What are the possible association statuses of a pointer? How can you determine the association status of a given pointer?

15-5. Is the program fragment shown below correct or incorrect? If it is incorrect, explain what is wrong with it. If it is correct, what is printed out by the WRITE statement?

```
REAL, POINTER :: p1, p2
REAL, TARGET :: x1 = 11.1, x2 = -3.2
p1 => x1
WRITE (*,*) ASSOCIATED(p1), ASSOCIATED(p2), ASSOCIATED(p1,x2)
```

15-6. What is the purpose of the function NULL()? What advantage does this function have over the nullify statement?

15-7. What are the proper Fortran statements to declare a pointer to an integer array, and then point that pointer to every tenth element in a 1000-element target array called my_data?

15-8. What is printed out by the program shown below?

```
PROGRAM ex15_8
IMPLICIT NONE
INTEGER :: i
REAL, DIMENSION(-25:25), TARGET :: info = [ (2.1*i, i=-25,25) ]
REAL, DIMENSION(:), POINTER :: ptr1, ptr2, ptr3
ptr1 => info(-25:25:5)
ptr2 => ptr1(1::2)
ptr3 => ptr2(3:5)
WRITE (*,'(A,11F6.1)') ' ptr1 = ', ptr1
WRITE (*,'(A,11F6.1)') ' ptr2 = ', ptr2
WRITE (*,'(A,11F6.1)') ' ptr3 = ', ptr3
WRITE (*,'(A,11F6.1)') ' ave of ptr3 = ', SUM(ptr3)/SIZE(ptr3)
END PROGRAM ex15_8
```

15-9. How is dynamic memory allocated and deallocated using pointers? How does memory allocation using pointers and allocatable arrays differ?

15-10. What is a memory leak? Why is it a problem, and how can it be avoided?

15-11. Is the program shown below correct or incorrect? If it is incorrect, explain what is wrong with it. If it is correct, what is printed out by the WRITE statement?

```
MODULE my_sub
CONTAINS
   SUBROUTINE running_sum (sum, value)
   REAL, POINTER :: sum, value
   ALLOCATE (sum)
   sum = sum + value
   END SUBROUTINE running_sum
END MODULE my_subs

PROGRAM sum_values
USE my_sub
IMPLICIT NONE
INTEGER :: istat
REAL, POINTER :: sum, value
ALLOCATE (sum, value, STAT=istat)
WRITE (*,*) 'Enter values to add: '
DO
   READ (*,*,IOSTAT=istat) value
   IF ( istat /= 0 ) EXIT
   CALL running_sum (sum, value)
   WRITE (*,*) ' The sum is ', sum
END DO
END PROGRAM sum_values
```

15-12. Is the program shown below correct or incorrect? If it is incorrect, explain what is wrong with it. If it is correct, what is printed out by the WRITE statements? What happens when this program is compiled and executed on your computer?

```
PROGRAM ex15_12
IMPLICIT NONE
INTEGER :: i, istat
INTEGER, DIMENSION(:), POINTER :: ptr1, ptr2

ALLOCATE (ptr1(1:10), STAT=istat)
ptr1 = [ (i, i = 1,10 ) ]
ptr2 => ptr1

WRITE (*,'(A,10I3)') ' ptr1 = ', ptr1
WRITE (*,'(A,10I3)') ' ptr2 = ', ptr2

DEALLOCATE(ptr1, STAT=istat)

ALLOCATE (ptr1(1:3), STAT=istat)
ptr1 = [ -2, 0, 2 ]

WRITE (*,'(A,10I3)') ' ptr1 = ', ptr1
WRITE (*,'(A,10I3)') ' ptr2 = ', ptr2

END PROGRAM ex15_12
```

15-13. Create a version of the insertion sort program that will sort a set of input character values in a case-insensitive manner (i.e., uppercase and lowercase are to be treated as equivalent). Ensure that the ASCII collating sequence is used regardless of the computer on which program is executed.

15-14. Insertion Sort Using a Binary Tree versus a Linked List (a) Create an insertion sort subroutine to sort an array of real data using a linked list. This subroutine will be similar to the program in Example 15-2, except that the input data will be presented all at once in an array instead of being read one value at a time from the disk. (b) Create a set of subroutines to perform an insertion sort on an array of real data using a binary tree structure. (c) Compare the two ways to perform insertion sorts by generating a set of 50,000 random numbers and sorting the list with both subroutines. Time both subroutines using the elapsed time subroutines developed in Exercise 7-29. Which sorting algorithm was fastest?

15-15. How can an array of pointers be generated in Fortran?

15-16. What is printed out by the following program?

```
PROGRAM ex15_16
TYPE :: ptr
   REAL, DIMENSION(:), POINTER :: p
END TYPE
TYPE (ptr), DIMENSION(4) :: p1
REAL, DIMENSION(4), TARGET :: a = [ 1., 2., 3., 4. ]
REAL, DIMENSION(2), TARGET :: b = [ 5., 6. ]
REAL, DIMENSION(3), TARGET :: c = [ 7., 8., 9. ]
REAL, DIMENSION(5), TARGET :: d = [ 10., 11., 12., 13., 14. ]
p1(1)%p => a
p1(2)%p => b
p1(3)%p => c
p1(4)%p => d
WRITE (*,'(F6.1,/)') p1(1)%p(2) + p1(4)%p(4) + p1(3)%p(3)
DO i = 1, 4
   WRITE (*,'(5F6.1)') p1(i)%p
END DO
END PROGRAM ex15_16
```

15-17. Write a function that accepts a real input array and returns a pointer to the largest value in the array.

15-18. Write a function that accepts a real input array and returns a pointer to the largest value in the array.

15-19. Write a function that accepts a pointer to a real input array and returns a pointer to the largest value in the array.

15-20. Linear Least-Squares Fit Write a program that reads in an unknown number of real (x, y) pairs from a file, and stores them in a linked list. When all of the values have all been read in, the list should be passed to a subroutine that will compute the linear least-squared fit of the data to a straight line. (The equations for the linear squares fit are introduced in Example 5-5.)

15-21. Doubly Linked Lists Linked lists have the limitation that in order to find a particular element in the list, it is always necessary to search the list from the top down. There is no way to work backward up the list to find a particular item. For example, suppose that a program had examined the 1000th item in a list and now wanted to examine the 999th item in the list. The only way to do so would be to go back to the top of the list and start over, working from item 1 down! We can get around this problem by creating a *doubly linked list*. A doubly linked list has pointers both to the next item in the list and to the previous item in the list, permitting searches to be conducted in either direction. Write a program that reads in an arbitrary number of real numbers, and adds them to a doubly linked list. Then, write out the numbers both in input order and in reverse input order using the pointers. Test the program by creating 20 random values between −100.0 and 100.0 and processing them with the program.

15-22. Insertion Sort with Doubly-Linked Lists Write a version of the insertion sort program that inserts the real input values into a doubly linked list. Test the program by creating 50 random values between −1000.0 and 1000.0, and sorting them with the program. Print out the sorted values in both ascending and descending order.

15-23. Manually reconstruct the binary tree created by the program in Example 15-4 for the given test data set. How many layers are there in the tree? Is the tree regular or irregular?

16

Object-Oriented Programming in Fortran

OBJECTIVES

- Understand the basics of objects and object-oriented programming.
- Understand the relationship between an object and a class.
- Understand inheritance in an object-oriented methodology.
- Understand the structure of a Fortran class.
- Be able to use the `CLASS` keyword, and understand how it differs from the `TYPE` keyword.
- Know how to create a class, including how to create methods bound to the class.
- Know how to control access to instance variable and methods within a class, and know *why* you should control such access.
- Understand what a finalizer is, and when it should be used.
- Understand how inheritance and polymorphism work.
- Understand what an abstract class is. Know how to declare one, and why you would wish to do so.

This chapter introduces the basic concepts of object-oriented programming (OOP) in Fortran.

Fortran is not fundamentally an object-oriented language, but some of the new features introduced in Fortran 2003 allow (but do not require) a programmer to write code in an object-oriented style. We have already met most of the features needed for object-oriented programming: extended data types, access controls, and bound methods. We will introduce one new concept (the `CLASS` keyword), and then combine them properly to produce Fortran object-oriented programming.

This chapter begins with an introduction to the basic concepts of object-oriented programming, and then shows how Fortran can be adapted to that approach.

Throughout this chapter, we will be using the standard terms of object-oriented programming, such as classes, objects, fields, methods, and so forth. Most of these terms are not a part of the official Fortran standard, but the basic functionality is all there. By using the standard terms, you will be better able to talk with and understand colleagues who were trained in object-oriented languages such as Java or C++.

16.1
AN INTRODUCTION TO OBJECT-ORIENTED PROGRAMMING

Object-oriented programming is the process of programming by modeling objects in software. The principal features of OOP are described in the following sections.

16.1.1 Objects

The physical world is full of objects: cars, pencils, trees, and so on. Any real object can be characterized by two different aspects: its *properties* and its *behavior*. For example, a car can be modeled as an object. A car has certain properties (color, speed, direction, fuel consumption) and certain behaviors (starting, stopping, turning, and so on).

In the software world, an **object** is a software component whose structure is like that of objects in the real world. Each object consists of a combination of data (called **properties**) and behaviors (called **methods**). The properties are variables describing the essential characteristics of the object, while the methods describe how the object behaves and how the properties of the object can be modified. Thus, an object is a software bundle of variables and related methods.

A software object is often represented as shown in Figure 16-1. The object can be thought of as a cell, with a central nucleus of variables (containing the object's properties) and an outer layer of methods that form an interface between the object's variables and the outside world. The nucleus of data is hidden from the outside world by the outer layer of methods. The object's variables are said to be *encapsulated* within the object, meaning that no code outside of the object can see or directly manipulate them. Any access to the object's data must be through calls to the object's methods.

The variables and methods in an object are known as **instance variables** and **instance methods.** Each object of a given type has its own copies of the instance variables, but all of the objects share the same instance methods.

Typically, encapsulation is used to hide the implementation details of an object from other objects in the program. If the other objects in the program cannot see the

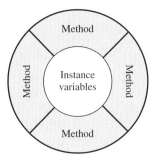

FIGURE 16-1
An object may be represented as a nucleus of data (instance variables) surrounded and protected by methods, which implement the object's behavior and form an interface between the variables and the outside world.

internal state of an object, they cannot introduce bugs by accidentally modifying the object's state. In addition, changes to the internal operation of the object will not affect the operation of the other objects in a program. As long as the interface to the outer world is unchanged, the implementation details of an object can change at any time without affecting other parts of the program.

Encapsulation provides two primary benefits to software developers:

1. **Modularity.** An object can be written and maintained independently of the source code for other objects. Therefore, the object can be easily reused and passed around in the system.
2. **Information Hiding.** An object has a public interface (the calling sequence of its methods) that other objects can use to communicate with it. However, the object's instance variables are not directly accessible to other objects. Therefore, if the public interface is not changed, an object's variables and methods can be changed at any time without introducing side effects in the other objects that depend on it.

Good Programming Practice
Always make instance variables private, so that they are hidden within an object. Such encapsulation makes your programs more modular and easier to modify.

16.1.2 Messages

In an object-oriented programming model, objects communicate by passing "*messages*" back and forth among themselves. These messages are really just method calls. For example, if Object A wants Object B to perform some action for it, it sends a message to Object B requesting the object to execute one of its methods (see Figure 16-2). The message causes Object B to execute the specified method.

Each message has three components, which provide all the information necessary for the receiving object to perform the desired action:

1. A reference pointing to the object to which the message is addressed.
2. The name of the method to perform on that object.
3. Any parameters needed by the method.

An object's behavior is expressed through its methods, so message passing supports all possible interactions between objects.

16.1.3 Classes

In object-oriented programming, **classes** are the software blueprints from which objects are made. A class is a software construct that specifies the number and type of variables to be included in an object, and the methods that will be defined for the object. Each component of a class is known as a **member.** The two types of members are **fields,** which specify the data types defined by the class, and **methods,** which specify the

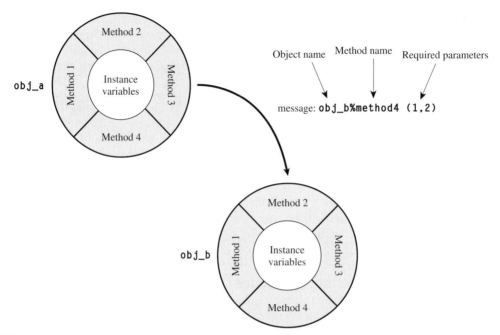

FIGURE 16-2
If object obj_a wants object obj_b to do some work for it, it sends a message to that object. The message contains three parts: A reference to the object to which it is addressed, the name of the method within the object that will do the work, and the required parameters. Note that the names of the object and method are separated by a % sign.

operations on those fields. For example, suppose that we wish to create an object to represent a complex number. Such an object would have two instance variables, one for the real part of the number (re) and one for the imaginary part of the number (im). In addition, it would have methods describing how to add, subtract, multiply, divide, etc., with complex numbers. To create such objects, we would write a class complex_ob that defines the required fields re and im, together with their associated methods.

Note that a class is a *blueprint* for an object, not an object itself. The class describes what an object will look and behave like once it is created. Each object is created or *instantiated* in memory from the blueprint provided by a class, and many different objects can be instantiated from the same class. For example, Figure 16-3 shows a class complex_ob, together with three objects a, b, and c created from that class. Each of the three objects has its own copies of the instance variables re and im, while sharing a single set of methods to modify them.

16.1.4 Class Hierarchy and Inheritance

The classes in an object-oriented language are organized in a **class hierarchy,** with the highest level classes being very general in behavior and lower-level ones becoming more specific. Each lower-level class is based on and derived from a higher-level class, and

Object-Oriented Programming in Fortran

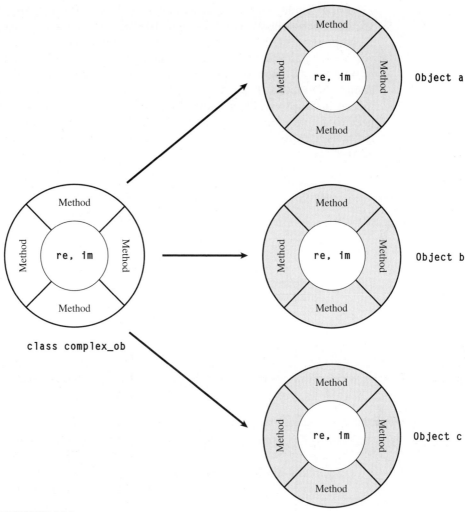

FIGURE 16-3
Many objects can be instantiated from a single class. In this example, three objects a, b, and c have been instantiated from class `complex_ob`.

the lower-level classes *inherit both the instance variables and the instance methods* of the class from which it is derived. A new class starts with all of the instance variables and methods of the class on which it is based, and the programmer then adds the additional variables and methods necessary for the new class to perform its function.

The class on which a new class is based is referred to as a **superclass**, and the new class is referred to as a **subclass**. The new subclass can itself become the superclass for another new subclass. A subclass normally adds instance variables and instance methods of its own, so a subclass is generally larger than its superclass. In addition, it can **override** some methods of its superclass, changing its behavior from that of its superclass. Because a subclass is more specific than its superclass, it represents a smaller group of objects.

For example, suppose that we define a class called vector_2d to contain 2D vectors. Such a class would have two instance variables x and y to contain the *x* and *y* components of the 2D vectors, and it would need methods to manipulate the vectors such as adding two vectors, subtracting two vectors, calculating the length of a vector, etc. Now suppose that we need to create a class called vector_3d to contain 3D vectors. If this class is based on vector_2d, then it will automatically inherit instance variables x and y from its superclass, so the new class will only need to define a variable z (see Figure 16-4). The new class will also override the methods used to manipulate 2D vectors to allow them to work properly with 3D vectors.

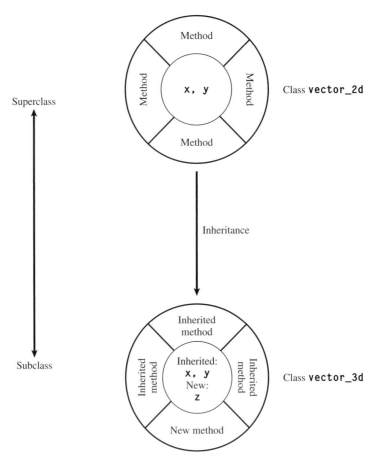

FIGURE 16-4
An example of inheritance. Class vector_2d has been defined to handle 2D vectors. When class vector_3d is defined as a subclass of vector_2d, it inherits the instance variables x and y, as well as many methods. The programmer then adds a new instance variable z and new methods to the ones inherited from the superclass.

The concepts of class hierarchy and inheritance are extremely important, since inheritance allows a programmer to define certain behaviors only once in a superclass, and to reuse those behaviors over and over again in many different subclasses. This reusability makes programming more efficient.

16.1.5 Object-Oriented Programming

Object-oriented programming (OOP) is the process of programming by modeling objects in software. In OOP, a programmer examines the problem to be solved, and tries to break it down into identifiable objects, each of which contains certain data and specific methods by which that data is manipulated. Sometimes these objects will correspond to physical objects in nature, and sometimes that will be purely abstract software constructs.

Once the objects making up the problem have been identified, the programmer identifies the type of data to be stored as instance variables in each object, and the exact calling sequence of each method needed to manipulate the data.

The programmer can then develop and test the classes in the model one at a time. As long as the *interfaces* between the classes (the calling sequence of the methods) are unchanged, each class can be developed and tested without needing to change any other part of the program.

16.2
THE STRUCTURE OF A FORTRAN CLASS

The remainder of this chapter shows how to implement object-oriented programming in Fortran, starting with the structure of a Fortran class. The major components (class members) of a Fortran class are (see Figure 16-5):

1. **Fields.** Fields define the instance variables that will be created when an object is instantiated from a class. Instance variables are the data encapsulated inside an object. A new set of instance variables is created each time that an object is instantiated from the class.
2. **Methods.** Methods implement the behaviors of a class. Some methods may be explicitly defined in a class, while other methods may be inherited from superclasses of the class.
3. **Constructor.** A constructor initializes the instance variables in an object when it is created. Fortran objects are either initialized using structure constructors, which were introduced in Section 12.1, or by special initializing methods.
4. **Finalizer.** Just before an object is destroyed, it makes a call to a special method called a **finalizer.** The method performs any necessary cleanup (releasing resources, etc.) before the object is destroyed. There can be at most one finalizer in a class, and many classes do not need a finalizer at all.

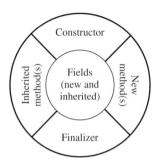

FIGURE 16-5
A class consists of fields (data), a constructor to initialize the data in an object, one or more methods to modify and manipulate the data, and up to one finalizer to cleanup before the object is destroyed. Note that both fields and methods may be inherited from a superclass.

The members of a class, whether variables or methods, are accessed by referring to an object created from the class using the **component selector,** the % symbol. For example, suppose that a class `my_class` contains an instance variable `a` and a method `process_a()`. If an object of this class is named `my_obj`, then the instance variable in `my_obj` would be accessed as `my_obj%a`, and the method would be accessed as `my_obj%process_a()`.

16.3
THE CLASS KEYWORD

The `CLASS` keyword is a variant of the `TYPE` keyword that adds special properties important for object-oriented programming.

In an ordinary Fortran, the type of each dummy argument in a procedure and the corresponding calling argument must match *exactly*, or there will be an error. Similarly, the type of a pointer and the corresponding target must match *exactly*, or there will be an error, and the type of an allocatable variable and the corresponding data must match *exactly*, or there will be an error.

The `CLASS` keyword relaxes this requirement in a special way. If an allocatable item, pointer, or dummy argument is declared with the `CLASS(type)` keyword, where `type` is a derived data type, then the item will match that data type *or any extension of that data type*.

For example, suppose that we declare the following two data types:

```
TYPE :: point
    REAL :: x
    REAL :: y
END TYPE

TYPE,EXTENDS(point) :: point_3d
    REAL :: z
END TYPE
```

Then a pointer declared as

```
TYPE(point),POINTER :: p
```

would only accept targets of type `point`, but a pointer declared as

```
CLASS(point),POINTER :: p
```

would accept targets of either type `point` or type `point_3d`, which is an extension of type `point`.

The type of a pointer or dummy argument declared with the `CLASS` keyword is known as the **declared type** of the pointer or dummy argument, and the type of the actual object assigned to the pointer or dummy argument at any time is known as the **dynamic type** of the pointer or dummy argument.

An item declared with the `CLASS` keyword is said to be **polymorphic** (meaning "many forms"), because it will match more than one data type.

Polymorphic pointers or dummy arguments have a special limitation: You can only access items of the *declared type* with them. Items defined in extensions are *not* accessible with the polymorphic pointer. For example, consider the following type definitions.

```
CLASS(point),POINTER :: p
TYPE(point),TARGET :: p1
TYPE(point_3d),TARGET :: p2
```

With these definitions, variables p1 and p2 can both be assigned to p, and the pointer p can be used to access components x and y within them. However, pointer p cannot be used to access component z, because that component is not defined in the declared type of the pointer.

To understand this more clearly, let's examine the code below. In line 1, pointer p is assigned to point to the target p1, and lines 2 and 3 access the components of p using the original variable name and the pointer, respectively. This all works fine. In line 4 below, pointer p is assigned to point to the target p2, which is of type `point_3d`. Lines 5 and 6 access the components of p using the original variable name and the pointer, respectively. Line 5 works fine, but line 6 produces an error, because we can't access component z using a pointer of class `point`, since component z is not defined in that derived type.

```
1 p => p1
2 WRITE (*,*) p1%x, p1%y         ! These two lines produce the same output
3 WRITE (*,*) p%x, p%y           ! These two lines produce the same output
4 p => p2
5 WRITE (*,*) p2%x, p2%y, p2%z   ! Legal
6 WRITE (*,*) p%x, p%y, p%z      ! Illegal-can't access z
```

There is a way around this limitation by using the `SELECT TYPE` construct, which we will meet later in the chapter.

It is also possible to define a pointer or dummy argument to be of `CLASS(*)`. Such a pointer or argument is said to be **unlimited polymorphic,** since it will match *any* derived type. However, you cannot directly access any components of the dynamic data type, since no components are defined in the declared type of the pointer or dummy argument.

16.4
IMPLEMENTING CLASSES AND OBJECTS IN FORTRAN

As we saw in Section 16.2, a Fortran class consists of instance variables, methods, a constructor, and possibly a finalizer. We will now learn how to create a simple Fortran class (without a finalizer), and how to instantiate objects from that class.

Each Fortran class should be placed in a separate module, so that we can control access to its components and have an explicit interface to the class via USE access.

16.4.1 Declaring Fields (Instance Variables)

The data fields (or instance variables) in a class are defined in a user-defined data type, and the name of that data type is the name of the class. In proper object-oriented programming, the data type should be declared with PUBLIC access, but the *components* of the data type should be declared PRIVATE. Thus, it will be possible to create objects of this type outside the module, but it will *not* be possible to read or modify the instance variables of the data type from outside the module.

In actual object-oriented Fortran programs, we often do *not* declare the components of the data type to be PRIVATE. If a Fortran object is to have subclasses that inherit the data from the superclass, then that data must be declared with PUBLIC access, or the subclasses (that are defined in different modules) will not be able to access the data. Also, the Fortran language does not allow constructors to be used if the data fields are declared to be PRIVATE. This is a limitation of the Fortran implementation of object-oriented programming.

As an example, suppose that we are defining a simple complex number class named complex_ob. This class will contain two instance variables, re and im, for the real and the imaginary components of the complex number. This can be accomplished as follows:

```
MODULE complex_class
IMPLICIT NONE

! Type definition
TYPE,PUBLIC :: complex_ob     ! This will be the name we instantiate
    PRIVATE                   ! (Should be used, but might not be)
    REAL :: re                ! Real part
    REAL :: im                ! Imaginary part
END TYPE complex_ob

! Now add methods
CONTAINS

    (Insert methods here)

END MODULE complex_class
```

The constructor for this class can be used to initialize the instance variables if the fields in the class are declared PUBLIC. The constructor consists of the data type name followed by the initial values of the data elements, in parentheses. For example, if the fields in the class are declared PUBLIC, then the following code creates a complex object in which initial x and y values are 1 and 2, and assigns it to pointer p.

```
CLASS(complex_ob),POINTER :: p
p = complex_ob(1.,2.)
```

If the fields in the class are declared PRIVATE, then the programmer will have to write a special method to initialize the data in the class.

16.4.2 Creating Methods

Object-oriented methods differ from ordinary Fortran procedures in that they are *bound* to a particular class, and can only work with data from that class. How do we bind Fortran procedures to a particular class (i.e., a defined data type), and so create methods in Fortran?

As we saw in Chapter 12, **type-bound Fortran procedures** are created by adding a CONTAINS statement to the type definition, and declaring the bindings after that statement. For example, suppose that we wanted to include a subroutine to add two items of type complex_ob in our class. Then we would declare the type definition as follows:

```
MODULE complex_class
IMPLICIT NONE

! Type definition
TYPE,PUBLIC :: complex_ob   ! This will be the name we instantiate
   PRIVATE
   REAL :: re               ! Real part
   REAL :: im               ! Imaginary part
CONTAINS
   PROCEDURE :: add => add_complex_to_complex
END TYPE complex_ob

! Declare access for methods
PRIVATE :: add_complex_to_complex

! Now add methods
CONTAINS

   ! Insert method add_complex_to_complex here:
   SUBROUTINE add_complex_to_complex( this, ... )
   CLASS(complex_ob) :: this
   ...
   END SUBROUTINE add_complex_to_complex

END MODULE complex_class
```

These statements declare that subroutine add_complex_to_complex is bound to this data type and only works with this data type, and that it will be accessed with the name "add". The subroutine itself must have an item of the type definition as its *first* argument, because the PASS attribute is the default for bound procedures. This means that the object to which it is bound will always be passed as the first argument to the subroutine whenever it is called.

Bindings can also be generic, with multiple procedures bound to the same name, as long as the procedures can be distinguished by their calling arguments. For example, we might want to add either a complex number or a real number to the object. In that case, the binding could be as follows:

```
MODULE complex_class
IMPLICIT NONE

! Type definition
TYPE,PUBLIC :: complex_ob      ! This will be the name we instantiate
    PRIVATE
    REAL :: re                 ! Real part
    REAL :: im                 ! Imaginary part
CONTAINS
    PRIVATE
    PROCEDURE :: ac => add_complex_to_complex
    PROCEDURE :: ar => add_real_to_complex
    GENERIC, PUBLIC :: add => ac, ar
END TYPE complex_ob

! Declare access for methods
PRIVATE :: add_complex_to_complex, add_real_to_complex

! Now add methods
CONTAINS

    ! Insert method add_complex_to_complex here:
    SUBROUTINE add_complex_to_complex( this, ... )
    CLASS(complex_ob) :: this
    ...
    END SUBROUTINE add_complex_to_complex

    ! Insert method add_real_to_complex here:
    SUBROUTINE add_real_to_complex( this, ... )
    CLASS(complex_ob) :: this
    ...
    END SUBROUTINE add_real_to_complex

END MODULE complex_class
```

This example defines a generic public binding `add`, and two private procedures `ac` and `ar` associated with the public binding. Note that `ac` and `ar` are mapped to subroutines with much longer names; the short forms are just for convenience. Also, note that `ac`, `ar`, `add_complex_to_complex`, and `add_real_to_complex` are all declared `PRIVATE`, so they cannot be accessed directly from outside the module.

As many methods as necessary can be created in this fashion, each one bound to the data object created from the class. All of the procedures would be accessed as `obj%add(...)`, where `obj` is the name of an object created from this class. The particular method that is invoked will be determined by the arguments of the `add` method.

16.4.3 Creating (Instantiating) Objects from a Class

Objects of type `complex_ob` can be instantiated in another procedure by USEing module `complex_class` in the procedure, and then declaring the object using the TYPE keyword.

```
USE complex_class
IMPLICIT NONE
TYPE(complex_ob) :: x, y, z
```

These statements have created (instantiated) three objects from the class `complex_ob`: x, y, and z. If the fields of the objects have *not* been declared PRIVATE, then they can also be initialized as they are created using constructors.

```
TYPE(complex_ob) :: x = complex_ob(1.,2.), y = complex_ob(3.,4.), z
```

Once they have been created, the methods in the objects can be accessed using the object name and the component selector. For example, the method `add` could be accessed for object x as follows:

```
z = x%add(...)
```

16.5
FIRST EXAMPLE: A timer CLASS

When developing software, it is often useful to be able to determine how long a particular part of a program takes to execute. This measurement can help us locate the "hot spots" in the code, the places where the program is spending most of its time, so that we can try to optimize them. This is usually done with an *elapsed time calculator*.

An elapsed time calculator makes a great first object, because it is so simple. It is analogous to a physical stopwatch. A stopwatch is an object that measures the elapsed time between a push on a start button and a push on a stop button (often they are the same physical button). The basic actions (methods) performed on a physical stopwatch are:

1. A button push to reset and start the timer.
2. A button push to stop the timer and display the elapsed time.

Internally, the stopwatch must remember the time of the first button push in order to calculate the elapsed time.

Similarly, an elapsed time class needs to contain the following components (members):

1. A method to store the start time of the timer (`start_timer`). This method will not require any input parameters from the calling program, and will not return any results to the calling program.
2. A method to return the elapsed time since the last start (`elapsed_time`). This method will not require any input parameters from the calling program, but it will return the elapsed time in seconds to the calling program.
3. A field (instance variable) to store the time that the timer started running, for use by the elapsed time method.

This class will not need a finalizer.

The timer class must be able to determine the current time whenever one of its methods is called. Fortunately, the intrinsic subroutine date_and_time (see Appendix B) provides this information. The optional argument values returns an array of eight integers, containing time information from the year all the way down to the current millisecond. These values can be turned into a current time in milliseconds since the start of the month as follows:

```
! Get time
CALL date_and_time ( VALUES=value )
time1 = 86400.D0 * value(3) + 3600.D0 * value(5) &
      + 60.D0 * value(6) + value(7) + 0.001D0 * value(8)
```

Be sure that variable time1 is a 64-bit REAL, or there will not be enough precision to save all of the time information.

16.5.1 Implementing The timer Class

We will implement the timer class in a series of steps, defining the instance variables, constructor, and methods in succession.

1. **Define Instance Variables.** The timer class must contain a single instance variable called saved_time, which contains the last time at which start_timer method was called. It must be a 64-bit real value (SELECTED_REAL_KIND(p=14)), so that it can hold fractional parts of seconds.

   Instance variables are declared after the class definition, and before the constructors and methods. Therefore, class timer will begin as follows:

```
MODULE timer_class
IMPLICIT NONE

! Declare constants
INTEGER,PARAMETER :: DBL = SELECTED_REAL_KIND(p=14)

! Type definition
TYPE,PUBLIC :: timer    ! This will be the name we instantiate
   PRIVATE
   REAL(KIND=DBL) :: saved_time
END TYPE timer
```

   Note that we are declaring the field saved_time to be PRIVATE, so it will not be possible to initialize the data value using a structure constructor. Instead, it must be initialized using a user-defined method.

2. **Create the Methods.** The class must also include two methods to start the timer and to read the elapsed time. Method start_timer() simply resets the start time in the instance variable. Method elapsed_time() returns the elapsed time since the start of the timer in seconds. Both of these methods must be bound to the class.

   The dummy arguments of the timer type that are declared in these methods should use the CLASS keyword, so that they will also work with any extensions of the timer class that might be defined later.

Object-Oriented Programming in Fortran

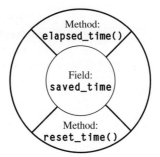

FIGURE 16-6
The timer class.

The resulting timer class is shown in Figure 16-6, and the source code for this class is shown in Figure 16-7.

FIGURE 16-7
The source code for the timer class.

```fortran
MODULE timer_class
!
!   This module implements a timer class.
!
!   Record of revisions:
!       Date        Programmer          Description of change
!       ====        ==========          =====================
!       01/06/16    S. J. Chapman       Original code
!
IMPLICIT NONE

! Declare constants
INTEGER,PARAMETER :: DBL = SELECTED_REAL_KIND(p=14)

! Type definition
TYPE,PUBLIC :: timer    ! This will be the name we instantiate

    ! Instance variables
    PRIVATE
    REAL(KIND=DBL) :: saved_time   ! Saved time in ms

CONTAINS

    ! Bound procedures
    PROCEDURE,PUBLIC :: start_timer => start_timer_sub
    PROCEDURE,PUBLIC :: elapsed_time => elapsed_time_fn

END TYPE timer

! Restrict access to the actual subroutine names
PRIVATE :: start_timer_sub, elapsed_time_fn

! Now add subroutines
CONTAINS
```

(continued)

(*concluded*)

```fortran
SUBROUTINE start_timer_sub(this)
!
! Subroutine to get and save the initial time
!
IMPLICIT NONE

! Declare calling arguments
CLASS(timer) :: this            ! Timer object

! Declare local variables
INTEGER,DIMENSION(8) :: value    ! Time value array

! Get time
CALL date_and_time ( VALUES=value )
this%saved_time = 86400.D0 * value(3) + 3600.D0 * value(5) &
              + 60.D0 * value(6) + value(7) + 0.001D0 * value(8)

END SUBROUTINE start_timer_sub

REAL FUNCTION elapsed_time_fn(this)
!
! Function to calculate elapsed time
!
IMPLICIT NONE

! Declare calling arguments
CLASS(timer) :: this            ! Timer object

! Declare local variables
INTEGER,DIMENSION(8) :: value    ! Time value array
REAL(KIND=DBL) :: current_time   ! Current time (ms)

! Get time
CALL date_and_time ( VALUES=value )
current_time = 86400.D0 * value(3) + 3600.D0 * value(5) &
           + 60.D0 * value(6) + value(7) + 0.001D0 * value(8)

! Get elapsed time in seconds
elapsed_time_fn = current_time - this%saved_time

END FUNCTION elapsed_time_fn
END MODULE timer_class
```

16.5.2 Using The `timer` Class

To use this class in a program, the programmer must first instantiate a `timer` object with a statement like

```fortran
TYPE(timer) :: t
```

This statement defines an object t of the timer class (see Figure 16-8). After this object has been created, t is a `timer` object, and the methods in the object can be called using that reference: `t%start_timer()` and `t%elapsed_time()`.

A program can reset the elapsed timer to zero at any time by calling method `start_timer()`, and can get the elapsed time by executing method `elapsed_time()`. An example program that uses the `timer` object is shown in Figure 16-9. The program

Object-Oriented Programming in Fortran

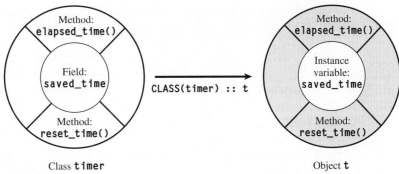

FIGURE 16-8
The statement "CLASS(timer) :: t" creates (instantiates) a new timer object from the template provided by the class definition, gives it the name t. This object has its own unique copy of the instance variable saved_time.

tests this class by measuring the time required to perform 100,000,000 iterations of a pair of nested DO loops.

FIGURE 16-9
A program to test the timer class.

```
PROGRAM test_timer
!
!   This program tests the timer class.
!
!   Record of revisions:
!       Date          Programmer              Description of change
!       ====          ==========              =====================
!     01/06/16      S. J. Chapman              Original code
!
USE timer_class                         ! Import timer class
IMPLICIT NONE

! Declare local variables
INTEGER :: i, j                         ! Loop index
INTEGER :: k                            ! Scratch variable
TYPE(timer) :: t                        ! Timer object

! Reset the timer
CALL t%start_timer()

! Waste some time
DO i = 1, 100000
   DO j = 1, 100000
      k = i + j
   END DO
END DO

! Get the elapsed time
WRITE (*,'(A,F8.3,A)') 'Time =', t%elapsed_time(), ' s'

END PROGRAM test_timer
```

When this program is executed on my Intel i7-class PC, the results are:

```
D:\book\fortran\chap16>test_timer
Time =    0.274 s
```

The measured time will of course differ on computers of different speeds, and will also vary depending on the compiler optimizations selected.

16.5.3 Comments on the `timer` Class

This section contains a few notes about the operation of our `timer` class, and of classes in general.

First, note that the `timer` class saves its start time in the instance variable `saved_time`. Each time that an object is instantiated from a class, it receives its *own copy* of all instance variables defined in the class. Therefore, many `timer` objects could be instantiated and used simultaneously in a program, and *they will not interfere with each other*, because each timer has its own private copy of the instance variable `saved_time`.

Also, notice that each class member in Figure 16-7 is declared with either a **PUBLIC** or **PRIVATE** keyword. Any instance variable or method definition declared with the PUBLIC keyword can be accessed by USE association from other parts of the program. Any instance variable or method declared with the PRIVATE keyword is only accessible to methods of the object in which it is defined.[1]

In this case, the instance variable `saved_time` is declared PRIVATE, so it cannot be seen or modified by any method outside of the object in which it is defined. Since no part of the program outside of `timer` can see `saved_time`, it is not possible for some other part of the program to accidentally modify the value stored there and so mess up the elapsed time measurement. The only way that a program can utilize the elapsed time measurement is through the PUBLIC bound methods `start_timer()` and `elapsed_time()`. You should always declare all instance variables within your classes to be PRIVATE.

Also, note that the actual method names `start_timer_sub` and `elapsed_time_fn` are declared PRIVATE. This means that the actual methods cannot be called directly from another part of the program. The *only* way to execute these methods is using the object name and the component selector (%).

16.6
CATEGORIES OF METHODS

Since instance variables are usually hidden within a class, the only way to work with them is through the interface formed by the class's methods. The methods are the public face of the class, providing a standard way to work with the information while hiding the unnecessary details of the implementation from the user.

---

[1] Actually, it is accessible to any other methods in the same module. Since we are putting each class in its own module, the PRIVATE keyword effectively restricts access to the object in which it is defined.

A class's methods must perform certain common "housekeeping" functions, as well as the specific actions required by the class. These housekeeping functions fall into a few broad categories, and they are common to most classes regardless of their specific purpose. A class must usually provide a way to store data into its instance variables, read data from its instance variables, test the status of its instance variables, and manipulate the instance variables as required to solve problems.

Since the instances variables in a class cannot be used directly, classes must define methods to store data into the instance variables and to read data from them. By convention among object-oriented programmers, the names of methods that store data begin with "set" and are called **set methods,** while the names of methods that read data begin with "get" and are called **get methods.**

Set methods take information from the outside world and store the data into the class's instance variables. In the process, they *should also check the data for validity and consistency*. This checking prevents the instance variables of the class from being set into an illegal state.

For example, suppose that we have created a class date containing instance variables day (with a range of 1–31), month (with a range of 1–12), and year (with a range of 1900–2100). If these instance variables were declared PUBLIC, then any part of the program that USEs the class could modify them directly. For example, assume that a date object was declared as

```
USE date_class
...
TYPE(date) :: d1
```

With this declaration, any method in the program could directly set the day to an illegal value.

```
d1%day = 32;
```

Set methods and private instance variables prevent this sort of illegal behavior by testing the input parameters. If the parameters are valid, the method stores them in the appropriate instance variables. If the parameters are invalid, the method either modifies the inputs to be legal or provides some type of error message to the caller.

Good Programming Practice
Use set methods to check the validity and consistency of input data before it is stored in an object's instance variables.

Get methods are used to retrieve information from the instance variables and to format it properly for presentation to the outside world. For example, our date class might include methods get_day( ), get_month( ), and get_year( ) to recover the day, month, and year, respectively.

Another category of method tests for the truth or falsity of some condition. These methods are called **predicate methods.** These methods typically begin with the word is, and they return a LOGICAL (true/false) result. For example, a date class might include a method is_leap_year(), which would return true if the specified year is a leap year, and false otherwise. It could also include methods like is_equal(), is_earlier(), and is_later() to compare two dates chronologically.

Good Programming Practice
Define predicate methods to test for the truth or falsity of conditions associated with any classes you create.

EXAMPLE 16-1

Creating a date Class:

We will illustrate the concepts described in this chapter by creating a date class designed to hold and manipulate dates on the Gregorian calendar.

This class should be able to hold the day, month, and year of a date in instance variables that are protected from outside access. The class must include set and get methods to change and retrieve the stored information, predicate methods to recover information about date objects and to allow two date objects to be compared, and a to_string method to allow the information in a date object to be displayed easily.

SOLUTION
The date class will need three instance variables, day, month, and year. They will be declared PRIVATE to protect them from direct manipulation by outside methods. The day variable should have a range of 1–31, corresponding to the days in a month. The month variable should have a range of 1–12, corresponding to the months in a year. The year variable will be greater than or equal to zero.

We will define a method set_date(day,month,year) to insert a new date into a date object, and three methods get_day(), get_month(), and get_year() to return the day, month, and year from a given date object.

The supported predicate methods will include is_leap_year() to test if a year is a leap year. This method will use the leap year test described in Example 4-3. In addition, we will create three methods is_equal(), is_earlier(), and is_later() to compare two date objects. Finally, method to_string() will format the date as a string in the normal US style: mm/dd/yyyy.

The resulting class is shown in Figure 16-10. Notice that we took advantage of the renaming capability of the bindings to give each procedure a name that identified whether it is a subroutine or a function. This is *not* required in OOP, but I find it convenient to help me keep subroutines and functions straight.

FIGURE 16-10
The date class.

```
MODULE date_class
!
!   This module implements a date class, which stores
!   and manipulates dates on the Gregorian calendar.
```

(continued)

(*continued*)

```fortran
!    It implements set methods, get methods, predicate
!    methods, and a "to_string" method for displays.
!
!   Record of revisions:
!      Date        Programmer         Description of change
!      ====        ==========         =====================
!    01/07/16    S. J. Chapman        Original code
!
IMPLICIT NONE

! Type definition
TYPE,PUBLIC :: date   ! This will be the name we instantiate

   ! Instance variables. Note that the default
   ! date is January 1, 1900.
   PRIVATE
   INTEGER :: year = 1900    ! Year (0 - xxxx)
   INTEGER :: month = 1      ! Month (1-12)
   INTEGER :: day = 1        ! Day (1-31)

CONTAINS

   ! Bound procedures
   PROCEDURE,PUBLIC :: set_date => set_date_sub
   PROCEDURE,PUBLIC :: get_day => get_day_fn
   PROCEDURE,PUBLIC :: get_month => get_month_fn
   PROCEDURE,PUBLIC :: get_year => get_year_fn
   PROCEDURE,PUBLIC :: is_leap_year => is_leap_year_fn
   PROCEDURE,PUBLIC :: is_equal => is_equal_fn
   PROCEDURE,PUBLIC :: is_earlier_than => is_earlier_fn
   PROCEDURE,PUBLIC :: is_later_than => is_later_fn
   PROCEDURE,PUBLIC :: to_string => to_string_fn

END TYPE date

! Restrict access to the actual procedure names
PRIVATE :: set_date_sub, get_day_fn, get_month_fn, get_year_fn
PRIVATE :: is_leap_year_fn, is_equal_fn, is_earlier_fn
PRIVATE :: is_later_fn, to_string_fn

! Now add methods
CONTAINS

   SUBROUTINE set_date_sub(this, day, month, year)
   !
   ! Subroutine to set the initial date
   !
   IMPLICIT NONE

   ! Declare calling arguments
   CLASS(date) :: this                  ! Date object
   INTEGER,INTENT(IN) :: day            ! Day (1-31)
   INTEGER,INTENT(IN) :: month          ! Month (1-12)
   INTEGER,INTENT(IN) :: year           ! Year (0 - xxxx)
```

(*continued*)

(*continued*)

```fortran
      ! Save date
      this%day   = day
      this%month = month
      this%year  = year

      END SUBROUTINE set_date_sub

      INTEGER FUNCTION get_day_fn(this)
      !
      ! Function to return the day from this object
      !
      IMPLICIT NONE

      ! Declare calling arguments
      CLASS(date),INTENT(IN) :: this    ! Date object

      ! Get day
      get_day_fn = this%day

      END FUNCTION get_day_fn

      INTEGER FUNCTION get_month_fn(this)
      !
      ! Function to return the month from this object
      !
      IMPLICIT NONE

      ! Declare calling arguments
      CLASS(date) :: this               ! Date object

      ! Get month
      gct_month_fn = this%month

      END FUNCTION get_month_fn

      INTEGER FUNCTION get_year_fn(this)
      !
      ! Function to return the year from this object
      !
      IMPLICIT NONE

      ! Declare calling arguments
      CLASS(date),INTENT(IN) :: this    ! Date object

      ! Get year
      get_year_fn = this%year

      END FUNCTION get_year_fn

      LOGICAL FUNCTION is_leap_year_fn(this)
      !
      ! Is this year a leap year?
      !
      IMPLICIT NONE
```

(*continued*)

(*continued*)

```fortran
    ! Declare calling arguments
    CLASS(date),INTENT(IN) :: this     ! Date object

    ! Perform calculation
    IF ( MOD(this%year, 400) == 0 ) THEN
       is_leap_year_fn = .TRUE.
    ELSE IF ( MOD(this%year, 100) == 0 ) THEN
       is_leap_year_fn = .FALSE.
    ELSE IF ( MOD(this%year, 4) == 0 ) THEN
       is_leap_year_fn = .TRUE.
    ELSE
       is_leap_year_fn = .FALSE.
    END IF

    END FUNCTION is_leap_year_fn

    LOGICAL FUNCTION is_equal_fn(this,that)
    !
    ! Are these two dates equal?
    !
    IMPLICIT NONE

    ! Declare calling arguments
    CLASS(date),INTENT(IN) :: this     ! Date object
    CLASS(date),INTENT(IN) :: that     ! Another date for comparison

    ! Perform calculation
    IF ( (this%year == that%year) .AND. &
         (this%month == that%month) .AND. &
         (this%day == that%day) ) THEN
       is_equal_fn = .TRUE.
    ELSE
       is_equal_fn = .FALSE.
    END IF

    END FUNCTION is_equal_fn

    LOGICAL FUNCTION is_earlier_fn(this,that)
    !
    ! Is the date in "that" earlier than the date
    ! stored in the object?
    !
    IMPLICIT NONE

    ! Declare calling arguments
    CLASS(date),INTENT(IN) :: this     ! Date object
    CLASS(date),INTENT(IN) :: that     ! Another date for comparison

    ! Perform calculation
    IF ( that%year > this%year ) THEN
       is_earlier_fn = .FALSE.
    ELSE IF ( that%year < this%year ) THEN
       is_earlier_fn = .TRUE.
```

(*continued*)

(*continued*)

```fortran
      ELSE
         IF ( that%month > this%month) THEN
            is_earlier_fn = .FALSE.
         ELSE IF ( that%month < this%month ) THEN
            is_earlier_fn = .TRUE.
         ELSE
            IF ( that%day >= this%day ) THEN
               is_earlier_fn = .FALSE.
            ELSE
               is_earlier_fn = .TRUE.
            END IF
         END IF
      END IF

      END FUNCTION is_earlier_fn

      LOGICAL FUNCTION is_later_fn(this,that)
      !
      ! Is the date in "that" later than the date
      ! stored in the object?
      !
      IMPLICIT NONE

      ! Declare calling arguments
      CLASS(date),INTENT(IN) :: this    ! Date object
      CLASS(date),INTENT(IN) :: that    ! Another date for comparison

      ! Perform calculation
      IF ( that%year > this%year ) THEN
         is_later_fn = .TRUE.
      ELSE IF ( that%year < this%year ) THEN
         is_later_fn = .FALSE.
      ELSE
         IF ( that%month > this%month ) THEN
            is_later_fn = .TRUE.
         ELSE IF ( that%month < this%month ) THEN
            is_later_fn = .FALSE.
         ELSE
            IF ( that%day > this%day ) THEN
               is_later_fn = .TRUE.
            ELSE
               is_later_fn = .FALSE.
            END IF
         END IF
      END IF

      END FUNCTION is_later_fn

      CHARACTER(len=10) FUNCTION to_string_fn(this)
      !
      ! Represent the date as a string: MM/DD/YYYY.
      !
      IMPLICIT NONE
```

(*continued*)

(*concluded*)

```
    ! Declare calling arguments
    CLASS(date),INTENT(IN) :: this      ! Date object

    ! Declare local variables
    CHARACTER(len=2) :: dd              ! Day
    CHARACTER(len=2) :: mm              ! Month
    CHARACTER(len=4) :: yy              ! Year

    ! Get components
    WRITE (dd,'(I2.2)') this%day
    WRITE (mm,'(I2.2)') this%month
    WRITE (yy,'(I4)') this%year

    ! Return string
    to_string_fn = mm // '/' // dd // '/' // yy

    END FUNCTION to_string_fn

END MODULE date_class
```

We must create a program to test the date class. Such a program is shown in Figure 16-11. Program test_date instantiates four date objects, and initializes them. It then exercises all of the methods defined in the class.

FIGURE 16-11
Program test_date to test the date class.

```
PROGRAM test_date
!
!   This program tests the date class.
!
!   Record of revisions:
!       Date         Programmer          Description of change
!       ====         ==========          =====================
!     01/07/16      S. J. Chapman         Original code
!
USE date_class                          ! Import date class
IMPLICIT NONE

! Declare local variables
TYPE(date) :: d1                        ! Date 1
TYPE(date) :: d2                        ! Date 2
TYPE(date) :: d3                        ! Date 3
TYPE(date) :: d4                        ! Date 4
CHARACTER(len=10) :: str1               ! Date strings
CHARACTER(len=10) :: str2               ! Date strings
CHARACTER(len=10) :: str3               ! Date strings
CHARACTER(len=10) :: str4               ! Date strings

! Initialize dates d1, d2, and d3 (d4 defaults)
CALL d1%set_date(4,1,2016)
```

(*continued*)

(*concluded*)

```fortran
CALL d2%set_date(1,3,2018)
CALL d3%set_date(3,1,2016)

! Write out the dates
str1 = d1%to_string()
str2 = d2%to_string()
str3 = d3%to_string()
str4 = d4%to_string()
WRITE (*,'(A,A)') 'Date 1 = ', str1
WRITE (*,'(A,A)') 'Date 2 = ', str2
WRITE (*,'(A,A)') 'Date 3 = ', str3
WRITE (*,'(A,A)') 'Date 4 = ', str4

! Check for leap years
IF ( d1%is_leap_year() ) THEN
   WRITE (*,'(I4,A)') d1%get_year(), ' is a leap year.'
ELSE
   WRITE (*,'(I4,A)') d1%get_year(), ' is a not leap year.'
END IF

IF ( d2%is_leap_year() ) THEN
   WRITE (*,'(I4,A)') d2%get_year(), ' is a leap year.'
ELSE
   WRITE (*,'(I4,A)') d2%get_year(), ' is a not leap year.'
END IF

! Check for equality
IF ( d1%is_equal(d3) ) THEN
   WRITE (*,'(3A)') str3, ' is equal to ', str1
ELSE
   WRITE (*,'(3A)') str3, ' is not equal to ', str1
END IF

! Check is_earlier
IF ( d1%is_earlier_than(d3) ) THEN
   WRITE (*,'(3A)') str3, ' is earlier than ', str1
ELSE
   WRITE (*,'(3A)') str3, ' is not earlier than ', str1
END IF

! Check is_later
IF ( d1%is_later_than(d3) ) THEN
   WRITE (*,'(3A)') str3, ' is later than ', str1
ELSE
   WRITE (*,'(3A)') str3, ' is not later than ', str1
END IF

END PROGRAM test_date
```

When this program is executed, the results are:

```
C:\book\fortran\chap16>test_date
Date 1 = 01/04/2016
Date 2 = 03/01/2018
```

```
                Date 3 = 01/03/2016
                Date 4 = 01/01/1900
                2016 is a leap year.
                2018 is a not leap year.
                01/03/2016 is not equal to 01/04/2016
                01/03/2016 is earlier than 01/04/2016
```

Note that the date strings are being written out in the order month/day/year. From the test results, this class appears to be functioning correctly.

This class works, but it could be improved. For example, there is no validity checking performed on the input values in the set_date() method, and the to_string() method could be modified to produce dates with explicit month names such as "January 1, 1900". In addition, the US order month/day/year is not used everywhere in the world. It would be possible to customize the to_string() method so that it writes out date strings in different orders in different parts of the world. You will be asked to improve this class as an end of chapter exercise.

16.7
CONTROLLING ACCESS TO CLASS MEMBERS

The instance variables of a class are normally declared PRIVATE and the methods of a class are normally declared PUBLIC, so that the methods form an interface with the outside world, hiding the internal behavior of the class from any other parts of the program. This approach has many advantages, since it makes programs more modular. For example, suppose that we have written a program that makes extensive use of timer objects. If necessary, we could completely redesign the internal behavior of the timer class, and the program will continue to work properly as long as we have not changed the parameters or returned values from methods start_timer() and elapsed_time(). This **public interface** isolates the internals of the class from rest of the program, making incremental modifications easier.

Good Programming Practice
The instance variables of a class should normally be declared PRIVATE, and the class methods should be used to provide a standard interface to the class.

There are some exceptions to this general rule. Many classes contain PRIVATE methods that perform specialized calculations in support of the PUBLIC methods of the class. These are called **utility methods;** since they are not intended to be called directly by users, they are declared with the PRIVATE access modifier.

16.8
FINALIZERS

Just before an object is destroyed, it makes a call to a special method called a **finalizer**, if one is defined. A finalizer performs any necessary cleanup (releasing resources, closing files, etc.) before the object is destroyed. There can be more than one finalizer in a class, but most classes do not need a finalizer at all.

A finalizer is bound to a class by adding a FINAL keyword in the CONTAINS section of the type definition. For example, the following data type contains a pair of pointers to arrays of *x* and *y* data points. When an object of this data type is created and used, arrays will be allocated and data will be assigned to the pointer v.

```
TYPE,PUBLIC :: vector
    PRIVATE
    REAL,DIMENSION(:),POINTER :: v
    LOGICAL :: allocated = .FALSE.
END TYPE
```

If the object of this data type were later deleted, the pointers would go away, but the allocated memory would remain, and the program would have a memory leak.

Now suppose that we declare a final subroutine called clean_vector for this data type.

```
TYPE,PUBLIC :: vector
    PRIVATE
    REAL,DIMENSION(:),POINTER :: v
    LOGICAL :: v_allocated = .FALSE.
CONTAINS
    FINAL :: clean_vector
END TYPE
```

When an item of the data type is destroyed, the final subroutine clean_vector is automatically called with the object as an argument, just before it is destroyed. This subroutine can deallocate any memory allocated on x or y, and thus avoid a memory leak.

Final subroutines are also used to close files that might be open in an object, and to release similar system resources.

EXAMPLE 16-2

Using Finalizers:

To illustrate the use of finalizers, we will create a simple class capable of storing an arbitrary-length vector of real data. Since we don't know how long the vector will be, we will declare this vector using a pointer and allocate an array of the proper size on the pointer.

This class will contain a set method to place the vector into the object, a put method to retrieve the data, and a final method to deallocate the data when the object is destroyed.

The resulting class is shown in Figure 16-12.

FIGURE 16-12
The vector class.

```fortran
MODULE vector_class
!
!   This module implements a vector class. This initial
!   version of the class holds an arbitrary-length rank 1
!   REAL vector. It includes procedures to put and gut
!   the data, as well as a finalizer to deallocate the
!   data before an object of this type is destroyed.
!
!   Record of revisions:
!       Date        Programmer       Description of change
!       ====        ==========       =====================
!     01/08/16    S. J. Chapman      Original code
!
IMPLICIT NONE

! Type definition
TYPE,PUBLIC :: vector   ! This will be the name we instantiate

    ! Instance variables.
    PRIVATE
    REAL,DIMENSION(:),POINTER :: v
    LOGICAL :: v_allocated = .FALSE.

CONTAINS

    ! Bound procedures
    PROCEDURE,PUBLIC :: set_vector => set_vector_sub
    PROCEDURE,PUBLIC :: get_vector => get_vector_sub
    FINAL :: clean_vector

END TYPE vector

! Restrict access to the actual procedure names
PRIVATE :: set_vector_sub, get_vector_sub, clean_vector_sub

! Now add methods
CONTAINS

    SUBROUTINE set_vector_sub(this, array)
    !
    ! Subroutine to store data in the vector
    !
    IMPLICIT NONE

    ! Declare calling arguments
    CLASS(vector) :: this                       ! Vector object
    REAL,DIMENSION(:),INTENT(IN) :: array       ! Input data

    ! Declare local variables
    INTEGER :: istat                            ! Allocate status

    ! Save data, for deleting any data that might have been
    ! stored in this object.
    IF ( this%v_allocated ) THEN
       DEALLOCATE(this%v,STAT=istat)
    END IF
```

(continued)

(*continued*)

```
   ALLOCATE(this%v(SIZE(array,1)),STAT=istat)
   this%v = array
   this%v_allocated = .TRUE.

   END SUBROUTINE set_vector_sub

   SUBROUTINE get_vector_sub(this, array)
   !
   ! Subroutine to get data in the vector
   !
   IMPLICIT NONE

   ! Declare calling arguments
   CLASS(vector) :: this                    ! Vector object
   REAL,DIMENSION(:),INTENT(OUT) :: array   ! Output data

   ! Declare local variables
   INTEGER :: array_length                  ! Length of array
   INTEGER :: data_length                   ! Length of data vector
   INTEGER :: istat                         ! Allocate status

   ! Retrieve data. If the size of the stored data does
   ! not match the array size, then return only a subset
   ! of the data or else pad the real data with zeros.
   IF ( this%v_allocated ) THEN

      ! Return as much data as possible, truncating or
      ! zero padding as necessary.
      array_length = SIZE(array,1)
      data_length  = SIZE(this%v,1)
      IF ( array_length > data_length ) THEN
         array(1:data_length) = this%v
         array(data_length+1:array_length) = 0
      ELSE IF ( array_length == data_length ) THEN
         array = this%v
      ELSE
         array = this%v(1:array_length)
      END IF

   ELSE
      ! No data--return zeros.
      array = 0

   END IF

   END SUBROUTINE get_vector_sub

   SUBROUTINE clean_vector_sub(this)
   !
   ! Subroutine to finalize the vector
   !
   IMPLICIT NONE

   ! Declare calling arguments
   CLASS(vector) :: this                    ! Vector object
```

(*continued*)

(*concluded*)

```
   ! Declare local variables
   INTEGER :: istat                  ! Allocate status

   ! Debugging message
   WRITE (*,*) 'In finalizer ...'

   ! Save data, for deleting any data that might have been
   ! stored in this object.
   IF ( this%v_allocated ) THEN
      DEALLOCATE(this%v,STAT=istat)
   END IF

   END SUBROUTINE clean_vector_sub

END MODULE vector_class
```

We must create a test driver program to test the `vector` class. Such a program is shown in Figure 16-13. This program creates a `vector` object by allocating it on a pointer. It stores and retrieves an array from the object, and then deallocates it. Note that when the object is deallocated, the final subroutine is automatically called to deallocate the instance variable v.

FIGURE 16-13
Test driver for the `vector` class.

```
PROGRAM test_vector
!
!   This program tests the vector class.
!
!   Record of revisions:
!       Date          Programmer          Description of change
!       ====          ==========          =====================
!     01/08/16      S. J. Chapman         Original code
!
USE vector_class                   ! Import vector class
IMPLICIT NONE

! Declare variables
REAL,DIMENSION(6) :: array         ! Array of data to load / save
INTEGER :: istat                   ! Allocate status
TYPE(vector),POINTER :: my_vec     ! Test object

! Create an object of type "vector" using the pointer
ALLOCATE( my_vec, STAT=istat )

! Save an array of data in this object.
array = [ 1., 2., 3., 4., 5., 6. ]
CALL my_vec%set_vector(array)

! Retrieve the data from this vector.
array = 0
CALL my_vec%get_vector(array)
WRITE (*,'(A,6F6.1)') 'vector = ', array
```

(*continued*)

(*concluded*)

```
! Destroy this object
WRITE (*,*) 'Deallocating vector object ...'
DEALLOCATE( my_vec, STAT=istat )

END PROGRAM test_vector
```

When this program was executed on my computer, the results were:

```
C:\book\fortran\chap16>test_vector
vector =    1.0   2.0   3.0   4.0   5.0   6.0
 Deallocating vector object ...
 In finalizer ...
```

Note that the data stored in the vector was recovered successfully. Also, note that the finalizer was called when the object was deallocated.

16.9

INHERITANCE AND POLYMORPHISM

In Section 16.1.4, we learned that classes could be organized in a class hierarchy with lower-level classes inheriting instance variables and methods from the higher-level classes that they were based on.

Any class above a specific class in the class hierarchy is known as a superclass of that class. The class just above a specific class in the hierarchy is known as the *immediate superclass* of the class. Any class below a specific class in the class hierarchy is known as a subclass of that class.

This section explains how inheritance allows Fortran to treat objects from different subclasses as a single unit by referring to them as objects of their common superclass. It also explains how, when working with a collection of superclass objects, Fortran is able to automatically apply the proper methods to each object, regardless of the subclass the object came from. This ability is known as **polymorphism.**

Inheritance is a major advantage of object-oriented programming; once a behavior (method) is defined in a superclass, that behavior is automatically inherited by all subclasses unless it is explicitly overridden with a modified method. Thus, behaviors only need to be coded *once*, and they can be used by all subclasses. A subclass need only provide methods to implement the *differences* between itself and its parent.

16.9.1 Superclasses and Subclasses

For example, suppose that we were to create a class `employee`, describing the characteristics of the employees of a company. This class would contain the name, social security number, address, etc., of the employee, together with pay information. However, most companies have two different types of employees, those on a salary and those paid by the hour. Therefore, we could create two subclasses of `employee`, `salaried_employee` and `hourly_employee`, with different methods for calculating monthly pay.

Object-Oriented Programming in Fortran

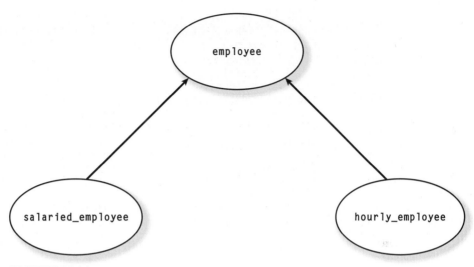

FIGURE 16-14
A simple inheritance hierarchy. Both salaried_employee and hourly_employee inherit from employee, and an object of either of their classes is also an object of the employee class.

Both of these subclasses would inherit all of the common information and methods from employee (name, etc.), but would override the method used to calculate pay.

Figure 16-14 shows this inheritance hierarchy. In object-oriented programming, the relationship between superclasses and subclasses is shown with arrows pointing from a subclass to the parent class. Here, class employee is the parent of both class salaried_employee and class hourly_employee.

Objects of either the salaried_employee *or* hourly_employee *classes may be treated as objects of the* employee *class*, and so forth for any additional classes up the inheritance hierarchy. This fact is very important since objects of the two subclasses can be grouped together and treated as a *single* collection of objects of the superclass employee.

Objects of either the salaried_employee or hourly_employee classes inherit all of the PUBLIC instance variables and methods of the employee class. This means that if an object is to work with instance variables or override methods defined in the parent class, those instance variables and/or methods must have been declared with PUBLIC access.

16.9.2 Defining and Using Subclasses

A class is declared as a subclass of another class by including an EXTENDS attribute in the type definition. For example, suppose that the instance variables and methods of class employee were declared as follows:

```
! Type definition
TYPE,PUBLIC :: employee    ! This will be the name we instantiate
    ! Instance variables.
    CHARACTER(len=30) :: first_name    ! First name
```

```
        CHARACTER(len=30) :: last_name      ! Last name
        CHARACTER(len=11) :: ssn            ! Social security number
        REAL :: pay = 0                     ! Monthly pay
    CONTAINS
        ! Bound procedures
        PROCEDURE,PUBLIC :: set_employee => set_employee_sub
        PROCEDURE,PUBLIC :: set_name => set_name_sub
        PROCEDURE,PUBLIC :: set_ssn => set_ssn_sub
        PROCEDURE,PUBLIC :: get_first_name => get_first_name_fn
        PROCEDURE,PUBLIC :: get_last_name => get_last_name_fn
        PROCEDURE,PUBLIC :: get_ssn => get_ssn_fn
        PROCEDURE,PUBLIC :: calc_pay => calc_pay_fn
    END TYPE employee
```

Then a subclass `salaried_employee` could be declared using the EXTENDS attribute as follows:

```
    ! Type definition
    TYPE,PUBLIC,EXTENDS(employee) :: salaried_employee
        ! Additional instance variables.
        PRIVATE
        REAL :: salary = 0                  ! Monthly salary
    CONTAINS
        ! Bound procedures
        PROCEDURE,PUBLIC :: set_salary => set_salary_sub
        PROCEDURE,PUBLIC :: calc_pay => calc_pay_fn
    END TYPE employee
```

This new subclass *inherits* all of the instance variables from class `employee`, and adds a new instance variable `salary` of its own. It also inherits the methods of the parent class, except that it *overrides* (replaces) method `calc_pay` with a new version of its own. This overridden method `calc_pay` will be used instead of the one defined in class `employee` for objects of this subclass. It also adds a unique method `set_salary` that did not exist in the parent class.

A similar definition could be created for subclass `hourly_employee`.

```
    ! Type definition
    TYPE,PUBLIC,EXTENDS(employee) :: hourly_employee
        ! Additional instance variables.
        PRIVATE
        REAL :: rate = 0                    ! Hourly rate
    CONTAINS
        ! Bound procedures
        PROCEDURE,PUBLIC :: set_pay_rate => set_pay_rate_sub
        PROCEDURE,PUBLIC :: calc_pay => calc_pay_fn
    END TYPE employee
```

This class also extends `employee`. This new subclass inherits all of the instance variables from class `employee`, and adds a new instance variable `rate` of its own.

It also inherits the methods of the parent class, except that it overrides method `calc_pay` with a new version of its own. This overridden method `calc_pay` will be used instead of the one defined in class `employee` for objects of this subclass. It also adds a unique method `set_pay_rate` that did not exist in the parent class.

For all practical purposes, any object of the subclass `salaried_employee` or subclass `hourly_employee` *is* an object of class `employee`. In object-oriented programming terms, we say that these classes have an "is a" relationship with `employee`, because an object of either class "is an" object of the parent class `employee`.

The Fortran code for the `employee` class is shown in Figure 16-15. This class includes four instance variables, `first_name`, `last_name`, `ssn`, and `pay`. The class also defines seven methods to manipulate the instance variables of the class.

FIGURE 16-15
The `employee_class` class.

```fortran
MODULE employee_class
!
!   This module implements an employee class.
!
!   Record of revisions:
!       Date         Programmer         Description of change
!       ====         ==========         =====================
!      01/09/16      S. J. Chapman      Original code
!
IMPLICIT NONE

! Type definition
TYPE,PUBLIC :: employee   ! This will be the name we instantiate

    ! Instance variables.
    CHARACTER(len=30) :: first_name    ! First name
    CHARACTER(len=30) :: last_name     ! Last name
    CHARACTER(len=11) :: ssn           ! Social security number
    REAL :: pay = 0                    ! Monthly pay
CONTAINS

    ! Bound procedures
    PROCEDURE,PUBLIC :: set_employee => set_employee_sub
    PROCEDURE,PUBLIC :: set_name => set_name_sub
    PROCEDURE,PUBLIC :: set_ssn => set_ssn_sub
    PROCEDURE,PUBLIC :: get_first_name => get_first_name_fn
    PROCEDURE,PUBLIC :: get_last_name => get_last_name_fn
    PROCEDURE,PUBLIC :: get_ssn => get_ssn_fn
    PROCEDURE,PUBLIC :: calc_pay => calc_pay_fn

END TYPE employee

! Restrict access to the actual procedure names
PRIVATE :: set_employee_sub, set_name_sub, set_ssn_sub
PRIVATE :: get_first_name_fn, get_last_name_fn, get_ssn_fn
PRIVATE :: calc_pay_fn
```

(*continued*)

(*continued*)

```fortran
! Now add methods
CONTAINS

   SUBROUTINE set_employee_sub(this, first, last, ssn)
   !
   ! Subroutine to initialize employee data.
   !
   IMPLICIT NONE

   ! Declare calling arguments
   CLASS(employee) :: this            ! Employee object
   CHARACTER(len=*) :: first          ! First name
   CHARACTER(len=*) :: last           ! Last name
   CHARACTER(len=*) :: ssn            ! SSN

   ! Save data in this object.
   this%first_name = first
   this%last_name  = last
   this%ssn        = ssn
   this%pay        = 0

   END SUBROUTINE set_employee_sub

   SUBROUTINE set_name_sub(this, first, last)
   !
   ! Subroutine to initialize employee name.
   !
   IMPLICIT NONE

   ! Declare calling arguments
   CLASS(employee) :: this                      ! Employee object
   CHARACTER(len=*),INTENT(IN) :: first         ! First name
   CHARACTER(len=*),INTENT(IN) :: last          ! Last name

   ! Save data in this object.
   this%first_name = first
   this%last_name  = last

   END SUBROUTINE set_name_sub

   SUBROUTINE set_ssn_sub(this, ssn)
   !
   ! Subroutine to initialize employee SSN.
   !
   IMPLICIT NONE

   ! Declare calling arguments
   CLASS(employee) :: this                      ! Employee object
   CHARACTER(len=*),INTENT(IN) :: ssn           ! SSN

   ! Save data in this object.
   this%ssn = ssn

   END SUBROUTINE set_ssn_sub

   CHARACTER(len=30) FUNCTION get_first_name_fn(this)
```

(*continued*)

(*concluded*)

```fortran
   !
   ! Function to return the first name.
   !
   IMPLICIT NONE

   ! Declare calling arguments
   CLASS(employee) :: this              ! Employee object

   ! Return the first name
   get_first_name_fn = this%first_name

   END FUNCTION get_first_name_fn
   CHARACTER(len=30) FUNCTION get_last_name_fn(this)
   !
   ! Function to return the last name.
   !
   IMPLICIT NONE

   ! Declare calling arguments
   CLASS(employee) :: this              ! Employee object

   ! Return the last name
   get_last_name_fn = this%last_name

   END FUNCTION get_last_name_fn

   CHARACTER(len=30) FUNCTION get_ssn_fn(this)
   !
   ! Function to return the SSN.
   !
   IMPLICIT NONE

   ! Declare calling arguments
   CLASS(employee) :: this              ! Employee object

   ! Return the last name
   get_ssn_fn = this%ssn

   END FUNCTION get_ssn_fn

   REAL FUNCTION calc_pay_fn(this,hours)
   !
   ! Function to calculate the employee pay. This
   ! function will be overridden by different subclasses.
   !
   IMPLICIT NONE

   ! Declare calling arguments
   CLASS(employee) :: this              ! Employee object
   REAL,INTENT(IN) :: hours             ! Hours worked

   ! Return pay
   calc_pay_fn = 0

   END FUNCTION calc_pay_fn
END MODULE employee_class
```

The method `calc_pay` in this class returns a zero instead of calculating a valid pay, since the method of calculating the pay will depend on the type of employee, and we don't know that information yet in this class.

Note that the instance variables in this class are *not* declared to be `PRIVATE`. These instance variables will need to be accessed by subclasses of the `employee` class. Since the subclasses are in different modules, they would not be able to access the instance variables if they were declared to be `PRIVATE`. This is a limitation of Fortran's object-oriented capabilities that is not found in proper object-oriented languages such as C++ and Java.

Note that the calling arguments in each bound method include the object itself as the first parameter. This is necessary, because whenever a bound method with the `PASS` attribute is referenced by an object using the format `obj%method()`, the object itself is passed to the method as its first argument. This allows the method to access or modify the contents of the object, if necessary. Furthermore, note that the object is declared using a `CLASS` keyword in each method call, for example,

```
SUBROUTINE set_name_sub(this, first, last)
!
! Subroutine to initialize employee name.
!
IMPLICIT NONE

! Declare calling arguments
CLASS(employee) :: this               ! Employee object
CHARACTER(len=*) :: first             ! First name
CHARACTER(len=*) :: last              ! Last name
```

The `CLASS` keyword in this list means that this subroutine will work with either an object of class `employee` or with an object of any subclass of `employee`. In Fortran terms, the **declared type** of the argument `this` is `employee`, but the **dynamic type** at runtime can be `employee` or any subclass of `employee`.

In contrast, if the calling argument were declared with a `TYPE` keyword instead

```
! Declare calling arguments
TYPE(employee) :: this                ! Employee object
```

then it would only work with an object of the `employee` class, *not* with any of the subclasses. In this case, the declared type and the dynamic type must be identical. To get polymorphic behavior, we must always declare the methods arguments with the `CLASS` keyword.

The Fortran code for the `salaried_employee` subclass is shown in Figure 16-16. This class *inherits* the four instance variables, `first_name`, `last_name`, `ssn`, and `pay`, and adds an additional instance variable `salary`. It also defines a new method `set_salary`, and overrides the method `calc_pay` from the superclass.

FIGURE 16-16
The `salaried_employee` class.

```
MODULE salaried_employee_class
!
!   This module implements a salaried employee class.
```

(continued)

(*continued*)

```fortran
!
!   Record of revisions:
!       Date          Programmer          Description of change
!       ====          ==========          =====================
!     01/09/16        S. J. Chapman       Original code
!
USE employee_class                      ! USE parent class
IMPLICIT NONE

! Type definition
TYPE,PUBLIC,EXTENDS(employee) :: salaried_employee

   ! Additional instance variables.
   PRIVATE
   REAL :: salary = 0                   ! Monthly salary

CONTAINS

   ! Bound procedures
   PROCEDURE,PUBLIC :: set_salary => set_salary_sub
   PROCEDURE,PUBLIC :: calc_pay => calc_pay_fn

END TYPE salaried_employee

! Restrict access to the actual procedure names
PRIVATE :: calc_pay_fn, set_salary_sub

! Now add methods
CONTAINS

   SUBROUTINE set_salary_sub(this, salary)
   !
   ! Subroutine to initialize the salary of the salaried
   ! employee. This is a new method.
   !
   IMPLICIT NONE

   ! Declare calling arguments
   CLASS(salaried_employee) :: this       ! Salaried employee object
   REAL,INTENT(IN) :: salary              ! Salary

   ! Save data in this object.
   this%pay    = salary
   this%salary = salary

   END SUBROUTINE set_salary_sub

   REAL FUNCTION calc_pay_fn(this,hours)
   !
   ! Function to calculate the salaried employee pay. This
   ! function overrides the one in the parent class.
   !
   IMPLICIT NONE

   ! Declare calling arguments
```

(*continued*)

(*concluded*)

```
        CLASS(salaried_employee) :: this    ! Salaried employee object
        REAL,INTENT(IN) :: hours             ! Hours worked

        ! Return pay
        calc_pay_fn = this%salary

        END FUNCTION calc_pay_fn

END MODULE salaried_employee_class
```

A class is declared as a subclass of another class by including an EXTENDS attribute in the type definition. In this case, class salaried_employee is a subclass of class employee because of the "EXTENDS(employee)" attribute in the type definition. Therefore, this class inherits all of the PUBLIC instance variables and methods from class employee.

The class adds one new instance variable salary and one new method set_salary to the ones inherited from the parent class. In addition, the class *overrides* method calc_pay_fn, changing the meaning of this method for objects of type salaried_employee.

The Fortran code for the hourly_employee subclass is shown in Figure 16-17. This class *inherits* the four instance variables, first_name, last_name, ssn, and pay, and adds an additional instance variable rate. It also defines a new method set_rate, and overrides the method calc_pay from the superclass.

FIGURE 16-17
The hourly_employee class.

```
MODULE hourly_employee_class
!
!   This module implements an hourly employee class.
!
!   Record of revisions:
!       Date        Programmer          Description of change
!       ====        ==========          =====================
!       01/09/16    S. J. Chapman       Original code
!
USE employee_class                      ! USE parent class
IMPLICIT NONE

! Type definition
TYPE,PUBLIC,EXTENDS(employee) :: hourly_employee

    ! Additional instance variables.
    PRIVATE
    REAL :: rate = 0                    ! Hourly rate

CONTAINS
```

(*continued*)

(*concluded*)

```
   ! Bound procedures
   PROCEDURE,PUBLIC :: set_pay_rate => set_pay_rate_sub
   PROCEDURE,PUBLIC :: calc_pay => calc_pay_fn

END TYPE hourly_employee

! Restrict access to the actual procedure names
PRIVATE :: calc_pay_fn, set_pay_rate_sub

! Now add methods
CONTAINS

   SUBROUTINE set_pay_rate_sub(this, rate)
   !
   ! Subroutine to initialize the pay rate of the hourly
   ! employee. This is a new method.
   !
   IMPLICIT NONE

   ! Declare calling arguments
   CLASS(hourly_employee) :: this        ! Hourly employee object
   REAL,INTENT(IN) :: rate               ! Pay rate ($/hr)

   ! Save data in this object.
   this%rate = rate

   END SUBROUTINE set_pay_rate_sub

   REAL FUNCTION calc_pay_fn(this,hours)
   !
   ! Function to calculate the hourly employee pay. This
   ! function overrides the one in the parent class.
   !
   IMPLICIT NONE

   ! Declare calling arguments
   CLASS(hourly_employee) :: this        ! Hourly employee object
   REAL,INTENT(IN) :: hours              ! Hours worked

   ! Return pay
   this%pay = hours * this%rate
   calc_pay_fn = this%pay

   END FUNCTION calc_pay_fn

END MODULE hourly_employee_class
```

Class `hourly_employee` is a subclass of class `employee` because of the "EXTENDS(employee)" attribute in the type definition. Therefore, this class inherits all of the instance variables and methods from class `employee`.

The class adds one new instance variable `rate` and one new method `set_rate` to the ones inherited from the parent class. In addition, the class *overrides* method `calc_pay_fn`, changing the meaning of this method for objects of type `hourly_employee`.

16.9.3 The Relationship between Superclass Objects and Subclass Objects

An object of a subclass inherits all of the instance variables and methods of its superclass. In fact, *an object of any subclass may be treated as ("is") an object of its superclass.* This fact implies that we can manipulate objects with either pointers to the subclass or pointers to the superclass. Figure 16-18 illustrates this point.

FIGURE 16-18
A program that illustrates the manipulation of objects with superclass pointers.

```
PROGRAM test_employee
!
!   This program tests the employee class and its subclasses.
!
!   Record of revisions:
!      Date         Programmer          Description of change
!      ====         ==========          =====================
!    01/09/16     S. J. Chapman        Original code
!
USE hourly_employee_class           ! Import hourly employee class
USE salaried_employee_class         ! Import salaried employee class
IMPLICIT NONE

! Declare variables
CLASS(employee),POINTER :: emp1, emp2       ! Employees
TYPE(salaried_employee),POINTER :: sal_emp  ! Salaried employee
TYPE(hourly_employee),POINTER :: hourly_emp ! Hourly employee
INTEGER :: istat                            ! Allocate status

! Create an object of type "salaried_employee"
ALLOCATE( sal_emp, STAT=istat )

! Initialize the data in this object
CALL sal_emp%set_employee('John','Jones','111-11-1111');
CALL sal_emp%set_salary(3000.00);

! Create an object of type "hourly_employee"
ALLOCATE( hourly_emp, STAT=istat )

! Initialize the data in this object
CALL hourly_emp%set_employee('Jane','Jones','222-22-2222');
CALL hourly_emp%set_pay_rate(12.50);

! Now create pointers to "employees".
emp1 => sal_emp
emp2 => hourly_emp

! Calculate pay using subclass pointers
WRITE (*,'(A)') 'Pay using subclass pointers:'
WRITE (*,'(A,F6.1)') 'Emp 1 Pay = ', sal_emp%calc_pay(160.)
WRITE (*,'(A,F6.1)') 'Emp 2 Pay = ', hourly_emp%calc_pay(160.)
```

(continued)

Object-Oriented Programming in Fortran

(*concluded*)

```
! Calculate pay using superclass pointers
WRITE (*,'(A)') 'Pay using superclass pointers:'
WRITE (*,'(A,F6.1)') 'Emp 1 Pay = ', emp1%calc_pay(160.)
WRITE (*,'(A,F6.1)') 'Emp 2 Pay = ', emp2%calc_pay(160.)

! List employee information using superclass pointers
WRITE (*,*) 'Employee information:'
WRITE (*,*) 'Emp1 Name / SSN = ', TRIM(emp1%get_first_name()) // &
            ' ' // TRIM(emp1%get_last_name()) // ' ', &
            TRIM(emp1%get_ssn())
WRITE (*,*) 'Emp 2 Name / SSN = ', TRIM(emp2%get_first_name()) // &
            ' ' // TRIM(emp2%get_last_name()) // ' ', &
            TRIM(emp2%get_ssn())
END PROGRAM test_employee
```

This test program creates one salaried_employee object and one hourly_employee object, and assigns them to pointers of the same types. Then it creates polymorphic pointers to employee objects, and assigns the two subtype objects to the employee pointers. Normally, it is illegal to assign an object of one type to a pointer of another type. However, it is ok here because *the objects of the subclassess* salaried_employee *and* hourly_employee *are also objects of the superclass* employee. The pointers were declared with the CLASS keyword, which allows them to match objects whose dynamic type is the declared type or any subclass of the declared type.

Once the program assigns the objects to the employee pointers, it uses both the original pointers and the employee pointers to access some methods. When this program executes, the results are:

```
D:\book\fortran\chap16>test_employee
Pay using subclass pointers:
Emp 1 Pay = 3000.0
Emp 2 Pay = 2000.0
Pay using superclass pointers:
Emp 1 Pay = 3000.0
Emp 2 Pay = 2000.0
 Employee information:
 Emp 1 Name / SSN = John Jones 111-11-1111
 Emp 2 Name / SSN = Jane Jones 222-22-2222
```

Notice that the pay calculated with the subclass pointers is identical to the pay calculated with the superclass pointers.

It is possible to freely assign an object of a subclass to a pointer of a superclass type, since the object of the subclass is also an object of the superclass. However, the converse is *not* true. An object of a superclass type is *not* an object of its subclass types. Thus, if e is a pointer to employee and s is a pointer to salaried_employee, then the statement

$$e => s$$

is perfectly legal. In contrast, the statement

$$s => e$$

is illegal and will produce a compile-time error.

16.9.4 Polymorphism

Let's look at the program in Figure 16-18 once more. Pay was calculated using superclass pointers, and employee information was displayed using superclass pointers. Note that the `calc_pay` method *differed* for `emp1` and `emp2`. The object referred to by `emp1` was really a `salaried_employee`, so Fortran used the `salaried_employee` version of `calc_pay()` to calculate the appropriate value for it. On the other hand, the object referred to by `emp2` was really an `hourly_employee`, so Fortran used the `hourly_employee` version of `calc_pay()` to calculate the appropriate value for it. The version of `calc_pay()` defined in class `employee` was never used at all.

Here, we were working with `employee` objects, but *this program automatically selected the proper method to apply to each given object based on the subclass that it also belonged to*. This ability to automatically vary methods depending on the subclass that an object belongs to is known as **polymorphism.**

Polymorphism is an incredibly powerful feature of object-oriented languages. It makes them very easy to change. For example, suppose that we wrote a program using arrays of `employees` to work out a company payroll, and then later the company wanted to add a new type of employee, one paid by the piece. We could define a new subclass called `piecework_employee` as a subclass of `employee`, overriding the `calc_pay()` method appropriately, and create employees of this type. *The rest of the program will not have to be changed*, since the program manipulates objects of class `employee`, and polymorphism allows Fortran to automatically select the proper version of a method to apply whenever an object belongs to a particular subclass.

Good Programming Practice
Polymorphism allows multiple objects of different subclasses to be treated as objects of a single superclass, while automatically selecting the proper methods to apply to a particular object based on the subclass that it belongs to.

Note that for polymorphism to work, the methods to be used must be *defined in the superclass and overridden in the various subclasses*. Polymorphism will *not* work if the method you want to use is only defined in the subclasses. Thus, a polymorphic method call like `emp1.calc_pay()` is legal, because method `calc_pay()` is defined in class `employee` and overridden in subclasses `salaried_employee` and `hourly_employee`. On the other hand, a method call like `emp1.set_rate()` is illegal, because method `set_rate()` is only defined in class `hourly_employee`, and we cannot use an `employee` pointer to refer to an `hourly_employee` method.

It is possible to access a subclass method or instance variable by using the SELECT TYPE construct, as we shall see in the next section.

Good Programming Practice
To create polymorphic behavior, declare all polymorphic methods in a common superclass, and then override the behavior of the methods in each subclass that inherits from the superclass.

16.9.5 The SELECT TYPE Construct

It is possible to explicitly determine which type of subclass a given object belongs to while it is being referenced with a superclass pointer. This is done using a SELECT TYPE construct. Once that information is known, a program can access the additional instance variables and methods that are unique to the subclass.

The form of a SELECT TYPE construct is

```
[name:] SELECT TYPE (obj)
TYPE IS ( type_1 ) [name]

    Block 1

TYPE IS ( type_2 ) [name]

    Block 2

CLASS IS ( type_3 ) [name]

    Block 3

CLASS DEFAULT [name]

    Block 4

END SELECT [name]
```

The declared type of *obj* should be a superclass of the other types in the construct. If the input object *obj* has the dynamic type *type_1*, then the statements in Block 1 will be executed, and *the object pointer will be treated as being type_1 during the execution of the block*. This means that the program can access the instance variables and methods unique to subclass *type_1*, even though the declared type of *obj* is of a superclass type.

Similarly, if the input object *obj* has the dynamic type *type_2*, then the statements in Block 2 will be executed, and *the object pointer will be treated as being type_2 during the execution of the block*.

If the dynamic type of the input object *obj* does not exactly match any of the "TYPE IS" clauses, then the structure will look at the "CLASS IS" clauses, and it will execute the code in the block that provides the best match to the dynamic type of the input object. The type of object will be treated as the type of the declared class during the execution of the statements in the block.

At most one block of statements will be executed by this construct. The rules for selecting the block to execute are:

1. If a TYPE IS block matches, execute it.
2. Otherwise, if a single CLASS IS block matches, execute it.
3. Otherwise, if several CLASS IS blocks match, one must be an extension of all the others, and it is executed.
4. Otherwise, if a CLASS DEFAULT block is defined, then it is executed.

An example program illustrating the use of this construct is shown in Figure 16-19. This program defines a 2D point type and two extensions of that type, one a 3D point and the other a 2D point with a temperature measurement. It then declares objects of each type and a pointer of class `point`, which can match any of the objects. In this case, the temperature point object is assigned to the pointer, and the `SELECT TYPE` construct will match the `TYPE IS ( point_temp )` clause. The program will then treat the `point` pointer as though it were a `point_temp` pointer, allowing access to the instance variable `temp` that is only found in that type.

FIGURE 16-19
Example program illustrating the use of the `SELECT TYPE` construct.

```
PROGRAM test_select_type
!
!   This program tests the select type construct.
!
!   Record of revisions:
!       Date        Programmer          Description of change
!       ====        ==========          =====================
!     01/09/16    S. J. Chapman         Original code
!
IMPLICIT NONE

! Declare a 2D point type
TYPE :: point
    REAL :: x
    REAL :: y
END TYPE point

! Declare a 3D point type
TYPE,EXTENDS(point) :: point3d
    REAL :: z
END TYPE point3d

! Declare a 2D point with temperature data
TYPE,EXTENDS(point) :: point_temp
    REAL :: temp
END TYPE point_temp

! Declare variables
TYPE(point),TARGET :: p2
TYPE(point3d),TARGET :: p3
TYPE(point_temp),TARGET :: pt
CLASS(point),POINTER :: p

! Initialize objects here...
p2%x = 1.
p2%y = 2.
p3%x = -1.
p3%y = 7.
p3%z = -2.
pt%x = 10.
pt%y = 0.
pt%temp = 700.

! Assign one of the objects to "p"
p => pt
```

(continued)

(concluded)

```
! Now access the data in that object
SELECT TYPE (p)
TYPE IS ( point3d )
   WRITE (*,*) 'Type is point3d'
   WRITE (*,*) p%x, p%y, p%z
TYPE IS ( point_temp )
   WRITE (*,*) 'Type is point_temp'
   WRITE (*,*) p%x, p%y, p%temp
CLASS IS ( point )
   WRITE (*,*) 'Class is point'
   WRITE (*,*) p%x, p%y
END SELECT

END PROGRAM test_select_type
```

When this program is executed, the results are:

```
D:\book\fortran\chap16>test_select_type
Type is point_temp
   10.00000        0.0000000E+00       700.0000
```

16.10

PREVENTING METHODS FROM BEING OVERRIDDEN IN SUBCLASSES

It is sometimes desirable to ensure that one or more methods are *not* modified in subclasses of a given superclass. This can be done by declaring them in the binding with the NON_OVERRIDABLE attribute, as shown below:

```
TYPE :: point
   REAL :: x
   REAL :: y
CONTAINS
   PROCEDURE,NON_OVERRIDABLE :: my_proc
   ...
END TYPE
```

With this attribute, procedure my_proc would be declared in the definition of the point class, and could not be modified in any subclasses of the point class.

16.11

ABSTRACT CLASSES

Look at the employee class again. Note that we defined method calc_pay() in that class, but *the method is never used*. Since we only ever instantiate members of the subclasses salaried_employee and hourly_employee, this method is *always* overridden polymorphically by the corresponding method in the two subclasses. If this method is never going to be used, why did we bother to write it at all? The answer is that, in order for polymorphism to work, the polymorphic methods must be bound to the parent class, and therefore inherited in all of the subclasses.

However, the actual *methods* in the parent class will never be used if no objects are ever instantiated from that class, so Fortran allows us to declare the bindings and interface definitions only without writing the actual methods. Such methods are called **abstract methods** or **deferred methods,** and types containing abstract methods are known as **abstract types,** as opposed to ordinary **concrete types.**

Abstract methods are declared using the DEFERRED attribute in the type definition, together with an ABSTRACT INTERFACE to define the calling sequence for the method. Any type containing a deferred method must be declared with the ABSTRACT attribute. It is illegal to create any objects directly from an abstract type, but it is legal to create pointers of that type that can be used to manipulate objects of various subtypes.

A deferred method is declared with the following statement

```
PROCEDURE(CALC_PAYX),PUBLIC,DEFERRED :: calc_pay
```

In this statement, the name in parentheses after the PROCEDURE declaration (CALC_PAYX here) is the name of the abstract interface that applies to this method, and calc_pay is the actual name of the method.

An abstract version of the Employee class is shown in Figure 16-20.

FIGURE 16-20
An abstract employee class.

```
MODULE employee_class
!
!   This module implements an abstract employee class.
!
!   Record of revisions:
!       Date        Programmer          Description of change
!       ====        ==========          =====================
!     01/11/16      S. J. Chapman       Original code
!
IMPLICIT NONE

! Type definition
TYPE,ABSTRACT,PUBLIC :: employee

    ! Instance variables.
    CHARACTER(len=30) :: first_name     ! First name
    CHARACTER(len=30) :: last_name      ! Last name
    CHARACTER(len=11) :: ssn            ! Social security number
    REAL :: pay = 0                     ! Monthly pay

CONTAINS

    ! Bound procedures
    PROCEDURE,PUBLIC :: set_employee => set_employee_sub
    PROCEDURE,PUBLIC :: set_name => set_name_sub
```

(continued)

(*continued*)

```
      PROCEDURE,PUBLIC :: set_ssn => set_ssn_sub
      PROCEDURE,PUBLIC :: get_first_name => get_first_name_fn
      PROCEDURE,PUBLIC :: get_last_name => get_last_name_fn
      PROCEDURE,PUBLIC :: get_ssn => get_ssn_fn
      PROCEDURE(CALC_PAYX),PUBLIC,DEFERRED :: calc_pay
   END TYPE employee

   ABSTRACT INTERFACE

      REAL FUNCTION CALC_PAYX(this,hours)
      !
      ! Function to calculate the employee pay. This
      ! function will be overridden by different subclasses.
      !
      IMPLICIT NONE

      ! Declare calling arguments
      CLASS(employee) :: this              ! Employee object
      REAL,INTENT(IN) :: hours             ! Hours worked

      END FUNCTION CALC_PAYX

   END INTERFACE

   ! Restrict access to the actual procedure names
   PRIVATE :: set_employee_sub, set_name_sub, set_ssn_sub
   PRIVATE :: get_first_name_fn, get_last_name_fn, get_ssn_fn

   ! Now add methods
   CONTAINS

      ! All methods are the same as before, except that there is
      ! no implementation of method calc_pay...

      SUBROUTINE set_employee_sub(this, first, last, ssn)
      !
      ! Subroutine to initialize employee data.
      !
      IMPLICIT NONE

      ! Declare calling arguments
      CLASS(employee) :: this              ! Employee object
      CHARACTER(len=*) :: first            ! First name
      CHARACTER(len=*) :: last             ! Last name
      CHARACTER(len=*) :: ssn              ! SSN

      ! Save data in this object.
      this%first_name = first
      this%last_name  = last
      this%ssn        = ssn
      this%pay        = 0

      END SUBROUTINE set_employee_sub
```

(*continued*)

(*continued*)

```fortran
SUBROUTINE set_name_sub(this, first, last)
!
! Subroutine to initialize employee name.
!
IMPLICIT NONE

! Declare calling arguments
CLASS(employee) :: this                   ! Employee object
CHARACTER(len=*),INTENT(IN) :: first      ! First name
CHARACTER(len=*),INTENT(IN) :: last       ! Last name

! Save data in this object.
this%first_name = first
this%last_name  = last

END SUBROUTINE set_name_sub

SUBROUTINE set_ssn_sub(this, ssn)
!
! Subroutine to initialize employee SSN.
!
IMPLICIT NONE

! Declare calling arguments
CLASS(employee) :: this                   ! Employee object
CHARACTER(len=*),INTENT(IN) :: ssn        ! SSN

! Save data in this object.
this%ssn = ssn

END SUBROUTINE set_ssn_sub

CHARACTER(len=30) FUNCTION get_first_name_fn(this)
!
! Function to return the first name.
!
IMPLICIT NONE

! Declare calling arguments
CLASS(employee) :: this                   ! Employee object

! Return the first name
get_first_name_fn = this%first_name

END FUNCTION get_first_name_fn

CHARACTER(len=30) FUNCTION get_last_name_fn(this)
!
! Function to return the last name.
!
IMPLICIT NONE

! Declare calling arguments
CLASS(employee) :: this                   ! Employee object
```

(*continued*)

(concluded)

```
  ! Return the last name
  get_last_name_fn = this%last_name

  END FUNCTION get_last_name_fn

  CHARACTER(len=30) FUNCTION get_ssn_fn(this)
  !
  ! Function to return the SSN.
  !
  IMPLICIT NONE

  ! Declare calling arguments
  CLASS(employee) :: this              ! Employee object

  ! Return the last name
  get_ssn_fn = this%ssn

  END FUNCTION get_ssn_fn
END MODULE employee_class
```

Abstract classes define the list of methods that will be available to subclasses of the class, and can provide partial implementations of those methods. For example, the abstract class employee in Figure 16-20 provides implementations of set_name and set_ssn that will be inherited by the subclasses of employee, but does *not* provide an implementation of calc_pay.

Any subclasses of an abstract class *must* override all abstract methods of the superclass, or they will be abstract themselves. Thus, classes salaried_employee and hourly_employee must override method calc_pay, or they will be abstract themselves.

Unlike concrete classes, *no objects may be instantiated from an abstract class*. Since an abstract class does not provide a complete definition of the behavior of an object, no object may be created from it. The class serves as a template for concrete subclasses, and objects may be instantiated from those concrete subclasses. An abstract class defines the types of polymorphic behaviors that can be used with subclasses of the class, but does *not* define the details of those behaviors.

Programming Pitfalls

Objects may not be instantiated from an abstract class.

Abstract classes often appear at the top of an object-oriented programming class hierarchy, defining the broad types of actions possible with objects of all subclasses of the class. Concrete classes appear at lower levels in a hierarchy, providing implementation details for each subclass.

Good Programming Practice
Use abstract classes to define broad types of behaviors at the top of an object-oriented programming class hierarchy, and use concrete classes to provide implementation details in the subclasses of the abstract classes.

In summary, to create polymorphic behavior in a program:

1. **Create a parent class containing all methods that will be needed to solve the problem.** The methods that will change in different subclasses can be declared DEFERRED, if desired, and we will not have to write a method for them in the superclass—just an interface. Note that this makes the superclass ABSTRACT—no objects may be instantiated directly from it.
2. **Define subclasses for each type of object to be manipulated.** The subclasses must implement a specific method for each abstract method in the superclass definition.
3. **Create objects of the various subclasses, and refer to them using superclass pointers.** When a method call appears with a superclass pointer, Fortran automatically executes the method in the object's actual subclass.

The trick to getting polymorphism right is to determine what behaviors objects of the superclass must exhibit, and to make sure that there is a method to represent every behavior in the superclass definition.

EXAMPLE 16-3

Putting it All Together—A Shape Class Hierarchy:

To illustrate the object-oriented programming concepts introduced in this chapter, let's consider generic 2D shapes. There are many types of shapes including circles, triangles, squares, rectangles, pentagons, and so forth. All of these shapes have certain characteristics in common, since they are closed 2D shapes having an enclosed area and a perimeter of finite length.

Create a generic shape class having methods to determine the area and perimeter of a shape, and then create an appropriate class hierarchy for the following specific shapes: circles, equilateral triangles, squares, rectangles, and pentagons. Then, illustrate polymorphic behavior by creating shapes of each type and determining their area and perimeter using references to the generic shape class.

SOLUTION
To solve this problem, we should create a general shape class and a series of subclasses below it.

The listed shapes fall into a logical hierarchy based on their relationships. Circles, equilateral triangles, rectangles, and pentagons are all specific types of shapes, so they should be subclasses of our general shape class. A square is a special kind of rectangle, so it should be a subclass of the rectangle class. These relationships are shown in Figure 16-21.

Object-Oriented Programming in Fortran

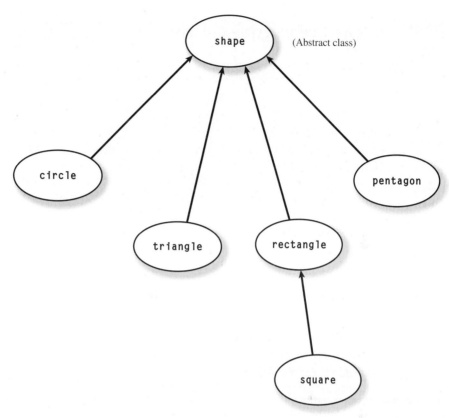

FIGURE 16-21
The shape class hierarchy.

A circle can be completely specified by its radius r, and the area A and perimeter (circumference) P of a circle can be calculated from the equations:

$$A = \pi r^2 \qquad (16\text{-}1)$$

$$P = 2\pi r \qquad (16\text{-}2)$$

An equilateral triangle can be completely specified by the length of one side s, and the area A and perimeter P of the equilateral triangle can be calculated from the equations:

$$A = \frac{\sqrt{3}}{4} s^2 \qquad (16\text{-}3)$$

$$P = 3s \qquad (16\text{-}4)$$

A rectangle can be completely specified by its length *l* and its width *w*, and the area *A* and perimeter *P* of the rectangle can be calculated from the equations:

$$A = lw \qquad (16\text{-}5)$$

$$P = 2(l + w) \qquad (16\text{-}6)$$

A square is a special rectangle whose length is equal to its width so it can be completely specified by setting the length and width of a rectangle to the same size *s*. The area *A* and perimeter *P* of the square can then be calculated from the Equations (16-5) and (16-6).

A pentagon can be completely specified by the length of one side *s*, and the area *A* and perimeter *P* of the pentagon can be calculated from the equations:

$$A = \frac{5}{4} s^2 \cot \frac{\pi}{5} \qquad (16\text{-}7)$$

$$P = 5s \qquad (16\text{-}8)$$

where cot is the cotangent, which is the reciprocal of the tangent.

1. **State the problem.**

   Define and implement a class `shape` with methods to calculate the area and perimeter of a specified shape. Define and implement appropriate subclasses for circles, equilateral triangles, rectangles, squares, and pentagons, with the area and perimeter calculations appropriate for each shape.

2. **Define the inputs and outputs.**

   The inputs to the various classes will be the radius *r* of the circles, the length of a side *s* for the equilateral triangles, the length *l* and width *w* for the rectangles, the length of a side *s* for the squares, and the length of a side *s* for the pentagons. The outputs will be the perimeters and areas of the various objects.

3. **Describe the algorithm.**

   Each class will need methods capable of initializing the appropriate objects. For circles, the initializing method will need the radius *r*. For equilateral triangles, the initializing method will need the length of a side *s*. For rectangles, the initializing method will need the length *l* and width *w*. For squares, the initializing method will need the length of a side *s*. For pentagons, the initializing method will need the length of a side *s*.

   Each of these classes will contain `area`, `perimeter`, and `to_string` methods, returning the area, perimeter, and a character representation of the shape, respectively. They will also contain methods to retrieve the key parameters for each type of shape (radius, etc.).

   The classes required for this problem are `shape`, `circle`, `triangle`, `rectangle`, `square`, and `pentagon`. Class `shape` is a superclass representing a closed, 2D object with a finite area and perimeter. Classes `circle`, `triangle`, `rectangle`, and `pentagon` are special kinds of shapes, so they should be subclasses of `shape`. Class `square` is special kind of rectangle, so it should be a subclass of `rectangle`. The methods in each class will be the class initializer, `area`, `perimeter`, `to_string`, and methods to recover the key parameters for the particular type of shape.

Object-Oriented Programming in Fortran

The pseudocode for the `area()` method in the `circle` class is:

```
get_area_fn = PI * this%r**2
```

The pseudocode for the `perimeter()` method in the `circle` class is:

```
get_perimeter_fn = 2.0 * PI * this%r
```

The pseudocode for the `area()` method in the `triangle` class is:

```
get_area_fn = SQRT(3.0) / 4.0 * this%s**2
```

The pseudocode for the `perimeter()` method in the `triangle` class is:

```
get_perimeter_fn = 3.0 * this%s
```

The pseudocode for the `area()` method in the `rectangle` class is:

```
get_area_fn = this%l * this%w
```

The pseudocode for the `perimeter()` method in the `rectangle` class is:

```
get_perimeter_fn = 2 * this%l + 2 * this%w
```

The pseudocode for the `area()` and `perimeter()` methods in the `square` class is the same as for the `rectangle` class. These methods may be directly inherited from the `rectangle` class.

The pseudocode for the `area()` method in the `pentagon` class is:

```
get_area_fn = 1.25 * this%s**2 / 0.72654253
```

The pseudocode for the `perimeter()` method in the `pentagon` class is:

```
get_perimeter_fn = 5.0 * this%s
```

4. **Turn the algorithm into Fortran statements.**

The abstract class `shape` is shown in Figure 16-22. Note that this class defines abstract methods `area()`, `perimeter()`, and `to_string()`, so that all subclasses will be required to implement these methods, and they may be used polymorphically with objects of type `shape`.

FIGURE 16-22
The parent class `shape`.

```
MODULE shape_class
!
!   This module implements a parent shape class.
!
!   Record of revisions:
!       Date        Programmer          Description of change
!       ====        ==========          =====================
!     01/13/16      S. J. Chapman       Original code
!
```

(continued)

(*continued*)

```fortran
IMPLICIT NONE

! Type definition
TYPE,PUBLIC :: shape

   ! Instance variables.
   ! <none>

CONTAINS

   ! Bound procedures
   PROCEDURE,PUBLIC :: area => calc_area_fn
   PROCEDURE,PUBLIC :: perimeter => calc_perimeter_fn
   PROCEDURE,PUBLIC :: to_string => to_string_fn

END TYPE shape

! Restrict access to the actual procedure names
PRIVATE :: calc_area_fn, calc_perimeter_fn, to_string_fn

CONTAINS

   REAL FUNCTION calc_area_fn(this)
   !
   ! Return the area of this object.
   !
   IMPLICIT NONE

   ! Declare calling arguments
   CLASS(shape) :: this              ! Shape object

   ! Return dummy area
   calc_area_fn = 0.

   END FUNCTION calc_area_fn

   REAL FUNCTION calc_perimeter_fn(this)
   !
   ! Return the perimeter of this object.
   !
   IMPLICIT NONE

   ! Declare calling arguments
   CLASS(shape) :: this              ! Shape object

   ! Return dummy perimeter
   calc_perimeter_fn = 0.

   END FUNCTION calc_perimeter_fn

   CHARACTER(len=50) FUNCTION to_string_fn(this)
   !
   ! Return the character description of this object.
   !
   IMPLICIT NONE
```

(*continued*)

Object-Oriented Programming in Fortran

(*concluded*)

```
    ! Declare calling arguments
    CLASS(shape) :: this              ! Shape object

    ! Return dummy string
    to_string_fn = ''

    END FUNCTION to_string_fn
END MODULE shape_class
```

The class `circle` is shown in Figure 16-23. This class defines an instance variable r for the radius of the circle, and provides concrete implementations of `area()`, `perimeter()`, and `to_string()`. It also defines a method `initialize` that is not inherited from the parent class.

FIGURE 16-23
Class `circle`.

```
MODULE circle_class
!
!   This module implements a circle class.
!
!   Record of revisions:
!       Date          Programmer          Description of change
!       ====          ==========          =====================
!     01/13/16       S. J. Chapman          Original code
!
USE shape_class                         ! USE parent class
IMPLICIT NONE

! Type definition
TYPE,PUBLIC,EXTENDS(shape) :: circle

    ! Additional instance variables.
    REAL :: r = 0                       ! Radius

CONTAINS

    ! Bound procedures
    PROCEDURE,PUBLIC :: initialize => initialize_sub
    PROCEDURE,PUBLIC :: area => get_area_fn
    PROCEDURE,PUBLIC :: perimeter => get_perimeter_fn
    PROCEDURE,PUBLIC :: to_string => to_string_fn

END TYPE circle

! Declare constant PI
REAL,PARAMETER :: PI = 3.141593

! Restrict access to the actual procedure names
PRIVATE :: initialize_sub, get_area_fn, get_perimeter_fn
```

(*continued*)

(*continued*)

```
PRIVATE :: to_string_fn

! Now add methods
CONTAINS
   SUBROUTINE initialize_sub(this,r)
   !
   ! Initialize the circle object.
   !
   IMPLICIT NONE

   ! Declare calling arguments
   CLASS(circle) :: this              ! Circle object
   REAL,INTENT(IN) :: r               ! Radius

   ! Initialize the circle
   this%r = r

   END SUBROUTINE initialize_sub

   REAL FUNCTION get_area_fn(this)
   !
   ! Return the area of this object.
   !
   IMPLICIT NONE

   ! Declare calling arguments
   CLASS(circle) :: this              ! Circle object

   ! Calculate area
   get_area_fn = PI * this%r**2

   END FUNCTION get_area_fn

   REAL FUNCTION get_perimeter_fn(this)
   !
   ! Return the perimeter of this object.
   !
   IMPLICIT NONE

   ! Declare calling arguments
   CLASS(circle) :: this              ! Circle object

   ! Calculate perimeter
   get_perimeter_fn = 2.0 * PI * this%r

   END FUNCTION get_perimeter_fn

   CHARACTER(len=50) FUNCTION to_string_fn(this)
   !
   ! Return the character description of this object.
   !
   IMPLICIT NONE
```

(*continued*)

(*concluded*)

```
    ! Declare calling arguments
    CLASS(circle) :: this              ! Circle object

    ! Return description
    WRITE (to_string_fn,'(A,F6.2)') 'Circle of radius ', &
       this%r

    END FUNCTION to_string_fn

END MODULE circle_class
```

The class `triangle` is shown in Figure 16-24. This class defines an instance variable `s` for the length of the side of the triangle, and provides concrete implementations of `area()`, `perimeter()`, and `to_string()`. It also defines a method `initialize` that is not inherited from the parent class.

FIGURE 16-24
Class `triangle`.

```
MODULE triangle_class
!
!   This module implements a triangle class.
!
!   Record of revisions:
!       Date            Programmer          Description of change
!       ====            ==========          =====================
!     01/13/16        S. J. Chapman          Original code
!
USE shape_class                     ! USE parent class
IMPLICIT NONE

! Type definition
TYPE,PUBLIC,EXTENDS(shape) :: triangle

    ! Additional instance variables.
    REAL :: s = 0                   ! Length of side

CONTAINS

    ! Bound procedures
    PROCEDURE,PUBLIC :: initialize => initialize_sub
    PROCEDURE,PUBLIC :: area => get_area_fn
    PROCEDURE,PUBLIC :: perimeter => get_perimeter_fn
    PROCEDURE,PUBLIC :: to_string => to_string_fn

END TYPE triangle

! Restrict access to the actual procedure names
PRIVATE :: initialize_sub, get_area_fn, get_perimeter_fn
PRIVATE :: to_string_fn
```

(*continued*)

(continued)

```
! Now add methods
CONTAINS

    SUBROUTINE initialize_sub(this,s)
    !
    ! Initialize the triangle object.
    !
    IMPLICIT NONE

    ! Declare calling arguments
    CLASS(triangle) :: this              ! Triangle object
    REAL,INTENT(IN) :: s                 ! Length of side

    ! Initialize the triangle
    this%s = s

    END SUBROUTINE initialize_sub

    REAL FUNCTION get_area_fn(this)
    !
    ! Return the area of this object.
    !
    IMPLICIT NONE

    ! Declare calling arguments
    CLASS(triangle) :: this              ! Triangle object

    ! Calculate area
    get_area_fn = SQRT(3.0) / 4.0 * this%s**2

    END FUNCTION get_area_fn

    REAL FUNCTION get_perimeter_fn(this)
    !
    ! Return the perimeter of this object.
    !
    IMPLICIT NONE

    ! Declare calling arguments
    CLASS(triangle) :: this              ! Triangle object

    ! Calculate perimeter
    get_perimeter_fn = 3.0 * this%s

    END FUNCTION get_perimeter_fn

    CHARACTER(len=50) FUNCTION to_string_fn(this)
    !
    ! Return the character description of this object.
    !
    IMPLICIT NONE

    ! Declare calling arguments
    CLASS(triangle) :: this              ! Triangle object
```

(continued)

Object-Oriented Programming in Fortran

(*concluded*)

```
      ! Return description
      WRITE (to_string_fn,'(A,F6.2)') 'Equilateral triangle of side ', &
         this%s

   END FUNCTION to_string_fn

END MODULE triangle_class
```

The class `rectangle` is shown in Figure 16-25. This class defines instance variables `l` and `w` for the length and width of the rectangle, and provides concrete implementations of `area()`, `perimeter()`, and `to_string()`. It also defines a method `initialize` that is not inherited from the parent class.

FIGURE 16-25
Class rectangle.

```
MODULE rectangle_class
!
!   This module implements a rectangle class.
!
!   Record of revisions:
!       Date            Programmer          Description of change
!       ====            ==========          =====================
!       01/13/16        S. J. Chapman       Original code
!
USE shape_class                             ! USE parent class
IMPLICIT NONE

! Type definition
TYPE,PUBLIC,EXTENDS(shape) :: rectangle

   ! Additional instance variables.
   REAL :: l = 0     ! Length
   REAL :: w = 0     ! Width

CONTAINS

   ! Bound procedures
   PROCEDURE,PUBLIC :: initialize => initialize_sub
   PROCEDURE,PUBLIC :: area => get_area_fn
   PROCEDURE,PUBLIC :: perimeter => get_perimeter_fn
   PROCEDURE,PUBLIC :: to_string => to_string_fn

END TYPE rectangle

! Restrict access to the actual procedure names
PRIVATE :: initialize_sub, get_area_fn, get_perimeter_fn
PRIVATE :: to_string_fn

! Now add methods
CONTAINS
```

(*continued*)

(*continued*)

```fortran
SUBROUTINE initialize_sub(this,l,w)
!
! Initialize the rectangle object.
!
IMPLICIT NONE

! Declare calling arguments
CLASS(rectangle) :: this           ! Rectangle object
REAL,INTENT(IN) :: l               ! Length
REAL,INTENT(IN) :: w               ! Width

! Initialize the rectangle
this%l = l
this%w = w

END SUBROUTINE initialize_sub

REAL FUNCTION get_area_fn(this)
!
! Return the area of this object.
!
IMPLICIT NONE

! Declare calling arguments
CLASS(rectangle) :: this           ! Rectangle object

! Calculate area
get_area_fn = this%l * this%w

END FUNCTION get_area_fn

REAL FUNCTION get_perimeter_fn(this)
!
! Return the perimeter of this object.
!
IMPLICIT NONE

! Declare calling arguments
CLASS(rectangle) :: this           ! Rectangle object

! Calculate perimeter
get_perimeter_fn = 2 * this%l + 2 * this%w

END FUNCTION get_perimeter_fn

CHARACTER(len=50) FUNCTION to_string_fn(this)
!
! Return the character description of this object.
!
IMPLICIT NONE

! Declare calling arguments
CLASS(rectangle) :: this           ! Rectangle object
```

(*continued*)

(*concluded*)

```
    ! Return description
    WRITE (to_string_fn,'(A,F6.2,A,F6.2)') 'Rectangle of length ', &
       this%l, ' and width ', this%w

    END FUNCTION to_string_fn

END MODULE rectangle_class
```

The class `square` is shown in Figure 16-26. Since a square is just a rectangle with its length equal to its width, this class *inherits* its instance variables l and w from class `rectangle`, as well as concrete implementations of `area()` and `perimeter()`. The class overrides method `to_string()`. It also defines a method `initialize` that is not inherited from the parent class.

FIGURE 16-26
Class `square`.

```
MODULE square_class
!
!   This module implements a square class.
!
!   Record of revisions:
!      Date         Programmer          Description of change
!      ====         ==========          =====================
!    01/13/16     S. J. Chapman         Original code
!
USE rectangle_class                ! USE parent class
IMPLICIT NONE

! Type definition
TYPE,PUBLIC,EXTENDS(rectangle) :: square

    ! Additional instance variables.
    !<none>

CONTAINS

    ! Bound procedures
    PROCEDURE,PUBLIC :: to_string => to_string_fn

END TYPE square

! Restrict access to the actual procedure names
PRIVATE :: to_string_fn

! Now add methods
CONTAINS

    CHARACTER(len=50) FUNCTION to_string_fn(this)
    !
    ! Return the character description of this object.
```

(*continued*)

(concluded)

```
!
IMPLICIT NONE

! Declare calling arguments
CLASS(square) :: this              ! Square object

! Return description
WRITE (to_string_fn,'(A,F6.2)') 'Square of length ', &
   this%l

END FUNCTION to_string_fn

END MODULE square_class
```

The class `pentagon` is shown in Figure 16-27. This class defines an instance variable `s` for the length of the side of the pentagon, and provides concrete implementations of methods `area()`, `perimeter()`, and `to_string()`. It also defines a method `initialize` that is not inherited from the parent class.

FIGURE 16-27
Class `pentagon`.

```
MODULE pentagon_class
!
!   This module implements a pentagon class.
!
!   Record of revisions:
!       Date          Programmer          Description of change
!       ====          ==========          =====================
!     01/13/16      S. J. Chapman          Original code
!
USE shape_class                         ! USE parent class
IMPLICIT NONE

! Type definition
TYPE,PUBLIC,EXTENDS(shape) :: pentagon

   ! Additional instance variables.
   REAL :: s = 0          ! Length of side

CONTAINS

   ! Bound procedures
   PROCEDURE,PUBLIC :: initialize => initialize_sub
   PROCEDURE,PUBLIC :: area => get_area_fn
   PROCEDURE,PUBLIC :: perimeter => get_perimeter_fn
   PROCEDURE,PUBLIC :: to_string => to_string_fn

END TYPE pentagon

! Restrict access to the actual procedure names
PRIVATE :: initialize_sub, get_area_fn, get_perimeter_fn
```

(continued)

(*continued*)

```
PRIVATE :: to_string_fn

! Now add methods
CONTAINS

   SUBROUTINE initialize_sub(this,s)
   !
   ! Initialize the pentagon object.
   !
   IMPLICIT NONE

   ! Declare calling arguments
   CLASS(pentagon) :: this              ! Pentagon object
   REAL,INTENT(IN) :: s                 ! Length of side

   ! Initialize the pentagon
   this%s = s

   END SUBROUTINE initialize_sub

   REAL FUNCTION get_area_fn(this)
   !
   ! Return the area of this object.
   !
   IMPLICIT NONE

   ! Declare calling arguments
   CLASS(pentagon) :: this              ! Pentagon object

   ! Calculate area [0.72654253 is tan(PI/5)]
   get_area_fn = 1.25 * this%s**2 / 0.72654253

   END FUNCTION get_area_fn

   REAL FUNCTION get_perimeter_fn(this)
   !
   ! Return the perimeter of this object.
   !
   IMPLICIT NONE

   ! Declare calling arguments
   CLASS(pentagon) :: this              ! Pentagon object

   ! Calculate perimeter
   get_perimeter_fn = 5.0 * this%s

   END FUNCTION get_perimeter_fn

   CHARACTER(len=50) FUNCTION to_string_fn(this)
   !
   ! Return the character description of this object.
   !
   IMPLICIT NONE
```

(*continued*)

(*concluded*)

```
    ! Declare calling arguments
    CLASS(pentagon) :: this            ! Pentagon object

    ! Return description
    WRITE (to_string_fn,'(A,F6.2)') 'Pentagon of side ', &
       this%s

    END FUNCTION to_string_fn

END MODULE pentagon_class
```

5. **Test the program.**

To test this program, we will calculate the area and perimeter of several shapes by hand, and compare the results with those produced by a test driver program.

Shape	Area	Perimeter
Circle of radius 2:	$A = \pi r^2 = 12.5664$	$P = 2\pi r = 12.5664$
Triangle of side 2:	$A = \dfrac{\sqrt{3}}{4}s^2 = 1.7321$	$P = 3s = 6$
Rectangle of length 2 and width 1:	$A = lw = 2$	$P = 2(l + w) = 6$
Square of side 2:	$A = lw = 2 \times 2 = 4$	$P = 2(l + w) = 8$
Pentagon of side 2:	$A = \dfrac{5}{4}s^2 \cot\dfrac{\pi}{5} = 6.8819$	$P = 5s = 10$

An appropriate test driver program is shown in Figure 16-28. Note that this program creates five objects of the various subclasses, and an array of pointers of type `shape` (as described in Section 15.6). It then assigns the objects to elements of the array. It then uses the methods `to_string()`, `area()`, and `perimeter()` on each object in the array `shapes`.

FIGURE 16-28
Program to test abstract class `shape` and its subclasses.

```
PROGRAM test_shape
!
!   This program tests polymorphism using the shape class
!   and its subclasses.
!
!   Record of revisions:
!       Date        Programmer          Description of change
!       ====        ==========          =====================
!       01/13/16    S. J. Chapman       Original code
!
```

(*continued*)

(continued)

```
    USE circle_class             ! Import circle class
    USE square_class             ! Import square class
    USE rectangle_class          ! Import rectangle class
    USE triangle_class           ! Import triangle class
    USE pentagon_class           ! Import pentagon class
    IMPLICIT NONE

    ! Declare variables
    TYPE(circle),POINTER :: cir       ! Circle object
    TYPE(square),POINTER :: squ       ! Square object
    TYPE(rectangle),POINTER :: rec    ! Rectangle object
    TYPE(triangle),POINTER :: tri     ! Triangle object
    TYPE(pentagon),POINTER :: pen     ! Pentagon object
    INTEGER :: i                      ! Loop index
    CHARACTER(len=50) :: id_string    ! ID string
    INTEGER :: istat                  ! Allocate status

    ! Create an array of shape pointers
    TYPE :: shape_ptr
       CLASS(shape),POINTER :: p       ! Pointer to shapes
    END TYPE shape_ptr
    TYPE(shape_ptr),DIMENSION(5) :: shapes

    ! Create and initialize circle
    ALLOCATE( cir, STAT=istat )
    CALL cir%initialize(2.0)

    ! Create and initialize square
    ALLOCATE( squ, STAT=istat )
    CALL squ%initialize(2.0,2.0)

    ! Create and initialize rectangle
    ALLOCATE( rec, STAT=istat )
    CALL rec%initialize(2.0,1.0)

    ! Create and initialize triangle
    ALLOCATE( tri, STAT=istat )
    CALL tri%initialize(2.0)

    ! Create and initialize pentagon
    ALLOCATE( pen, STAT=istat )
    CALL pen%initialize(2.0)

    ! Create the array of shape pointers
    shapes(1)%p => cir
    shapes(2)%p => squ
    shapes(3)%p => rec
    shapes(4)%p => tri
    shapes(5)%p => pen

    ! Now display the results using the array of
    ! shape pointers.
    DO i = 1, 5
```

(continued)

(*concluded*)

```
        ! Get ID string
        id_string = shapes(i)%p%to_string()
        WRITE (*,'(/A)') id_string

        ! Get the area and perimeter
        WRITE (*,'(A,F8.4)') 'Area      = ', shapes(i)%p%area()
        WRITE (*,'(A,F8.4)') 'Perimeter = ', shapes(i)%p%perimeter()
     END DO

END PROGRAM test_shape
```

When this program is executed, the results are:

```
C:\book\fortran\chap16>test_shape
Circle of radius    2.00
Area      =   12.5664
Perimeter =   12.5664

Square of length    2.00
Area      =    4.0000
Perimeter =    8.0000

Rectangle of length   2.00 and width    1.00
Area      =    2.0000
Perimeter =    6.0000

Equilateral triangle of side    2.00
Area      =    1.7321
Perimeter =    6.0000

Pentagon of side    2.00
Area      =    6.8819
Perimeter =   10.0000
```

The results of the program agree with our hand calculations to the number of significant digits that we performed the calculation. Note that the program called the correct polymorphic version of each method.

Quiz 16-1

This quiz provides a quick check to see if you have understood the concepts introduced in Sections 16.1 through 16.9. If you have trouble with the quiz, reread the section, ask your instructor, or discuss the material with a fellow student. The answers to this quiz are found in the back of the book.

1. What are the principal advantages of object-oriented programming?
2. Name the major components of a class, and describe their purposes.

(*continued*)

> (*concluded*)
>
> 3. What types of access modifiers may be defined in Fortran, and what access does each type give? What access modifier should normally be used for instance variables? for methods?
> 4. How are type-bound methods created in Fortran?
> 5. What is a finalizer? Why is a finalizer needed? How do you create one?
> 6. What is inheritance?
> 7. What is polymorphism?
> 8. What are abstract classes and abstract methods? Why would you wish to use abstract classes and methods in your programs?

16.12
SUMMARY

An object is a self-contained software component that consists of properties (variables) and methods. The properties (variables) are usually hidden from the outside world, and are only modified through the methods that are associated with them. Objects communicate with each other via messages (that are really method calls). An object uses a message to request another object to perform a task for it.

Classes are the software blueprints from which objects are made. The members of a class are instance variables, methods, and possibly a finalizer. The members of a class are accessed using the object name and the access operator—the % operator.

A finalizer is a special method used to release resources just before an object is destroyed. A class can have at most one finalizer, but most classes do not need one.

When an object is instantiated from a class, a separate copy of each instance variable is created for the object. All objects derived from a given class share a single set of methods.

When a new class is created from some other class ("extends" the class), it inherits the instance variables and methods of its parent class. The class on which a new class is based is called the superclass of the new class, and the new class is a subclass of the class on which it is based. The subclass only needs to provide instance variables and methods to implement the *differences* between itself and its parent.

An object of a subclass may be treated as an object of its corresponding superclass. Thus, an object of a subclass may be freely assigned to a superclass pointer.

Polymorphism is the ability to automatically vary methods depending on the subclass that an object belongs to. To create polymorphic behavior, define all polymorphic methods in the common superclass, and override the behavior of the methods in each subclass that inherits from the superclass. All pointers and dummy arguments manipulating the objects must be declared to be the superclass type using the CLASS keyword.

An abstract method is a method whose interface is declared without an associated method being written. An abstract method is declared by adding the DEFERRED attribute to the binding, and by providing an abstract interface for the method. A class containing one or more abstract methods is called an abstract class. Each subclass of an abstract class must provide an implementation of all abstract methods, or the subclass will remain abstract.

16.12.1 Summary of Good Programming Practice

The following guidelines introduced in this chapter will help you to develop good programs:

1. Always make instance variables private, so that they are hidden within an object. Such encapsulation makes your programs more modular and easier to modify.
2. Use set methods to check the validity and consistency of input data before it is stored in an object's instance variables.
3. Define predicate methods to test for the truth or falsity of conditions associated with any classes you create.
4. The instance variables of a class should normally be declared PRIVATE, and the class methods should be used to provide a standard interface to the class.
5. Polymorphism allows multiple objects of different subclasses to be treated as objects of a single superclass, while automatically selecting the proper methods to apply to a particular object based on the subclass that it belongs to.
6. To create polymorphic behavior, declare all polymorphic methods in a common superclass, and then override the behavior of the methods in each subclass that inherits from the superclass.
7. Use abstract classes to define broad types of behaviors at the top of an object-oriented programming class hierarchy, and use concrete classes to provide implementation details in the subclasses of the abstract classes.

16.12.2 Summary of Fortran Statements and Structures

ABSTRACT Attribute:

```
TYPE,ABSTRACT :: type_name
```

Examples:

```
TYPE,ABSTRACT :: test
   INTEGER :: a
   INTEGER :: b
CONTAINS
   PROCEDURE(ADD_PROC),DEFERRED :: add
END TYPE
```

(continued)

Object-Oriented Programming in Fortran

(*concluded*)

Description:
The ABSTRACT attribute declares that a data type is abstract, meaning that no objects of this type can be created, because one or more of the bound methods are deferred.

ABSTRACT INTERFACE Construct:

```
ABSTRACT INTERFACE
```

Examples:

```
TYPE,ABSTRACT :: test
    INTEGER :: a
    INTEGER :: b
CONTAINS
    PROCEDURE(ADD_PROC),DEFERRED :: add
END TYPE
ABSTRACT INTERFACE
    SUBROUTINE add_proc (this, b)
    ...
    END SUBROUTINE add_proc
END INTERFACE
```

Description:
The ABSTRACT INTERFACE construct declares the interface of a deferred procedure, so that the Fortran compiler will know the required calling sequence of the procedure.

CLASS Keyword:

```
CLASS(type_name) :: obj1, obj2, ...
```

Examples:

```
CLASS(point) :: my_point
CLASS(point),POINTER :: p1
CLASS(*),POINTER :: p2
```

Description:
The CLASS keyword defines a pointer or dummy argument that can accept a target of the specified type, or of any type that extends the specified type. In other words, the pointer or dummy argument will work with targets of the specified class or of any subclass of the specified class.

The last form of the CLASS keyword creates an unlimited polymorphic pointer that can match an object of any class, but the fields and methods of the object can only be accessed using the SELECT TYPE structure.

DEFERRED Attribute:

```
PROCEDURE,DEFERRED :: proc_name
```

Examples:

```
TYPE,ABSTRACT :: test
   INTEGER :: a
   INTEGER :: b
CONTAINS
   PROCEDURE(ADD_PROC),DEFERRED :: add
END TYPE
```

Description:
The DEFERRED attribute declares that a procedure bound to a derived data type is not defined in the data type, making the type abstract. No object can be created with this data type. A concrete implementation must be defined in a subclass before objects of that type can be created.

EXTENDS Attribute:

```
TYPE,EXTENDS(parent_type) :: new_type
```

Example:

```
TYPE,EXTENDS(point2d) :: point3d
   REAL :: z
END TYPE
```

Description:
The EXTENDS attribute indicates that the new type being defined is an extension of the type specified in the EXTENDS attribute. The new type inherits all the instance variables and methods of the original type, except for ones explicitly overridden in the type definition.

NON_OVERRIDABLE Attribute:

```
PROCEDURE,NON_OVERRIDABLE :: proc_name
```

Example:

```
TYPE :: point
   REAL :: x
   REAL :: y
CONTAINS
   PROCEDURE,NON_OVERRIDABLE :: my_proc
END TYPE
```

Description:
The NON_OVERRIDABLE attribute indicates that a bound procedure cannot be overridden in any subclasses derived from this class.

SELECT TYPE Construct:

```
[name:] SELECT TYPE (obj)
TYPE IS ( type_1 ) [name]

   Block 1

TYPE IS ( type_2 ) [name]

   Block 2

CLASS IS ( type_3 ) [name]

   Block 3

CLASS DEFAULT [name]

   Block 4

END SELECT [name]
```

Example:

```
SELECT TYPE (obj)
TYPE IS (class1)

CLASS DEFAULT

END SELECT
```

Description:
The SELECT TYPE construct selects a block of code to execute depending on the particular subclass of *obj*. If the type in a TYPE IS block matches the type of the object exactly, that block will be executed. Otherwise, if the type in a CLASS IS block is a superclass of the object, that block will be executed. If more than one CLASS IS block is a superclass of the object, then the block with the highest superclass will be executed.

16.12.3 Exercises

16-1. List and describe the major components of a class.

16-2. Enhance the date class created in this chapter by adding:

1. A method to calculate the day-of-year for the specified date.
2. A method to calculate the number of days since January 1, 1900, for the specified date.
3. A method to calculate the number of days between the date in the current date object and the date in another date object.

Also, convert the to_string method to generate the date string in the form Month dd, yyyy. Generate a test driver program to test all of the methods in the class.

16-3. Create a new class called salary_plus_employee as a subclass of the employee class created in this chapter. A salary-plus employee will receive a fixed salary for his/

her normal work week, plus bonus overtime pay at an hourly rate for any hours greater than 42 in any given week. Override all of the necessary methods for this subclass. Then modify program `test_employee` to demonstrate the proper operation of all three subclasses of `employee`.

16-4. General Polygons Create a class called `point`, containing two instance variables x and y, representing the *(x, y)* location of a point on a Cartesian plane. Then, define a class `polygon` as a subclass of the `shape` class developed in Example 16-3. The polygon should be specified by an ordered series of *(x, y)* points denoting the ends of each line segment forming the polygon. For example, a triangle is specified by three *(x, y)* points, a quadrilateral is specified by three *(x, y)* points, and so forth.

The initializing method for this class should accept the number of points used to specify a particular polygon, and should allocate an array of `point` objects to hold the *(x, y)* information. The class should implement `set` and `get` methods to allow the locations of each point to be set and retrieved, as well as area and perimeter calculations.

The area of a general polygon may be found from the equation

$$A = \frac{1}{2}(x_1 y_2 + x_2 y_3 + \cdots + x_{n-1} y_n + x_n y_1 - y_1 x_2 - y_2 x_3 - \cdots - y_{n-1} x_n - y_n x_1) \quad (16\text{-}9)$$

where x_i and y_i are *(x, y)* values of the *i*th point. The perimeter of the general polygon will be the sum of the lengths of each line segment, where the length of segment *i* is found from the equation:

$$length = \sqrt{(x_{i+1} - x_i)^2 + (y_{i+1} - y_i)^2} \quad (16\text{-}10)$$

Once this class is created, write a test program that creates an array of `shapes` of various sorts including general polygons, and sorts the shapes into ascending order of area.

16-5. Create an abstract class called `vec`, which includes instance variables x and y, and abstract methods to add and subtract two vectors. Create two subclasses, `vec2d` and `vec3d`, that implement these methods for 2D and 3D vectors, respectively. Class `vec3d` must also define the additional instance variable z. Write a test program to demonstrate that the proper methods are called polymorphically when `vec` objects are passed to the addition and subtraction methods.

16-6. In Chapter 15, we learned how to create a linked list. Write a program that creates and manipulates linked lists of `Employee` objects, as defined in this chapter.

16-7. Generalize the linked list program created in Exercise 16-6 to work with any type of object. (*Hint:* Use the unlimited polymorphic version of the `CLASS` keyword to create program.)

17

Coarrays and Parallel Processing

OBJECTIVES

- Understand the advantages and disadvantages of parallel processing on modern computers.
- Understand the single program multiple data (SPMD) approach to parallel processing.
- Understand how to create a program with multiple images.
- Understand how to create **coarrays,** which are arrays of data that are shared between images in a multiimage program.
- Learn how to synchronize communications and data transfer among images operating in parallel.
- Learn about the problems associated with race conditions and deadlocks in parallel programs.

This chapter introduces the basic concepts of parallel processing and coarrays in Fortran.

Modern computers now have many **cores,** which are separate processing units that run in parallel. For example, the computer that I am writing this book on contains eight cores, so it can do eight different things simultaneously.

In earlier times, a computer would have one computational unit, and computers became faster by running the computational unit at higher and higher clock speeds.[1] Unfortunately, clock speeds cannot increase forever, because the physical design of a semiconductor chip has limits on how fast signals can propagate. In addition, power requirements (and heat dissipation) increase dramatically with higher clock speeds. As a result, the past decade has seen increases in computer performance by placing more and more computational units (cores) in parallel on a single silicon chip instead of significant increases in throughput for a single core.

These extra cores make a single computer much more powerful than before, but *only* if it is doing more than one thing at a time to run on separate cores. If we create a

---

[1] And by designing more efficient computer instructions.

classic computer program that runs instructions one at a time, it can only run on a single core, and it will only be as fast as a single core of the computer. Classic Fortran programs operate like this, and they have not speeded up very much when running on a modern CPU compared to previous generations. Such programs are called **single-threaded programs.**

To make our modern programs faster, we need to be able to divide the work of the program up so that it can run on more than one core in a computer, with each core doing a part of the work in parallel. Such programs are **parallel programs,** and they are *much* more complicated than the simple programs that we have discussed so far in this book. In a sequential program such as the ones we have seen so far, the program can always know that the results of previous lines in the program have already been calculated before the next line is executed. In a parallel program, this is not true unless the programmer makes special efforts to ensure that the calculations on the multiple cores are coordinated.

For example, suppose that we wanted to perform some calculation on a very large array, and to speed up the calculation we split it up among multiple cores. If each calculation on an array element depended on the values of the neighboring elements, then the results of a calculation in one core would depend on whether another core had updated the neighboring values or not before the calculation was performed. The parallel program would produce different results depending on the relative timing of each core's calculations. This sort of situation is called a **race condition,** and it must be avoided in parallel programming.

To produce reliable results, a parallel program must have mechanisms to synchronize the different parallel parts so that the prerequisite data required by one core has been computed before a given calculation starts. These synchronizing statements are absolutely essential for the proper operation of parallel programs.

Fortran was originally designed to be a single-threaded language, in which every statement is executed sequentially. However, a parallel processing extension was added to the language in Fortran 2008. This option is called **Coarray Fortran.** It consists of new data structures called **coarrays** that allow data to be shared between multiple cores working together to solve a problem, and a set of synchronizing statements to coordinate the operation of the programs on the parallel cores.

Coarray Fortran is designed to allow parallel processing with a relatively simple interface, and therefore to be relatively easy to use. It was designed to preserve all of the efficiencies of existing Fortran, with a simple syntax that directly extends the existing Fortran design. Users of Coarray Fortran do not need to know all of the dirty details of sharing memory and processing between images—it is a hidden behind a simple façade.

17.1
PARALLEL PROCESSING IN COARRAY FORTRAN

Parallel processing in Coarray Fortran works on the single-program, multiple data (SPMD) model. Many copies of a single program can be started up in parallel, with each copy having its own data, and being able to share certain data from other copies. Each copy of the program is referred to as an **image.** The number of parallel images employed can be specified at compilation and/or runtime, depending on the compiler.

The multiple images of a program can be run on multiple cores in a single computer (called the **host**), or sometimes on multiple cores on different computers connected together by a network. Some compilers only support parallel processing for images running on cores within a single host, while others can support spreading the images across multiple hosts connected by networks. Check your compiler documentation to see the type of support that it provides.

We will divide the discussion about parallel processing in Fortran into two sections. First we will learn how to create programs with multiple images, and then we will learn how to synchronize and share data between the images.

17.2
CREATING A SIMPLE PARALLEL PROGRAM

A Coarray Fortran program consists of *n* parallel copies of a single program. Each copy is called an image, and the language provides intrinsic functions to let each image know what its copy number is, and how many total images are being used. The function `this_image()` returns the image number of a particular image, and the function `num_images()` returns the total number of images that are running in parallel.

Figure 17-1 shows a simple Hello World program that can be run in parallel. Each copy will write out a string identifying its image number, and then shut down.

FIGURE 17-1
The source code for the `timer` class.

```
PROGRAM hello_world

WRITE (*,*) 'Hello from image ', this_image(), ' out of ', &
    num_images(), ' images.'

END PROGRAM hello_world
```

There are special compiler switches required to compile this program for parallel operation, and those switches are compiler dependent. In the case of Intel Fortran running on Windows, the option `/Qcoarray:shared` specifies that the program should be run in parallel with shared memory, and the option `/Qcoarray-num-images:n` specifies that there should be *n* parallel images of the program.

This program can be compiled from the command line as follows:

```
C:\book\fortran\chap17>ifort /Qcoarray:shared /Qcoarray-num-images:4
hello_world.f90 /Fehello_world.exe
Intel(R) Visual Fortran Intel(R) 64 Compiler for applications running on
Intel(R) 64, Version 16.0.3.207 Build 20160415
Copyright (C) 1985-2016 Intel Corporation.  All rights reserved.

Microsoft (R) Incremental Linker Version 12.00.40629.0
Copyright (C) Microsoft Corporation.  All rights reserved.

-out:hello_world.exe
-subsystem:console
hello_world.obj
```

When this program is executed, the result are as follows:

```
C:\book\fortran\chap17>hello_world
 Hello from image    4 out of    4 images.
 Hello from image    1 out of    4 images.
 Hello from image    2 out of    4 images.
 Hello from image    3 out of    4 images.

C:\book\fortran\chap17>hello_world
 Hello from image    2 out of    4 images.
 Hello from image    3 out of    4 images.
 Hello from image    4 out of    4 images.
 Hello from image    1 out of    4 images.
```

Note that the order in which the images write out their data varies from execution to execution. The order is not deterministic. Instead, it depends on which image happens to reach the WRITE statement first during a given execution. This is an example of a race condition. In a later section, we will learn about special synchronization commands that can be added to a program to resolve race conditions.

The first image of a coarray program is special, and it is usually referred to as the **master image.** For example, only the master image can read data from the standard input device. Any image can write data out, but only the master image can read data in. The master image is usually used to coordinate the functions of the other images, which are often called **worker images.**

Good Programming Practice
Use the master image (image 1) to coordinate and control the function of the various worker images in a coarray Fortran program.

Good Programming Practice
Only the master image can read data from the standard input device. If that data is to be made available to the worker images, the master image must copy it to them.

An image will terminate when it reaches the end of the program, or when it executes a STOP statement, whichever happens first. If one image terminates, all of the remaining ones continue to run until they also reach the end of the program or a STOP statement.

If you want all of the images in a program to stop, then use the STOP ALL statement. When this statement is executed in any image, it will force all images to terminate. The STOP ALL statement has the same syntax as the STOP statement, so a number or a string can be printed out when the program executes the STOP ALL.

Good Programming Practice
Use the STOP ALL statement to force all images in a coarray program to terminate.

17.3
COARRAYS

A **coarray** is a scalar or array that is allocated separately in every image, but the copy in any image can be accessed from any other image. There will be a separate copy of the data in each image, and a special kind of addressing allows any image to use either the copy of the data in its local memory or the copy of data stored in any other image.

A coarray is declared using a special `CODIMENSION` attribute in the type declaration statement, or by using the `[]` syntax with the variable name. For example, the following statements declare a scalar a and an array b as coarrays using the `CODIMENSION` attribute.

```
INTEGER,CODIMENSION[*] :: a
REAL, DIMENSION(3,3), CODIMENSION[*] :: b
```

The first statement declares that each image of the program will have an integer scalar called a, and that each variable a in each image can be accessed from any other image. The second statement declares that each image of the program will have a 3 × 3 real array called b, and that each array b in each image can be accessed from any other image. Note that a codimension is always declared and accessed using square brackets, and also that the declaration uses the *, representing an indefinite size. The last dimension of any coarray declaration must *always* be *, because the actual number of images to use is not known until compile time (or until runtime, depending on the compiler).

Coarrays can also be declared using an alternate syntax without the `CODIMENSION` keyword.

```
INTEGER :: a[*]
REAL :: b(3,3)[*]
```

These declarations are identical to the ones shown above.

Coarrays must always be implicitly or explicitly declared as having the `SAVE` attribute, because the values in the coarray must be maintained even if they go out of scope in a particular image. If a coarray is declared in a `PROGRAM` or `MODULE`, a `SAVE` attribute is automatically implied. If a coarray is declared in a subroutine or function, it must explicitly have a `SAVE` attribute declared so that the coarray never goes out of scope.

A coarray value can be used just like any ordinary variable or array. You can add, subtract, multiply, divide, and so forth just like any other array element or scalar. The value in the brackets corresponds to the image that the data is to be read from or written to. For example, the following code adds b(3,1) from image 2 to b(1,3) from image 1, and stores the result in the scalar a declared in image 3.

```
a[3] = b(3,1)[2] + b(1,3)[1]
```

The coarray syntax automatically handles the communication between different images without the programmer having to do any special manipulations. The Fortran hides all of the complex software required to connect the different images together with a very simple syntax.

Good Programming Practice
The coarray syntax allows very easy communication between data stored in different executing images of a program.

If a variable is declared as a coarray, it can be used as either a coarray or normal memory in an image. If the variable b is declared in an image as

```
REAL, DIMENSION(3,3), CODIMENSION[*] :: b
```

then the local copy of the array can be used in an image by addressing it as a normal array. For example, the first element of the array in the current images can be accessed as b(1,1). To address the first element of the array from image 3, we would use the syntax b(1,1)[3]. Note that b(1,1) and b(1,1)[this_image()] refer to exactly the same memory location, but the memory access is faster and more efficient if you refer to memory on the local machine without coarray subscript.

Coarrays can also be declared with a multidimensional syntax. For example, the array b could be declared as

```
REAL :: b(3,3)[2,*]
```

This is known as a **corank 2 coarray.** In this case, the images are addressed in column major order, like arrays. The first image would be addressed as [1,1], the second image would be addressed as [2,1], the third image would be addressed as [1,2], and so forth. Note that not all coarray addresses may be defined, because there are only a finite number of images when the program executes. For example, if the program is executed with five images, then the valid images would be addressed as [1,1], [2,1], [1,2], [2,2], and [1,3]. Other values would not correspond to existing images in that particular execution.

Fortran includes two functions co_lbound and co_ubound that return the lowest and highest cobounds of a particular coarray. They function just like lbound and ubound, except that they return coarray sizes instead of array sizes.

There is an intrinsic function image_index() that can be used to determine the index of a particular image containing a coarray variable. This function returns a zero if the coarray index does not match any existing image. Thus, image_index(b, [1,1]) would return an image index of 1, image_index(b, [1,3]) would return an image index of 5, and image_index(b, [2,3]) would return an image index of 0, because that image does not exist in the specified execution.

The maximum number of subscripts plus the maximum number of codimensions on an array must be less than or equal to 15.

Good Programming Practice
The total number of dimensions plus codimensions on an array must be less than or equal to 15.

If any image terminates at the end of the program or at a STOP statement, the coarrays declared in the terminated image remain allocated and available for use by all of the other images. If any image terminates on a STOP ALL statement, all images are terminated and all coarrays are deallocated.

17.4
SYNCHRONIZATION BETWEEN IMAGES

We have learned that any image can access data from any other image, and also that only the master image (image 1) can read data from the standard input stream. Thus, we could create a program that uses image 1 to read an input value and use that value to initialize a coarray in all of the images. After that, each image will write out the local copy of the data that exists in its own image. Such a program is shown in Figure 17-2. Note that image 1 *only* prompts the user for an integer, and then uses that value to initialize coarray a in each executing image. Then, each image writes out the value of a present in its local image.

FIGURE 17-2
This program uses the master image 1 to initialize the coarray a in all images, and then prints out the data present in each image.

```
PROGRAM initialize_image
IMPLICIT NONE

! Declare variables
INTEGER ::a[*]            ! Coarray
INTEGER :: i              ! Loop index
INTEGER :: m              ! Seed

IF ( this_image() == 1 ) THEN

   ! Get the seed value using image 1
   WRITE(*,'(A)') 'Enter an integer:'
   READ(*,*) m

   ! Use it to initialize the other images
   DO i = 1, num_images()
      a[i] = i*m
   END DO

END IF

! Now write out the results from each image.
WRITE (*,'(A,I0,A,I0)4') 'The result from image ', &
      this_image(), ' is ', a

END PROGRAM initialize_image
```

If this program is compiled and executed with eight images, the results will be something like:

```
C:\book\fortran\chap17>initialize_image
The result from image 4 is 0
```

```
The result from image 2 is 0
The result from image 6 is 0
Enter an integer:
The result from image 3 is 0
The result from image 5 is 0
The result from image 7 is 0
The result from image 8 is 0
4
The result from image 1 is 4
```

This is not right! The first image should have produced a 4, the second image should have produced an 8, the third image should have produced a 12, and so forth. What went wrong here?

The problem is that each image is running *independently* in parallel, and images 2 through 8 finished before image 1 ever got around to initializing the value of a in those images. We must do something to hold off the execution of all images until after initialization is complete. The simplest way to do this is to use the SYNC ALL statement. A SYNC ALL statement produces a **synchronization point** in the program. If a SYNC ALL statement is executed in *any* image, then execution will halt in that image until *all other images* have also reached a SYNC ALL statement. When all other images have reached a SYNC ALL statement, then program execution will resume in all images.

The format of a SYNC ALL statement is

```
SYNC ALL
SYNC ALL( [sync_stat_list] )
```

The sync_stat_list can contain optional STAT= and ERMSG= clauses. If the clauses are missing and an error occurs, the executing image will terminate. If the clauses are present, execution will continue so that the user can try to handle the situation.

The value returned by the STAT= clause will be zero if the statement is successful. It will be the constant STAT_STOPPED_IMAGE (defined in module ISO_FORTRAN_ENV) if one or more images has stopped, and will be some other positive number if another error occurs. The string returned by the ERMSG= clause will be a description of the error.

If we add the SYNC ALL statement to the program right after the initialization block (see Figure 17-3), then all images will halt at that point until the initialization has been completed by image 1. Image 1 will be the last one to reach the synchronization point, and after it reaches that point all images will resume executing.

FIGURE 17-3

A modified form of program initialize_image that synchronizes execution of all images right after initialization, holding up execution until the initialization is complete.

```
PROGRAM initialize_image2
IMPLICIT NONE

! Declare variables
INTEGER ::a[*]              ! Coarray
INTEGER :: i                ! Loop index
INTEGER :: m                ! Seed
```

(*continued*)

Coarrays and Parallel Processing

(*concluded*)

```
IF ( this_image() == 1 ) THEN
   ! Get the seed value using image 1
   WRITE(*,'(A)') 'Enter an integer:'
   READ(*,*) m

   ! Use it to initialize the other images
   DO i = 1, num_images()
      a[i] = i*m
   END DO

END IF

! Synchronize all images before continuing
SYNC ALL

! Now write out the results from each image.
WRITE (*,'(A,I0,A,I0)4') 'The result from image ', &
      this_image(), ' is ', a
END PROGRAM initialize_image2
```

If this program is compiled and executed with eight images, the results will be something like:

```
C:\book\fortran\chap17>initialize_image2
Enter an integer:
4
The result from image 1 is 4
The result from image 4 is 16
The result from image 3 is 12
The result from image 7 is 28
The result from image 8 is 32
The result from image 2 is 8
The result from image 5 is 20
The result from image 6 is 24
```

This time, the coarray was initialized before the images continued executing.

Good Programming Practice

Use the SYNC ALL statement to ensure that all images in a program reach a common point before allowing any of them to continue execution.

There is also a SYNC IMAGES command. This command allows a program to synchronize a named list of images instead of every image. The format of a SYNC IMAGES statement is

```
SYNC IMAGES(* [, sync_stat_list] )
SYNC IMAGES(int[, sync_stat_list])
SYNC IMAGES(int array[, sync_stat_list])
```

If the first argument is *, then the image stops until every other image executes a `SYNC IMAGES` that specified the original image. If the first argument is an integer, then the image stops until the specified image executes a `SYNC IMAGES` that specified the original image. If the first argument is an integer array, then the image stops until all of the specified images in the array execute a `SYNC IMAGES` that specified the original image. The `STAT=` and `ERMSG=` clauses are both optional in the statement, and they have the same meaning as in the `SYNC ALL` statement.

Then when a `SYNC IMAGES` command is executed, the image issuing the command is frozen until the corresponding synchronization is complete. If not done carefully, this can cause a program to freeze and never recover. For example, examine the program in Figure 17-4. This program is compiled with three images. Image 1 attempts to sync with images 2 and 3, and image 2 attempts to sync with image 1. Image 3 does not do a synchronization at all.

FIGURE 17-4
Program illustrating possible problems with image synchronization.

```
PROGRAM test_sync_image
! Image 1
IF ( this_image() == 1 ) THEN
   WRITE (*,'(A)') 'Image 1 syncing with images 2 and 3.'
   SYNC IMAGES( [2,3])
   WRITE (*,'(A)') 'Image 1 after the sync point'
END IF

! Image 2
IF ( this_image() == 2 ) THEN
   WRITE (*,'(A)') 'Image 2 syncing with image 1'
   SYNC IMAGES (1)
   WRITE (*,'(A)') 'Image 2 after the sync point'
END IF

! Image 3
IF ( this_image() == 3 ) THEN
   WRITE (*,'(A)') 'Image 3 not syncing with Image 1'
END IF

! All
WRITE (*,'(A,I0,A)') 'Image ', this_image(), ' reached end.'

END PROGRAM test_sync_image
```

When this program executes, the results are indeterminate. If image 1 should happen to execute first, then image 1 will hang waiting for images 2 and 3 to synchronize. Image 2 will synchronize with image 1, and then execute to the end, and image 3 will execute to the end without synchronization. However, image 1 will *never* finish, because it is held up waiting for image 3 to synchronize with it. This program will hang forever.

```
C:\book\fortran\chap17>test_sync_image
Image 1 syncing with images 2 and 3.
```

```
Image 2 syncing with image 1
Image 3 not syncing with Image 1
Image 3 reached end.
Image 2 after the sync point
Image 2 reached end.
```

On the other hand, if image 3 finished before image 1 tries to synchronize with it, the program will crash because an image that we are trying to synchronize with has terminated before the SYNC IMAGES command was ever issued.

```
C:\book\fortran\chap17>test_sync_image

Image 2 syncing with image 1
Image 1 syncing with images 2 and 3.
Image 3 not syncing with Image 1
Image 3 reached end.
forrtl: severe (778): One of the images to be synchronized with has terminated.
In coarray image 1
Image                PC                    Routine     Line      Source
libicaf.dll          00007FF8DA49719B      Unknown     Unknown   Unknown
test_sync_image.e    00007FF645DF128E      Unknown     Unknown   Unknown
test_sync_image.e    00007FF645E4154E      Unknown     Unknown   Unknown
test_sync_image.e    00007FF645E41D10      Unknown     Unknown   Unknown
KERNEL32.DLL         00007FF8F4C68102      Unknown     Unknown   Unknown
ntdll.dll            00007FF8F503C5B4      Unknown     Unknown   Unknown

application called MPI_Abort(comm=0x84000000, 3) - process 0
```

Both of these possibilities can happen, depending on the timing of the images. This is another example of a race condition, which is the absolute bane of parallel programming. Unless the programmer is *very* careful, race conditions will produce programs whose results are not repeatable.

Programming Pitfalls
Race conditions in a parallel program can produce nonrepeatable results, and programs must be coded as to avoid them.

Also, a failure to synchronize all images in a reciprocal fashion can cause a program image to hang forever. This is known as a **deadlock** condition, and it is a major problem in large parallel programs. It is easy to see and predict deadlocks in a simple program such as the one we are working with now, but it can be very hard to spot all possible deadlock conditions in large parallel programs.

Programming Pitfalls
If one image calls SYNC IMAGE with another image, then the other image must call SYNC IMAGE with the first one, or the first image will hang indefinitely waiting for the synchronization. This is known as a deadlock.

The SYNC IMAGES functions can be used to force particular images to execute in a particular order. For example, examine the modified program in Figure 17-5. The block of code in bold face makes image 2 synch with image 1, image 3 sync with images 2, and so forth. Image 1 can run freely, which unblocks image 2 to write its output, and that unblocks image 3 to write its output, and so forth. The result is that the output lines are written in consecutive order.

FIGURE 17-5
A modified form of program initialize_image that forces the output statements to be in order.

```
PROGRAM initialize_image3
IMPLICIT NONE

! Declare variables
INTEGER ::a[*]             ! Coarray
INTEGER :: i               ! Loop index
INTEGER :: m               ! Seed

IF ( this_image() == 1 ) THEN

   ! Get the seed value using image 1
   WRITE(*,'(A)') 'Enter an integer:'
   READ(*,*) m

   ! Use it to initialize the other images
   DO i = 1, num_images()
      a[i] = i*m
   END DO

END IF

! Synchronize all images before continuing
SYNC ALL

! Now write out the results from each image
! in sequence.
me = this_image()
IF ( me > 1 ) SYNC IMAGES(me - 1)
WRITE (*,'(A,I0,A,I0)') 'The result from image ', &
  this_image(), ' is ', a
IF ( me < NUM_IMAGES() ) SYNC IMAGES(me + 1)

END PROGRAM initialize_image3
```

If this program is compiled and executed with eight images, the results will be:

```
C:\book\fortran\chap17>initialize_image3
Enter an integer:
4
The result from image 1 is 4
The result from image 2 is 8
The result from image 3 is 12
```

```
The result from image 4 is 16
The result from image 5 is 20
The result from image 6 is 24
The result from image 7 is 28
The result from image 8 is 32
```

The output statements are now in consecutive order.

Note that we do *not* want to constrain the images like this in general. If we do, most of the images will be blocked waiting for others to execute, and we would lose the whole advantage of parallel processing! When the images are doing calculations, we want them to be able to run as freely as possible for maximum speed. You should only use synchronization when the result of one image depends on having the output of another image.

Good Programming Practice

Only use synchronization points when the calculations of one image depend on the input from another image. In that case, synchronization guarantees that the required data from the other image is present before execution starts. If not required, do not use extra synchronization points, as they block parallel execution and slow the overall program down.

Another type of synchronization is SYNC MEMORY. If a SYNC MEMORY command is issued, then all executions in all images will halt until any pending memory writes have propagated to all remote images. All other synchronization statements in Coarray Fortran automatically perform a SYNC MEMORY when they are executed.

To understand SYNC MEMORY, consider the following lines of code:

```
REAL :: var
...
IF ( this_image() == 1 ) THEN
   var[2] = -6
   ! At this point, we have started to send a 6 to var in image 2,
   ! but it may not have arrived yet
   SYNC MEMORY
   ! Now we are sure that var[2] has the new value, and we can use it...
   a = 6 * var[2]
END IF
```

In the first line of the IF block, we assign the copy of variable var in image 2 to have a value of −6. That value is used in the last line of the IF block, but without the SYNC MEMORY statement, we would not have a guarantee that the data in var[2] has been updated by the time we wanted to use it. This is another example of a race condition. Adding the SYNC MEMORY statement halts image execution until the memory update is complete, ensuring that var[2] will be −6 when we access it.

17.5

EXAMPLE: SORTING A LARGE DATA SET

To illustrate the advantages of parallel processing, we will create a large data set to sort, and compare the time required to sort it in one image versus the time required to sort it in two images operating in parallel. The sorting will be done by the selection sort subroutine (`sort`) developed in Chapter 13. Note that this is not a very efficient sort, but it will be suitable for illustrating the advantages of parallel processing. We will sort the data twice, once in a single image, and once using two parallel images. We will time the operation of the serial and parallel sorts using the `timer_class` object developed in Chapter 16.

EXAMPLE 17-1 *Performing a Sort with Parallel Processing*

Create a program that sorts a large real data set into ascending order, comparing the execution time if it is run on one image versus the time if it is run on two images in parallel.

SOLUTION

To perform this calculation, we will create a master image and a second worker image. The master image will prompt the user for the number of samples to sort, and create an array of random data samples using the intrinsic subroutine `random_number`. Then to solve the problem in parallel, we will pass half of the data to one image and the other half to the other image. Each worker image will sort its half of the data using the `sort` subroutine, and then wait for the other worker image to complete. When they have completed, the master image will run a merge subroutine to combine the data from the two sorted arrays into a single output. Image 1 will then display the first few and last few samples in the sorted array, and display the time required for the calculation.

Next, the same sorts and merge will be performed using image 1 only. Image 1 will then display the first few and last few samples in the sorted array, and display the time required for the calculation. The output data displayed should be identical, but the time required for the calculation will differ.

1. **State the problem.**

    Sort an array of real numbers in ascending order using two parallel images, and using a single image. Compare the time to perform the sort using those two approaches.

2. **Define the inputs and outputs.**

    The input to the program is the number of samples to sort. The output from the program is the first few and last few sorted values, along with the time required to perform each sort.

3. **Describe the algorithm.**

    The approach is:

    (*a*) Create an array containing the random samples to sort.
    (*b*) Start the timer running.

Coarrays and Parallel Processing

(c) Pass half of the samples to image 1 and half of the samples to image 2, and sort each set.
(d) When the data is sorted, merge the outputs in a merge subroutine.
(e) Get the elapsed time from the timer object, and display the first and last few values plus the sort time.
(f) Reset the timer.
(g) Sort half of the original data in image 1, and then sort the other half of the data in image 1.
(h) When the data is sorted, merge the outputs in a merge subroutine.
(i) Get the elapsed time from timer object, and display the first and last few values plus the sort time.

4. **Turn the algorithm into Fortran statements.**

The program to test sorting with multiple images is shown in Figure 17-6:

FIGURE 17-6
Program test_sort.

```
PROGRAM test_sort
!
!   This module tests sorting with parallel images.
!
!   Record of revisions:
!      Date          Programmer         Description of change
!      ====          ==========         =====================
!     05/06/16     S. J. Chapman         Original code
!
USE merge_module                       ! Merge
USE timer_class                        ! Timer class
IMPLICIT NONE

! Parameters
INTEGER,PARAMETER :: N_SAMPLES = 100000

! Declare variables
REAL,DIMENSION(N_SAMPLES) :: a         ! Input values to sort
REAL,DIMENSION(N_SAMPLES/2) :: b[*]    ! Coarray for parallel sorting
REAL,DIMENSION(N_SAMPLES/2) :: b1      ! Arrays for sequential sorting
REAL,DIMENSION(N_SAMPLES/2) :: b2      ! Arrays for sequential sorting
REAL :: elapsed_time                   ! Elapsed time (s)
INTEGER :: i                           ! Loop index
INTEGER :: m                           ! Number of values to sort
REAL,DIMENSION(N_SAMPLES) :: out       ! Sorted output values
TYPE(timer) :: t                       ! Timer object

!**************************************************
!**************************************************
! Now sort the data using two images
!**************************************************
!**************************************************

!**************************************************
```

(continued)

(*continued*)

```fortran
! Create the input array using Image 1
!****************************************************
IF ( this_image() == 1 ) THEN

   ! Allocate the data to sort
   CALL random_number(a)

   ! Start the timer
   CALL t%start_timer()

   ! Copy the data into the working arrays
   ! for each image.
   b[1] = a(1:N_SAMPLES/2)
   b[2] = a(N_SAMPLES/2+1:N_SAMPLES)
END IF

!****************************************************
! Synchronize all images during the creation of
! the input data
!****************************************************
SYNC ALL

!****************************************************
! Now all images can run in parallel.
! Sort the data in each image
!****************************************************
CALL sort( b, N_SAMPLES/2 )

!****************************************************
! Wait until all images are finished
!****************************************************
SYNC ALL

!****************************************************
! Now merge the data back into a common output
! array using image 1, and display the results.
!****************************************************
IF ( this_image() == 1 ) THEN

   ! Merge the data
   CALL merge ( b[1], N_SAMPLES/2, b[2], N_SAMPLES/2, out, N_SAMPLES )

   ! Stop the timer
   elapsed_time = t%elapsed_time()

   ! Display the elapsed time
   WRITE (*,'(A,F8.3,A)') &
      'Parallel sort elapsed time =', elapsed_time, ' s'

   ! Display first 5 samples
   WRITE (*,'(A)') 'First 5 samples:'
   DO i = 1, 5
      WRITE (*,'(F10.6)') out(i)
   END DO
```

(*continued*)

(*concluded*)

```fortran
      ! Display last 5 samples
      WRITE (*,'(A)') 'Last 5 samples:'
      DO i = N_SAMPLES-4, N_SAMPLES
         WRITE (*,'(F10.6)') out(i)
      END DO

   ELSE

      ! Stop other images--they won't be used again
      STOP

   END IF
   !***************************************************
   !***************************************************
   ! Now sort the data using a single image
   !***************************************************
   !***************************************************
   IF ( this_image() == 1 ) THEN

      ! Start the timer
      CALL t%start_timer()

      ! Copy the data into the working arrays
      b1 = a(1:N_SAMPLES/2)
      b2 = a(N_SAMPLES/2+1:N_SAMPLES)

      ! Sort the data in the single image
      CALL sort( b1, N_SAMPLES/2 )
      CALL sort( b2, N_SAMPLES/2 )

      ! Merge the data
      CALL merge ( b1, N_SAMPLES/2, b2, N_SAMPLES/2, out, N_SAMPLES )

      ! Stop the timer
      elapsed_time = t%elapsed_time()

      ! Display the elapsed time
      WRITE (*,'(/A,F8.3,A)') &
         'Sequential sort elapsed time =', elapsed_time, ' s'

      ! Display first 5 samples
      WRITE (*,'(A)') 'First 5 samples:'
      DO i = 1, 5
         WRITE (*,'(F10.6)') out(i)
      END DO

      ! Display last 5 samples
      WRITE (*,'(A)') 'Last 5 samples:'
      DO i = N_SAMPLES-4, N_SAMPLES
         WRITE (*,'(F10.6)') out(i)
      END DO

   END IF
   END PROGRAM test_sort
```

The subroutine to merge the two data sets together is placed in a module so that it has an explicit interface. This subroutine is shown in Figure 17-7:

FIGURE 17-7
The merge subroutine to combine the sorted data from two sorts into a single output array.

```
MODULE merge_module
!
!   This module implements a merge subroutine.
!
IMPLICIT NONE

! Now add methods
CONTAINS

SUBROUTINE merge(b1, size1, b2, size2, out, size_out)
!
!   Subroutine to merge two sorted arrays together
!   in increasing order.
!
!   Record of revisions:
!       Date            Programmer          Description of change
!       ====            ==========          =====================
!     05/06/16        S. J. Chapman         Original code
!
IMPLICIT NONE

! Declare calling arguments
INTEGER :: size1                        ! Size of array b1
REAL,DIMENSION(size1) :: b1             ! Input array b1
INTEGER :: size2                        ! Size of array b2
REAL,DIMENSION(size1) :: b2             ! Input array b1
INTEGER :: size_out                     ! Size of array out
REAL,DIMENSION(size_out) :: out         ! Output array b1

! Declare local variables
INTEGER :: i1                           ! Pointer in b1
INTEGER :: i2                           ! Pointer in b2
INTEGER :: iout                         ! Pointer in out

! Initialize pointers
i1   = 1
i2   = 1
iout = 1

! Now do the merge, putting the smaller value
! from either input into the output array at
! each step.
DO
   IF ( iout > size_out ) THEN

      ! All done, get out.
      EXIT

   ELSE IF ( i1 > size1 ) THEN
```

(continued)

(*concluded*)

```
         ! If b1 is finished, use b2
         out(iout) = b2(i2)
         iout = iout + 1
         i2   = i2 + 1

      ELSE IF ( i2 > size2 ) THEN

         ! If b2 is finished, use b1
         out(iout) = b2(i1)
         iout = iout + 1
         i1   = i1 + 1

      ELSE IF ( b1(i1) <= b2(i2) ) THEN

         ! If b1 is smaller, use it
         out(iout) = b1(i1)
         iout = iout + 1
         i1   = i1 + 1

      ELSE IF ( b1(i1) > b2(i2) ) THEN

         ! If b2 is smaller, use it
         out(iout) = b2(i2)
         iout = iout + 1
         i2   = i2 + 1

      END IF

   END DO

END SUBROUTINE merge

END MODULE merge_module
```

When this program is executed, the results are:

```
C:\book\fortran\chap17>test_sort
Parallel sort elapsed time =    1.232 s
First 5 samples:
  0.000000
  0.000007
  0.000010
  0.000021
  0.000029
Last 5 samples:
  0.999927
  0.999945
  0.999979
  0.999983
  0.999998

Sequential sort elapsed time =    2.467 s
First 5 samples:
  0.000000
  0.000007
  0.000010
  0.000021
  0.000029
```

```
Last 5 samples:
  0.999927
  0.999945
  0.999979
  0.999983
  0.999998
```

The results of sort are identical when the data is sorted in parallel images and in a single image. However, it is twice as fast if there are two images doing the work!

In some end-of-chapter problems, you will be asked to generalize this sort to support arbitrary amounts of data, and to support more images in parallel.

17.6
ALLOCATABLE COARRAYS AND DERIVED DATA TYPES

Coarrays can be allocatable as well as static. To declare an allocatable coarray, simply add the `CODIMENSION` attribute to the type declaration statement (or add the `[]` syntax to the declaration, which is the same thing). For example, the following two statements each declare a 2D allocatable real array `arr` that is also a coarray:

```
REAL,ALLOCATABLE,DIMENSION(:,:),CODIMENSION[:] :: var
REAL,ALLOCATABLE :: var(:,:)[:]
```

Note that *all* the dimensions in the array and coarray of the type declaration statement must be declared as deferred (:). This variable would be allocated in an image using the `ALLOCATE` statement very much like any other variable:

```
ALLOCATE( var(10,20)[*], STAT=istat )
```

If an array is allocated in one image, it *must* immediately be allocated in all other images to be the same size as the first allocation. All allocations must be finished before any code is executed, because any image could try to access the data in any other image. Therefore, it is customary to put a `SYNC ALL` statement after the allocation to ensure that all images finish the allocation before proceeding.

```
ALLOCATE( var(10,20)[*], STAT=istat )
SYNC ALL
```

Allocatable coarrays are deallocated using the `DEALLOCATE` statement, just like ordinary allocatable arrays.

```
DEALLOCATE( var, STAT=istat )
SYNC ALL
```

Allocatable coarrays must have the `SAVE` attribute. Allocatable arrays declared in a program or module automatically have an implied `SAVE` attribute. Allocatable arrays declared in a subroutine or function must have an explicit `SAVE` attribute included in the declaration statement.

Allocatable arrays allow a program to be dynamically resized to support different data sets. In an end-of-chapter problem, you will be asked to modify Example 17-1 so that it works with data set of any size.

Good Programming Practice
Allocatable arrays can be used as coarrays. When they are allocated, they must be allocated in all images at the same time.

Derived data types can also be coarrays. They can be declared as scalars, static arrays, or allocatable arrays. The derived data types can also contain pointers as elements within them. As with allocatable arrays, all allocatable derived data types must be declared in every image before execution can advance. For example, the following code declares a derived data type `my_type`, and then allocates an array of that data type for each image. As before, all of the images must synchronize before execution continues.

```
TYPE :: my_type
   REAL :: a
   REAL,POINTER,DIMENSION(:) :: b
   LOGICAL :: valid = .FALSE.
END TYPE my_type

TYPE(my_type),ALLOCATABLE,DIMENSION(:),CODIMENSION[:] :: arr
ALLOCATE( arr(10)[*], STAT=istat )
DO i = 1, num_images()
   ALLOCATE (arr(i)%b(100), STAT=istat)
   arr(i)%b = this_image()
END DO

SYNC ALL
```

A pointer itself cannot be a coarray, but a pointer can exist inside a derived data type which itself is a coarray.

Good Programming Practice
Pointers cannot be used as coarrays. However, a pointer can exist inside a derived data type that is a coarray.

17.7
PASSING COARRAYS TO PROCEDURES

Coarrays can be passed to subroutines or functions as long as the procedure has an explicit interface. If an argument is declared with the coarray syntax, then the procedure can access both the local copy and the copies in other images. If the argument is declared without the coarray syntax, then the procedure can only access the local copy of the data. For example, the module below declares two subroutines `sub1` and `sub2`.

```
MODULE test_module
CONTAINS
SUBROUTINE sub1(b)
REAL,DIMENSION(:),CODIMENSION(*) :: b
...
...
...
END SUBROUTINE sub1

SUBROUTINE sub2(b)
REAL,DIMENSION(:) :: b
...
...
...
END SUBROUTINE sub2

END MODULE test_module
```

These subroutines are called by a main program as follows:

```
PROGRAM test
USE test_module
...
...
CALL SUBROUTINE sub1(b)
CALL SUBROUTINE sub2(b)
...
...
END PROGRAM test
```

Subroutine sub1 can use both the local and remote copies of array b, but subroutine sub2 can use only the local copy of array b.

Good Programming Practice

Procedures can use both local and remote copies of coarrays if the dummy argument is declared with a coarray syntax and the procedure has an explicit interface.

17.8
CRITICAL SECTIONS

Critical sections are another feature of parallel programming. Sometimes there are calculations in an image that depend on a set of input values. If these input values were modified by another image *during the calculation*, the inconsistent input data would produce invalid results. For example, suppose that a parallel program is calculating the angle of an incoming ray using the expression

```
angle = ATAN2D(y,x)
```

where x is the horizontal length and y is the vertical height of the ray. If the values of x and y can be modified by other images, then they could change *in the middle of the calculation*, and the resulting angle would be invalid. In cases like this, the user could place the calculations in a critical section. A critical section is a block of code that *only one image could enter at a time*. If multiple images want to modify this code, then they will queue up and take turns. The second image would start to execute this code when the first image is finished with it, and so forth.

The format of a critical section is

```
CRITICAL
    angle = ATAN2D(y,x)
END CRITICAL
```

All of the code between the `CRITICAL` and `END CRITICAL` statements is accessible to only one image at a time.

Good Programming Practice
Use critical sections to protect pieces of code and data that should only be accessed by one image at a time.

17.9
THE PERILS OF PARALLEL PROGRAMMING

Parallel programming has many problems not evident in ordinary sequential programming, which must be addressed by the developer. In general, parallel programs should be faster than sequential programs because parts of the work are being done on separate cores. However, the results of the program are *not* deterministic. They can differ depending on which image finishes first in a given execution. In addition, parallel programs can suffer **deadlocks,** in which one image waits forever to synchronize with another image (we saw an example of that in Section 17.4). These problems are often not repeatable, happening on some executions and not on other ones depending on the order in which things happen.

For example, consider the following simple program.

```
PROGRAM test_race

INTEGER,CODIMENSION[*] :: i_sum = 0

i_sum[1] = i_sum[1] + this_image()

WRITE (*,'(A,I0,A,I0)') 'Image ', this_image(), &
                        ' finishing: i_sum = ', i_sum[1]

END PROGRAM test_race
```

This program looks like it should add the image number to the value in `i_sum[1]`. If the images always executed in order, we would expect the result to be 1 after image 1 executed, 3 after image 2 executed, and 6 after image 3 executed. In fact, the results are *very* different:

```
C:\book\fortran\chap17\test_race>test_race
Image 1 finishing: i_sum = 1
```

```
Image 2 finishing: i_sum = 3
Image 3 finishing: i_sum = 4

C:\book\fortran\chap17\test_race>test_race
Image 1 finishing: i_sum = 1
Image 3 finishing: i_sum = 3
Image 2 finishing: i_sum = 3

C:\book\fortran\chap17\test_race>test_race
Image 1 finishing: i_sum = 1
Image 2 finishing: i_sum = 4
Image 3 finishing: i_sum = 4

C:\book\fortran\chap17\test_race>test_race
Image 1 finishing: i_sum = 3
Image 3 finishing: i_sum = 3
Image 2 finishing: i_sum = 5
```

Notice that the images are not executed in the same order every time, and even when they are executed in the same order, the answers are different! What is going on here?

This is a classic example of the race conditions that plague parallel programming. The statement "i_sum[1] = i_sum[1] + this_image()" is a major source of the problem. Each image reads the value of i_sum from image 1, adds the image number to it, and saves it back into i_sum[1]. However, multiple images are running in parallel, and a second image is reading the value of i_sum[1] before the first image finishes updating it. As a result, both images started with the same input value for i_sum[1], and added their different image numbers to it. Whichever image was the last to finish would be the one whose calculation was saved in the variable and printed out. For example, we sometimes see [1 3 3]. This happens if the first image executes and saves its result in i_sum[1], and then the second and third images read that value and add their image number to it. If the *second* image is the last one to update the variable, the results are [1 3 3]. If the third image is the last one to update the variable, the results would be [1 4 4].

This is obviously not an acceptable way to program. What can we do to ensure that results are more repeatable? One possibility is to place the summation into a CRITICAL block so that no more than one image can access it at a time.

```
PROGRAM test_race2

INTEGER,CODIMENSION[*] :: i_sum = 0

CRITICAL
   i_sum[1] = i_sum[1] + this_image()
END CRITICAL

WRITE (*,'(A,I0,A,I0)') 'Image ', this_image(), &
                        ' finishing: i_sum = ', i_sum[1]

END PROGRAM test_race2
```

When this program executes, the results are a bit better:

```
C:\book\fortran\chap17\test_race>test_race2
Image 1 finishing:, i_sum = 1
Image 2 finishing:, i_sum = 3
Image 3 finishing:, i_sum = 6
```

Coarrays and Parallel Processing

```
C:\book\fortran\chap17\test_race>test_race2
Image 1 finishing:, i_sum = 1
Image 2 finishing:, i_sum = 3
Image 3 finishing:, i_sum = 6

C:\book\fortran\chap17\test_race>test_race2
Image 1 finishing:, i_sum = 4
Image 3 finishing:, i_sum = 4
Image 2 finishing:, i_sum = 6

C:\book\fortran\chap17\test_race>test_race2
Image 1 finishing:, i_sum = 1
Image 3 finishing:, i_sum = 4
Image 2 finishing:, i_sum = 6
```

Each image can only access the critical section one at a time, so the image numbers always add to 6 (1 + 2 + 3). However, the numbers printed out by the WRITE statements vary because the values of i_sum[1] differ at the time each image writes its output.

We could go further and put the WRITE statement in the CRITICAL block as well. Now things will look even better:

```
PROGRAM test_race3

INTEGER,CODIMENSION[*] :: i_sum = 0

CRITICAL
   i_sum[1] = i_sum[1] + this_image()
   WRITE (*,'(A,I0,A,I0)') 'Image ', this_image(), &
                           ' finishing: i_sum = ', i_sum[1]
END CRITICAL

END PROGRAM test_race3
```

Now the results only depend on the order in which the images enter the critical section:

```
C:\book\fortran\chap17\test_race>test_race3
Image 1 finishing:, i_sum = 1
Image 3 finishing:, i_sum = 4
Image 2 finishing:, i_sum = 6

C:\book\fortran\chap17\test_race>test_race3
Image 1 finishing:, i_sum = 1
Image 3 finishing:, i_sum = 4
Image 2 finishing:, i_sum = 6

C:\book\fortran\chap17\test_race>test_race3
Image 1 finishing:, i_sum = 1
Image 2 finishing:, i_sum = 3
Image 3 finishing:, i_sum = 6
```

If we went further and constrained the order in which the images executed, the results would become deterministic.

```
PROGRAM test_race4

INTEGER,CODIMENSION[*] :: i_sum = 0
INTEGER :: me
```

```
me = this_image()
IF ( me > 1 ) SYNC IMAGES(me - 1)
i_sum[1] = i_sum[1] + this_image()
WRITE (*,'(A,I0,A,I0)') 'Image ', this_image(), &
                       ' finishing: i_sum = ', i_sum[1]
IF ( me < NUM_IMAGES() ) SYNC IMAGES(me + 1)

END PROGRAM test_race4
```

When this program executes, the results are a bit better:

```
C:\book\fortran\chap17\test_race>test_race4
Image 1 finishing: i_sum = 1
Image 2 finishing: i_sum = 3
Image 3 finishing: i_sum = 6

C:\book\fortran\chap17\test_race>test_race4
Image 1 finishing: i_sum = 1
Image 2 finishing: i_sum = 3
Image 3 finishing: i_sum = 6

C:\book\fortran\chap17\test_race>test_race4
Image 1 finishing: i_sum = 1
Image 2 finishing: i_sum = 3
Image 3 finishing: i_sum = 6
```

In this last case, the results are deterministic, but we have lost the advantage of parallel processing because we have forced the images to run sequentially.

It can be very hard to avoid race conditions and deadlocks in parallel code, depending on the algorithms that you are trying to implement. Be careful in your design, and try to run simple cases before you do complex ones to identify any problems associated with race conditions or deadlocks. Write statements can be very useful in this process, as they can show the contents of specific variables at each point in each image's execution.

Learning to write good parallel programs is a challenge, and it requires extra skills that we have not taught in this text. If you are going to do extensive work in this area, be sure to get a text that specializes in this area.

Quiz 17-1

This quiz provides a quick check to see if you have understood the concepts introduced in Sections 17.1 through 17.9. If you have trouble with the quiz, reread the section, ask your instructor, or discuss the material with a fellow student. The answers to this quiz are found in the back of the book.

1. How can we create a Fortran program that supports multiple images?
2. What is the SPMD model?
3. In a Coarray Fortran program, how does the program distinguish one image from another one? How does each image know what code to execute?

(continued)

(*concluded*)

4. What is a coarray?
5. How do images in Coarray Fortran communicate with each other?
6. What is a race condition? What can we do in a program to minimize the occurrence of race conditions?
7. What is a critical section?
8. The program shown below has four images, and it is designed to calculate and print out the values of $\sin 0$, $\sin\frac{\pi}{2}$, $\sin \pi$, and $\sin\frac{3\pi}{2}$. Will it work properly? If it will work, how many images are required to print out the desired result?

```
PROGRAM test
REAL,PARAMETER :: PI = 3.141593
REAL :: in_val[*]
REAL :: sin_val[*]
INTEGER :: i

IF ( this_image() == 1 ) THEN
   DO i = 1, num_images()-1
      in_val[i+1] = i * PI/2
      SYNC MEMORY
      IF ( i > 0 ) SYNC IMAGES([i+1])
   END DO
ELSE
   SYNC IMAGES([1])
END IF
sin_val = SIN(in_val)
IF ( this_image() > 1 ) THEN
    WRITE (*,'(A,F9.5,A,F9.5)') 'sin(', in_val, ') = ', sin_val
END IF

END PROGRAM test
```

17.10
SUMMARY

Coarray Fortran is an extension to Fortran that performs parallel processing. It is a direct extension to Fortran using the single-program, multiple data (SPMD) model, designed to work with the minimum possible number of changes to standard Fortran. In Coarray Fortran, every possible parallel program is called an image, and every image is identical. However, different images can perform different functions by using the `this_image()` function to distinguish one image from another.

Images communicate with each other through synchronization commands, and they share data through coarrays. Coarrays are scalars or arrays that are declared identically in every image, but a special syntax allows any image to access the copy of the memory in any other image.

The images in a program can be forced to synchronize with each other using `SYNC ALL`, `SYNC IMAGES`, `NOTIFY`, or `QUERY` statements. If an image executes the `SYNC ALL` command, then that image will freeze, and it will remain frozen until every other image also executes the `SYNC ALL` command. This can be done to ensure that all images have the required information before they start executing. The `SYNC IMAGES` function allows specific images to be synchronized instead of all of them. The `NOTIFY` or `QUERY` statements perform a function like `SYNC IMAGES`, but more flexibly. Specific images can synchronize without freezing until the synchronization is complete.

Parallel processing is relatively hard compared to normal programming because of race conditions and deadlocks. Race conditions are situations in which the value of a variable depends on the order in which the images ran. Deadlocks are situations in which one image tries to synchronize with another image, and the other image doesn't respond. This causes a particular image to freeze forever. It requires very careful programming to avoid problems from race conditions and deadlocks.

17.10.1 Summary of Good Programming Practice

The following guidelines introduced in this chapter will help you to develop good programs:

1. Use the master image (image 1) to coordinate and control the function of the various worker images in a coarray Fortran program.
2. Only the master image can read data from the standard input device. If that data is to be made available to the worker images, the master image must copy it to them.
3. Use the `STOP ALL` statement to force all images in a coarray program to terminate.
4. The coarray syntax allows very easy communication between data stored in different executing images of a program.
5. The total number of dimensions plus codimensions on an array must be less than or equal to 15.
6. Use the `SYNC ALL` statement to ensure that all images in a program reach a common point before allowing any of them to continue execution.
7. Only use synchronization points when the calculations of one image depend on the input from another image. In that case, synchronization guarantees that the required data from the other image is present before execution starts. If not required, do not use extra synchronization points, as they block parallel execution and slow the overall program down.
8. Allocatable arrays can be used as coarrays. When they are allocated, they must be allocated in all images at the same time.
9. Pointers cannot be used as coarrays. However, a pointer can exist inside a derived data type that is a coarray.

10. Procedures can use both local and remote copies of coarrays if the dummy argument is declared with a coarray syntax and the procedure has an explicit interface.
11. Use NOTIFY and QUERY to provide more flexible synchronization between images in a parallel program.
12. Use critical sections to protect pieces of code and data that should only be accessed by one image at a time.

17.10.2 Summary of Fortran Statements and Structures

CODIMENSION Attribute::

```
TYPE,CODIMENSION[*] :: type_name
```

Examples:

```
REAL,CODIMENSION[*] :: value
REAL,DIMENSION(4,4),CODIMENSION[2,*] :: array
INTEGER :: i(2,2)[*]
```

Description:
The CODIMENSION attribute declares that a variable or array is shared across multiple images. The syntax can be either in the form of a CODIMENSION attribute or as a set of square brackets after the variable declaration.

CO_LBOUND Function:

```
co_lbound(coarray)
```

Examples:

```
co_lbound(coarray)
```

Description:
The CO_LBOUND function returns the lowest value of each dimension in a coarray.

CO_UBOUND Function:

```
co_ubound(coarray)
```

Examples:

```
co_ubound(coarray)
```

Description:
The CO_UBOUND function returns the highest value of each dimension in a coarray.

CRITICAL Section:

```
CRITICAL
    ...
END CRITICAL
```

Examples:
```
CRITICAL
    ival[2] = ival[1] + ival[2]
END CRITICAL
```

Description:
The CRITICAL section marks a part of the code where only one image can execute at a time. If more than one image tries to execute this code, all the remaining ones will wait until the currently executing images leave the critical section.

NUM_IMAGES Function:

```
num_images()
```

Examples:
```
num_images()
```

Description:
This function returns the total number of images in a program.

SYNC ALL Statement:

```
SYNC ALL
```

Example:
```
SYNC ALL
```

Description:
The SYNC ALL statement causes images that executes it to stop and wait, until every image in the program has synchronized. At that point, all images start executing again.

SYNC IMAGES Statement:

```
SYNC IMAGES()
```

Example:
```
SYNC IMAGES(*)
SYNC IMAGES(1)
SYNC IMAGES([2,3,4])
```

Description:
The SYNC IMAGES statement causes the calling image to stop until the specified images in the list call SYNC IMAGES with original caller as an argument. At that point, all images start executing again. If the argument is *, then the calling image waits on every other image. If the argument is a list of specific images, then the calling image waits on every one of the numbered images in the list.

Coarrays and Parallel Processing

SYNC MEMORY Statement:

Example:
```
        SYNC MEMORY

        SYNC MEMORY
```

Description:
The `SYNC MEMORY` statement causes all images to stop until any pending memory writes have finished posting. At that point, execution resumes on any images that are not otherwise blocked.

THIS_IMAGE Function:

Examples:
```
        this_image()

        this_image()
```

Description:
This function returns the number of the current image.

17.10.3 Exercises

17-1. Modify the sort program in Example 17-1 so that it works with an arbitrary number of images. Perform the final sort by calling the `merge` subroutine repeatedly.

17-2. Modify the sort program in Example 17-1 so that it works with an arbitrary number of data samples. Prompt the user for the number of samples to sort, and then use allocatable arrays to create arrays of the size required to do the sort.

17-3. Modify the sort program in Example 17-1 so that it uses image 1 as a master image, and images $2-n$ as worker images. The data to sort should be created by image 1, and the final merge should also be performed by image 1. Worker images $2-n$ should do the actual sorting of subsets of the data.

17-4. Calculating π Figure 17-8 shows a square whose sides are of length 2, enclosing a circle of radius 1. The area of the square is given by

$$A = l^2 \qquad (17\text{-}1)$$

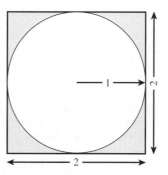

FIGURE 17-8
A circle of radius 0.5 touching the sides of a square of side 1.0.

where l is the length of the side of the square. The area of the enclosed circle is given by

$$A = \pi r^2 \quad (17\text{-}2)$$

where r is the radius of the circle. In this case, where side $l = 2$ and the radius $r = 1$, the area of the square is 4 and the area of the circle is π. Therefore, the ratio of the area of the circle to the area of the square is $\pi/4$.

This relationship can provide an interesting approach to calculating the value of π. Suppose that we drew two random numbers x and y from a uniform distribution over the range $-1 \leq x < 1$ and $-1 \leq y < 1$. Then each possible (x, y) point will fall inside the area of the square (or exactly on its border). If the points also satisfy the constraint

$$\sqrt{x^2 + y^2} < r \quad (17\text{-}3)$$

then they will fall within the area of the circle as well. This leads to a way to estimate π.

1. Initialize variables counting the number of points falling inside the square (N_{sq}) and the number of points falling inside the circle (N_{cir}).

2. Select a random x and y from a uniform distribution over the range $-1 \leq x < 1$ and $-1 \leq y < 1$. This point will be inside the square, so increment the number N_{sq} by 1.

3. If $\sqrt{x^2 + y^2} < 1$, this point is also inside the circle, so increment the number N_{cir} by 1. Otherwise, leave this value unchanged.

4. After very many examples,

$$\frac{N_{cir}}{N_{sq}} \approx \frac{\pi}{4} \quad (17\text{-}4)$$

or

$$\pi \approx 4 \frac{N_{cir}}{N_{sq}} \quad (17\text{-}5)$$

The more samples we include in the test, the better the approximation is to the value of π.

Create a program that contains a subroutine to return a uniform random value in the range $-1 \leq x < 1$, and then use that random value subroutine to write a parallel processing program that estimates the value π by performing millions of trials. Use eight parallel images in the calculation, and determine the calculation when the result is accurate to eight significant digits. Be sure to write out the elapsed time required for the calculation.

17-5. Perform the calculation in Exercise 17-5 with 1, 2, 4, and 8 images running in parallel. How does the time taken to perform the calculation vary with the number of images?

17-6. In the calculation in Exercise 17-5, it is necessary to increment the number of samples that fall in the square and the number of samples that fall in the circle at separate times. If the coarrays containing sums of those two values are read at different times, then the data could be incorrect—one of the values could have been increment and the other might not have had a chance to be incremented yet. Use a critical section to prevent the sum data from being read by another image while it is being modified.

Execute the modified program. Is it faster or slower than before, and if so, by how much?

17-7. Can you come up with another way to synchronize images so that the data access will not occur during the time that the sums are being generated? How fast is the resulting code?

18

Redundant, Obsolescent, and Deleted Fortran Features

OBJECTIVES

- Be able to look up and understand redundant, obsolescent, and deleted Fortran features when you encounter them.
- Understand that these features should *never* be used in any new program.

There are a number of odds and ends in the Fortran language that have not fit logically into our discussions in the previous chapters. These miscellaneous features of the language are described here.

Many of the features we will be describing in this chapter date from the early days of the Fortran language. They are the skeletons in Fortran's closet. For the most part, they are either incompatible with good structured programming or are obsolete and have been replaced by better methods. As such, *they should not be used in new programs that you write*. However, you may see them in existing programs that you are required to maintain or modify, so you should be familiar with them.

Many of these features are classified as either **obsolescent** or **deleted** in Fortran 2008. An *obsolescent* feature is one that has been declared undesirable, and that has been replaced in good usage by better methods. It is still supported by all compilers, but it should not be used in any new code. Obsolescent features are candidates for deletion in future versions of Fortran as their use declines. A *deleted* feature is one that has officially been removed from the Fortran language. It may be supported by your Fortran compiler for backward compatibility reasons, but there is no guarantee that it will work with all compilers.

Because the features described in this chapter are generally undesirable, there are no examples or quizzes featuring them. The contents of the chapter may be used as a cross-reference to help you understand (and possibly replace) older features found in existing programs.

18.1
PRE-FORTRAN 90 CHARACTER RESTRICTIONS

Before Fortran 90, the Fortran character set for naming variables officially included only the uppercase letters A–Z and the digits 0–9. The lowercase letters were undefined in the standard, but were usually made equivalent to the corresponding uppercase ones if they were supported at all by a particular compiler. In addition, the underscore character (_) was not legal in a variable name.

All Fortran names (procedure names, variable names, etc.) were restricted to a maximum of six characters. Because of these restrictions, you may encounter strange and hard-to-read names in older programs.

18.2
OBSOLESCENT SOURCE FORM

As we mentioned in Chapter 1, Fortran was one of the first major computer languages to be developed. It originated in the days before video displays and keyboards, when the punched card was the major form of input to the computer. Each punched card had a fixed length of 80 columns, and one character, number, or symbol could be typed in each column. The structure of statements in earlier versions of Fortran reflected this fixed limitation of 80 characters per line. By contrast, Fortran 90 and later versions were developed in the age of the video display and keyboard, so it allows free entry of statements in any column. For backward compatibility, Fortran 90 and later also supports the old fixed form used by earlier versions of Fortran.

A fixed source form Fortran statement still reflects the structure of the punched computer card. Each card has 80 columns. Figure 18-1 shows the use of these 80 columns in a fixed form Fortran statement.

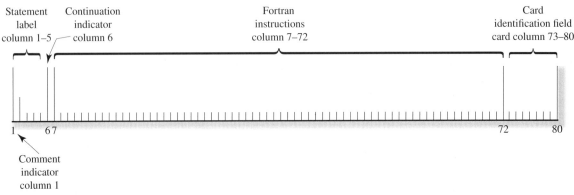

FIGURE 18-1
An object may be represented as a nucleus of data (instance variables) surrounded and protected by methods, which implement the object's behavior and form an interface between the variables and the outside world.

Columns 1 through 5 are reserved for statement labels. A statement label may be located anywhere within columns 1 through 5, with either leading or trailing blanks. For example, the label 100 could be placed in columns 1 to 3, 2 to 4, or 3 to 5, and it would still be the same label.

A letter C or an asterisk (*) placed in column 1 indicates that the statement is a **comment.** The Fortran compiler completely ignores any statement beginning with these characters.

Column 6 is normally blank. If any character other than a blank or a zero is placed in that column, then the statement is interpreted as a continuation of the statement immediately preceding it.

Columns 7 to 72 contain the Fortran instructions that are interpreted by the compiler. The instructions may be freely placed anywhere within this area. Programmers typically take advantage of this freedom to indent certain instructions (loops and branches) to make their code more readable.

Columns 73 to 80 are sometimes called the **card identification field.** This field is totally ignored by the compiler, and may be used by the programmer for any desired purpose. In the days when programs were saved on decks of punched cards, this field was used to number the cards in consecutive order. If someone accidentally dropped a numbered card deck, it was possible to reconstruct the order of the statements in the program from the numbers on the cards. Today, these columns are usually blank.

Figure 1-8 shows a sample Fortran program using the fixed source form. Note that the statement label 100 falls in columns 1 to 5, and the Fortran instructions begin in column 7.

It is easy to convert a fixed source form program into free source form. A Fortran program to accomplish this conversion is freely available on the Internet. It was written by Michael Metcalf at CERN in Geneva, and is named `convert.f90`. It is freely available from many sources on the Internet.

The fixed source form has been declared obsolescent as of Fortran 95, which means that it is a candidate for deletion in future versions of Fortran. All new programs should use the free source form.

18.3
REDUNDANT DATA TYPE

In versions of Fortran before Fortran 90, there were two types of real variables: REAL and DOUBLE PRECISION. Double-precision variables were defined as having higher precision than real variables, but the exact precision and range of each data type varied from computer to computer. Thus, double precision on a VAX computer was a 64-bit variable, while double precision in a Cray supercomputer was a 128-bit variable. This difference made programs that depended on having variables with a certain minimum range and precision inherently less portable.

These older data types have been replaced by the parameterized real data type, in which it is possible to explicitly specify the range and/or precision required for a given data item. The DOUBLE PRECISION data type should never be used in new Fortran programs.

18.4

OLDER, OBSOLESCENT, AND/OR UNDESIRABLE SPECIFICATION STATEMENTS

The syntax of many specification statements was different before Fortran 90. In addition, there are five obsolete and/or undesirable Fortran statements that may appear in the declaration section of a Fortran program. They are

1. The `IMPLICIT` statement
2. The `DIMENSION` statement
3. The `EQUIVALENCE` statement
4. The `DATA` statement
5. The `PARAMETER` statement

These statements are described below.

18.4.1 Pre-Fortran 90 Specification Statements

The form of many specification statements was different before Fortran 90. It was not possible to declare attributes in a type declaration statement, and the double colons (::) were not used. It was also not possible to initialize variables in a type declaration statement.

In addition, the lengths of character variables were declared using an asterisk followed by the length in characters. The asterisk and length could be attached to either the `CHARACTER` statement, in which case it applied to all variables in the statement, or it could be attached to a specific variable name. If it were attached to a specific variable name, the length applied only to that variable.

Pre-Fortran 90 type specification statements took one of the following forms:

```
INTEGER list of integer variables
REAL list of real variables
DOUBLE PRECISION list of double precision variables
COMPLEX list of complex variables
LOGICAL list of logical variables
CHARACTER list of character variables
CHARACTER*<len> list of character variables
```

Examples of some pre-Fortran 90 type specification statements are shown below:

```
INTEGER I, J, K
DOUBLE PRECISION BIGVAL
CHARACTER*20 FILNM1, FILNM2, YN*1
```

The `CHARACTER*<len>` form of the character type declaration statement has been declared obsolescent in Fortran 95, which means that it is a candidate for deletion in future versions of Fortran.

18.4.2 The `IMPLICIT` Statement

By default, named constants and variables whose names begin with the letters I through N are integers, while all other named constants and variables are of type real. The `IMPLICIT` statement permits us to override these defaults.

The general form of the IMPLICIT statement is

   IMPLICIT *type1* (a_1, a_2, a_3, ...), *type2* (b_1, b_2, b_3, ...), ...

where *type1*, *type2*, etc., are any legal data types: INTEGER, REAL, LOGICAL, CHARACTER, DOUBLE PRECISION, or COMPLEX. The letters a_1, a_2, a_3, etc., are the first letters whose type will be *type1*, and so forth for the other types. If a range of letters is to be declared as the same type, then the range may be indicated by the first and last letters separated by a dash (–). For example, the following statements declare that variables starting with the letters a, b, c, i, and z will be COMPLEX, and variables beginning with the letter d will be DOUBLE PRECISION. Variables beginning with other letters will retain their default types. Finally, the variables i1 and i2 are explicitly declared to be integers, overriding the IMPLICIT statement.

   IMPLICIT COMPLEX (a-c, i, z), DOUBLE PRECISION d
   INTEGER :: i1, i2

The IMPLICIT NONE statement was described in Chapter 2 and has been used throughout the book. It cancels all default types. When the IMPLICIT NONE statement is used in a program, every named constant, variable, and function name in the program must be declared explicitly. Since every named constant and variable in your program should be declared explicitly, there is no need for the standard IMPLICIT statement in any well-designed program. Only the IMPLICIT NONE statement should be used. However, you must be familiar with it, since you will encounter it in older Fortran programs.

Good Programming Practice
Do not use IMPLICIT statements in your programs, except for IMPLICIT NONE. All of your programs should include the IMPLICIT NONE statement, and all named constants, variables, and functions in your programs should be explicitly typed.

18.4.3 The DIMENSION Statement

The DIMENSION statement is a declaration statement used to declare the *length* of arrays. The general form of a DIMENSION statement is

   DIMENSION *array*([i1:]i2, [j1:]j2, ...), ...

where *array* is an array name, and i_1, i_2, j_1, j_2, etc. are the dimensions of the arrays. For example, a 6-element array array1 could be declared with the following statement:

   DIMENSION array1(6)

Notice that the DIMENSION statement declares the length of an array, but not its type. If array1 is not included in any type specification statement, then its type will default to real because the name begins with the letter A. If we wish to declare both the type and the length of the array, then we would have to use one of the following sets of statements.

```
        REAL, DIMENSION(6) :: array1
```
or
```
        REAL :: array1
        DIMENSION array1(6)
```

The `DIMENSION` statement is only needed when we declare the length of an array while using default typing. Since we never use default typing in good Fortran programs, there is no need to ever use this statement. It is a holdover from earlier versions of Fortran.

Good Programming Practice

Do not use `DIMENSION` statements in your programs. Since all variables and arrays in your programs will be explicitly typed, the lengths of the arrays can be declared in the type declaration statements with the `DIMENSION` attribute. There is never a need for `DIMENSION` statements in well-designed programs.

18.4.4 The DATA Statement

Before Fortran 90, it was not possible to initialize variables in a type declaration statement. Instead, the variables were initialized by a separate DATA statement, which took the form

```
        DATA var_names/values/, var_names/values/, ...
```

where *var_names* are a list of variable names and *values* are the values to be assigned to those variables. There must be a one-to-one correspondence between the number of variables in the data statements and the number of values to initialize them. A single value could be repeated several times by preceding it with a repeat count followed by an asterisk. For example, the following statement initializes variables a1, b1, and c1 to 1.0, 0.0, and 0.0, respectively.

```
        DATA a1, b1, c1 / 1.0, 2*0.0 /
```

Arrays may also be initialized in DATA statements. If an array is mentioned in a DATA statement, then there must be enough data values to initialize all of the elements in the array. The values are assigned to the array elements in *column order*. The following DATA statement initializes the 2 × 2 array a1.

```
        REAL a1(2,2)
        DATA a1 / 1., 2., 3., 4. /
```

Since values are assigned to the array elements in column order, this initializes a1(1,1) to 1.0, a1(2,1) to 2.0, a1(1,2) to 3.0, and a1(2,2) to 4.0.

It is possible to change the order in which values are assigned to array elements by using an implied DO loop. Thus, the following DATA statement initializes the 2 × 2 array a2.

```
REAL a2(2,2)
DATA ((a2(i,j), j=1,2), i=1,2) / 1., 2., 3., 4. /
```

This implied DO loop initializes the array elements in the order $a2(1,1)$, $a2(1,2)$, $a2(2,1)$, and $a2(2,2)$, so the values become $a(1,1) = 1.0$, $a(1,2) = 2.0$, $a(2,1) = 3.0$, and $a(2,2) = 4.0$.

The DATA statement is redundant, since data initializations can be made directly in type declaration statements. It should not be used in new programs.

Good Programming Practice

Do not use DATA statements in your programs. Instead, initialize your variables in their type declaration statements.

18.4.5 The PARAMETER Statement

The parameter or named constant was introduced in FORTRAN 77. At that time, parameters were declared in a PARAMETER statement of the form:

```
INTEGER SIZE
PARAMETER ( SIZE = 1000 )
```

The PARAMETER attribute was introduced in Fortran 90, so that the same parameter is now declared as

```
INTEGER, PARAMETER :: size = 1000
```

The older PARAMETER statement was retained for backward compatibility, but it should never be used. The syntax of that statement is not consistent with other Fortran statements, and it is simpler to declare a parameter value in its type declaration statement anyway.

Good Programming Practice

Do not use PARAMETER statements in your programs. Instead, use the PARAMETER attribute of a type declaration statement.

18.5

SHARING MEMORY LOCATIONS: COMMON AND EQUIVALENCE

Fortran includes two statements that permit different variables to physically share the same memory locations, either between program units or within a single program unit: the COMMON statement and the EQUIVALENCE statement. Both of these statements have been replaced by better methods in Fortran 90 and later versions.

18.5.1 COMMON Blocks

We saw in Chapter 7 that modules may be used to share data between program units. If a data item is defined in a module and has the PUBLIC attribute, then any program unit that uses the module can access the data item. This is the standard way to share data among program units in modern Fortran. However, modules did not exist before Fortran 90, and a totally different mechanism was used to share data.

Before Fortran 90, information was shared among program units through COMMON blocks. A COMMON block is a declaration of a region of computer memory that is accessible to every program unit containing the common block. The structure of a COMMON block is

```
COMMON / name / var1, var2, var3, ...
```

where *name* is the name of the COMMON block, and *var1, var2*, etc., are variables or arrays allocated in successive memory locations starting at the beginning of the block. Before Fortran 90, a COMMON block could contain any mixture of real, integer, and logical variables and arrays, or it could contain character data, but noncharacter and character data could not be mixed in the same COMMON block. This restriction was removed in Fortran 90.

A procedure can have as many COMMON blocks as the programmer wishes to declare, so it is possible to create separate COMMON blocks for logical groupings of data that need to be shared. Each separate COMMON block must have a unique name. The names have global scope, so they must be unique within the entire program.

When an array appears in a COMMON block, the size of the array may be declared in either the COMMON block or the type declaration statement, but *not* in both places. The following pairs of statements are legal and completely equivalent:

```
REAL, DIMENSION(10) :: a      ! Preferred
COMMON / data1 / a

REAL :: a
COMMON / data1 / a(10)
```

while the following statements are illegal and will produce an error at compilation time:

```
REAL, DIMENSION(10) :: a
COMMON / data1 / a(10)
```

COMMON blocks permit procedures to share data by sharing a common region of memory. The Fortran compiler allocates all COMMON blocks with the same name in any program unit to the *same region of memory*, so that any data stored there by one procedure may be read and used by any of the other ones. The COMMON blocks with a given name do not all have to be the same length in every procedure, since the Fortran compiler and linker are smart enough to allocate enough memory to hold the largest block declared in any of the procedures.

A sample pair of routines with COMMON blocks are shown in Figure 18-2.

FIGURE 18-2
A main program and subroutine sharing data through a COMMON block.

```
PROGRAM test_common
IMPLICIT NONE
REAL :: a, b
REAL, DIMENSION(5) :: c
INTEGER :: i
COMMON / common1 / a, b, c, i
...
CALL sub11
END PROGRAM

SUBROUTINE sub11
REAL :: x
REAL,DIMENSION(5) :: y
INTEGER :: i, j
COMMON / common1 / x, y, i, j
...
END SUBROUTINE
```

Variables and arrays are allocated in a COMMON block in the order in which they are declared in the COMMON statement. In the main program, variable a occupies the first word in the block, variable b occupies the second word, etc. In the subroutine, variable x occupies the first word in the block, and array element y(1) occupies the second word, etc. Therefore, *variable* a *in the main program is really the same as variable* x *in the subroutine.* They are two different ways to refer to identically the same memory location. Note that the variables in the main program and the subroutine are related by their relative positions in their common blocks. This is known as **storage association,** because the variables are associated by sharing a single physical storage location.

Both COMMON blocks and modules are convenient ways to share large volumes of data between procedures. However, COMMON blocks must be used carefully to avoid problems, since they are subject to two types of errors that don't affect modules. Both of these errors are illustrated in Figure 18-3. Note that the 5-element array c in the main program and the corresponding 5-element array y in the subroutine are misaligned, because there is one fewer value declared in the block before the array in the subroutine than in the main program. Therefore, c(1) in the main program will be the same variable as y(2) in the subroutine. If arrays c and y are supposed to be the same, this misalignment will cause severe problems. Also, note that real array element c(5) in the main program is identical to integer variable i in the subroutine. It is extremely unlikely that the real variable stored in c(5) will be usable as an integer in subroutine sub1. This type mismatch must also be prevented. Neither the array misalignment nor the type mismatch can occur when using modules to share data between program units, so modules are the best way to share data in all modern Fortran programs.

To properly use a COMMON block, we must ensure that all variables in the block appear *in the same order* and have *the same type and size* in every program unit containing the block. In addition, it is good programming practice to keep the *same names* for each of the variables in every program unit containing the block. The program will

Memory address	Program main	Subroutine sub1
0000	a	x
0001	b	y(1)
0002	c(1)	y(2)
0003	c(2)	y(3)
0004	c(3)	y(4)
0005	c(4)	y(5)
0006	c(5)	i
0007	i	j

FIGURE 18-3
Memory allocation in COMMON block /common1/, showing the misalignment between arrays c and y.

be much more understandable if the same names apply to the same variables in all procedures.

Common blocks have been declared obsolete, and should never be used in any new program.

Good Programming Practice
Use modules rather than COMMON blocks to share data between program units. If you do use COMMON blocks, you should be sure to declare the blocks identically in every procedure containing them, so that the variables always have the same name, type, and order in each procedure.

18.5.2 Initializing Data in COMMON Blocks: The BLOCK DATA Subprogram

The DATA statement was introduced above. It may be used to initialize the values associated with local variables in a main program or subprogram. However, it may *not* be used in a main program or procedure to initialize variables in COMMON blocks. The reason for this restriction is illustrated in the following example program:

```
PROGRAM test
CALL sub1
CALL sub2
END PROGRAM
```

```
        SUBROUTINE sub1
        INTEGER ival1, ival2
        COMMON / mydata / ival1, ival2
        DATA ival1, ival2 /1, 2/
        ...
        END SUBROUTINE sub1

        SUBROUTINE sub2
        INTEGER ival1, ival2
        COMMON / mydata / ival1, ival2
        DATA ival1, ival2 /3, 4/
        ...
        END SUBROUTINE sub2
```

Here, COMMON block /mydata/ is exchanged between subroutines sub1 and sub2. Subroutine sub1 attempts to initialize ival1 and ival2 to 1 and 2, respectively, while subroutine sub2 attempts to initialize ival1 and ival2 to 3 and 4, respectively. And yet, they are the same two variables! How could the Fortran compiler possibly make sense of this situation? The simple answer is that it can't.

To guarantee that there is only one set of initial values for the variables in a COMMON block, the Fortran language prohibits the use of DATA statements with common variables in any Fortran main program or procedure. Instead, it includes a special type of program unit *whose only function is to initialize the variables in a* COMMON *block*: the BLOCK DATA subprogram. Since there is one and only one place where COMMON variables may be initialized, there is no ambiguity about what values to assign to them.

A BLOCK DATA subprogram begins with a BLOCK DATA statement, and may contain any number of type definition statements, COMMON statements and DATA statements. *It must not contain any executable statements.* An example BLOCK DATA subprogram is shown below:

```
        BLOCK DATA initial
        INTEGER ival1, ival2
        COMMON / mydata / ival1, ival2
        DATA ival1, ival2 /1, 2/
        END BLOCK DATA
```

The name of this BLOCK DATA subprogram is initial. (BLOCK DATA names are optional: This subprogram would have worked equally well with no name.) The subprogram initializes the variables ival1 and ival2 in COMMON block /mydata/ to 1 and 2, respectively.

BLOCK DATA subprograms have been declared obsolescent, and should never be used in any new program.

18.5.3 The Unlabeled COMMON Statement

There is an alternate form of the COMMON statement that is called the unlabeled COMMON statement. An unlabeled COMMON statement has the form

```
        COMMON var1, var2, var3, ...
```

where `var1, var2,` etc., are variables or arrays allocated in successive memory locations starting at the beginning of the common block. The unlabeled `COMMON` statement is exactly like an ordinary `COMMON` block, except that this block has no name.

The unlabeled `COMMON` statement is a relic left over from earlier versions of Fortran. Before FORTRAN 66, it was only possible to declare one `COMMON` area in any given program. The unlabeled `COMMON` statement should never be used in any modern Fortran program.

18.5.4 The EQUIVALENCE Statement

In the past, it was sometimes useful to refer to a particular location in computer memory by more than one name. Computer memory was a limited and very expensive resource. Because computer memory was so expensive, it was common for large computer programs to reuse portions of memory for scratch calculations in different procedures within the program. Since dynamic memory allocation did not exist before Fortran 90, a single fixed block of scratch memory would be declared that would be large enough for any temporary calculations within the program. This block of memory would be used over and over wherever scratch memory was needed. The scratch memory would often be referred to by different names in different portions of the program, but the same physical memory would be used each time.

To support such applications, Fortran provided a mechanism for assigning two or more names to the same physical memory location: the `EQUIVALENCE` statement. The `EQUIVALENCE` statement appears in the declaration section of a program after all type declaration statements and before any `DATA` statements. The form of the `EQUIVALENCE` statement is

```
EQUIVALENCE ( var1, var2, var3, ... )
```

where `var1, var2,` etc., are variables or array elements. *Every variable appearing within the parentheses in an* `EQUIVALENCE` *statement is assigned to the same memory location by the Fortran compiler.* If some of the variables are array elements, then this statement also fixes the relative relationships of all elements within the arrays. Consider the following example:

```
INTEGER, DIMENSION(2,2) :: i1
INTEGER, DIMENSION(5) :: j1
EQUIVALENCE ( i1(2,1), j1(4) )
```

Here, `i1(2,1)` and `j1(4)` occupy the same memory location. Because of the way arrays are laid out in memory, `i1(1,2)` and `j1(5)` will also occupy a single memory location (see Figure 18-4).

`EQUIVALENCE` statements are inherently quite dangerous. A common problem occurs when we first perform some calculation using an equivalenced array under one name (say, `array1`) in a program, and then perform a different calculation using the same equivalenced array under another name (say, `array2`) in another part of the program. If we then try to access values in `array1`, we will find that they have all been destroyed by the operations on `array2`. This can be an especially big problem if the

Memory address	Name 1	Name 2
001		j1(1)
002		j1(2)
003	i1(1,1)	j1(3)
004	i1(2,1)	j1(4)
005	i1(1,2)	j1(5)
006	i1(2,2)	
007		
008		

Effect of the statement

EQUIVALENCE (I1(2,1), J1(4))

on memory allocation in a Fortran program

FIGURE 18-4
The effect of the EQUIVALENCE statement on memory allocation in a Fortran program. Because i1(2,1) and j1(4) must be the same physical location, arrays i1 and j1 will overlap in the computer's memory.

program is being modified by some person other than the original programmer. Since the data in array1 has been destroyed without array1 ever appearing in an assignment statement, it can be very hard to track down this bug.

Since computer memory has gotten both cheaper and more plentiful over the years, the need for equivalencing arrays has decreased dramatically. You should not equivalence variable names in your programs unless you have a very good reason to do so. If you need to reuse scratch memory arrays in your program, *it is better to use allocatable arrays or pointers to allocate and deallocate the scratch memory dynamically.*

Another use of the EQUIVALENCE statement was to assign the same memory address to variables of *different types* so that the same bit pattern could be examined in different ways. For example, a real variable and an integer variable could be assigned to the same location. When a real variable was stored in that location, the integer variable could be used to examine the bit patterns. If you have any older code that uses EQUIVALENCE statements in this fashion, it can be replaced by the TRANSFER intrinsic function. For example, the following code takes the exact bit pattern of real variable value and stores it in integer variable ivalue:

```
INTEGER :: ivalue
REAL :: value
...
ivalue = TRANSFER(value, 0)
```

Finally, note that the EQUIVALENCE statement effectively assigns two or more different names to the same *memory location*. From this statement, it follows that names must be associated with memory locations, or they may not be equivalenced. Names that are not associated with a specific memory location (e.g., dummy argument names) may not be used in an EQUIVALENCE statement.

EQUIVALENCE statements have been declared obsolescent, and should never be used in any new program.

Good Programming Practice
Do not use EQUIVALENCE statements in your programs. If you need to reuse scratch memory arrays in your program, it is better to allocate and deallocate them dynamically with allocatable arrays or pointers.

18.6
UNDESIRABLE SUBPROGRAM FEATURES

There are four subprogram features that are undesirable and should never be used in modern Fortran programs. They are:

1. Alternate subroutine returns
2. Alternate entry points
3. The statement function
4. Passing intrinsic functions as arguments

18.6.1 Alternate Subroutine Returns

When a Fortran program calls a normal subroutine, the subroutine is executed, and then control returns to the first executable statement following the subroutine call.

It is sometimes useful to execute different code in the calling procedure depending on the results of the subroutine call. Earlier version of Fortran supported such operation by providing alternate subroutine returns. Alternate subroutine returns are statement labels passed as calling arguments to the subroutine. When the subroutine executes, it can decide to return control to any of the statement labels specified in the argument list. Alternate subroutine returns are specified in the following manner:

1. The statement labels associated with all possible alternate returns are specified as arguments in the CALL statement by preceding each label with an asterisk:

    CALL SUB1 (a, b, c, *n1, *n2, *n3)

where n1, n2, and n3 are the statement numbers to which execution may be transferred.

2. The alternate returns are specified in the SUBROUTINE statement by asterisks:

   SUBROUTINE SUB1 (a, b, c, *, *, *)

   where the asterisks correspond to the locations of the alternate returns in the calling statement.

3. The particular alternate return to be executed is specified by a parameter on the RETURN statement:

   RETURN k

   where *k* is the *position* of the alternate return to be executed. In the above example, there are three possible alternate returns, so *k* could take on a value from 1 to 3.

In the example in Figure 18-5, there are two possible returns. The first return is for normal completion, and the second one is for error conditions.

FIGURE 18-5
A program fragment illustrating the use of alternate subroutine returns.

```
CALL calc ( a1, a2, result, *100, *999 )
! Normal return--continue execution.
100 ...
    ...
STOP

! Error in subroutine call--process error and stop.
999 WRITE (*,*) 'Error in subroutine calc.  Execution aborted.'
STOP 999
END PROGRAM

SUBROUTINE calc ( a1, a2, result, *, * )
REAL a1, a2, result, temp
IF ( a1 * a2 >= 0. ) THEN
   result = SQRT(a1 * a2)
ELSE
   RETURN 2
END IF

RETURN 1
END SUBROUTINE
```

Alternate subroutine returns should *never* be used in modern Fortran code. They make program maintenance and debugging much harder by making it difficult to follow the execution path through the program. They contribute to the "spaghetti code" so commonly found in older programs. There are other, much better ways to provide for different program execution paths depending on the results of a subroutine call. The simplest and best approach is to include a logical IF construct that tests the subroutine return parameters immediately after the subroutine call, and then takes action depending on the status returned by the subroutine.

Alternate subroutine returns have been declared obsolescent in Fortran 95, which means that they are a candidate for deletion in future versions of Fortran.

> **Good Programming Practice**
> Do not use alternate subroutine returns in your programs. They make programming debugging and maintenance much harder, and simple, structured alternatives are available.

18.6.2 Alternate Entry Points

The normal entry point for a Fortran procedure is the first executable statement in the procedure. However, it is possible to get program execution to start at a different point within the procedure if that point is specified with an ENTRY statement. An ENTRY statement has the form

```
ENTRY name ( arg1, arg2, ... )
```

where *name* is the name of the entry point, and *arg1*, *arg2*, etc., are the dummy arguments passed to the procedure at the entry point. When a subprogram is invoked by the name specified in the ENTRY statement, execution begins at the first executable statement following the ENTRY statement instead of the first executable statement in the subprogram.

A common use of the ENTRY statement occurs when a subprogram must be initialized the first time it is used but not thereafter. In that case, a special initialization entry point is sometimes included in the subprogram. For example, consider the subroutine in Figure 18-6, which evaluates a third order polynomial for a specific input value x. Before the polynomial can be evaluated, the coefficients of the polynomial must be specified. If the coefficients of the polynomial change infrequently, we could specify them in a special ENTRY to the subroutine.

FIGURE 18-6
A subroutine illustrating the use of multiple entry points.

```
PROGRAM test_entry
REAL :: a = 1., b = 2., c = 1., d = 2.
CALL init1 ( a, b, c, d )
DO I = 1, 10
   CALL eval3 ( REAL(i), result )
   WRITE (*,*) 'EVAL3(', i, ') = ', result
END DO
END PROGRAM

SUBROUTINE eval3 ( x, result )
!
!    Evaluates a third order polynomial of the form:
!       RESULT = A + B*X + C*X**2 + D*X**3
!
!    Declare calling arguments
IMPLICIT NONE
REAL :: a1, b1, c1, d1, x, result
```

(continued)

Redundant, Obsolescent, and Deleted Fortran Features

(*concluded*)

```
! Declare local variables
REAL, SAVE :: a, b, c, d

! Calculate result
result = a + b**x + c*x**2 + d*x**3

RETURN

! Entry INITL specifies the values of a, b, c, and d
! to be used when evaluating the polynomial.

ENTRY initl(a1, b1, c1, d1)
a = a1
b = b1
c = c1
d = d1

RETURN
END SUBROUTINE
```

Note from the above example that the various entry points in a subroutine do not have to have the same calling sequence. However, we must be sure to call each entry point with the proper argument list for that particular entry point.

The use of entry points should be discouraged. A major disadvantage of ENTRY statements occurs when we need to modify the code of a procedure containing multiple entry points. If there are any code segments or variables in common to the different entry points, we can get in serious trouble. In the process of changing the procedure to make one entry point work correctly, we can inadvertently screw up the operation of another entry point. After a procedure containing multiple entry points is modified, it must be tested *very* carefully, both the entry point being modified and all other entry points.

The original reason for using multiple entry points in a procedure was to share segments of code for multiple purposes, thus reducing the size of the completed program. This reason no longer makes sense today. As cheap as memory is now, there is no good reason to *ever* use an entry point. If you write separate procedure for each function you need, your code will be much more maintainable.

If you need to share data among multiple procedures, the data (and possibly the procedures themselves) should be placed in a module. The previous example can be rewritten without entry points as shown in Figure 18-7. Variables a, b, c, and d are made available to both subroutines eval3 and initl through host association, and the subroutines are made available to the main program through USE association.

FIGURE 18-7
The previous example program rewritten without multiple entry points.

```
MODULE evaluate
IMPLICIT NONE
PRIVATE
PUBLIC eval3, initl
```

(*continued*)

(*concluded*)

```fortran
! Declare shared data.
REAL, SAVE :: a, b, c, d

! Declare procedures
CONTAINS
   SUBROUTINE eval3 ( x, result )
   !
   !      Evaluates a third order polynomial of the form:
   !         RESULT = A + B*X + C*X**2 + D*X**3
   !
   ! Declare calling arguments
   REAL, INTENT(IN) :: x
   REAL, INTENT(OUT) :: result

   ! Calculate result
   result = a + b**x + c*x**2 + d*x**3

   END SUBROUTINE eval3

   SUBROUTINE initl (a1, b1, c1, d1 )
   !
   ! Subroutine INITL specifies the values of a, b, c, and d
   ! to be used when evaluating the polynomial.
   !
   REAL, INTENT(IN) :: a1, b1, c1, d1
   a = a1
   b = b1
   c = c1
   d = d1
   END SUBROUTINE initl
END MODULE evaluate

PROGRAM test_noentry
USE evaluate
REAL :: a = 1., b = 2., c = 1., d = 2.
CALL initl ( a, b, c, d )
DO i = 1, 10
   CALL eval3 ( REAL(i), result )
   WRITE (*,*) 'EVAL3(', i, ') = ', result
END DO
END PROGRAM test_noentry
```

ENTRY statements have been declared obsolescent, and should never be used in any new program.

Good Programming Practice

Avoid alternate entry points in your programs. There is no good reason to use them in a modern Fortran program.

18.6.3 The Statement Function

In Chapter 7, we introduced the external functions. An external function is a procedure that returns a single value to the invoking program unit. Its input values are passed via an argument list. An external function is invoked by being named as a part of a Fortran expression.

In Chapter 9, we introduced the internal functions. Internal functions are similar to external functions, except they are entirely contained within another program unit, and may only be invoked from that program unit.

There is a third type of Fortran function: the *statement function*. A statement function consists of a *single statement*. It must be defined in the declaration section of a Fortran program unit before the first executable statement in the program unit. An example of a statement function is shown in Figure 18-8.

FIGURE 18-8
A program using a statement function.

```
PROGRAM polyfn
!
! This program evaluates a third order polynomial
! of the form:
!     RES = A + B*X + C*X**2 + D*X**3
! using a statement function.

IMPLICIT NONE

! Declare local variables.
REAL :: a, b, c, d, x, y
INTEGER :: i

! Declare dummy arguments of the statement function.
REAL :: a1, b1, c1, d1, x1, res

! Declare statement function res.
res(a1,b1,c1,d1,x1) = a1 + b1**x1 + c1*x1**2 + d1*x1**3

! Set up coefficients of polynomial res.
a = 1.
b = 2.
c = 1.
d = 2.

! Evaluate polynomial for x values of 1 through 10.
DO i = 1, 10
   x = REAL(i)
   y = res(a,b,c,d,x)
   WRITE (*,*) 'y(',i,') = ', y
END DO
END PROGRAM polyfn
```

In this example, real statement function res is defined as

$$res(a1,b1,c1,d1,x1) = a1 + b1**x1 + c1*x1**2 + d1*x1**3$$

where a1, b1, c1, d1, and x1 are *dummy arguments*. The types of the function and its dummy arguments must all be declared or defaulted before the function is defined. The dummy arguments are placeholders for the actual values that are used when the function is executed later in the program. The dummy arguments must agree in type and number with the actual arguments that are used when the function is executed. At execution time, the value in the first argument of the function will be used instead of `a1` wherever `a1` appears in the statement function, and so forth for all other arguments.

If you take a close look, you will notice that a statement function looks exactly like an assignment statement that assigns a value to an array element. Since this is so, how can the Fortran compiler tell the difference between them? To make it possible to tell the difference between them, Fortran requires that *all statement functions must be defined in the declaration section of a program*, before the first executable statement.

Like internal functions, statement functions can only be used in the program unit in which they are declared. They are limited to functions that can be evaluated in a single expression with no branches or loops. In addition, the calling arguments must be variables, constants, or array elements. Unlike external or internal functions, it is not possible to pass whole arrays to a statement function.

Statement functions are a very old feature of Fortran, dating all the way back to Fortran 1 in 1954. They have been replaced by internal functions. Internal functions can do anything a statement function can do, and much more besides. There is no reason to ever use statement functions in your programs.

Statement functions have been declared obsolescent as of Fortran 95, which means that they are a candidate for deletion in future versions of Fortran.

Good Programming Practice

Never use statement functions in your programs. Use internal functions instead.

18.6.4 Passing Intrinsic Functions as Arguments

It is possible to pass a **specific intrinsic function** as a calling argument to another procedure. If the name of a specific intrinsic function is included as an actual argument in a procedure call, then a pointer to that function is passed to the procedure. If the corresponding dummy argument in the procedure is used as a function, then when the procedure is executed, the intrinsic function in the calling argument list will be used in place of the dummy function in the procedure. *Generic intrinsic functions may not be used as calling arguments*—only specific intrinsic functions may be used.

Before it can be used as a calling argument to a procedure, a specific intrinsic function must be declared in an `INTRINSIC` statement in the calling program. The `INTRINSIC` statement is a specification statement of the form

```
INTRINSIC name1, name2, ...
```

It states that `name1, name2,` etc., are names of intrinsic functions. The `INTRINSIC` statement must appear in the declaration section of a procedure, before the first executable statement. The reason that an `INTRINSIC` statement is required is the same as the reason that an `EXTERNAL` statement is required: It permits the compiler to distinguish between a variable name and an intrinsic function of the same type.

An example program illustrating the passing of a specific intrinsic function as an argument is shown in Figure 18-9. This program is a modification of the test driver program in Figure 6-25. It calculates the average value of the intrinsic function SIN(X) over 101 samples in the interval [0, 2π], and the result is printed out.

FIGURE 18-9
Program illustrating the passing of an intrinsic function as a calling argument.

```
PROGRAM test_ave_value2
!
!  Purpose:
!    To test function ave_value by calling it with the intrinsic
!    function sin.
!
!  Record of revisions:
!      Date        Programmer        Description of change
!      ====        ==========        =====================
!    2/26/16       S. J. Chapman     Original code
!
IMPLICIT NONE

! Declare functions:
REAL :: ave_value                ! Average value of function
INTRINSIC sin                    ! Function to evaluate

! Declare parameters,:
REAL, PARAMETER :: TWOPI = 6.283185 ! 2 * Pi

! Declare local variables:
REAL :: ave                      ! Average of my_function

! Call function with func=sin.
ave = ave_value ( sin, 0., TWOPI, 101 )
WRITE (*,1000) 'SIN', ave
1000 FORMAT ('The average value of ',A,' between 0. and twopi is ', &
             F16.6,'.')

END PROGRAM test_ave_value2
```

When program `test_ave_value2` is executed, the results are

```
C:\book\fortran\chap18>test_ave_value2
The average value of SIN between 0. and TWOPI is          .000000.
```

The passing of intrinsic functions as calling arguments is very confusing, and is only possible for specific intrinsic functions. It should never be done in modern Fortran programs.

18.7

MISCELLANEOUS EXECUTION CONTROL FEATURES

There are two statements that pause or stop the execution of a program: the PAUSE and STOP statements. The PAUSE statement is rarely used in a modern Fortran program, since the same function can be done more flexibly with a combination of WRITE and READ statements. The STOP statement is more common, but it is often not necessary either, since program execution will terminate when the END statement is reached. However, it is occasionally useful to have multiple stopping points in a program. In that case, each stopping point will need a STOP statement. If there are multiple STOP statements in a program, each one should be labeled with a unique argument (as explained below) so that the user can tell which STOP statement was executed.

Finally, there is an older form of the END statement to indicate the end of a separately-compiled program unit.

18.7.1 The PAUSE Statement

When we write Fortran programs whose results are meant to be viewed from a terminal, it is necessary to pause the program at certain points while the user examines the results displayed on the terminal. Otherwise, the information may scroll off the top of the display before it can be read. After the user reads output data on the terminal, he or she can either continue the program or abort it.

Earlier versions of Fortran included a special statement designed to pause the execution of a program until the user starts it up again: the PAUSE statement. The general form of the PAUSE statement is

   PAUSE *prompt*

where *prompt* is an optional value to be displayed when the PAUSE statement is executed. The prompt may be either a character constant or an integer between 0 and 99999. When the PAUSE statement is executed, the value of *prompt* is displayed on the terminal, and execution stops until the user restarts the program. When the program is restarted, execution will begin at the statement following the PAUSE statement.

The PAUSE statement was never particularly common, since it is possible to perform the same function with WRITE and READ statements with much more flexibility.

The PAUSE statement has been deleted from the language as of Fortran 95, which means that it is no longer an official part of the Fortran language.

18.7.2 Arguments Associated with the STOP Statement

Like the PAUSE statement described above, it is possible to include an argument with the STOP statement. The general form of the STOP statement is

   STOP *stop_value*

where *stop_value* is an optional value to be displayed when the STOP statement is executed. The *stop_value* may be either a character constant or an integer between 0 and 99999. It is mainly used when there are multiple STOP statements in a program. If there are multiple STOP statements and a separate *stop_value* is associated with each one, then the programmer and user can tell which of the STOP statements was executed when the program quit.

If there are multiple STOP statements in a program, it is a good idea to use either a separate argument on each one or a separate WRITE statement before each one, so that a user can tell which STOP a program halted on. An example program with multiple STOP statements is shown in Figure 18-8. The first STOP occurs if the file specified by the user does not exist. It is clearly marked by the WRITE statement that occurs just before it. The second STOP occurs when the program completes normally. If this stop is executed, the message 'Normal Completion.' will be printed out when the program terminates.

FIGURE 18-10
A program to illustrate the use of multiple STOP statements in a single program unit.

```
PROGRAM stop_test
!
!  Purpose:
!    To illustrate multiple STOP statements in a program.
!
IMPLICIT NONE

! Declare parameters:
INTEGER, PARAMETER :: lu = 12       ! I/O unit

! Declare variables:
INTEGER :: error                     ! Error flag
CHARACTER(len=20) :: filename        ! File name

! Prompt user and get the name of the input file.
WRITE (*,*) 'Enter file name: '
READ (*,'(A)') filename

! Open the input file
OPEN (UNIT=lu, FILE=filename, STATUS='OLD', IOSTAT=error )

! Check to see of the OPEN failed.
IF ( error > 0 ) THEN
   WRITE (*,1020) filename
   1020 FORMAT ('ERROR: File ',A,' does not exist!')
   STOP
END IF

! Normal processing...
...
! Close input file, and quit.
CLOSE (lu)

STOP 'Normal completion.'
END PROGRAM stop_test
```

As Fortran has improved over the years, the use of multiple STOP statements has declined. Modern structured techniques usually result in programs with a single starting point and a single stopping point. However, there are still occasions when multiple stopping points might occur in different error paths. If you do have multiple stopping points, be sure that each one is labeled distinctively so that they can be easily distinguished.

Also, if a modern program does have multiple stops in it, the programmer will normally use the ERROR STOP statement for error stops instead of an ordinary STOP, because that statement returns an error message to the operating system when it is executed.

18.7.3 The END Statement

Before Fortran 90, all program units terminated with an `END` statement instead of separate `END PROGRAM`, `END SUBROUTINE`, `END FUNCTION`, `END MODULE`, `END SUBMODULE`, or `END BLOCK DATA` statements. The `END` statement is still accepted for backward compatibility in independently-compiled program units such as main programs, external subroutines, and external functions.

However, internal procedures and module procedures *must* end with an `END SUBROUTINE` or `END FUNCTION` statement—the older form won't work in these new types of procedures that did not exist before Fortran 90.

18.8
OBSOLETE BRANCHING AND LOOPING STRUCTURES

In Chapter 3, we described the logical `IF` structure and the `CASE` structure, which are the standard ways to implement branches in modern Fortran. In Chapter 4, we described the various forms of the `DO` loop, which are the standard iterative and while loops in modern Fortran. This section describes several additional ways to produce branches, and older forms of `DO` loops. They are all archaic survivals from earlier versions of Fortran that are still supported for backward compatibility. These features should *never* be used in any new Fortran program. However, you may run into them if you ever have to work with old Fortran programs. They are described here for possible future reference.

18.8.1 The Arithmetic IF Statement

The Arithmetic `IF` statement goes all the way back to the origins of Fortran in 1954. The structure of an arithmetic `IF` statement is

```
IF (arithmetic_expression) label1, label2, label3
```

where *arithmetic_expression* is any integer, real, or double-precision arithmetic expression, and *label1, label2,* and *label3* are labels of executable Fortran statements. When the arithmetic `IF` statement is executed, the arithmetic expression is evaluated. If the resulting value is negative, execution transfers to the statement at *label1*. If the value is zero, execution transfers to the statement at *label2*. If the value is positive, execution transfers to the statement at *label3*. An example statement is

```
         IF ( x - y ) 10, 20, 30
     10  (code for negative case)
         ...
         GO TO 100
     20  (code for zero case)
         ...
         GO TO 100
     40  (code for positive case)
         ...
    100  CONTINUE
         ...
```

The arithmetic IF statement has been deleted from the language as of Fortran 2008, and it should never be used in any modern Fortran program.

Good Programming Practice
Never use arithmetic IF statement in your programs. Use the logical IF structure instead.

18.8.2 The Unconditional GO TO Statement

The GO TO statement has the form

```
GO TO label
```

where *label* is the label of an executable Fortran statement. When this statement is executed, control jumps unconditionally to the statement with the specified label.

In the past, GO TO statements were often combined with IF statements to create loops and conditional branches. For example, a while loop could be implemented as

```
10 CONTINUE
   ...
   IF ( condition ) GO TO 20
   ...
   GO TO 10
20 ...
```

There are better ways to create loops and branches in modern Fortran, so the GO TO statement is now rarely used. The excessive use of GO TO statements tends to lead to "spaghetti code", so their use should be discouraged. However, there may be some rare occasions (such as exception handling) when the statement will prove useful.

Good Programming Practice
Avoid the use of GO TO statements whenever possible. Use structured loops and branches instead.

18.8.3 The Computed GO TO Statement

The computed GO TO statement has the form

```
GO TO (label1, label2, label3,..., labelk), int_expr
```

where *label1* through *labelk* are labels of executable Fortran statements, and the *int_expr* evaluates to an integer between 1 and *k*. If the integer expression evaluates to 1, then the statement at *label1* is executed. If the integer expression evaluates to 2, then the statement at *label2* is executed, and so forth up to *k*. If the integer expression is less than 1 or greater than *k*, this is an error condition, and the behavior of the statement will vary from processor to processor.

An example of a computed GO TO statement is shown below. In this example, the number 2 would be printed out when the program is executed.

```
PROGRAM test
i = 2
GO TO (10, 20), i
10 WRITE (*,*) '1'
GO TO 30
20 WRITE (*,*) '2'
30 STOP
END PROGRAM
```

The computed GO TO should never be used in any modern Fortran program. It has been entirely replaced by the CASE structure.

The computed GO TO statement has been declared obsolescent as of Fortran 95, which means that they are a candidate for deletion in future versions of Fortran.

Good Programming Practice

Never use the computed GO TO statement in your programs. Use the CASE structure instead.

18.8.4 The Assigned GO TO Statement

The assigned GO TO statement has two possible forms:

GO TO *integer variable*, (*label1*, *label2*, *label3*,..., *labelk*)

or

GO TO *integer variable*

where *integer variable* contains the statement number of the statement to be executed next, and *label1* through *labelk* are labels of executable Fortran statements. Before this statement is executed, a statement label must be assigned to the integer variable using the ASSIGN statement:

ASSIGN *label* TO *integer variable*

When the first form of the assigned GO TO is executed, the program checks the value of the integer variable against the list of statement labels. If the value of the variable is in the list, then execution branches to the statement with that label. If the value of the variable is not in the list, an error occurs.

When the second form of the assigned GO TO is executed, no error checking is done. If the value of the variable is a legal statement label in the program, control branches to the statement with that label. If the value of the variable is not a legal statement label, execution continues with the next executable statement after the assigned GO TO.

An example of an assigned GO TO statement is shown below. In this example, the number 1 would be printed out when the program is executed.

```
PROGRAM test
ASSIGN 10 TO i
GO TO i (10, 20)
10 WRITE (*,*) '1'
GO TO 30
20 WRITE (*,*) '2'
30 END PROGRAM
```

The assigned GO TO should never be used in any modern Fortran program.

The ASSIGN statement and the assigned GO TO statement have been deleted from the language as of Fortran 95, which means that they are no longer an official part of the Fortran language.

Good Programming Practice

Never use the assigned GO TO statement in your programs. Use the logical IF structure instead.

18.8.5 Older Forms of DO Loops

Before Fortran 90, DO loops had a different form than the one taught in this book. Modern counting DO loops have the structure

```
DO i = istart, iend, incr
   ...
END DO
```

where istart is the starting value of the loop, iend is the ending value of the loop, and incr is the loop increment.

Early FORTRAN DO loops had the structure

```
DO 100 i = istart, iend, index
   ...
100    ...
```

A statement label is included in this form of the loop, and all of the code from the DO statement until the statement containing that statement label is included in the loop. An example of the earlier loop structure is:

```
    DO 100 i = 1, 100
       a(i) = REAL(i)
100    b(i) = 2. * REAL(i)
```

This was the standard form of the DO loop used by most programmers from the beginning of FORTRAN until about the mid-1970s.

Because the end of this earlier form of the DO loop is so hard to recognize, many programmers developed the habit of always ending DO loops on a CONTINUE statement, which is a statement that does nothing. In addition, they indented all of the statements between the `DO` and the `CONTINUE` statement. An example of a "good" FORTRAN 77 DO loop is:

```
      DO 200 i = 1, 100
         a(i) = REAL(i)
         b(i) = 2. * REAL(i)
  200 CONTINUE
```

As you can see, this form of the loop is much easier to understand.

The termination of a `DO` loop on any statement other than an `END DO` or a `CONTINUE` has been deleted from the language as of Fortran 2008.

Another feature of older `DO` loops was the ability to terminate more than one loop on a single statement. For example, in the following code, two `DO` loops terminate on a single statement.

```
      DO 10 i = 1, 10
      DO 10 j = 1, 10
   10 a(i,j) = REAL(i+j)
```

This sort of structure was terribly confusing, and it should not be used in any modern Fortran program.

The termination of more than one `DO` loop on a single statement has also been deleted from the language as of Fortran 2008.

Finally, FORTRAN 77 added the ability to use single-precision or double-precision real numbers as `DO` loop indices. This was a terrible decision, since the behavior of `DO` loops with real indices varied from processor to processor (this was explained in Chapter 4). Fortran 90 deleted the use of real numbers as loop indices.

Good Programming Practice

Never use any of these older forms of the `DO` loop in any new program.

18.9

REDUNDANT FEATURES OF I/O STATEMENTS

A number of features of I/O statements have become redundant and should not be used in modern Fortran programs. The `END=` and `ERR=` clauses in I/O statements have been largely replaced by the `IOSTAT=` clause. The `IOSTAT=` clause is more flexible and more compatible with modern structured programming techniques than the older clauses, and only the `IOSTAT=` clause should be used in new programs.

Similarly, three format descriptors have been made redundant and should no longer be used in modern Fortran programs. The `H` format descriptor was an old way to specify character strings in a `FORMAT` statement. It was briefly mentioned in Table 10-1.

It has been completely replaced by the use of character strings surrounded by single or double quotes.

The H format descriptor has been deleted from the language as of Fortran 95, which means that it is no longer an official part of the Fortran language.

The P scale factor was used to shift the decimal point in data displayed with the E and F format descriptors. It has been made redundant by the introduction of the ES and EN format descriptors, and should never be used in any new programs.

The D format descriptor was used to input and output double-precision numbers in earlier versions of Fortran. It is now identical to the E descriptor, except that on output a D instead of an E may appear as the marker of the exponent. There is no need to ever use the D descriptor in a new program.

The BN and BZ format descriptors control the way blanks are interpreted when reading fields from card-image files. By default, modern Fortran ignores blanks in input fields. In FORTRAN 66 and earlier, blanks were treated as zeros. These descriptors are provided for backward compatibility with very early version of Fortran; they should never be needed in any new program.

The S, SP, and SS format descriptors control the display of positive signs in a format descriptor. These descriptors are completely unnecessary and should never be used.

■ 18.10
SUMMARY

In this chapter, we introduced a variety of miscellaneous Fortran features. Most of these features are either redundant, obsolescent, or incompatible with modern structured programming. They are maintained for backward compatibility with older versions of Fortran.

None of the features described here should be used in any new programs, except possibly for arguments on multiple STOP statements. Since modern programming practices greatly reduce the need for STOP statements, they will not be used very often. However, if you do write a program that contains multiple STOP statements, you should make sure that you use WRITE statements or arguments on STOP statements to distinguish each of the possible stopping points in the program.

COMMON blocks may occasionally be needed in procedures that must work with older Fortran code, but completely new programs should use modules for data sharing instead of COMMON blocks.

There may also be rare circumstances in which the unconditional GO TO statement is useful, such as for exception handling. Most of the traditional uses of the GO TO statement have been replaced by the modern IF, CASE, and DO constructs, so they will be very rare in any modern Fortran program.

Table 18-1 summarizes the features of Fortran that should not be used in new programs, and gives suggestions as to how to replace them if you run into them in older code.

TABLE 18-1
Summary of older Fortran features

Feature	Status	Comment
Source form		
Fixed source form	Obsolescent in Fortran 95	Use free form.
Specification statements		
CHARACTER*<len> statement	Obsolescent in Fortran 95	Use CHARACTER(len=<len>) form.
COMMON blocks	Redundant	Use modules to exchange data.
DATA statement	Redundant	Use initialization in type declaration statements.
DIMENSION statement	Redundant	Use dimension attribute in type declaration statements.
EQUIVALENCE statement	Unnecessary and confusing	Use dynamic memory allocation for temporary memory. Use the TRANSFER function to change the type of a particular data value.
IMPLICIT statement	Confusing but legal	Do not use. Always use IMPLICIT NONE and explicit type declaration statements.
PARAMETER statement	Redundant, and confusing syntax	Use parameter attribute in type declaration statements.
Unlabeled COMMON	Redundant	Use modules to exchange data.
Undesirable subprogram features		
Alternate entry points	Unnecessary and confusing	Share data between procedures in modules, and do not share code between procedures.
Alternate subroutine returns	Obsolescent in Fortran 95	Use status variable, and test status of variable after subroutine call.
Statement function	Obsolescent in Fortran 95	Use internal procedures.
Execution control statement		
PAUSE statement	Deleted in Fortran 95	Use WRITE statement followed by a READ statement.
Branching and looping control statements		
Arithmetic IF statement	Deleted in Fortran 2008	Use logical IF.
Assigned GO TO Statement	Deleted in Fortran 95	Use block IF or CASE construct.
Computed GO TO statement	Obsolescent in Fortran 2008	Use CASE construct.
GO TO statement	Rarely needed	Largely replaced by IF, CASE, and DO constructs with CYCLE and EXIT statements.
DO 100 . . .	Redundant	Use DO . . .
. . .		. . .
100 CONTINUE		END DO
DO loops terminating on executable statement	Deleted in Fortran 2008	Terminate loops on END DO statements.
Multiple DO loops terminating on same statement	Deleted in Fortran 2008	Terminate loops on separate statements.
I/O features		
H format descriptor	Deleted in Fortran 95	Use single or double quotes to delimit strings.
D format descriptor	Redundant	Use E format descriptor.
P scale factor	Redundant and confusing	Use ES or EN format descriptors.
BN and BZ format descriptors	Unnecessary	Blanks should always be nulls, which is the default case.
S, SP, and SS format descriptors	Unnecessary	Processor's default behavior is acceptable.
ERR= clause	Redundant and confusing	Use IOSTAT= and IOMSG= clauses.
END= clause	Redundant and confusing	Use IOSTAT= and IOMSG= clauses.

18.10.1 Summary of Good Programming Practice

None of the features described in this chapter should be used in any new programs, except possibly for arguments on multiple STOP or ERROR STOP statements. Since modern programming practices greatly reduce the need for multiple STOP statements, they will not be used very often. However, if you do write a program that contains multiple STOP or ERROR STOP statements, you should make sure that you use WRITE statements or arguments on STOP statements to distinguish each of the possible stopping points in the program.

18.10.2 Summary of Fortran Statements and Structures

Arithmetic IF Statement:

```
IF (arithmetic expression) label1, label2, label3
```

Example:

```
IF (b**2-4.*a*c) 10, 20, 30
```

Description:
The arithmetic IF statement is an obsolete conditional branching statement. If the arithmetic expression is negative, control will be transferred to statement with label *label1*. If the arithmetic expression is zero, control will be transferred to statement with label *label2*, and if the arithmetic expression is positive, control will be transferred to statement with label *label3*.
    The arithmetic IF statement has been declared obsolescent in Fortran 95.

Assigned GO TO Statement:

```
ASSIGN label TO int_var
GO TO int_var
```

or

```
GO TO int_var, (label1, label2, ... labelk)
```

Example:

```
ASSIGN 100 TO i
...
GO TO i
...
100... (execution continues here)
```

Description:
The assigned GO TO statement is an obsolete branching structure. A statement label is first assigned to an integer variable using the ASSIGN statement. When the assigned GO TO statement is executed, control branches to the statement whose label was assigned to the integer variable.
    The assigned GO TO statement has been deleted in Fortran 95.

COMMON Block:

```
COMMON / name / var1, var2, ...
COMMON var1, var2, ...
```

Example:

```
COMMON / shared / a, b, c
COMMON a, i(-3:3)
```

Description:
This statement defines a COMMON block. The variables declared in the block will be allocated consecutively starting at a specific memory location. They will be accessible to any program unit in which the COMMON block is declared. The COMMON block has been replaced by data values declared in modules.

Computed GO TO Statement:

```
GO TO (label1, label2, ... labelk), int_var
```

Example:

```
GO TO (100, 200, 300, 400), i
```

Description:
The computed GO TO statement is an obsolete branching structure. Control is transferred to one of the statements whose label is listed, depending upon the value of the integer variable. If the variable is 1, then control is transferred to the first statement in the list, etc.

The computed GO TO statement has been declared obsolescent in Fortran 95.

CONTINUE Statement:

```
CONTINUE
```

Description:
This statement is a placeholder statement that does nothing. It is sometimes used to terminate DO loops, or as a location to attach a statement label.

DIMENSION Statement:

```
DIMENSION array( [i1:]i2, [j1:]j2, ... ), ...
```

Example:

```
DIMENSION a1(100), a2(-5:5), i(2)
```

Description:
This statement declares the size of an array but *not* its type. Either the type must be declared in a separate type declaration statement, or else it will be defaulted. DIMENSION statements are not required in well-written code, since type declaration statements will perform the same purpose.

Redundant, Obsolescent, and Deleted Fortran Features

DO Loops (old versions):

```
DO k index = istart, iend, incr
    ...
k   CONTINUE
```

or

```
DO k index = istart, iend, incr
    ...
k   Executable statement
```

Examples:

```
DO 100 index = 1, 10, 3
    ...
100 CONTINUE
```

or

```
DO 200 i = 1, 10
200 a(i) = REAL(i**2)
```

Description:
These forms of the DO loop repeatedly execute the block of code from the statement immediately following the DO up to and including the statement whose label appears in the DO. The loop control parameters are the same in these loops as they are in modern DO constructs.

Only the versions of the DO loop that end in an END DO statement should be used in new programs. DO loops that terminate in a CONTINUE statement are legal but redundant, and should not be used. DO loops that terminate on other statements (such as the one in the second example) have been deleted from the language as of Fortran 2008.

ENTRY Statement:

```
ENTRY name( arg1, arg2, ... )
```

Example:

```
ENTRY sorti ( num, data1 )
```

Description:
This statement declares an entry point into a Fortran subroutine or function subprogram. The entry point is executed with a CALL statement or function reference. The dummy arguments $arg1$, $arg2$, ... are placeholders for the calling arguments passed when the subprogram is executed. This statement should be avoided in modern programs.

EQUIVALENCE Statement:

```
EQUIVALENCE ( var1, var2, ...)
```

Example:

```
EQUIVALENCE ( scr1, iscr1 )
```

Description:
The EQUIVALENCE statement is a specification statement that specifies that all of the variables in the parentheses occupy the same location in memory.

GO TO Statement:

                GO TO *label*

Example:

                GO TO 100

Description:
The GO TO statement transfers control unconditionally to the executable statement that has the specified statement label.

IMPLICIT Statement:

                IMPLICIT *type1* (a_1, a_2, a_3, ...), *type2* (b_1, b_2, b_3, ...), ...

Example:

                IMPLICIT COMPLEX (c,z), LOGICAL (l)

Description:
The IMPLICIT statement is a specification statement that overrides the default typing built into Fortran. It specifies the default type to assume for parameters and variables whose names begin with the specified letters. This statement should never be used in any modern program.

PAUSE Statement:

                PAUSE *prompt*

Example:

                PAUSE 12

Description:
The PAUSE statement is an executable statement that temporarily stops the execution of the Fortran program, until the user resumes it. The prompt is either an integer between 0 and 99999 or a character constant. It is displayed when the PAUSE statement is executed.
    The PAUSE statement has been deleted from the language as of Fortran 95.

Statement Function:

        name(*arg1*,*arg2*,...) = expression containing *arg1*, *arg2*, ...

Example:

    Definition: quad(a,b,c,x) = a * x**2 + b * x + c
    Use: result = 2. * pi * quad(a1,b1,c1,1.5*t)

Description:
The statement function is an older structure that has been replaced by the internal function. It is defined in the declaration section of a program, and may be used only within that program. The arguments *arg1*, *arg2*, etc., are dummy arguments that are replaced by actual values when the function is used.
    Statement functions have been declared obsolescent as of Fortran 95. They should never be used in any modern program.

APPENDIX A

The ASCII Character Set

Each character in the default Fortran character set is stored in 1 byte of memory, so there are 256 possible values for each character variable. The table shown below shows the ASCII character set, with the first two decimal digits of the character number defined by the row, and the third digit defined by the column. Thus, the letter 'R' is on row 8 and column 2, so it is character 82 in the ASCII character set.

	0	1	2	3	4	5	6	7	8	9
0	nul	soh	stx	etx	eot	enq	ack	bel	bs	ht
1	nl	vt	ff	cr	so	si	dle	dc1	dc2	dc3
2	dc4	nak	syn	etb	can	em	sub	esc	fs	gs
3	rs	us	sp	!	"	#	$	%	&	'
4	(	)	*	+	,	-	.	/	0	1
5	2	3	4	5	6	7	8	9	:	;
6	<	=	>	?	@	A	B	C	D	E
7	F	G	H	I	J	K	L	M	N	O
8	P	Q	R	S	T	U	V	W	X	Y
9	Z	[	\	]	^	_	`	a	b	c
10	d	e	f	g	h	i	j	k	l	m
11	n	o	p	q	r	s	t	u	v	w
12	x	y	z	{	|	}	~	del		

APPENDIX B

Fortran/C Interoperability

While Fortran is an excellent language for scientific computation, other languages like C and C++ are better for functions such as connecting to sockets interfaces, manipulating data on a display, working with low-level bit manipulations on interfaces, and calling Web services. Because C and Fortran have different strengths, Fortran 2003 and later introduced a *standard* mechanism to allow Fortran to interoperate with C/C++, so that Fortran could call C/C++ functions and C/C++ could call Fortran subroutines and functions. This had previously been done for many years on an *ad hoc* basis, different for each compiler and operating system. Program written for one computer/compiler combination would not be portable to other systems without rewriting. With the new Fortran/C interoperability features, it is now possible for Fortran to call C and vice versa in a *standard* way that is transportable across different compilers and operating systems.

The key to Fortran/C Interoperability is the ability to make calls between the languages reliably. This requires us to ensure that the two languages use the same calling sequences, to pass data using the same method (value or reference), and to use the same data types for each item in the calling sequence. A new intrinsic module called iso_c_binding has been introduced to Fortran to provide names and structures to guarantee this interoperability.

The intrinsic module iso_c_binding declares all of the features required to make interoperability easy. They include:

1. Fortran kind types to match each interoperable type of C data.[1]
2. Fortran named constants corresponding to common nonprintable C characters, such as the null character (\0), the new line character (\n), and the horizontal tab character (\t).
3. Fortran names for key C procedures and pointers, such as C_F_POINTER to convert a C pointer to a Fortran pointer and C_LOC to return the memory address of a variable.

Table B-1 contains a partial list of the data types and kinds declared in module iso_c_binding. Table B-2 contains a partial list of the constants and generic procedures declared in module iso_c_binding.

---

[1] Only *interoperable* data types are defined in the module and can be used to share data between languages. For example, the unsigned integer data types exist in C, and there is no equivalent in Fortran, so they cannot be used in a mixed language interface. Similarly, Fortran strings contain a hidden length variable. There is no equivalent to the hidden length variable in C string, so it cannot be used in a mixed language interface.

Fortran/C Interoperability

Table B-1:
Selected data types declared in module `iso_c_binding`

Fortran type	Fortran kind	C type
INTEGER	C_INT	int
	C_SHORT	short int
	C_LONG	long inrt
	C_LONG_LONG	long long int
	C_SIGNED_CHAR	signed char
		unsigned char
	C_SIZE_T	size_t
	C_INT8_T	int8_t
	C_INT16_T	int16_t
	C_INT32_T	int32_t
	C_INT64_T	int64_t
REAL	C_FLOAT	float
	C_DOUBLE	double
	C_LONG_DOUBLE	long double
COMPLEX	C_FLOAT_COMPLEX	float _Complex
	C_DOUBLE_COMPLEX	double _Complex
	C_LONG_DOUBLE_COMPLEX	long double _Complex
LOGICAL	C_BOOL	_Bool
CHARACTER	C_CHAR	char

Table B-2:
Constants and procedures declared in module `iso_c_binding`

Name	Description
Named constants	
C_NULL_CHAR	Null character ('\0')
C_ALERT	Alert ('\a')
C_BACKSPACE	Backspace ('\b')
C_FORMFEED	Form feed ('\f')
C_NEW_LINE	New Line ('\n')
C_CARRIAGE_RETURN	Carriage return ('\r')
C_HORIZONTAL_TAB	Horizontal tab ('\t')
C_VERTICAL_TAB	Vertical tab ('\v')
Procedures	
C_ASSOCIATED	Function to test if a C pointer is associated.
C_F_POINTER	Function to convert a C pointer to a Fortran pointer.

(continued)

Name	Description
C_F_PROCPOINTER	Function to convert a C function pointer to a Fortran procedure pointer.
C_FUNLOC	Returns the address in memory of a C function.
C_LOC	Returns the address in memory of a C data item.
C_SIZEOF	Returns the size of a C data item in bytes.
Types to interoperate with C pointers	
C_PTR	Derived type representing any C pointer type.
C_FUNPTR	Derived type representing any C function pointer type.
C_NULL_PTR	The value of a null C pointer.
C_NULL_FUNPTR	The value of a null C function pointer.

Fortran programs wishing to use the iso_c_binding module must declare it using a USE statement with an INTRINSIC attribute, so that the compiler knows that the module is built in and not externally defined:

```
USE, INTRINSIC :: iso_c_binding
```

B.1

DECLARING INTEROPERABLE DATA TYPES

A Fortran data type that is to be interoperable with C must be declared using one of the KINDs defined in Table B-1. For example, an integer variable that is to be exchanged between a Fortran program and a C function might be declared as follows:

```
INTEGER(KIND=C_INT) :: ival
```

Similarly, floating-point variable that is to be exchanged between a Fortran program and a C function might be declared as follows:

```
REAL(KIND=C_FLOAT) :: value
```

Note that the LOGICAL and CHARACTER data types present special problems in interoperable programs. In Fortran, a LOGICAL value is true if the most significant bit is 1, which means that the value is negative. Normally, logical true values are represented by a −1, which has all bits set to 1, and logical false values are represented by a 0, which has all bits set to 0. In contrast, a C _Bool data type is defined as true when it has the value 1, and false when it has the value 0. Some compilers have a compile-time switch that can change the way Fortran logicals are represented to make them compatible with C _Bool values.[2] Otherwise, you must be careful how you interpret logical values that are passed between languages.

---

[2] In Intel Fortran, the compile-time option "fpscomp logicals" changes the definition of a Fortran logical to make it compatible with C.

Character data also has a problem when being passed between languages. There is no equivalent to a Fortran character string in C. The Fortran data type has a hidden length argument, so that the language knows how long a string is. Since hidden arguments cannot be passed between languages, Fortran character strings cannot be used in calls between languages.

The C language uses an array of characters (type `char`) to represent a string, and the language knows where the string ends by looking for a null character ('\0') at the end of the string. A Fortran string can be passed to a C program by adding a null character at the end of the data and calling the C function with that string. A C string can only be passed to a Fortran program by declaring an array of characters, and having the C array map into the Fortran character array. The Fortran procedure must then look for the null character to know where the string ends.

Arrays in Fortran are 1-based by default, but this can be overridden in `DIMENSION` attribute. Arrays in C are always 0-based. Therefore, it is simpler and more understandable if arrays to be passed between Fortran and C are declared as starting from 0 instead of 1:

```
REAL(KIND=C_FLOAT),DIMENSION(0:10) :: array
```

Derived data types can also be passed between the languages, as long as the type structure is declared with the special `BIND(C)` attribute. This attribute ensures that the data in the Fortran structure is laid out exactly the same way as the data in corresponding C `struct`, so that the data can be read properly in the other language. For example, suppose that we wanted to include one integer and two single-precision real values in a derived data type. This type can be made compatible with C by declaring it as follows:

```
TYPE, BIND(C) :: my_type
   INTEGER(KIND=C_INT) :: count
   REAL(KIND=C_FLOAT) :: data1
   REAL(KIND=C_FLOAT) :: data2
END TYPE
```

This data structure would map exactly to a C `struct` declared as follows:

```
typedef struct {
   int count;
   float data1;
   float data2;
} MyType
```

B.2

DECLARING INTEROPERABLE PROCEDURES

A Fortran procedure can be declared as interoperable with C by using the `BIND(C)` attribute. This attribute should appear after the calling arguments in the subroutine or function. Similarly, a Fortran interface describing a C function should be declared with the `BIND(C)` attribute, so that the compiler knows that the function being called has

C syntax. In either case, a procedure declared with the BIND(C) attribute has the following special characteristics:

1. The external name for the procedure is what a C compiler would use, with the Fortran procedure name converted to lowercase.
2. Arguments are passed and received by *reference*, which is the Fortran standard, instead of the C default of passing by value.
3. Only interoperable arguments are allowed, which means that all variables must have one of the kinds defined in Table B-1.
4. No hidden arguments are allowed, so a Fortran character string passed to a C function will not have a hidden length, and must be terminated with a null character.

Fortran supports two type of procedures: subroutines and functions. In contrast, C only supports functions. However, C functions that have been declared void do not return a value, so they are like subroutines. Fortran subroutines correspond to void C functions, and Fortran functions correspond to C functions that return a value.

B.3

SAMPLE PROGRAMS—FORTRAN CALLING C

A simple example of Fortran program calling a C function is shown below. The Fortran program defines an interface for the C function, and declares all calling arguments as kinds found in module iso_c_binding.

Figure B-1 shows a simple Fortran program that reads in two floating-point values and calls a C function to calculate the sum of the two values. The resulting sum is returned to the Fortran program and printed out.

FIGURE B-1
A simple Fortran program that calls a C function to perform a calculation, and displays the result of the calculation.

```
PROGRAM fortran_calls_c
USE, INTRINSIC :: iso_c_binding

! Declare an interface for the C function
INTERFACE

    SUBROUTINE calc(a, b, c) BIND(C)
    USE, INTRINSIC :: iso_c_binding
    REAL(KIND=C_FLOAT) :: a, b, c
    END SUBROUTINE calc

END INTERFACE

! Get data
WRITE(*,*) 'Enter a:'
READ (*,*) a
WRITE(*,*) 'Enter b:'
READ (*,*) b
```

(*continued*)

(*concluded*)

```fortran
! Call C function
CALL calc(a, b, c)

! Write output
WRITE (*,*) 'In Fortran: a + b = ', c

END PROGRAM fortran_calls_c
```

Note that the Fortran program defines an interface for the C function. Because the function is void, the Fortran interface is a subroutine. The subroutine is declared as a C function by the BIND(C) clause on the SUBROUTINE statement.

The corresponding C function is shown in Figure B-2. This function passes the command line arguments by reference, and adds variables a and b, storing and returning the result in variable c. It also prints out the result of the calculation in C using a printf statement.

FIGURE B-2
The C function calc.

```c
void calc (float *a, float *b, float *c)
{
    // Sum a and b

    *c = *a + *b;
    // Tell user
    printf(" In C: a + b = %f\n", *c);

}
```

The way that the Fortran program and the C function are compiled and linked is compiler and operating system dependent. For Intel Fortran and Microsoft C running on Windows, we first compile the C function into an object file, and then compile the Fortran program, supplying the C function object file name on the command line.

```
C:\Data\book\fortran\appB\fortran_calls_c>cl /c calc.c
Microsoft (R) C/C++ Optimizing Compiler Version 18.00.40629 for x86
Copyright (C) Microsoft Corporation. All rights reserved.

calc.c

C:\Data\book\fortran\appB\fortran_calls_c>ifort /standard-semantics
fortran_calls_c.f90 calc.obj
Intel(R) Visual Fortran Compiler for applications running on IA-32,
Version 16.0.3.207 Build 20160415
Copyright (C) 1985-2016 Intel Corporation. All rights reserved.

Microsoft (R) Incremental Linker Version 12.00.40629.0
Copyright (C) Microsoft Corporation. All rights reserved.

-out:fortran_calls_c.exe
-subsystem:console
fortran_calls_c.obj
calc.obj
```

When the program is executed, the results are:

```
C:\Data\book\fortran\appB\fortran_calls_c>fortran_calls_c
 Enter a:
2
 Enter b:
4
 In C: a + b = 6.000000
 In Fortran: a + b =  6.000000
```

A second example program is shown below. This program illustrates passing user-defined data types and strings between Fortran and C. As before, the Fortran program defines an interface for the C function, and declares all calling arguments as kinds found in module iso_c_binding.

Figure B-3 shows a simple Fortran program that passes a data structure and a string to a C function. The data structure is called my_type, and it is defined in a module, which is used in the main program. The program initializes the values in the data type and the string, and then calls the C function c_sub. It prints out the data before and after the call to c_sub.

FIGURE B-3
A simple Fortran program that calls a C function with a derived data type and a string.

```
MODULE data_types
USE, INTRINSIC :: iso_c_binding
IMPLICIT NONE

TYPE,BIND(C) :: my_type
   INTEGER(KIND=C_INT) :: n
   REAL(KIND=C_FLOAT) :: data1
   REAL(KIND=C_FLOAT) :: data2
END TYPE my_type

END MODULE data_types

PROGRAM fortran_calls_c2
USE, INTRINSIC :: iso_c_binding
USE data_types

TYPE(my_type) :: my_struct
CHARACTER(KIND=C_CHAR),DIMENSION(20) :: c

! Declare an interface for the C function
INTERFACE

    SUBROUTINE c_sub(my_struct, msg) BIND(C)
    USE, INTRINSIC :: iso_c_binding
    USE data_types
    TYPE(my_type) :: my_struct
    CHARACTER(KIND=C_CHAR),DIMENSION(20) :: msg
    END SUBROUTINE c_sub

END INTERFACE

! Initialize data
my_struct%n = 3
```

(*continued*)

(*concluded*)

```fortran
my_struct%data1 = 6
my_struct%data2 = 0
c(1) = 'H'
c(2) = 'e'
c(3) = 'l'
c(4) = 'l'
c(5) = 'o'
c(6) = C_NULL_CHAR

! Write output before the call
WRITE (*,*) 'Output before the call:'
WRITE (*,*) 'my_struct%n     = ', my_struct%n
WRITE (*,*) 'my_struct%data1 = ', my_struct%data1
WRITE (*,*) 'my_struct%data2 = ', my_struct%data2

! Call C function
CALL c_sub(my_struct, c)

! Write output after the call
WRITE (*,*) 'Output after the call:'
WRITE (*,*) 'my_struct%n     = ', my_struct%n
WRITE (*,*) 'my_struct%data1 = ', my_struct%data1
WRITE (*,*) 'my_struct%data2 = ', my_struct%data2

END PROGRAM fortran_calls_c2
```

The C function multiplies the values n and data1 in the structure, and stores the result in data2. It also prints out the string passed to the C function. The C function is shown in Figure B-4.

FIGURE B-4
The C function c_sub.

```c
typedef struct {
   int n;
   float data1;
   float data2;
} MyType;

void c_sub (MyType *my_struct, char c[])
{
   // Multiply n * data1 and store in data2
   my_struct->data2 = my_struct->n * my_struct->data1;

   // Print the character string
   printf(" String = %s\n", c);
}
```

For Intel Fortran and Microsoft C running on Windows, the Fortran main program and C function are compiled as shown below:

```
C:\Data\book\fortran\appB\fortran_calls_c2>cl /c c_sub.c
Microsoft (R) C/C++ Optimizing Compiler Version 18.00.40629 for x86
Copyright (C) Microsoft Corporation.  All rights reserved.
```

c_sub.c

```
C:\Data\book\fortran\appB\fortran_calls_c2>ifort /standard-semantics
data_types.f90 fortran_calls_c2.f90 c_sub.obj /Fefortran_calls_c2.exe
Intel(R) Visual Fortran Compiler for applications running on IA-32, Version
16.0.3.207 Build 20160415
Copyright (C) 1985-2016 Intel Corporation.  All rights reserved.

Microsoft (R) Incremental Linker Version 12.00.40629.0
Copyright (C) Microsoft Corporation.  All rights reserved.

-out:fortran_calls_c2.exe
-subsystem:console
fortran_calls_c2.obj
calc.obj
```

When the program is executed, the results are:

```
C:\Data\book\fortran\appB\fortran_calls_c2>fortran_calls_c2
Output before the call:
 my_struct%n     =    3
 my_struct%data1 =    6.000000
 my_struct%data2 =    0.000000
 String = Hello
Output after the call:
 my_struct%n     =    3
 my_struct%data1 =    6.000000
 my_struct%data2 =    9.000000
```

B.4

SAMPLE PROGRAM—C CALLING FORTRAN

C main programs can also call Fortran subroutines or functions using the Fortran/C interoperability features. The following program illustrates a C main function calling a Fortran subroutine. The Fortran subroutine accepts three arguments, and multiplies the first two together, saving the result in the third argument, as shown below:

FIGURE B-5
A Fortran subroutine that can be called by a C main program.

```fortran
SUBROUTINE my_sub(a, b, c) BIND(C)
USE, INTRINSIC :: iso_c_binding
IMPLICIT NONE
REAL(KIND=C_FLOAT) :: a, b, c

c = a * b

END SUBROUTINE my_sub
```

The C main program that calls the subroutine is shown below:

FIGURE B-6
The C main program that calls the Fortran subroutine.

```
int main ()
{
   float a = 3;
   float b = 4;
   float c;

   /* Call the Fortran subroutine */
   my_sub( &a, &b, &c );

   printf("a = %f\n", a);
   printf("b = %f\n", b);
   printf("c = %f\n", c);

   return 0;
}
```

As we have stated before, the steps to compile and execute a combined C/Fortran program vary depending on compiler and operating system. For Intel Fortran and Microsoft C, the Fortran and C are compiled as shown below:

```
C:\Data\book\fortran\appB\c_calls_fortran>ifort /c my_sub.f90
Intel(R) Visual Fortran Compiler for applications running on IA-32, Version
16.0.3.207 Build 20160415
Copyright (C) 1985-2016 Intel Corporation.  All rights reserved.

C:\Data\book\fortran\appB\c_calls_fortran>cl cmain.c my_sub.obj
Microsoft (R) C/C++ Optimizing Compiler Version 18.00.40629 for x86
Copyright (C) Microsoft Corporation.  All rights reserved.

cmain.c
Microsoft (R) Incremental Linker Version 12.00.40629.0
Copyright (C) Microsoft Corporation.  All rights reserved.

/out:cmain.exe
cmain.obj
my_sub.obj
```

When the program is executed, the results are:

```
C:\Data\book\fortran\appB\c_calls_fortran>cmain
a = 3.000000
b = 4.000000
c = 12.000000
```

APPENDIX C

Fortran Intrinsic Procedures

This appendix describes the intrinsic procedures built into the Fortran language, and provides some suggestions for their proper use. The majority of Fortran intrinsic procedures are functions, although there are a few intrinsic subroutines.

C.1
CLASSES OF INTRINSIC PROCEDURES

Fortran intrinsic procedures can be broken down into three classes: elemental, inquiry, or transformational. An **elemental function**[1] is one that is specified for scalar arguments, but that may also be applied to array arguments. If the argument of an elemental function is a scalar, then the result of the function will be a scalar. If the argument of the function is an array, then the result of the function will be an array of the same shape as the input argument. If there is more than one input argument, all of the arguments must have the same shape. If an elemental function is applied to an array, the result will be the same as if the function were applied to each element of the array on an element-by-element basis.

An **inquiry function** or **inquiry subroutine** is a procedure whose value depends on the properties of an object being investigated. For example, the function `PRESENT(A)` is an inquiry function that returns a true value if the optional argument A is present in a procedure call. Other inquiry functions can return properties of the system used to represent real numbers and integers on a particular processor.

A **transformational function** is a function that has one or more array-valued arguments or an array-valued result. Unlike elemental functions that operate on an element-by-element basis, transformational functions operate on arrays as a whole. The output of a transformational function will often not have the same shape as the input arguments. For example, the function `DOT_PRODUCT` has two vector input arguments of the same size, and produces a scalar output.

C.2
ALPHABETICAL LIST OF INTRINSIC PROCEDURES

Table C-1 contains an alphabetical listing of the intrinsic procedures included in Fortran. The table is organized into five columns. The first column of the table

---

[1] One intrinsic subroutine is also elemental.

contains the *generic name* of each procedure, and its calling sequence. The calling sequence is represented by the keywords associated with each argument. Mandatory arguments are shown in roman type, and optional arguments are shown in italics. The use of keywords is optional, but they must be supplied for optional arguments if earlier optional arguments in the calling sequence are missing, or if the arguments are specified in a nondefault order (see Section 13.3). For example, the function SIN has one argument, and the keyword of the argument is X. This function can be invoked either with or without the keyword, so the following two statements are equivalent.

```
result = sin(X=3.141593)
result = sin(3.141593)
```

Another example is the function MAXVAL. This function has one required argument and two optional arguments:

```
MAVXAL ( ARRAY, DIM, MASK )
```

If all three calling values are specified in that order, then they may be simply included in the argument list without the keywords. However, if the MASK is to be specified without DIM, then keywords must be used.

```
value = MAVXAL ( array, MASK=mask )
```

The types of the most common argument keywords are as shown below (any kind of the specified type may be used):

A	Any
ARRAY	Any array
BACK	Logical
CHAR	Character
COARRAY	Any coarray
DIM	Integer
I	Integer
KIND	Integer
MASK	Logical
SCALAR	Any scalar
STRING	Character
X, Y	Numeric (integer, real, or complex)
Z	Complex

For the types of other keywords, refer to the detailed procedure descriptions below.

The second column contains the *specific name* of an intrinsic function, which is the name by which the function must be called if it is to appear in an INTRINSIC statement and be passed to another procedure as an actual argument. If this column is blank, then the procedure does not have a specific name, and so may not be used as a calling argument. The types of arguments used with the specific functions are:

c, c1, c2, ...	Default complex
d, d1, d2, ...	Double-precision real
i, i1, i2, ...	Default integer
r, r1, r2, ...	Default real

```
1, 11, 12, ...                Logical
str1, str2, ...               Character
```

The third column contains the type of the value returned by the procedure if it is a function. Obviously, intrinsic subroutines do not have a type associated with them. The fourth column is a reference to the section of this Appendix in which the procedure is described, and the fifth column is for notes that are found at the end of the Table.

Table C-1:
Specific and generic names for all Fortran intrinsic procedures

Generic name, keyword(s), and calling sequence	Specific name	Function type	Section	Notes
ABS(A)		Argument type	C.3	
	ABS(r)	Default real		
	CABS(c)	Default real		2
	DABS(d)	Double precision		
	IABS(i)	Default integer		
ACHAR(I,*KIND*)		Character(1)	C.7	
ACOS(X)		Argument type	C.3	
	ACOS(r)	Default real		
	DACOS(d)	Double precision		
ACOSH(X)		Argument type	C.3	
ADJUSTL(STRING)		Character	C.7	
ADJUSTR(STRING)		Character	C.7	
AIMAG(Z)	AIMAG(c)	Real	C.3	
AINT(A,*KIND*)		Argument type	C.3	
	AINT(r)	Default real		
	DINT(d)	Double precision		
ALL(MASK,*DIM*)		Logical	C.8	
ALLOCATED(SCALAR)		Logical	C.9	
ANINT(A,*KIND*)		Argument type	C.3	
	ANINT(r)	Real		
	DNINT(d)	Double precision		
ANY(MASK,*DIM*)		Logical	C.8	
ASIN(X)	ASIN(r)	Argument type		
	ASIN(r)	Real		
	DASIN(d)	Double precision		
ASINH(X)		Argument type	C.3	

(continued)

Fortran Intrinsic Procedures

Generic name, keyword(s), and calling sequence	Specific name	Function type	Section	Notes
ASSOCIATED(POINTER,*TARGET*)		Logical	C.9	
ATAN(X,*Y*)		Argument type	C.3	
	ATAN(r)	Real		
	DATAN(d)	Double precision		
ATAN2(Y,X)		Argument type	C.3	
	ATAN2(r2,r1)	Real		
	DATAN2(d2,d1)	Double precision		
ATANH(X)		Argument type	C.3	
BESSEL_J0(X)		Argument type	C.3	
BESSEL_J1(X)		Argument type	C.3	
BESSEL_JN(N,X)		Argument type	C.3	
BESSEL_JN(N1,N2,X)		Argument type	C.3	
BESSEL_Y0(X)		Argument type	C.3	
BESSEL_Y1(X)		Argument type	C.3	
BESSEL_YN(N,X)		Argument type	C.3	
BESSEL_YN(N1,N2,X)		Argument type	C.3	
BGE(I,J)		Logical	C.6	
BGT(I,J)		Logical	C.6	
BIT_SIZE(I)		Integer	C.4	
BLE(I,J)		Logical	C.6	
BLT(I,J)		Logical	C.6	
BTEST(I,POS)		Logical	C.6	
CEILING(A,*KIND*)		Integer	C.3	
CHAR(I,*KIND*)		Character(1)	C.7	
CMPLX(X,*Y*,*KIND*)		Complex	C.3	
COMMAND_ARGUMENT_COUNT()		Integer	C.5	
CONGJ(Z)	CONJG(c)	Complex	C.3	
COS(X)		Argument type	C.3	
	CCOS(c)	Complex		
	COS(r)	Real		
	DCOS(d)	Double precision		
COSH(X)		Argument type	C.3	
	COSH(r)	Real		
	DCOSH(d)	Double precision		

(continued)

Generic name, keyword(s), and calling sequence	Specific name	Function type	Section	Notes
COSHAPE(COARRAY,*KIND*)		Integer	C.11	
COUNT(MASK,*DIM*)		Integer	C.8	
CPU_TIME(TIME)		Subroutine	C.5	
CSHIFT(ARRAY,*SHIFT,DIM*)		Array type	C.8	
DSHIFTL(I,J,SHIFT)		Integer	C.6	
DSHIFTR(I,J,SHIFT)		Integer	C.6	
DATE_AND_TIME(*DATE,TIME,ZONE,VALUES*)		Subroutine	C.5	
DBLE(A)		Double precision	C.3	
DIGITS(X)		Integer	C.4	
DIM(X,Y)		Argument type	C.3	
	DDIM(d1,d2)	Double precision		
	DIM(r1,r2)	Real		
	IDIM(i1,i2)	Integer		
DOT_PRODUCT(VECTOR_A,VECTOR_B)		Argument type	C.3	
DPROD(X,Y)	DPROD(x1,x2)	Double precision	C.3	
EOSHIFT(ARRAY,SHIFT,*BOUNDARY,DIM*)		Array type	C.8	
EPSILON(X)		Real	C.4	
ERF(X)		Argument type	C.3	
ERFC(X)		Argument type	C.3	
ERFC_SCALED(X)		Argument type	C.3	
EXECUTE_COMMAND_LINE(COMMAND, *WAIT,EXITSTAT,CMDSTAT,CMDMSG*)		Argument type	C.3	
EXP(X)		Argument type	C.3	
	CEXP(c)	Complex		
	DEXP(d)	Double precision		
	EXP(r)	Real		
EXPONENT(X)		Integer	C.4	
FINDLOC(ARRAY,VALUE,*DIM,MASK, KIND,BACK*)		Integer	C.8	
FLOOR(A,*KIND*)		Integer	C.3	4
FRACTION(X)		Real	C.4	
GAMMA(X)		Argument type	C.3	
GET_COMMAND(*COMMAND,LENGTH,STATUS*)			C.5	
GET_COMMAND_ARGUMENT(NUMBER, *COMMAND,LENGTH,STATUS*)			C.5	

(*continued*)

Fortran Intrinsic Procedures

Generic name, keyword(s), and calling sequence	Specific name	Function type	Section	Notes
GET_ENVIRONMENT_VARIABLE(NAME, *VALUE*, *LENGTH*, *STATUS*, *TRIM_NAME*)			C.5	
HUGE(X)		Argument type	C.4	
HYPOT(X,Y)		Argument type	C.3	
IACHAR(C)		Integer	C.7	
IALL(ARRAY, *DIM*, *MASK*)		Integer	C.6	
IANY(ARRAY, *DIM*, *MASK*)		Integer	C.6	
IAND(I,J)		Integer	C.6	
IBCLR(I,POS)		Argument type	C.6	
IBITS(I,POS,LEN)		Argument type	C.6	
IBSET(I,POS)		Argument type	C.6	
ICHAR(C)		Integer	C.7	
IEOR(I,J)		Argument type	C.6	
IMAGE_INDEX(COARRAY, *SUB*)		Integer	C.11	
INDEX(STRING,SUBSTRING, *BACK*)	INDEX(str1,str2)	Integer	C.7	
INT(A, *KIND*)		Integer	C.3	
	IDINT(i)	Integer		1
	IFIX(r)	Integer		1
IOR(I,J)		Argument type	C.6	
IPARITY(ARRAY, *DIM*, *MASK*)		Argument type	C.6	
IS_IOSTAT_END(I)		Logical	C.5	
IS_IOSTAT_EOR(I)		Logical	C.5	
ISHFT(I,SHIFT)		Argument type	C.6	
ISHFTC(I,SHIFT, *SIZE*)		Argument type	C.6	
KIND(X)		Integer	C.4	
LBOUND(ARRAY, *DIM*, *KIND*)		Integer	C.8	
LCOBOUND(COARRAY, *DIM*, *KIND*)		Integer	C.11	
LEADZ(I,J,SHIFT)		Integer	C.6	
LEN(STRING)	LEN(str)	Integer	C.7	
LEN_TRIM(STRING)		Integer	C.7	
LGE(STRING_A,STRING_B)		Logical	C.7	
LGT(STRING_A,STRING_B)		Logical	C.7	
LLE(STRING_A,STRING_B)		Logical	C.7	
LLT(STRING_A,STRING_B)		Logical	C.7	
LOG(X)		Argument type	C.3	
	ALOG(r)	Real		
	CLOG(c)	Complex		

(*continued*)

Generic name, keyword(s), and calling sequence	Specific name	Function type	Section	Notes
	DLOG(d)	Double precision		
LOG10(X)		Argument type	C.3	
	ALOG10(r)	Real		
	DLOG10(d)	Double precision		
LOG_GAMMA(X)		Argument type	C.3	
LOGICAL(L,*KIND*)		Logical	C.3	
MASKL(I)		Integer	C.6	
MASKR(I)		Integer	C.6	
MATMUL(MATRIX_A,MATRIX_B)		Argument type	C.3	
MAX(A1,A2,*A3*,...)		Argument type	C.3	
	AMAX0(i1,i2,...)	Real		1
	AMAX1(r1,r2,...)	Real		1
	DMAX1(d1,d2,...)	Double precision		1
	MAX0(i1,i2,...)	Integer		1
	MAX1(r1,r2,...)	Integer		1
MAXEXPONENT(X)		Integer	C.4	
MAXLOC(ARRAY,*DIM*,*MASK*,*KIND*,*BACK*)		Integer	C.8	
MAXVAL(ARRAY,*DIM*,*MASK*)		Argument type	C.8	
MERGE(TSOURCE,FSOURCE,MASK)		Argument type	C.8	
MERGE_BITS(I,J,MASK)		Integer	C.6	
MIN(A1,A2,*A3*,...)		Argument type	C.3	
	AMIN0(i1,i2,...)	Real		1
	AMIN1(r1,r2,...)	Real		1
	DMIN1(d1,d2,...)	Double precision		1
	MIN0(i1,i2,...)	Integer		1
	MIN1(r1,r2,...)	Integer		1
MINEXPONENT(X)		Integer	C.4	
MINLOC(ARRAY,*DIM*,*MASK*,*KIND*,*BACK*)		Integer	C.8	
MINVAL(ARRAY,*DIM*,*MASK*)		Argument type	C.8	
MOD(A,P)		Argument type	C.3	
	AMOD(r1,r2)	Real		
	MOD(i,j)	Integer		
	DMOD(d1,d2)	Double precision		

(*continued*)

Generic name, keyword(s), and calling sequence	Specific name	Function type	Section	Notes
MODULO(A,P)		Argument type	C.3	
MOVE_ALLOC(FROM,TO)		Subroutine	C.10	
MVBITS(FROM,FROMPOS,LEN,TO,TOPOS)		Subroutine	C.6	
NEAREST(X,S)		Real	C.3	
NEW_LINE(CHAR)		Character	C.7	
NINT(A,*KIND*)		Integer	C.3	
	IDNINT(i)	Integer		
	NINT(x)	Integer		
NORM2(X,Y)		Argument type	C.3	
NOT(I)		Argument type	C.6	
NULL(*MOLD*)		Pointer	C.8	
NUM_IMAGES		Integer	C.11	
PACK(ARRAY,MASK,*VECTOR*)		Argument type	C.8	
PARITY(MASK,*DIM*)		Argument type	C.8	
POPCNT(I)		Integer	C.6	
POPPAR(I)		Integer	C.6	
PRECISION(X)		Integer	C.4	
PRESENT(A)		Logical	C.9	
PRODUCT(ARRAY,*DIM,MASK*)		Argument type	C.8	
RADIX(X)		Integer	C.4	
RANDOM_NUMBER(HARVEST)		Subroutine	C.3	
RANDOM_SEED(*SIZE,PUT,GET*)		Subroutine	C.3	
RANGE(X)		Integer	C.4	
REAL(A,*KIND*)		Real	C.3	
	FLOAT(i)	Real		1
	SNGL(d)	Real		1
REPEAT(STRING,NCOPIES)		Character	C.7	
RESHAPE(SOURCE,SHAPE,*PAD,ORDER*)		Argument type	C.8	
RRSPACING(X)		Argument type	C.4	
SCALE(X,I)		Argument type	C.4	
SCAN(STRING,SET,*BACK*)		Integer	C.7	
SELECTED_CHAR_KIND(NAME)		Integer	C.4	
SELECTED_INT_KIND(R)		Integer	C.4	
SELECTED_REAL_KIND(*P,R*)		Integer	C.4	3
SET_EXPONENT(X,I)		Argument type	C.4	
SHAPE(SOURCE,*KIND*)		Integer	C.8	
SIGN(A,B)		Argument type	C.3	

(*continued*)

Generic name, keyword(s), and calling sequence	Specific name	Function type	Section	Notes
	DSIGN(d1,d2)	Double precision		
	ISIGN(i1,i2)	Integer		
	SIGN(r1,r2)	Real		
SIN(X)		Argument type	C.3	
	CSIN(c)	Complex		
	DSIN(d)	Double precision		
	SIN(r)	Real		
SINH(X)		Argument type	C.3	
	DSINH(d)	Double precision		
	SINH(r)	Real		
SIZE(ARRAY, DIM)		Integer	C.8	
SHIFTA(I,SHIFT)		Integer	C.6	
SHIFTL(I,SHIFT)		Integer	C.6	
SHIFTR(I,SHIFT)		Integer	C.6	
SPACING(X)		Argument type	C.4	
SPREAD(SOURCE, DIM, NCOPIES)		Argument type	C.8	
SQRT(X)		Argument type	C.3	
	CSQRT(c)	Complex		
	DSQRT(d)	Double precision		
	SQRT(r)	Real		
STORAGE_SIZE(X,KIND)		Argument type	C.9	
SUM(ARRAY, DIM, MASK)		Argument type	C.8	
SYSTEM_CLOCK(COUNT, COUNT_RATE, COUNT_MAX)		Subroutine	C.5	
TAN(X)		Argument type	C.3	
	DTAN(d)	Double precision		
	TAN(r)	Real		
TANH(X)		Argument type	C.3	
	DTANH(d)	Double precision		
	TANH(r)	Real		
THIS_IMAGE(COARRAY,DIM)		Integer	C.11	
TINY(X)		Real	C.4	
TRAILZ(I,J,SHIFT)		Integer	C.6	

(*continued*)

Generic name, keyword(s), and calling sequence	Specific name	Function type	Section	Notes
TRANSFER(SOURCE, MOLD, *SIZE*)		Argument type	C.8	
TRANSPOSE(MATRIX)		Argument type	C.8	
TRIM(STRING)		Character	C.7	
UBOUND(ARRAY, *DIM, KIND*)			C.8	
UCOBOUND(COARRAY, *DIM, KIND*)		Integer	C.11	
UNPACK(VECTOR, MASK, FIELD)		Argument type	C.8	
VERIFY(STRING, SET, *BACK*)		Integer	C.7	

1. These intrinsic functions cannot be passed to procedures as calling arguments.
2. The result of function CABS is real with the same kind as the input complex argument.
3. At least one of P and R must be specified in any given call.

These intrinsic procedures are divided into broad categories based on their functions. Refer to Table C-1 to determine which of the following sections will contain a description of any particular function of interest.

The following information applies to all of the intrinsic procedure descriptions:

1. All arguments of all intrinsic functions have INTENT(IN). In other words, all of the functions are pure. The intent of subroutine arguments are specified in the description of each subroutine.
2. Optional arguments are shown in italics in all calling sequences.
3. When a function has an optional KIND dummy argument, then the function result will be of the kind specified in that argument. If the KIND argument is missing, then the result will be of the default kind. If the KIND argument is specified, it must correspond to a legal kind on the specified processor, or the function will abort. The KIND argument is always an integer.
4. When a procedure is said to have two arguments of the same type, it is understood that they must also be of the same kind. If this is not true for a particular procedure, the fact will be explicitly mentioned in the procedure description.
5. The lengths of arrays and character strings will be shown by an appended number in parentheses. For example, the expression

    Integer(*m*)

    implies that a particular argument is an integer array containing m values.

C.3

MATHEMATICAL AND TYPE CONVERSION INTRINSIC PROCEDURES

ABS(A)
- Elemental function of the same type and kind as A.
- Returns the absolute value of A, |A|.
- If A is complex, the function returns $\sqrt{\text{real}^2 + \text{imag}^2}$.

ACOS(X)
- Elemental function of the same type and kind as X.
- Returns the inverse cosine of X in radians.
- Argument is real of any kind, with $|X| \leq 1.0$, and $0 \leq ACOS(X) \leq \pi$.
- Argument can be complex.

ACOSH(X)
- Elemental function of the same type and kind as X.
- Returns the inverse hyperbolic cosine of X.

AIMAG(Z)
- Real elemental function of the same kind as Z.
- Returns the imaginary part of complex argument Z.

AINT(A, *KIND*)
- Real elemental function.
- Returns A truncated to a whole number. AINT(A) is the largest integer that is smaller than $|A|$, with the sign of A. For example, AINT(3.7) is 3.0, and AINT(-3.7) is -3.0.
- Argument A is real; optional argument *KIND* is integer.

ANINT(A, *KIND*)
- Real elemental function.
- Returns the nearest whole number to A. For example, ANINT(3.7) is 4.0, and AINT(-3.7) is -4.0.
- Argument A is real; optional argument *KIND* is integer.

ASIN(X)
- Elemental function of the same type and kind as X.
- Returns the inverse sine of X in radians.
- Argument is real of any kind, with $|X| \leq 1.0$, and $-\pi/2 \leq ASIN(X) \leq \pi/2$.
- Argument can be complex.

ASINH(X)
- Elemental function of the same type and kind as X.
- Returns the inverse hyperbolic sine of X.

ATAN(X, *Y*)
- Elemental function of the same type and kind as X.
- Returns the inverse tangent of X in radians.
- Argument is real of any kind, with $-\pi/2 \leq ATAN(X) \leq \pi/2$.
- Argument can be complex.
- If the optional *Y* argument is present, this function is the same as ATAN2 (see next function).

ATAN2(Y, X)
- Elemental function of the same type and kind as X.
- Returns the four-quadrant inverse tangent of Y/X in the range $-\pi < ATAN2(Y, X) \leq \pi$.
- X, Y are real of any kind, and must be of same kind.
- Both X and Y cannot be simultaneously 0.

ATANH(X)
- Elemental function of the same type and kind as X.
- Returns the inverse hyperbolic tangent of X.

BESSEL_J0(X)
- Elemental function of the same type and kind as X.
- Returns Bessel function of the first kind, order 0.

BESSEL_J1(X)
- Elemental function of the same type and kind as X.
- Returns Bessel function of the first kind, order 1.

BESSEL_JN(N,X)
- Elemental function of the same type and kind as X.
- Returns Bessel function of the first kind, order n.

BESSEL_JN(N1,N2,X)
- Elemental function of the same type and kind as X.
- Returns Bessel functions of the first kind.

BESSEL_Y0(X)
- Elemental function of the same type and kind as X.
- Returns Bessel function of the second kind, order 0.

BESSEL_Y1(X)
- Elemental function of the same type and kind as X.
- Returns Bessel function of the second kind, order 1.

BESSEL_YN(N,X)
- Elemental function of the same type and kind as X.
- Returns Bessel function of the second kind, order n.

BESSEL_YN(N1,N2,X)
- Elemental function of the same type and kind as X.
- Returns Bessel functions of the second kind.

CEILING(A,*KIND*)
- Integer elemental function.
- Returns the smallest integer $\geq$ A. For example, CEILING(3.7) is 4, and CEILING(-3.7) is −3.
- Argument A is real of any kind; optional argument *KIND* is integer.

CMPLX(X,Y,*KIND*)
- Complex elemental function.
- Returns a complex value as follows:
    1. If X is complex, then Y must not exist, and the value of X is returned.
    2. If X is not complex, and Y doesn't exist, then the returned value is (X,0).
    3. If X is not complex and Y exists, then the returned value is (X,Y).
- X is complex, real, or integer, Y is real or integer, and *KIND* is an integer.

CONJG(Z)
- Complex elemental function of the same kind as Z.
- Returns the complex conjugate of Z.
- Z is complex.

COS(X)
- Elemental function of the same type and kind as X.
- Returns the cosine of X, where X is in radians.
- X is real or complex.

COSH(X)
- Elemental function of the same type and kind as X.
- Returns the hyperbolic cosine of X.
- X is real or complex.

DIM(X,Y)
- Elemental function of the same type and kind as X.
- Returns X-Y if > 0; otherwise returns 0.
- X and Y are integer or real; both must be of the same type and kind.

DBLE(A)
- Double-precision real elemental function.
- Converts value of A to double-precision real.
- A is numeric. If A is complex, then only the real part of A is converted.

DOT_PRODUCT(VECTOR_A,VECTOR_B)
- Transformational function of the same type as VECTOR_A.
- Returns the dot product of numeric or logical vectors.
- Arguments are numeric or logical vectors. Both vectors must be of the same type, kind, and length.

DPROD(X,Y)
- Double-precision real elemental function.
- Returns the double-precision product of X and Y.
- Arguments X and Y are default real.

ERF(X)
- Elemental function of the same type and kind as X.
- Returns error function.
- X is real.

ERFC(X)
- Elemental function of the same type and kind as X.
- Returns complementary error function.
- X is real.

ERFC_SCALED(X)
- Elemental function of the same type and kind as X.
- Returns scaled complementary error function.
- X is real.

EXP(X)
- Elemental function of the same type and kind as X.
- Returns e^x.
- X is real or complex.

FLOOR(A, *KIND*)
- Integer elemental function.
- Returns the largest integer $\leq$ A. For example, FLOOR(3.7) is 3, and FLOOR(-3.7) is −4.
- Argument A is real of any kind; optional argument *KIND* is integer.

GAMMA(X)
- Elemental function of the same type and kind as X.
- Returns the gamma function.
- X is real.

HYPOT(X,Y)
- Elemental function of the same type and kind as X.
- Returns the Euclidean distance function.
- X and Y are real.

INT(A, *KIND*)
- Integer elemental function.
- This function truncates A and converts it into an integer. If A is complex, only the real part is converted. If A is integer, this function changes the kind only.
- A is numeric; optional argument *KIND* is integer.

LOG(X)
- Elemental function of the same type and kind as X.
- Returns $\log_e(x)$.
- X is real or complex. If real, X > 0. If complex, X $\neq$ 0.

LOG10(X)
- Elemental function of the same type and kind as X.
- Returns $\log_{10}(x)$.
- X is real and positive.

LOG_GAMMA(X)
- Elemental function of the same type and kind as X.
- Returns the logarithm of the absolute value of the gamma function.
- X is real and positive.

LOGICAL(L, *KIND*)
- Logical elemental function.
- Converts the logical value L to the specified kind.
- L is logical, and *KIND* is integer.

MATMUL(MATRIX_A,MATRIX_B)
- Transformational function of the same type and kind as MATRIX_A.

- Returns the *matrix product* of numeric or logical matrices. The resulting matrix will have the same number of rows as MATRIX_A and the same number of columns as MATRIX_B.
- Arguments are numeric or logical matrices. Both matrices must be of the same type and kind, and of compatible sizes. The following constraints apply:
  1. In general, both matrices are of rank 2.
  2. MATRIX_A may be rank 1. If so, MATRIX_B must be rank 2 with only one column.
  3. In all cases, the number of columns in MATRIX_A must be the same as the number of rows in MATRIX_B.

MAX(A1,A2,A3,...)
- Elemental function of same kind as its arguments.
- Returns the maximum value of A1, A2, etc.
- Arguments may be real, integer, or character; all must be of the same type.

MIN(A1,A2,A3,...)
- Elemental function of same kind as its arguments.
- Returns the minimum value of A1, A2, etc.
- Arguments may be real or integer, or character; all must be of the same type.

MOD(A1,P)
- Elemental function of same kind as its arguments.
- Returns the value MOD(A,P) = A - P*INT(A/P) if $P \neq 0$. Results are processor dependent if $P = 0$.
- Arguments may be real or integer; they must be of the same type.
- Examples:

Function	Result
MOD(5,3)	2
MOD(-5,3)	-2
MOD(5,-3)	2
MOD(-5,-3)	-2

MODULO(A1,P)
- Elemental function of same kind as its arguments.
- Returns the modulo of A with respect to P if $P \neq 0$. Results are processor dependent if $P = 0$.
- Arguments may be real or integer; they must be of the same type.
- If $P > 0$, then the function determines the positive difference between A and then next lowest multiple of P. If $P < 0$, then the function determines the negative difference between A and then next highest multiple of P.
- Results agree with the MOD function for two positive or two negative arguments; results disagree for arguments of mixed signs.

- Examples:

Function	Result	Explanation
MODULO(5,3)	2	5 is 2 up from 3
MODULO(-5,3)	1	−5 is 1 up from −6
MODULO(5,-3)	−1	5 is 1 down from 6
MODULO(-5,-3)	−2	−5 is 2 down from −3

NEAREST(X,S)
- Real elemental function.
- Returns the nearest machine-representable number different from X in the direction of S. The returned value will be of the same kind as X.
- X and S are real, and $S \neq 0$.

NINT(A,*KIND*)
- Integer elemental function.
- Returns the nearest integer to the real value A.
- A is real.

NORM2(X,Y)
- Elemental function of the same type and kind as X.
- Returns the L_2 norm.
- X and Y are real.

RANDOM_NUMBER(HARVEST)
- Intrinsic subroutine.
- Returns pseudorandom number(s) from a uniform distribution in the range $0 \leq$ HARVEST < 1. HARVEST may be either a scalar or an array. If it is an array, then a separate random number will be returned in each element of the array.
- Arguments:

| HARVEST | Real | OUT | Holds random numbers. May be scalar or array. |

RANDOM_SEED(*SIZE,PUT,GET*)
- Intrinsic subroutine.
- Performs three functions: (1) restarts the pseudorandom number generator used by subroutine RANDOM_NUMBER, (2) gets information about the generator, and (3) puts a new seed into the generator.
- Arguments:

SIZE	Integer	OUT	Number of integers used to hold the seed (n)
PUT	Integer(m)	IN	Set the seed to the value in *PUT*. Note that $m \geq n$.
GET	Integer(m)	OUT	Get the current value of the seed. Note that $m \geq n$.

- *SIZE* is an integer, and *PUT* and *GET* are integer arrays. All arguments are optional, and at most one can be specified in any given call.

- Functions:
  1. If no argument is specified, the call to RANDOM_SEED restarts the pseudorandom number generator.
  2. If *SIZE* is specified, then the subroutine returns the number of integers used by the generator to hold the seed.
  3. If *GET* is specified, then the current random generator seed is returned to the user. The integer array associated with keyword *GET* must be at least as long as *SIZE*.
  4. If *PUT* is specified, then the value in the integer array associated with keyword *PUT* is set into the generator as a new seed. The integer array associated with keyword *PUT* must be at least as long as *SIZE*.

REAL(A, *KIND*)
- Real elemental function.
- This function converts A into a real value. If A is complex, it converts the real part of A only. If A is real, this function changes the kind only.
- A is numeric; *KIND* is integer.

SIGN(A,B)
- Elemental function of same kind as its arguments.
- Returns the value of A with the sign of B.
- Arguments may be real or integer; they must be of the same type.

SIN(X)
- Elemental function of the same type and kind as X.
- Returns the sine of X, where X is in radians.
- X is real or complex.

SINH(X)
- Elemental function of the same type and kind as X.
- Returns the hyperbolic sine of X.
- X is real or complex.

SQRT(X)
- Elemental function of the same type and kind as X.
- Returns the square root of X.
- X is real or complex.
- If X is real, X must be ≥ 0. If X is complex, then the real part of X must be ≥ 0. If X is purely imaginary, then the imaginary part of X must be ≥ 0.

TAN(X)
- Elemental function of the same type and kind as X.
- Returns the tangent of X, where X is in radians.
- X is real or complex.

TANH(X)
- Elemental function of the same type and kind as X.
- Returns the hyperbolic tangent of X.
- X is real or complex.

C.4
KIND AND NUMERIC PROCESSOR INTRINSIC FUNCTIONS

Many of the functions in this section are based on the Fortran models for integer and real data. These models must be understood in order to make sense of the values returned by the functions.

Fortran uses **numeric models** to insulate a programmer from the physical details of how bits are laid out in a particular computer. For example, some computers use two's complement representations for numbers, while other computers use sign-magnitude representations for numbers. Approximately the same range of numbers can be represented in either case, but the bit patterns are different. The numeric models tell the programmer what range and precision can be represented by a given type and kind of numbers without requiring a knowledge of the physical bit layout on a particular machine.

The Fortran model for an integer i is

$$i = s \times \sum_{k=0}^{q-1} w_k \times r^k \tag{C-1}$$

where r is an integer exceeding one, q is a positive integer, each w_k is a nonnegative integer less than r, and s is $+1$ or -1. The values of r and q determine the set of model integers for a processor. They are chosen to make the model fit as well as possible to the machine on which the program is executed. Note that this model is independent of the actual bit pattern used to store integers on a particular processor.

The value r in this model is the **radix** or base of the numbering system used to represent integers on a particular computer. Essentially all modern computers use a base 2 numbering system, so r is 2. If r is 2, then the value q is one less than the number of bits used to represent an integer (1 bit is used for the sign of the number). For a typical 32-bit integer on a base 2 computer, the model of an integer becomes

$$i = \pm \sum_{k=0}^{30} w_k \times 2^k \tag{C-2}$$

where each w_k is either 0 or 1.

The Fortran model for a real number x is

$$x = \begin{cases} 0 & \text{or} \\ s \times b^e \times \sum_{k=1}^{p} f_k \times b^{-k} \end{cases} \tag{C-3}$$

where b and p are integers exceeding one, each f_k is a nonnegative integer less than b (and f_1 must not be zero), s is $+1$ or -1, and e is an integer that lies between some integer maximum e_{max} and some integer minimum e_{min}. The values of b, p, e_{min}, and e_{max} determine the set of model floating-point numbers. They are chosen to make the model fit as well as possible to the machine on which the program is executed. This model is independent of the actual bit pattern used to store floating-point numbers on a particular processor.

The value b in this model is the **radix** or base of the numbering system used to represent real numbers on a particular computer. Essentially all modern computers use a base 2 numbering system, so b is 2, and each f_k must be either 0 or 1 (f_1 must be 1).

The bits that make up a real or floating-point number are divided into two separate fields, one for the mantissa (the fractional part of the number) and one for the exponent. For a base 2 system, p is the number of bits in the mantissa, and the value of e is stored in a field that is one less than the number of bits in the exponent.[2] Since the IEEE single-precision standard devotes 24 bits to the mantissa and 8 bits to the exponent, p is 24, $e_{max} = 2^7 = 127$, and $e_{min} = -126$. For a typical 32-bit single-precision real number on a base 2 computer, the model of the number becomes

$$x = \begin{cases} 0 & \text{or} \\ \pm 2^e \times \left(\frac{1}{2} + \sum_{k=2}^{24} f_k \times 2^{-k} \right), & -126 \leq e \leq 127 \end{cases} \tag{C-4}$$

The inquiry functions DIGITS, EPSILON, HUGE, MAXEXPONENT, MINEXPONENT, PRECISION, RANGE, RADIX, and TINY all return values related to the model parameters for the type and kind associated with the calling arguments. Of these functions, only PRECISION and RANGE matter to most programmers.

BIT_SIZE(I)
- Integer inquiry function.
- Returns the number of bits in integer I.
- I must be integer.

DIGITS(X)
- Integer inquiry function.
- Returns the number of significant digits in X. (This function returns q from the integer model in Equation C-1, or p from the real model in Equation C-3.)
- X must be integer or real.
- **Caution:** This function returns the number of significant digits in the base of the numbering system used on the computer. For most modern computers, this is base 2, so this function returns the number of significant bits. If you want the number of significant decimal digits, use PRECISION(X) instead.

EPSILON(X)
- Integer inquiry function of the same type as X.
- Returns a positive number that is almost negligible compared to 1.0 of the same type and kind as X. (The returned value is b^{1-p}, where b and p are defined in Equation C-3.)
- X must be real.
- Essentially, EPSILON(X) is the number that, when added to 1.0, produces the next number representable by the given KIND of real number on a particular processor.

---

[2] It is one less than the number of bits in the exponent because 1 bit is reserved for the sign of the exponent.

Fortran Intrinsic Procedures

EXPONENT(X)
- Integer inquiry function of the same type as X. Returns the exponent of X in the base of the computer numbering system. (This is e from the real number model as defined in Equation C-3.)
- X must be real.

FRACTION(X)
- Real elemental function of same kind as X.
- Returns the mantissa or the fractional part of the model representation of X. (This function returns the summation term from Equation C-3.)
- X must be real.

HUGE(X)
- Integer inquiry function of the same type as X.
- Returns the largest number of the same type and kind as X.
- X must be integer or real.

KIND(X)
- Integer inquiry function.
- Returns the kind value of X.
- X may be any intrinsic type.

MAXEXPONENT(X)
- Integer inquiry function.
- Returns the maximum exponent of the same type and kind as X. (The returned value is e_{max} from the model in Equation C-3.)
- X must be real.
- **Caution:** This function returns the maximum exponent in the base of the numbering system used on the computer. For most modern computers, this is base 2, so this function returns the maximum exponent as a base 2 number. If you want the maximum exponent as a decimal value, use RANGE(X) instead.

MINEXPONENT(X)
- Integer inquiry function.
- Returns the minimum exponent of the same type and kind as X. (The returned value is e_{min} from the model in Equation C-3.)
- X must be real.

PRECISION(X)
- Integer inquiry function.
- Returns the number of significant *decimal digits* in values of the same type and kind as X.
- X must be real or complex.

RADIX(X)
- Integer inquiry function.
- Returns the base of the mathematical model for the type and kind of X. Since most modern computers work on a base 2 system, this number will almost always be 2. (This is r in Equation C-1, or b in Equation C-3.)
- X must be integer or real.

RANGE(X)
- Integer inquiry function.
- Returns the *decimal* exponent range for values of the same type and kind as X.
- X must be integer, real, or complex.

RRSPACING(X)
- Elemental function of the same type and kind as X.
- Returns the reciprocal of the relative spacing of the numbers near X. (The result has the value $|x \times b^{-e}| \times b^p$, where b, e, and p are defined as in Equation C-3.)
- X must be real.

SCALE(X,I)
- Elemental function of the same type and kind as X.
- Returns the value $x \times b^I$, where b is the base of the model used to represent X. The base b can be found with the RADIX(X) function; it is almost always 2.
- X must be real, and I must be integer.

SELECTED_CHAR_KIND(STRING)
- Integer transformational function.
- Returns the kind number associated with the character input argument.
- STRING must be character.

SELECTED_INT_KIND(R)
- Integer transformational function.
- Returns the kind number for the smallest integer kind that can represent all integers n whose values satisfy the condition ABS(n) < 10**R. If more than one kind satisfies this constraint, then the kind returned will be the one with the smallest decimal range. If no kind satisfies the requirement, the value -1 is returned.
- R must be an integer.

SELECTED_REAL_KIND(P,R)
- Integer transformational function.
- Returns the kind number for the smallest real kind that has a decimal precision of at least P digits and an exponent range of at least R powers of 10. If more than one kind satisfies this constraint, then the kind returned will be the one with the smallest decimal precision.
- If no real kind satisfies the requirement, a -1 is returned if the requested precision was not available, a -2 is returned if the requested range was not available, and a -3 is returned if neither was available.
- P and R must be integers.

SET_EXPONENT(X,I)
- Elemental function of the same type as X.
- Returns the number whose fractional part is the fractional part of the number X, and whose exponent part is I. If $X = 0$, then the result is 0.
- X is real, and I is integer.

SPACING(X)
- Elemental function of the same type and kind as X.

- Returns the absolute spacing of the numbers near X in the model used to represent real numbers. If the absolute spacing is out of range, then this function returns the same value as TINY(X). (This function returns the value b^{e-p}, where b, e, and p are as defined in Equation C-3, as long as that value is in range.)
- X must be real.
- The result of this function is useful for establishing convergence criteria in a processor-independent manner. For example, we might conclude that a root-solving algorithm has converged when the answer gets within 10 times the minimum representable spacing.

TINY(X)
- Elemental function of the same type and kind as X.
- Returns the smallest positive number of the same type and kind as X. (The returned value is $b^{e_{min}-1}$, where b and e_{min} are as defined in Equation C-3.)
- X must be real.

C.5

SYSTEM ENVIRONMENT PROCEDURES

COMMAND_ARGUMENT_COUNT()
- Intrinsic function.
- Returns the number of command line arguments.
- Arguments:
  None
- The purpose of this is to return the number of command line arguments. Argument 0 is the name of the program being executed, and arguments 1 to n are the actual arguments on the command line.

CPU_TIME(TIME)
- Intrinsic subroutine.
- Returns processor time expended on current program in seconds.
- Arguments:
  TIME   Real   OUT   Processor time
- The purpose of this subroutine is to time sections of code by comparing the processor time before and after the code is executed.
- The definition of the time returned by this subroutine is processor dependent. On most processors, it is the CPU time spent executing the current program.
- On computers with multiple CPUs, TIME may be implemented as an array containing the times associated with each processor.

DATE_AND_TIME(DATE,TIME,ZONE,VALUE)
- Intrinsic subroutine.
- Returns date and time.
- All arguments are optional, but at least one must be included:
  DATE   Character(8)   OUT   Returns a string in the form CCYYMMDD, where CC is century, YY is year, MM is month, and DD is day.

TIME	Character(10)	OUT	Returns a string in the form HHMMSS.SSS, where HH is hour, MM is minute, SS is second, and SSS is millisecond.
ZONE	Character(5)	OUT	Returns a string in the form ±HHMM, where HHMM is the time difference between local time and Coordinated Universal Time (UCT, or GMT).
VALUES	Integer(8)	OUT	See table below for values.

- If a value is not available for *DATE*, *TIME*, or *ZONE*, then the string is blank.
- The information returned in array *VALUES* is:

VALUES(1)	Century and year (e.g., 1996)
VALUES(2)	Month (1–12)
VALUES(3)	Day (1–31)
VALUES(4)	Time zone difference from UTC in minutes.
VALUES(5)	Hour (0–23)
VALUES(6)	Minutes (0–59)
VALUES(7)	Seconds (0–60)
VALUES(8)	Milliseconds (0–999)

- If no information is available for one of the elements of array *VALUES*, that element is set to the most negative representable integer (-HUGE(0)).
- Note that the seconds field ranges from 0 to 60. The extra second is included to allow for leap-seconds.

EXECUTE_COMMAND_LINE(COMMAND, *WAIT, EXITSTAT, CMDSTAT, CMDMSG*)
- Intrinsic subroutine.
- The COMMAND argument is passed to the shell and executed, using the C library's system call. If wait is present and has the value .FALSE., the execution of the command is asynchronous if the system supports it; otherwise, the command is executed synchronously.
- The remaining arguments are optional:

WAIT	Logical	IN	If true, wait for command completion. If false, continue executing immediately while the system command runs.
EXITSTAT	Default integer	OUT	Returns the exit code after the command is executed, as returned by the system library call.
CMDSTAT	Default integer	OUT	Returns 0 if the command line was executed.
CMDMSG	Character	OUT	Return a character error message if an error occurred.

GET_COMMAND(*COMMAND, LENGTH, STATUS*)
- Intrinsic subroutine.
- Returns the entire command line used to start the program.
- All arguments are optional:

COMMAND	Character(*)	OUT	Returns a string containing the command line.
LENGTH	Integer	OUT	Returns the length of the command line.
STATUS	Integer	OUT	Status: 0 = success; −1 = command line present but COMMAND is too short to hold it all; other value = retrieval failed.

GET_COMMAND_ARGUMENT(*NUMBER, VALUE, LENGTH, STATUS*)
- Intrinsic subroutine.
- Returns a specified command argument.
- Argument list:

NUMBER	Integer	IN	Argument number to return, in the range 0 to COMMAND_ARGUMENT_COUNT()
VALUE	Character(*)	OUT	Returns the specified argument.
LENGTH	Integer	OUT	Returns the length of the argument.
STATUS	Integer	OUT	Status: 0 = success; −1 = command line present but COMMAND is too short to hold it all; other value = retrieval failed.

GET_ENVIRONMENT_VARIABLE(*NAME, VALUE, LENGTH, STATUS, TRIM_NAME*)
- Intrinsic subroutine.
- Returns a specified command argument.
- All arguments are optional:

NAME	Character(*)	IN	Name of environment variable to retrieve.
VALUE	Character(*)	OUT	Returns the value of the specified environment variable.
LENGTH	Integer	OUT	Returns the length of the value in characters.
STATUS	Integer	OUT	Status: 0 = success; −1 = command line present but COMMAND is too short to hold it all; 2 = processor does not support environment

			variables; other value = retrieval failed.
`TRIM_NAME`	Logical	IN	If true, ignore trailing blanks in `NAME` when matching to an environment variable; otherwise, include the blanks. If this argument is missing, trailing blanks are ignored.

`IS_IOSTAT_END(I)`
- Intrinsic function.
- Returns true if the value of I is equal to the `IOSTAT_END` flag.
- Arguments:

I	Integer	IN	This is the result of a `READ` operation returned by the `IOSTAT=` clause.

- The purpose of this is to provide a simple way to test for the end-of-file condition during a read operation.

`IS_IOSTAT_EOR(I)`
- Intrinsic function.
- Returns true if the value of I is equal to the `IOSTAT_EOR` flag.
- Arguments:

I	Integer	IN	This is the result of a `READ` operation returned by the `IOSTAT=` clause.

- The purpose of this is to provide a simple way to test for the end-of-record condition during a read operation with `ADVANCE='NO'`.

`SYSTEM_CLOCK(COUNT,COUNT_RATE,COUNT_MAX)`
- Intrinsic subroutine.
- Returns raw counts from the processor's real-time clock. The value in `COUNT` is increased by one for each clock count until `COUNT_MAX` is reached. When `COUNT_MAX` is reached, the value in `COUNT` is reset to 0 on the next clock count. Variable `COUNT_RATE` specifies the number of real-time clock counts per second, so it tells how to interpret the count information.
- Arguments:

`COUNT`	Integer	OUT	Number of counts of the system clock. The starting count is arbitrary.
`COUNT_RATE`	Integer, or Real	OUT	Number of clock counts per second.
`COUNT_MAX`	Integer	OUT	The maximum value for `COUNT`.

- If there is no clock, `COUNT` and `COUNT_RATE` are set to `-HUGE(0)` and `COUNT_MAX` is set to 0.

C.6
BIT INTRINSIC PROCEDURES

The layout of bits within an integer varies from processor to processor. For example, some processors place the most significant bit of a value at the bottom of the memory representing that value, while other processors place the least significant bit of a value at the top of the memory representing that value. To insulate programmers from these machine dependencies, Fortran defines a bit to be a binary digit w located at position k of a nonnegative integer based on a model nonnegative integer defined by

$$j = \sum_{k=0}^{z-1} w_k \times 2^k \tag{C-5}$$

where w_k can be either 0 or 1. Thus, bit 0 is the coefficient of 2^0, bit 1 is the coefficient of 2^1, etc. In this model, z is the number of bits in the integer, and the bits are numbered $0, 1, \ldots, z-1$, regardless of the physical layout of the integer. The least significant bit is considered to be at the right of the model and the most significant bit is considered to be at the left of the model, regardless of the actual physical implementation. Thus, shifting a bit left increases its value, and shifting a bit right decreases its value.

Fortran includes elemental functions and one elemental subroutine that manipulate bits according to this model. Logical operations on bits are performed by the elemental functions IOR, IAND, NOT, and IEOR. Shift operations are performed by the elemental functions ISHFT and ISHFTC. Bit subfields may be referenced by the elemental function IBITS and the elemental subroutine MVBITS. Finally, single-bit processing is performed by the elemental functions BTEST, IBSET, and IBCLR.

BGE(I,J)
- Logical elemental function.
- Determines whether an integer is a bitwise greater than or equal to another integer.
- I and J must be integers of the same kind.

BGT(I,J)
- Logical elemental function.
- Determines whether an integer is a bitwise greater than another integer.
- I and J must be integers of the same kind.

BLE(I,J)
- Logical elemental function.
- Determines whether an integer is a bitwise less than or equal to another integer.
- I and J must be integers of the same kind.

BLT(I,J)
- Logical elemental function.
- Determines whether an integer is a bitwise less than another integer.
- I and J must be integers of the same kind.

BTEST(I,POS)
- Logical elemental function.

- Returns true if bit POS of I is 1, and false otherwise.
- I and POS must be integers, with $0 \le POS < BIT\_SIZE(I)$.

DSHIFTL(I,J,SHIFT)
- Integer elemental function.
- DSHIFTL(I,J,SHIFT) combines bits of I and J. The rightmost SHIFT bits of the result are the leftmost SHIFT bits of J, and the remaining bits are the rightmost bits of I.
- I and J must be integers of the same kind.

DSHIFTR(I,J,SHIFT)
- Integer elemental function.
- DSHIFTR(I,J,SHIFT) combines bits of I and J. The leftmost SHIFT bits of the result are the rightmost SHIFT bits of J, and the remaining bits are the rightmost bits of I.
- I and J must be integers of the same kind.

IALL(ARRAY,*DIM*,*MASK*)
- Transformational function of the type and kind as ARRAY.
- ARRAY is an array of type integer.
- *DIM* is a scalar of type integer, with a value from $1 - n$, where n is the rank of the ARRAY.
- *MASK* is a logical scalar, or a logical array of the same shape as ARRAY.
- The value returned is 1 in any given bit if the corresponding bit in every element of the array was 1; otherwise, it is zero.
- The result is an integer of the same type as ARRAY.

IAND(I,J)
- Elemental function of the same type and kind as I.
- Returns the bit by bit logical AND of I and J.
- I and J must be integers of the same kind.

IANY(ARRAY,*DIM*,*MASK*)
- Transformational function of the type and kind as ARRAY.
- ARRAY is an array of type integer.
- *DIM* is a scalar of type integer, with a value from $1 - n$, where n is the rank of the ARRAY.
- *MASK* is a logical scalar, or a logical array of the same shape as ARRAY.
- The value returned is 1 in any given bit if the corresponding bit in any element of the array was 1; otherwise, it is zero.
- The result is an integer of the same type as ARRAY.

IBCLR(I,POS)
- Elemental function of the same type and kind as I.
- Returns I with bit POS set to 0.
- I and POS must be integers, with $0 \le POS < BIT\_SIZE(I)$.

IBITS(I,POS,LEN)
- Elemental function of the same type and kind as I.

- Returns a right-adjusted sequence of bits extracted from I of length LEN starting at bit POS. All other bits are zero.
- I, POS, and LEN must be integers, with POS + LEN < BIT_SIZE(I).

IBSET(I,POS)
- Elemental function of the same type and kind as I.
- Returns I with bit POS set to 1.
- I and POS must be integers, with $0 \leq$ POS $<$ BIT_SIZE(I).

IEOR(I,J)
- Elemental function of the same type and kind as I.
- Returns the bit by bit exclusive OR of I and J.
- I and J must be integers of the same kind.

IOR(I,J)
- Elemental function of the same type and kind as I.
- Returns the bit by bit inclusive OR of I and J.
- I and J must be integers of the same kind.

IPARITY(ARRAY,*DIM*,*MASK*)
- Transformational function of the type and kind as ARRAY.
- ARRAY is an array of type integer.
- *DIM* is a scalar of type integer, with a value from $1 - n$, where n is the rank of the ARRAY.
- *MASK* is a logical scalar, or a logical array of the same shape as ARRAY.
- The value returned in any given bit is the exclusive OR of that bit in all elements of the input array.
- The result is an integer of the same type as ARRAY.

ISHFT(I,SHIFT)
- Elemental function of the same type and kind as I.
- Returns I logically shifted to the left (if SHIFT is positive) or right (if SHIFT is negative). The empty bits are filled with zeros.
- I must be an integer.
- SHIFT must be an integer, with ABS(SHIFT) <= BIT_SIZE(I).
- A shift to the left implies moving the bit in position i to position $i+1$, and a shift to the right implies moving the bit in position i to position $i-1$.

ISHFTC(I,SHIFT,*SIZE*)
- Elemental function of the same type and kind as I.
- Returns the value obtained by shifting the *SIZE* rightmost bits of I circularly by SHIFT bits. If SHIFT is positive, the bits are shifted left, and if SHIFT is negative, the bits are shifted right. If the optional argument *SIZE* is missing, all BIT_SIZE(I) bits of I are shifted.
- I must be an integer.
- SHIFT must be an integer, with ABS(SHIFT) <= *SIZE*.
- *SIZE* must be a positive integer, with $0 <$ *SIZE* $<=$ BIT_SIZE(I).

LEADZ(I)
- Elemental function of the type default integer.

- Returns the number of leading zeros in the bit pattern in I.
- I must be an integer.

MASKL(I,*KIND*)
- Elemental function of the type *KIND*. If *KIND* is not present, the type is default integer.
- Returns a mask with the leftmost I bits set.
- I must be an integer.

MASKR(I,*KIND*)
- Elemental function of the type *KIND*. If *KIND* is not present, the type is default integer.
- Returns a mask with the rightmost I bits set.
- I must be an integer.

MERGE_BITS(I,J.MASK)
- Elemental function of the type integer with the same kind as I.
- MERGE_BITS(I,J.MASK) merges the bits of I and J as determined by the mask. The *k*th bit of the result is equal to the *kk*th bit of I if the *k*th bit of mask is 1; it is equal to the *kk*th bit of J otherwise.
- I and J must be integers of the same kind.

MVBITS(FROM,FROMPOS,LEN,TO,TOPOS)
- Elemental subroutine.
- Copies a sequence of bits from integer FROM to integer TO. The subroutine copies a sequence of LEN bits starting at FROMPOS in integer FROM, and stores them starting at TOPOS in integer TO. All other bits in integer TO are undisturbed.
- Note that FROM and TO can be the same integer.
- Arguments:

FROM	Integer	IN	The object from which the bits are to be moved.
FROMPOS	Integer	IN	Starting bit to move; must be ≥ 0.
LEN	Integer	IN	Number of bits to move; FROMPOS+LEN must be $\leq$ BIT_SIZE(FROM).
TO	Integer, same kind as FROM	INOUT	Destination object.
TOPOS	Integer	IN	Starting bit in destination; $0 \leq$ TOPOS+LEN $\leq$ BIT_SIZE(TO).

NOT(I)
- Elemental function of the same type and kind as I.
- Returns the logical complement of the bits in I.
- I must be an integer.

POPCNT(I)
- Elemental function of the type default integer.

- Returns the number of bits set in I.
- I must be an integer.

POPPAR(I)
- Elemental function of the type default integer.
- Returns the parity of the number of bits set in I.
- I must be an integer.

SHIFTA(I,SHIFT)
- Elemental function of the type integer with the same kind as I.
- Returns a value with all bits shifted right by SHIFT places, with the value of the sign bit filling in for each new value on the left. Bits shifted out on the right end are lost.
- I must be an integer.

SHIFTL(I,SHIFT)
- Elemental function of the type integer with the same kind as I.
- Returns a value with all bits shifted left by SHIFT places, with zeros filling in for each new value on the left. Bits shifted out on the left end are lost.
- I must be an integer.

SHIFTR(I,SHIFT)
- Elemental function of the type integer with the same kind as I.
- Returns a value with all bits shifted right by SHIFT places, with zeros filling in for each new value on the left. Bits shifted out on the right end are lost.
- I must be an integer.

TRAILZ(I)
- Elemental function of the type default integer.
- Returns the number of trailing zeros in the bit pattern in I.
- I must be an integer.

C.7
CHARACTER INTRINSIC FUNCTIONS

These functions produce, manipulate, or provide information about character strings.

ACHAR(I,*KIND*)
- Character(1) elemental function.
- Returns the character in position I of the ASCII collating sequence.
- If $0 \leq I \leq 127$, the result is the character in position I of the ASCII collating sequence. If $I \geq 128$, the results are processor-dependent.
- I must be an integer.
- *KIND* must be an integer whose value is a legal kind of character for the particular computer; if it is absent, the default kind of character is assumed.
- IACHAR is the inverse function of ACHAR.

ADJUSTL(STRING)
- Character elemental function.

- Returns a character value of the same length as STRING, with the nonblank contents left justified. That is, the leading blanks of STRING are removed and the same number of trailing blanks are added at the end.
- STRING must be character.

ADJUSTR(STRING)
- Character elemental function.
- Returns a character value of the same length as STRING, with the nonblank contents right justified. That is, the trailing blanks of STRING are removed and the same number of leading blanks are added at the beginning.
- STRING must be character.

CHAR(I,*KIND*)
- Character(1) elemental function.
- Returns the character in position I of the processor collating sequence associated with the specified kind.
- I must be an integer in the range $0 \leq I \leq n-1$, where n is the number of characters in the processor-dependent collating sequence.
- *KIND* must be an integer whose value is a legal kind of character for the particular computer; if it is absent, the default kind of character is assumed.
- ICHAR is the inverse function of CHAR.

IACHAR(C)
- Integer elemental function.
- Returns the position of a character in the ASCII collating sequence. A processor-dependent value is returned if C is not in the collating sequence.
- C must be character(1).
- ACHAR is the inverse function of IACHAR.

ICHAR(C)
- Integer elemental function.
- Returns the position of a character in the processor collating sequence associated with the kind of the character.
- C must be character(1).
- The result is in the range $0 \leq ICHAR(C) \leq n-1$, where n is the number of characters in the processor-dependent collating sequence.
- CHAR is the inverse function of ICHAR.

INDEX(STRING,SUBSTRING,*BACK*)
- Integer elemental function.
- Returns the starting position of a substring within a string.
- STRING and SUBSTRING must be character values of the same kind, and *BACK* must be logical.
- If the substring is longer than the string, the result is 0. If the length of the substring is 0, then the result is 1. Otherwise, if *BACK* is missing or false, the function returns the starting position of the *first* occurrence of the substring within the string, searching from left to right through the string. If *BACK* is true, the function returns the starting position of the *last* occurrence of the substring within the string.

Fortran Intrinsic Procedures

LEN(STRING)
- Integer inquiry function.
- Returns the length of STRING in characters.
- STRING must be character.

LEN_TRIM(STRING)
- Integer inquiry function.
- Returns the length of STRING in characters, less any trailing blanks. If STRING is completely blank, then the result is 0.
- STRING must be character.

LGE(STRING_A,STRING_B)
- Logical elemental function.
- Returns true if STRING_A $\geq$ STRING_B in the ASCII collating sequence.
- STRING_A and STRING_B must be of type default character.
- The comparison process is similar to that used by the >= relational operator, except that the comparison always uses the ASCII collating sequence.

LGT(STRING_A,STRING_B)
- Logical elemental function.
- Returns true if STRING_A > STRING_B in the ASCII collating sequence.
- STRING_A and STRING_B must be of type default character.
- The comparison process is similar to that used by the > relational operator, except that the comparison always uses the ASCII collating sequence.

LLE(STRING_A,STRING_B)
- Logical elemental function.
- Returns true if STRING_A $\leq$ STRING_B in the ASCII collating sequence.
- STRING_A and STRING_B must be of type default character.
- The comparison process is similar to that used by the <= relational operator, except that the comparison always uses the ASCII collating sequence.

LLT(STRING_A,STRING_B)
- Logical elemental function.
- Returns true if STRING_A < STRING_B in the ASCII collating sequence.
- STRING_A and STRING_B must be of type default character.
- The comparison process is similar to that used by the < relational operator, except that the comparison always uses the ASCII collating sequence.

NEW_LINE(CHAR)
- Inquiry function.
- Returns the newline character for the KIND of the input character string.

REPEAT(STRING,NCOPIES)
- Character transformational function.
- Returns a character string formed by concatenating NCOPIES copies of STRING one after another. If STRING is zero length or if NCOPIES is 0, the function returns a zero length string.
- STRING must be of type character; NCOPIES must be a nonnegative integer.

SCAN(STRING,SET,*BACK*)
- Integer elemental function.
- Scans STRING for the first occurrence of any one of the characters in SET, and returns the position of that occurrence. If no character of STRING is in set, or if either STRING or SET is zero length, the function returns a zero.
- STRING and SET must be of type character and the same kind, and *BACK* must be of type logical.
- If *BACK* is missing or false, the function returns the position of the *first* occurrence (searching left to right) of any of the characters contained in SET. If *BACK* is true, the function returns the position of the *last* occurrence (searching right to left) of any of the characters contained in SET.

TRIM(STRING)
- Character transformational function.
- Returns STRING with trailing blanks removed. If STRING is completely blank, then a zero length string is returned.
- STRING must be of type character.

VERIFY(STRING,SET,*BACK*)
- Integer elemental function.
- Scans STRING for the first occurrence of any one of the characters *not* in SET, and returns the position of that occurrence. If all characters of STRING are in SET, or if either STRING or SET is zero length, the function returns a zero.
- STRING and SET must be of type character and the same kind, and *BACK* must be of type logical.
- If *BACK* is missing or false, the function returns the position of the *first* occurrence (searching left to right) of any of the characters not contained in SET. If *BACK* is true, the function returns the position of the *last* occurrence (searching right to left) of any of the characters not in SET.

C.8
ARRAY AND POINTER INTRINSIC FUNCTIONS

This section describes the 24 standard array and pointer intrinsic functions. Because certain arguments appear in many of these functions, they will be described in detail before we examine the functions themselves.

1. The rank of an array is defined as the number of dimensions in the array. It is abbreviated as r throughout this section.
2. A scalar is defined to be an array of rank 0.
3. The optional argument *MASK* is used by some functions to select the elements of another argument to operate on. When present, *MASK* must be a logical array of the same size and shape as the target array; if an element of *MASK* is true, then the corresponding element of the target array will be operated on.
4. The optional argument *DIM* is used by some functions to determine the dimension of an array along which to operate. When supplied, *DIM* must be a number in the range $1 \leq DIM \leq r$.

5. In the functions ALL, ANY, LBOUND, MAXVAL, MINVAL, PRODUCT, SUM, and UBOUND, the optional argument *DIM* affects the type of argument returned by the function. If the argument is absent, then the function returns a scalar result. If the argument is present, then the function returns a vector result. Because the presence or absence of *DIM* affects the type of value returned by the function, the compiler must be able to determine whether or not the argument is present when the program is compiled. Therefore, *the actual argument corresponding to DIM must not be an optional dummy argument in the calling program unit*. If it were, the compiler would be unable to determine whether or not DIM is present at compilation time. This restriction does not apply to functions CSHIFT, EOSHIFT, SIZE, and SPREAD, since the argument *DIM* does not affect the type of value returned from these functions.

To illustrate the use of *MASK* and *DIM*, let's apply the function MAXVAL to a 2×3 real array array1 ($r = 2$) and two masking arrays mask1 and mask2 defined as follows:

$$\text{array1} = \begin{bmatrix} 1. & 2. & 3. \\ 4. & 5. & 6. \end{bmatrix}$$

$$\text{mask1} = \begin{bmatrix} .\text{TRUE.} & .\text{TRUE.} & .\text{TRUE.} \\ .\text{TRUE.} & .\text{TRUE.} & .\text{TRUE.} \end{bmatrix}$$

$$\text{mask2} = \begin{bmatrix} .\text{TRUE.} & .\text{TRUE.} & .\text{FALSE.} \\ .\text{TRUE.} & .\text{TRUE.} & .\text{FALSE.} \end{bmatrix}$$

The function MAXVAL returns the maximum value(s) along the dimension *DIM* of an array corresponding to the true elements of *MASK*. It has the calling sequence

```
result = MAXVAL(ARRAY,DIM,MASK)
```

If *DIM* is not present, the function returns a scalar equal to the largest value in the array for which *MASK* is true. Therefore, the function

```
result = MAXVAL(array1,MASK=mask1)
```

will produce a value of 6, while the function

```
result = MAXVAL(array1,MASK=mask2)
```

will produce a value of 5. If *DIM* is present, then the function will return an array of rank $r-1$ containing the maximum values along dimension *DIM* for which *MASK* is true. That is, the function will hold the subscript in the specified dimension constant while searching along all other dimensions to find the masked maximum value in that subarray, and then repeat the process for every other possible value of the specified dimension. Since there are three elements in each row of the array, the function

```
result = MAXVAL(array1,DIM=1,MASK=mask1)
```

will search along the *columns* of the array at each row position, and will produce the vector [4. 5. 6.], where 4. is the maximum value in column 1, 5. is the maximum value in column 2, and 6. is the maximum value in column 3. Similarly, there are two elements in each column of the array, so the function

```
result = MAXVAL(array1,DIM=2,MASK=mask1)
```

will search along the *rows* of the array at each column position, and will produce the vector [3. 6.], where 3. is the maximum value in row 1, and 6. is the maximum value in row 2.

ALL(MASK,*DIM*)
- Logical transformational function.
- Returns true if all MASK values are true along dimension *DIM*, or if MASK has zero size. Otherwise, it returns false.
- MASK is a logical array. *DIM* is an integer in the range $1 \leq DIM \leq r$. The corresponding actual argument must not be an optional argument in the calling procedure.
- The result is a scalar if *DIM* is absent. It is an array of rank $r-1$ and shape (d(1),d(2),...,d(DIM-1),d(DIM+1),...,d(r)) where the shape of MASK is (d(1),d(2),...,d(r)). In other words, the shape of the returned vector is the same as the shape of the original mask with dimension *DIM* deleted.

ANY(MASK,*DIM*)
- Logical transformational function.
- Returns true if any MASK value is true along dimension *DIM*. Otherwise, it returns false. If MASK has zero size, it returns false.
- MASK is a logical array. *DIM* is an integer in the range $1 \leq DIM \leq r$. The corresponding actual argument must not be an optional argument in the calling procedure.
- The result is a scalar if *DIM* is absent. It is an array of rank $r-1$ and shape (d(1),d(2),...,d(DIM-1),d(DIM+1),...,d(r)) where the shape of MASK is (d(1),d(2),...,d(r)). In other words, the shape of the returned vector is the same as the shape of the original mask with dimension *DIM* deleted.

COUNT(MASK,*DIM*)
- Logical transformational function.
- Returns the number of true elements of MASK along dimension *DIM*, and returns 0 if MASK has zero size.
- MASK is a logical array. *DIM* is an integer in the range $1 \leq DIM \leq r$. The corresponding actual argument must not be an optional argument in the calling procedure.
- The result is a scalar if *DIM* is absent. It is an array of rank $r-1$ and shape (d(1),d(2),...,d(DIM-1),d(DIM+1),...,d(r)) where the shape of MASK is (d(1),d(2),...,d(r)). In other words, the shape of the returned vector is the same as the shape of the original mask with dimension *DIM* deleted.

CSHIFT(ARRAY,SHIFT,*DIM*)
- Transformational function of the same type as ARRAY.
- Performs a circular shift on an array expression of rank 1, or performs circular shifts on all the complete rank 1 sections along a given dimension of an array expression of rank 2 or greater. Elements shifted out at one end of a section are shifted in at the other end. Different sections may be shifted by different amounts and in different directions.

- ARRAY may be an array of any type and rank, but not a scalar. SHIFT is a scalar if ARRAY is rank 1. Otherwise, it is an array of rank $r-1$ and of shape (d(1),d(2),...,d(DIM-1),d(DIM+1),...,d(r)) where the shape of ARRAY is (d(1),d(2),...,d(r)). DIM is an optional integer in the range $1 \leq DIM \leq r$. If DIM is missing, the function behaves as though DIM were present and equal to 1.

EOSHIFT(ARRAY,SHIFT,BOUNDARY,DIM)
- Transformational function of the same type as ARRAY.
- Performs an end-off shift on an array expression of rank 1, or performs end-off shifts on all the complete rank 1 sections along a given dimension of an array expression of rank 2 or greater. Elements are shifted off at one end of a section and copies of a boundary value are shifted in at the other end. Different sections may have different boundary values and may be shifted by different amounts and in different directions.
- ARRAY may be an array of any type and rank, but not a scalar. SHIFT is a scalar if ARRAY is rank 1. Otherwise, it is an array of rank $r-1$ and of shape (d(1),d(2),...,d(DIM-1),d(DIM+1),...,d(r)) where the shape of ARRAY is (d(1),d(2),...,d(r)). BOUNDARY is the value to be shifted in when the old values are shifted. It must be of the same type and type and kind as ARRAY. It may either be a scalar or a rank n−1 array of shape ($d1, d2, d_{DIM-1}, d_{DIM+1}, dn$). If BOUNDARY is missing, the default value for the given data type will be used (0 for integer, 0.0 for real, etc.). DIM is an optional integer in the range $1 \leq DIM \leq r$. If DIM is missing, the function behaves as though DIM were present and equal to 1.

FINDLOC(ARRAY,VALUE,DIM,MASK,KIND,BACK)
- Integer transformational function, returning a rank 1 array of size r.
- Returns the location of the specified VALUE in the elements of ARRAY along dimension DIM (if present) corresponding to the true elements of MASK (if present). If more than one element has the same maximum value, the location of the first one found is returned.
- ARRAY is an array of type integer, real, or character. DIM is an integer in the range $1 \leq DIM \leq r$. The corresponding actual argument must not be an optional argument in the calling procedure.
- MASK is a logical scalar or a logical array conformable with ARRAY.
- BACK is a logical scalar.
- If DIM is not present and MASK is not present, the result is a rank 1 array containing the subscripts of the first element found in ARRAY having the specified value. If DIM is not present and MASK is present, the search is restricted to those elements for which MASK is true. If DIM is present, the result is an array of rank $r-1$ and of shape (d(1),d(2),...,d(DIM-1), d(DIM+1),...,d(r)) where the shape of ARRAY is (d(1),d(2),...,d(r)). This array contains the subscripts of the largest values found along dimension DIM.
- For example, if

$$\text{ARRAY} = \begin{bmatrix} 1 & 3 & -9 \\ 2 & 2 & 6 \end{bmatrix}$$

then the result of the function FINDLOC(ARRAY,2) is (/2,1/). Note that the search is in column major order: the first subscript, then the second subscript, etc. In that order, (/2,1/) is the location of the first value of 2 detected.
- If more than one element has the same maximum value, the location of the first such element will be returned. If *BACK* is present and has a true value, the location of the last such element will be returned. Thus, the result of FINDLOC(ARRAY,2,BACK=.TRUE.) is (/2,2/), since that is the first value of 2 encountered when running backward.
- If *KIND* is present, the result is an integer of that kind. If it is absent, the result is of type default integer.

LBOUND(ARRAY,DIM,KIND)
- Integer inquiry function.
- Returns all of the lower bounds or a specified lower bound of ARRAY.
- ARRAY is an array of any type. It must not be an unassociated pointer or an unallocated allocatable array.
- *DIM* is an integer in the range $1 \le DIM \le r$. The corresponding actual argument must not be an optional argument in the calling procedure.
- If *DIM* is present, the result is a scalar. If the actual argument corresponding to ARRAY is an array section or an array expression, or if dimension *DIM* has zero size, then the function will return 1. Otherwise, it will return the lower bound of that dimension of ARRAY. If *DIM* is not present, then the function will return an array whose ith element is LBOUND(ARRAY,i) for $i=1,2,\ldots,r$.
- The returned value will be of the kind specified in the *KIND* parameter. If no *KIND* parameter is supplied, it will be of type default integer.

MAXLOC(ARRAY,DIM,MASK,KIND,BACK)
- Integer transformational function, returning a rank 1 array of size r.
- Returns the location of the maximum value of the elements in ARRAY along dimension *DIM* (if present) corresponding to the true elements of *MASK* (if present). If more than one element has the same maximum value, the location of the first one found is returned.
- ARRAY is an array of type integer, real, or character. *DIM* is an integer in the range $1 \le DIM \le r$. The corresponding actual argument must not be an optional argument in the calling procedure.
- *MASK* is a logical scalar or a logical array conformable with ARRAY.
- *BACK* is a logical scalar.
- If *DIM* is not present and *MASK* is not present, the result is a rank 1 array containing the subscripts of the first element found in ARRAY having the maximum value. If *DIM* is not present and *MASK* is present, the search is restricted to those elements for which *MASK* is true. If *DIM* is present, the result is an array of rank $r-1$ and of shape (d(1),d(2),...,d(DIM-1), d(DIM+1),...,d(r)) where the shape of ARRAY is (d(1),d(2),...,d(r)). This array contains the subscripts of the largest values found along dimension *DIM*.
- For example, if

$$\text{ARRAY} = \begin{bmatrix} 1 & 3 & -9 \\ 2 & 2 & 6 \end{bmatrix} \quad \text{and} \quad \text{MASK} = \begin{bmatrix} \text{TRUE} & \text{FALSE} & \text{FALSE} \\ \text{TRUE} & \text{TRUE} & \text{FALSE} \end{bmatrix}$$

then the result of the function MAXLOC(ARRAY) is (/2,3/). The result of MAXLOC(ARRAY,MASK) is (/2,1/). The result of MAXLOC(ARRAY,DIM=1) is (/2,1,2/), and the result of MAXLOC(ARRAY,DIM=2) is (/2,3/).
- If more than one element has the same maximum value, the location of the first such element will be returned. If *BACK* is present and has a true value, the location of the last such element will be returned.
- If *KIND* is present, the result is an integer of that kind. If it is absent, the result is of type default integer.

MAXVAL(ARRAY, *DIM*, *MASK*)
- Transformational function of the same type as ARRAY.
- Returns the maximum value of the elements in ARRAY along dimension *DIM* (if present) corresponding to the true elements of *MASK* (if present). If ARRAY has zero size, or if all the elements of *MASK* are false, then the result is the largest possible negative number of the same type and kind as ARRAY.
- ARRAY is an array of type integer, real, or character. *DIM* is an integer in the range $1 \leq DIM \leq r$. The corresponding actual argument must not be an optional argument in the calling procedure. *MASK* is a logical scalar or a logical array conformable with ARRAY.
- If *DIM* is not present, the result is a scalar containing the maximum value found in the elements of ARRAY corresponding to true elements of *MASK*. If *MASK* is absent, the search is over all of the elements in ARRAY. If *DIM* is present, the result is an array of rank $r-1$ and of shape (d(1),d(2),...,d(DIM-1), d(DIM+1),...,d(r)) where the shape of ARRAY is (d(1),d(2),...,d(r)).
- For example, if

$$\text{ARRAY} = \begin{bmatrix} 1 & 3 & -9 \\ 2 & 2 & 6 \end{bmatrix} \quad \text{and} \quad \text{MASK} = \begin{bmatrix} \text{TRUE} & \text{FALSE} & \text{FALSE} \\ \text{TRUE} & \text{TRUE} & \text{FALSE} \end{bmatrix}$$

then the result of the function MAXVAL(ARRAY) is 6. The result of MAXVAL(ARRAY,MASK) is 2. The result of MAXVAL(ARRAY,DIM = 1) is (/2,3,6/), and the result of MAXLOC(ARRAY,DIM = 2) is (/3,6/).

MERGE(TSOURCE, FSOURCE, MASK)
- Elemental function of the same type as TSOURCE.
- Selects one of two alternative values according to MASK. If a given element of MASK is true, then the corresponding element of the result comes from array TSOURCE. If a given element of MASK is false, then the corresponding element of the result comes from array FSOURCE. MASK may also be a scalar, in which case either all of TSOURCE or all of FSOURCE is selected.
- TSOURCE is any type of array; FSOURCE is the same type and kind as TSOURCE. MASK is a logical scalar, or a logical array conformable with TSOURCE.

MINLOC(ARRAY, *DIM*, *MASK*, *KIND*, *BACK*)
- Integer transformational function, returning a rank 1 array of size r.
- Returns the *location* of the minimum value of the elements in ARRAY along dimension *DIM* (if present) corresponding to the true elements of *MASK* (if present). If more than one element has the same minimum value, the location of the first one found is returned.

- ARRAY is an array of type integer, real, or character. *DIM* is an integer in the range $1 \leq DIM \leq r$. The corresponding actual argument must not be an optional argument in the calling procedure. *MASK* is a logical scalar, or a logical array conformable with ARRAY. *BACK* is a logical scalar.
- If *DIM* and *MASK* are not present, the result is a rank 1 array containing the subscripts of the first element found in ARRAY having the minimum value. If *DIM* is not present and *MASK* is present, the search is restricted to those elements for which *MASK* is true. If *DIM* is present, the result is an array of rank $r-1$ and of shape (d(1),d(2),...,d(DIM-1), d(DIM+1),...,d(r)) where the shape of ARRAY is (d(1),d(2),...,d(r)). This array contains the subscripts of the smallest values found along dimension *DIM*.
- For example, if

$$\text{ARRAY} = \begin{bmatrix} 1 & 3 & -9 \\ 2 & 2 & 6 \end{bmatrix} \quad \text{and} \quad \text{MASK} = \begin{bmatrix} \text{TRUE} & \text{FALSE} & \text{FALSE} \\ \text{TRUE} & \text{TRUE} & \text{FALSE} \end{bmatrix}$$

then the result of the function MINLOC(ARRAY) is (/1,3/). The result of MINLOC(ARRAY,MASK) is (/1,1/). The result of MINLOC(ARRAY,DIM = 1) is (/1,2,1/), and the result of MINLOC(ARRAY,DIM = 2) is (/3,1/).
- If more than one element has the same maximum value, the location of the first such element will be returned. If *BACK* is present and has a true value, the location of the last such element will be returned.
- If *KIND* is present, the result is an integer of that kind. If it is absent, the result is of type default integer.

MINVAL(ARRAY,*DIM*,*MASK*)
- Transformational function of the same type as ARRAY.
- Returns the minimum value of the elements in ARRAY along dimension *DIM* (if present) corresponding to the true elements of *MASK* (if present). If ARRAY has zero size, or if all the elements of *MASK* are false, then the result is the largest possible positive number of the same type and kind as ARRAY.
- ARRAY is an array of type integer, real, or character. *DIM* is an integer in the range $1 \leq DIM \leq r$. The corresponding actual argument must not be an optional argument in the calling procedure. *MASK* is a logical scalar, or a logical array conformable with ARRAY.
- If *DIM* is not present, the result is a scalar containing the minimum value found in the elements of ARRAY corresponding to true elements of *MASK*. If *MASK* is absent, the search is over all of the elements in ARRAY. If *DIM* is present, the result is an array of rank $r-1$ and of shape (d(1),d(2),...,d(DIM-1), d(DIM+1),...,d(r)) where the shape of ARRAY is (d(1),d(2),...,d(r)).
- For example, if

$$\text{ARRAY} = \begin{bmatrix} 1 & 3 & -9 \\ 2 & 2 & 6 \end{bmatrix} \quad \text{and} \quad \text{MASK} = \begin{bmatrix} \text{TRUE} & \text{FALSE} & \text{FALSE} \\ \text{TRUE} & \text{TRUE} & \text{FALSE} \end{bmatrix}$$

then the result of the function MINVAL(ARRAY) is -9. The result of MINVAL(ARRAY,MASK) is 1. The result of MINVAL(ARRAY,DIM = 1) is (/1,2,-9/), and the result of MINLOC(ARRAY,DIM = 2) is (/-9,2/).

NULL(*MOLD*)
- Transformational function.
- Returns a disassociated pointer of the same type as *MOLD*, if present. If *MOLD* is not present, the pointer type is determined by context. (For example, if NULL() is being used to initialize an integer pointer, the returned value will be a disassociated integer pointer.)
- *MOLD* is a pointer of any type. Its pointer association status may be undefined, disassociated, or associated.
- This function is useful for initializing the status of a pointer at the time it is declared.

PACK(ARRAY,MASK,*VECTOR*)
- Transformational function of the same type as ARRAY.
- Packs an array into an array of rank 1 under the control of a mask.
- ARRAY is an array of any type. MASK is a logical scalar, or a logical array conformable with ARRAY. *VECTOR* is a rank 1 array of the same type as ARRAY. It must have at least as many elements as there are true values in the mask. If MASK is a true scalar with the value true, then it must have at least as many elements as there are in ARRAY.
- This function packs the elements of ARRAY into an array of rank 1 under the control of MASK. An element of ARRAY will be packed into the output vector if the corresponding element of MASK is true. If MASK is a true scalar value, then the entire input array will be packed into the output array. The packing is done in column order.
- If argument *VECTOR* is present, then the length of the function output will be the length of *VECTOR*. This length must be greater than or equal to the number of elements to be packed.
- For example, if

$$\text{ARRAY} = \begin{bmatrix} 1 & -3 \\ 4 & -2 \end{bmatrix} \quad \text{and} \quad \text{MASK} = \begin{bmatrix} \text{FALSE} & \text{TRUE} \\ \text{TRUE} & \text{TRUE} \end{bmatrix}$$

then the result of the function PACK(ARRAY,MASK) will be [4 -3 -2].

PRODUCT(ARRAY,*DIM*,*MASK*)
- Transformational function of the same type as ARRAY.
- Returns the product of the elements in ARRAY along dimension *DIM* (if present) corresponding to the true elements of *MASK* (if present). If ARRAY has zero size, or if all the elements of *MASK* are false, then the result has the value one.
- ARRAY is an array of type integer, real, or complex. *DIM* is an integer in the range $1 \leq DIM \leq r$. The corresponding actual argument must not be an optional argument in the calling procedure. *MASK* is a logical scalar or a logical array conformable with ARRAY.
- If *DIM* is not present or if ARRAY has rank 1, the result is a scalar containing the product of all the elements of ARRAY corresponding to true elements of *MASK*. If *MASK* is also absent, the result is the product of all of the elements in ARRAY. If *DIM* is present, the result is an array of rank $r-1$ and of shape (d(1), d(2),...,d(DIM-1),d(DIM+1),...,d(r)) where the shape of ARRAY is (d(1),d(2),...,d(r)).

RESHAPE(SOURCE,SHAPE,*PAD*,*ORDER*)
- Transformational function of the same type as SOURCE.
- Constructs an array of a specified shape from the elements of another array.
- SOURCE is an array of any type. SHAPE is a one- to seven-element integer array containing the desired extent of each dimension of the output array. PAD is a rank 1 array of the same type as SOURCE. It contains elements to be used as a pad at the end of the output array if there are not enough elements in SOURCE. ORDER is an integer array of the same shape as SHAPE. It specifies the order in which dimensions are to be filled with elements from SOURCE.
- The result of this function is an array of shape SHAPE constructed from the elements of SOURCE. If SOURCE does not contain enough elements, the elements of PAD are used repeatedly to fill out the remainder of the output array. ORDER specifies the order in which the dimensions of the output array will be filled; by default, they fill in the order $(1, 2, \ldots, n)$ where n is the size of SHAPE.
- For example, if SOURCE= [1 2 3 4 5 6], SHAPE=[2 5], and PAD = [0 0], then

$$\text{RESHAPE(SOURCE,SHAPE,PAD)} = \begin{bmatrix} 1 & 3 & 5 & 0 & 0 \\ 2 & 4 & 6 & 0 & 0 \end{bmatrix}$$

and

$$\text{RESHAPE(SOURCE,SHAPE,PAD,(/2,1/))} = \begin{bmatrix} 1 & 2 & 3 & 4 & 5 \\ 6 & 0 & 0 & 0 & 0 \end{bmatrix}$$

SHAPE(SOURCE,*KIND*)
- Integer inquiry function.
- Returns the shape of SOURCE as a rank 1 array whose size is r and whose elements are the extents of the corresponding dimensions of SOURCE. If SOURCE is a scalar, a rank 1 array of size zero is returned.
- SOURCE is an array or scalar of any type. It must not be an unassociated pointer or an unallocated allocatable array.
- If KIND is present, the rank 1 array has the specified kind. Otherwise, it is of the default integer type.

SIZE(ARRAY,*DIM*)
- Integer inquiry function.
- Returns either the extent of ARRAY along a particular dimension if DIM is present; otherwise, it returns the total number of elements in the array.
- ARRAY is an array of any type. It must not be an unassociated pointer or an unallocated allocatable array. DIM is an integer in the range $1 \leq DIM \leq r$. If ARRAY is an assumed-size array, DIM must be present, and must have a value less than r.

SPREAD(SOURCE,DIM,NCOPIES)
- Transformational function of the same type as SOURCE.
- Constructs an array of rank $r+1$ by copying SOURCE along a specified dimension (as in forming a book from copies of a single page).
- SOURCE is an array or scalar of any type. The rank of SOURCE must be less than 7. DIM is an integer specifying the dimension over which to copy SOURCE. It must

satisfy the condition $1 \leq DIM \leq r+1$. NCOPIES is the number of copies of SOURCE to make along dimension DIM. If NCOPIES is less than or equal to zero, a zero-sized array is produced.
- If SOURCE is a scalar, each element in the result has a value equal to SOURCE. If source is an array, the element in the result with subscripts $(s_1, s_2, \ldots, s_{n+1})$ has the value SOURCE$(s_1, s_2, \ldots, s_{DIM-1}, s_{DIM+1}, \ldots, s_{n+1})$.
- For example, if SOURCE= [1 3 5], then the result of function SPREAD (SOURCE, DIM = 1, NCOPIES = 3) is the array $\begin{bmatrix} 1 & 3 & 5 \\ 1 & 3 & 5 \\ 1 & 3 & 5 \end{bmatrix}$.

SUM(ARRAY, *DIM, MASK*)
- Transformational function of the same type as ARRAY.
- Returns the sum of the elements in ARRAY along dimension *DIM* (if present) corresponding to the true elements of *MASK* (if present). If ARRAY has zero size, or if all the elements of *MASK* are false, then the result has the value zero.
- ARRAY is an array of type integer, real, or complex. *DIM* is an integer in the range $1 \leq DIM \leq r$. The corresponding actual argument must not be an optional argument in the calling procedure. *MASK* is a logical scalar or a logical array conformable with ARRAY.
- If *DIM* is not present or if ARRAY has rank 1, the result is a scalar containing the sum of all the elements of ARRAY corresponding to true elements of *MASK*. If *MASK* is also absent, the result is the sum of all of the elements in ARRAY. If *DIM* is present, the result is an array of rank $r-1$ and of shape (d(1),d(2),..., d(DIM-1),d(DIM+1),...,d(r)) where the shape of ARRAY is (d(1),d(2),...,d(r)).

TRANSFER(SOURCE, MOLD, *SIZE*)
- Transformational function of the same type as MOLD.
- Returns either a scalar or a rank 1 array with a physical representation identical to that of SOURCE, but interpreted with the type and kind of MOLD. Effectively, this function takes the bit patterns in SOURCE and interprets them as though they were of the type and kind of MOLD.
- SOURCE is an array or scalar of any type. MOLD is an array or scalar of any type. *SIZE* is a scalar integer value. The corresponding actual argument must not be an optional argument in the calling procedure.
- If MOLD is a scalar and *SIZE* is absent, the result is a scalar. If MOLD is an array and *SIZE* is absent, the result has the smallest possible size that makes use of all of the bits in SOURCE. If *SIZE* is present, the result is a rank 1 array of length *SIZE*. If the number of bits in the result and in SOURCE are not the same, then bits will be truncated or extra bits will be added in an undefined, processor-dependent manner.
- Example 1: TRANSFER(4.0,0) has the integer value 1082130432 on a PC using IEEE Standard floating-point numbers, because the bit representations of a floating point 4.0 and an integer 1082130432 are identical. The transfer function has caused the bit associated with the floating point 4.0 to be reinterpreted as an integer.

- Example 2: In the function TRANSFER((/1.1,2.2,3.3/),(/(0.,0.)/)), the SOURCE is three real values long. The MOLD is a rank 1 array containing a complex number, which is two real values long. Therefore, the output will be a complex rank 1 array. In order to use all of the bits in SOURCE, the result of the function is a complex rank 1 array with two elements. The first element in the output array is (1.1,2.2), and the second element has a real part of 3.3 together with an unknown imaginary part.
- Example 3: In the function TRANSFER((/1.1,2.2,3.3/),(/(0.,0.)/),1), the SOURCE is three real values long. The MOLD is a rank 1 array containing a complex number, which is two real values long. Therefore, the output will be a complex rank 1 array. Since the SIZE is specified to be 1, only one complex value is produced. The result of the function is a complex rank 1 array with one element: (1.1,2.2).

TRANSPOSE(MATRIX)
- Transformational function of the same type as MATRIX.
- Transposes a matrix of rank 2. Element (i, j) of the output has the value of MATRIX(j,i).
- MATRIX is a rank 2 matrix of any type.

UBOUND(ARRAY,DIM,KIND)
- Integer inquiry function.
- Returns all of the upper bounds or a specified upper bound of ARRAY.
- ARRAY is an array of any type. It must not be an unassociated pointer or an unallocated allocatable array.
- DIM is an integer in the range $1 \leq DIM \leq r$. The corresponding actual argument must not be an optional argument in the calling procedure.
- If DIM is present, the result is a scalar. If the actual argument corresponding to ARRAY is an array section or an array expression, or if dimension DIM has zero size, then the function will return 1. Otherwise, it will return the upper bound of that dimension of ARRAY. If DIM is not present, then the function will return an array whose ith element is UBOUND(ARRAY,i) for $i = 1,2,\ldots,r$.
- The returned value will be of the kind specified in the KIND parameter. If no KIND parameter is supplied, it will be of type default integer.

UPACK(VECTOR,MASK,FIELD)
- Transformational function of the same type as VECTOR.
- Unpacks a rank 1 array into an array under the control of a mask. The result is an array of the same type and type parameters as VECTOR and the same shape as MASK.
- VECTOR is a rank 1 array of any type. It must be at least as large as the number of true elements in MASK. MASK is a logical array. FIELD is of the same type as VECTOR and comfortable with MASK.
- This function produces an array with the shape of MASK. The first element of the VECTOR is placed in the location corresponding to the first true value in MASK, the second element of VECTOR is placed in the location corresponding to the second true value in MASK, etc. If a location in MASK is false, then the

corresponding element from FIELD is placed in the output array. If FIELD is a scalar, the same value is placed in the output array for all false locations.
- This function is the inverse of the PACK function.
- For example, suppose that $V = [1 \quad 2 \quad 3]$, $M = \begin{bmatrix} \text{TRUE} & \text{FALSE} & \text{FALSE} \\ \text{FALSE} & \text{FALSE} & \text{FALSE} \\ \text{TRUE} & \text{FALSE} & \text{TRUE} \end{bmatrix}$, and $F = \begin{bmatrix} 0 & 0 & 0 \\ 1 & 1 & 1 \\ 0 & 0 & 0 \end{bmatrix}$. Then the function UNPACK(V,MASK=M,FIELD=0) would have the value $\begin{bmatrix} 1 & 0 & 0 \\ 0 & 0 & 0 \\ 2 & 0 & 3 \end{bmatrix}$, and the function UNPACK(V,MASK=M,FIELD=F) would have the value $\begin{bmatrix} 1 & 0 & 0 \\ 1 & 1 & 1 \\ 2 & 0 & 3 \end{bmatrix}$.

C.9
MISCELLANEOUS INQUIRY FUNCTIONS

ALLOCATED(ARRAY)
- Logical inquiry function.
- Returns true if ARRAY is currently allocated, and false if ARRAY is not currently allocated. The result is undefined if the allocation status of ARRAY is undefined.
- ARRAY is any type of allocatable array.

ASSOCIATED(POINTER,*TARGET*)
- Logical inquiry function.
- There are three possible cases for this function:
  1. If *TARGET* is not present, this function returns true if POINTER is associated, and false otherwise.
  2. If *TARGET* is present and is a target, the result is true if TARGET does not have size zero and POINTER is currently associated with TARGET. Otherwise, the result is false.
  3. If *TARGET* is present and is a pointer, the result is true if both POINTER and TARGET are currently associated with the same nonzero-sized target. Otherwise, the result is false.
- POINTER is any type of pointer whose pointer association status is not undefined. *TARGET* is any type of pointer or target. If it is a pointer, its pointer association status must not be undefined.

PARITY(MASK,*DIM*)
- Logical inquiry function.
- MASK is an array of type logical.

- DIM is an integer in the range $1 \leq DIM \leq r$. The corresponding actual argument must not be an optional argument in the calling procedure.
- Returns true if an odd number of elements in MASK are true, and false otherwise.

PRESENT(A)
- Logical inquiry function.
- Returns true if optional argument A is present, and false otherwise.
- A is any optional argument.

STORAGE_SIZE(A, KIND)
- Integer inquiry function.
- A is a scalar or integer of any type.
- This function returns an integer containing the number of bits in scalar A, or in one element of array A. If KIND is present, the integer will be of the specified kind.

C.10
MISCELLANEOUS PROCEDURES

MOVE_ALLOC(FROM, TO)
- Pure subroutine.
- Arguments:

FROM	Any	INOUT	Allocatable scalar or array of any type and rank.
TO	Same as FROM	OUT	Allocatable scalar or array compatible with the FROM argument.

- Transfers the current allocation from the FROM object to the TO object.
- The FROM object will be unallocated at the end of this subroutine.
- If the FROM object is unallocated at the time of the call, the TO object becomes unallocated.
- If the FROM object is allocated at the time of the call, the TO object becomes allocated with the type, type parameters, array bounds, and value originally in the FROM object.
- If the TO object has the TARGET attribute, then any pointers that used to point to the FROM object will now point to the TO object.
- If the TO object does not have the TARGET attribute, then any pointers that used to point to the FROM object will become undefined.

C.11
COARRAY FUNCTIONS

COSHAPE(COARRAY, KIND)
- Integer inquiry function.
- Returns the coshape of COARRAY as a rank 1 array whose size is r and whose elements are the extents of the corresponding codimensions of COARRAY.

- SOURCE is an array or scalar of any type. It must not be an unassociated pointer or an unallocated allocatable array.
- If KIND is present, the rank 1 array has the specified kind. Otherwise, it is of the default integer type.

IMAGE_INDEX(COARRAY,SUB)
- Integer inquiry function.
- COARRAY is any coarray.
- SUB is a rank 1 array of cosubscripts for the coarray.
- If the set of cosubscripts corresponds to a valid coarray address, the function returns the image containing the specified coarray data.
- SOURCE is an array or scalar of any type. It must not be an unassociated pointer or an unallocated allocatable array.
- If KIND is present, the rank 1 array has the specified kind. Otherwise, it is of the default integer type.

LCOBOUND(COARRAY,DIM,KIND)
- Integer inquiry function.
- Returns all of the lower coarray bounds or a specified lower bound of COARRAY.
- COARRAY is a coarray of any type.
- DIM is an integer in the range $1 \leq DIM \leq r$. The corresponding actual argument must not be an optional argument in the calling procedure.
- If DIM is present, the result is a scalar. If the actual argument corresponding to ARRAY is an array section or an array expression, or if dimension DIM has zero size, then the function will return 1. Otherwise, it will return the lower bound of that dimension of ARRAY. If DIM is not present, then the function will return an array whose ith element is LBOUND(ARRAY,i) for $i=1,2,\ldots,r$.
- The returned value will be of the kind specified in the KIND parameter. If no KIND parameter is supplied, it will be of type default integer.

NUM_IMAGES()
- Transformational function.
- Returns the number of images currently running.

THIS_IMAGE(COARRAY,DIM)
- Transformational function.
- COARRAY is the name of a coarray.
- DIM is an integer in the range $1 \leq DIM \leq r$. The corresponding actual argument must not be an optional argument in the calling procedure.
- If the function has no arguments, it returns the number of the current image as a default integer.
- If the function has the COARRAY argument, it returns the sequence of cosubscript values for the COARRAY that would invoke to specify the invoking image.
- If the function has the COARRAY and DIM arguments, it returns the cosubscript value for the COARRAY corresponding to DIM that would invoke to specify the invoking image.

UCOBOUND(COARRAY,*DIM,KIND*)
- Integer inquiry function.
- Returns all of the upper coarray bounds or a specified lower bound of COARRAY.
- COARRAY is a coarray of any type.
- *DIM* is an integer in the range $1 \leq DIM \leq r$. The corresponding actual argument must not be an optional argument in the calling procedure.
- If *DIM* is present, the result is a scalar. If the actual argument corresponding to ARRAY is an array section or an array expression, or if dimension *DIM* has zero size, then the function will return 1. Otherwise, it will return the lower bound of that dimension of ARRAY. If *DIM* is not present, then the function will return an array whose *i*th element is LBOUND(ARRAY,*i*) for $i=1,2,\ldots,r$.
- The returned value will be of the kind specified in the *KIND* parameter. If no *KIND* parameter is supplied, it will be of type default integer.

APPENDIX D

Order of Statements in a Fortran Program

Fortran programs consist of one or more program units, each of which contains at least two legal Fortran statements. Any number and type of program units may be included in the program, with the exception that one and only one main program may be included.

All Fortran statements may be grouped into one of 17 possible categories, which are listed below. (In this list, all undesirable, obsolescent, or deleted Fortran statements are shown in small type.)

1. Initial statements (PROGRAM, SUBROUTINE, FUNCTION, MODULE, SUBMODULE, and BLOCK DATA)
2. Comments
3. USE statements
4. IMPLICIT NONE statement
5. Other IMPLICIT statements
6. PARAMETER statements
7. DATA statements
8. Derived type definitions
9. Type declaration statements
10. Interface blocks
11. Statement function declarations
12. Other specification statements (PUBLIC, PRIVATE, SAVE, etc.)
13. FORMAT statements
14. ENTRY statements
15. Executable statements and constructs
16. CONTAINS statement
17. END statements (END PROGRAM, END FUNCTION, END MODULE, etc.)

The order in which these statements may appear in a program unit is specified in Table D-1. In this table, horizontal lines indicate varieties of statements that may not be mixed, while vertical lines indicate types of statements that may be interspersed.

Note from this table that nonexecutable statements generally precede executable statements in a program unit. The only nonexecutable statements that may be legally mixed with executable statements are FORMAT statements, ENTRY statements, and DATA statements.

Table D-1
Requirements on Statement Ordering

PROGRAM, FUNCTION, MODULE, SUBROUTINE, or BLOCK DATA statement	
USE statements	
IMPORT statements	
IMPLICIT NONE statement	
FORMAT and ENTRY statements	PARAMETER statements / IMPLICIT statements
	PARAMETER and DATA statements / Derived type definitions, interface blocks, type declaration statements, enumeration definitions, procedure declarations, specification statements, and statement function statements
	DATA statements / Executable statements and constructs
CONTAINS statement	
Internal subprograms or module subprograms	
END statement	

In addition to the above constraints, not every type of Fortran statement may appear in every type of Fortran scoping unit. Table D-2 shows which types of Fortran statements are allowed in which scoping units.

Table D-2
Statements Allowed in Scoping Units

Kind of scoping unit	Main program	Module	Block Data	External subprog	Module subprog	Internal subprog	Interface Body
USE statement	Yes	Yes	Yes	Yes	Yes	Yes	Yes
ENTRY statement	No	No	No	Yes	Yes	No	No
FORMAT statement	Yes	No	No	Yes	Yes	Yes	No
Misc. declarations (see notes)	Yes	Yes	Yes	Yes	Yes	Yes	Yes
DATA statement	Yes	Yes	Yes	Yes	Yes	Yes	No
Derived-type definition	Yes	Yes	Yes	Yes	Yes	Yes	Yes
Interface block	Yes	Yes	No	Yes	Yes	Yes	Yes
Executable statement	Yes	No	No	Yes	Yes	Yes	No
CONTAINS statement	Yes	Yes	No	Yes	Yes	No	No
Statement function statement	Yes	No	No	Yes	Yes	Yes	No

Notes:

1. Miscellaneous declarations are PARAMETER statements, IMPLICIT statements, type declaration statements, and specification statements such as PUBLIC, SAVE, etc.
2. Derived type definitions are also scoping units, but they do not contain any of the above statements, and so have not been listed in the table.
3. The scoping unit of a module does not include any module subprograms that the module contains.

APPENDIX E

Glossary

This appendix contains a glossary of Fortran terms. Many of the definitions here are paraphrased from the definitions of terms in the current Fortran Standard, ISO/IEC 1539: 2010 (Fortran 2008).

abstract type A derived type that has the ABSTRACT attribute. It can only be used as a basis for type extension—no objects of this type can be defined.

actual argument An expression, variable, or procedure that is specified in a procedure invocation (a subroutine call or a function reference). It is associated with the dummy argument in the corresponding position of procedure definition, unless keywords are used to change the order of arguments.

algorithm The "formula" or sequence of steps used to solve a specific problem.

allocatable array An array specified as ALLOCATABLE with a certain type and rank. It can be allocated to a certain extent with the ALLOCATE statement. The array cannot be referenced or defined until it has been allocated. When no longer needed, the corresponding storage area can be released with the DEALLOCATE statement.

allocatable variable A variable, either intrinsic or user defined, specified as ALLOCATABLE. It can be allocated with the ALLOCATE statement. The variable cannot be referenced or defined until it has been allocated. When no longer needed, the corresponding storage area can be released with the DEALLOCATE statement.

allocation statement A statement that allocates memory for an allocatable array or a pointer.

allocation status A logical value indicating whether or not an allocatable array is currently allocated. It can be examined using the ALLOCATED intrinsic function.

alpha release The first completed version of a large program. The alpha release is normally tested by the programmers themselves and a few others very close to them, in order to discover the most serious bugs present in the program.

argument A placeholder for a value or variable name that will be passed to a procedure when it is invoked (a dummy argument), or the value or variable name that is actually passed to the procedure when it is invoked (an actual argument). Arguments appear in parentheses after a procedure name both when the procedure is declared and when the procedure is invoked.

argument association The relationship between an actual argument and a dummy argument during the execution of a procedure reference. Argument association is performed either by the relative position of actual and dummy arguments in the procedure reference and the procedure definition, or by means of argument keywords.

argument keyword A dummy argument name. It may be used in a procedure reference followed by the equals symbol provided the procedure has an explicit interface.

argument list A list of values and variables that is passed to a procedure when it is invoked. Argument lists appear in parentheses after a procedure name both when the procedure is declared and when the procedure is invoked.

array A set data items, all of the same type and kind, which are referred to by the same name. Individual elements within an array are accessed by using the array name followed by one or more subscripts.

array constant A constant that creates an array.

array constructor An array-valued constant.

array element An individual data item within an array.

array element order The order in which the elements of an array appear to be stored. The physical storage arrangement within a computer's memory may be different, but any reference to the array will make the elements appear to be in this order.

array overflow An attempt to use an array element with an index outside the valid range for the array; an out-of-bounds reference.

array pointer A pointer to an array.

array section A subset of an array, which can be used and manipulated as an array in its own right.

array specification A means of defining the name, shape, and size of an array in a type declaration statement.

array variable An array-valued variable.

array-valued Having the property of being an array.

array-valued function A function whose result is an array.

ASCII The American Standard Code for Information Interchange (ISO/IEC 646:1991), a widely used internal character coding set. This set is also known as ISO 646 (International Reference Version).

ASCII collating sequence The collating sequence of the ASCII character set.

assignment Storing the value of an expression into a variable.

assignment operator The equal (=) sign, which indicates that the value of the expression to the right of the equal sign should be assigned to the variable named on the left of the sign.

assignment statement A Fortran statement that causes the value of an expression to be stored into a variable. The form of an assignment statement is "variable = expression".

associated A pointer is associated with a target if it currently points to that target.

association status A logical value indicating whether or not a pointer is currently associated with a target. The possible pointer association status values are: undefined, associated, and unassociated. It can be examined using the ASSOCIATED intrinsic function.

assumed length character declaration The declaration of a character dummy argument with an asterisk for its length. The actual length is determined from the corresponding actual argument when the procedure is invoked. For example:

```
CHARACTER(len=*) :: string
```

assumed length character function A character function whose *return length* is specified with an asterisk. These functions must have an explicit interface. They have been declared obsolescent in Fortran 95. In the example below, my_fun is an assumed length character function:

```
FUNCTION my_fun ( str1, str2 )
CHARACTER(len=*), INTENT(IN) :: str1, str2
CHARACTER(len=*) :: my_fun
```

assumed-shape array A dummy array argument whose bounds in each dimension are represented by colons, with the actual bounds being obtained from the corresponding actual argument when the procedure is invoked. An assumed-shape array has a declared data type and rank, but its size is unknown until the procedure is actually executed. It may only be used in procedures with explicit interfaces. For example:

```
SUBROUTINE test( a, ... )
REAL, DIMENSION(:,:) :: a
```

Glossary

assumed-size array An older Pre-Fortran 90 mechanism for declaring dummy arrays in procedures. In an assumed size array, all of the dimensions of a dummy array are explicitly declared except for the last dimension, which is declared with an asterisk. Assumed-size arrays have been superseded by assumed-shape arrays.

asynchronous input/output Input or output operations that can occur simultaneously with other Fortran statement executions.

attribute A property of a variable or constant that may be declared in a type declaration statement. Examples are PARAMETER, DIMENSION, SAVE, ALLOCATABLE, ASYNCHRONOUS, VOLATILE, and POINTER.

automatic array An explicit-shape array that is local to a procedure, some or all of whose bounds are provided when the procedure is invoked. The array can have a different size and shape each time the procedure is invoked. When the procedure is invoked, the array is automatically allocated with the proper size, and when the procedure terminates, the array is automatically deallocated. In the example below, scratch is an automatic array:

```
SUBROUTINE my_sub ( a, rows, cols )
INTEGER :: rows, cols
...
REAL, DIMENSION(rows,cols) :: scratch
```

automatic length character function A character function whose return length is specified when the function is invoked either by a dummy argument or by a value in a module or COMMON block. These functions must have an explicit interface. In the example below, my_fun is an automatic length character function:

```
FUNCTION my_fun ( str1, str2, n )
INTEGER, INTENT(IN) :: n
CHARACTER(len=*), INTENT(IN) :: str1, str2
CHARACTER(len=n) :: my_fun
```

automatic character variable A local character variable in a procedure whose length is specified when the procedure is invoked either by a dummy argument or by a value in a module or COMMON block. When the procedure is invoked, the variable is automatically created with the proper size, and when the procedure terminates, the variable is automatically destroyed. In the example below, temp is an automatic character variable:

```
SUBROUTINE my_sub ( str1, str2, n )
CHARACTER(len=*) :: str1, str2
...
CHARACTER(len=n) :: temp
```

batch processing A mode of processing in which a program is compiled and executed without input or interaction with a user.

beta release The second completed version of a large program. The beta release is normally given to "friendly" outside users who have a need for the program in their day-to-day jobs. These users exercise the program under many different conditions and with many different input data sets, and they report any bugs that they find to the program developers.

binary digit A 0 or 1, the two possible digits in a base 2 system.

binary operator An operator that is written between two operands. Examples include +, −, *, /, >, <, .AND., etc.

binary tree A tree structure that splits into two branches at each node.

bit A binary digit.

binding   The process of associating a procedure with a particular derived data type.
block   A sequence of executable statements embedded in an executable construct, bounded by statements that are particular to the construct, and treated as an integral unit. For example, the statements between IF and END IF below are a block.

```
IF ( x > 0. ) THEN
   ...
   (code block)
   ...
END IF
```

BLOCK DATA program unit   A program unit that provides initial values for variables in named COMMON blocks.
block IF construct   A program unit in which the execution of one or more blocks of statements is controlled by an IF statement, and optionally by one or more ELSE IF statements and up to one ELSE statement.
bound   An upper bound or a lower bound; the maximum or minimum value permitted for a subscript in an array.
bound procedure   A procedure that is bound to a derived data type, and that is accessible through the component selection syntax (i.e., using a variable name followed by the % component selector: a%proc()).
bounds checking   The process of checking each array reference before it is executed to ensure that the specified subscripts are within the declared bounds of the array.
branch   (*a*) A transfer of control within a program, as in an IF or CASE structure. (*b*) A linked list that forms part of a binary tree.
bug   A programming error that causes a program to behave improperly.
byte   A group of 8 bits.

card identification field   Columns 73 to 80 of a fixed source form line. These columns are ignored by the compiler. In the past, these columns were used to number the individual cards in a source card deck.
central processing unit   The part of the computer that carries out the main data processing functions. It usually consists of one or more *control units* to select the data and the operations to be performed on it, and *arithmetic logic units* to perform arithmetic calculations.
character   (*a*) A letter, digit, or other symbol. (*b*) An intrinsic data type used to represent characters.
character constant   A constant that contains a character string between single or double quotes.
character constant edit descriptor   An edit descriptor that takes the form of a character constant in an output format. For example, in the statement

```
100 FORMAT (" X = ", x)
```

the " X = " is a character constant edit descriptor.
character context   Characters that form a part of a character literal constant or a character constant edit descriptor. Any legal character in a computer's character set may be used in a character context, not just those in the Fortran character set.
character data type   An intrinsic data type used to represent characters.
character expression   A combination of character constants, character variables, and character operators that calculates a result.
character length parameter   The type parameter that specifies the number of characters for an entity of type character.
character operator   An operator that operates on character data.

Glossary

character set A collection of letters, numbers, and symbols that may be used in character strings. These common character sets are ASCII and Unicode.

character storage unit The unit of storage that can hold a single character of the default type.

character string A sequence of one or more characters.

character variable A variable that can be used to store one or more characters.

child A derived data type extended from a parent data type. It is defined with an `EXTENDS` clause.

class The set of defined data types all extended from a single prototype, which is declared with the `CLASS` statement instead of the `TYPE` statement.

class hierarchy An ordering of classes, indicting which classes inherit from other classes. Parent classes are at the top of the hierarchy, with subclasses that inherit from them below.

close The process of terminating the link between a file and an input/output unit.

coarray A coarray is a type of array that is allocated across all images running a coarray Fortran program. Any image can access the data in any part of the coarray on any image using coarray syntax.

Coarray Fortran A form of Fortran program in which multiple identical copies of a program run in parallel, sharing data and computational tasks.

corank The number of dimensions of a coarray. The maximum rank plus corank of a Fortran coarray must be less than or equal to 15.

corank 2 coarray A coarray whose images are organized in a 2D structure.

collating sequence The order in which a particular character set is sorted by relational operators.

column major order The way multidimensional Fortran arrays are allocated in memory. In column major order, subscript 1 runs through all of its values before subscript 2 is incremented, and so forth for higher subscripts. For example, if an array `a` is 2×3, then the array elements will be allocated in the order `a(1,1), a(2,1), a(1,2), a(2,2), a(1,3), a(2,3)`.

combinational operator An operator whose operand(s) are logical values, and whose result is a logical value. Examples include `.AND.`, `.OR.`, `.NOT.`, etc.

comment Text within a program unit that is ignored by a compiler, but provides information for the programmer. In free source form, comments begin with the first exclamation point (!) on a line that is not in a character context, and continue to the end of the line. In fixed source form, comments begin with a C or * in column 1, and continue to the end of the line.

COMMON block A block of physical storage that may be accessed by any of the scoping units in a program. The data in the block is identified by its relative position, regardless of the name and type of the variable in that position.

compilation error An error that is detected by a Fortran compiler during compilation.

compiler A computer program that translates a program written in a computer language such as Fortran into the machine code used by a particular computer. The compiler usually translates the code into an intermediate form called object code, which is then prepared for execution by a separate linker.

complex An intrinsic data type used to represent complex numbers.

complex constant A constant of the complex type, written as an ordered pair of real values enclosed in parentheses. For example, (3.,-4.) is a complex constant.

complex number A number consisting of a real part and an imaginary part.

component One of the elements of a derived data type.

component order The order of components in a derived data type.

component selector The method of addressing a specific component within a structure. It consists of the structure name and the component name, separated by a percent (%) sign. For example, `student%age`.

computer A device that stores both information (data) and instructions for modifying that information (programs). The computer executes programs to manipulate its data in useful ways.

concatenation   The process of attaching one character string to the end of another one, by means of a concatenation operator.

concatenation operator   An operator (//) that combines two character strings to form a single character string.

concrete type   A derived type that does *not* have the ABSTRACT attribute. It is possible to create objects from classes of a concrete type.

conformable   Two arrays are said to be conformable if they have the same shape. A scalar is conformable with any array. Intrinsic operations are only defined for conformable data items.

constant   A data object whose value is unchanged throughout the execution of a program. Constants may be named (i.e., parameters) or unnamed.

construct   A sequence of statements starting with a DO, IF, SELECT CASE, FORALL, ASSOCIATE, or WHERE statement and ending with the corresponding terminal statement.

construct association   The association between the selector of an ASSOCIATE or SELECT TYPE construct and the associated construct entity.

control character   The first character in an output buffer, which is used to control the vertical spacing for the current line.

control mask   In a WHERE statement or construct, an array of type logical whose value determines which elements of an array will be operated on. This definition also applies to the MASK argument in many array intrinsic functions.

core   A core is an individual processing unit on a CPU chip that contains more than one core. Each core can perform independent calculations in parallel with the other cores on the chip.

coshape   The corank and extent of a coarray in each of its codimensions. The coshape can be stored in a rank 1 array, with each element of the array containing the extent of one codimension.

counting loop   A DO loop that executes a specified number of times, based on the loop control parameters (also known as an iterative loop).

CPU   See central processing unit.

critical section   A section of code in a parallel program whose results are indeterminate unless only one image executes the code at a time. The CRITICAL ... END CRITICAL structure prevents more than one image from executing that code at any given time.

data   Information to be processed by a computer.

data abstraction   The ability to create new data types, together with associated operators, and to hide the internal structure and operations from the user.

data dictionary   A list of the names and definitions of all named variables and constants used in a program unit. The definitions should include both a description of the contents of the item and the units in which it is measured.

data hiding   The idea that some items in a program unit may not be accessible to other program units. Local data items in a procedure are hidden from any program unit that invokes the procedure. Access to the data items and procedures in a module may be controlled using PUBLIC and PRIVATE statements.

data object   A constant or a variable.

data type   A named category of data that is characterized by a set of values, together with a way to denote these values and a collection of operations that interpret and manipulate the values.

deadlock   A condition in which one image is waiting for another one to synchronize, while the second image is waiting for the first one to synchronize. In this case, the program will hang forever.

deallocation statement   A statement that frees memory previously allocated for an allocatable array or a pointer.

Glossary

debugging Locating and eliminating bugs from a program.

decimal symbol The character that separates the whole and fractional parts of a real number. This is a period in the United States, United Kingdom, and many other countries, and a comma in Spain, France, and some other parts of Europe.

declared type The type declared for an argument in a procedure. The dynamic type of an actual argument can either be the declared type or some subclass of the declared type.

default character set The set of characters available for use by programs on a particular computer if no special action is taken to select another character set.

default complex The kind of complex value used when no kind type parameter is specified.

default integer The kind of integer value used when no kind type parameter is specified.

default kind The kind type parameter used for a specific data type when no kind is explicitly specified. The default kinds of each data type are known as default integer, default real, default complex, etc. Default kinds vary from processor to processor.

default real The kind of real value used when no kind type parameter is specified.

default typing The type assigned to a variable when no type declaration statement is present in a program unit, based on the first letter of the variable name.

deferred-shape array An allocatable array or a pointer array. The type and rank of these arrays are declared in type declaration statements, but the shape of the array is not determined until memory is allocated in an `ALLOCATE` statement.

defined assignment A user-defined assignment that involves a derived data type. This is done with the `INTERFACE ASSIGNMENT` construct.

defined operation A user-defined operation that either extends an intrinsic operation for use with derived types or defines a new operation for use with either intrinsic types or derived types. This is done with the `INTERFACE OPERATOR` construct.

deleted feature A feature of older versions of Fortran that has been deleted from later versions of the language. An example is the Hollerith (H) format descriptor.

dereferencing The process of accessing the corresponding target when a reference to a pointer appears in an operation or assignment statement.

derived type (or **derived data type**) A user-defined data type consisting of components, each of which is either of intrinsic type or of another derived type.

dimension attribute An attribute of a type declaration statement used to specify the number of subscripts in an array, and the characteristics of those subscripts such as their bounds and extent. This information can also be specified in a separate `DIMENSION` statement.

direct access Reading or writing the contents of a file in arbitrary order.

direct access file A form of file in which the individual records can be written and read in any order. Direct access files must have records of fixed length so that the location of any particular record can be quickly calculated.

disassociated A pointer is disassociated if it is not associated with a target. A pointer can be disassociated using the `NULLIFY()` statement or the `null()` intrinsic function.

DO construct A loop that begins with a `DO` statement and ends with an `END DO` statement.

DO loop A loop that is controlled by a `DO` statement.

DO loop index The variable that is used to control the number of times the loop is executed in an iterative `DO` loop.

double precision A method of storing floating-point numbers on a computer that uses twice as much memory as single precision, resulting in more significant digits and (usually) a greater range in the representation of the numbers. Before Fortran 90, double-precision variables were declared with a `DOUBLE PRECISION` type declaration statement. In Fortran 95 / 2003, they are just another kind of the real data type.

dummy argument An argument used in a procedure definition that will be associated with an actual argument when the procedure is invoked.

dynamic memory allocation   Allocating memory for variables or arrays at execution time, as opposed to static memory allocation, which occurs at compilation time.

dynamic type   The type of a data entity during execution. For polymorphic entities, it will be of the parent data type or a child of the parent type. For nonpolymorphic entities, it is the same as the declared data type.

dynamic variable   A variable that is created when it is needed during the course of a program's execution, and that is destroyed when it is no longer needed. Examples are automatic arrays and character variables, allocatable arrays, and allocated pointer targets.

edit descriptor   An item in a format that specifies the conversion between the internal and external representations of a data item. (Identical to format descriptor.)

elemental   An adjective applied to an operation, procedure, or assignment that is applied independently to the elements of an array or corresponding elements of a set of conformable arrays and scalars. Elemental operations, procedures, or assignments may be easily partitioned among many processors in a parallel computer.

elemental function   A function that is elemental.

elemental subroutine   A subroutine that is elemental.

elemental intrinsic function   An intrinsic function that is defined for scalar inputs and outputs, but that can accept an array-valued argument or arguments and will deliver an array-valued result obtained by applying the procedure to the corresponding elements of the argument array(s) in turn.

elemental intrinsic procedure   An intrinsic procedure that is defined for scalar inputs and outputs, but that can accept an array-valued argument or arguments and will deliver an array-valued result obtained by applying the procedure to the corresponding elements of the argument array(s) in turn.

elemental procedure (user-defined)   A user-defined procedure that is defined with only scalar dummy arguments (no pointers or procedures) and with a scalar result (not a pointer). An elemental function must have no side effects, meaning that all arguments are INTENT(IN). An elemental subroutine must have no side effects except for arguments explicitly specified with INTENT(OUT) or INTENT(INOUT). If the procedure is declared with the ELEMENTAL prefix, it will be able to accept an array-valued argument or arguments and will deliver an array-valued result obtained by applying the procedure to the corresponding elements of the argument arrays in turn. User-defined elemental procedures are available in Fortran 95 only.

end-of-file condition   A condition set when an endfile record is read from a file, which can be detected by an IOSTAT clause in a READ statement.

endfile record   A special record that only occurs at the end of a sequential file. It can be written by an ENDFILE statement.

error flag   A variable returned from a subroutine to indicate the status of the operation performed by the subroutine.

executable statement   A statement that causes the computer to perform some action during the execution of a program.

execution error   An error that occurs during the execution of a program (also called a runtime error).

explicit interface   A procedure interface that is known to the program unit that will invoke the procedure. An explicit interface to an external procedure may be created by an interface block, or by placing the external procedures in modules and then accessing them by USE association. An explicit interface is automatically created for any internal procedures, or for recursive procedures referencing themselves. (Compare with implicit interface, below.)

explicit-shape array   A named array that is declared with explicit bounds in every dimension.

Glossary

explicit typing   Explicitly declaring the type of a variable in a type declaration statement (as opposed to default typing).

exponent   (*a*) In a binary representation, the power of 2 by which the mantissa is multiplied to produce a complete floating-point number. (*b*) In a decimal representation, the power of 10 by which the mantissa is multiplied to produce a complete floating-point number.

exponential notation   Representing real or floating-point numbers as a mantissa multiplied by a power of 10.

expression   A sequence of operands, operators, and parentheses, where the operands may be variables, constants, or function references.

extent   The number of elements in a particular dimension of an array.

external file   A file that is stored on some external medium. This contrasts with an internal file, which is a character variable within a program.

external function   A function that is not an intrinsic function or an internal function.

external procedure   A function subprogram or a subroutine subprogram, which is not a part of any other program unit.

external unit   An I/O unit that can be connected to an external file. External units are represented by numbers in Fortran I/O statements.

field   A description of a data type defined in a class.

field width   The number of characters available for displaying an output formatted value, or reading an input formatted value.

file   A unit of data that is held on some medium outside the memory of the computer. It is organized into records, which can be accessed individually using READ and WRITE statements.

file storage unit   The basic unit of storage for an unformatted or stream file.

final subroutine   A subroutine that is called automatically by the processor during the finalization of a derived data entity.

finalizer   A method that is called just before an object is destroyed to allow the object to clean up any resources it has allocated. In Fortran, a finalizer is a final subroutine.

finalizable   A derived data type that has final subroutine, or that has a finalizable component. Also, any object of a finalizable type.

finalization   The process of calling a final subroutine before an object is destroyed.

fixed source form   An obsolescent method of writing Fortran programs in which fixed columns were reserved for specific purposes. (Compare with free source form.)

floating-point   A method of representing numbers in which the memory associated with the number is divided into separate fields for a mantissa (fractional part) and an exponent.

floating-point arithmetic   Arithmetic calculations performed with real or floating-point constants and variables.

format   A sequence of edit descriptors that determine the interpretation of an input data record, or that specify the form of an output data record. A format may be found in a FORMAT statement, or in a character constant or variable.

format descriptor   An item in a format that specifies the conversion between the internal and external representations of a data item. (Identical to edit descriptor.)

format statement   A labeled statement that defines a format.

formatted file   A file containing data stored as recognizable numbers, characters, etc.

formatted output statement   A formatted WRITE statement or a PRINT statement.

formatted READ statement   A READ statement that uses format descriptors to specify how to translate the data in the input buffer as it is read.

formatted WRITE statement   A WRITE statement that uses format descriptors to specify how to format the output data as it is displayed.

Fortran Character Set The 86 characters that can be used to write a Fortran program.

free format List-directed I/O statements, which do not require formats for either input or output.

free source form The newer and preferred method of writing Fortran programs, in which any character position in a line can be used for any purpose. (Compare with fixed source form.)

function A procedure that is invoked in an expression, and that computes a single result that is then used in evaluating the expression.

function pointer A type of pointer that points to the location of a function instead of a data item.

function reference The use of a function name in an expression, which invokes (executes) the function to carry out some calculation, and returns the result for use in evaluating the expression. A function is invoked or executed by naming it in an expression.

function subprogram A program unit that begins with a `FUNCTION` statement and ends with an `END FUNCTION` statement.

function value The value that is returned when the function executes.

generic function A function that can be called with different types of arguments. For example, the intrinsic function `ABS` is a generic function, since it can be invoked with integer, real, or complex arguments.

generic interface block A form of interface block used to define a generic name for a set of procedures.

generic name A name that is used to identify two or more procedures, with the required procedure being determined by the compiler determined at each invocation from the types of the nonoptional arguments in the procedure invocation. A generic name is defined for a set of procedures in a generic interface block.

get methods Methods that access and return the values of data stored in an object.

global accessibility The ability to directly access data and derived type definitions from any program unit. This capability is provided by `USE` association of modules.

global entity An entity whose scope is that of the whole program. It may be a program unit, a common block, or an external procedure.

global storage A block of memory accessible from any program unit—a `COMMON` block. Global storage in `COMMON` blocks has largely been replaced by global accessibility through modules.

guard digits Extra digits in a mathematical calculation that are beyond the precision of the kind of real values used in the calculation. They are used to minimize truncation and round-off errors.

hard disk (or hard disk drive) A data storage device made of rigid magnetic platters, capable of storing large amounts of data.

head The first item in a linked list.

hexadecimal The base 16 number system, in which the legal digits are 0 through 9 and A through F.

high-level language A computer language with a more English-like syntax and more complex programming constructs, as opposed to machine language or assembly language.

host A main program or subprogram that contains an internal subprogram is called the host of the internal subprogram. A module that contains a module subprogram is called the host of the module subprogram.

host association The process by which data entities in a host scoping unit are made available to an inner scoping unit.

host scoping unit A scoping unit that surrounds another scoping unit.

Glossary

ill-conditioned system   A system of equations whose solution is highly sensitive to small changes in the values of its coefficients, or to truncation and round-off errors.

image   One of the multiple copies of a parallel program that execute simultaneously in Coarray Fortran.

imaginary part   The second of the two numbers that make up a COMPLEX data value.

implicit type declaration   Determining the type of a variable from the first letter of its name Implicit type declaration should never be used in any modern Fortran program.

implicit interface   A procedure interface that is not fully known to the program unit that invokes the procedure. A Fortran program cannot detect type, size, or similar mismatches between actual arguments and dummy arguments when an implicit interface is used, so some programming errors will not be caught by the compiler. All pre-Fortran 90 interfaces were implicit. (Compare with explicit interface, above.)

implied DO loop   A shorthand loop structure used in input/output statements, array constructors, and DATA statements, which specifies the order in which the elements of an array are used in that statement.

implied DO variable   A variable used to control an implied DO loop.

impure elemental procedure   An elemental procedure that modifies one or more of its calling arguments.

index array   An array containing indices to other arrays. Index arrays are often used in sorting to avoid swapping large chunks of data.

Inf   Infinite value returned by IEEE 754 arithmetic. It represents an infinite result.

infinite loop   A loop that never terminates, typically because of a programming error.

initial statement   The first statement of a program unit: a PROGRAM, SUBROUTINE, FUNCTION, MODULE, or BLOCK DATA statement.

initialization expression   A restricted form of constant expression that can appear as an initial value in a declaration statement. For example, the initialization expression in the following type declaration statement initializes pi to 3.141592.

```
REAL :: pi = 3.141592
```

input buffer   A section of memory used to hold a line of input data as it is entered from an input device such as a keyboard. When the entire line has been input, the input buffer is made available for processing by the computer.

input device   A device used to enter data into a computer. A common example is a keyboard.

input format   A format used in a formatted input statement.

input list   The list of variable, array, and/or array element names in a READ statement into which data is to be read.

input statement   A READ statement.

input/output unit   A number, asterisk, or name in an input/output statement referring to either an external unit or an internal unit. A number is used to refer to an external file unit, which may be connected to a specific file using an OPEN statement and disconnected using a CLOSE statement. An asterisk is used to refer to the standard input and output devices for a processor. A name is used to refer to an internal file unit, which is just a character variable in the program's memory.

inquiry intrinsic function   An intrinsic function whose result depends on properties of the object being investigated, other than the value of the argument. Other inquiry functions can return properties related to the number system on a particular computer.

inquiry subroutine   A subroutine whose result depends on properties of the object being investigated, other than the value of the argument.

instance method   A bound procedure associated with an object, which can modify the instance variables in the object.

instance variable A variable stored in an object, where each object instantiated has a different copy of the variable.
integer An intrinsic data type used to represent whole numbers.
integer arithmetic Mathematical operations involving only data of the integer data type.
integer constant A numeric constant that does not contain a decimal point.
integer division Division of one integer by another integer. In integer division, the fractional part of the result is lost. Thus, the result of an integer 7 by an integer 4 is 1.
interactive processing A mode of processing in which a user enters data into a program from the keyboard during execution.
integer variable A variable that stores integer data.
interface The name of a procedure, the names and characteristics of its dummy arguments, and (for functions) the characteristics of the result variable.
interface assignment block An interface block used to extend the meaning of the assignment operator (=).
interface block (*a*) A means of making an interface to a procedure explicit (*b*) A means of defining a generic procedure, operator, or assignment.
interface body A sequence of statements in an interface block from a `FUNCTION` or `SUBROUTINE` statement to the corresponding `END` statement. The body specifies the calling sequence of the function or subroutine.
interface function A function used to isolate calls to processor-specific procedures from the main portion of a program.
interface operator block An interface block used to define a new operator or to extend the meaning of a standard Fortran operator (+, −, *, /, >, etc.).
internal file A character variable that can be read from and written to by normal formatted `READ` and `WRITE` statements.
internal function An internal procedure that is a function.
internal procedure An subroutine or function that is contained within another program unit, and that can only be invoked from within that program unit.
intrinsic data type One of the pre-defined data types in Fortran: integer, real, double precision, logical, complex, and character.
intrinsic function An intrinsic procedure that is a function.
intrinsic module A module that is defined as a part of the standard Fortran language.
intrinsic procedure A procedure that is defined as a part of the standard Fortran language (see Appendix B).
intrinsic subroutine An intrinsic procedure that is a subroutine.
I/O unit See input/output unit.
invoke To `CALL` a subroutine, or to reference a function in an expression.
iteration count The number of times that an iterative `DO` loop is executed.
iterative DO loop A `DO` loop that executes a specified number of times, based on the loop control parameters (also known as a counting loop).

keyword A word that has a defined meaning in the Fortran language.
keyword argument A method of specifying the association between dummy arguments and actual arguments of the form: "DUMMY_ARGUMENT=actual_argument". Keyword arguments permit arguments to be specified in any order when a procedure is invoked, and are especially useful with optional arguments. Keyword arguments may only be used in procedures with explicit interfaces. An example of the use of a keyword argument is:

```
kind_value = SELECTED_REAL_KIND(r=100)
```

Glossary

kind All intrinsic data types except for DOUBLE PRECISION may have more than one, processor-dependent representation. Each representation is known as a different kind of that type, and is identified by a processor-dependent integer called a kind type parameter.

kind selector The means of specifying the kind type parameter of a variable or named constant.

kind type parameter An integer value used to identify the kind of an intrinsic data type.

language extension The ability to use the features of a language to extend the language for other purposes. The principal language extension features of Fortran are derived types, user-defined operations, and data hiding.

lexical functions Intrinsic functions used to compare two character strings in a character-set-independent manner.

librarian A program that creates and maintains libraries of compiled object files.

library A collection of procedures that is made available for use by a program. They may be in the form of modules or separately linked object libraries.

line printer A type of printer used to print Fortran programs and output on large computer systems. It got its name from the fact that large line printers print an entire line at a time.

link The process of combining object modules produced from program units to form an executable program.

linked list A data structure in which each element contains a pointer that points to the next element in the structure. (It sometimes contains a pointer to the previous element as well.)

list-directed input A special type of formatted input in which the format used to interpret the input data is selected by the processor in accordance with the type of the data items in the input list.

list-directed I/O statement An input or output statement that uses list-directed input or output.

list-directed output A special type of formatted output in which the format used to display the output data is selected by the processor in accordance with the type of the data items in the output list.

literal constant A constant whose value is written directly, as opposed to a named constant. For example, 14.4 is a literal constant.

local entity An entity defined within a single scoping unit.

local variable A variable declared within a program unit, which is not also in a COMMON block. Such variables are local to that scoping unit.

logical A data type that can have only two possible values: TRUE or FALSE.

logical constant A constant with a logical value: TRUE or FALSE.

logical error A bug or error in a program caused by a mistake in program design (improper branching, looping, etc.).

logical expression An expression whose result is either TRUE or FALSE.

logical IF statement A statement in which a logical expression controls whether or not the rest of the statement is executed.

logical operator An operator whose result is a logical value. There are two type of logical operators: combinational (.AND., .OR., .NOT., etc.) and relational (>, < ==, etc.)

logical variable A variable of type LOGICAL.

loop A sequence of statements repeated multiple times, and usually controlled by a DO statement.

loop index An integer variable that is incremented or decremented each time an iterative DO loop is executed.

lower bound The minimum value permitted for a subscript of an array.

machine language The collection of binary instructions (also called op codes) actually understood and executed by a particular processor.

main memory The computer memory used to store programs that are currently being executed and the data associated with them. This is typically semiconductor memory. Main memory is typically much faster than secondary memory, but it is also much more expensive.

main program A program unit that starts with a PROGRAM statement. Execution begins here when a program is started. There can be only one main program unit in any program.

mantissa (*a*) In a binary representation, the fractional part of a floating-point number that, when multiplied by a power of 2, produces the complete number. The power of 2 required is known as the exponent of the number. The value of the mantissa is always between 0.5 and 1.0. (*b*) In a decimal representation, the fractional part of a floating-point number that, when multiplied by a power of 10, produces the complete number. The power of 10 required is known as the exponent of the number. The value of the mantissa is always between 0.0 and 1.0.

many-one array section An array section with a vector subscript having two or more elements with the same value. Such an array section cannot appear on the left side of an assignment statement.

mask (*a*) A logical expression that is used to control assignment of array elements in a masked array assignment (a WHERE statement or a WHERE construct). (*b*) A logical argument in several array intrinsic functions that determines which array elements will be included in the operation.

masked array assignment An array assignment statement whose operation is controlled by a logical MASK that is the same shape as the array. The operation specified in the assignment statement is only applied to those elements of the array corresponding to true elements of the MASK. Masked array assignments are implemented as WHERE statements or WHERE constructs.

master image Image number 1 in a Coarray Fortran program.

matrix A rank 2 array.

member A component of a class, either a field or a method.

method A procedure that is bound to an object. Most methods access or modify the data stored in the object.

mixed-mode expression An arithmetic expression involving operands of different types. For example, the addition of a real value and an integer is a mixed-mode expression.

module A program unit that allows other program units to access constants, variables, derived type definitions, interfaces, and procedures declared within it by USE association.

module procedure A procedure contained within a module.

name A lexical token consisting of a letter followed by up to 30 alphanumeric characters (letters, digits, and underscores). The named entity could be a variable, a named constant, a pointer, or a program unit.

name association Argument association, USE association, host association, or construct association.

named constant A constant that has been named by a PARAMETER attribute in a type declaration statement, or by a PARAMETER statement.

NAMELIST input/output A form of input or output in which the values in the data are accompanied by the names of the corresponding variables, in the form "NAME=*value*". NAMELISTs are defined once in each program unit, and can be used repeatedly in many I/O statements. NAMELIST input statements can be used to update only a portion of the variables listed in the NAMELIST.

NaN not-a-number value returned by IEEE 754 arithmetic. It represents an undefined value or the result of an illegal operation.

nested The inclusion of one program construct as a part of another program construct, such as nested DO loops or nested block IF constructs.

node An element in a linked list or binary tree.

Glossary

nonadvancing input/output A method of formatted I/O in which each READ, WRITE, or PRINT statement does not necessarily begin a new record.

nonexecutable statement A statement used to configure the program environment in which computational actions take place. Examples include the IMPLICIT NONE statement and type declaration statements.

nonvolatile memory Memory that preserves its data when power is turned off.

numeric model A model that describes the range and precision that can be achieved for a given type and kind of numbers, without going down to the physical details of how bits are laid out in memory on a particular machine.

numeric type Integer, real or complex data type.

object A data object.

object designator A designator for a data object.

object module The file output by most compilers. Multiple object modules are combined with libraries in a linker to produce the final executable program.

obsolescent feature A feature from earlier versions of Fortran that is considered to be redundant but that is still in frequent use. Obsolescent features have been replaced by better methods in later versions of Fortran. An example is the fixed source form, which has been replaced by free form. Obsolescent features are candidates for deletion in future version of Fortran as their use declines.

octal The base 8 number system, in which the legal digits are 0 through 7.

one-dimensional array A rank 1 array, or vector.

operand An expression that precedes or follows an operator.

operation A computation involving one or two operands.

operator A character or sequence of characters that defines an operation. There are two kinds: unary operators, which have one operand, and binary operators, which have two operands.

optional argument A dummy argument in a procedure that does not need to have a corresponding actual argument every time that the procedure is invoked. Optional arguments may only exist in procedures with an explicit interface.

out-of-bounds reference A reference to an array using a subscript either smaller than the lower bound or larger than the upper bound of the corresponding array dimension.

output buffer A section of memory used to hold a line of output data before it is sent to an output device.

output device A device used to output data from a computer. Common examples are printers and CRT displays.

output format A format used in a formatted output statement.

output statement A statement that sends formatted or unformatted data to an output device or file.

override Method overriding is a language feature that allows a subclass to provide a specific version of a method that is already defined its parent classes. The method in the subclass overrides the method in the superclass as long as it has the same name and signature.

parallel program A program containing multiple images that execute in parallel; a Coarray Fortran program.

parameter attribute An attribute in a type declaration statement that specifies that the named item is a constant instead of a variable.

parameterized variable A variable whose kind is explicitly specified.

parent The type being extended in an extended derived data type. This type appears in the parentheses after the EXTENDS(*parent_type*) clause.

pass-by-reference A scheme in which arguments are exchanged between procedures by passing the memory locations of the arguments, instead of the values of the arguments.

pointer A variable that has the `POINTER` attribute. A pointer may not be referenced or defined unless it is pointer associated with a target. If it is an array, it does not have a shape until it is associated, although it does have a rank. When a pointer is associated with a target, it contains the memory address of the target, and thus "points" to it.

pointer array An array that is declared with the `POINTER` attribute. Its rank is determined in the type declaration statement, but its shape and size are not known until memory is allocated for the array in an `ALLOCATE` statement.

pointer assignment statement A statement that associates a pointer with a target. Pointer assignment statement takes the form "`pointer => target`".

pointer association The process by which a pointer becomes associated with a target. The association status of a pointer can be checked with the `ASSOCIATED` intrinsic function.

pointer attribute An attribute in a type declaration statement that specifies that the named item is a pointer instead of a variable.

polymorphic Able to be of different types during program execution. A derived data type declared with the `CLASS` keyword is polymorphic.

pre-connected An input or output unit that is automatically connected to the program and does not require an `OPEN` statement. Examples are the standard input and standard output units.

precision The number of significant decimal digits that can be represented in a floating-point number.

present A dummy argument is present in a procedure invocation if it is associated with an actual argument, and the corresponding actual argument is present in the invoking program unit. The presence of a dummy argument can be checked with the `PRESENT` intrinsic function.

printer control character The first character of each output buffer. When it is sent to the printer, it controls the vertical movement of the paper before the line is written.

private An entity in a module that is not accessible outside the module by `USE` association; declared by a `PRIVATE` attribute or in a `PRIVATE` statement.

procedure A subroutine or function.

procedure interface The characteristics of a procedure, the name of the procedure, the name of each dummy argument, and the generic identifiers (if any) by which it may be referenced.

processor A processor is the combination of a specific computer with a specific compiler. Processor-dependent items can vary from computer to computer, or from compiler to compiler on the same computer.

program A sequence of instructions on a computer that causes the computer to carry out some specific functions.

program unit A main program, a subroutine, a function, a module, or a block data subprogram. Each of these units is separately compiled.

properties The data stored in an object.

pseudocode A set of English statements structured in a Fortran-like manner, and used to outline the approach to be taken in solving a problem without getting buried in the details of Fortran syntax.

public An entity in a module that is accessible outside the module by `USE` association; declared by a `PUBLIC` attribute or in a `PUBLIC` statement. An entity in a module is public by default.

pure function A pure procedure that is a function.

pure procedure A pure procedure is a procedure without side effects. A pure function must not modify its dummy arguments in any fashion, and all arguments must be `INTENT(IN)`. A pure subroutine must have no side effects except for arguments explicitly specified with `INTENT(OUT)` or `INTENT(INOUT)`. Such a procedure is declared with a `PURE` prefix,

Glossary

and pure functions may be used in specification expressions to initialize data in type declaration statements. Note that all elemental procedures are also pure.

pure subroutine A pure procedure that is a subroutine.

race condition A situation in which the results of a calculation depend on the speed at which multiple parallel calculations are performed. If different calculations finish in different orders, the final results of the calculation will differ.

random access Reading or writing the contents of a file in arbitrary order.

random access file Another name for a direct access file: A form of file in which the individual records can be written and read in any order. Direct access files must have records of fixed length so that the location of any particular record can be quickly calculated.

random access memory (RAM) The semiconductor memory used to store the programs and data that are actually being executed by a computer at a particular time.

range The difference between the largest and smallest numbers that can be represented on a computer with a given data type and kind. For example, on most computers a single precision real number has a range of 10^{-38} to 10^{38}, 0, and -10^{-38} to -10^{38}.

rank The number of dimensions of an array. The rank of a scalar is zero. The maximum rank of a Fortran array is 15.

rank 1 array An array having only one dimension, where each array element is addressed with a single subscript.

rank 2 array An array having two dimensions, where each array element is addressed with two subscripts.

rank n array An array having n dimensions, where each array element is addressed with n subscripts.

real An intrinsic data type used to represent numbers with a floating-point representation.

real arithmetic Arithmetic calculations performed with real or floating-point constants and variables.

real constant A numeric constant that contains a decimal point.

real number A number of the REAL data type.

real part The first of the two numbers that make up a COMPLEX data value.

real variable A variable that stores real (floating-point) data.

record A sequence of values or characters that is treated as a unit within a file. (A record is a "line" or unit of data from a file.)

record number The index number of a record in a direct access (or random access) file.

recursion The invocation of a procedure by itself, either directly or indirectly. Recursion is only allowed if the procedure is declared with the RECURSIVE keyword.

recursive Capable of being invoked recursively.

reference The appearance of a data object name in a context requiring the value at that point during execution, the appearance of a procedure name, its operator symbol, or a defined assignment statement in a context requiring execution of the procedure at that point, or the appearance of a module name in a USE statement. Neither the act of defining a variable nor the appearance of the name of a procedure as an actual argument is regarded as a reference.

relational expression A logical expression in which two nonlogical operands are compared by a relational operator to give a logical value for the expressions.

relational operator An operator that compares two nonlogical operands and returns a TRUE or FALSE result. Examples include >, >=, <. <=, ==, and /=.

repeat count The number before a format descriptor or a group of format descriptors, which specifies then the number of times that they are to be repeated. For example, the descriptor 4F10.4 is used four times.

root (*a*) The solution to an equation of the form $f(x) = 0$; (*b*) The node from which a binary tree grows.

round-off error The cumulative error that occurs during floating-point operations when the result of each calculation is rounded off to the nearest value representable with a particular kind of real values.

result variable The variable that returns the value of a function.

runtime error An error that only manifests itself when a program is executed.

SAVE attribute An attribute in the type declaration statement of a local variable in a procedure that specifies that value of the named item is to be preserved between invocations of the procedure. This attribute can also be specified in a separate SAVE statement.

scalar variable A variable that is not an array variable. The variable name refers to a single item of an intrinsic or derived type, and no subscripts are used with the name.

scope The part of a program in which a name or entity has a specified interpretation. There are three possible scopes in Fortran: global scope, local scope, and statement scope.

scoping unit A scoping unit is a single region of local scope within a Fortran program. All local variables have a single interpretation throughout a scoping unit. The scoping units in Fortran are: (*a*) a derived type definition, (*b*) an interface body, excluding any derived-type definitions and interface bodies within it, or (*c*) a program unit or subprogram, excluding derived-type definitions, interface bodies, and subprograms within it.

scratch file A temporary file that is used by a program during execution, and that is automatically deleted when it is closed. A scratch file may not be given a name.

secondary memory The computer memory used to store programs that are not currently being executed and the data that is not currently needed. This is typically a disk. Secondary memory is typically much slower than main memory, but it is also much cheaper.

separate procedure A procedure that is defined in a submodule.

sequential access Reading or writing the contents of a file in sequential order.

sequential file A form of file in which each record is read or written in sequential order. Sequential files do not require a fixed record length. They are the default file type in Fortran.

set methods Methods that modify the values of data stored in an object.

shape The rank and extent of an array in each of its dimensions. The shape can be stored in a rank 1 array, with each element of the array containing the extent of 1D.

side effects The modification by a function of the variables in its input argument list, or variables in modules made available by USE association, or variables in COMMON blocks.

single-precision A method of storing floating-point numbers on a computer that uses less memory than double precision, resulting in fewer significant digits and (usually) a smaller range in the representation of the numbers. Single-precision numbers are the "default real" type, the type of real number that results if no kind is specified.

single-threaded program A program that performs one calculation at a time sequentially from the time it starts until it ends.

size The total number of elements in an array.

solid state disk (or solid state drive, SSD) A data storage device that stores large amounts of data in solid-state nonvolatile memory.

source form The style in which a Fortran program is written—either free form or fixed form.

specific function A function that must always be called with a single type of argument. For example, the intrinsic function IABS is a specific function, while the intrinsic function ABS is a generic function.

specific intrinsic function An intrinsic function that is specific.

specification expression A restricted form of scalar integer constant expression that can appear in a type specification statement as a bound in an array declaration or as the length in a character declaration.

specifier   An item in a control list that provides additional information for the input/output statement in which it appears. Examples are the input/output unit number and the format specification for `READ` and `WRITE` statements.

statement entity   An entity whose scope is a single statement or part of a statement, such as the index variable in the implied `DO` loop of an array constructor.

standard error stream   This is output stream reserved for error messages.

standard input stream   The I/O unit accessed by the `READ (*,...)` statement. This is usually the keyboard.

standard output stream   The I/O unit accessed by the `WRITE (*,...)` statement. This is usually display.

statement label   A number preceding a statement that can be used to refer to that statement.

static memory allocation   Allocating memory for variables or arrays at compilation time, as opposed to dynamic memory allocation, which occurs during program execution.

static variable   A variable allocated at compilation time, and remaining in existence throughout the execution of a program.

storage association   A method of associating two or more variables or arrays by aligning their physical storage in a computer's memory. This was commonly achieved with `COMMON` blocks and `EQUIVALENCE` statements, but is not recommended for new programs.

stride   The increment specified in a subscript triplet.

structure   (*a*) An item of a derived data type. (*b*) An organized, standard way to describe an algorithm.

structure constructor   An unnamed (or literal) constant of a derived type. It consists of the name of the type followed by the components of the type in parentheses. The components appear in the order in which they were declared in the definition of the derived type. For example, the following line declares a constant of type `person`:

```
john = person('John','R','Jones','323-6439',21,'M','123-45-6789')
```

structure component   A part of an object of derived type that may be referenced by a component selector. A component selector consists of the object's name followed by the component's name, separated by a percent sign (%).

structured program   A program designed using a structured manner.

submodule   A program unit that extends a module or another submodule. Submodules provide a way to implement procedures whose interface is declared in a module, thus separating the interface of the procedure from its implementation.

subroutine   A procedure that is invoked by a `CALL` statement, and that returns its result through its arguments.

subscript   One of the integer values in parentheses following an array name, which are used to identify a particular element of the array. There is one subscript value for each dimension of the array.

subscript triplet   A method of specifying 1D of an array section by means of the initial and final values and a stride. The three components of the subscript triplet are written separated by colons, and some of them may be defaulted. For example, the following array section contains two subscript triplets: `array(1:3:2,2:4)`.

substring   A contiguous portion of a scalar character string.

substring specification   The specification of a substring of a character string. The specification takes the form `char_var(istart:iend)`, where `char_var` is the name of a character variable, `istart` is the first character in `char_var` to include in the substring, and `iend` is the first character in `char_var` to include in the substring.

subclass   A class that inherits data and methods from a parent class.

superclass   A class upon which a subclass is based.

synchronization point A point within a coarray Fortran program where two or more images wait for each other to arrive before execution continues.

synchronous input/output With synchronous input/output operations, program execution stops and waits until the input/output operations are completed.

syntax error An error in the syntax of a Fortran statement, detected by the compiler during compilation.

tail The last item in a linked list.

target A variable that has the TARGET attribute, and that can be the destination of a pointer.

test driver program A small program that is written specifically to invoke a procedure for the purpose of testing it.

top-down design The process of analyzing a problem by starting with the major steps, and successively refining each step until all of the small steps are easy to implement in Fortran code.

transformational intrinsic function An intrinsic function that is neither an elemental function nor an inquiry function. It usually has array arguments and an array result whose elements have values that depend on the values of many of the elements of the arguments.

tree A form of linked list in which each node points to two or more other nodes. If each node points to two other nodes, the structure is a binary tree.

truncation (*a*) The process in which the fractional part of a real number is discarded before the number is assigned to an integer variable. (*b*) The process in which excess characters are removed from the right-hand side of a character string before it is assigned to a character variable of shorter length.

truncation error (*a*) The error caused by terminating a calculation before it is complete. (*b*) The cumulative error that occurs during floating-point operations when the result of each calculation is truncated to the next lower value representable with a particular kind of real values.

truth table A table showing the result of a combinational logic expression for all possible combinations of operand values.

two-dimensional array A rank 2 array.

type-bound procedure A type-bound procedure is a procedure declared in a derived data type, which can only be accessed by reference to the data type.

type declaration statement A statement that specifies the type and optionally the attributes of one or more variables or constants: An INTEGER, REAL, DOUBLE PRECISION, COMPLEX, CHARACTER, LOGICAL, or TYPE (type-name) statement.

type parameter A parameter of an intrinsic data type. KIND and LEN are the type parameters.

ultimate type A structure component that is of intrinsic type. Structure components of derived data types are *not* ultimate types.

unary operator An operator that has only one operand, such as .NOT. or the unary minus.

undefined A data entity that does not have a defined value.

unformatted file A file containing data stored in a sequence of bit patterns that are a direct copy of a portion of the computer's memory. Unformatted files are processor dependent, and can only be produced by unformatted WRITE statements and read by unformatted READ statements on the particular type of processor that produced them.

unformatted input statement An unformatted READ statement.

unformatted output statement An unformatted WRITE statement.

unformatted READ statement A READ statement that does not contain a format specifier. Unformatted READ statements transfer bit patterns directly from an external device into memory without interpretation.

Glossary

unformatted record A record consisting of a sequence of bit patterns that are a direct copy of a portion of the computer's memory. Unformatted records are processor dependent, and can only be produced by unformatted `WRITE` statements and read by unformatted `READ` statements on the particular type of processor that produced them.

unformatted WRITE statement A `WRITE` statement that does not contain a format specifier. Unformatted `WRITE` statements transfer bit patterns directly from a processor's memory to an external device without interpretation.

Unicode An internal character coding scheme that uses 2 bytes to represent each character. The Unicode system can represent 65,536 possible different characters. The first 128 Unicode characters are identical to the ASCII character set, and other blocks of characters are devoted to various languages such as Chinese, Japanese, Hebrew, Arabic, Hindi, etc.

uninitialized array An array, some or all of whose elements have not been initialized.

uninitialized variable A variable that has been defined in a type declaration statement, but for which no initial value has been assigned.

unit An input/output unit.

unit specifier A specifier that specifies the unit on which input or output is to occur.

unit testing The process of testing individual procedures separately and independently before they are combined into a final program.

unlimited polymorphic A pointer or dummy argument declared to be of type `CLASS(*)` is unlimited polymorphic, because the pointer or dummy argument will work with objects of any class.

upper bound The maximum value permitted for a subscript of an array.

USE association The manner in which the contents of a module are made available for use in a program unit.

USE statement A statement that references a module in order to make the contents of the module available for use in the program unit containing it.

user-defined function A function written by a user.

utility method A method inside an object that performs some function, but is not intended to be called directly by a user.

value separator A comma, a space, a slash, or end of record that separates two data values in a list-directed input.

variable A data object whose value may be changed during program execution.

variable declaration The declaration of the type and, optionally, the attributes of a variable.

vector A rank 1 array.

vector subscript A method of specifying an array section by a rank 1 array containing the subscripts of the elements to include in the array section.

volatile memory Memory that is erased when power is turned off.

well-conditioned system A system of equations whose solution is relatively insensitive to small changes in the values of its coefficients, or to truncation and round-off errors.

while loop A loop that executes indefinitely until some specified condition is satisfied.

whole array An array that has a name.

WHERE construct The construct used in a masked array assignment.

word The fundamental unit of memory on a particular computer. The size of a word varies from processor to processor, but it is typically 16, 32, or 64 bits.

work array A temporary array used for the storage of intermediate results. This can be implemented as an automatic array in modern Fortran.

worker image Images 2 – n in a Coarray Fortran program.

APPENDIX F

Answers to Quizzes

QUIZ 1–1

1. (a) 11011_2 (b) 1011_2 (c) 100011_2 (d) 1111111_2
2. (a) 14_{10} (b) 85_{10} (c) 9_{10}
3. (a) 162655_8 or $E5AD_{16}$ (b) 1675_8 or $3BD_{16}$ (c) 113477_8 or $973F_{16}$
4. $131_{10} = 10000011_2$, so the fourth bit is a zero.
5. (a) ASCII: M   (b) ASCII: {   (c) ASCII: (unused)
6. (a) –32768   (b) 32767
7. Yes, a 4-byte variable of the real data type can be used to store larger numbers than a 4-byte variable of the integer data type. The 8 bits of exponent in a real variable can represent values as large as 10^{38}. A 4-byte integer can only represent values as large as 2,147,483,647 (about 10^9). To do this, the real variable is restricted to 6 or 7 decimal digits of precision, while the integer variable has 9 or 10 decimal digits of precision.

QUIZ 2–1

1. Valid real constant.
2. Invalid—commas not permitted within constants.
3. Invalid—real constants must have a decimal point.
4. Invalid—single quotes within a character string delimited by single quotes must be doubled. Correct forms are: `'That"s ok!'` or `"That's ok!"`.
5. Valid integer constant.
6. Valid real constant.
7. Valid character constant.
8. Valid character constant.
9. Invalid—character constants must be enclosed by symmetrical single or double quotes.
10. Valid character constant.
11. Valid real constant.
12. Invalid—real exponents are expressed using the E symbol instead of ^.
13. Same.
14. Same.
15. Different.
16. Different.
17. Valid program name.
18. Invalid—program name must begin with a letter.
19. Valid integer variable.
20. Valid real variable.
21. Invalid—name must begin with a letter.

22. Valid real variable.
23. Invalid—name must begin with a letter.
24. Invalid—no double colons (::) present.
25. Valid.

QUIZ 2–2

1. The order is (1) exponentials, working from right to left; (2) multiplications and divisions, working from left to right; (3) additions and subtractions, working from left to right. Parentheses modify this order—terms in parentheses are evaluated first, starting from the innermost parentheses and working outward.
2. (a) Legal: Result = 12; (b) Legal: Result = 42; (c) Legal: Result = 2; (d) Legal: Result = 2; (e) Illegal: Division by 0; (f) Legal: Result = −40.5 Note that this result is legal because exponentiation precedes negation in operator precedence. It is equivalent to the expression: -(3.**(4./2.)), and does *not* involve taking the real power of a negative number; (g) Legal: Result = 0.111111; (h) Illegal: Two adjacent operators.
3. (a) 7; (b) −21; (c) 7; (d) 9
4. (a) Legal: Result = 256; (b) Legal: Result = 0.25; (c) Legal: Result = 4; (d) Illegal: Negative real number raised to a real power; (e) Legal: Result = 0.25; (f) Legal: Result = −0.125.
5. The statements are illegal, since they try to assign a value to named constant k.
6. result = 43.5.
7. a = 3.0; b = 3.333333; n = 3.

QUIZ 2–3

1. r_eq = r1 + r2 + r3 + r4
2. r_eq = 1. / (1./r1 + 1./r2 + 1./r3 + 1./r4)
3. t = 2. * pi * SQRT(l / g)
4. v = v_max * EXP(- alpha * t) * COS(omega * t)
5. $d = \frac{1}{2}at^2 + v_o t + x_o$
6. $f = \dfrac{1}{2\pi\sqrt{LC}}$
7. $E = \frac{1}{2}Li^2$
8. The results are
    126   5.000000E-02

    Make sure that you can explain why a is equal to 0.05!
9. The results are shown below. Can you explain why each value was assigned to a given variable by the READ statements?
    1   3   180   2.000000   30.000000   3.4899499E-02

QUIZ 3–1

1. (a) Legal: Result = .FALSE.; (b) Illegal: .NOT. only works with logical values; (c) Legal: Result = .TRUE.; (d) Legal: Result = .TRUE. (because the .NOT. is evaluated before the .AND.) (e) Legal: Result = .TRUE.; (f) Legal: Result = .TRUE.; (g) Legal: Result = .FALSE.; (h) Illegal: .OR. only works with logical values.

2. An "F" (for false) will be printed, because i + j = 4 while k = 2, so that the expression i + j == k evaluates to be false.

QUIZ 3–2

1.
```
IF ( x >= 0. ) THEN
    sqrt_x = SQRT( x )
    WRITE (*,*) 'The square root of x is ', sqrt_x
ELSE
    WRITE (*,*) 'Error--x < 0!'
    sqrt_x = 0.
END IF
```
2.
```
IF ( ABS(denominator) < 1.0E-10 ) THEN
    WRITE (*,*) 'Divide by zero error!'
ELSE
    fun = numerator / denominator
    WRITE (*,*) 'FUN = ', fun
END IF
```
3.
```
IF ( distance > 300. ) THEN
    cost = 70. + 0.15 * ( distance - 300. )
ELSE IF ( distance > 100. ) THEN
    cost = 30. + 0.20 * ( distance - 100. )
ELSE
    cost = 0.30 * distance
END IF
average_cost = cost / distance
```
4. These statements are incorrect. There is no ELSE in front of IF (VOLTS < 105.).
5. These statements are correct. They will print out the warning because warn is .TRUE., even though the speed limit is not exceeded.
6. These statements are incorrect, since a real value is used to control the operation of a CASE statement.
7. These statements are correct. They will print out the message 'Prepare to stop.'
8. These statements are technically correct, but they are unlikely to do what the user intended. If the temperature is greater than 100°, then the user probably wants 'Boiling point of water exceeded' to be printed out. Instead, the message 'Human body temperature exceeded' will be printed out, since the IF structure executes the first true branch that it comes to. If the temperature is greater than 100°, it is also greater than 37°.

QUIZ 4–1

1. 6
2. 0
3. 1
4. 7
5. 6
6. 0
7. ires = 10
8. ires = 55

9. ires = 10 (Note that once ires = 10, the loop will begin to cycle, and ires will never be updated again no matter how many times the loop executes!)
10. ires = 100
11. ires = 60
12. Invalid: These statements redefine DO loop index i within the loop.
13. Valid.
14. Illegal: DO loops overlap.

QUIZ 4–2

1. (*a*) Legal: Result = .FALSE. (*b*) Legal: Result = .TRUE. (*c*) Legal: Result = 'Hello there' (*d*) Legal: Result = 'Hellothere'
2. (*a*) Legal: Result = 'bcd' (*b*) Legal: Result = 'ABCd' (*c*) Legal: Result = .FALSE. (*d*) Legal: Result = .TRUE. (*e*) Illegal: Can't compare character strings and integers (*f*) Legal: Result = .TRUE. (*g*) Legal: Result = .FALSE.
3. The length of str3 is 20, so the first WRITE statement produces a 20. The contents of str3 are 'Hello   World' (with five blanks in the middle), so the trimmed length of the string is 15. After the next set of manipulations, the contents of str3 are 'HelloWorld', so the third WRITE statement prints out 20 and the fourth one prints out 10.

QUIZ 5–1

Note: There are more than one way to write each of the FORMAT statements in this Quiz. The answers shown below represent one of many possible correct answers to these questions.

1. WRITE (*,100)
   100 FORMAT (24X,'This is a test!')
2. WRITE (*,110) i, j, data1
   100 FORMAT (/,2I10,F10.2)
3. WRITE (*,110) result
   110 FORMAT (T12,'The result is ',ES12.4)
4.   -.0001*********   3.1416
   ----|----|----|----|----|----|
      5   10 15 20 25 30
5.   .000   .602E+24   3.14159
   ----|----|----|----|----|----|----|
      5   10 15 20 25 30 35
6. ********* 6.0200E+23   3.1416
   ----|----|----|----|----|----|-
      5   10 15 20 25 30
7. 32767
      24
   *****
   ----|----|----|----|----|----|
      5   10 15 20 25 30
8.   32767 00000024 -1010101
   ----|----|----|----|----|----|
      5   10 15 20 25 30

9. ```
   ABCDEFGHIJ    12345
   ----|----|----|----|----|----|
       5   10   15   20   25   30
   ```
10. ```
                         ABC12345IJ
   ----|----|----|----|----|----|
       5   10   15   20   25   30
   ```
11. ```
   ABCDE 12345
   ----|----|----|----|----|----|
       5   10   15   20   25   30
   ```
12. Correct—all format descriptors match variable types.
13. Incorrect. Format descriptors do not match variable types for `test` and `ierror`.
14. This program writes the following data.

```
         Output Data
         ===========

   POINT( 1) =      1.200000      2.400000
   POINT( 2) =      2.400000      4.800000


   ----|----|----|----|----|----|----|----|
       5   10   15   20   25   30   25   40
```

QUIZ 5-2

Note: There are more than one way to write each of the FORMAT statements in this Quiz. The answers shown below represent one of many possible correct answers to these questions.

1. ```
   READ (*,100) amplitude, count, identity
   100 FORMAT (9X,F11.2,T30,I6,T60,A13)
   ```
2. ```
   READ (*,110) title, i1, i2, i3, i4, i5
   110 FORMAT (T10,A25,/(4X,I8))
   ```
3. ```
   READ (*,120) string, number
   120 FORMAT (T11,A10,///,T11,I10)
   ```
4. a = 1.65 x 10^{-10}, b = 17., c = -11.7
5. a = -3.141593, b = 2.718282, c = 37.55
6. i = -35, j = 6705, k = 3687
7. string1 = 'FGHIJ', string2 = 'KLMNOPQRST', string3 = 'UVWXYZ0123', string4 = ' _TEST_ 1'
8. Correct.
9. Correct. These statements read integer `junk` from columns 60 to 74 of one line, and then read real variable `scratch` from columns 1 to 15 of the next line.
10. Incorrect. Real variable `elevation` will be read with an I6 format descriptor.

QUIZ 5-3

1. ```
   OPEN (UNIT=25, FILE='IN052691', ACTION='READ', IOSTAT=ierror, &
         IOMSG=msg)
      IF ( istat /= 0 ) THEN
   ```
   ```

Answers to Quizzes

```
            WRITE (*,'(A,I6)') 'Open error on file. IOSTAT = ', ierror
            WRITE (*,'(A)') msg
         ELSE
            ...
         END IF
```
2. `OPEN (UNIT=4, FILE=out_name, STATUS='NEW', ACTION='WRITE', &`
      `IOSTAT=istat, IOMSG=msg)`
3. `CLOSE (UNIT=24)`
4. `READ (8,*,IOSTAT=istat) first, last`
   `IF ( istat < 0 ) THEN`
      `WRITE (*,*) 'End of file encountered on unit 8.'`
   `END IF`
5. `DO i = 1, 8`
      `BACKSPACE (UNIT=13)`
   `END DO`
6. Incorrect. File `data1` has been replaced, so there is no data to read.
7. Incorrect. You cannot specify a file name with a scratch file.
8. Incorrect. There is nothing in the scratch file to read, since the file was created when it was opened.
9. Incorrect. You cannot use a real value as an i/o unit number.
10. Correct.

QUIZ 6–1

1. 15
2. 256
3. 41
4. Valid. The array will be initialized with the values in the array constructor.
5. Valid. All 10 values in the array will be initialized to 0.
6. Valid. Every tenth value in the array will be initialized to 1000, and all other values will be initialized to zero. The values will then be written out 10 per line.
7. Invalid. The arrays are not conformable, since `array1` is 11 elements long and `array2` is 10 elements long.
8. Valid. Every tenth element of array `in` will be initialized to 10, 20, 30, etc. All other elements will be zero. The 10-element array `sub1` will be initialized to 10, 20, 30, ..., 100, and the 10-element array `sub2` will be initialized to 1, 2, 3, ..., 10. The multiplication will work because arrays `sub1` and `sub2` are conformable.
9. Mostly valid. The values in array `error` will be printed out. However, since `error(0)` was never initialized, we don't know what will be printed out, or even whether printing that array element will cause an I/O error.
10. Valid. Array `ivec1` will be initialized to 1, 2, ..., 10, and array `ivec2` will be initialized to 10, 9, ..., 1. Array `data1` will be assigned the values 1., 4., 9., ..., 100. The `WRITE` statement will print out 100., 81., 64., ..., 1., because of the vector subscript.
11. Probably invalid. These statements will compile correctly, but they probably do *not* do what the programmer intended. A 10-element integer array `mydata` will be created. Each `READ` statement reads values into the entire array, so array `mydata` will be initialized 10 times over (using up 100 input values!). The user probably intended for each array element to be initialized only once.

QUIZ 7-1

1. The call to `ave_sd` is incorrect. The second argument is declared as an integer in the calling program, but it is a real within the subroutine.
2. These statements are valid. When the subroutine finishes executing, `string2` contains the mirror image of the characters in `string1`.
3. These statements are incorrect. Subroutine `sub3` uses 30 elements in array `iarray`, but there are only 25 values in the array passed from the calling program. Also, the subroutine uses an assumes-size dummy array, which should not be used in any new programs.

QUIZ 7-2

1. If data values are defined in a module, and then two or more procedures USE that module, they can all see and share the data. This is a convenient way to share private data among a group of related procedures, such as `random0` and `seed` in Example 7-4.
2. If procedures are placed in a module and accessed by USE association, then they will have explicit interfaces, allowing the compiler to catch many errors in calling sequence.
3. There is no error in this program. The main program and the subroutine share data using module `mydata`. The output from the program is `a(5) = 5.0`.
4. This program is invalid. Subroutine `sub2` is called with a constant as the second argument, which is declared to be INTENT(OUT) in the subroutine. The compiler will catch this error because the subroutine is inside a module accessed by USE association.

QUIZ 7-3

1. ```
   REAL FUNCTION f2(x)
   IMPLICIT NONE
   REAL, INTENT(IN) :: x
   f2 = (x -1.) / (x + 1.)
   END FUNCTION f2
   ```
2. ```
   REAL FUNCTION tanh(x)
   IMPLICIT NONE
   REAL, INTENT(IN) :: x
   tanh = (EXP(x)-EXP(-x)) / (EXP(x)+EXP(-x))
   END FUNCTION tanh
   ```
3. ```
   FUNCTION fact(n)
   IMPLICIT NONE
   INTEGER, INTENT(IN) :: n
   INTEGER :: fact
   INTEGER :: i
   fact = 1.
   DO i = n, 1, -1
      fact = fact * i
   END DO
   END FUNCTION fact
   ```
4. ```
   LOGICAL FUNCTION compare(x,y)
   IMPLICIT NONE
   REAL, INTENT(IN) :: x, y
   ```

```
        compare = (x**2 + y**2) > 1.0
        END FUNCTION compare
```
5. This function is incorrect because `sum` is never initialized. The sum must be set to zero before
   the DO loop is executed.
6. This function is invalid. Argument a is INTENT(IN), but its value is modified in the function.
7. This function is valid.

QUIZ 8–1

1. 645 elements. The valid range is `data_input(-64,0)` to `data_input(64,4)`.
2. 213 elements. The valid range is `filenm(1,0)` to `filenm(3,70)`.
3. 294 elements. The valid range is `in(-3,-3,1)` to `in(3,3,6)`.
4. Invalid. The array constructor is not conformable with array `dist`.
5. Valid. `dist` will be initialized with the values in the array constructor.
6. Valid. Arrays `data1`, `data2`, and `data_out` are all conformable, so this addition is valid. The first WRITE statement prints the five values: 1., 11., 11., 11., 11., and the second WRITE statement prints the two values: 11., 11.
7. Valid. These statements initialize the array, and then select the subset specified by `list1` = [1,4,2,2], and `list2` = [1,2,3]. The resulting array section is

$$\text{array(list1,list2)} = \begin{bmatrix} \text{array}(1,1) & \text{array}(1,2) & \text{array}(1,3) \\ \text{array}(4,1) & \text{array}(4,2) & \text{array}(4,3) \\ \text{array}(2,1) & \text{array}(2,2) & \text{array}(2,3) \\ \text{array}(2,1) & \text{array}(2,2) & \text{array}(2,3) \end{bmatrix}$$

$$\text{array(list1,list2)} = \begin{bmatrix} 11 & 21 & 31 \\ 14 & 24 & 34 \\ 12 & 22 & 32 \\ 12 & 22 & 32 \end{bmatrix}$$

8. Invalid. There is a many-one array section of the left-hand side of an assignment statement.
9. The data on the first three lines would be read into array `input`. However, the data is read in column order, so `mydata(1,1)` = 11.2, `mydata(2,1)` = 16.5, `mydata(3,1)` = 31.3, etc. `mydata(2,4)` = 15.0.
10. The data on the first three lines would be read into array `input`. The data is read in column order, so `mydata(0,2)` = 11.2, `mydata(1,2)` = 16.5, `mydata(2,2)` = 31.3, etc. `mydata(2,4)` = 17.1.
11. The data on the first three lines would be read into array input. This time, the data is read in row order, so `mydata(1,1)` = 11.2, `mydata(1,2)` = 16.5, `mydata(1,3)` = 31.3, etc. `mydata(2,4)` = 17.1.
12. The data on the first three lines would be read into array input. The data is read in row order, but only the first five values on each line are read by each READ statement. The next READ statement begins with the first value on the next input line. Therefore, `mydata(2,4)` = 11.0.

13. −9.0.
14. The rank of array `mydata` is 2.
15. The shape of array `mydata` is 3 × 5.
16. The extent of the first dimension of array `data_input` is 129.
17. 15.

QUIZ 8–2

1. LBOUND(values,1) = -3, UBOUND(values,2) = 50, SIZE(values,1) = 7,
   SIZE(values) = 357, SHAPE(values) = [7,51]
2. UBOUND(values,2) = 4, SIZE(values) = 60, SHAPE(values) = [3,4,5]
3. MAXVAL(input1) = 9.0, MAXLOC(input1) = [5,5]
4. SUM(arr1) = 5.0, PRODUCT(arr1) = 0.0,
   PRODUCT(arr1, MASK=arr1 /= 0.) = -45.0, ANY(arr1>0) = T,
   ALL(arr1>0) = F
5. The values printed out are: SUM(arr2, MASK=arr2 > 0.) = 20.0
6. REAL, DIMENSION(5,5) :: input1
   FORALL (i=1:5, j=1:5)
      input1(i,j) = i+j-1
   END FORALL
   WRITE (*,*) MAXVAL(input1)
   WRITE (*,*) MAXLOC(input1)
7. Invalid. The expression in the WHERE structure is not conformable with the masking statement.
8. Invalid. Array `time` must be allocated before it is initialized.
9. Valid. Since the array is not allocated, the result of the ALLOCATED function is .FALSE., and output of the WRITE statement is F.

QUIZ 9–1

1. The SAVE statement or the SAVE attribute should be used in any procedure that depends on local data values being unchanged between invocations of the procedure. All local variables that must remain constant between invocations should be declared with the SAVE attribute.
2. An automatic array is a local array in a procedure whose extent is specified by variables passed to the procedure when it is invoked. The array is automatically created each time procedure is invoked, and is automatically destroyed each time the procedures exit. Automatic arrays should be used for temporary storage within a procedure. An allocatable array is an array declared with the ALLOCATABLE attribute, and allocated with an ALLOCATE statement. It is more general and flexible than an automatic array, since it may appear in either main programs or procedures. Allocatable arrays can create memory leaks if misused. Allocatable arrays should be used to allocate memory in main programs.
3. Assumed-shape dummy arrays have the advantage (compared to assumed-size arrays) that they can be used with whole array operations, array intrinsic functions, and array sections. They are simpler than explicit-shape dummy arrays because the bounds of each array do not have to be passed to the procedure. The only disadvantage associated with them is that they must be used with an explicit interface.

4. This program will work on many processors, but it has two potentially serious problems. First, the value of variable `isum` is never initialized. Second, `isum` is not saved between calls to `sub1`. When it works, it will initialize the values of the array to 1, 2, ..., 10.
5. This program will work. When array `b` is written out, it will contain the values:

$$b = \begin{bmatrix} 2. & 8. & 18. \\ 32. & 50. & 72. \\ 98. & 128. & 162. \end{bmatrix}$$

6. This program is invalid. Subroutine `sub4` uses assumed-shape arrays but does not have an explicit interface.

QUIZ 10-1

1. False.
2. False.
3. False.
4. These statements are legal.
5. This function is legal, provided that it has an explicit interface. Automatic length character functions must have an explicit interface.
6. Variable `name` will contain the string:
   `'JOHNSON        ,JAMES              R'`.
7. `a = '123'; b = 'ABCD23   IJKL'`
8. `ipos1 = 17, ipos2 = 0, ipos3 = 14, ipos4 = 37`

QUIZ 10-2

1. Valid. The result is -1234, because `buff1(10:10)` is `'J'`, not `'K'`.
2. Valid. After these statements `outbuf` contains
   `'    123        0        -11            '`
3. The statements are valid. `ival1 = 456789, ival2 = 234, rval3 = 5678.90`

QUIZ 11-1

1. This answer to this question is processor-dependent. You must consult the manuals for your particular compiler.
2. `(-1.980198E-02,-1.980198E-01)`
3. ```
   PROGRAM complex_math
   !
   !   Purpose:
   !     To perform the complex calculation:
   !        D = ( A + B ) / C
   !     where A = ( 1., -1.)
   !           B = (-1., -1.)
   !           C = (10.,  1.)
   ```
   ```

```fortran
!      without using the COMPLEX data type.
!
IMPLICIT NONE
!
REAL :: ar = 1.,  ai = -1.
REAL :: br = -1., bi = -1.
REAL :: cr = 10., ci = 1.
REAL :: dr, di
REAL :: tempr, tempi

CALL complex_add ( ar, ai, br, bi, tempr, tempi )
CALL complex_divide ( tempr, tempi, cr, ci, dr, di )

WRITE (*,100) dr, di
100 FORMAT (1X,'D = (',F10.5,',',F10.5,')' )

END PROGRAM complex_math

SUBROUTINE complex_add ( x1, y1, x2, y2, x3, y3 )
!
! Purpose:
!   Subroutine to add two complex numbers (x1, y1) and
!   (x2, y2), and store the result in (x3, y3).
!
IMPLICIT NONE

REAL, INTENT(IN) :: x1, y1, x2, y2
REAL, INTENT(OUT) :: x3, y3

x3 = x1 + x2
y3 = y1 + y2

END SUBROUTINE complex_add

SUBROUTINE complex_divide ( x1, y1, x2, y2, x3, y3 )
!
! Purpose:
!   Subroutine to divide two complex numbers (x1, y1) and
!   (x2, y2), and store the result in (x3, y3).
!
IMPLICIT NONE

REAL, INTENT(IN) :: x1, y1, x2, y2
REAL, INTENT(OUT) :: x3, y3
REAL :: denom

denom = x2**2 + y2**2
x3 = (x1 * x2 + y1 * y2) / denom
y3 = (y1 * x2 - x1 * y2) / denom

END SUBROUTINE complex_divide
```

It is much easier to use the complex data type to solve the problem than it is to use the definitions of complex operations and real numbers.

QUIZ 12-1

1. ```
   WRITE (*,100) points(7)%plot_time%day, points(7)%plot_time%month, &
                 points(7)%plot_time%year, points(7)%plot_time%hour, &
                 points(7)%plot_time%minute, points(7)%plot_time%second
   100 FORMAT (I2.2,'/',I2.2,'/',I4.4,' ',I2.2,':',I2.2,':',I2.2)
   ```
2. ```
   WRITE (*,110) points(7)%plot_position%x, &
                 points(7)%plot_position%y, &
                 points(7)%plot_position%z
   110 FORMAT (' x = ',F12.4,' y = ',F12.4,' z = ',F12.4 )
   ```
3. To calculate the time difference, we must subtract the times associated with the two points, taking into account the different scales associated with hours, minutes, seconds, etc. The code below converts the times to seconds before subtracting them, and also assumes that both points occur on the same day, month, and year. (It is easy to extend this calculation to handle arbitrary days, months, and years as well, but double-precision real arithmetic must be used for the calculations.) To calculate the position difference, we use the equation

$$dpos = \sqrt{(x_2 - x_1)^2 + (y_2 - y_1)^2 + (z_2 - z_1)^2}$$

```
time1 = points(2)%plot_time%second + 60.*points(2)%plot_time%minute &
      + 3600.*points(2)%plot_time%hour
time2 = points(3)%plot_time%second + 60.*points(3)%plot_time%minute &
      + 3600.*points(3)%plot_time%hour
dtime = time2 - time1

dpos = SQRT ( &
       (points(3)%plot_position%x - points(2)%plot_position%x )**2 &
     + (points(3)%plot_position%y - points(2)%plot_position%y )**2 &
     + (points(3)%plot_position%z - points(2)%plot_position%z )**2 )

rate = dpos / dtime
```
4. Valid. This statement prints out all of the components of the first element of array points.
5. Invalid. The format descriptors do not match the order of the data in `points(4)`.
6. Invalid. Intrinsic operations are not defined for derived data types, and component `plot_position` is a derived data type.

QUIZ 13-1

1. The scope of an object is the portion of a Fortran program over which the object is defined. The four levels of scope are global, local, block, and statement.
2. Host association is the process by which data entities in a host scoping unit are made available to an inner scoping unit. If variables and constants are defined in a host scoping unit, then those variables and constants are inherited by any inner scoping units *unless* another object with the same name is explicitly defined in the inner scoping unit.
3. When this program is executed z = 3.666667. Initially, z is set to 10.0, and then function `fun1(z)` is invoked. The function is an internal function, so it inherits the values of derived type variable xyz by host association. Since xyz%x = 1.0 and xyz%z = 3.0, the function evaluates to (10. + 1.)/3. = 3.666667. This function result is then stored in variable z.

4. i = 20. The first executable statement changes i to 27, and the fourth executable statement subtracts 7 from it to produce the final answer. (The i in the third statement has statement scope only, and so does not affect the value of i in the main program.)
5. This program is illegal. The program name abc must be unique within the program.
6. Recursive procedures are procedures that can call themselves. They are declared using the RECURSIVE keyword in SUBROUTINE or FUNCTION statement. If the recursive procedure is a function, then the FUNCTION statement should also include a RESULT clause.
7. This function is illegal. The type of the *function name* is declared. The type of the *function result* sum should be declared instead.
8. Keyword arguments are calling arguments of the form KEYWORD=value, where KEYWORD is the name used to declare the dummy argument in the procedure definition, and value is the value to be passed to that dummy argument when the procedure is invoked. Keyword arguments may only be used if the procedure being invoked has an explicit interface. Keyword arguments may be used to allow calling arguments to be specified in a different order, or to specify only certain optional arguments.
9. Optional arguments are arguments that do not have to be present when a procedure is invoked, but which will be used if they are present. Optional arguments may only be used if the procedure being invoked has an explicit interface. They may be used for input or output data that is not needed every time a procedure is invoked.

QUIZ 13–2

1. An interface block is a way to specify an explicit interface for a separately-compiled external procedure. It consists of an INTERFACE statement and an END INTERFACE statement. Between these two statements are statements declaring the calling sequence of the procedure, including the order, type, and intent of each argument. Interface blocks may be placed in the declaration section of an invoking program unit, or else they may be placed in a module, and that module may be accessed by the invoking program unit via USE association.
2. A programmer might choose to create an interface block for a procedure because the procedure may be written in a language other than Fortran, or because the procedure must work with both Fortran 90 (and later) and older FORTRAN 77 applications.
3. The interface body contains a SUBROUTINE or FUNCTION statement declaring the name of the procedure and its dummy arguments, followed by type declaration statements for each of the dummy arguments. It concludes with an END SUBROUTINE or END FUNCTION statement.
4. This program is valid. The multiple definitions for x1 and x2 do not interfere with each other because they are in different scoping units. When the program is executed, the results are:
   This is a test.   613.000   248.000
5. A generic procedure is defined using a named interface block. The name of the generic procedure is specified in the INTERFACE statement, and the calling sequences of all possible specific procedures are specified in the body of the interface block. Each specific procedure must be distinguishable from all of the other specific procedures by some combination of its nonoptional calling arguments. If the generic interface block appears in a module and the corresponding specific procedures are also defined in the module, then they are specified as being a part of the generic procedure with MODULE PROCEDURE statements.

6. A generic bound procedure is defined using a GENERIC statement in the type definition. The GENERIC statement will declare the generic name of the procedure, followed by the list of specific procedures associated with it:
```
TYPE :: point
   REAL :: x
   REAL :: y
CONTAINS
   GENERIC :: add => point_plus_point, point_plus_scalar
END TYPE point
```
7. This generic interface is illegal, because the number, types, and order of the dummy arguments for the two specific procedures are identical. There must be a difference between the two sets of dummy arguments so that the compiler can determine which one to use.
8. A MODULE PROCEDURE statement is used to specify that a specific procedure is a part of a generic procedure (or operator definition) when both the specific procedure and the generic procedure (or operator definition) appear within the same module. It is used because any procedure in a module automatically has an explicit interface. Respecifying the interface in a generic interface block would involve declaring the explicit interface of the procedure twice, which is illegal.
9. A user-defined operator is declared using the INTERFACE OPERATOR block, while a user-defined assignment is declared using the INTERFACE ASSIGNMENT block. A user-defined operator is implemented by a one- or two-argument function (for unary and binary operators, respectively). The arguments of the function must have INTENT(IN), and the result of the function is the result of the operation. A user-defined assignment is implemented using a two-argument subroutine. The first argument must be INTENT(OUT) or INTENT(INOUT), and the second argument must be INTENT(IN). The first argument is the result of the assignment operation.
10. Access to the contents of a module may be controlled using PUBLIC, PRIVATE, and PROTECTED statements or attributes. It might be desirable to restrict access to the internal components of some user-defined data types, or to restrict direct access to procedures used to implement user-defined operators or assignments, so these items can be declared to be PRIVATE. The PROTECTED access allows a variable to be used but not modified, so it is effective read-only outside of the module in which it is defined.
11. The default type of access for items in a module is PUBLIC.
12. A program unit accessing items in a module by USE association can limit the items in the module that it accesses by using the ONLY clause in the USE statement. A programmer might wish to limit access in this manner to avoid conflicts if a public item in the module has the same name as a local item in the programming unit.
13. A program unit accessing items in a module by USE association can rename the items in the module that it accesses by using the => option in the USE statement. A programmer might wish to rename an item in order to avoid conflicts if an item in the module has the same name as a local item in the programming unit.
14. This program is illegal, because the program attempts to modify the protected value t1%z.

QUIZ 14–1

1.
```
       4096.1  4096.07  .40961E+04  4096.1      4096.
---|----|----|----|----|----|----|----|----|----|----|----|
    5   10   15   20   25   30   35   40   45   50   55   60
```

2. Data1(1) = -17.2000,   Data1(2) = 4.0000,
    Data1(3) =   4.0000,   Data1(4) =   .3000,
    Data1(5) = -2.2200
```
---|----|----|----|----|----|----|----|----|----|----|----|
    5   10   15   20   25   30   35   40   45   50   55   60
```
3. 12.200000E-06 12.345600E+06
    1.220000E-05   1.234560E+07
```
---|----|----|----|----|----|
    5   10   15   20   25   30
```
4. i =   -2002 j =   -1001 k =   -3
```
---|----|----|----|----|----|----|----|----|
    5   10   15   20   25   30   35   40   45
```

QUIZ 14–2

1. A formatted file contains information stored as ASCII or Unicode characters. The information in a formatted file can be read with a text editor. By contrast, an unformatted file contains information stored in a form that is an exact copy of the bit patterns in the computer's memory. Its contents cannot be easily examined. Formatted files are portable between processors, but they occupy a relatively large amount of space and require extra processor time to perform the translations on input and output. Unformatted files are more compact and more efficient to read and write, but they are not portable between processors of different types.

2. A direct access file is a file whose records can be read and written in any arbitrary order. A sequential access file is a file whose records must be read and written sequentially. Direct access files are more efficient for accessing data in random order, but every record in a direct access file must be the same length. Sequential access files are efficient for reading and writing data in sequential order, but are very poor for random access. However, the records in a sequential access file may have variable lengths.

3. The INQUIRE statement is used to retrieve information about a file. The information may be retrieved by (1) file name or (2) I/O unit number. The third form of the INQUIRE statement is the IOLENGTH form. It calculates the length of a record in an unformatted direct-access file in processor-dependent units.

4. Invalid. It is illegal to use a file name with a scratch file.

5. Invalid. The RECL= clause must be specified when opening a direct access file.

6. Invalid. By default, direct access files are opened unformatted. Formatted I/O cannot be performed to unformatted files.

7. Invalid. By default, sequential access files are opened formatted. Unformatted I/O cannot be performed to formatted files.

8. Invalid. Either a file name or an i/o unit may be specified in an INQUIRE statement, but not both.

9. The contents of file 'out.dat' will be:
```
&LOCAL_DATA
A =   -200.000000  -17.000000  0.000000E+00  100.000000   30.000000
B =    -37.000000
C =      0.000000E+00
/
```

Answers to Quizzes

QUIZ 15-1

1. A pointer is a Fortran variable that contains the *address* of another Fortran variable or array. A target is an ordinary Fortran variable or array that has been declared with the TARGET attribute, so that a pointer can point to it. The difference between a pointer and an ordinary variable is that a pointer contains the address of another Fortran variable or array, while an ordinary Fortran variable contains data.
2. A pointer assignment statement assigns the address of a target to a pointer. The difference between a pointer assignment statement and an ordinary assignment statement is that a pointer assignment statement assigns the address of a Fortran variable or array to a pointer, while an ordinary assignment statement assigns the value of an expression to the target pointed to by the pointer.

    ```
    ptr1 => var    ! Assigns address of var to ptr1
    ptr1 = var     ! Assigns value of var to target of ptr1
    ```

3. The possible association statuses of a pointer are: associated, disassociated, and undefined. When a pointer is first declared, its status is undefined. It may be associated with a target using a pointer assignment statement or an ALLOCATE statement. The pointer may be disassociated from a target by the NULLIFY statement, the DEALLOCATE statement, by assigning a null pointer to it in a pointer assignment statement, or by using the NULL( ) function (Fortran 95 only).
4. Dereferencing is the process of accessing the corresponding target when a reference to a pointer appears in an operation or assignment statement.
5. Memory may be dynamically allocated with pointers using the ALLOCATE statement. Memory may be deallocated using the DEALLOCATE statement.
6. Invalid. This is an attempt to use ptr2 before it is associated with a target.
7. Valid. This statement assigns the address of the target variable value to pointer ptr2.
8. Invalid. A pointer must be of the same type as its target.
9. Valid. This statement assigns the address of the target array array to pointer ptr4. It illustrates the use of POINTER and TARGET statements.
10. Valid, but with a memory leak. The first WRITE statement will print out an F, because pointer ptr is not associated. The second WRITE statement will print out a T followed by the value 137, because a memory location was allocated using the pointer, and the value 137 was assigned to that location. The final statement nullifies the pointer, leaving the allocated memory location inaccessible.
11. Invalid. These statements allocate a 10-element array using ptr1 and assign values to it. The address of the array is assigned to ptr2, and then the array is deallocated using ptr1. This leaves ptr2 pointing to an invalid memory location. When the WRITE statement is executed, the results are unpredictable.
12. Valid. These statements define a derived data type containing a pointer, and then declare an array of that derived data type. The pointer contained in each element of the array is then used to allocate an array, and each array is initialized. Finally, the entire array pointed to by the pointer in the fourth element is printed out, and the first element of the array pointed to by the pointer in the seventh element is printed out. The resulting output is:

    ```
    31 32 33 34 35 36 37 38 39 40
    61
    ```

QUIZ 16–1

1. Object-oriented programming provides a number of advantages:
   - **Encapsulation and data hiding.** Data inside an object cannot be accidentally or deliberately modified by other programming modules. The other modules can only communicate with the object through the defined interfaces, which are the objects public method calls. This allows a user to modify the internals of an object without affecting any other part of the code, as long as the interfaces are not changed.
   - **Reuse.** Since objects are self-contained, it is easy to reuse them in other projects.
   - **Reduced effort.** Methods and behaviors can be coded once only in a superclass and inherited by all subclasses of that superclass. Each subclass only has to code the *differences* between it and its parent class.
2. The principal components of a class are:
   - **Fields.** Fields define the instance variables that will be created when an object is instantiated from a class. Instance variables are the data encapsulated inside an object. A new set of instance variables is created each time that an object is instantiated from the class.
   - **Methods.** Methods implement the behaviors of a class. Some methods may be explicitly defined in a class, while other methods may be inherited from superclasses of the class.
   - **Finalizer.** Just before an object is destroyed, it makes a call to a special method called a **finalizer.** The method performs any necessary cleanup (releasing resources, etc.) before the object is destroyed. There can be at most one finalizer in a class, and many classes do not need a finalizer at all.
3. The three types of access modifiers are PUBLIC, PRIVATE, and PROTECTED. PUBLIC instance variables and methods may be accessed from any procedure that USEs the module containing the definitions. PRIVATE instance variables and methods may *not* be accessed from any procedure that USEs the module containing the definitions. PROTECTED instance variables may be read but not written from any procedure that USEs the module containing the definitions. The PRIVATE access modifier should normally be used for instance variables, so that they are not visible from outside the class. The PUBLIC access modifier should normally be used for methods, so that they can be used from outside the class.
4. Type-bound methods are created using the CONTAINS clause in a derived type definition.
5. A finalizer is a special method that is called just before an object is destroyed. A finalizer performs any necessary cleanup (releasing resources, etc.) before the object is destroyed. There can be more than one finalizer in a class, but most classes do not need a finalizer at all. A finalizer is declared by adding a FINAL keyword in the CONTAINS section of the type definition.
6. Inheritance is the process by which a subclass receives all of the instance variables and bound methods from its parent class. If a new class extends an existing class, then all of the instance variables and bound methods from its parent class will automatically be included in the child class.
7. Polymorphism is the ability to work with objects of many different subclasses as though they were all objects of a common superclass. When a bound method is called on one of the objects, the program will automatically pick the proper version of the method for an object of that particular subclass.
8. Abstract methods are methods whose interface is declared in a superclass, but whose implementation is deferred until subclasses are derived from the superclass. Abstract methods can be used where you want to achieve polymorphic behavior, but the specific method will always be overridden in subclasses derived from the method. Any class with one or more abstract methods will be an abstract class. No objects can be derived from an abstract class, but pointers and dummy arguments can be of that type.

QUIZ 17–1

1. The way to create a parallel program (a Coarray Fortran program) differs from compiler to compiler. For Intel Fortran, the compiler option `/Qcoarray:shared` specifies that the program should be run in parallel with shared memory, and the option `/Qcoarray-num-images:n` specifies that there should be `n` parallel images of the program. For GNU Fortran, the compilation option is `-fcoarray=lib`, where `lib` is the library to link to.
2. The SPMD (Single Program Multiple Data) program model is a multiprocessing model in which every image runs exactly the same program, but different images can run different parts of the program in parallel.
3. In a Coarray Fortran program, each image can determine its image number using the `this_image()` function. The particular code that an image executes can be controlled using `IF` statements by specifying a particular image or range of images.
4. A coarray is an array in which an equal-sized array is allocated in each image, and the memory in each image can be accessed by every other image. The data in a particular image can be accessed from any other image by specifying the desired image in square brackets, for example, element `a(3,3)` from image 2 could be accessed as `a(3,3)[2]`.
5. The images in Coarray Fortran can communicate with each other through the `SYNC` statements. Each image can synchronize calculations with other ones using these commands.
6. A race condition is a condition in which two or more images are calculating simultaneously, and the final result depends on which image finishes first. A program can minimize race conditions by designing algorithms so that as much as possible the function of each image does not depend on the function of the other images. When communication is required between images, synchronization statements and critical sections can be used to ensure that the data being exchanged is consistent.
7. A critical section is a part of the code that can only be entered by one image at a time. If there are sections of code that could produce invalid results if they were accessed by more than one image at a time, that code can be placed in a critical section to avoid the possible collision. For example, suppose that two variables must both be changed by a calculation for the results to be consistent, and an incorrect result would occur if another image read the two variables after the first one had been modified and before the second one had been modified. If that calculation were placed in a critical section, then no other image could access the data until both variables had been updated properly.
8. This program will print out the desired result, as long as there are five images: one master to control the execution, and four slaves to each print out one calculation. Note the `SYNC MEMORY` statement ensures that the input values reach the slave images before they perform their calculations. Try it with that statement out and see what happens.

INDEX

A

ABS function, 48, 49, 518, 581
Absolute value function, 517
Abstract classes
　principles of, 809–814
　shape class hierarchy example, 814–830
Abstract methods, 810
Abstract types, 810
Acceleration due to gravity, 167, 238
ACCESS= clause
　in INQUIRE statements, 656
　in OPEN statements, 646, 648–649
Access restrictions to modules, 607–613
ACHAR function, 48, 157, 463, 465
ACOS function, 48
ACTION= clause
　in INQUIRE statements, 656
　in OPEN statements, 646, 650
　purpose, 214, 215, 216
Actual arguments, dummy arguments versus, 300
add_arrays program, 262
Addition
　in hierarchy of operations, 39
　in two's complement arithmetic, 6
　vector quantities, 540–543, 550–552, 597
A descriptor, 189, 207–208, 634
ADVANCE= clause, 660, 662
Algorithms
　constructs for describing, 86–88
　defined, 84
　heapsort, 359
　for random number generators, 323
　for selection sorts, 272
　tracking, 559
ALL function, 377
allocatable_arguments program, 431–432
allocatable_function program, 433
Allocatable arrays
　automatic arrays versus, 422–423
　basic rules for using, 384–386, 428
　Fortran 2003 support, 430–433
　in procedures, 421
　sample programs demonstrating, 386–389
ALLOCATABLE attribute, 432, 544–545
Allocatable coarrays, and derived data types, 856–857
Allocatable functions, 432–433
ALLOCATED function, 376
ALLOCATE statements, 385, 856
　for deferred-shape arrays, 428
　form in Fortran, 384
　for pointers, 709, 711, 727
　required when SAVE attribute absent, 421

Alpha-beta tracker, 559–560
Alphabetic characters, ASCII collating sequence, 156, 158–159
Alphabetization program, 459–462. *See also* Sorting data
Alpha releases, 85
Alternate entry points, 884–886
Alternate subroutine returns, 882–884
American Standard Code for Information Interchange, 8–9. *See also* ASCII character set
Amperes, 61
Ampersand, 23, 668, 669
ANY function, 377
Apogee, 242
Apostrophe, 32, 651
APOSTROPHE option, DELIM= clause, 651
Apparent power program, 60–63
APPEND option, POSITION= clause, 650
Area equations, 815–816, 828
Argument lists
　of functions, 332, 334
　passing procedures to other procedures in, 339–344
　role in subroutines, 298, 299
Arguments
　defined, 47
　passing intrinsic functions as, 888–889
　passing procedures as, 339–344
　types of most common keywords in, 915
　types used with specific functions, 915–916
　with STOP statements, 890–891
Arithmetic calculations
　hierarchy of, 39–41
　on integer and real data, 37–38
　mixed mode, 41–45
　standard operators, 36–37
Arithmetic IF statement, 892–893
Arithmetic logic unit, 3
Arithmetic mean. *See* Mean
Arithmetic operators, 36–37
array_io program, 269–270
array_ptr program, 708
array2 program, 311–312
Array constants, 247, 249–250
Array constructors, 247, 363–365
Array elements
　changing subscripts, 251–252
　defined, 245
　initializing, 248–251
　input/output operations on, 265–269
　in namelists, 671
　as ordinary variables, 247–248
　out-of-bounds subscripts, 252–256, 278–282
Array overflow conditions, 278–282

Index

Arrays
　addressing derived data components in, 529
　allocatable, 383–391, 421, 422–423, 430–433
　automatic, 422–426, 427
　basic features, 245–246
　changing subscript ranges, 251–252
　in COMMON blocks, 876–878
　declaring, 246–247, 361
　declaring sizes with named constants, 256
　derived data type resemblance to, 527–528
　DIMENSION statement, 873–874
　in elemental functions, 434–435
　elements as ordinary variables, 247–248
　FORALL construct with, 381–383
　initializing, 248–251
　input/output operations on, 269–270
　intrinsic functions with, 375–378, 946–957
　linked lists versus, 712–713
　masked assignment, 378–381
　multidimensional array overview, 372
　obsolescent specification statements, 873–875
　operations on subsets of, 263–265, 269–270
　out-of-bounds subscripts, 252–256
　passing to subroutines, 305–305, 309–310, 404–406
　of pointers, 725
　pointers to, 699–700, 707–709
　rank 1 versus rank 2, 360–361
　rank 2 declaration and storage, 361–362
　rank 2 initialization, 362–366
　sample program for data sorting, 271–282
　sample program for median calculation, 282–287
　sample programs to find largest and smallest values, 256–261
　saving between procedure calls, 417–421
　summary of types, 426–428
　when to use, 287–288
　whole array operations, 261–263, 269–270, 371
Array sections
　operations on, 263–265, 269–270
　pointers to, 707–708
　selection from rank 2 arrays, 371
Arrows in flowcharts, 87
ASCII character set, 903
　collating sequence, 156, 158–159, 159
　overview, 457
ASCII coding system, 8–9
ASIN function, 48
ASIS option, POSITION= clause, 650
Assigned GO TO statement, 894–895
Assignment operator, 36
Assignment statements
　basic features, 36–37, 89
　for character expressions, 154
　checking for errors, 67
　creating for pointers, 701–703
　equal signs in, 91
　of functions, 332
　initializing arrays with, 248–249, 363–365
　initializing complex variables in, 515
　masked array, 378–381
　using pointers in, 705–707
　variable initialization in, 56
ASSIGN statement, 894–895
ASSOCIATE construct, 552–553
Association status of pointers, 704–705, 727, 731
assumed_shape program, 416
Assumed length character functions, 471
Assumed-shape dummy arrays
　multidimensional, 405
　one-dimensional, 310, 311, 330
　overview, 427
　programming examples using, 415–416
Assumed-size dummy arrays, 310, 406, 427
Asterisks, 191
　to declare character function length, 471
　to declare character variables, 313
　in fields, 50
　for real output, 186, 187
　for standard input and output devices, 212
ASYNCHRONOUS= clause
　in INQUIRE statements, 657
　in OPEN statements, 646, 649, 688
　in READ statements, 660, 663, 687, 688
　in WRITE statements, 687–688
ASYNCHRONOUS attribute, 688–689
Asynchronous I/O mode, 687–688
ATAN function, 48
Automatic arrays, 422–426, 427
Automatic character variables, 466
Automatic length character functions, 471–472
ave_sd subroutine, 319–320
ave_value function, 340–341
Average. *See* Mean
Avogadro's number, 186–187

B

BACKSPACE statements, 213, 222, 666, 667
Backward compatibility, 16
bad_argument subroutine, 329–330
bad_call2 program, 329–330
bad_call program, 308
bad_ptr program, 711–712
bad program, 708–709
Balanced parentheses, 41
Ball, writing program to calculate maximum range, 162–168
ball program, 166–167
Base 10 system, 4
Batch mode, 28
B descriptor, 634, 638
Best-fit lines, 226–231
Beta releases, 85
binary_tree program, 753–754
Binary data format descriptors, 638
Binary digits, 4, 939
Binary number systems, 4–8
Binary operators, 37, 595
Binary trees
　basic features, 736–740
　creating, 742–743
　importance, 740–742
　programming examples using, 743–756
Binding procedures. *See* Bound procedures
Bit intrinsic procedures, 939–943
Bits, 4, 939
BLANK= clause
　in INQUIRE statements, 656
　in OPEN statements, 647, 651
Blank characters
　adding with format descriptors, 190–191, 192
　as control characters, 183, 192
　no characters versus, 31
　variables padded with, 154–155, 208, 651
Blank descriptors, 641

BLOCK DATA subprogram, 878–879
Block IF constructs
    common errors, 118–119
    with ELSE clause, 96–98
    form and actions of, 95–96
    introduction in FORTRAN 76, 15
    logical IF versus, 110
    naming, 106–107, 108–109
    nested loops within, 152
    nesting, 107–110
    sample programs demonstrating, 99–106
Blocks, 567–568
BN descriptor, 635, 641, 896–897
Body of loop, 135
Bound assignments and operators, 607
Bound procedures
    declaring in CONTAINS statements, 548, 550, 773
    to enable user-defined I/O operations, 679
    generic, 591–594, 774
    procedure pointers versus, 736
Bounds checking, 253–256, 311–312
bounds program, 254–256
Branches
    debugging programs with, 118–119
    defined, 81, 94
    from DO loops, 144
    naming, 287
    obsolescent structures, 892–895
    using block IF, 95–110
    using SELECT CASE, 110–116
btree module, 748–753
Buffers, input. *See* Input buffers
Buffers, output. *See* Output buffers
Bugs, 66. *See also* Debugging
Builds, 85
Bytes, 4
BZ descriptor, 635, 641, 896–897

C

CABS function, 517, 518, 581
Cache memory, 3
calc_hypotenuse program, 300
CALL statements, 299, 300
capacitor program, 200–202
Capacitors, 197–203
Capitalization, 23, 27, 34
Carbon 14 dating program example, 63–66
Card identification field, 871
Cards, 871
C language, 15
Cartesian planes, 557
CASE constructs, 110–116, 894
Case conversion programs, 158–161, 468–470
CASE DEFAULT block, 111, 112–113
Case insensitivity, 23
CEILING function, 43
Central processing unit (CPU), 2, 3
Centripetal acceleration equation, 79
C function
    calling Fortran subroutine, 912–913
    Fortran program calling, 908–912
CHARACTER_STORAGE_SIZE constant, ISO_FORTRAN_ENV module, 689
character_subs module, 471
Character constants, 31–32, 34, 181
Character context, 31

Character data. *See also* Character variables
    assignment statements, 154–155
    basic features, 8–9
    case conversion programs, 158–161, 468–470
    concatenation, 156
    constants and variables, 31–32
    converting real data to, 474–478
    converting to numeric, 473–474
    declaration statements, 33, 361
    early length declarations, 872
    format descriptors, 189, 207–208, 634
    intrinsic functions, 157, 463–465, 943–946
    kinds, 511–512
    relational operators for, 156, 458
    substring specifications for, 155
Character expressions, 154
Character functions
    defined, 154
    intrinsic, 157, 463–465, 943–946
    lexical, 462–463
    variable-length, 471–472
Character limitations in PROGRAM statement, 25
Character operators, 154–156
Character sets
    ASCII upper- versus lowercase offset, 158, 159, 458, 468
    collating sequence variations, 156, 458, 468
    Fortran, 23
    major versions, 457
    prior to Fortran 90, 870
    specifying in OPEN statements, 649
    supporting multiple versions, 511–512
CHARACTER statements, 465–467
Character variables. *See also* Character data
    assignment statements, 154–155
    comparison operations, 458–463
    declaration statements, 33
    defined, 32, 457
    early length declarations, 872
    format descriptors, 207–208
    format specifications in, 181
    intrinsic functions, 157, 463–465
    for OPEN statement error messages, 213–215
    passing to subroutines, 313, 465–470
CHAR function, 463
check_array program, 376–377
circle_class module, 819–821
Circles
    area and circumference equations, 815
    area and circumference program, 819–821
    in flowcharts, 87
Circumference equations, 815, 828
C language, 15
Classes
    abstract, 809–830
    hierarchy of, 766–769
    implementing in Fortran, 772–775
    in object-oriented programming model, 765–766, 766
    structure in Fortran, 769–770
Class hierarchy, 766–769
CLASS keyword, 549, 770–771, 800
CLOSE statements, 213, 216, 653–655
CMPLX function, 518–519
Coarray Fortran, 838
    parallel processing in, 838–839
Coarrays, 837, 838, 841–843
    allocatable, and derived data types, 856–857
    functions, 958–960
    passing to procedures, 857–858

Index

Code pages, 9
CODIMENSION attribute, 841
Coding examples. *See* Programming examples
co_lbound function, 842
Collating sequences
    of upper- and lowercase ASCII characters, 158, 159
    variations, 156, 458–459, 468
Colons
    for array dimensions, 384, 405
    as format descriptors, 639–640
Column major order, 362
Combinational logic operators, 91–93
COMMAND_ARGUMENT_COUNT function, 616
Command line, 60, 615–617
Commas
    prohibited in constants, 29, 30
    to separate format descriptors, 181
    in type declaration statements, 34
Comments, 24, 871
COMMON blocks, 876–879
compare program, 160–161
Comparison operations (character), 458–463
COMPATIBLE option, ROUND= clause, 649
Compilers, 12, 27–28, 618
complex_class module, 772, 773, 774
Complex constants, 514
Complex numbers, 512–519, 630
COMPLEX statements, 514
Complex variables, 514, 515–516
Compilers, 184
Component selectors, 529, 770
Computations, indicating in flowcharts, 87
Computed GO TO statement, 893–894
Computer languages, 12–13
Computer programs, 1
Computers
    data representation systems, 4–11
    importance, 1–2
    major components, 2–4
    Type 704, 13–14
Concatenation, 156
Concrete types, 810, 813–814
Conditional stopping points, 639–640
Conformable arrays, 262, 363
Constants
    capitalizing, 27
    complex, 514
    conversion factors as, 49
    declared in iso_c_binding module, 905
    declaring kind, 487–488, 511
    defined, 28
    of derived data type, 528
    maintaining consistency, 34
    major types, 29–32
    in output lists, 53
Constructors, 528, 769, 772–773
Constructs, 86
CONTAINS statements
    for internal procedures, 436
    for procedures in modules, 328
    for type-bound procedures, 548–552, 773
CONTINUE statement, 896
Control characters
    blanks in, 192
    in printer output, 182–184
    in output buffer, 182, 183
Control statements, 81
Control unit of CPU, 3
Conversion functions

    intrinsic, 49, 923–930
    between real and integer data, 43–44
Conversion of temperature units, 58–60
convert.f90 program, 871
Corank 2 coarray, 842
Cores, 837
Correlation coefficient, 243
COS function, 48
co_ubound function, 842
COUNT function, 377
Counting loops, 87. *See also* Iterative loops
Cray supercomputers, 489, 510
Critical sections, 858–859
Cross product, 295, 352, 598
Current flow equation, 174
CYCLE statements
    actions in DO loops, 145–147
    naming, 148, 152
    in nested loops, 151

D

DABS function, 581
Databases, 741, 742
Data dictionaries, 29
Data files. *See* Files
Data hiding, 299, 608–611
Data representation
    basic principles, 4
    binary numbers, 4–6
    data types, 8–11
    octal and hexadecimal numbers, 7–8
Data sharing
    basic features of modules for, 321–323
    modules versus entry points, 885
    obsolescent features, 876–879
DATA statement, 874–875, 878–879
Data types, 89–93
    declared in iso_c_binding module, 905
    interoperable, declaring, 906–907
date_class module, 782–787
Date class creation example, 782–789
Day-of-week selection, 113–116
Day-of-year calculation, 137–140
DBLE function, 494, 518
DC descriptor, 635, 642
D descriptor, 634, 636, 897
Deadlock condition, 847
Deadlocks, 859
DEALLOCATE statements, 856
    form in Fortran, 386
    for pointers, 709, 710–711, 727
Debugging. *See also* Errors; Testing
    bounds checking with, 253
    branches, 118–119
    development time spent on, 86
    loops, 168–169
    overview, 66–67
Decibels, 124–125, 176
DECIMAL= clause
    in OPEN statements, 646, 649
    in READ statements, 660, 663
Decimal descriptors, 635
Decimal points
    in constants, 29, 30
    in real input fields, 206–207

Declaration sections, 25, 33
Decomposition, 84
Default case clauses, 111, 112–113
Default character sets, 458
Default complex kind, 515
Defaulting values in list-directed input, 642–643
Default real variables, 485, 487
Default variables, 32–33
DEFERRED attribute, 810
Deferred methods, 810
Deferred-shape arrays, 384, 428, 699–700
Deleted Fortran features, 16, 869
DELETE option, STATUS= clause, 654
DELIM= clause
  in INQUIRE statements, 657
  in OPEN statements, 646, 651
  in READ statements, 661, 663
Delimiters for array constants, 247
Dereferencing pointers, 705–706
Derivative of a function, 353–354, 495–499
Derived data types
  access restrictions for, 610–611
  allocatable coarrays and, 856–857
  in binary trees, 736. *See also* Binary trees
  binding new operators to, 607
  declaring in modules, 531–539
  defining new operators for, 594–595
  dynamic allocation, 544–545
  extension of, 546–547
  Fortran 2003 enhancement of, 15
  input/output operations on, 529–531, 678–680
  overview, 29, 527–529
  parameterized, 545–546
  pointers in, 712–725
  pointers to, 699
  procedures bound to, 548–552
  returning from functions, 540–543
Derived type definitions. *See* Type definitions
Design process. *See* Program design
Diagonal elements, extracting from matrices, 728–731
Diamonds in flowcharts, 87
Dielectrics, 197
diff program, 497–498
DIM argument, 946–947
DIMENSION attribute, 725
Dimensions, representing arrays in, 247
DIMENSION statement, 873–874
direct_access_formatted program, 674–675
direct_access program, 675–677
Direct access mode
  default file format, 650
  defined, 212, 648
  file creation for, 673–677
DIRECT clause, in INQUIRE statements, 656
Disassociating pointers, 704, 711
Discriminants of equations, 96
Divide-by-zero errors, 132, 336, 337, 408
Division, 37, 39, 598
DO loops
  common errors, 168–169
  CYCLE statements, 145–147, 148
  EXIT statements, 146, 148
  implied, 250, 265–268, 365–366, 370–371, 875
  implied versus standard, 268–269
  initializing arrays with, 248–249
  iterative, 134–145, 164–165
  nesting, 148–151, 382
  obsolescent forms, 895–896

DO statements, in while loops, 127
DOT_PRODUCT function, 377
Dot products of vectors, 294, 598
Double-lined rectangles in flowcharts, 87
Double-precision complex values, 515, 518–519
Double-precision real values
  determining kind numbers associated with, 488
  format descriptors, 636
  mixing with single, 492–494
  obsolescent data type, 871
  origins of term, 486
  selecting, 489
  when to use, 494–495, 498, 501
Double slashes, 156
Doubly linked lists, 762
DO WHILE loops, 134
DOWN option, ROUND= clause, 649
doy program, 138–139
DP descriptor, 635, 642
dsimul subroutine, 502–504
DT descriptor, 635, 679
Dummy arguments
  allocatable, 430–432
  associating INTENT attribute with, 306–307
  for character variables in procedures, 465–467
  constraints for elemental functions, 435
  declared and dynamic types, 771
  defined, 299
  in interface blocks, 580
  keyword arguments with, 571–573
  pointers as, 727
  in statement functions, 887
  with user-defined operators, 595, 596
Dummy arrays
  passing to subroutines, 310, 330, 404–406
  summary of types, 427
Dynamic memory allocation
  defined, 384
  for derived data, 544–545
  with pointers, 709–712

E

Echoing values from keyboard input, 51
E descriptor
  actions in FORMAT statements, 186–188
  actions in READ statements, 207
  F descriptor output versus, 637
  optional forms, 636
  P descriptor with, 641
  usage, 634
Eighty-column cards, 870–871
Elapsed-time calculator example, 775–780
Electric power generation sample program, 366–370
Elemental functions, 434–435, 914
Elemental intrinsic functions, 263, 375
Elemental subroutines, 436
ELSE clauses, 96–98, 107
ELSE IF clauses, 96–98, 107, 109
employee_class module, 797–799, 810–813
Encapsulation of object variables, 764, 765
ENCODING= clause
  in INQUIRE statements, 657
  in OPEN statements, 646, 649
END= clause
  in READ statements, 661, 663

Index

END DO statements, 148, 149, 150
EN descriptor, 634, 636
ENDFILE statements, 667
END FORALL statements, 381
END IF statement, 95–96, 107
END PROGRAM statements, 26
END statements, 892
END WHERE statements, 379
Engineering notation, 636
Entry points, alternate, 884–886
ENTRY statement, 884–886
Environment variables, 615, 617–618
EOR= clause, in READ statements, 660, 662–663
Equal sign, 36, 91
Equations
   acceleration due to gravity, 238
   area, 815–816, 828
   binary digit, 939
   capacitance, 198
   centripetal acceleration, 79
   complex number form, 512
   correlation coefficient, 243
   cross product of two vectors, 295
   current flow, 174
   decibels, 124–125, 176
   derivative of a function, 353–354, 495
   discriminants, 96
   E descriptor width, 187
   electrical power, 61, 63
   energy, 76
   escape velocity, 80
   ES descriptor width, 188
   Euler's equation, 526
   factorial function, 567
   flight of thrown ball, 162, 163, 164
   future value, 78, 240
   gain, 241
   Gauss-Jordan elimination, 406–408
   geometric mean, 177
   gravitational force between two bodies, 358–359
   harmonic mean, 178
   ideal gas law, 179
   integer model in Fortran, 931
   kinetic energy, 76, 240
   least-squares method, 226, 243
   leverage, 179
   line in Cartesian plane, 557
   loop iterations, 135
   magnitude of a complex number, 517
   mean, 128, 288, 420
   mean time between failures, 178
   output power, 241
   pendulum period, 76–77
   perimeter, 815–816, 828
   potential energy, 76, 240
   radial acceleration, 244
   radioactive decay, 63–64
   random number generators, 323
   real numbers in Fortran, 931, 932
   resonant frequency, 79
   root-mean-square average, 177
   satellite orbit, 174, 242
   sinc function, 336
   smallest and largest integer value, 9–10
   Snell's law, 125
   standard deviation, 128, 288, 420–421
   standardized normal distribution, 358
   temperature conversion, 59
   truncated infinite series, 176–177
   two-dimensional vectors, 292
   velocity of falling object, 226
   velocity of orbiting object, 295
   velocity of thrown ball, 163
Equilateral triangles, 815
Equivalence relational operator, 91
EQUIVALENCE statement, 880–882
ERR= clause
   in CLOSE statements, 654
   in file positioning statements, 667
   in INQUIRE statements, 657
   in OPEN statements, 647, 651
   in READ statements, 661, 663–664
ERROR_UNIT constant, ISO_FORTRAN_ENV module, 689
Error flags, 314
Error messages, 214–215
ERROR STOP statement, 26, 891
Errors. *See also* Debugging
   with COMMON blocks, 876–878
   debugging basics, 66–67
   handling in subroutines, 313–314
   handling with IOSTAT5 and IOMSG= clauses, 651–652
   reducing with IMPLICIT NONE, 57–58
   sensitivity of systems of linear equations to, 499–501
   from single-precision math, 495
Escape velocity equation, 80
ES descriptor
   actions in FORMAT statements, 188–189
   actions in READ statements, 207
   basic usage, 634
   optional forms, 636
Euler's equation, 526
evaluate module, 885–886
every_fifth function, 732
Examples. *See* Programming examples
Exclamation points, 24
Executable statements, 23
Executing programs, 27–28
Execution sections, 25, 26
EXIST clause, in INQUIRE statements, 655
EXIT statements, 152
   actions in DO loops, 148
   naming, 148, 152
   in nested loops, 151
   in while loops, 127
EXP function, 48
Explicit interfaces
   defined, 329
   for functions of derived data types, 540
   keyword arguments with, 571–572
   obstacles to using, 577
   optional arguments with, 572
   in pointer-valued functions, 732
   for user-defined generic procedures, 582–583
Explicit-shape arrays, 384, 426
Explicit-shape dummy arrays
   multidimensional, 404–405
   one-dimensional, 309–310, 311, 330
   overview, 426–427
   programming examples using, 406–415
Explicit variables, 32–33
Exponential notation, formatting output for, 186–189
Exponentiation operator, 37
Exponents
   in constants, 30
   formatting output for, 186–189

Exponents—*Cont.*
  in hierarchy of operations, 39
  in mixed-mode arithmetic, 44–45
  in REAL data type, 485–486
  in scientific notation, 10, 11
Expressions
  for character data manipulation, 154
  conversion factors as, 49
  mixed mode, 42, 43
  in program design, 93
Extended Binary Coded Decimal Interchange Code.
    See EBCDIC system
EXTENDS attribute, 795–797, 802
Extension
  of derived data types, 546–547
  of operator meanings, 595–596
Extents of arrays, 247, 262
EXTERNAL attribute, 339–340
External functions, intrinsic functions versus, 47
External procedures. *See* Procedures
EXTERNAL statements, 340, 342
Extrapolation, 456
extremes program, 258–260
extremes subroutine, 574–575

F

fact function, 570
Factorial function, 136–137, 568–571
factorial subroutine, 569
Fahrenheit-to-kelvin conversion program, 58–60
F descriptor
  actions in FORMAT statements, 186
  actions in READ statements, 206–207
  basic usage, 634
  E descriptor output versus, 637
  optional forms, 636
  P descriptor with, 641
Fields
  as class members, 765, 769
  declaring in Fortran, 772–773
FILE= clause
  actions of, 214
  in INQUIRE statements, 655
  in OPEN statements, 646, 647
FILE_STORAGE_SIZE constant, ISO_FORTRAN_ENV
    module, 689
File positioning statements, 222, 667
Files
  basic concepts, 211–213
  direct access, 673–677
  formatted versus unformatted, 650, 671–672
  internal, 473–474
  major options for opening and closing, 214–216
  noisy data example, 226–231
  OPEN statement options for, 646–653
  positioning, 223–225
  reading and writing to, 216–222
Finalizers, 769, 790–794
FINAL keyword, 769
Fixed-source form statements, 870–871
Flight of a ball, 162–168
Floating-point data
  arithmetic operation rules, 38
  basic features, 10–11
  format descriptors, 636, 642

  range and precision, 30–31
  REAL data type overview, 485–486
Floating-point operations, computer capacities, 14
FLOOR function, 43
Flowcharts, 87, 88
FLUSH statements, 668
FMT= clause, in READ statements, 660, 661–662
FORALL construct, 15, 381–383, 434
FORALL statements, 383
FORM= clause, in OPEN statements, 646, 650
Format descriptors
  actions of, 181
  basic types in FORMAT statements, 184–192
  basic types in READ statements, 206–209
  complete listing, 633–635
  processing with READ statements, 209–210
  processing with WRITE statements, 192–203
  varying to match output data, 474–478
Formats
  descriptor types, 184–192, 633–642
  output devices and, 182–184
  overview, 180–181
  processing with READ statements, 205–210
  processing with WRITE statements, 192–196
  varying to match output data, 474–478
FORMAT statements, 191
  actions on output devices, 182–184
  descriptor types, 184–192
  sample programs demonstrating, 198–203
  WRITE statements versus, 203
FORMATTED clause, in INQUIRE statements, 656
Formatted files
  advantages and disadvantages, 671, 672
  defined, 650
  as direct access files, 673–677
FORM clause, in INQUIRE statements, 656
Formulas. *See* Equations
Fortran, history and continued development, 13–19
Fortran 90 introduction, 15
Fortran allocatable arrays, 384–386
FORTRAN 77, 15
Fortran 2008, 15–16
  access to command line and environment variables,
      615–618, 689
  allocatable arrays in, 430–433
  array constructor options, 247
  ASSOCIATE construct, 552–553
  asynchronous I/O mode, 687–688
  bound assignments and operators, 607
  data hiding attributes, 608–609, 611
  derived data type extension, 546–547
  dynamic allocation of derived data in, 544–545
  exclusive CLOSE statement options listed, 653, 655
  exclusive format descriptors listed, 634, 635, 641–642
  exclusive INQUIRE statement options listed, 655,
      656–657
  exclusive OPEN statement options listed, 646–647,
      648–650, 678
  exclusive READ statement options listed, 659, 660, 663
  finding standard I/O units in, 213
  generic bound procedures in, 591–594
  IMPORT statement, 581
  INTENT attribute with pointers, 731
  intrinsic modules, 615
  IOMSG = clause, 214, 217, 222.
  major improvements of, 15
  naming complex constants in, 514
  procedure pointers, 733–736
  RANDOM_NUMBER subroutine, 328

Index

type-bound procedures, 548–552
volatile variables, 618
WAIT and FLUSH statements, 667–668
Fortran character set, 23
fortran_calls_c program, 908–912
Fortran/C interoperability, 904–913
Fortran programs. *See* Programs
Fortran Working Group, 16
Fractions, 10, 37–38
Free-format output, 180
Functions, 471–472. *See also* Intrinsic functions
   allocatable, 432–433
   basic types and requirements for, 331–332
   as basis of user-defined operators, 595, 598
   character, defined, 154
   defined, 47, 298, 331
   derivatives, 353–354
   elemental, 434–435
   generic versus specific, 49, 581
   lexical, 462–463
   within modules, 328
   obsolescent features, 886–889
   passing as arguments, 339–342, 888–889
   passing character variables to, 465–470
   pointer-valued, 732
   pure, 15, 434
   recursive, 570
   returning derived data types from, 540–543
   unintended effects, 334
Functions (mathematical), 103–106
   with deliberate side effects, 334–335
Function subprograms. *See* User-defined functions
Future value equation, 78, 240

G

Gain equation, 241
Gaussian distribution, 358
Gauss-Jordan elimination
   principles of, 406–408, 499–501
   sample programs using, 408–415, 501–509
G descriptor, 634, 637–638
G0 format descriptor, 638
GENDAT file, 369
Generalized format descriptor, 637–638
Generator output sample program, 366–370
generic_maxval module, 586–589
generic_procedure_module module, 592–593
Generic functions, 49
Generic interface blocks, 582–583
Generic procedures
   bound, 591–594, 774
   overview, 581
   user-defined, 582–591
GENERIC statement, 591, 607
Geometric mean equation, 177
GET_COMMAND_ARGUMENT subroutine, 616
get_command_line program, 616
GET_COMMAND subroutine, 616
get_diagonal subroutine, 728–729
GET_ENVIRONMENT_VARIABLE subroutine, 617–618
get_env program, 617
Get methods, 781
Global objects, 562, 565
GO TO statements, 893–895

Graphical symbols for flowcharts, 87, 88
Gravitational force between two bodies, 358–359
Gravity, acceleration due to, 167, 238
Grouped format descriptors, repeating, 191, 194

H

Half-life, 66
Hard disks, 4
Harmonic mean equation, 178
Hashing techniques, 742
H descriptor, 897
Heapsort algorithm, 359
hello_world program, 839–840
Hexadecimal numbers, 7–8, 638
Hierarchy of classes, 766–769, 794. *See also* Inheritance
Hierarchy of operations
   in arithmetic calculations, 39–41
   combinational logic operators in, 92
   relational operators in, 91
Higher-order least-squares fits, 454
High-level languages, 13–14
High-precision real values, 494–495. *See also* Double-precision real values
Histograms, 483–484
Horizontal positioning, 190–191, 208
Host, 839
Host association, 437–438, 563, 565
Host program units, 436, 437–438
Host scoping units, 563, 564–565
hourly_employee_class module, 802–803
Hypotenuse calculations, 300

I

IABS function, 581
IACHAR function, 48, 157, 463, 465
IBM Type 704 computer, 13–14
ICHAR function, 463, 465
ID= clause
   in INQUIRE statements, 657
   in READ statements, 661, 663
Ideal gas law equation, 179
I descriptor, 184–185, 206, 634
IEEE modules, 615
IEEE Standard 755, 30–31, 485
IF (…) THEN statement, 95–96
IF statements
   block IF actions, 15, 95–96. *See also* Block IF constructs
   common errors, 118–119
   in while loops, 127
Ill-conditioned systems, 501, 508
Images
   defined, 838
   master, 840
   synchronization between, 843–849
   worker, 840
Imaginary part of complex number, 512
Immediate superclasses, 794
Implicit interfaces, 329
IMPLICIT NONE statement, 57–58, 332, 873
IMPLICIT statements, 872–873

Implied DO loops
    basic actions of, 250
    initializing arrays with, 365–366, 370–371, 875
    input/output operations with, 265–269
    nesting, 267–268
Implied multiplication, 37
IMPORT statement, 581
Impure elemental procedures, 435–436
Indention, 95, 96, 143
INDEX function, 463, 464
index variables (DO loop)
    for nested DO loops, 150
    problems of modifying, 143–144
    purpose, 134–135
    real type, 144
    value upon loop completion, 144–145
Infinite loops, 143
Infinite series, 176–177
Information hiding, 765
Inheritance
    benefits of, 794
    class hierarchy overview, 766–769
    defining and using subclasses, 795–803
    in derived data type extension, 546–547
Initialization of array elements, 248–251
Initialization of variables, 55–57
initialize_image program, 843–844
initialize_image2 program, 844–845
initialize_image3 program, 848–849
INPUT_UNIT constant, ISO_FORTRAN_ENV module, 689
Input arguments, accidental modification, 306–307, 334
Input buffers, 205, 208, 662
Input data
    checking for errors, 66–67
    defining in top-down design approach, 84
    file concepts, 211–213
    formatted, 205–210
    with interactive mode, 27–28
    noisy, 226–231
    opening a file for, 215
Input devices, 2, 4, 205
Input operations
    on array elements, 265
    on arrays and sections, 269–270
    on derived data types, 529–530
    format descriptors listed, 634–635
    with implied DO loops, 265–269
    indicating in flowcharts, 87
Input/output statements
    CLOSE options, 653–655
    defined, 49–50
    file positioning with, 223, 667
    INQUIRE options, 655–659
    list-directed input, 49–53, 642–643
    logical values in, 93
    namelist, 669–671
    OPEN options, 644–653
    READ options, 659–664
    redundant features of, 896–897
    unformatted, 672
    user-defined, 678–680
    WRITE and PRINT options, 665–666
Input/output units
    connecting to disk files, 644–645, 647
    definition of, 212
    predefined, 213
    specifying in file positioning statements, 666
    specifying in OPEN statements, 647
    specifying in READ statements, 661
Input parameter files, 482–483
INQUIRE statements, 655–659, 672
Inquiry intrinsic functions, 375, 376, 914
insertion_sort program, 722–724
Insertion sorts, 719–724
Instance methods, 764, 767
Instance variables, 800
    declaring in Fortran, 772–773, 776
    defined, 764
    for each instantiated object, 780
    in Fortran classes, 769
    inheritance in class hierarchies, 767
Integer arithmetic, 37–38
Integer constants, 29–30, 511
Integer data
    arithmetic operation rules, 37–38
    basic features, 9–10
    format descriptors, 184–185, 206, 634
    kinds, 509–511
    operations with real data, 41–45
    overview of constants and variables, 29–30
    sample array declarations, 361
Integers for i/o unit numbers, 212
Integers (Fortran model), 931
INTEGER type declaration statement, 25
Integer variables, 30, 214
Intel Itanium chip, 489
Intel Visual Fortran, 490, 510
INTENT(IN) attribute
    in allocatable dummy arguments, 430
    to prevent accidental input argument changes, 334
    in pure functions, 434, 923
INTENT(INOUT) attribute, 430
INTENT(OUT) attribute, 430
INTENT attribute
    actions of, 306–307
    in allocatable dummy arguments, 430
    with pointers, 731
    to prevent accidental input argument changes, 334
Interactive mode, 27–28
Interface, 439
interface_example program, 578–579
Interface assignment blocks, 596
Interface blocks, 577–584
Interface operator blocks, 595, 746
Internal files, 473–474
Internal functions, 47, 437, 888
Internal memory, 3
Internal procedures, 436–438
International Organization for Standardization (ISO), 16
Interpolation, 455
INT function, 43, 518
Intrinsic data types, 527
Intrinsic functions
    array and pointer, 946–957
    with arrays, 375–378
    bit, 939–943
    character, 157, 463–465, 943–946
    for command line access, 615–616
    for complex numbers, 517, 517–519
    elemental, 263
    generic versus specific, 49, 581
    kind and numeric processor, 931–935
    lexical, 462–463
    overview, 47–49, 332
    passing as arguments, 888–889
    for selecting character kind, 512

Index

for selecting integer kind, 490, 510–511
for selecting real variable precision, 489–491, 494
Intrinsic procedures
　alphabetical list of, 914–923
　array and pointer, 946–957
　bit, 939–943
　character, 943–946
　classes of, 914
　kind and numeric processor, 931–935
　mathematical and type conversion, 923–930
　system environment, 935–938
Intrinsic modules, 615, 689
INTRINSIC statement, 888–889
Intrinsic subroutines
　for command line access, 615–616
　current time, 357
　for matrix multiplication, 451
　for random number generation, 328, 452, 455
IOLENGTH= clause, in INQUIRE statements, 657
IOMSG= clause
　in CLOSE statements, 654
　in file positioning statements, 666
　in INQUIRE statements, 655
　in OPEN statements, 214, 222, 646, 648, 651
　in READ statements, 217, 660, 662, 664
IOSTAT= clause
　basic actions of, 214
　in CLOSE statements, 654
　in file positioning statements, 666
　in INQUIRE statements, 655
　in OPEN statements, 646, 648, 651–652
　preferred over END5 and ERR5, 896
　in READ statements, 217, 660, 662, 664
IOSTAT_END constant, ISO_FORTRAN_ENV module, 689
IOSTAT_EOR constant, ISO_FORTRAN_ENV module, 689
IOSTAT errors, 222
ISO_C_BINDING module, 615, 904
　data types declared in, 905
ISO_FORTRAN_ENV module, 492, 615, 689
ISO-8859 standard series, 9
Iterative loops
　actions of, 134–136
　details for using, 143–145
　in flowcharts, 87
　sample programs demonstrating, 136–143, 164–165
　while loops versus, 126

J

Java programming style, 27
J3 Committee (INCITS), 16

K

KEEP option, STATUS= clause, 654
Kelvins, converting Fahrenheit degrees to, 58–60
Keyword arguments, 571–573, 915
Keywords, 27, 915
Kind and numeric processor intrinsic functions, 931–935
KIND function, 488–489, 490
Kinds of character data, 511–512

Kinds of derived data, 545–546
Kinds of integers, 30, 509–511
Kinds of real numbers, 31
Kinds of real variables and constants
　declaring, 486–488
　determining, 488–489
　selecting in code, 489–491, 493
kinds program, 488
Kind type parameters, 486–489
Kinetic energy, 76, 240

L

Lahey Fortran, 510
Languages (computer), 12–13
Languages (human), 9
Largest data values, program for finding, 256–261
Laser printers, 182
LBOUND function, 376
L descriptor, 189, 207, 634
Leap years, 137–138
least_squares_fit program, 228–229
Least-squares method, 226–231, 452–456
Left arrows in pseudocode, 88
LEN_TRIM function, 157, 463, 465
len field, 33
LEN function, 157, 464, 465
Length limits
　of lines in statements, 23
　in pre-Fortran 90 statement lines, 870
　of program names, 25
　of variables, 28
Lengths of character strings, 464, 471–472, 872
Lengths of integers, 509–511
Lengths of records, 650
Letter grade assignment program, 109–110
Leverage equation, 179
Lexical functions, 462–463
LGE function, 465
LGT function, 465
Linear equations, 499–509
Linear regression, 226–231
Line in Cartesian plane, 557
Line printers, 182
Lines in Fortran statements, 23–24
linked_list program, 717–718
Linked lists
　creating, 715–719
　doubly linked, 762
　insertion sort implementation with, 719–724
　using pointers in, 712–715
Linkers, 12
Linking programs, 27–28
List-directed input, 50–51, 642–643
List-directed statements, 49–53
LLE function, 465
LLT function, 465, 533
Local objects, scoping units and, 562–563, 565
Local variables
　defined, 301
　saving between procedure calls, 417–421
　temporary length declarations for subroutines, 466
LOG function, 48
LOG10 function, 48
Logical calculations, assignment statements for, 89
Logical constants, 89

Logical errors, 66
Logical expressions
  in block IF, 95–96
  common errors, 118
  in program design, 93
Logical IF statements, 110
Logical operators, 89–93, 118
LOGICAL statement, 89
Logical variables
  defined, 89
  format descriptors, 189, 207, 634
  in program design, 93
Loop indexes, 134–135
Loops. *See also* DO loops
  CYCLE and EXIT statements with, 145–147
  debugging, 168–169
  defined, 81, 126
  DO WHILE, 134
  iterative, 134–145
  named, 146–148, 151
  nesting, 148–151
  obsolescent structures, 893–896
  while loops, 127–134
Lowercase letters
  ASCII offset from uppercase, 158, 159, 458
  equivalent to uppercase in Fortran, 23
  programs shifting to uppercase, 158–161, 468–470
lt_city function, 538
lt_last function, 537
lt_zip function, 538

M

Machine language, 12, 15
Magnetic tapes, 212
Magnitude of a complex number, 517
Main memory, 2, 3
Maintenance. *See* Program maintenance
Mantissa
  in REAL data type, 485–486
  in scientific notation, 10, 11, 30
Many-one array sections, 265
MASK argument, 946–947
Masked array assignment, 378–381
Massively parallel computers, 382–383, 434–436
Master image, 840
Mathematical and type conversion intrinsic procedures, 923–930
Mathematical functions, 517
MATMUL function, 377, 451
Matrices, 360–361, 728–731
MAX function, 48
Maximum pivot technique, 408
Maximum value
  calculating with arrays, 258–260
  calculating with subroutines, 316–320
  calculating with subroutines and optional output arguments, 574–575
  generic subroutines to find, 584–591
MAXLOC function, 378
MAXVAL function, 378
Mean
  arrays not required for, 288
  equation for, 128, 288, 420
  median versus, 282
  programs to calculate with arrays, 282–288
  programs to calculate with counting loops, 140–143
  programs to calculate with subroutines, 316–320
  programs to calculate with while loops, 129–134
  running averages program example, 418–421
Mean time between failures equation, 178
Measurement, noisy, 226–231, 323
Median
  arrays required for, 288
  defined, 282
  programs to calculate with arrays, 282–288
  programs to calculate with subroutines, 316–320
median subroutine, 320
mem_leak program, 710
Members of classes, 765
Memory
  allocating for derived data types, 531
  array demands on, 288, 384
  common data types in, 8–11
  dynamic allocation, 384–391
  major computer components, 2, 3
  multidimensional array allocation, 372
  obsolescent sharing methods, 875–882
  primary and secondary, 212
Memory bus, 3
Memory leaks, 709–710, 790
Memory registers, 3
Messages, 765, 766
Metcalf, Michael, 871
Method of least squares, 226–231
Methods (object)
  abstract, 809–810
  categories in Fortran, 780–789
  as class members, 765, 769
  creating in Fortran, 773–774
  defined, 764
  protecting from modification, 809
Microfarads, 198
MIN function, 48
Minimum value
  calculating with arrays, 258–260
  calculating with subroutines, 316–320
  calculating with subroutines and optional output arguments, 574–575
MINLOC function, 378
MINVAL function, 378
Mixed-mode arithmetic, 41–45, 492–494, 516
Mixed-mode expressions, 42, 43
MOD function, 48, 138
Modularity, as benefit of encapsulation, 765
module_example module, 564
Module procedures, 328–330, 438
MODULE PROCEDURE statement, 584
Modules
  basic features, 320–323
  declaring derived data types with, 531–539
  external procedures within, 328–330
  functions of derived data types in, 540
  intrinsic, 615, 689
  restricting access to, 607–613
MODULE statements, 321
Modulo function, 284
Multidimensional arrays. *See also* Arrays
  with assumed-shape dummy arrays, 415–416
  with explicit-shape dummy arrays, 406–415
  overview, 372
  passing to subroutines and functions, 404–406
Multiple entry points, 884–886
Multiplication, 37, 39, 597
The Mythical Man-Month, 86

Index

N

NAGWare Fortran, 510
NAME clause, in INQUIRE statements, 656
NAMED clause, in INQUIRE statements, 656
Named constants
 complex, 514
 conversion factors as, 49
 to declare array sizes, 256
 defined, 34
 for kind numbers, 487, 488
Namelist I/O, 669–671
Names
 for block IF constructs, 106–107, 108–109
 for branches, 287
 for CASE constructs, 111
 changing for data items or procedures, 612–613
 for constants, 34, 49
 for FORALL constructs, 381
 for loops, 146–148, 151, 287
 scope and, 562–567
 for WHERE constructs, 380
n descriptor, 635
Near equality, testing for, 119
NEAREST option, ROUND= clause, 649
Negative numbers, 5–6, 45
Nesting
 block IF construct, 107–110
 format descriptors, 191
 implied DO loops, 267–268
 loops, 148–151, 382
NEW file status, 647
Newline character, 678
NEWUNIT= clause, 647
NEXTREC clause, in INQUIRE statements, 656
NINT function, 43
NML= clause, in READ statements, 660, 662
Noisy measurements, fitting line to, 226–231
NON_OVERRIDABLE attribute, 809
NONE option, DELIM= clause, 651
Nonexecutable statements, 23, 25, 57–58
Nonunique solutions, 408
Nonvolatile memory, 3–4
NOPASS attribute, 549
Notional allocation of multidimensional arrays, 372
.NOT. operator, 93
NULL() function, 705
NULLIFY statements, 704, 711
Null values in list-directed READ statements, 642
NUMBER clause, in INQUIRE statements, 656
NUMERIC_STORAGE_SIZE constant, ISO_FORTRAN_ENV module, 689
Numerical analysis, 499
Numeric data, converting to character, 473–478
Numeric models, 931
Numeric processor and kind intrinsic functions, 931–935

O

Object-oriented programming
 abstract classes, 809–830
 basic concepts, 764–769, 794
 CLASS keyword, 770–771
 class member access controls, 789
 defining and using subclasses, 795–803
 finalizers, 769, 790–794
 Fortran 2003 support, 15, 763
 Fortran class structure, 769–770
 implementing in Fortran, 772–775
 method categories, 780–789
 polymorphism, 771, 794, 806
 SELECT TYPE construct, 807–809
 superclass/subclass object relations, 804–805
 timer class example, 775–780
Objects
 basic concepts, 764–765
 instantiating in Fortran, 774–775, 778
 scope levels of, 562
Obsolescent features
 alternative READ form, 665
 assumed character length functions, 471
 branching and looping structures, 892–896
 COMMON blocks, 876–879
 D descriptor, 636
 DOUBLE PRECISION data type, 871
 fixed-source form, 870–871
 identifying and deleting, 16, 869
 for index variables and control parameters, 144
 specification statements, 872–875
 subprogram, 882–889
Octal numbers, 7–8, 638
O descriptor, 634, 638
OLD file status, 647
1 (one) character, as control character, 182–183
One-dimensional arrays, 360, 361. *See also* Arrays
ONLY clauses, with USE statements, 612, 613
open_file program, 658–659
OPENED clause, in INQUIRE statements, 655
OPEN statements
 basic purpose, 213
 common clauses in, 214–216
 complete listing of clauses, 644–653
 for direct access files, 673
 for stream access, 648, 678
Operation codes, 12
Operators
 character, 154–156
 standard, 36–37
 user-defined, overview, 594–596
 user-defined examples, 597–607
Optimizers, 618, 700
Optional arguments, 572
Order of statements in Fortran, 961–962
Out-of-bounds array subscripts, 252–256, 278–282
OUTPUT_UNIT constant, ISO_FORTRAN_ENV module, 689
Output buffers
 basic features, 182
 forcing writes to disk, 668
 sending to printer with slash descriptor, 191
Output devices, 2, 4, 182–184
Output files, 215
Output operations
 on array elements, 265
 on arrays and sections, 269–270
 on derived data types, 529–531
 format descriptors listed, 634–635
 with implied DO loops, 265–269
 indicating in flowcharts, 87
Output power equation, 241
Outputs, defining, 83–84. *See also* Formats

Output statements, defined, 49–50. *See also* Input/output statements
Ovals in flowcharts, 87
Overflow condition, 10
Overriding, protecting methods from, 809

P

PAD= clause
  in INQUIRE statements, 657
  in OPEN statements, 647, 651
Parabolic flight paths, 162
Parallelograms in flowcharts, 87
Parallel processing
  in Coarray Fortran, 838–839
  sorting with, 850–856
Parallel programs/programming, 838, 859–862
  creation of, 839–840
PARAMETER attributes, 34
Parameterized variables, 487
PARAMETER statement, 875
Parentheses
  for arguments, 47
  arithmetic operation rules for, 37, 39, 41, 92
  checking for errors, 67
  in list-directed statements, 50, 52
PASS attribute, 549
Pass-by-reference scheme, 307–309
Patterns, searching character strings for, 464
PAUSE statement, 890
P descriptor, 635, 641
PENDING= clause, in INQUIRE statements, 657
pentagon_class module, 826–828
Pentagons, 816
Percent sign, 529, 770
Perigee, 242
Perimeter equations, 815–816, 828
Periods, 89, 91
Personal computers, 14
Picofarads, 198
Plan position indicator displays, 558
Plus character, 183
PLUS option, SIGN= clause, 649
Pointer assignment statements, 701–703
POINTER attribute, 435
Pointers
  arrays of, 725
  in binary trees, 736–739
  creating and associating with targets, 699–705
  declared and dynamic types, 771
  in derived data types, 712–725
  dynamic allocation with, 709–712
  intrinsic functions for, 946–957
  ordinary variables versus, 698–699
  in procedures, 727–732
  to procedures, 733–736
  in superclass/subclass object relations, 804–805
  using in assignment statements, 705–707
  using with arrays, 707–709
Pointer-valued functions, 732
Poisson distribution, 352
Polar complex numbers, 630
Polar coordinates, 293, 512–513
polyfn program, 887
Polymorphism
  abstract classes required, 809–814

  benefits of, 806
  defined, 771, 794
POS= clause
  in INQUIRE statements, 657
  in READ statements, 661, 663
POSITION= clause
  in INQUIRE statements, 656
  in OPEN statements, 646, 650
Positioning descriptors, 635
Positioning files, 222–225
Potential energy, 76, 240
Power calculation program example, 60–63
Power generation sample program, 366–370
Precision
  defined for real numbers, 10
  factors determining, 30–31
  limitations in arithmetic operations, 38
  maximizing for constants, 34
  obsolescent features, 871
  selecting, 489–491, 493
  when to increase, 494–495
PRECISION function, 490
Predefined units, 213
Predicate methods, 781–782
PRESENT function, 573
Primary memory, 212
Printers, 182–184
PRINT statements, 665–666
PRIVATE attributes and statements, 608–609, 610–611
PRIVATE keyword
  in timer class example, 780
  typical use of, 772–773, 789
Probability distributions, 352
Problems, articulating for top-down program design, 82–83, 99
Procedure pointers, 733–736
Procedures. *See also* Intrinsic procedures
  allocatable arrays in, 422, 430–433
  automatic arrays in, 422–426
  benefits of, 298–299
  declared in iso_c_binding module, 905
  defined, 298
  generic bound, 591–594
  generic procedures overview, 581
  generic user-defined, 582–591
  impure elemental, 435–436
  internal, 436–438
  interoperable, declaring, 907–908
  miscellaneous, 958
  within modules, 328–330, 438, 571, 579, 584
  passing coarrays to, 857–858
  passing to other procedures, 339–344
  pure and elemental, 434–436
  recursive, 568–571
  renaming, 612–613
  type-bound, 548–552
  use of SAVE with, 417–421
  using pointers in, 727–732
PROCESSOR DEFINED option, 650
PRODUCT function, 378
Program design
  basic data types, 89–93
  pseudocode and flowchart use, 86–88
  top-down technique overview, 82–86
Program maintenance, benefits of external procedures for, 298–299
Programming examples
  add_arrays, 262
  allocatable_arguments, 431–432

Index

allocatable_function, 433
array_io, 269–270
array_ptr, 708
array2, 311–312
assumed_shape, 416
ave_sd subroutine, 319–320
ave_value function, 340–341
bad, 708–709
bad_argument subroutine, 329–330
bad_call, 308
bad_call2, 329–330
bad_ptr, 711–712
ball, 166–167
binary_tree, 753–754
bounds, 254–256
btree module, 748–753
calc_hypotenuse, 300
capacitor, 200–202
character_subs module, 471
check_array, 376–377
circle_class module, 819–821
compare, 160–161
complex_class module, 772, 773, 774
customer_database, 534–536
date_class module, 782–787
diff, 497–498
direct_access, 675–677
direct_access_formatted, 674–675
doy, 138–139
dsimul subroutine, 502–504
employee_class module, 797–799, 809–813
evaluate module, 885–886
every_fifth function, 732
extremes, 258–260
extremes subroutine, 574–575
fact function, 570
factorial subroutine, 569
fortran_calls_c, 908–912
generate, 367–369
generic_maxval module, 586–589
generic_procedure_module module, 592–593
get_command_line, 616
get_diagonal subroutine, 728–729
get_env, 617
hello_world, 839–840
hourly_employee_class module, 802–803
insertion_sort, 722–724
interface_example, 578–579
initialize_image, 843–844
initialize_image2, 844–845
initialize_image3, 848–849
kinds, 488
least_squares_fit, 228–229
linked_list, 717–718
lt_city function, 538
lt_last function, 537
lt_zip function, 538
median subroutine, 320
mem_leak, 710
module_example, 564
open_file, 658–659
pentagon_class module, 826–828
polyfn, 887
ptr_array, 725
quadf function, 333
ran001 module, 325
random0 subroutine, 325–326
read_file, 220–221
read_namelist, 670
real_to_char function, 476
rectangle_class module, 823–825
rmax subroutine, 317–318
rmin subroutine, 318–319
roots_2, 520
running_average subroutine, 418–419
salaried_employee_class module, 800–802
scoping_test, 565
scratch_file, 223–225
seed subroutine, 326
select_kinds, 491
shape_class module, 817–819
shared_data module, 321–322
simul2 subroutine, 424–426
simul subroutine, 410–412
sinc function, 337
sort_database subroutine, 536–537
sort1, 276–278
sort2, 280–281
sort3, 302–305
sort4, 459–461
sortc subroutine, 467–468
square_and_cube_roots, 266–267
square_class module, 825–826
square_roots, 250–251
squares, 249
squares_2, 252
stats_1, 131
stats_2, 133–134
stats_3, 142–143
stats_4, 285–287
stats_5, 386–389
stock, 683–685
stop_test, 891
subs_as_arguments subroutine, 342
table, 196
test_abc, 471–472
test_alloc_fun function, 432–433
test_alloc subroutine, 430–432
test_array subroutine, 416
test_ave_value, 341–342
test_ave_value2, 889
test_char1, 155
test_date, 787–788
test_diagonal, 729–731
test_dsimul, 504–507
test_employee, 804–805
test_entry, 884
test_generic_procedures, 593–594
test_hypotenuse, 301–302
test_internal, 436–437
test_io, 530
test_keywords, 572
test_ptr, 701
test_ptr2, 703
test_ptr3, 705
test_ptr4, 706
test_quadf, 333
test_random0, 326–327
test_real_to_char, 477
test_running_average, 419–421
test_select_type, 808–809
test_shape, 828–830
test_simul, 412–414
test_sinc, 337–338
test_sort, 851–853
test_subs_as_arguments, 343–344
test_timer, 779
test_type_extension, 547

Programming examples—*Cont.*
  test_ucase, 470
  test_vector, 793–794
  test_vectors, 542–543, 551–552, 605–606
  text_maxval, 590–591
  timer_class module, 777–778
  triangle_class module, 821–823
  types module, 534
  ucase subroutine, 469–470
  vector_class module, 791–793
  vector_module module, 541–542, 550–552
  vectors module, 601–604, 609–610
  write_namelist, 669
Programming styles, 27
Programs
  basic structure, 24–25
  compiling, linking, and executing, 27–28
  debugging, 66–67. *See also* Debugging
  defined, 1
  parallel (see Parallel programs)
  single-threaded, 838
  styles for writing, 27
  testing, 84–86
PROGRAM statement, 25
Program units, 298
Properties (object), 764
PROTECTED attributes and statements, 608–609
P scale factor, 897
Pseudocode, 84, 86–88
ptr_array program, 725
PUBLIC attributes and statements, 608–609, 610–611
PUBLIC keyword
  in timer class example, 780
  typical use of, 772–773, 789
Punched cards, 870–871
Pure functions, 15, 434, 923
Pure procedures, 434
Pure subroutines, 434

Q

quadf function, 333
Quadratic equations
  solving and evaluating with block IF, 96, 99–103
  solving in different Fortran versions, 17–19
  solving with complex variables, 519–521
Quote marks, 31, 32, 651
QUOTE option, DELIM= clause, 651

R

Race condition, 838
Radar trackers, 557–560s
Radial acceleration equation, 244
Radioactive isotopes, 63–66
Radix, 931, 932
ran001 module, 325
RANDOM_NUMBER subroutine, 328
random0 subroutine, 325–326
Random data, binary tree advantages, 742
Random number generators, 323–328

Range
  defined for real numbers, 11
  factors determining, 30–31
  selecting, 489–491
RANGE function, 490, 491
Rank 1 arrays, 360, 361. *See also* Arrays
Rank 2 arrays
  basic features, 360–361, 362
  declaration, 361
  initializing, 362–366
  sample programs using, 366–370
  storage, 362
  whole array operations, 371
Rank n arrays, 372
Ranks of arrays, 247, 262
RC descriptor, 635, 642
RD descriptor, 635, 642
Reactive power, 60–63
READ= clause, in INQUIRE statements, 656
read_file program, 220–221
read_namelist program, 670
READ statements
  alternative form, 665
  array elements in, 265
  asynchronous, 687, 688
  basic purpose, 26
  complete listing of clauses, 659–664
  for data files, 216–217
  formatted, 205–210
  initializing arrays with, 251, 365–366
  initializing complex variables in, 515
  from internal files, 474
  with list-directed input, 51, 642–643
  namelist directed, 669
  sample programs demonstrating, 218–222
  variable initialization in, 56
READWRITE= clause, in INQUIRE statements, 657
real_to_char function, 476
Real arithmetic, 38
Real constants
  arithmetic operation rules, 38
  declaring kind, 487–488
  overview, 30–31
Real data
  approaches to representing, 485–486
  arithmetic operation rules, 38
  basic features, 10–11
  converting complex data to, 517
  converting to character, 474–478
  format descriptors, 186–188, 206–207, 634
  kinds of variables and constants, 486–489
  mixed-mode arithmetic, 492–494
  operations with integer data, 41–45
  sample array declarations, 361
  selecting precision, 489–491, 493
  solving large systems of linear equations, 499–509
REAL data type, 485–489
REAL function, 43, 518
Real numbers, Fortran model, 931, 932
Real part of complex number, 512
Real power, 60–63
Real variables
  arithmetic operation rules, 38
  default, 485
  index variables as, 144
  kinds, 486–489
  overview, 30–31
  testing for equality, 119
REC= clause, in READ statements, 660, 662

Index

RECL= clause
  in INQUIRE statements, 656
  in OPEN statements, 646, 650
Record length, 673
Records, 212
rectangle_class module, 823–825
Rectangle area and perimeter equations, 816
Rectangles in flowcharts, 87
Rectangular coordinates, 293, 512–513
Recursion
  binary tree advantages, 742–743
  binary tree examples, 743–756
  overview, 300, 568–571
RECURSIVE keyword, 569
Redundant I/O statement features, 896–897
Refinement, stepwise, 84
Relational logic operators
  for character strings, 156, 458–462
  common errors, 118
  with complex numbers, 516–517
  using, 90–91
Renaming data items and procedures, 612–613
Repeating groups of format descriptors, 191, 194
Repetition counts, 191, 194–195
REPLACE file status, 647
RESHAPE function, 364–365, 378
Resizing allocatable arrays, 423
Resonant frequency equation, 79
Restricting access to modules, 607–613
RESULT clauses, 570
RETURN statements, 301, 332
Reusable code, 298
REWIND option, POSITION= clause, 650
REWIND statements, 213, 222, 666, 667
Right justification, 185, 186, 189
rmax subroutine, 317–318
rmin subroutine, 318–319
RN descriptor, 635, 642
Root-mean-square average equation, 177
Root nodes in binary trees, 737
roots_2 program, 520
ROUND= clause
  in INQUIRE statements, 657
  in OPEN statements, 646, 649
  in READ statements, 661, 663
Rounding descriptors, 635, 641–642
Round-off errors. *See also* Precision
  avoiding in Gauss-Jordan elimination, 408, 499–501
  common problems with, 119
  defined, 11
RP descriptor, 635, 642
RU descriptor, 635, 642
running_average subroutine, 418–419
Running averages, 418–421
Run-time errors, 66, 193
RZ descriptor, 635, 642

S

salaried_employee_class module, 800–802
Sample programs. *See* Programming examples
Satellite orbit equation, 174, 242
SAVE attribute
  for allocatable arrays, 427
  illegal for automatic arrays, 422, 427
  illegal in pure procedures, 434
  purpose, 417
SAVE statements, 321, 323, 417
Scalar values
  applied to arrays, 262–263, 292
  multiplying and dividing vectors by, 597–598
Scale factors, 641
Scanning control descriptor, 635
Scientific notation, 10, 186–189
Scope, 562–567
scoping_test program, 564
Scoping units, 562–567
scratch_file program, 223–225
Scratch files, 216, 223–225, 648
SCRATCH file status, 648
S descriptor, 635, 641, 897
Searches, binary tree advantages, 741
Secondary memory, 2, 3–4, 212
Second-order least-squares fits, 452–454
Seed of a sequence, 325
seed subroutine, 326
select_kinds program, 491
SELECT CASE constructs
  actions of, 110–113
  sample programs demonstrating, 113–116, 138, 139
SELECTED_CHAR_KIND function, 512
SELECTED_INT_KIND function, 490, 510–511
SELECTED_REAL_KIND function, 489–491, 573
Selection sorts
  defined, 272
  sample programs demonstrating, 273–282, 302–305
SELECT TYPE construct, 807–809
SEQUENCE statements, 532
Sequential access mode
  default file format, 650
  defined, 212, 648
  file positioning with, 222, 667
SEQUENTIAL clause, in INQUIRE statements, 656
Sequential programs, 81
Set methods, 781
shape_class module, 817–819
Shape class hierarchy example, 814–830
SHAPE function, 376
Shapes of arrays
  changing, 364–365
  defined, 247
  whole array operations and, 261, 262
shared_data module, 321–322
Sharing data. *See* Data sharing
Side effects, 334, 434
SIGN= clause
  in INQUIRE statements, 657
  in OPEN statements, 646, 649–650
  in READ statements, 661, 663
Sign bit, 5
SIGN descriptors, 641
Significant digits of real numbers, 10
simul2 subroutine, 424–426
Simulations, 323
simul subroutine, 410–412
Simultaneous linear equations, 406
Sinc function, 336–338
SIN function, 48
Single-precision complex values, 515
Single-precision real values
  determining kind numbers associated with, 488
  mixing with double, 492–494
  origins of term, 486

Single-precision real values—*Cont.*
  selecting, 489
  when to use, 494–495, 498
Single-threaded programs, 838
SIZE= clause, in READ statements, 660, 662
SIZE function, 376
Sizes of arrays, 247
Slash character
  actions in FORMAT statements, 191–192
  actions in READ statements, 208–209
  basic purpose, 635
  for concatenation, 156
  to terminate namelists, 668, 669
Smallest data values, program for finding, 256–261
Snell's law, 125
sort1 program, 276–278
sort2 program, 280–281
sort3 program, 302–305
sort4 program, 459–461
sortc subroutine, 467–468
Sorted data, creating binary trees from, 742
Sorting data
  basic concepts, 271–272
  in binary trees, 736–740
  insertion sort program, 719–724
  with parallel processing, 850–856
  selection sort of derived data, 532–539
  selection sort program, 273–282
  selection sort subroutine, 302–305
Spare parts database example, 681–686
SP descriptor, 635, 641, 897
Specification statements, obsolescent, 872–875
Specific functions, 49, 581
Specific intrinsic functions, 888–889
Specific procedures, 582
SQRT function, 48
square_and_cube_roots program, 266–267
square_class module, 825–826
square_roots program, 250–251
Square brackets, 247
Squares, 816
squares_2 program, 252
squares program, 249
SS descriptor, 635, 641, 897
Standard deviation
  arrays not required for, 288
  defined, 128–129
  equation for, 420–421
  programs to calculate with arrays, 282–288
  programs to calculate with counting loops, 140–143
  programs to calculate with subroutines, 316–320
  programs to calculate with while loops, 129–134
  in running averages program example, 418–421
Standard error device, 213
Standard input devices, 213
Standardized normal distribution, 358
Standard output devices, 213
STAT= clause, 385
Statement function, 886–888
Statement labels, 24, 871, 882
Statement numbers, 97
Statements
  assignment, 36–37
  basic structure, 23–24
  fixed-source form, 870–871
  IMPLICIT NONE, 57–58
  list-directed, 49–53
  order in Fortran, 961–962

  in program sections, 25
  specifying variables in, 33
Statement scope, 562, 565
Static memory allocation, 383
Static variables, 698
stats_1 program, 131
stats_2 program, 133–134
stats_3 program, 142–143
stats_4 program, 285–287
stats_5 program, 386–389
STATUS= clause
  basic actions of, 214, 215
  in CLOSE statements, 654
  in OPEN statements, 646, 647–648
Stepwise refinement, 84
stock program, 683–685
STOP ALL statement, 840
stop_test program, 891
STOP statements, 26
  arguments with, 890–891
  avoiding in subroutines, 314
  basic purpose, 26
  limiting use of, 686
Stopwatches, 775
Storage association, 877
Storage devices, 212
Stream access mode, 648, 678
STREAM clause, in INQUIRE statements, 656
Strings. *See* Character data
Structure constructors, 528
Structured programs, 86
Structures (derived data), 528
Stubs, 84
Subclasses
  abstract, 813, 814
  basic principles, 767, 794
  defining and using, 795–803
  relationship to superclass objects, 804–805
  SELECT TYPE construct in, 807–809
Submodules, 438–446
Subprogram features, undesirable, 882–889
Subroutines
  alternate returns, 882–884
  basic features, 299–302
  bounds checking with, 311–312
  for command line access, 615–616
  defined, 298
  elemental, 436
  error handling, 313–314
  INTENT attribute, 306–307
  within modules, 328
  pass-by-reference scheme for, 307–309
  passing arrays to, 309–310, 404–406
  passing as arguments, 342–344
  passing character variables to, 313, 465–470
  pure, 434
  recursive, 300, 569, 743
  in sample program to sort data, 302–305
  in sample random number generator program, 326
  samples to calculate basic statistics, 316–320
  scope concepts, 564–566
  for user-defined operators, 596
SUBROUTINE statements, 299
Subroutines
  merging, 854–856
subs_as_arguments subroutine, 342
Subscripts (array)
  changing, 251–252
  out-of-bounds, 252–256

Index

purpose, 245
replacing to create sections, 263–265, 371
Subscript triplets, 263–264, 371, 381
Substring specifications, 154, 155
Subtasks
  breaking into procedures, 297–298
  testing separately, 84–85
  top-down design approach, 82, 84
Subtraction
  in hierarchy of operations, 39
  vector quantities, 540–543, 550–552, 597
SUM function, 378
Superclasses, 767, 794, 804–805
SUPPRESS option, SIGN= clause, 650
Symbolic debuggers, 67, 118, 168
Symbols
  for flowcharts, 87, 88
  to represent character data, 8–9
  used with format descriptors, 185
SYNC ALL statement, 844, 845
Synchronization point, 844
Synchronous I/O, 687
SYNC IMAGES statement, 845–846
SYNC MEMORY command, 849
Syntax errors, 66
System environment intrinsic procedures, 935–938

T

Tab format descriptor, 190–191, 197
TAB format descriptors, 638–639
table program, 196
Tables of information, sample programs creating, 196–197
TAN function, 48
TARGET attribute, 700
Targets, associating pointers with, 699–705
T descriptor
  actions in FORMAT statements, 190–191
  actions in READ statements, 208
  basic actions of, 635, 638–639
Temperature conversion program example, 58–60
Termination sections, 25, 26
test_abc program, 471–472
test_alloc_fun function, 432–433
test_alloc subroutine, 431
test_array subroutine, 416
test_ave_value2 program, 889
test_ave_value program, 341–342
test_char1 program, 155
test_date program, 787–788
test_diagonal program, 729–731
test_dsimul program, 504–507
test_employee program, 804–805
test_entry program, 884
test_generic_procedures program, 593–594
test_hypotenuse program, 301–302
test_internal program, 436–437
test_io program, 530
test_keywords program, 572
test_ptr2 program, 703
test_ptr3 program, 705
test_ptr4 program, 706
test_ptr program, 701
test_quadf program, 333

test_random0 program, 326–327
test_real_to_char program, 477
test_running_average program, 419–421
test_select_type program, 808–809
test_shape program, 828–830
test_simul program, 412–414
test_sinc program, 337–338
test_sort program, 851–853
test_subs_as_arguments program, 343–344
test_timer program, 779
test_type_extension program, 547
test_ucase program, 470
test_vector program, 793–794
test_vectors programs, 542–543, 551–552, 605–606
Test drivers, 84
Testing. *See also* Debugging
  for all possible inputs, 132
  of sample programs, 102–103, 106
  of subtasks, 298
  in top-down design, 84–86
text_maxval program, 590–591
Three-component vectors, 597–598
timer_class module, 777–778
timer class example, 775–780
TL descriptor, 635, 638–639
Top-down program design, 82–86, 297
Track while scan radars, 559
Transformational intrinsic functions, 377, 914
TRANSPOSE function, 378
TR descriptor, 635, 639
triangle_class module, 821–823
Triangles, 815, 821–823
TRIM function, 157, 463, 465
Truncation, 37–38, 176–177
Truth tables, 91, 92
Two-dimensional arrays. *See* Rank 2 arrays
Two-dimensional vectors, 292
Two's complement arithmetic, 5–6
Type 704 computer, 13–14
Type-bound procedures, 548–552, 773. *See also* Bound procedures
Type conversion functions, 43–44, 517
Type conversion and mathematical intrinsic procedures, 923–930
Type declaration statements
  access control attributes in, 608
  ALLOCATABLE attribute, 384
  associating INTENT attribute with, 306
  avoiding with USE association, 322
  defined, 33
  for derived data types, 528, 545–546
  for dummy character arguments in procedures, 465–467
  EXTERNAL attribute in, 340
  of functions, 332
  initializing complex variables in, 514, 515
  local variables saved from, 417
  PARAMETER attributes, 34
  pre-Fortran 90 limitations, 872, 874
  for rank 1 arrays, 246–247, 249–251
  for rank 2 arrays, 361, 365
  variable initialization in, 56–57
Type definitions
  bound procedures in, 548, 773
  CLASS keyword in, 771
  data item status in, 608
  declaring abstract methods in, 810
  declaring pointers in, 699

Type definitions—Cont.
  declaring targets in, 700
  extending, 546, 795–797, 802
  form of, 528
  order of components in, 529–530, 532
  as scoping units, 563
TYPE keyword, 800
Type mismatches, 308, 329–330
types module, 534
Types of variables, 32–33
Type specification statements, obsolescent, 872–875
Typographical errors, 66

U

UBOUND function, 376
ucase subroutine, 469–470
Unary operators, 37, 595
Unconditional GO TO statement, 893
Undefined pointer association status, 704
Underscore character, 27, 29
UNFORMATTED clause, in INQUIRE statements, 656
Unformatted files
  advantages and disadvantages, 671–672
  defined, 650
  as direct access files, 673–677
Unicode system, 9, 15, 457
Uninitialized arrays, 248
Uninitialized variables, 55–56
UNIT= clause
  actions of, 214
  in CLOSE statements, 654
  in file positioning statements, 666
  in INQUIRE statements, 655
  in OPEN statements, 646, 647
  in READ statements, 660, 661
Unit testing, 84–85, 298
UNKNOWN file status, 648
Unlabeled COMMON statement, 879–880
Unlimited polymorphic pointers and arguments, 771
UP option, ROUND= clause, 649
Uppercase letters
  ASCII offset from lowercase, 158, 159, 458
  equivalent to lowercase in Fortran, 23
  programs shifting strings to, 158–161, 468–470
USB memory, 4
USE association, 321, 563, 566
USE statements
  to access derived data in modules, 540, 543
  advanced options, 611–613
  command line access via, 689
  to extend object scope, 563, 566
  form of, 321
User-defined functions
  overview, 332
  passing as arguments, 339–342
  sample programs, 333
  type declarations for, 334
User-defined generic procedures, 582–591
User-defined I/O operations, 678–680
User-defined operators
  overview, 594–596
  programming examples, 597–607
UTF-8 encoding scheme, 458
Utility methods, 789

V

Variable-length character functions, 471–472
Variable names
  basic requirements, 28–29
  lowercase for, 27
  typographical errors in, 66
Variables
  access restrictions for, 611
  array elements as, 247–248
  assignment statements, 36–37
  in COMMON blocks, 877
  conflicts with descriptors, 192–193, 209–210
  conversion factors as, 49
  default and explicit typing, 32–33
  defined, 28
  initializing, 55–57, 874. *See also* Type declaration statements
  in list-directed input, 49–51
  local, 301
  logical, 89
  major types, 29–32
  namelist, 669–671
  naming, 28–29
  in objects, 764–765. *See also* Instance variables
  passing to subroutines, 307–309
  pointer versus ordinary, 698–699
  sample program to evaluate functions of, 103–106
  saving between procedure calls, 417–421
  scope concepts, 562–563, 565–566
  showing with WRITE statements, 67, 118
  volatile, 618
vector_class module, 791–793
vector_module module, 541–542, 550–552
Vectors
  adding and subtracting, 540–543, 550–552, 597
  with bound generic procedures, 592–594
  creating classes for storing, 790–793
  dot products, 294, 598
  most common operations on, 597–598
  one-dimensional arrays as, 360
  scalar quantities versus, 292
  user-defined operators with, 598–607
vectors module, 601–604, 609–610
Vector subscripts, 263, 264–265, 371
Velocity of falling object, 226
VOLATILE attributes or statements, 618
Volt-amperes, 61
Volt-amperes-reactive, 61
Volts, 61

W

WAIT statements, 667–668
Watts, 61
Weekday/weekend program example, 115–116
Well-conditioned systems, 501, 507
WG5, 16
WHERE construct, 379–380
WHERE statements, 380–381
While loops
  actions of, 127–134
  common errors, 169
  in interactive programs, 478

Index

iterative loops versus, 126
for reading data sets, 218
Whole array operations, 261–263, 269–270, 371
Words, 4, 30
Worker images, 840
WRITE= clause, in INQUIRE statements, 656
write_namelist program, 669
WRITE statements
  array elements in, 265, 266
  asynchronous, 687–688
  basic purpose, 26
  character constants in, 32
  clauses available for, 665
  conditional stopping points in, 639–640
  for data files, 217
  formatted, 181, 192–196, 203, 639–640
  to internal files, 474
  list-directed, 52–53
  namelist directed, 669
  sample programs demonstrating, 196–197
  to show intermediate calculations, 67, 118, 168

X

(x, y) data, 226–231
X descriptor, 190–191, 208

Z

Z descriptor, 634, 638
0 (zero) character, 183
Zero length descriptor, 185
ZERO option, ROUND= clause, 649

Summary of Selected Fortran Statements and Structures

This table presents a quick summary of some common Fortran statements and constructs. Less common and/or obsolete statements are not included here.

Statement	Description (page in text)	Example of Usage
ALLOCATE	Allocate memory to an allocatable array or pointer (384, 709)	`ALLOCATE ( x(100,100) )`
Assignment Statement	Assigns a value to a variable (36)	`pi = 3.141593` `name = 'James'`

ASSOCIATE construct	Allow variables with long names to be addressed by shorter names within the construct (552)	`ASSOCIATE (x => target(i)%x, &`
		`            y => target(i)%y )`
		`   dist(i) = SQRT(x**2 + y**2)`
		`END ASSOCIATE`

BACKSPACE	Backspace one record in a file (222)	`BACKSPACE (UNIT=9)`

Block IF construct	Branching construct (95)	`test: IF ( x > 0. ) THEN`
		`   res = SQRT(x)`
		`ELSE IF ( x == 0. ) THEN`
		`   res = 0.`
		`ELSE`
		`   res = SQRT(-x)`
		`END IF test`

CALL	Call a subroutine (300)	`CALL sort ( array, n )`

CASE construct	Branching among mutually exclusive choices (110)	`SELECT CASE ( ii )`
		`CASE (selector_1)`
		`   block 1`
		`CASE (selector_2)`
		`   block 2`
		`CASE DEFAULT`
		`   block 3`
		`END SELECT`

CHARACTER	Declares variables or named constants of type CHARACTER (33, 457)	`CHARACTER(len=12) :: surname`
CLOSE	Closes a file (216, 653)	`CLOSE (UNIT=1)`
COMPLEX	Declares variables or named constants of type COMPLEX (512)	`COMPLEX(KIND=sgl) :: cval` `COMPLEX,DIMENSION(10) :: array`
CONTAINS	Specifies that a module or procedure contains internal procedures (328)	`CONTAINS`
CRITICAL	Marks start of a critical section, in which only one image is allowed to execute at at time (859)	`CRITICAL`
CYCLE	Branch to top of loop (145)	`CYCLE`
DEALLOCATE	Deallocate memory associated with an allocatable array or pointer (384, 709)	`DEALLOCATE ( x )`

DO (counting loop) construct	A loop that repeats a block of statements a specified number of times (134)	`DO i = 1, 6, 2`
		`   sqr(i) = i**2`
		`END DO`

DO (while loop) construct	A loop that repeats a block of statements a specified number of times (127)	`DO`
		`   IF ( condition ) EXIT`
		`   ...`
		`END DO`

END CRITICAL	Last statement of a critical section (859)	`END CRITICAL`
END FUNCTION	Last statement of a function (332)	`END FUNCTION myfun`
END MODULE	Last statement of a module (321)	`END MODULE modulename`
END PROGRAM	Last statement of a program (25)	`END PROGRAM progname`
END SUBROUTINE	Last statement of a subroutine (299)	`END SUBROUTINE mysub`
ENDFILE	Writes an end-of-file marker in a file (667)	`ENDFILE (UNIT=lu)`

Summary of Selected Fortran Statements and Structures

Statement	Description (page in text)	Example of Usage
EXIT	Branch to first statement after end of loop (145)	`IF ( value < 0 ) EXIT`
FLUSH	Flush output buffers to disk (668)	`FLUSH (UNIT=8)`

FORALL construct	Execute statements based on a mask and index values (381)	`FORALL (i=1:3, j=1:3, i > j)`
		`    arr1(i,j) = ABS(i-j) + 3`
		`END FORALL`

FORMAT	Defines descriptors used to format input or output data (181)	`5 FORMAT (' I = ',I6)`
FUNCTION	Declares the start of a function subprogram (332)	`INTEGER FUNCTION fact(n)`
IF Statement	Executes or skips a statement, depending on whether a logical expression is true or false (110)	`IF ( x < 0. ) x = -x / 2.`
IMPORT	Imports type definitions into an interface block from the containing procedure (581)	`IMPORT :: a, b`
IMPLICIT NONE	Cancels default typing (57)	`IMPLICIT NONE`
INQUIRE	Used to learn information about a file either by name or logical unit (655)	`INQUIRE (NAME='x', EXIST=flag)`
INTEGER	Declares variables or named constants of type INTEGER (33)	`INTEGER :: i, j, k`

INTERFACE	Creates an explicit interface, a generic procedure, or a user-defined operator (557)	`INTERFACE :: sort`
		`    MODULE PROCEDURE sort_i`
		`    MODULE PROCEDURE sort_r`
		`END INTERFACE`

LOGICAL	Declares variables or named constants of type LOGICAL (89)	`LOGICAL :: test1, test2`
MODULE	Declares the start of a module (321)	`MODULE mysubs`
OPEN	Opens a file (213, 644)	`OPEN (UNIT=10,FILE='x')`
PRIVATE	Declares that the specified items in a module are not accessible outside the module (608)	`PRIVATE :: internal_data` `PRIVATE`
PROTECTED	Declares that an object in a module is protected, meaning that it can be used but not modified outside the module in which it is defined (608)	`PROTECTED :: x`
PROGRAM	Defines the start of a program, and gives it a name (25)	`PROGRAM my_program`
PUBLIC	Declares that the specified items in a module are accessible outside the module (608)	`PUBLIC :: proc1, proc2`

READ	Read in data (50)	`READ (12,100) rate, time`
		`READ (unit,'(I6)') count`
		`READ (*,*) nvals`

REAL	Declares variables or named constants of type REAL (33)	`REAL(KIND=sgl) :: value`
RETURN	Returns control from a procedure to the invoking routine (299)	`RETURN`
REWIND	Position file pointer at first record in a file (222)	`REWIND (UNIT=3)`
SAVE	Preserve local variables in a subprogram between calls to the subprogram (417)	`SAVE ncalls, iseed` `SAVE`
STOP	Stop program execution (26)	`STOP`
SUBROUTINE	Declares the start of a subroutine (299)	`SUBROUTINE sort (array, n)`
SYNC IMAGES() SYNC ALL	Synchronise one or more images in a Coarray Fortran program (844)	`SYNC ALL` `SYNC IMAGES(*)`
TYPE	Declares a derived data type (528)	`TYPE (point) :: x, y`
USE	Makes the contents of a module available to a program unit (321)	`USE mysubs`
VOLATILE	Declares that the value of a variable might be changed at any time by some source external to the program (618)	`VOLATILE :: val1`

WHERE construct	Masked array assignment (378)	`WHERE ( x > 0. )`
		`    x = SQRT(x)`
		`END WHERE`

Statement	Description (page in text)	Example of Usage
WRITE	Write out data (180, 665)	WRITE (12,100) rate, time WRITE (unit,'(1X,I6)') count WRITE (*,*) nvals

This table presents a quick summary of common attributes used in type declaration statements.

Attribute	Description (page in text)	Example of Usage
ALLOCATABLE	Declares that an array is allocatable (384)	REAL,ALLOCATABLE,DIMENSION(:) :: a
DIMENSION	Declares the rank and shape of an array (246)	REAL,DIMENSION(10,10) :: matrix
CODIMENSION	Declares the rank and shape of a coarray, which is allocated across multiple executing images (841)	REAL,CODIMENSION(*) :: a
EXTERNAL	Declares that a name is a function external to a program unit (339)	REAL,EXTERNAL :: fun1
INTENT	Specifies the intended use of a dummy argument (300)	INTEGER,INTENT(IN) :: ndim
INTRINSIC	Declares that a name is a specific intrinsic function (888)	REAL,INTRINSIC :: sin
NOPASS	Declares that the derived data type variable used to invoke a bound procedure will *not* be passed to it as its first calling argument (549)	PROCEDURE,NOPASS :: add
OPTIONAL	Declares that a dummy argument is optional (572)	REAL,OPTIONAL,INTENT(IN) :: maxval
NON_OVERRIDABLE	Declares a bound procedure cannot be overridden in a subclass of this class (809)	PROCEDURE, NON_OVERRIDABLE :: pr
PARAMETER	Defines named constant (34)	REAL,PARAMETER :: pi = 3.141593
PASS	Declares that the derived data type variable used to invoke a bound procedure will be passed to it as its first calling argument (548)	PROCEDURE,PASS :: add
POINTER	Declares that a variable is a pointer (699)	INTEGER,POINTER :: ptr
PRIVATE	Declares that an object is private to a module (608)	REAL,PRIVATE :: internal_data
PROTECTED	Declares that an object in a module is protected, meaning that it can be used but not modified outside the module in which it is defined (608)	REAL,PROTECTED :: x
PUBLIC	Declares that an object in a module is visible outside the module (608)	REAL,PUBLIC :: pi = 3.141593
SAVE	Preserve local variables in a procedure between invocations of the procedure (417)	REAL,SAVE :: sum SAVE
TARGET	Declares that a variable may be pointed to by a pointer (700)	INTEGER,TARGET :: val1
VOLATILE	Declares that the value of a variable might be changed at any time by some source external to the program (618)	REAL,VOLATILE :: val1